Biofuel and Biorefinery Technologies

Volume 17

This book series provides detailed information on recent developments in biofuels & bioenergy and related research. The individual volumes highlight all relevant biofuel production technologies and integrated biorefinery methods, describing the merits and shortcomings of each, including cost-efficiency. All volumes are written and edited by international experts, academics and researchers in the respective research areas.

Biofuel and Biorefinery Technologies will appeal to researchers and post-graduates in the fields of biofuels & bioenergy technology and applications, offering not only an overview of these specific fields of research, but also a wealth of detailed information.The ideas and technologies presented in this book series contribute to the UN Sustainable Development Goal 7: *Affordable and Clean Energy.*

Adenise Lorenci Woiciechowski
Arion Zandona Filho • Satinder Kaur Brar
Carlos Ricardo Soccol
Editors

Biomass for Bioenergy and Bioproducts

Fundamentals, Processes and Advances

Springer

Editors
Adenise Lorenci Woiciechowski
Department of Bioprocess Engineering and Biotechnology
Federal University of Paraná
Curitiba, Paraná, Brazil

Arion Zandona Filho
Chemical Engineering Department
Federal University of Paraná
Curitiba, Paraná, Brazil

Satinder Kaur Brar
Department of Civil Engineering
York University
Woodbridge, ON, Canada

Carlos Ricardo Soccol
Department of Bioprocess Engineering and Biotechnology
Federal University of Paraná
Curitiba, Paraná, Brazil

ISSN 2363-7609 ISSN 2363-7617 (electronic)
Biofuel and Biorefinery Technologies
ISBN 978-3-032-19897-6 ISBN 978-3-032-19898-3 (eBook)
https://doi.org/10.1007/978-3-032-19898-3

This Springer imprint is published by the registered company Springer Nature Switzerland AG
The registered company address is: Gewerbestrasse 11, 6330 Cham, Switzerland

Preface

Natural and planted forests are fundamental sources of biomass, a power plant, along with crops and industrial waste, which play an indispensable role in a low-carbon future. Plants and the organisms that have chlorophyll use the carbon from the CO_2 of the air and through photosynthesis, produce their carbonaceous complex structural molecules synthetizing their own biomass, using the sun's light as energy source. This process converts solar energy into chemical bond energy, storing solar energy at the chemical bonds of their biomass structural molecules. When this biomass is used for energy, the carbon is released back into the atmosphere, establishing a closed-loop system. Biomass is vital to balance global energy sources, enhancing energy security. This form of carbon is referred to as *green carbon*, synthetized from CO_2, present at the atmosphere, a component of the air, different from the *fossil carbon*, which is sourced from subterranean deposits. When this fossil fuel undergoes combustion to produce energy, the carbon that was underground is released as CO_2, increasing the greenhouse gases at the atmosphere.

Several efforts are being made over the years to replace part of the fossil fuels with green fuels and biofuel, and gradually introducing them into the countries' energetic matrix as energy from renewable sources. The Conference of the Parties (COP) summit, held in different countries around the world, has taken small steps to accomplish this goal. To replace, even in part, the energetic matix of the countries based on the petrochemicl industry is important but it is not easy, due to the efficiency of the petrochemical industry and the large range of petrochemical products at the world market, since solid, liquid, and gaseous fuels, until soil fertilizers, chemical for many products, such as tissue, paints, polymers, plastic and others.

In this sense, biomass is a versatile resource that generates various products and forms of energy, including electricity, heat, and liquid or gaseous fuels such as bioethanol and biogas. Biomass can be processed in biorefineries, where multiple fuels and high-value bioproducts are produced, thereby improving economic and operational efficiency. Lignocellulosic biomass structure disruption involves chemical, biochemical, and thermochemical processes, resulting in diverse outputs.

Despite its availability, challenges such as maintaining feedstock quality, avoiding competition with food production, minimizing land-use change, and improving supply chains efficiency must be addressed to ensure sustainable biomass utilization.

Biomass is a vital renewable resource essential for facilitating the transition toward sustainable energy and bioproducts. It comprises organic materials such as sugar- and starch-rich biomass (e.g., sugarcane, corn, wheat), oilseed biomass (soybeans, canola, sunflower), lignocellulosic mass (crop residues, forestry waste, energy crops), and algae, each serving specific applications. Crops high in sugar and starch contents are particularly suitable for bioethanol production. Oilseed crops are used to produce bio-oil and biodiesel, and are also employed in the manufacture of bioplastics and in biorefineries. Lignocellulosic biomass can be converted into biofuels, biochemicals, and biomaterials through pretreatment and conversion processes. Studies on advanced methodologies such as, physical, chemical, and thermochemical treatments along with enzymatic hydrolysis support the development of sustainable biorefineries to break down biomass into sugars and convert them into valuable products.

Biofuels are classified into three generations: first from food crops, second from non-food biomass, and third from algae and microbes. Biochar, produced through pyrolysis, enhances soil quality and carbon sequestration and assists in pollution remediation. Bio-oil, similarly derived from pyrolysis, functions as a renewable fuel or chemical feedstock. Industrial waste can be processed via anaerobic digestion to generate biogas for heating, electricity, or vehicular use. The pulp industry can adopt biorefinery technologies to convert waste into valuable products, thereby fostering a circular economy and minimizing environmental impact. The emphasis on biomass conversion studies aims to optimize feedstock selection, pretreatment procedures, and technological methodologies and strategies to maximize the pretreatments yield and quality of bioproducts, which are subsequently transformed into biofuels, biochemicals, and biomaterials.

Therefore, effective utilization of this vital energy resource is crucial for maintaining a sustainable planet. Research on energy transition, changing fossil fuels to biofuels, plays a key role in this effort, considering technical and economic aspects.

The COP30 climate summit in Belém, Brazil, held in November 2025, failed to incorporate language on the exclusion of fossil fuels despite robust evidences of the importance. Nevertheless, it underscored the need to implement practical solutions at the local and national levels to protect the forests.

Curitiba, Paraná, Brazil — Adenise Lorenci Woiciechowski
Curitiba, Paraná, Brazil — Arion Zandona Filho
Woodbridge, ON, Canada — Satinder Kaur Brar
Curitiba, Paraná, Brazil — Carlos Ricardo Soccol

Contents

Chapter 1
Importance of Biomass: Overview

Adenise Lorenci Woiciechowski, Júlio Cesar de Carvalho, Arion Zandoná Filho, Luiz Alberto Júnior Letti, Leonardo José Duda, and Carlos Ricardo Soccol

Abstract Biomass, more specifically phytobiomass, can be defined as all materials, molecules, or structures produce by photosynthesis, using carbon dioxide (CO_2) and water (H_2O) as substrates, processed using solar light as the energy source, producing all the diversity of molecules that compose plants, including forest trees, fruit trees, cereals, vegetables, grains, flowers, planktons, micro- and macroalgae, and all organisms that contain chlorophyll. This chapter brings basic informations to provide a brief description of the photosynthesis process, the main energy molecule that is accumulated by these organisms, the structural molecules of the lignocellulosic vegetal structure, and its composition, along with the importance of the photosynthesis process that changes light energy into chemical energy and stores it at the chemical linkage C–C in the synthetized molecules that form the chlorophyll organisms available to be used by other organisms, including human beings, at its own metabolisms, metabolic energy of vital importance to the survival of life on Earth. Besides, as phytobiomass structural molecules are basically composed of carbonaceous matter, the photosynthetic process throught phytobiomass synthesis contributes to preserving atmospheric quality in terms of its characteristic composition, which is essential for human breathing. This occurs due to the carbon from CO_2 storage capacity of phytobiomass. Thus, CO_2 is removed from the atmosphere and incorporated into its structure.

A. L. Woiciechowski (✉) · J. C. de Carvalho · L. A. J. Letti · L. J. Duda · C. R. Soccol
Biotechnololgy and Bioprocess Engineering Department, Federal University of Paraná – Centro Politécnico, Curitiba, Paraná, Brazil
e-mail: adenise@ufpr.br

A. Z. Filho
Chemical Engineering Department, Federal University of Paraná – Centro Politécnico, Curitiba, Paraná, Brazil

A. L. Woiciechowski et al. (eds.), *Biomass for Bioenergy and Bioproducts*, Biofuel and Biorefinery Technologies 17, https://doi.org/10.1007/978-3-032-19898-3_1

1.1 Introduction

Energy sources available are extremely important to life preservation on the planet. Among the kinds of energy are metabolic, thermic, mechanical, chemical, and others. For the Earth and the planets of the solar system, the sun is the only form Pof primary energy. Light comes through the space and reaches the Earth in the form of electromagnetic energy, with a range of wavelengths, including the infrared, visible light, and ultraviolet ranges. As benefit of the solar energy (light and heat) for our planet can be included maintaining the temperature within a range suitable for life and enabling the photosynthesis process in organisms that contain chlorophyll in their cells. These organisms are able to synthesize their own biomass from simple molecules, carbon dioxide (CO_2) and water (H_2O) using energy from the sun. These organisms can be, more precisely, named phytobiomass. So the photosynthesis process is responsible for converting light energy into chemical energy stored in the atoms' linkages of the molecules' constituents in the biomass cells.

The chemical energy stored in phytobiomass is vital to life on Earth due to the wide variety of feed, food products and industrial raw material derived from it, including wood, vegetables, grains, fruits, and cereals, as well as products obtained from raw material processing, such as processed foods, meals, beverages and also, animal protein is produced from livestock fed such as grass and silage. Some animal species have the ability to convert vegetal biomass into animal protein, which is an important component of the human daily diet. Additionally, residues from some processed food production chains still contain energetic molecules within their cellular structures.

The use of phytobiomass (firewood, animal residues, and agricultural and forest residues) for thermic energy generation is as ancient as the discovery of fire, and nowadays, it is even used as the primary source of energy for warmth, food cooking, and more recently, industrial steam generation (Carioca and Arora 1984).

Biofuels such as bioethanol and biodiesel are also produced from phytobiomass, and biofuels are becoming more and more important day by day in the energy matrix of countries in many forms.

Oil discovery and the development of the oil industry has hardly affected this kind of phytobiomass use, but more recently, oil crisis has imposed a more important participation of phytobiomass in energetic purposes.

1.2 Phytobiomass Synthesis

The photosynthetic process is the most practical, natural, and widespread way of solar energy utilization, which happens in even minimal areas of the plant that has chlorophyll since the first instant it receives solar light.

Phytobiomass, the living biomass derived from photosynthesis in plants, algae, and photosynthetic microorganisms, is by far the most important fraction of the biosphere. With an estimated 450 Gt of carbon, plant biomass makes up 82% of the biosphere (Bar-On et al. 2018). In contrast, marine phytobiomass makes up 20% of the 6.5 Gt of marine biomass (Bar-On and Milo 2019), showing its fast turnover in the food webs. While light harvesting is believed to be as old as 3.4 billion years, oxygenic photosynthesis (OP) appeared only 2.7 billion years ago and reshaped the atmosphere during the next 400 million years. Biomass productivity increased approximately 1000 times after the emergence of OP in cyanobacteria with the evolution of algae and plants (Crockford et al. 2023).

Sunlight is the energy source used by photosynthetic organisms to synthesize organic compounds from simple inorganic substances, carbon dioxide, and water, using these organic compounds as sources of energy and cellular building blocks. Although there are important classes of organisms that perform non-oxygenic photosynthesis and also microorganisms that derive their energy from the oxidation of inorganic chemical species—a reminder of the way life first thrived on Earth—OP is today the most important mechanism for biomass synthesis, as detailed in the next section.

1.2.1 Photosynthesis Process

Photosynthesis is a complex cascade of reactions that uses light energy to convert water and carbon dioxide into organic molecules while storing energy. The process relies on photosynthetic pigments—light-sensitive molecules such as chlorophyll—that absorb photons, rise to an excited state, and release the absorbed energy while splitting water into protons, electrons, and oxygen in the photosystem II (PSII) complex.

In photosynthesis, two photons in PSII and two photons in PSI are required to excite two electrons taken from water. These electrons move through the electron transport chain and drive the Q-cycle in cytochrome b_6f, which translocates four protons (H^+) from the stroma to the lumen. In addition, two protons are released directly into the lumen from water splitting in PSII. The same electrons, after excitation in PSI, are ultimately used to reduce $NADP^+$ to one molecule of NADPH. The proton gradient created across the thylakoid membrane is used by ATP synthase to create ATP, requiring about 4.67 H^+ per ATP synthesized (Hahn et al. 2018). This constitutes the "light-dependent" phase of photosynthesis.

$$6\,H_2O \rightarrow 3\,O_2 + 12\,{H^+}_t + 12{e^-}_{PSII}$$

$$12{e^-}_{PSII} + 24{H^+}_s \rightarrow 24\,{H^+}_t + 12{e^-}_{PSI} \quad \text{(through b6F)}$$

$$12{e^-}_{PSI} + 6NADP^+ + 6{H^+}_s \rightarrow 6NADPH \quad \text{(stroma side)}$$

In total, 24 H^+t are translocated, plus 12 H^+t released from water, yielding 36 H^+t. These protons can produce 36/4.67 $\approx$ 7.7 ATP, which is less than the

9 ATP needed for the fixation of 3 CO_2. Therefore, cyclic photophosphorylation (CP) is employed to provide additional protons without producing NADPH:

$$3e^-{}_{PSI} + 6H^+{}_s \rightarrow 6\,H^+{}_t + 3e^-{}_{PSI} \quad \text{(electrons } \textit{cycled} \text{ via } b_6F).$$

Then, 42 $H^+{}_t$ + 9ADP + 9Pi $\rightarrow$ 9ATP + 42 $H^+{}_s$ (translocated through ATP synthase)

Overall light-dependent equation:

$$6\,H_2O + 6\,NADP^+ + 9\,ADP + 9\,Pi \rightarrow 6\,NADPH + 3\,O_2 + 9\,ATP$$

where the subscripts "s" and "t" indicate that the protons are on the stroma or thylakoid side. One proton is consumed in the stroma during $NADP^+$ reduction, which slightly increases the effective number of H^+ needed per ATP, so the effective H^+/ATP ratio is ~4.67 instead of 4. These "extra" protons can be provided by cyclic photophosphorylation.

Another crucial step in photosynthesis is the reduction of carbon dioxide in the Calvin cycle. Carbon atoms at an oxidation state of +4 in CO_2 require four electrons to be reduced to C at an average oxidation state of 0. This is achieved with 6 NADPH and 9 ATP, which together form one molecule of glyceraldehyde 3-phosphate, $C_3H_7O_6P$, or G3P (where Pi is the phosphate group). This carbon reduction is carried out in the chloroplast stroma through the Calvin cycle, a complex cascade of "light-independent" reactions, summarized as:

$$3CO_2 + 6NADPH + 9ATP \rightarrow C_3H_7O_6P + 6\,NADP^+ + 9ADP + 8Pi$$

Balancing both light-dependent and light-independent reactions,

$$6\,H_2O + 6\,NADP^+ + 9\,ADP + 9\,Pi \rightarrow 6\,NADPH + 3\,O_2 + 9\,ATP$$

$$3CO_2 + 6NADPH + 9ATP \rightarrow C_3H_7O_6P + 6\,NADP^+ + 9ADP + 8Pi$$

$$3H_2O + 3CO_2 \rightarrow C_3H_7O_6P + 3\,O_2$$

While it is common to describe photosynthesis as producing one hexose (e.g. glucose) from 6 CO_2 and 6 CO_2, photosynthesis actually produces G3P, which may be used directly or converted into hexoses. In these reactions, O_2 comes from water during light reactions, while a part of the oxygen in CO_2 remains in G3P, and a part is used in the hydrolysis of ATPs. Photosynthesis is the most widespread and fundamental way of capturing solar energy, occurring in every chlorophyll-containing part of the plant upon receiving light.

1.2.2 Green Carbon Versus Fossil Carbon

It is believed that the early Earth atmosphere was primarily composed of nitrogen, carbon dioxide, some methane, and hydrogen. The atmosphere changed slowly, with H_2 steadily lost to space, even with some production through CH_2 photolysis. Carbon dioxide was consumed through carbonate formation for about two billion years, until the Great Oxygenation Event (2.4 Ga ago) began to modestly contribute to CO_2 reduction while producing O_2 (Catling and Zahnle 2020). However, when land plants spread, about 400 Ma ago, oxygen concentration in the atmosphere rose markedly, while copious amounts of CO_2 were converted into biomass, in the aptly called Carboniferous Period. Coal deposits date from this era, and in that sense, coal and other fossil fuels are *green* in origin. However, while the drop from thousands of ppm of carbon dioxide in the atmosphere to about 300 was fast in geological terms, human contribution to the rising CO_2 levels by burning fossil fuels is orders of magnitude faster, leading to environmental changes that can be hard for nature to cope with. Therefore, carbon capture is essential, as is the use of carbon-neutral fuels. However, it is important to note that biomass is not a carbon sink unless it is maintained as forests, soil carbon, or permanent materials such as wood or bioplastics. Biofuels, for example, are solely energy vectors where sunlight is stored for a while—better than fossil fuels but not carbon-negative.

1.2.3 CO_2: Bad or Good?

As with any aspect of nature, CO_2 is neither good nor bad. On the one hand, it is essential for photosynthesis and for the greenhouse effect that maintains Earth at its current average temperature. While water vapor is responsible for about two-thirds of the greenhouse effect, CO_2 is responsible for another third (Kasting and Catling 2003). It can be estimated that Earth would be at least 10 °C colder without the radiative forcing provided by CO_2—probably more because of the consequent reduction in water content in the colder atmosphere. But also, as with anything in nature, too much of a good thing can be bad, and the increasing CO_2 concentrations—nowadays, at about 425 ppm and rising by 0.6% per year, per NOAA data (NOAA 2025)—seem to be causing a temperature rise of 0.13% per year. As a consequence, there is an increase in the amount of water in the air, ocean acidification, and coral reef dissolution.

For land plants, however, more CO_2 means an easier photosynthesis—the Calvin cycle is a CO_2-dependent process that uses RuBisCo, an enzyme complex that operates more efficiently with a higher CO_2-to-O_2 ratio (Raven and Beardall 2016). While C4 plants (from about 35 Ma ago), such as grasses, cope with "low" CO_2 by concentrating it around RuBisCo, C3 plants use the older mechanism that worked better when CO_2 concentrations were higher; these plants may thrive in a richer and warmer environment. Global warming and rising CO_2 awareness are not about "saving the planet" but about avoiding changes that are too fast for environmental adaptation without loss of diversity and the increase of extreme climatic events.

1.3 Phytobiomass Energy Reserve

As a result of the photosynthesis process driven by sunlight, solar energy is converted into chemical energy and stored in plants and vegetation, at the chemical linkage of the biomolecules synthetized, forming carbon hydrates (carbohydrates), in various structural forms, including mono-, di-, oligo-, and polysaccharides, as well as in the form of proteins, lipids, and other biomolecules:

$nCO_2 + nH_2O \rightarrow (CH_2O)_n + nO_2$, in the presence of chlorophyll and with the input of energy coming from the solar light. Thus, the molecules in the form of $(CH_2O)_n$ store energy in their atoms' linkages. This organic matter can be used as food, and at the digestion process, convert chemical energy into metabolic energy, as far as can be used as raw material for many industrial processes and fuels production.

1.3.1 Phytobiomass Sources

Phytobiomass can be classified, according to Carioca and Arora (1984), into groups related to its structural matrix, composed of cellulose, hemicellulose, and lignin, and the chemical species that is mainly accumulated in its vegetal cells, as follows:

- Saccharide energetic cultures
- Starch energetic cultures
- Oilseed energetic cultures
- Agropastoral lignocellulosic residue cultures
- Lignocellulosic natural forest cultures
- Lignocellulosic planted forest cultures
- Aquatic phytobiomass cultures
- Other solid and liquid, organic, industrial and urban residues

According to Woiciechowski et al. (2014), each phytobiomass component presents distinct and specific characteristics, as well as an specific composition that must be studied and carefully analyzed before being used or reused. This is particularly important in the case of residues applied in the development of industrial processes or new production routes that use industrial or urban residues as raw materials, within the concepts of biorefinery or circular economy.

The composition in terms of structural and energetic reserve of each phytobiomass depends not only on the phytobiomass species but also on the variety, the cultivar, the regional climate, the rainfall incidence, the stage of harvesting, and others. Each industrial process is defined to extract from the phytobiomass a specific

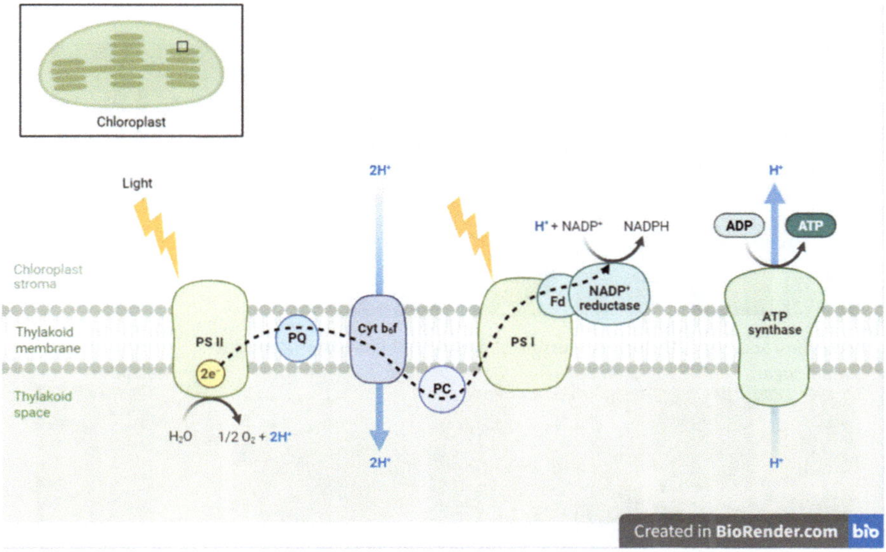

Fig. 1.1 The photosynthesis process

industrial product (juice, sugar, starch, oil, or protein), generating a residue (pulp, bagasse, husk, comb, straw, sawdust, and others), mainly of a lignocellulosic characteristic and still with a very energetic composition. Figure 1.1 represents the phytobiomass sources with their main reserve molecule and their industrial processing residues.

1.3.2 Saccharide Phytobiomass

Saccharide Phytobiomass is a very wide group of phytobiomass, including sugarcane, fruits in general, and sugar beet, with a cell structure composed of cellulose, hemicellulose, and lignin, which accumulates into the vegetal cells, a large amounts of a simple sugar, mono or disaccharide, directly fermentable or a polysaccharide easily hydrolysed. The kind of accumulated sugar depends on the species and can be a monosaccharide, such as glucose, fructose, or galactose; a disaccharide, such as sucrose formed by glucose and fructose linked by an α(1-4) glycosidic bond; a polysaccharide, such as pectin, which is a heteropolysaccharide composed mainly of galacturonic acid; and inulin, a complex polysaccharide of fructose. This kind of biomass is important because the simple sugars industrially extracted, normally as a juice, can be directly consumed or used as a carbon source for many fermentative processes, generating important industrial and commercial products, such as wine or liquor, or bioethanol through alcoholic fermentation; also, the polysaccharides can

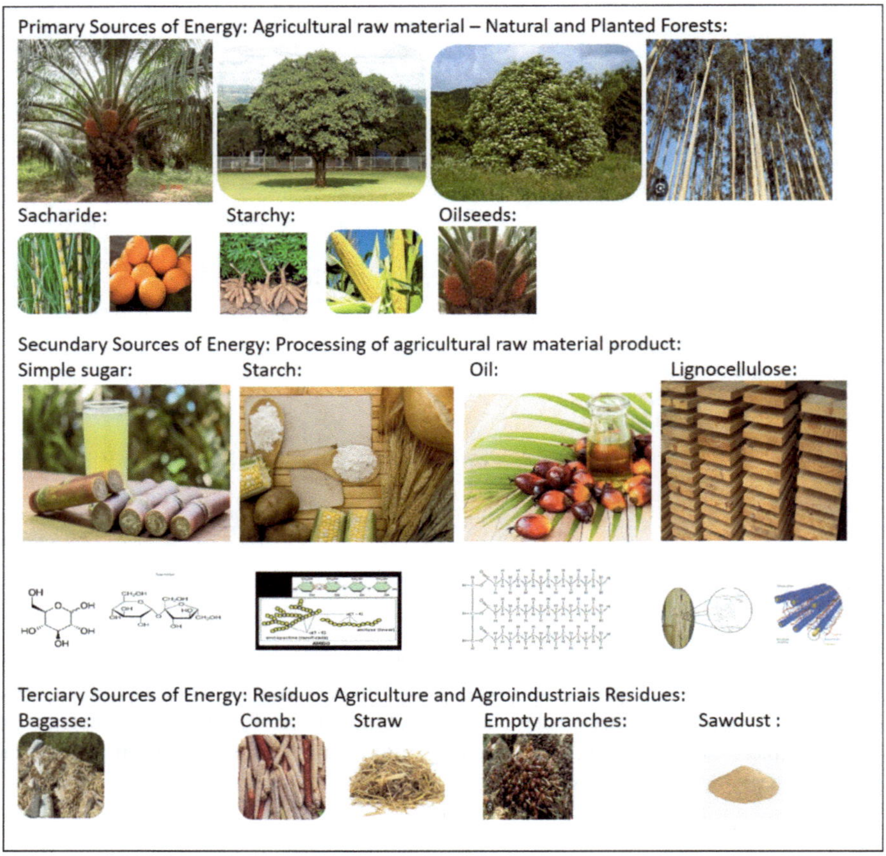

Fig. 1.2 Agricultural and agro-industrial phytobiomasses: products and subproducts

be easily hydrolysed, chemically or enzymatically, producing a sugary broth with their monomers, which can easily be fermented (Fig. 1.2).

The process of extracting the simple sugar can generate a residue with lignocellulosic characteristics that can be conveniently pretreated to recover its constituents' fractions, cellulose, hemicellulose, and lignin, which are energetic molecules with many applications in new industrial processes, such as second-generation biofuels.

Some examples of saccharide species that can be cited are (1) sugarcane: the juice is composed of sucrose, glucose, and fructose; the juice, concentrated till the crystallization point, produces sugar, which is also used as a substrate to produce bioethanol through alcoholic fermentation; the residue of the sugarcane juice extraction is the lignocellulosic sugarcane bagasse; (2) grapefruit: the juice is composed mainly of fructose used in the wine industry, which is very important economically, and the residue of the juice extraction, the grapefruit peels, is easily fermented or used to produce jams and sweets; (3) orange: orange juice is an important commercial product, and the residue of orange juice extraction is the orange bagasse composed mainly of pectin, which is easily hydrolysed into its simple sugar constituents; (4) apple: it is used for direct consumption or industrially processed to juice production; pomace, the residue of apple juice extraction, is a very rich source of pectin, vegetal proteins, and antioxidants; and (5) sweet sorghum: the sweet sorghum juice, which is rich in sugar and composed of sucrose, glucose, and fructose, can be used as a substrate for fermentation processes, and the residue of sweet sorghum juice extraction is the sweet sorghum bagasse of lignocellulosic characteristics. Sweet sorghum bagasse, like sugarcane bagasse, due to its lignocellulosic structure, can be pretreated to separate and recover cellulose, hemicellulose, and lignin. After chemical or enzymatic hydrolysis, these fractions can be reused in various applications, such as fermentation processes, as additives, or for the production of substances with biological activity, among others. This approach adds value to the lignocellulosic fraction of phytobiomass within the biorefinery concept (Woiciechowski et al. 2014; Woiciechowski et al. 2016).

1.3.3 Starch Phytobiomass

This group of phytobiomass uses starch as its source of energy, accumulating the starch in the form of insoluble and compact grains, namely crude starch, in certain parts of the plant, normally in the root. So crude starch is the main and particular form of energy source of many vegetal species, such as wheat, cassava, potato, sweet potato, and corn, among others. Starch is a polymer of glucose, which forms two structures in proportions depending on the starch source: amylose, linear, with glucose linked by glucoside linkages α(1-4), and amylopectin, branched, with chains like amylose with glycosidic linkages α(1-6) at the branches. Crude starch grains are insoluble at room temperature, but with heating, depending on the starch source and the proportions of amylose and amylopectin, around 60–70 °C, they reach the gelatinization temperature, where the grain slowly swells up, and the chains of amylose and amylopectin are released from the starch grain to the media, producing viscous solutions; with the media cooling, the chains of amylose and amylopectin remain organized and solvated, producing gels, which are very useful in the food industry. Also, the glycosidic linkages at the α position generate molecules of amylose and amylopectin with a structure that permits easy access of water into its structure, at relatively low temperatures, and an easy chemical or enzymatic hydrolysis, liberating glucose in the medium, available to be a carbon source for fermentative processes. Crude or gelatinized starch can be used as a substrate for the production and recovery of amylases using amylolytic microorganisms such as filamentous fungi (*Aspergillus*, *Pleurotus*, and *Rhizopus*), some yeasts, and bacteria (Woiciechowski et al. 2014).

The most important starch-producing plant is cassava (*Manihot esculenta*), which is directly consumed as a human and animal meal around the world and is a raw material for the industry to produce cassava flour and other industrial products.

Cassava processing residues are manipueira, the liquid residue, and the bagasse, bark, and skin, the solid residue, which are both highly biodegradable, with high organic contents, that can be used in submerged or solid fermentative processes (Woiciechowski et al. 2016).

1.3.4 Oilseed Phytobiomass

Oilseed is the phytobiomass that produces and accumulates an oil as its energetic molecule reserve, normally at its pulp, seed, or grain, in the form of triacylglycerol, triester of a fatty acid. The oil extraction generates a solid residue, a cake, formed by the solid part of the pulp, seed, or grain, along with many energetic molecules, proteins, sugar, and many bioactive molecules, which can be processed to recover these products. The refined oil extracted from them is commercially very important and is largely used by consumers and is used industrially as frying oil. Soya, oil palm, sunflower, and olive are examples of this phytobiomass, cultivated all around the world. Each oil, depending on the origin, has properties with its own composition and specific characteristics. While sunflower and olive oils are used mainly for edible purposes, more recently, oil palm and soya oils are also being used to produce biodiesel through the transesterification process, generating a large amount of glycerol as the residue of the process. Many studies are being done all around the world to use glycerol in many chemical or biological processes (Woiciechowski et al. 2016).

1.3.5 Lignocellulosic Forest Phytobiomass

Natural and planted forests, due to the worldwide cultivation and availability all over the world, are the most important phytobiomass. They are composed mainly of cellulose, hemicellulose, and lignin in a very typical and stable structure, where molecules of crystalline cellulose are juxtaposed in layers, forming stable fibres, involved by molecules of amorphous hemicellulose and the entire structure is immersed in a matrix of lignin. This structure is responsible for the chemical, mechanical, and biological properties, giving recalcitrance to the lignocellulosic complex. This phytobiomass is the classic raw material for the pulp and paper industry to produce cellulose pulp and paper. Depending on the species, a specific extract, a component of low molecular weight, and ashes are present. This kind of phytobiomass can be categorized into two groups: hardwood, including those from trees with broad leaves (mahogany, oak, and eucalyptus), and softwood, which includes those from coniferous trees (pines, cedar, and ipê). Due to the amount present all around the world and the diversity, forest phytobiomasses represent a great source of energy for many applications. It is also the classical raw material for the important industrial sector of paper and cellulose pulp production, carried out through well-established thermochemical processes, such as alkaline sulphite, kraft, Stora, and bisulphite, among others. These processes differ in processing time, temperature, and the chemical agents used. In the industrial processing of phytobiomass for cellulose pulp production, due to the thermochemical conditions used, the hemicellulose, lignin and extractive of the wood are solubilized while the cellulose fraction remains insoluble. However, different cellulose yields can be achieved depending on the process conditions used. The liquid residue from the process is the black liquor, which is a very aggressive effluent that is burned to produce steam, an alternative fuel for the process.

Many alternatives of thermochemical pretreatments in mild conditions are being studied to efficiently disrupt the lignocellulosic structure and recover the three main fractions of the phytobiomass, involving a first thermal pretreatment step in acid conditions to hydrolyse and recover the sugars of hemicellulose and then a second thermal pretreatment step in basic conditions to soften and solubilize the lignin, preserving the cellulosic fraction at the solid phase, which can be enzymatically hydrolysed to glucose. So the syrup with the sugars of hemicellulose and the glucose from cellulose can be used for the development of bioprocess media, even to reach an economical second-generation biofuel production; Lignin, due to its biological properties, can be used to control the growth of bacteria, fungi, and parasites, as well as in other applications ranging from rubber to plastic additives.

1.3.6 Lignocellulosic Residue Phytobiomass

According to Fig. 1.1, agricultural residues, those generated in the field after the culture harvest, such as leaves, branches, and straw, are in part left in the field to protect the soil and in part can be collected to reuse, while the agro-industrial residues, generated in the industry during the industrial processing of the agricultural raw material to extract the simple sugar, the starch, or the oil are seen not as a residue but as a subproduct, still very important energetically, because it is composed mainly of very important molecules, cellulose, hemicellulose, and lignin.

As said above, despite their energetic characteristics, to recover those main phytobiomass components, due to their recalcitrant structure, a suitable pretreatment must be defined, and each industrial processing by-product will have a pretreatment that is more efficient technically and economically than the other, so each process must be studied to define the best operational conditions of time, temperature, and chemical catalyst for the pretreatment.

1.3.7 Aquatic Phytobiomass

Microalgae (e.g. *Chlorella* and *Nannochloropsis*) are unicellular, fast-growing phytobiomass with efficient CO_2 fixation. They often attain a very high protein content (50–70% DW) and can accumulate substantial lipids under stress, making them oleaginous feedstocks (García-Encinas et al. 2025), and the cultivation on non-arable land or in wastewater allows nutrient recycling. Microalgal biomass can be converted to biodiesel from the lipids, bioethanol from carbohydrates, and methane or hydrogen via dark fermentation. It is also used as a protein-rich feed and as a source of nutraceuticals (omega-3 fatty acids and pigments) (Peter et al. 2021).

Macroalgae or seaweeds, e.g. *Ulva*, *Gracilaria,* and *Kappaphycus*, are multicellular aquatic plants with polysaccharide-rich cell walls (alginate, agar, and carrageenan) and virtually no lignin (Sharma et al. 2020). They are carbohydrate- and mineral-rich (median ~48.5% CHO and 31.8% ash), with modest protein (~10% DW) and low lipids (~3%). For example, *Kappaphycus* biomass contains ~6.4–54.3% carbohydrates, ~2.7–23.6% proteins, and only ~0.19–0.91% lipids (Ajik and Tahiluddin 2024; Duda et al. 2024; Kumar et al. 2014). These polysaccharides can be hydrolysed and fermented to bioethanol, such as from *Gracilaria* agar, or fed to anaerobic digesters for biogas, and extracted lipids can be converted to biodiesel. Seaweeds also furnish bioplastic precursors and industrial biopolymers, for example, alginate, agar, or cellulose (Elkaliny et al. 2024), and are exploited as feed supplements and in wastewater bioremediation.

Free-floating aquatic plants, such as macrophytes, like water hyacinth (*Eichhornia crassipes*) and duckweed (*Lemna minor*), grow rapidly, yielding abundant biomass. Duckweed is notable for high protein (~20–45% DW) and starch, and it efficiently removes nutrients in wastewater (Xu et al. 2023). This biomass can be anaerobically digested or fermented to biogas and bioethanol. Aquatic plants are also used as animal feed and can be processed for biohydrogen or bioplastic production. In sum, aquatic phytobiomass provides carbohydrate- and lipid-rich biomass for a range of biofuels, such as biodiesel, bioethanol, biogas, or biohydrogen, and bioproducts, like feed, bioplastics, and nutraceuticals (Wang et al. 2021).

1.3.8 Cellulose, Hemicellulose, and Lignin

Due to the importance related to the amount available all around the world and the energetic power stored in the molecular structures in the form of chemical energy at the linkage between the atoms, it is worth briefly remembering the structure of these three important molecules.

Cellulose is a crystalline structural homopolysaccharide of glucose found in the vegetal wall cells. Without a bone skeleton, all the vegetal weight is supported by its wall cells, from grasses to big trees. Unlike the α(1–4) and α(1–6) linkages present in starch, that form spiral and branched structures, the β(1–4) glycosidic linkages form a linear molecule. Together with the axial position of the glucose hydroxyl groups, this structure enables cellulose molecules to align closely and form highly recalcitrant cellulose fibers. These fibers are stabilized by intra- and intermolecular hydrogen bonding, making them highly resistant to chemical and biological degradation (Conn and Stumpf 1980; Woiciechowski et al. 2014; Lima et al. 2016).

Hemicellulose is an heteropolysaccharide with amorphous character, non-starchy and non-cellulosic, of low molecular weight, at the vegetal cell is tightly connected to the cellulose by hydrogen bonding, also known as a structural polysaccharide in plants, formed by pentoses (arabinose and xylose), hexoses (mannose, galactose, and glucose), deoxyhexoses (fucose and rhamnose), and uronic acids (glucuronic, galacturonic, and methyl glucuronic), which, due to its branched structure, can be easily hydrolysed, producing a sugar syrup that can be used as a substrate for fermentative processes using a suitable microorganism for the simple sugar present (Woiciechowski et al. 2014; Lima et al. 2016).

Lignin is a complex and amorphous polymer formed by units derived from phenyl propane (sinapyl, coniferyl, and p-coumaryl alcohols), linked by C–C and ether linkages, with different degrees of methylation and acetylation, producing a very stable polymer that coats the structural polymer of the vegetal cells, improving the stability and recalcitrance of the vegetal cell structure (Woiciechowski et al. 2014; Lima et al. 2016).

1.4 Importance of Phytobiomass

Due to its complex and diverse structure, phytobiomass is essential for the survival of life on Earth. Food and energy are supplied mainly by phytobiomass. It is continually being produced by photosynthesis due to these two important factors, light and chlorophyll, which is seen as fundamental phytobiomass synthesis. Besides, the use of CO_2 as a carbon source to synthesize the complex molecules, such as simple sugar, polysaccharides, starch, lipids, oil, cellulose, hemicellulose, and lignin, that will compose the structure of phytobiomass is essential to preserve the atmospheric quality, allowing the survival of human life.

Continuous CO_2 removal from the atmosphere by photosynthetic and primitive microorganisms since the most rePmote times of the Earth's evolution has slowly changed the Earth's atmosphere, from an inappropriate mixture of gases, unfeasible for human life, to a breathable atmosphere.

So it is undeniable that, over time, since the most remote ages, photosynthetic organisms have been producing an atmosphere that could permit life as we know it nowadays. Besides, these organisms kept evolving, improving, and producing tissue differentiation from a simple and primitive cell until the huge diversity of plants that we know nowadays, giving important raw materials for many industrial processes, providing us a great diversity of food, energy, fuel, light, heat, clothing, shelter, and so on.

1.5 Conclusions

When it comes to food, feed, raw industrial material, energy, and biofuel, seen from any point of view or any perspective, it is impossible to think of the world nowadays, the life as we know it, and the survival of humanity without biomass, more specifically, phytobiomass utilization. The great amount and diversity of components can provide the living organisms and microorganisms with all carbon and energy sources for all their metabolic needs.

Produced through photosynthetic processes, phytobiomass is essential for life and its basic necessities. It can be generated wherever sunlight is available even on a minimal leaf surface, in organisms or microorganisms containing chlorophyll, and is distributed worldwide.

Besides, as phytobiomass components are produced from CO_2 and H_2O, they are continuously removing a great amount of carbonic gas from the atmosphere and acting as a natural reservoir of this gas, which is important to maintain the quality of the atmosphere.

References

Ajik KO, Tahiluddin AB (2024) Proximate composition and heavy metal contents of edible seaweeds *Kappaphycus alvarezii* and *Caulerpa cf. macrodisca ecad corynephora*. Akademik Gıda 22(1):43–50

Bar-On YM, Milo R (2019) The biomass composition of the oceans: a blueprint of our blue planet. Cell 179(7):1451–1454. https://doi.org/10.1016/j.cell.2019.11.018

Bar-On YM, Phillips R, Milo R (2018) The biomass distribution on Earth. Proc Natl Acad Sci 115(25):6506–6511. https://doi.org/10.1073/pnas.1711842115

Carioca JOB, Arora HL (1984) BIOMASSA: Fundamentos e Aplicações Tecnologicas. Federal University of Ceará. Brazil. 644 p.

Catling DC, Zahnle KJ (2020) The Archean atmosphere. Sci Adv 6(9):eaax1420. https://doi.org/10.1126/sciadv.aax1420

Conn EE, Stumpf PK (1980) Traduction from the original in English: Outlines of Biochemistry. John Wiley ans Sons, Inc. USA. 1976: Introdução à Bioquimica. Edgard Blucher Editor Ltda. São Paulo, 525 p.

Crockford PW, Bar-On YM, Ward LM, Milo R, Halevy I (2023) The geologic history of primary productivity. Curr Biol 33(21):4741–4750.e5. https://doi.org/10.1016/j.cub.2023.09.040

Duda LJ, Woicichowski AL, Medeiros MC, Scapini T, Soccol CR (2024) Optimizing hydrolysis of Kappaphycus alvarezii for fermentable sugars: a potential feedstock for bioethanol production. In: Proceedings of the 1st Brazilian Congress in industrial biotechnology, Florianópolis

Elkaliny NE, Alzamel NM, Moussa SH, Elodamy NI, Madkor EA, Ibrahim EM, Elshobary ME, Ismail GA (2024) Macroalgae bioplastics: a sustainable shift to mitigate the ecological impact of petroleum-based plastics. Polymers 16(9):1246. https://doi.org/10.3390/polym16091246

García-Encinas JP, Ruiz-Cruz S, Juárez J, Ornelas-Paz JDJ, Del Toro-Sánchez CL, Márquez-Ríos E (2025) Proteins from microalgae: nutritional, functional and bioactive properties. Foods 14(6): 921. https://doi.org/10.3390/foods14060921

Hahn A, Hummer G, Kuhlbrandt W (2018) Structure, mechanism, and regulation of the chloroplast ATP synthase. Science 360(6389):eaat4318. https://doi.org/10.1126/science.aat4318

Kasting JF, Catling DC (2003) Evolution of a habitable planet. Annu Rev Astron Astrophys 41: 429–463. https://doi.org/10.1146/annurev.astro.41.071601.170049

Kumar S, Kumari P, Reddy CRK, Jha B (2014) Seasonal variation in nutritional composition of *Kappaphycus alvarezii* (Doty) Doty – an edible seaweed. Food Chem 157:368–374

Lima DR, Silveira MHL, Rio LD, Ramos LP (2016) Pretreatment processes for cellulosic ethanol production: processes integration and modelling for the utilization of lignoellulosics such as sugarcane straw. In: Soccol CR, Brar SK, Faulds C, Ramos LP (eds) Green fuels technology. Springer, Cham, pp 107–131

NOAA – National Oceanic and Atmospheric Administration (NOAA) (2025) Trends in atmospheric carbon dioxide. NOAA Global Monitoring Laboratory. https://www.esrl.noaa.gov/gmd/ccgg/trends/. Accessed 19 Aug 2025

Peter AP, Khoo KS, Chew KW, Ling TC, Ho S-H, Chang J-S, Show PL (2021) Microalgae for biofuels, wastewater treatment and environmental monitoring. Environ Chem Lett 19(4): 2891–2904. https://doi.org/10.1007/s10311-021-01219-6

Raven JA, Beardall J (2016) The ins and outs of CO_2. J Exp Bot 67(1):1–13. https://doi.org/10.1093/jxb/erv451

Sharma S, Sharma A, Mulla SI, Pant D, Sharma T, Kumar A (2020) Lignin as potent industrial biopolymer: an introduction. In: Sharma S, Kumar A (eds) Lignin. Springer, pp 1–15. https://doi.org/10.1007/978-3-030-40663-9_1

Wang K, Khoo KS, Chew KW, Selvarajoo A, Chen W-H, Chang J-S, Show PL (2021) Microalgae: the future supply house of biohydrogen and biogas. Front Energy Res 9:660399. https://doi.org/10.3389/fenrg.2021.660399

Woiciechowski AL, Medeiros ABP, Rodrigues C, Vandenberghe LPS, Tanobe VOA, Dall'Agnol A, Gazzoni DL, Soccol CR (2016) Feedstocks for biofuels. In: Soccol CR, Brar SK, Faulds C, Ramos LP (eds) Green fuels technology. Springer, Cham, pp 15–39. https://doi.org/10.1007/978-3-319-30205-8

Woiciechowski AL, Karp SG, Sobral K Carvalho JC, Letti LAJ, Soccol VT, Soccol CR (2014) Pretreatment Strategies to Enhance Value Addition of Agro-industrial Wastes. In: Biotransformation of Waste Biomass into High Value Biochemicals. Edited by: Satinder Kaur Brar; Gurpreet Singh and Carlos Ricardo Soccol. Springer, New York, pp 29–49

Xu J, Shen Y, Zheng Y, Smith G, Sun XS, Wang D, Zhao Y, Zhang W, Li Y (2023) Duckweed (Lemnaceae) for potentially nutritious human food: a review. Food Rev Int 39(7):3620–3634. https://doi.org/10.1080/87559129.2021.2012800

Chapter 2
Saccharide Biomasses

Rafaela de Oliveira Penha, Walter José Martinez-Burgos, Jessica Carolina Medina Gallardo, Willian Pereira Gomes, Desireé Soares da Silva, and Leonardo Wedderhoff Herrmann

Abstract Saccharide biomasses that produce mono- and disaccharides play a crucial role in food, biofuel, and biochemical industries. Understanding the structure and properties of these saccharides is essential to optimizing their utilization. Monosaccharides, such as glucose and fructose, and disaccharides, like sucrose, have distinct physicochemical characteristics that influence their behavior in biological and industrial processes. Several plant species, including sugarcane, sugar beet, and sweet sorghum, are primary sources of these sugars, each with unique accumulation mechanisms. Traditional processing methods, such as extraction and purification, have been refined over time to improve efficiency and sustainability. The economic aspects of saccharide production highlight its significance in global markets, with key players like Brazil, India, and the United States. In this context, saccharide biomasses remain vital resources for multiple industries, reinforcing the need for continued research and innovation to enhance their production and application. This chapter focuses on mono- and disaccharide-rich biomasses, providing a comprehensive overview of the structure and properties of these saccharides, the main plant species that accumulate these sugars and the techniques of processing them, the economic aspects, and finally, the bio- and by-products originating from saccharide biomasses.

R. de Oliveira Penha (✉) · W. J. Martinez-Burgos · J. C. M. Gallardo
Bioprocess Engineering and Biotechnology Department, Federal University of Paraná, Centro Politécnico, Curitiba, Paraná, Brazil
e-mail: rafaelapenha@ufpr.br

W. P. Gomes · D. S. da Silva
SENAI Innovation Institute for Biomass, Três Lagoas, Mato Grosso do Sul, Brazil

L. W. Herrmann
Embrapa Food Agroindustry, Guaratiba, Rio de Janeiro, Brazil

A. L. Woiciechowski et al. (eds.), *Biomass for Bioenergy and Bioproducts*, Biofuel and Biorefinery Technologies 17, https://doi.org/10.1007/978-3-032-19898-3_2

2.1 Introduction

Increased anthropogenic greenhouse gas (GHG) emissions are the main driver of global environmental change and its impacts on human populations and biodiversity. Yet the world's energy portfolio remains dominated by fossil fuels. The Paris Agreement was established as a global strategy in the 21st Conference of the Parties (COP21) by the United Nations Framework Convention on Climate Change (UNFCCC) in order to deal with climate change and limit the rise in temperature due to greenhouse gases. The terms of this agreement required each nation to commit to specific reductions in CO_2 emissions and individual strategies for actually reducing those outputs (Vandenberghe et al. 2022). In this scenario, biomass plays an important role.

Biomass is all organic matter derived from organisms, including plants, microorganisms, and animals. It serves as a renewable resource, being a part of the natural and recurring cycles of ecological processes. Specifically, plant biomass, named phytobiomass, originates through the mechanism of photosynthesis, converting solar energy into chemical energy stored within cellular structures as organic matter. This mechanism is crucial for sustaining ecological balance and promoting the preservation and enhancement of both biological and soil diversity (Flórez Pardo et al. 2018). In contrast to other sustainable energy sources like wind, solar, geothermal, marine, and hydropower, biomass has the unique capability to directly generate fuel, in addition to chemicals (Osman et al. 2021).

The composition of plant biomasses includes saccharides, polysaccharides, lignin, and other substances, including fats and proteins. Saccharides are important feedstocks utilized in different industries, such as energy, food, cosmetics, and pharmaceuticals, and fermentation processes. These compounds are categorized into four types: monosaccharides (degree of polymerization) ($DP = 1$), disaccharides ($DP = 2$), oligosaccharides ($DP = 3$ to 9), and polysaccharides ($DP > 10$) (Li et al. 2021).

This chapter focuses on mono- and disaccharide-rich biomasses, providing a comprehensive overview of the structure and properties of these saccharides, the main plant species that accumulate these sugars and the techniques of processing them, the economic aspects, and finally, the bio- and by-products originating from saccharide biomasses.

2.2 Structure and Properties of Saccharides

The saccharides, also known as carbohydrates or sugars, comprise a group of chemical compounds that contain almost exclusively carbon, oxygen, and hydrogen. These compounds are presented in nature both in the form of small, separated units and the polymerized union of those units into long and complex structures. A single monomer of a carbohydrate is named a monosaccharide, with glucose, fructose, and galactose being the main representatives. Combining two monosaccharides with a specific type of binding named glycosidic bond results in a disaccharide, such as sucrose, lactose, and maltose. Both monosaccharides and disaccharides are among the oligosaccharides, which are composed of a few units of the same monomer or a few different monomers. When several monosaccharides are incorporated into a large chain with sequential glycosidic bonds, usually more than 20 units, they become a polysaccharide, such as the commonly found cellulose, starch, and glycogen. While polysaccharides are excellent for energy storage and structural purposes, oligosaccharides are the preferred option in terms of fast energy usage and transportation. The main focus of this chapter is oligosaccharides and how they can be directed toward biotechnological purposes and value-added products (Nelson and Cox 2017; Seeberger 2017; Niaz et al. 2020).

The general appearance of sugars is as a colorless crystalline solid, highly soluble in water and having a neutral pH. Regarding their chemical properties, monosaccharides are essentially one string of three to eight carbons connected by single covalent bonds without ramifications, each of them having a hydroxyl group, classifying the structure as a polyhydroxy aldehyde or ketone, or polyalcohol. Among the carbons, one of them presents an aldehyde or ketone (the carbonyl group) instead of a hydroxyl, which divides the monosaccharides into aldoses and ketoses. For the same number of carbon atoms, there is both an aldose and a ketose, and these functional groups are especially important for forming a cycle in the organic structure. The unit can also be named after the number of carbons it contains as tetrose, pentose, hexose, and heptose containing four to seven carbons, respectively (Nelson and Cox 2017; Gunasekara and Zhao 2017).

Across the central scaffold, as almost every covalent bond of the four that the carbon can make is connected to a different radical, each carbon atom is chiral or asymmetric, except the carbonyl one. This confers two different tridimensional conformations, called enantiomers, for every asymmetric carbon that is in the compound. For instance, one aldose with six carbons has four chiral centers, allowing 16 different enantiomers for the same structure. As a standard, the classification method in these situations refers to the carbon that is most distant from the aldehyde/ketone group and compares the enantiomer to glyceraldehyde, which has only one asymmetric carbon. If the carbon of the compound is similar to that of the D-glyceraldehyde, it is classified as D (dextrorotatory), and if it is similar to that of the L-glyceraldehyde, it is classified as L (levorotatory). The specificity of the tridimensional conformation of each monosaccharide provides more selectivity to the enzymes that manage sugars, turning the biological processing systems especially precise in terms of chirality (Zhang et al. 2018; Yao et al. 2023).

When the monosaccharide has more than four carbons, instead of remaining in an open chain, it forms a predominantly cyclic structure in aqueous solution. This results from a reaction between the carbonyl and one of the alcoholic hydroxyls, removing a water molecule and binding both extremities into a ring. The ring that is similar to a hexagon is called pyranose, and the one related to a pentagon is named furanose. The bond, identified as hemiacetal or hemiketal reaction, creates a new asymmetric carbon where the carbonyl was previously located, the anomeric carbon, having two possible enantiomeric configurations, α and β, according to the position of the remaining hydroxyl. Both the carbonyl group in the open chain form and the anomeric carbon in the cyclic form are capable of being oxidized by other chemicals, which classify those carbohydrates as reducing sugars. Simultaneously, whenever in the cyclic form, the pyranose or furanose do not remain flat due to conformational stability, causing them to slightly fold into a more stable chair-like tridimensional figure (Solomons and Fryhle 2011; Nelson and Cox 2017).

The hemiacetal or hemiketal bond is also capable of performing another reaction, which can occur between the α or β hydroxyl and another molecule of the monosaccharide, regardless of the carbon of the second unit. The new bond connecting two monosaccharides is named O-glycosidic and forms a disaccharide. The α or β nomenclature comes from the monosaccharide, where the anomeric carbon was used for the reaction. As the reducing terminal of hemiacetal or hemiketal is now bound to another unit, the first monosaccharide is not capable of oxidizing, and if the second unit also utilizes the anomeric carbon for the bond, the resulting disaccharide is no longer a reducing sugar. If other glycosidic bonds are added to the disaccharide, and the chain continues growing, it forms a polysaccharide. Usually, the polysaccharide has only one reducing extremity, and the longer the number of monomers, the higher the effect on properties such as solubility, crystallinity, and reactivity (Schindler et al. 2017; Liu et al. 2021).

Additionally, the monosaccharide structure can have some of the covalent bonds replaced by other organic groups, such as amine, amide, carboxylic acid, or other carbonyls. Glucosamine, glucose-6-phosphate, muramic acid, glucuronate, and galactosamine are examples of modifications made in the organic groups across the scaffold, each of them with different metabolic purposes. The absence of a hydroxyl group also gives rise to other compounds, such as fucose, rhamnose, and deoxyribonucleic acid, part of the DNA strand. Furthermore, when the aldehyde or the last carbon is oxidized into a carboxylic acid, aldonic acids and uronic acids are formed, respectively, with acid properties and several pharmacological applications (Wang et al. 2017; Elferink et al. 2019).

Regarding biological properties, monosaccharides are the main source of energy for all living beings, especially glucose. The anaerobic metabolic pathway for energy consists basically of glycolysis, followed by alcoholic fermentation, lactic acid fermentation, or acetic acid fermentation, wherein glycolysis plays a pivotal role in breaking glucose and liberating energy in the form of 2 ATP, while fermentation balances the electrons and cofactors. For aerobic beings, glycolysis is also the starting point, followed by the Krebs cycle and the electron transport chain, extracting more than 30 ATP per glucose. Complementary pathways, such as the phosphate pentose pathway and the feeder pathways for glycolysis, allow the monosaccharides to be interconvertible among themselves, giving rise to precursors for other biological substances and/or transforming other units, for example, galactose, into intermediates of glycolysis, mainly glucose. Besides energy, oligosaccharides are the main source of carbon for all biological molecules, from proteins, lipids, and genetic material to other carbohydrates, being essential for supplementation in all heterotrophic diets. In microbial systems, monosaccharides are considerably faster in terms of assimilation and bioavailability, having their own transport through the membrane, as an example of the phosphotransferase system (PTS), which imports and phosphorylates monosaccharides simultaneously. The use of polysaccharides requires extracellular enzymes to break the complex chain into oligomers before being transported into the cell (Nelson and Cox 2017; Wang et al. 2024).

2.3 Main Plant Species That Accumulate Simple Sugars

Monosaccharides, such as glucose and fructose, and disaccharides, like sucrose, hold critical and essential functions in plant metabolism. These sugars are not only important sources of energy but also perform as signaling molecules and carbon reserves, which regulate plant growth, development, and responses to both abiotic and biotic stress. The ability of plants to accumulate sugars is controlled by enzymatic pathways regulating sugar synthesis, transport, and storage. Moreover, genetic factors, environmental conditions, and developmental aspects dictate the type and amount of monosaccharide that will be accumulated in any given plant. It is also the case, however, that some plants have evolved to accumulate and store sugars as part of their strategies for surviving in certain ecological niches. Others have been selectively bred for higher sugar content to meet human consumption requirements and also industrial demands, like commercial sugar extraction.

Below, we provide a detailed overview of the main plant species that accumulate simple sugars, their mechanisms of sugar accumulation, their role as primary energy sources, and their economic and ecological significance.

*2.3.1 Sugarcane (*Saccharum officinarum*)*

Sugarcane is one of the most efficient plants for sucrose production, which is the main sugar present in this plant. In fact, sucrose content in sugarcane typically ranges between 10% and 15% of the mature plant's weight. In addition to sucrose, it also contains smaller amounts of monosaccharides, including glucose and fructose. It is a tropical grass that belongs to the C4 group, ensuring efficient photosynthesis even at high temperatures and under intense light conditions. The sucrose accumulation in sugarcane is regulated by the activities of several enzymes, mainly by sucrose phosphate synthase (SPS), sucrose phosphate phosphatase (SPP), and invertase, which regulates sucrose breakdown (Wang et al. 2013). Uptake of sucrose into the parenchyma cells of stalks is also facilitated by sucrose transporters (SUTs), which maintain a high gradient between storage tissue and cell wall sugars (Rae et al. 2005). In sugarcane, it is sucrose that serves as the main energy reserve to fuel metabolic activities like cell division, growth, and the development of structure. It reaches its maximum sucrose content during the later stages of development and just before harvest.

Sugarcane is thought to have originated in Southeast Asia and the Pacific and South Pacific islands, where it was first domesticated thousands of years ago. Most of the sugarcane is produced in more than 100 countries throughout the world, with the present global annual output exceeding 2.0 billion tons, according to the Food and Agriculture Organization (FAO 2023). The main producers are Brazil, India, China, Thailand, and Pakistan. Brazil, for example, is responsible for 40% of the world's total sugarcane production, followed by India with 23% and China with 5%. Sugarcane growing is largely confined to tropical and subtropical regions, where warm weather, sunlight, and water resources are all in abundant supply, allowing for high yields per hectare and, consequently, high production of simple sugars, such as sucrose.

Currently, sugarcane represents a major driver of the world economy and daily life for many countries. By number alone, it provides jobs for millions of people, including in farming, processing, and other derived industries. It is a key raw material in the production of sugar, ethanol, and biofuels. Consequently, to develop these sectors, sugarcane clones with desirable industrial traits have emerged during the past decades from modern biotechnological tools. Genetic engineering, CRISPR-based genome editing, and marker-assisted breeding have made it possible to select traits that make sugarcane varieties more profitable for use in industrial applications (Li and Iqbal 2024).

Studies have shown that genetic manipulation, such as overexpression of SPS genes, sucrose transporter genes, and downregulation of invertase genes, could lead to higher sucrose accumulation in commercial sugarcane clones (Khan et al. 2023; Mehdi et al. 2024b). These advanced tools help select traits that enhance not just the quantity but also the quality of biomass for industrial applications. Higher sucrose content, enhanced stress and disease resistance, and improved processing efficiency are some of the characteristics that ensure sustainable production.

Regarding environmental conditions, sugar accumulation in stalks is influenced by sunlight exposure, temperature, soil moisture and fertility, water availability, and the timing of harvest. While optimal temperatures around 25–35 °C are favorable for photosynthesis and sugar production, excessive heat increases respiration in plants, and consequently, it can reduce sugar production and accumulation. Although sufficient irrigation guarantees growth, controlled drought stress at ripening improves sucrose concentration by directing carbohydrate allocation away from vegetation into sugar storage. Sugar content is strongly influenced by light intensity; high solar radiation stimulates photosynthesis and sugar synthesis, while cloudy conditions reduce carbon fixation and, thus, sugar accumulation. Soil nutrient composition and management, mainly nitrogen and potassium levels, optimize sugar storage, as shown by Desalegn et al. (2023). Last, the timing of harvesting the plant is extremely important and can significantly affect the overall sugar yield; in fact, harvesting at peak maturity ensures optimal sucrose formation. On the other hand, harvesting too late significantly degrades the sugar content through both enzymatic activity and microbial activity.

2.3.2 *Sugar Beet (*Beta vulgaris*)*

Sugar beet is an important C3 plant and the primary source of sucrose in temperate climates. Roots of sugar beet contain an average of 18%–20% sucrose on a fresh weight basis (Panella et al. 2015). In addition to sucrose, beets contain glucose and fructose; nevertheless, these sugars are present in much lower concentrations compared to sucrose. It is predominantly grown for sugar extraction, and it is a key component in the worldwide sugar industry, which contributes significantly to food production and the agricultural economy. Unlike sugarcane, sugar beet stores simple sugars such as sucrose in its taproots, which develop through active sugar partitioning during vegetative growth. Sugar beet primarily utilizes its stored sucrose to maintain metabolic homeostasis and cell energy and respond to environmental stresses. It also provides energy reserves necessary for regrowth when harvested incompletely. Genetic characteristics of sugar beet have a key effect on sucrose accumulation. Recent advances in breeding techniques, including conventional crossbreeding and molecular breeding methods, have led to the establishment of high-yielding and high-sucrose-type varieties. The key enzymes involved include the regulation of sucrose phosphate synthase (SPS) and the activity of fructokinase, while sucrose degradation by invertase is downregulated to maintain high sucrose concentrations (Jammer et al. 2020). Characteristics such as leaf area index, root morphology, and photosynthetic efficiency play an essential role in sucrose yield in sugar beet. Research in recent years has identified hundreds of quantitative trait loci (QTLs) associated with sugar content so that they can be used for genetic selection (Schneider et al. 2002).

Physical and chemical properties of soil, especially the nutrient availability, are important factors in determining sugar beet growth. Well-drained, rich soils promote the growth of the plant and the development of its roots. Sugar beets are generally grown in a temperate climate with moderate rainfall. Moreover, cold temperatures during late growth stages enhance sucrose accumulation. By contrast, high temperatures will lead to premature bolting of the sugar beet, which has a negative effect on both sucrose content and its quality, reducing the productivity of the crop. In relation to the water supply, drought stress or waterlogging would have a negative effect on sucrose concentration since it directs the sugar beet's energy toward survival (Kroupin et al. 2023). However, studies have shown that water stress can slightly increase the sugar content as a physiological adaptation response, but this potential benefit is often outweighed by the overall negative impact of water stress on root yield and quality (Ebmeyer and Hoffmann 2020).

Sucrose content of sugar beet is also greatly influenced by the developmental stage of the plant (Jammer et al. 2020). The accumulation of sugars takes place mainly during the late phase of the growth cycle, when plants switch from vegetative development to storage organ formation (Kroupin et al. 2023). Harvest timing is critical; harvesting too early will reduce the sucrose content; nevertheless, harvesting too late leads to sugar degradation and reduces the quality of the crop.

Sugar beet is a major agricultural commodity, especially in the European and North American context, and is one of the key raw materials in the global sugar value chain. Most of the economic effects of sugar beet production come not from the sugar produced but from the number of jobs created in rural communities and the development of other sectors due to sugar beet production, such as bioethanol and livestock feed from sugar beet residues.

2.3.3 *Sweet Sorghum (***Sorghum bicolor***)*

Sweet sorghum is a high-yielding biomass crop species generally known for its high accumulation of fermentable sugars in stalks. Due to its capacity to adapt to different environmental conditions and shorter growth cycle than sugarcane and relatively lower inputs, it is nowadays considered an important and ideal feedstock for biofuel production, especially in areas unsuitable for sugarcane production (Rooney et al. 2007; Mathur et al. 2017).

Sweet sorghum is distinct from grain sorghum, where the grain is the primary focus of the crop; rather, sweet sorghum is selected for the juice of its stalks, which is mainly rich in sucrose, glucose, and fructose. Its juice generally contains 10%–20% of fermentable sugars, with a Brix value that has been reported to range from 16 to 23, which compares to the sugar content found in sugarcane (Kawahigashi et al. 2013). The main sugars in sweet sorghum juice are sucrose, comprising 50%–70% of the total sugar content, and glucose and fructose with 20%–30% and 10%–20%, respectively. Different from sugarcane, sweet sorghum contains a higher fraction of reducing sugars, which makes it more fermentable for ethanol production, as reducing sugars can be directly fermented by yeast.

As in the other crops studied here, the sugar content in sweet sorghum also varies depending on the genetic, environmental, and developmental conditions in which it is grown. Genetic variation for sugar yield, juice, and biomass production among sorghum cultivars is particularly large. Some cultivars are specifically bred for their high sugar content, and the translocation and storage of sugar are regulated by key metabolic enzymes, including sucrose phosphate synthase (SPS), sucrose synthase (SUS), and invertases (Siddique et al. 2018). As a C4 plant, sweet sorghum has high photosynthetic efficiency, allowing for greater carbon fixation and sugar production relative to C3 crops such as sugarcane or wheat.

From an environmental perspective, many factors have a great impact on sugar accumulation in sweet sorghum. Temperature is one of the most important factors that must be controlled, as storage and syrup yields have optimal temperature ranges between 25 and 35 °C. Thus, heat stress can decrease photosynthetic efficiency, induce senescence of the plant, and consequently, lead to much lower sugar yields. Besides, water availability plays a complex role, as moderate drought stress can increase sugar accumulation by preventing excessive vegetative growth and directing more of this carbon into storage in the stalks (Mastrorilli et al. 1999). However, severe and prolonged droughts negatively affect the overall biomass and sugar yields. Soil fertility is also essential; nitrogen fertilization supports vegetative growth, but when applied in excess, it can accentuate the deviation of carbohydrates from storage as sugars into the leaves and grains. Research has shown that increased soil salinity and reduced irrigation have a direct impact on sweet sorghum's growth; they directly impact photosynthetic efficiency and sugar accumulation, especially in arid regions (Sun et al. 2024b). Moreover, photoperiod sensitivity influences flowering time and, consequently, sugar accumulation. Genotypes with delayed flowering typically accumulate higher sugar concentrations due to a prolonged vegetative phase, which enhances carbon assimilation and storage. The peak accumulation of sugars within the stalk occurs during the reproductive phase, particularly around the soft dough stage of grain development (Mathur et al. 2017). Therefore, harvesting at this point ensures maximum sugar yield and juice quality.

2.3.4 Fruits

Fruits are vital foods for humans, as they are often rich sources of vitamins, minerals, fiber, and sugars. They are important for nutrition and energy supply, which are essential for general health and well-being. The composition of fruits is highly species-dependent, with some fruits having a particularly high sugar content; fruits are therefore important not only for direct consumption but also for industrial applications, such as juice, wine, and dried fruit production.

Among the vast diversity of fruits, apples (*Malus domestica*), oranges (*Citrus sinensis*), and grapes (*Vitis vinifera*) play a vital role in humans' daily life, offering essential sugars that contribute to their taste, energy content, and market value. There are big differences in sugar concentrations between different species and cultivars of these fruits. Apples, for instance, are composed of 10%–16% of total sugars; 'Royal Gala' contains up to 11%, while the 'Golden Delicious' variety contains between 13% and 16% (Temma et al. 2002). 2002). The main sugars in apples are fructose, glucose, and sucrose, with fructose typically being the most abundant. Oranges usually have less sugar than apples, ranging from 8% to 12% of the fruit's weight. The main sugar is sucrose, followed by fructose and glucose. Grapes, by contrast, tend to store higher sugar concentrations, typically ranging from 15% to 25% of the fruit's weight (Liu et al. 2006). The main sugars found in grapes are glucose and fructose, with sucrose present in much smaller concentrations.

The presence of sugars in these fruits depends on the genetic species, environmental conditions, and developmental stage. The individual cultivars may show different sugar profiles and, thus, different sweetness and flavor profiles, based on differences in carbohydrate metabolism, sugar transport, and key enzyme activity. For instance, the composition of sugars in grapes shows a wide variation in the relative percentage of both glucose, varying from 42% to 46%, and fructose, ranging from 42% to 51%, in total sugars (Zhong et al. 2023). Different genetic traits affect not only the fruit sweetness but also how suitable they are for fresh consumption or later processing into juices and wines.

Regarding environmental factors, sugar synthesis is extremely influenced by light intensity, temperature, and water supply. Normally, high solar irradiation promotes an increase in photosynthetic activity, which, in turn, enhances carbohydrate synthesis, while temperature affects all enzymatic functions relating to sugar metabolism. Higher temperatures negatively affect the timing and duration of sugar accumulation, especially in grapes, with subsequent effects on fruit quality, as has been demonstrated in studies that correlate temperature to fruit quality (Venios et al. 2020; Stanfield et al. 2024).

Water availability significantly affects sugar levels in grapevines, particularly during controlled drought periods, which can also positively affect the sugar concentration. Regulated deficit irrigation is a technique employed with wine grapes in order to apply water stress to vines, as a result of which osmotic adjustments lead to higher concentration of sugars and positive flavor components (Hewitt et al. 2023; Lakatos and Mitre 2023).

The sugars found in these fruits are also influenced by their developmental stage, as during ripening, there is a transition from organic acid accumulation to sugar storage (Lu et al. 2024). The breakdown of starch reserves into simple sugars takes place progressively in apples, whereas in grapes, the process of sugar accumulation, primarily glucose and fructose, occurs rapidly during the *veraison*, the stage of the onset of ripening, marked by softening and a change in berry color (Suter et al. 2021). Thus, the process is decisive for fruit sweetness, consumer acceptance, and industrial processing potential.

Sugar accumulation in these fruits is significant in terms of both the economy and ecology. High sugar is especially desired in fresh and processed fruits, which is directly related to the marketability and profitability of these fruits for producers. Furthermore, many fruits high in sugar are crucial for interactions with pollinators and seed dispersers, some of which further support biodiversity and stability in ecosystems. However, climate change emerges as an increasing target due to rising temperatures and modified precipitation regimes disrupting sugar metabolism, potentially impacting fruit yield and quality (Venios et al. 2020).

Due to their significant sugar content, these three fruits are also important, as they play a key role in various agricultural and food industries. Among these crops, grape shows the greatest worldwide consumption, and therefore, understanding the components that drive the maximization of sugar storage in grape berries could enhance the fruit quality and optimize productivity. Advanced biotechnological strategies, genetic engineering, and agronomic practices will further optimize their sugar accumulation, enhancing yield, sustainability, and economic viability. Additionally, more efficient breeding strategies and improved adaptation to growth conditions will help sugarcane and sugar beet crops better contribute to global food security and renewable energy supply.

Table 2.1 provides a comparative overview of sugar accumulation in different plant species. Sugarcane and sugar beet are known as the greatest sucrose accumulators, reaching up to 20% of sugar in fresh weight. Sweet sorghum is a drought-tolerant, multipurpose crop with lower sucrose content, but it is a key biofuel crop. Major fruit crops (e.g., apples, oranges, and grapes) are popular fruits consumed around the world, with several by-products such as juice, pulp, and essential oils that

Table 2.1 Comparison of sugar content, by-products, and industrial applications in major plant species that accumulate simple sugars

Plant species	Main sugar present	Other sugars	High sugar content (% fresh weight)	Low sugar content (% fresh weight)	By-products	Industrial applications
Sugarcane (*Saccharum officinarum*)	Sucrose	Glucose, fructose	18%–20%	12%	Bagasse, molasses, filter cake	Sugar, ethanol, bioenergy, animal feed
Sugar beet (*Beta vulgaris*)	Sucrose	Glucose, fructose	18%–20%	15%	Pulp, molasses	Sugar, ethanol, animal feed
Sweet sorghum (*Sorghum bicolor*)	Sucrose	Glucose, fructose	10%–20%	8%	Bagasse, syrup	Bioethanol, sweeteners, forage
Apple (*Malus domestica*)	Fructose	Sucrose, glucose	10%–16%	8%	Apple juice, pulp, cider	Juice, cider, vinegar, pectin
Orange (*Citrus sinensis*)	Sucrose	Fructose, glucose	8%–12%	6%	Peel, essential oils, pulp	Juice, essential oils, citric acid
Grape (*Vitis vinifera*)	Glucose, fructose	Sucrose	15%–25%	12%	Pomace	Wine making, juice, raisins, tartaric acid

contribute significantly to the food industry. Their sugar composition influences their use in industries ranging from juice and vinegar production, wine making, and citric acid extraction.

The ability of plants to accumulate simple sugars is determined by their metabolic pathways, environmental adaptations, and genetic makeup. Species such as sugarcane, sugar beet, and sweet sorghum are pivotal for industrial sugar production and bioenergy, while fruits and trees like grapes, maple, and date palms play specialized roles in diets and industries. Advances in molecular biology and genomics offer promising avenues for enhancing sugar accumulation in these species, particularly in the context of climate change and sustainable agriculture. Further research on genetic modifications and adaptive agricultural practices can improve sugar yields and optimize the economic viability of sugar-producing crops.

2.3.4.1 Grapes (*Vitis vinifera*)

Grapevine (*Vitis vinifera*) is one of the world's most important fruit crops. According to the 2019 Statistical Report on World Vitiviniculture, the total global grape production is on the order of 75–80 million tons annually (approx. 77.8 Mt. in

2018). Grapes are cultivated on 7–8 million hectares worldwide, with leading producers including China, Italy, Spain, the United States, and France. Notably, a majority of grapes (>50%) are used in wine making: Roughly 57% of the world grape production is converted to wine, about 36% is sold as table grapes for fresh consumption, and the remainder (~7%) is dried into raisins (OIV 2019). Regions such as Europe, the Americas, and Asia all contribute to grape production; for example, Europe accounts for about 60% of the world's wine output, whereas China and other countries have expanded table grape acreage in recent years. Grapes are valued both economically and culturally, with thousands of cultivars grown for different uses (Vilela and Pinto 2021).

The main industrial products from grapes reflect these uses. Wine (red, white, sparkling, and fortified) is by far the dominant industrial product, made by fermenting grape juice. Table grapes are sold fresh; raisins are produced by sun-drying or mechanically dehydrating grapes. Grape juice and concentrate are also significant products for direct consumption or use in soft drinks. Other grape-derived products include grapevine leaves (used in some cuisines), grape molasses, jellies, jams, and vinegar. The organic acid, tartaric acid, abundant in grapes, is recovered industrially: It crystallizes from wine-making residues and is used as a food acidulant. Grape seeds yield a small but valuable amount of grapeseed oil, rich in unsaturated fats and vitamin E (Khan et al. 2020). Table 2.1 lists these main uses: Grapes yield juice and wine, with pomace (skins, seeds, and stems) as a by-product, and industrial outputs include wine, juice, raisins, and tartaric acid.

Grape processing generates substantial by-products. In wine making, pressing the berries leaves behind grape pomace (also called marc): the mixed solids of skins, seeds, and stems. Pomace is rich in dietary fiber, lignin, and polyphenolic compounds (resveratrol, anthocyanins, and tannins). Although pomace is fibrous (making direct animal feed use limited), it is increasingly valorized as a source of natural antioxidants and other bioactives. For example, phenolic extracts from grape pomace are used as antioxidant additives in foods or cosmetics, and natural colorants (anthocyanins) can be obtained. Grape pomace is also fermented to produce bioethanol or used as a substrate for secondary fermentation (e.g., in grappa or brandy). Some pomace is composted or used for biogas. Grape seeds (from pomace or fresh grapes) are processed to extract grapeseed oil and phenolic oils. Overall, the concept of circular bioeconomy is applied to grape residues: Various technologies recover tannins, fibers, and other compounds, reducing waste and creating value-added ingredients (Lopes et al. 2025).

2.3.4.2 Orange (*Citrus sinensis*)

Orange is the most widely produced citrus fruit, accounting for roughly 58% of the global citrus output (Pineda-Lozano et al. 2022). In 2021, the global orange harvest reached the order of 49–50 million tons, led by Brazil (the world's largest producer), followed by China, the United States, and Mexico. Oranges are primarily consumed fresh or as juice, with only a small fraction (~4%) of harvested oranges processed into industrial juice and the majority marketed fresh (Afifi et al. 2023).

The main industrial products from oranges are orange juice (both ready-to-drink and concentrate) and essential oils extracted from the peel. Orange juice (fresh or concentrated) is a major commercial product, often used in beverages or as a food ingredient. The flavedo (colored outermost layer of the peel) yields an essential oil rich in limonene and other terpenes, used in flavorings, fragrances, and cleaning products (Gavahian et al. 2019). Citric acid, a major organic acid component of oranges, is also recovered industrially from citrus residues (Pineda-Lozano et al. 2022). Orange peel is likewise processed into pectin (a gelling agent) (Reichembach and de Oliveira Petkowicz 2021) and candied peel for confectionery (Mohanta et al. 2021). Other products include marmalades or jams and beverages like orange vinegar. These uses are summarized in Table 2.1: Oranges supply juice and oil, and their processing can yield citric acid and other value-added products.

Orange processing generates large volumes of by-products (peel, pulp, and seeds). Approximately half of an orange's fresh weight becomes waste after juice extraction (Pineda-Lozano et al. 2022). The peel (flavedo plus albedo) is the largest by-product (≈50%–65% of fruit mass) and is rich in sugars, cellulose, pectin, flavonoids (e.g., hesperidin and naringin), carotenoids, and oil (Pineda-Lozano et al. 2022; Afifi et al. 2023). These peels can be revalorized: For instance, dried citrus peel is used as animal feed or fertilizer or treated to extract pectin, flavonoids, and essential oil (Afifi et al. 2023). D-limnonene, a major component of orange oil, is distilled from peel waste for use as a solvent or precursor to fine chemicals. Orange seeds and pulp residues (orange pulp) also contain oils, proteins, and sugars; they are used in some regions for animal feed or fermented to produce ethanol. In summary, citrus processing residues (especially peel) are a rich source of natural compounds (fiber, antioxidants, and aroma compounds) that are being explored for food, nutraceutical, cosmetic, and bioenergy applications (Pineda-Lozano et al. 2022; Afifi et al. 2023).

2.4 Saccharide-Processing Techniques

Saccharides are vital components of various biomass sources such as sugarcane, beetroot, sweet sorghum, and fruits. These sugars are relevant to numerous industrial applications, including biofuel production, food processing, and biomolecule production (Flórez Pardo et al. 2018). Though individual steps in the extraction process will differ depending on which type of biomass it is, most extractions entail milling or slicing to aid extraction and/or solubilization and juice pressing, followed by juice purification before concentrating out desired saccharides. The extraction efficiency and yield vary with the sugar content of the delivered biomass feedstock, processing methods used, and the desired end products (Regassa and Wortmann 2014; Sunori et al. 2015; Duraisam et al. 2017; Mushtaq 2018).

2.4.1 Traditional Methods of Saccharide Processing

2.4.1.1 Sugarcane

2.4.1.1.1 Pretreatment of Biomass

Sugarcane biomass is a primary source of simple sugar, having a productivity between 52.2 and 121.8 t/ha, depending on where it is cultivated (Ram and Ramaiyan 2021). Its processing can be broken down into three main stages: cane reception, cane preparation, and sugar extraction. Once the sugarcane, harvested mechanically, arrives at the factory, it is placed on tables before moving them either to the cleaning system or directly to the feeding tables that guide it to the cane preparation area. For chopped sugarcane, a dry-cleaning system is advisable to avoid sugar loss. During this stage, some mineral and vegetable impurities are removed, resulting in cleaned cane ready for preparation (Dias et al. 2015).

Mechanically harvested sugarcane usually already comes in forms of billets (stalk sections) (Kishore et al. 2017); however, the billets also need to be shredded into smaller pieces that are fibers of around 1–2 cm in size (Sunori et al. 2015). These fibers then pass through a set of mills containing five to six crushing mills, separating the raw sugarcane juice from the bagasse (Dias et al. 2015; Sunori et al. 2015; Karp et al. 2021). The raw sugarcane juice can be utilized directly to produce many biomolecules, such as ethanol, or undergo processing, where the sugar can be recovered (Karp et al. 2021). Sugarcane juice, abundant in sucrose, passes through a complex multistep process, including clarification, evaporation, crystallization, centrifugation, and dry product packaging (Fig. 2.1) (Meng et al. 2021).

2.4.1.1.2 Juice Clarification

The juice constitutes approximately 80%–90% of the mass of the sugarcane stalks and contains high levels of sugar, including 12%–18% of sucrose, glucose, and fructose (Karp et al. 2022a). The suspended material is primarily made up of bagacillo (a by-product of lignocellulosic material), as well as soil, sand, clay, starch, waxes, fats, and gums (Doherty 2011). The rapid darkening of the raw sugarcane juice after extraction is attributed to the oxidation of certain constituents, which compromises its commercial viability and sugar production (Prati and Moretti 2010). Based on this, effective clarification of sugarcane juice is critical to optimizing sugar production. If the clarification process is not properly controlled, it can result in issues such as sugar loss, undesirable color quality, and elevated sulfur levels in the clarified juice. These issues significantly affect the subsequent stages of sugar manufacturing (Meng et al. 2021). In this process, it is essential to remove unwanted components with minimal sucrose and sugar decomposition and prevent the formation of new non-sugar compounds. Three clarification methods of industrial significance are defecation, sulfitation, and carbonation (Rodrigues et al. 2018).

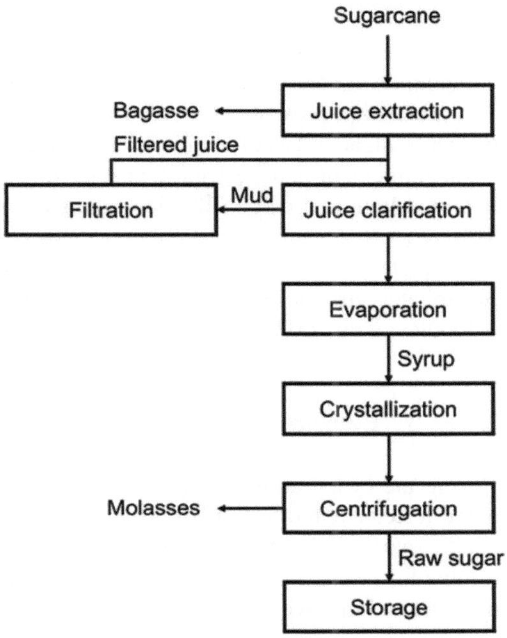

Fig. 2.1 Processes involved in sugarcane sugar production

In many sugar cane factories, the clarification method is often simple defecation. Lime is added as milk of lime or lime saccharate (lime dissolved in juice or syrup) after the raw juice is initially heated to 72–76 °C, raising its pH to around 7.8–8.0. The juice is then further heated to 102–105 °C, flashed for air removal, and sent to the clarifier. Then a flocculant polymer, such as polyacrylamide, is introduced in the clarifier, aiding in the formation of large flocs by precipitating calcium phosphates (Doherty 2011; Dias et al. 2015). The mud collected from the clarifier undergoes filtration to enhance sugar recovery (Dias et al. 2015). A notable drawback of using lime or lime saccharate to raise juice pH for calcium phosphate precipitation is the high level of CaO in the clarified juices, originating from both the added lime and the natural calcium in the raw juice. Excess CaO negatively impacts clarified juice quality, leading to scaling in evaporators, pans, and centrifugals and reduced molasses exhaustion (Doherty 2011).

The sulfitation process involves the introduction of gaseous sulfur dioxide (SO_2) into the juice until it reaches a pH of 3.8–4.2. Subsequently, the juice is neutralized to a pH of 7.0–7.2 through the addition of hydrated lime ($Ca(OH)_2$). This alkalinization results in the formation of calcium sulfite, which precipitates and adsorbs undesirable compounds. The treated juice is then heated to 100–105 °C and passed through sedimentation tanks to remove the precipitates (Hamerski et al. 2012). Also, some factories add phosphate in the form of superphosphate fertilizer or phosphoric acid to facilitate enhanced removal of colloids, resulting in lighter juice and expedited sedimentation (Prati and Moretti 2010; Doherty 2011). However, challenges have limited the application of the sulfitation method, especially in relation to food safety standards, which typically mandate lower residual levels of sulfur-based compounds in consumables (Hamerski et al. 2012).

Carbonation presents as an alternative to sulfur usage, facilitating CO_2 sequestration in sugar and alcohol industries (Saska et al. 2010). Primarily employed in raw sugar refinement and beet juice purification, carbonation involves the addition of hydrated lime and carbon dioxide to sugarcane juice under controlled conditions. This process generates crystalline calcium carbonate precipitates that adsorb and incorporate colloidal and insoluble matter, inorganic compounds, and dyes. The precipitates are subsequently removed from the clarified juice through filtration (Moodley et al. 2002; Hamerski et al. 2012).

2.4.1.1.3 Evaporation

During the evaporation process, diluted sugar juice is transformed into a concentrated juice. The thermal energy necessary for this evaporation is supplied by steam condensation. High-pressure steam generated by a boiler is directed to the steam turbine for electricity production. Traditional cogeneration systems employ a back-pressure turbine, where steam is released at a lower pressure. In contrast, modern systems use an extraction-condensing steam turbine, where a portion of the steam is extracted at a lower pressure, and the rest is sent to the condenser (Chantasiriwan 2020). The extent and speed of sucrose inversion during evaporation are affected by the juice's temperature, flow rate, and time spent in the evaporator. The product of evaporation is a syrup with 78%–86% of sugar purity and dissolved solids between 60 and 65°Brix, containing 3.5%–4.5% of invert sugars. The sucrose content of this syrup is favorable for sugar crystallization (Delgado and Casanova 2001).

2.4.1.1.4 Crystallization and Centrifugation

Crystallization is a process in which molecules transition from a solute dissolved in the liquid phase to a solid phase. This transformation occurs through two primary mechanisms: nucleation and crystal growth (Velazquez-Camilo et al. 2010). The first step in sugar crystallization relies on boiling the syrup in vacuum pans to increase its supersaturation (Delgado and Casanova 2001). Vacuum pans receive the syrup derived from the evaporation process and the molasses recirculated from the centrifugation process. Heating steam is supplied to the vacuum pan, where it flows around its tubes within a sealed chamber. This heating steam facilitates the evaporation of water from the solution and increases its concentration, forming massecuite, a thick mixture of sugar crystals and syrup (Morales et al. 2024). Finally, a centrifugation process is applied to separate the molasses from the crystalline sugar (Delgado and Casanova 2001).

2.4.1.2 Sweet Sorghum

The average productivity of sweet sorghum is over 35 t/ha (dry weight basis) (Rooney et al. 2007), with its fresh juice comprising 65%–80% of the total weight (Gnansounou et al. 2005; Eggleston et al. 2013). In traditional methods, the process of harvesting sweet sorghum for syrup production involves the removal of the plant tops and the stripping of the leaves (Regassa and Wortmann 2014). Mechanically harvested sweet sorghum is presented in the form of billets, from which the raw juice will later be extracted. The most commonly applied method for sweet sorghum juice extraction is roller-milling through mills constituted of two to four rollers (Gnansounou et al. 2005; Eggleston et al. 2013). After that, the extracted juice goes through similar steps as sugarcane juice, including clarification using lime and heat and concentration. However, its final product is a syrup instead of a dry product. This is attributable to its comparatively lower purity (~75%) relative to sugarcane (~85%), in conjunction with elevated levels of reducing sugars, starch, and aconitic acid (Eggleston et al. 2010, 2013).

2.4.1.3 Sugar Beet

Almost all industrial raw and refined crystal sugar originates from sugar beet and sugar cane, and their processing procedures are largely alike, with only a few differences (Zicari et al. 2019). Sugar beet is an alternative to sugarcane for producing sugar in temperate climates, being specially produced in developed countries (Duraisam et al. 2017). These plants have the capacity to store between 15% and 20% of their root mass as sucrose, with typical biomass yields varying from 40 to 100 metric tons per hectare (Panella and Kaffka 2010; Zicari et al. 2019). The beets are harvested mechanically with well-established equipment, and the pretreatment consists of washing and slicing the sugar beets to produce beet cossettes (thin strips) (Duraisam et al. 2017). The cossettes undergo a continuous counter-current extraction using hot water, a process referred to as diffusion. After diffusion, beet contents are pressed to generate a raw juice stream. This raw juice undergoes purification through several stages involving lime and carbon dioxide addition, followed by filtration, resulting in a clarified liquid known as thin juice. Thin juice is then evaporated to create thick juice, which contains over 60% sucrose. Thick juice can be either stored or sent to crystallizers for further dewatering and centrifugation under partial vacuum and high temperatures, ultimately producing white sugar that can be dried and packaged (Zicari et al. 2019).

2.4.1.4 Fruits

Although it is possible to extract sugar from fruits, this practice is not commonly employed on a large scale. This is because sugar cane and sugar beets, which contain much higher concentrations of natural sugars, offer a more efficient and cost-effective means of production. Consequently, producing sugar from fruits on an industrial scale is impractical and not commercially viable. On the other hand, fruit juice extraction is a very common process.

Generally, the fruits are washed to remove or minimize any physical or chemical contamination. In modern juice manufacturing facilities, washing conveyor belts are employed. Typically, a washing conveyor consists of rows (10–20) of nylon brushes rotating in the direction of the fruit to save energy and shower lines for water, detergent, and cleaning solutions. After that, the fruit juice can be extracted using different types of equipment, such as grinding mills (mainly used for apple, pear, pineapple, guava, and mango) and finger cup extractors (frequently used for citrus juice extraction) (Mushtaq 2018). Fruit juices can be used in a variety of applications, such as food ingredients, for flavor, color, sweetness, and texture (Priyadarshini and Priyadarshini 2018), and for fermentation processes, such as the production of alcoholic beverages (Beech 1972).

2.4.2 Advanced Methods of Saccharide Modification

The mechanical methods for sugar extraction, such as slicing, milling, and pressing, usually retain the saccharides that are more present in the structure. Their less expensive approach and high yield are ideal for plants with a high content of the desired sugar, as mentioned previously. However, when the interest is on a specific monosaccharide that is not the most abundant, those classical methods are not efficient, especially if the sugar has too low a concentration or if there are too many varieties in the composition. Therefore, complementary techniques are necessary in order to obtain high purity and better yields, such as membrane filtration and chromatography (Harrison et al. 2003).

The filtration alternative is one of the most commonly applied processes in the industry, yet it greatly varies in terms of range and cost. The method applies a porous system through which the liquid passes, with the particles larger than the pore being retained and the smaller ones passing through the device. The pore size can be altered according to the type of molecule, with sugar separation typically occurring within the nanofiltration and reverse osmosis range, from 0.9 to 25 nm approximately. Oligosaccharides can easily be purified from cells, proteins, and major molecule mixtures through filtration, yet they are more difficult to be separated from themselves and from salts and small organic molecules due to their size. However, in industrial processes, a sequence of filtrations is necessary, from conventional to micro-, ultra-, and nanofiltration, avoiding the obstruction of the membranes (Luo et al. 2015; Li et al. 2020).

Another alternative is chromatography, which consists of a separation technique based on two phases, one stationary phase and one mobile phase. The stationary phase, usually a solid, gel, resin, or liquid, remains static during the process in a column, while the mobile phase, a liquid or a gas, flows through the stationary phase continuously. If the compound of interest has more affinity to the mobile phase, it elutes faster, and if the compound has more affinity to the stationary phase, it takes longer to elute. High-performance liquid chromatography (HPLC), a laboratory-scale precision equipment, is capable of separating monosaccharides with a quantitative and qualitative approach. The columns utilized are versatile, as they classify alcohols and organic acids, together with oligosaccharides, due to a similar affinity to the resin (Niaz et al. 2020). For instance, a resin based on cross-linked hydrogen or calcium and a mobile phase of diluted sulfuric acid can efficiently separate citric acid, lactic acid, glucose, fructose, maltose, mannitol, and ethanol (Ball and Lloyd 2011). Another example is the use of amino and amide columns with an acetonitrile mobile phase, capable of isolating glucose, galactose, arabinose, fucose, and xylose (Yan et al. 2016). The combination of adequate phases will provide better resolution of the carbohydrates. In industry, the high cost of HPLC and its low sample throughput limit its large-scale application. As an alternative, chromatographic separation is implemented in large adsorption columns designed for higher processing volumes. While less precise and selective for only certain monosaccharides, these systems are still useful for purification purposes (Harrison et al. 2003).

2.4.3 Saccharide Modification Methods

Some saccharides, even after purification processes, undergo further modification methods to alter specific properties of the compound. The most famous one is the production of inverted sugar syrup, which has a slightly stronger sweetening power, retains more moisture, and avoids crystallization better than common sucrose. The inverted sugar comes from the sucrose itself, when an additional step to break the glycosidic bond and increase the glucose and fructose content occurs. This reaction is catalyzed by an enzyme called invertase, in a proportion of around 0.15% in pH 5.0 and 60.0 °C temperature, commonly applying citric acid, lemon juice, or tartar cream as catalysts in small proportions. While sucrose is being hydrolyzed, the optical rotation of its asymmetrical carbons occurs in the opposite direction as glucose and fructose are liberated in the solution, which is the reason for the "inverted" denomination. The resulting syrup is more bioavailable for general microorganisms for presenting monosaccharides instead of disaccharides. It also finds applications in the sweet food sector, contributing functional properties to products such as honey, jam, beverages, and creams (Aider et al. 2007; Obodovych et al. 2021).

2.5 Economic Aspects and Global Production of Saccharides

Saccharides, especially sugars and their derivatives, are essential commodities in the world economy (Niaz et al. 2020). These play a fundamental role in human and animal nutrition, in the production of bioenergy, and also in its applications in other industries, such as chemistry (Zaitoun et al. 2018; Deeba et al. 2021).

The main saccharides of economic importance are sucrose, obtained from sugar cane and sugar beet; starch from corn, potatoes, wheat, and cassava; syrup (glucose and fructose) obtained from corn processing; and cellulose, obtained from wood and vegetable fibers (Dwivedi 2022). However, the latter will not be discussed in this chapter.

In the food industry, sugars are used as sweeteners for juices, jams, sweets, bakery products, etc. In the case of starches, they are used as thickeners or binders in sauces, creams, sausages, ice cream, etc. Bioenergy is another important area of application for saccharides, which is generated through the conversion of fermentable carbohydrates, such as glucose and sucrose, into biofuels and other forms of energy. This process is widely used in the production of ethanol and biogas, which are renewable alternatives to fossil fuels (Karp et al. 2022b; Mello et al. 2022; Woiciechowski et al. 2022).

Among the main producers of saccharides are Brazil and India, especially in the production of sucrose; the former uses sugar cane as a source, while the latter uses sugar beet as a source. In Fig. 2.2, it can be seen that the sucrose production of these countries together is approximately 50% of the world production. Other countries with significant production are the European Union, which together produces around 15 million tons, followed by China, Thailand, and the United States with a production of 9.9, 8.8, and 8.3 million tons, respectively. Furthermore, the figure also shows that global sugar production has increased every year, with the exception of 2019 (Statista 2024a, b). The increase in global sugar production can be explained by the expansion of planted areas in key countries such as Brazil, India, and Thailand, as well as technological improvements in agricultural practices, such as the use of genetically modified seeds, effective fertilizers, and better mechanization systems.

Furthermore, production is mainly influenced by high demand; currently, sucrose is the main substrate used in the production of bioethanol (Sydney et al. 2021). The global sugar market is a sector of broad interest in the world economy, as it reached US$ 51.6 billion in 2023 and is projected to reach US$ 98 billion by 2033 (Spherical-Insights. 2023).

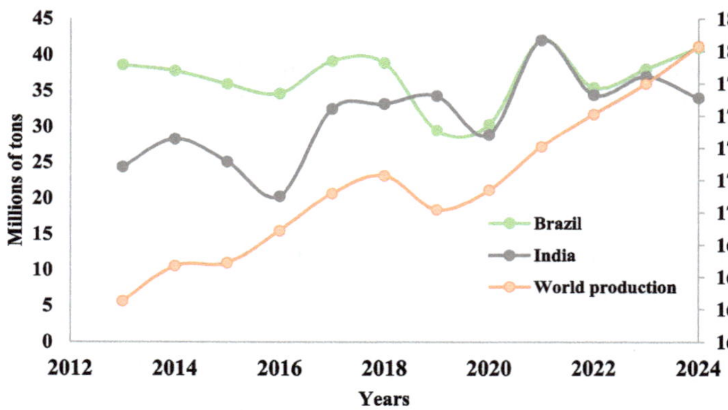

Fig. 2.2 Behavior of world sugar production. (Source: Statista 2024c; Statista 2024b; Statista 2024a)

Another important saccharide in the world market is starch (native starch and modified starch), which has been constantly increasing. In 2022, starch production was estimated at 134.5 million tons, and it is projected that by 2030, production will reach approximately 200 million tons, which means an annual growth rate of 5.1% (Globenewswire 2023). The increase is due to the growing demand for food products that use starch in their formulations, which is the main sector for starch applications, accounting for 55% of the total production volume. Other important applications are in pharmaceutical products and retail, which consume 2.7% of the total volume of starch.

The largest starch market is China, which is expected to reach 75 million tons by 2030, representing an annual growth of 5.9%. The second largest starch market is the United States, which reached approximately 31 million tons in 2022 and is projected to grow steadily over the forecast period (Globenewswire 2023).

In 2024, the industrial starch market size reached US$57.2 billion, and it is estimated to reach US$74.4 billion by 2030. The main raw materials used to obtain starch are corn (75%), cassava (14%), wheat (7.0%), and potato (4.0%) (Vilpoux and Junior 2023). Starch obtained from corn is one of the most sought-after, mainly due to the textural properties it provides to foods, specifically in dairy products and beverages. In addition, this type of starch also has a marked advantage over wheat starch, and that is the absence of gluten, which is a challenge for wheat starches, considering the traces of this protein in the raw material (Zhang et al. 2021). In the case of cassava starch, its production involves more complex logistics, particularly during cultivation and post-harvest handling. Additionally, cassava roots are highly perishable due to their elevated moisture content.

It is worth noting that the global market for modified starches has also shown steady growth in recent years. The size of this market in 2023 reached US$ 13.8 billion and is projected to reach approximately US$ 20 billion by 2030, that is, a compound annual growth rate of 5.4% (Verified-Market-Reports 2024). Imports of these starches reached US$ 6.7 billion, with the main importing countries being China, Germany, Japan, and the United States, with 11, 8.6, 8.3, and 7.3%, respectively.

The production of saccharides is influenced by several factors, such as the following: (i) Climatic conditions affect the yield or production of carbohydrates and, therefore, the market, that is, the supply and demand of the material. (ii) The policies and regulations of each country also affect it since government subsidies for the production of raw materials or derivatives can make production more viable, or otherwise, environmental and regulatory restrictions can increase costs. (iii) Available technologies are another factor to consider; genetic improvement and more efficient industrial processes improve yield and reduce production costs. (iv) Competition with products that can perform the same function and are healthier has a negative effect, for example, the market for sugar for human consumption (Dimopoulou et al. 2021; Mehdi et al. 2024a).

2.6 Main Energy and Biological Products Derived from the Use of Natural Simple Sugars

Natural simple sugars such as glucose, fructose, and sucrose play an essential role as substrates in the production of energy and high-value-added biotechnological compounds. These sugars are rapidly metabolized by microorganisms and used in fermentative and enzymatic processes for the synthesis of a variety of industrial products. Commercially important products in this energy scenario include bioethanol, organic acids, bioplastics, and pharmaceuticals.

2.6.1 High-Value Biological Products Derived from Simple Sugars and Conversion Mechanisms

Simple sugars are essential precursors in the synthesis of several industrial bioproducts through fermentation processes and enzymatic biotransformation. Converting sugars into highly valuable bioproducts requires the integration of biotechnology, metabolic engineering, and sustainable industrial process concepts. The following section summarizes the main mechanisms involved in sugar conversion, including fermentative processing, enzymatic processing, and genetic engineering developments. Additionally, economic data and sector growth projections are presented.

Microbial fermentation is one of the most used techniques for converting sugars into various bioproducts. Through metabolic activities, bacteria and yeast metabolize sugars present in renewable feedstocks, resulting in the production of bioethanol, organic acids, biopolymers, and bioinoculants.

2.6.1.1 Bioethanol

Bioethanol is one of the main energy products obtained through the fermentation process of sugars by several microorganisms, such as *Saccharomyces cerevisiae* yeast, which is one of the most used and genetically studied microorganisms for industrial applications. Sugarcane is a common source of sucrose used for the production of bioethanol, with Brazil being one of the world leaders in this sector. The fermentation process occurs in the presence of yeasts that convert sucrose into ethanol and carbon dioxide, reaching a theoretical yield of 90% (Basso et al. 2011). Bioethanol can be used as pure fuel or blended with gasoline in specific ratios such as 85:15 to reduce greenhouse gas emissions, promoting a low-carbon economy. Yeast strains play a significant role in bioethanol production; thus, their efficiency is influenced by factors like temperature, sugar concentration, pH, fermentation time, agitation rate, and inoculum size. In recent decades, many studies have explored alternative microorganisms, such as bacteria, to enhance ethanol yields. Moreover, advanced biotechnological tools have demonstrated a huge potential in order to achieve higher conversion efficiencies and increased productivity using engineered microorganisms (Dien et al. 2003; Sun et al. 2024a).

2.6.1.2 Biogas

The anaerobic conversion of simple and more complex sugars results in the production of biogas, a mixture of mostly methane. This process is carried out by a series of microorganisms, including hydrolytic, acidogenic, acetogenic, and methanogenic bacteria. Residues rich in carbohydrates, such as sugarcane bagasse, vinasse, and filter cake from sugar and ethanol production, along with milk by-products, fruit and vegetable wastes, and food processing wastes, may be used for biogas production in biodigesters (Janke et al. 2015; Almeida et al. 2024).

2.6.1.3 Organic Acids

Organic acids, such as lactic, acetic, and succinic acids, are produced through fermentation and find vast application in the food, pharmaceutical, and biodegradable materials industries. For instance, lactic acid is a precursor for polylactic acid (PLA), a biodegradable bioplastic used in sustainable packaging. PLA is synthesized through the polymerization of lactic acid monomers, which are derived from renewable resources such as sugarcane, corn, cassava roots, and sugar beet (Balla et al. 2021). This biopolymer offers a reduced carbon footprint compared to traditional fossil-based plastics, making it an environmentally friendly alternative in the packaging industry.

2.6.1.4 Biopolymers

Microorganisms are capable of synthesizing natural polymers, such as polyhydroxybutyrate (PHB), from sugars. PHB is a biodegradable polyester with similar properties to those of conventional plastic, which offers an attractive solution for the material industry, as there is a need for a more ecologically friendly option. Bacteria and archaea produce PHB as carbon and energy reserves, and due to its rapid degradation in natural environments, PHB is considered a sustainable substitute for synthetic plastics (Adnan et al. 2022).

2.6.1.5 Bioinoculants

Bioinoculants are produced through submerged liquid- and solid-state fermentation processes using isolated beneficial microorganisms, commonly bacteria and fungi, in a controlled environment where sugars are the main carbon source. When added to

Table 2.2 Main biotechnological products obtained from simple sugars

Product	Process	Microorganism	Application
Ethanol	Fermentation	*Saccharomyces cerevisiae*	Biofuel production
Lactic acid	Fermentation	*Lactobacillus* spp., *Escherichia coli*	Food, pharmaceutical, bioplastics industry
Succinic acid	Fermentation	*Actinobacillus succinogenes*, *Corynebacterium glutamicum*	Solvents, biodegradable plastics
Bioplastics (PHA and PLA)	Fermentation	*Cupriavidus necator*, *Lactobacillus* spp.	Bioplastics
Antibiotics (erythromycin and streptomycin)	Fermentation	*Saccharopolyspora erythraea*, *Streptomyces griseus*	Medical use
Industrial enzymes	Fermentation	*Aspergillus* spp., *Bacillus* spp.	Food industry, detergents
Bioinoculants	Fermentation	*Aspergillus* spp., *Bacillus* spp., *Pseudomonas* spp., *Rhizobium* spp., *Trichoderma* spp.	Growth promoters, biocontrol, disease resistance

the substrates, simple sugars such as glucose, sucrose, or molasses supply a more accessible metabolizable form of energy, which stimulates the microbial population and consequently increases the liquid-phase extraction of bioactive compounds, such as enzymes, phytohormones, and secondary metabolites (Vassileva et al. 2021). Factors like pH, temperature, oxygen availability, and nutrient composition are controlled during the fermentation process to optimize microbial viability and efficacy. After achieving the desired concentration of microorganisms, the bioinoculant formulation is stabilized (for instance, dried or encapsulated) to extend shelf life.

Table 2.2 summarizes the main biotechnological products derived from simple sugars through microbial fermentation processes, highlighting the associated microorganisms and their applications.

2.6.1.6 Enzymatic Processes for Sugar Conversion

Enzymes play a fundamental role in converting complex matrixes into fermentable sugars, which are then fermented into bioproducts. Enzymatic hydrolysis is a biological process in which specific enzymes break down polysaccharides into simple sugars (monomers). In the past decade, enzymatic hydrolysis of lignocellulosic biomass has been widely studied for the industrialization of second-generation ethanol in order to make its commercial production viable. Lignocellulosic biomass, which contains cellulose, hemicellulose, and lignin, is a rich source of sugars. Enzymes such as cellulases and hemicellulases catalyze the breakdown of these macromolecules into simple sugars, such as glucose and xylose. Recent research has also identified novel, effective enzymes that degrade lignin, allowing better access to cellulose and hemicellulose fractions and increasing efficiency (Janusz et al. 2017).

Conversion of intermediary metabolites by species-specific enzymes can convert sugars into valuable industrial compounds. For example, glucose can be isomerized into fructose by the enzyme glucose isomerase to produce high fructose corn syrup, a common sweetener in the food industry.

2.6.2 Advances in Metabolic Engineering and Gene Editing

Metabolic engineering and gene editing have revolutionized the production of bioproducts, allowing the modification of metabolic pathways in microorganisms to increase efficiency and productivity. Through metabolic engineering, it is possible to redirect metabolic flows for the preferential production of certain compounds. For example, overexpression of genes involved in the glycolytic pathway can increase the production of precursors necessary for the synthesis of biopolymers. Techniques such as CRISPR–Cas9 allow precise editing of genomes, eliminating unwanted genes or inserting new ones that confer specific metabolic abilities. In the biofuel production context, this approach has been used to develop strains capable of metabolizing different types of sugars or resistant to inhibitors present in pretreated biomass (Lakhawat et al. 2022).

To conclude, natural and simple sugars have been shown to be sustainable and versatile substrates in the production of high-value energy and bioproducts. By enabling their exploitation biotechnologically, the portfolio of sugar-derived products has been expanded to drive the bioeconomy, reduce the reliance on fossil fuels, and achieve environmental sustainability. Furthermore, the recent development of innovative biotechnological platforms, such as metabolic engineering and biocatalysis, suggests an even greater expansion of the sugar-derived product landscape, firmly establishing their strategic role in the transition to a circular economy.

2.7 By-Products Generated by Simple Sugar Processing and Their Applications in a Circular Economy Approach

Circular bioeconomy represents an integrated approach that promotes the rational use of renewable resources and the minimization of waste while generating added value through the conversion of by-products. In this context, the theoretical framework of biocircularity provides a basis for evaluating and planning systems that transform biomass residues—such as those derived from sugarcane, sugar beet, sweet sorghum, and various fruits—into sustainable bioproducts. This approach emphasizes not only efficiency in the utilization of natural resources but also the preservation of natural capital stocks, guiding the development of processes that prioritize maximum reuse and recycling of materials (Igbokwe et al. 2022; Holden et al. 2023).

The advancement of emerging technologies has enabled the optimization of conversion strategies, broadening the spectrum of applications for the obtained bioproducts. In particular, the integration of innovative methods for converting materials, previously considered waste, into high-value products plays a critical role in reducing environmental impacts and strengthening the transition to sustainable economic models (Swaminaathan et al. 2024; Upadhyay et al. 2024). This view is strengthened by a range of initiatives focused on boosting energy efficiency, cutting down waste, and enhancing industrial competitiveness, ultimately making the circular bioeconomy a smart choice for sustainable development.

Within this framework, various techniques—physical, chemical, biological, and thermochemical—work together to transform waste into high-value products. This integrated approach not only maximizes resource recovery but also drives innovation, helping industries become more sustainable and competitive (Ashokkumar et al. 2022). This integration optimizes the extraction and conversion of complex biomolecules into renewable inputs, effectively closing the production cycle and minimizing waste. Such technological convergence reinforces the principles of waste reduction and resource regeneration while positively impacting the economic viability and environmental sustainability of industrial processes through the systematic reintegration of by-products into the production system (Begum et al. 2024; Swaminaathan et al. 2024).

The sugarcane industry is responsible for generating a significant amount of by-products during processing, notably bagasse, molasses, and press mud. Historically regarded as valueless waste, these materials have undergone a conceptual transformation (Ungureanu et al. 2022). Nowadays, in light of the principles of circular economy and bioeconomy, they are viewed as renewable resources with a high potential for generating value-added products. This paradigm shift not only minimizes the environmental impacts associated with improper disposal but also fosters new economic opportunities and innovation within the sector.

Within the context of a circular economy, the reutilization of these by-products is systematically carried out using innovative technologies, markedly deviating from the traditional linear “extract–produce–discard” model. By integrating the principles of reduce, reuse, and recycle, every component of the sugarcane processing chain is reintegrated into the production system, maximizing its utility and contributing to the efficient use of natural resources. This approach promotes a closed-loop production cycle, wherein what was once considered waste becomes a feedstock for the production of high-value bioproducts such as biofuels, bioplastics, and organic fertilizers (Begum et al. 2024; Hassan et al. 2024). Such an integrative strategy not only reinforces environmental sustainability but also transforms traditional challenges into business opportunities, elevating the competitive level of the sugar energy industry. By redefining the concept of waste, a transition toward a regenerative and resilient production model is driven, one that aligns with the demands of the modern bioeconomy and global sustainability targets.

Bagasse, the product obtained after juice extraction, is composed mainly of cellulose, hemicellulose, and lignin. Its efficient utilization can significantly contribute to the circular economy and sustainability (Begum et al. 2024). The composition of bagasse favors its use as a raw material for the generation of bioenergy through direct combustion or its use in boilers and the production of solid biofuels, such as pellets or briquettes (Lee et al. 2023). Following pretreatment processes such as delignification, hydrolysis, and purification, the cellulosic fibers extracted from bagasse can be transformed into high-value products, including biocomposites (Mahmud et al. 2023), biodegradable packaging materials (Singh et al. 2022), hydrogels (Ban et al. 2022), and even cellulose nanocrystals (CNCs) (Tang et al. 2023), which exhibit exceptional mechanical properties and are being explored in emerging areas such as flexible electronics, filtration membranes, and controlled drug delivery systems.

Although bagasse can be transformed into advanced cellulosic fibers, its processing faces technological difficulties unlike that of traditional raw materials such as eucalyptus or pine. Delignification, alongside acid or enzymatic hydrolysis and mechanical treatments, needs substantial energy input and large financial investments for infrastructure development (Mahmud and Anannya 2021; El Sayed et al. 2022). The feasibility of these applications requires cost-effective production processes because conventional raw materials already enjoy established economies of scale. The utilization of sugarcane bagasse stands as a distinctive chance to drive the transformation toward a circular economy despite existing challenges. Utilizing bagasse in various production networks enables the establishment of a closed-loop system where waste transforms into valuable resources, which reduces the strain on finite natural resources and aids both climate change mitigation and ecosystem conservation.

Sugarcane molasses, which is the liquid derivative from sugar production, contains fermentable sugars, such as sucrose, glucose, and fructose, and it is a simple source of raw material for most biotechnological applications. Perhaps the most valuable application of this material is in the production of ethanol owing to its considerably high concentration of active soluble carbohydrates in relation to its low cost as a valuable substrate for fermentation. In this process, the sugars are converted by microorganisms, such as *Saccharomyces cerevisiae*, into ethanol and carbon dioxide (Parascanu et al. 2021; Mustafa et al. 2023). In addition to producing renewable biofuel, this process reduces greenhouse gas emissions due to a lesser dependence on fossil fuels. Molasses is also advantageous from strong economic points since it represents another purpose for a by-product from the sugar alcohol industry, which otherwise would be wasted (Mahmud and Anannya 2021).

Molasses plays a significant role in the nutrition of animals as an energy-enhanced supplement carrying simple carbohydrates plus essential minerals. When added to livestock feed, it enhances both the nutritional quality and palatability of the diet, supporting healthier and more efficient growth. Its natural prebiotic and antimicrobial properties further benefit gut health, improving feed conversion rates and reducing the need for synthetic additives (Jamir et al. 2021). This versatility highlights the principles of the circular economy, where a single by-product meets multiple industrial needs, linking agriculture, bioenergy, and livestock production in a way that minimizes waste and maximizes resource efficiency.

Another valuable by-product of sugar production is press mud, also known as filter cake, which is produced during the clarification of sugarcane juice. Press mud is rich in organic matter and active nutrients like nitrogen, phosphorus, potassium, iron, zinc, and manganese, which act as natural soil conditioners. Its use in farming teaches sustainable methods that enhance soil fertility across a crop cycle by providing nutrients on a slow-release basis (Caione et al. 2015; Jamir et al. 2021). Additionally, the organic matter in press mud enhances soil structure, increases water retention, and stimulates microbial activity, further enriching soil health. The phosphorus content is particularly beneficial, as it is highly bioavailable to plants, reducing reliance on synthetic phosphate fertilizers, whose extraction is often associated with negative environmental impacts (Almeida Júnior et al. 2011).

Beyond its immediate benefits for plant nutrition, press mud also contributes to long-term soil health, reinforcing the sustainability of agricultural systems. By repurposing an industrial by-product as a natural fertilizer, its use aligns with the circular economy model, reducing dependency on synthetic inputs and lowering the environmental impact of conventional fertilizer production.

Building on this resource efficiency, the integration of biorefinery processes and advanced bioprocessing technologies, such as thermomechanical, chemical, and biological processes, has further enhanced the valorization of these sugarcane by-products. These innovations have contributed to cost reduction and increased processing efficiency, enabling the production of a range of high-value bioproducts, including ethanol, biogas, nanocellulose, and biopolymers. What was once considered waste is now a key resource in a sustainable and economically viable production cycle. Achieving this requires the perfect alignment of bioprocess conditions and parameters with the specific characteristics of the biomass and the desired bioproducts. For example, enzymatic hydrolysis, microbial fermentation, and thermochemical conversions demand distinct operational parameters to maximize efficiency and minimize resource waste (Chandel et al. 2024).

The production efficiency of these bioproducts depends on the optimization of bioprocesses through advanced and intelligent bioreactors designed to process complex biomass feedstocks while maintaining precise control over critical parameters such as temperature, pH, dissolved oxygen, and substrate concentration (Jofre et al. 2025). Furthermore, the integration of real-time monitoring technologies and continuous improvements in bioreactor design are essential to achieving high yields and consistent product quality, ensuring optimal performance with superior conversion rates and reduced energy consumption and production costs—even under variable conditions. This, in turn, enhances the economic viability of the processes and renewable bioproducts, guaranteeing the consistency and reproducibility of industrial-scale production processes.

The process of turning sugarcane processing by-products into valuable resources demonstrates how industries can shift from linear production to circular systems that convert waste into useful products. By extracting and processing components like cellulose, hemicellulose, and simple sugars from bagasse, molasses, and press mud, production processes show maximum efficiency through the combined application of thermomechanical, chemical, and biological technologies.

The transformation establishes environmental protection from disposal issues while advancing technology and creating market opportunities, leading to enhanced sustainability and resilience for the agribusiness sector. The effectiveness of this strategy requires ongoing advancements in optimizing bioprocesses and bioreactors that can monitor and adjust vital parameters in real time.

To compete globally as viable inputs, sugarcane coproducts need manufacturers to address operational challenges while implementing cost reductions. When public policies and economic incentives support sustainability goals, the integration of these new processes will strengthen both natural resource preservation and climate change mitigation while encouraging regenerative production models that create a resilient and self-sustaining production cycle.

2.8 Conclusion

Saccharide biomasses that produce mono- and disaccharides are fundamental to various industries, from food to biofuels. Key plant species, such as sugarcane and sugar beet, efficiently accumulate these sugars, supporting large-scale production. Traditional methods of saccharide biomass processing to obtain sugar are still very much applied; however, more advanced methods, such as chromatography and ultrafiltration, are used to obtain highly pure saccharides. The economic relevance of saccharides is evident in their global production, with major producers shaping market dynamics. As demand for renewable and sustainable sugar sources grows, technological advancements will be essential to optimizing production and reducing environmental impact. Future research must focus on enhancing yield, refining processing techniques, and exploring new applications for its products and by-products, guaranteeing a sustainable production cycle.

Compliance with Ethical Standards The authors declare that they have no conflicts of interest. This chapter does not contain any studies involving animals performed by any of the authors. This chapter does not contain any studies involving human participants performed by any of the authors.

References

Adnan M, Siddiqui AJ, Ashraf SA, Snoussi M, Badraoui R, Alreshidi M, Elasbali AM, Al-Soud WA, Alharethi SH, Sachidanandan M, Patel M (2022) Polyhydroxybutyrate (PHB)-based biodegradable polymer from *Agromyces indicus*: enhanced production, characterization, and optimization. Polymers 14(19):3982. https://doi.org/10.3390/polym14193982

Afifi SM, Gök R, Eikenberg I, Krygier D, Rottmann E, Stübler AS, Aganovic K, Hillebrand S, Esatbeyoglu T (2023) Comparative flavonoid profile of orange (*Citrus sinensis*) flavedo and albedo extracted by conventional and emerging techniques using UPLC-IMS-MS, chemometrics and antioxidant effects. Front Nutr 10:1158473. https://doi.org/10.3389/FNUT.2023.1158473/BIBTEX

Aider M, de Halleux D, Belkacemi K (2007) Production of granulated sugar from maple syrup with high content of inverted sugar. J Food Eng 80(3):791–797. https://doi.org/10.1016/J.JFOODENG.2006.07.008

Almeida Júnior ABD, do Nascimento CWA, Sobral MF, da Silva FBV, Gomes WA (2011) Fertilidade do solo e absorção de nutrientes em cana-de-açúcar fertilizada com torta de filtro. Rev Bras Eng Agríc Ambiental 15(10):1004–1013. https://doi.org/10.1590/S1415-43662011001000003

Almeida PDS, de Menezes CA, Cavalcante WDA, Pinheiro NF, da Silva Júnior FDCG, Viana MB, Leitão RC (2024) Mesophilic anaerobic digestion of fruit and vegetable waste: single- versus two-stage reactor, and Modeling of the liquid and solid fractions of the residue. Ind Biotechnol 20(4):139–146. https://doi.org/10.1089/ind.2024.0007

Ashokkumar V, Flora G, Venkatkarthick R, SenthilKannan K, Kuppam C, Mary Stephy G, Kamyab H, Chen W-H, Thomas J, Ngamcharussrivichai C (2022) Advanced technologies on the sustainable approaches for conversion of organic waste to valuable bioproducts: emerging circular bioeconomy perspective. Fuel 324:124313. https://doi.org/10.1016/j.fuel.2022.124313

Ball S, Lloyd L (2011) Agilent Hi-Plex columns for carbohydrates, alcohols, and acids organic acids on Hi-Plex H. Agil Technol Inc 4:1–4

Balla E, Daniilidis V, Karlioti G, Kalamas T, Stefanidou M, Bikiaris ND, Vlachopoulos A, Koumentakou I, Bikiaris DN (2021) Poly(lactic acid): a versatile biobased polymer for the future with multifunctional properties—from monomer synthesis, polymerization techniques and molecular weight increase to PLA applications. Polymers 13(11):1822. https://doi.org/10.3390/polym13111822

Ban MT, Mahadin N, Abd Karim KJ (2022) Synthesis of hydrogel from sugarcane bagasse extracted cellulose for swelling properties study. Mater Today Proc 50:2567–2575. https://doi.org/10.1016/j.matpr.2021.08.342

Basso LC, Basso TO, Rocha SN (2011) Ethanol production in Brazil: the industrial process and its impact on yeast fermentation. In: Biofuel production-recent developments and prospects, Bernardes. InTech, pp 85–100

Beech FW (1972) Cider making and cider research: a review. J Inst Brew 78(6):477–491. https://doi.org/10.1002/j.2050-0416.1972.tb03485.x

Begum YA, Kumari S, Jain SK, Garg MC (2024) A review on waste biomass-to-energy: integrated thermochemical and biochemical conversion for resource recovery. Environ Sci Adv 3(9): 1197–1216. https://doi.org/10.1039/D4VA00109E

Caione G, de Mello Prado R, Campos CNS, Rodrigues M, Pavinato PS, Agostinho FB (2015) Phosphorus fractionation in soil cultivated with sugarcane fertilized by filter cake and phosphate sources. Commun Soil Sci Plant Anal 46(19):2449–2459. https://doi.org/10.1080/00103624.2015.1081926

Chandel H, Kumar P, Chandel AK, Verma ML (2024) Biotechnological advances in biomass pretreatment for bio-renewable production through nanotechnological intervention. Biomass Convers Biorefinery 14(3):2959–2981. https://doi.org/10.1007/s13399-022-02746-0

Chantasiriwan S (2020) Modification of conventional sugar juice evaporation process for increasing energy efficiency and decreasing sucrose inversion loss. Processes 8(7):765. https://doi.org/10.3390/pr8070765

Deeba F, Kumar KK, Gaur NA (2021) Saccharide to biodiesel. In: Biomass for bioenergy and biomaterials. CRC Press, pp 23–46

Delgado AV, Casanova CA (2001) Sugar processing and by-products of the sugar industry. FAO Agriculture. FAO of the United Nations, Rome

Desalegn B, Kebede E, Legesse H, Fite T (2023) Sugarcane productivity and sugar yield improvement: selecting variety, nitrogen fertilizer rate, and bioregulator as a first-line treatment. Heliyon 9(4):e15520. https://doi.org/10.1016/j.heliyon.2023.e15520

Dias MODS, Maciel Filho R, Mantelatto PE, Cavalett O, Rossell CEV, Bonomi A, Leal MRLV (2015) Sugarcane processing for ethanol and sugar in Brazil. Environ Dev 15:35–51. https://doi.org/10.1016/j.envdev.2015.03.004

Dien BS, Cotta MA, Jeffries TW (2003) Bacteria engineered for fuel ethanol production: current status. Appl Microbiol Biotechnol 63(3):258–266. https://doi.org/10.1007/s00253-003-1444-y

Dimopoulou M, Offiah V, Falade K, Smith AM, Kontogiorgos V, Angelis-Dimakis A (2021) Techno-economic assessment of polysaccharide extraction from baobab: a scale up analysis. Sustainability 13(17):1–13. https://doi.org/10.3390/su13179915

Doherty WOS (2011) Improved sugar cane juice clarification by understanding calcium oxide-phosphate-sucrose systems. J Agric Food Chem 59(5):1829–1836. https://doi.org/10.1021/jf1043212

Duraisam R, Salelgn K, Berekete AK (2017) Production of beet sugar and bio-ethanol from sugar beet and it bagasse: a review. Int J Eng Trends Technol 43(4):222–233. https://doi.org/10.14445/22315381/IJETT-V43P237

Dwivedi RS (2022) Saccharide sweet (SS) principles, classification and structural and functional details of SS sweeteners and plants. Alternative. In: Alternative sweet and supersweet principles. Springer, pp 113–223

Ebmeyer H, Hoffmann CM (2020) Early drought stress: effects on yield formation and quality of sugar beet. Sugar Ind 145:104–113. https://doi.org/10.36961/si24062

Eggleston G, Tew T, Panella L, Klasson T (2010) Ethanol from sugar crops. In: Singh B (ed) Industrial crops and uses. CABI, Wallingford, pp 60–83

Eggleston G, Cole M, Andrzejewski B (2013) New commercially viable processing technologies for the production of sugar feedstocks from sweet sorghum (*Sorghum bicolor* L. Moench) for manufacture of biofuels and bioproducts. Sugar Tech 15(3):232–249. https://doi.org/10.1007/s12355-013-0229-6

El Sayed A, Hamad K, Eid Y, Alaa A, Hassan A, Mohsen A, Safwat A, Atta K, Benbella M, Abdel Tawab S (2022) Production of paper using chemical pulping process of sugarcane bagasse. Int J Ind Sustain Dev 3:57–65. https://doi.org/10.21608/ijisd.2022.145854.1013

Elferink H, Mensink RA, Castelijns WWA, Jansen O, Bruekers JPJ, Martens J, Oomens J, Rijs AM, Boltje TJ (2019) The glycosylation mechanisms of 6,3-uronic acid lactones. Angew Chem Int Ed 58(26):8746–8751. https://doi.org/10.1002/ANIE.201902507

FAO (2023) Agricultural production statistics 2010–2023. https://openknowledge.fao.org/server/api/core/bitstreams/df90e6cf-4178-4361-97d4-5154a9213877/content

Flórez Pardo LM, López Galán JE, Lozano Ramírez T (2018) Saccharide biomass for biofuels, biomaterials, and chemicals. In: Biomass and green chemistry. Springer International Publishing, Cham, pp 11–30

Gavahian M, Chu YH, Mousavi Khaneghah A (2019) Recent advances in orange oil extraction: an opportunity for the valorisation of orange peel waste a review. Int J Food Sci Technol 54(4):925–932. https://doi.org/10.1111/IJFS.13987

Globenewswire (2023) Starch global market to reach 199.8 million metric tons by 2030: use of starch as a fat replacer drives growth. Retrieved from: https://www.globenewswire.com/news-release/2023/02/28/2617539/28124/en/Starch-Global-Market-to-Reach-199-8-Million-Metric-Tons-by-2030-Use-of-Starch-as-a-Fat-Replacer-Drives-Growth.html

Gnansounou E, Dauriat A, Wyman CE (2005) Refining sweet sorghum to ethanol and sugar: economic trade-offs in the context of North China. Bioresour Technol 96(9):985–1002. https://doi.org/10.1016/j.biortech.2004.09.015

Gunasekara RW, Zhao Y (2017) A general method for selective recognition of monosaccharides and oligosaccharides in water. J Am Chem Soc 139(2):829–835. https://doi.org/10.1021/JACS.6B10773/SUPPL_FILE/JA6B10773_SI_001.PDF

Hamerski F, da Silva VR, Corazza ML, Ndiaye PM, de Aquino AD (2012) Assessment of variables effects on sugar cane juice clarification by carbonation process. Int J Food Sci Technol 47(2):422–428. https://doi.org/10.1111/j.1365-2621.2011.02857.x

Harrison RG, Todd P, Rudge SR, Petrides DP (2003) Bioseparations science and engineering. Oxford University Press

Hassan S, Ngo T, Ball AS (2024) Valorisation of sugarcane bagasse for the sustainable production of polyhydroxyalkanoates. Sustainability 16(5):2200. https://doi.org/10.3390/su16052200

Hewitt S, Hernández-Montes E, Dhingra A, Keller M (2023) Impact of heat stress, water stress, and their combined effects on the metabolism and transcriptome of grape berries. Sci Rep 13(1):9907. https://doi.org/10.1038/s41598-023-36160-x

Holden NM, Neill AM, Stout JC, O'Brien D, Morris MA (2023) Biocircularity: a framework to define sustainable, circular bioeconomy. Circ Econ Sustain 3(1):77–91. https://doi.org/10.1007/s43615-022-00180-y

Igbokwe VC, Ezugworie FN, Onwosi CO, Aliyu GO, Obi CJ (2022) Biochemical biorefinery: a low-cost and non-waste concept for promoting sustainable circular bioeconomy. J Environ Manag 305:114333. https://doi.org/10.1016/j.jenvman.2021.114333

Jamir L, Kumar V, Kaur J, Kumar S, Singh H (2021) Composition, valorization and therapeutical potential of molasses: a critical review. Environ Technol Rev 10(1):131–142. https://doi.org/10.1080/21622515.2021.1892203

Jammer A, Albacete A, Schulz B, Koch W, Weltmeier F, van der Graaff E, Pfeifhofer HW, Roitsch TG (2020) Early-stage sugar beet taproot development is characterized by three distinct physiological phases. Plant Direct 4(7):e00221. https://doi.org/10.1002/pld3.221

Janke L, Leite A, Nikolausz M, Schmidt T, Liebetrau J, Nelles M, Stinner W (2015) Biogas production from sugarcane waste: assessment on kinetic challenges for process designing. Int J Mol Sci 16(9):20685–20703. https://doi.org/10.3390/ijms160920685

Janusz G, Pawlik A, Sulej J, Świderska-Burek U, Jarosz-Wilkołazka A, Paszczyński A (2017) Lignin degradation: microorganisms, enzymes involved, genomes analysis and evolution. FEMS Microbiol Rev 41(6):941–962. https://doi.org/10.1093/femsre/fux049

Jofre FM, Prado CA, Shibukawa VP, Rodrigues BG, Sarangi PK, Chandel AK (2025) Critical analysis of process parameters towards smart bioreactors development in biorefinery for biorenewables production. Int J Biol Macromol 305:140957. https://doi.org/10.1016/j.ijbiomac.2025.140957

Karp SG, Medina JDC, Letti LAJ, Woiciechowski AL, de Carvalho JC, Schmitt CC, de Oliveira PR, Kumlehn GS, Soccol CR (2021) Bioeconomy and biofuels: the case of sugarcane ethanol in Brazil. Biofuels Bioprod Biorefin 15(3):899–912. https://doi.org/10.1002/bbb.2195

Karp SG, Burgos WJM, Vandenberghe LPS, Diestra KV, Torres LAZ, Woiciechowski AL, Letti LAJ, Pereira GVM, Thomaz-Soccol V, Rodrigues C, de Carvalho JC, Soccol CR (2022a) Sugarcane: a promising source of green carbon in the circular bioeconomy. Sugar Tech 24(4): 1230–1245. https://doi.org/10.1007/s12355-022-01161-z

Karp SG, de Mello AFM, Herrmann LW, Vale ADS, Martinez-Burgos WJ, Savian CM, Soccol CR (2022b) New technologies for bioethanol production: patents and innovation. In: Liquid biofuels: bioethanol. Springer, pp 489–515

Kawahigashi H, Kasuga S, Okuizumi H, Hiradate S, Yonemaru J (2013) Evaluation of Brix and sugar content in stem juice from sorghum varieties. Grassl Sci 59(1):11–19. https://doi.org/10.1111/grs.12006

Khan N, Fahad S, Naushad M, Faisal S (2020) Grape production critical review in the world. SSRN Electron J. https://doi.org/10.2139/SSRN.3595842

Khan Q, Qin Y, Guo D-J, Yang L-T, Song X-P, Xing Y-X, Li Y-R (2023) A review of the diverse genes and molecules involved in sucrose metabolism and innovative approaches to improve sucrose content in sugarcane. Agronomy 13(12):2957. https://doi.org/10.3390/agronomy13122957

Kishore N, Gayathri D, Venkatesh J, Rajeswari V, Sangeeta B, Chandrika A (2017) Present mechanization status in sugarcane—a review. Int J Agric Sci 9(22):4247–4253

Kroupin PY, Kroupina AY, Karlov GI, Divashuk MG (2023) Root causes of flowering: two sides of bolting in sugar beet. Agronomy 13(11):2671. https://doi.org/10.3390/agronomy13112671

Lakatos L, Mitre Z (2023) Effect of drought on the future sugar content of wine grape varieties till 2100: possible adaptation in the Hungarian Eger wine region. Biomolecules 13(7):1143. https://doi.org/10.3390/biom13071143

Lakhawat SS, Malik N, Kumar V, Kumar S, Sharma PK (2022) Implications of CRISPR-Cas9 in developing next generation biofuel: a mini-review. Curr Protein Pept Sci 23(9):574–584. https://doi.org/10.2174/1389203723666220907110310

Lee H, Jung Sohn Y, Jeon S, Yang H, Son J, Jin Kim Y, Jae Park S (2023) Sugarcane wastes as microbial feedstocks: a review of the biorefinery framework from resource recovery to production of value-added products. Bioresour Technol 376:128879. https://doi.org/10.1016/j.biortech.2023.128879

Li C, Iqbal MA (2024) Leveraging the sugarcane CRISPR/Cas9 technique for genetic improvement of non-cultivated grasses. Front Plant Sci 15:1369416. https://doi.org/10.3389/fpls.2024.1369416

Li Y, Qi B, Wan Y (2020) Separation of monosaccharides from pretreatment inhibitors by nanofiltration in lignocellulosic hydrolysate: fouling mitigation by activated carbon adsorption. Biomass Bioenergy 136:105527. https://doi.org/10.1016/J.BIOMBIOE.2020.105527

Li X, Tan S, Luo J, Pinelo M (2021) Nanofiltration for separation and purification of saccharides from biomass. Front Chem Sci Eng 15(4):837–853. https://doi.org/10.1007/s11705-020-2020-z

Liu Y, Wen J, Ouyang A, Ying Y (2006) Rapid analysis of sugar content of intact orange fruit using ultraviolet and visible transmittance techniques. In: Chen Y-R, Meyer GE, Tu S-I (eds) Proceedings. SPIE, p 638115

Liu D, Tang W, Yin JY, Nie SP, Xie MY (2021) Monosaccharide composition analysis of polysaccharides from natural sources: hydrolysis condition and detection method development. Food Hydrocoll 116:106641. https://doi.org/10.1016/J.FOODHYD.2021.106641

Lopes JDC, Madureira J, Margaça FMA, Cabo Verde S (2025) Grape pomace: a review of its bioactive phenolic compounds, health benefits, and applications. Molecules 30(2):362. https://doi.org/10.3390/MOLECULES30020362

Lu L, Delrot S, Liang Z (2024) From acidity to sweetness: a comprehensive review of carbon accumulation in grape berries. Mol Hortic 4(1):22. https://doi.org/10.1186/s43897-024-00100-8

Luo J, Zeuner B, Morthensen ST, Meyer AS, Pinelo M (2015) Separation of phenolic acids from monosaccharides by low-pressure nanofiltration integrated with laccase pre-treatments. J Memb Sci 482:83–91. https://doi.org/10.1016/J.MEMSCI.2015.02.022

Mahmud MA, Anannya FR (2021) Sugarcane bagasse—a source of cellulosic fiber for diverse applications. Heliyon 7(8):e07771. https://doi.org/10.1016/j.heliyon.2021.e07771

Mahmud MA, Belal SA, Gafur MA (2023) Development of a biocomposite material using sugarcane bagasse and modified starch for packaging purposes. J Mater Res Technol 24: 1856–1874. https://doi.org/10.1016/j.jmrt.2023.03.083

Mastrorilli M, Katerji N, Rana G (1999) Productivity and water use efficiency of sweet sorghum as affected by soil water deficit occurring at different vegetative growth stages. Eur J Agron 11(3–4):207–215. https://doi.org/10.1016/S1161-0301(99)00032-5

Mathur S, Umakanth AV, Tonapi VA, Sharma R, Sharma MK (2017) Sweet sorghum as biofuel feedstock: recent advances and available resources. Biotechnol Biofuels 10(1):146. https://doi.org/10.1186/s13068-017-0834-9

Mehdi F, Cao Z, Zhang S, Gan Y, Cai W, Peng L, Wu Y, Wang W, Yang B (2024a) Factors affecting the production of sugarcane yield and sucrose accumulation: suggested potential biological solutions. Front Plant Sci 15(May):1–21. https://doi.org/10.3389/fpls.2024.1374228

Mehdi F, Galani S, Wickramasinghe KP, Zhao P, Lu X, Lin X, Xu C, Liu H, Li X, Liu X (2024b) Current perspectives on the regulatory mechanisms of sucrose accumulation in sugarcane. Heliyon 10(5):e27277. https://doi.org/10.1016/j.heliyon.2024.e27277

Mello AFMD, Vandenberghe LPDS, Diestra KKV, Bittencourt GA, WJM B, Soccol CR (2022) Corn first-generation bioethanol unities with energy and dried grains with solubles (DDGS) production. Angew Chem Int Ed 6(11):109–132

Meng Y, Yu S, Qiu Z, Zhang J, Wu J, Yao T, Qin J (2021) Modeling and optimization of sugarcane juice clarification process. J Food Eng 291:110223. https://doi.org/10.1016/j.jfoodeng.2020.110223

Mohanta V, Mukherjee I, Chottopadhyay JP (2021) Waste product utilization: preparation of candy from orange (*Citrus sinensis*) peel. Int J Agric Appl Sci 2(2):114–119. https://doi.org/10.52804/ijaas2021.2217

Moodley M, Schorn PM, Walthew DC, Masinga P (2002) Optimising the carbonation process. Int Sugar J 105:469–476

Morales H, di Sciascio F, Aguirre-Zapata E, Amicarelli A (2024) Crystallization process in the sugar industry: a discussion on fundamentals, industrial practices, modeling, estimation and control. Food Eng Rev 16(3):441–469. https://doi.org/10.1007/s12393-024-09377-3

Mushtaq M (2018) Extraction of fruit juice: an overview. In: Rajauria G, Tiwari B (eds) Fruit juices: extraction composition quality and analysis. Elsevier, pp 131–159

Mustafa G, Arshad M, Bano I, Abbas M (2023) Biotechnological applications of sugarcane bagasse and sugar beet molasses. Biomass Convers Biorefinery 13(2):1489–1501. https://doi.org/10.1007/s13399-020-01141-x

Nelson DL, Cox MM (2017) Lehninger principles of biochemistry. W. H. Freeman

Niaz K, Khan F, Shah MA (2020) Analysis of carbohydrates (monosaccharides, polysaccharides). In: Recent advances in natural products analysis. Elsevier, pp 621–633

Obodovych O, Shevchenko O, Myronchuk V, Lymar A, Sydorenko V, Yakobchuk R (2021) Intensification of the inverted sugar syrup production using the rotor-pulsation processing. Ukr Food J 10(3):605–614. https://doi.org/10.24263/2304-974X-2021-10-3-13

OIV (2019) Statistical Report on World Vitiviniculture. Retrieved July 15, 2025, from https://www.oiv.int/public/medias/6782/oiv-2019-statistical-report-on-world-vitiviniculture.pdf#:~:text=Grapes%20intended%20for%20all%20uses,of%20dried%20grape

Osman AI, Mehta N, Elgarahy AM, Al-Hinai A, Al-Muhtaseb AH, Rooney DW (2021) Conversion of biomass to biofuels and life cycle assessment: a review. Environ Chem Lett 19(6): 4075–4118. https://doi.org/10.1007/s10311-021-01273-0

Panella L, Kaffka SR (2010) Sugar beet (*Beta vulgaris* L) as a biofuel feedstock in the United States. ACS Publications, pp 163–175

Panella L, Kaffka SR, Lewellen RT, Mitchell McGrath J, Metzger MS, Strausbaugh CA (2015) Sugarbeet. In: Smith S, Diers B, Specht J, Carver B. (eds) Yield gains in major U.S. field crops. CSSA, Madison, pp. 357–395

Parascanu MM, Sanchez N, Sandoval-Salas F, Carreto CM, Soreanu G, Sanchez-Silva L (2021) Environmental and economic analysis of bioethanol production from sugarcane molasses and agave juice. Environ Sci Pollut Res 28(45):64374–64393. https://doi.org/10.1007/s11356-021-15471-4

Pineda-Lozano JE, Fonseca-Bustos V, Martinez-Moreno AG, Virgen-Carrillo CA (2022) The biological effect of orange (*Citrus sinensis* L.) by-products on metabolic biomarkers: a systematic review. Front Sustain Food Syst 6:1003144. https://doi.org/10.3389/FSUFS.2022.1003144/BIBTEX

Prati P, Moretti RH (2010) Study of clarification process of sugar cane juice for consumption. Ciência Tecnol Aliment 30(3):776–783. https://doi.org/10.1590/S0101-20612010000300033

Priyadarshini A, Priyadarshini A (2018) Market dimensions of the fruit juice industry. In: Rajauria G, Tiwari BK (eds) Fruit juices. Elsevier, pp 15–32

Rae AL, Perroux JM, Grof CPL (2005) Sucrose partitioning between vascular bundles and storage parenchyma in the sugarcane stem: a potential role for the ShSUT1 sucrose transporter. Planta 220(6):817–825. https://doi.org/10.1007/s00425-004-1399-y

Ram B, Ramaiyan K (2021) Current status of sugarcane agriculture and sugar industry in the world with special reference to India. In: Ram K, Hemaprabha G, Nashine R, Mohanraj K, Murali P, Shah RA (eds) Recent scientific advances in sugarcane cultivation for doubling farmers' income. Dilpreet Publishers and Distributors, New Delhi

Regassa TH, Wortmann CS (2014) Sweet sorghum as a bioenergy crop: literature review. Biomass Bioenergy 64:348–355. https://doi.org/10.1016/j.biombioe.2014.03.052

Reichembach LH, de Oliveira Petkowicz CL (2021) Pectins from alternative sources and uses beyond sweets and jellies: an overview. Food Hydrocoll 118:106824. https://doi.org/10.1016/J.FOODHYD.2021.106824

Rodrigues R, Sperandio LCC, Andrade CMG (2018) Investigation of color and turbidity in the clarification of sugarcane juice by ozone. J Food Process Eng 41(3). https://doi.org/10.1111/jfpe.12661

Rooney WL, Blumenthal J, Bean B, Mullet JE (2007) Designing sorghum as a dedicated bioenergy feedstock. Biofuels Bioprod Biorefin 1(2):147–157. https://doi.org/10.1002/bbb.15

Saska M, Zossi S, Liu H (2010) Removal of colour in sugar cane juice clarification by defecation sulfitation and carbonation. Int Sugar J 112:258–264

Schindler B, Barnes L, Renois G, Gray C, Chambert S, Fort S, Flitsch S, Loison C, Allouche AR, Compagnon I (2017) Anomeric memory of the glycosidic bond upon fragmentation and its consequences for carbohydrate sequencing. Nat Commun 8(1):1–7. https://doi.org/10.1038/s41467-017-01179-y

Schneider K, Schäfer-Pregl R, Borchardt D, Salamini F (2002) Mapping QTLs for sucrose content, yield and quality in a sugar beet population fingerprinted by EST-related markers. Theor Appl Genet 104(6):1107–1113. https://doi.org/10.1007/s00122-002-0890-8

Seeberger PH (2017) Chapter 2: Monosaccharide diversity. In: Essentials glycobiol, 3rd edn. Cold Spring Harbor Laboratory Press, Cold Spring Harbor

Siddique A, Upadhyay H, Sharma M, Kumar A (2018) Regulation of sugar in sweet sorghum crop—a review. J Pure Appl Microbiol 12(1):355–359. https://doi.org/10.22207/JPAM.12.1.41

Singh RV, Sharma P, Sambyal K (2022) Application of sugarcane bagasse in chemicals and food packaging industry: potential and challenges. Circ Econ Sustain 2(4):1479–1500. https://doi.org/10.1007/s43615-022-00167-9

Solomons TWG, Fryhle CB (2011) Organic chemistry. Wiley

Spherical-Insights (2023) Global industrial sugar market size insights forecasts to 2032. Retrieved from: https://www.sphericalinsights.com/reports/industrial-sugar-market (1 Feb 2017), p 249677. https://doi.org/10.1007/978-3-031-01241-9

Stanfield RC, Forrestel EJ, Elmendorf KE, Bagshaw SB, Bartlett MK (2024) Phloem anatomy predicts berry sugar accumulation across 13 wine-grape cultivars. Front Plant Sci 15:1360381. https://doi.org/10.3389/fpls.2024.1360381

Statista (2024a) Production volume of sugar across India from 2018/2019 to 2023/2024. Retrieved from: https://www.statista.com/statistics/761818/india-sugar-production-volume/ (1 Feb 2017)

Statista (2024b) Sugar production in Brazil from 2009/2010 to 2023/24. Retrieved from: https://www.statista.com/statistics/249677/production-of-sugar-in-brazil/ (1 Feb 2017)

Statista (2024c) Total sugar production worldwide from 2010/11 to 2024/25 (in million metric tons), vol 25. Retrieved from: https://www.statista.com/statistics/249679/total-production-of-sugar-worldwide/ (1 Feb 2025) 2023–2024. https://doi.org/10.1007/978-3-031-01241-9

Sun M, Gao AX, Liu X, Bai Z, Wang P, Ledesma-Amaro R (2024a) Microbial conversion of ethanol to high-value products: progress and challenges. Biotechnol Biofuels Bioprod 17(1): 115. https://doi.org/10.1186/s13068-024-02546-w

Sun W, He Z, Liu B, Ma D, Si R, Li R, Wang S, Malekian A (2024b) Changes in photosynthetic efficiency, biomass, and sugar content of sweet sorghum under different water and salt conditions in arid region of Northwest China. Agriculture 14(12):2321. https://doi.org/10.3390/agriculture14122321

Sunori SK, Shree S, Juneja PK (2015) Control of sugarcane crushing mill process: a comparative analysis. In: 2015 International conference on soft computing techniques and implementations (ICSCTI). IEEE, pp 1–5

Suter B, Destrac Irvine A, Gowdy M, Dai Z, van Leeuwen C (2021) Adapting wine grape ripening to global change requires a multi-trait approach. Front Plant Sci 12:624867. https://doi.org/10.3389/fpls.2021.624867

Swaminaathan P, Saravanan A, Thamarai P (2024) Utilization of bioresources for high-value bioproducts production: sustainability and perspectives in circular bioeconomy. Sustain Energy Technol Assess 63:103672. https://doi.org/10.1016/j.seta.2024.103672

Sydney EB, de Carvalho JC, Letti LAJ, Magalhães AI, Karp SG, Martinez-Burgos WJ, Candeo EDS, Rodrigues C, Vandenberghe LPDS, Neto CJD, Torres LAZ, Medeiros ABP, Woiciechowski AL, Soccol CR (2021) Current developments and challenges of green technologies for the valorization of liquid, solid, and gaseous wastes from sugarcane ethanol production. J Hazard Mater 404(September 2020):124059. https://doi.org/10.1016/j.jhazmat.2020.124059

Tang S, Chen Z, Chen F, Lai X, Wei Q, Chen X, Jiang C (2023) Extraction and surface functionalization of cellulose nanocrystals from sugarcane bagasse. Molecules 28(14):5444. https://doi.org/10.3390/molecules28145444

Temma T, Hanamatsu K, Shinoki F (2002) Measuring the sugar content of apples and apple juice by near infrared spectroscopy. Opt Rev 9(2):40–44. https://doi.org/10.1007/s10043-002-0040-1

Ungureanu N, Vlăduț V, Biriș S-Ș (2022) Sustainable valorization of waste and by-products from sugarcane processing. Sustainability 14(17):11089. https://doi.org/10.3390/su141711089

Upadhyay SK, Singh G, Rani N, Rajput VD, Seth CS, Dwivedi P, Minkina T, Wong MH, Show PL, Khoo KS (2024) Transforming bio-waste into value-added products mediated microbes for enhancing soil health and crop production: perspective views on circular economy. Environ Technol Innov 34:103573. https://doi.org/10.1016/j.eti.2024.103573

Vandenberghe LPS, Valladares-Diestra KK, Bittencourt GA, Zevallos Torres LA, Vieira S, Karp SG, Sydney EB, de Carvalho JC, Thomaz Soccol V, Soccol CR (2022) Beyond sugar and ethanol: the future of sugarcane biorefineries in Brazil. Renew Sust Energ Rev 167:112721. https://doi.org/10.1016/j.rser.2022.112721

Vassileva M, Malusà E, Sas-Paszt L, Trzcinski P, Galvez A, Flor-Peregrin E, Shilev S, Canfora L, Mocali S, Vassilev N (2021) Fermentation strategies to improve soil bio-inoculant production and quality. Microorganisms 9(6):1254. https://doi.org/10.3390/microorganisms9061254

Velazquez-Camilo O, Bolaños-Reynoso E, Rodriguez E, Alvarez-Ramirez J (2010) Characterization of cane sugar crystallization using image fractal analysis. J Food Eng 100(1):77–84. https://doi.org/10.1016/j.jfoodeng.2010.03.030

Venios X, Korkas E, Nisiotou A, Banilas G (2020) Grapevine responses to heat stress and global warming. Plants 9(12):1754. https://doi.org/10.3390/plants9121754

Verified-Market-Reports (2024) Modified starch market insights. Retrieved from: https://www.verifiedmarketreports.com/product/global-modified-starch-market-size-and-forecast-to-2025 (1 Feb 2025). https://doi.org/10.1007/978-3-031-01241-9

Vilela A, Pinto T (2021) Grape infusions: between nutraceutical and green chemistry. Sustain Chem 2(3):441–466. https://doi.org/10.3390/SUSCHEM2030025

Vilpoux OF, Junior JFSS (2023) Chapter 3—Global production and use of starch. In: Starchy crops morphology, extraction, properties and applications. Elsevier, pp 43–66

Wang J, Nayak S, Koch K, Ming R (2013) Carbon partitioning in sugarcane (Saccharum species). Front Plant Sci 4:201. https://doi.org/10.3389/fpls.2013.00201

Wang X, Zhang L, Wu J, Xu W, Wang X, Lü X (2017) Improvement of simultaneous determination of neutral monosaccharides and uronic acids by gas chromatography. Food Chem 220:198–207. https://doi.org/10.1016/J.FOODCHEM.2016.10.008

Wang Z, Zheng Y, Lai Z, Hu X, Wang L, Wang X, Li Z, Gao M, Yang Y, Wang Q, Li N (2024) Effect of monosaccharide composition and proportion on the bioactivity of polysaccharides: a review. Int J Biol Macromol 254:127955. https://doi.org/10.1016/J.IJBIOMAC.2023.127955

Woiciechowski AL, Letti LAJ, Karp SG, Filho AZ, Torres LAZ, Burgos WJM, Savian CM, Soccol CR (2022) Enzymatic hydrolysis of feedstocks for 1G bioethanol production. In: Liquid biofuels: bioethanol. Springer, pp 61–83

Yan J, Shi S, Wang H, Liu R, Li N, Chen Y, Wang S (2016) Neutral monosaccharide composition analysis of plant-derived oligo- and polysaccharides by high performance liquid chromatography. Carbohydr Polym 136:1273–1280. https://doi.org/10.1016/J.CARBPOL.2015.10.028

Yao Y, Meng X, Li C, Bernaerts KV, Zhang K (2023) Tuning the chiral structures from self-assembled carbohydrate derivatives. Small 19(25):2208286. https://doi.org/10.1002/SMLL.202208286

Zaitoun M, Ghanem M, Harphous S (2018) Sugars: types and their functional properties in food and human health. Int J Public Health Res 6(4):93–99

Zhang N, Huang W, Xia G, Oppong MB, Ding L, Li P, Qiu F (2018) Methods for determination of absolute configuration of monosaccharides. Chin Herb Med 10(1):14–22. https://doi.org/10.1016/J.CHMED.2017.12.009

Zhang R, Ma S, Li L, Zhang M, Tian S, Wang D, Liu K, Liu H, Zhu W, Wang X (2021) Comprehensive utilization of corn starch processing by-products: a review. Grain Oil Sci Technol 4(3):89–107. https://doi.org/10.1016/j.gaost.2021.08.003

Zhong H, Yadav V, Wen Z, Zhou X, Wang M, Han S, Pan M, Zhang C, Zhang F, Wu X (2023) Comprehensive metabolomics-based analysis of sugar composition and content in berries of 18 grape varieties. Front Plant Sci 14:1200071. https://doi.org/10.3389/fpls.2023.1200071

Zicari S, Zhang R, Kaffka S (2019) Sugar Beet. In: Pan Z, Zhang R, Zicari S (eds) Integrated processing technologies for food and agricultural by-products. Elsevier, pp 331–351

Chapter 3
Amylaceous Biomasses

Luiz Alberto Junior Letti, Adenise Lorenci Woiciechowski, Fernando Enrique Rosas Vega, Agnes de Freitas Diniz de Souza, Isabela Panato Mücke, Felipe Marcondes Stival, Maísa Emanuelle Colodel, and Carlos Ricardo Soccol

Abstract Starch is a fundamental polysaccharide widely distributed in amylaceous biomasses such as corn, cassava, potato, wheat, and rice. This chapter explores the structural, biochemical, and technological aspects of starch, emphasizing its role in food, bioenergy, and industrial applications. This chapter describes the biosynthesis of amylose and amylopectin, the diversity of starch granules, and the influence of environmental and genetic factors on starch composition. Extraction and purification methods are discussed, along with physical, chemical, enzymatic, and genetic modification techniques that enhance starch functionality. Economic aspects, including global production, trade policies, and cost structures, are analyzed in light of the growing demand for starch in bioplastics and biofuels. Special attention is given to innovative, sustainable processes that align with circular bioeconomy principles. This chapter also highlights the role of starch in biorefineries and its potential in emerging applications such as ethanol-to-jet (ETJ) fuels and biodegradable materials. By integrating scientific, technological, and economic perspectives, this work offers a comprehensive framework for understanding and advancing the use of amylaceous biomasses in sustainable development.

3.1 Introduction

Starch is one of the most abundant and versatile polysaccharides in nature, serving as a primary energy reserve in plants and as a key raw material for numerous industrial applications. Its wide availability, biodegradability, and functional diversity make starch an essential component in the food, pharmaceutical, textile, packaging, and bioenergy sectors. The increasing global demand for renewable, low-impact

L. A. J. Letti (✉) · A. L. Woiciechowski · F. E. R. Vega · A. de Freitas Diniz de Souza · I. P. Mücke · F. M. Stival · M. E. Colodel · C. R. Soccol
DDepartment of Bioprocess Engineering and Biotechnology, Federal University of Paraná, Centro Politécnico, Curitiba, Paraná, Brazil
e-mail: letti@ufpr.br

A. L. Woiciechowski et al. (eds.), *Biomass for Bioenergy and Bioproducts*, Biofuel and Biorefinery Technologies 17, https://doi.org/10.1007/978-3-032-19898-3_3

materials further amplifies the strategic value of starch as a cornerstone of the bioeconomy.

Amylaceous biomasses, such as corn, cassava, potato, rice, and wheat, are major agricultural sources of starch. These biomasses accumulate starch in specialized organelles called amyloplasts, which synthesize and store it in granule form. The molecular structure of starch—comprising amylose and amylopectin—determines its physicochemical properties and, thus, its suitability for specific industrial uses. Understanding the variation in starch content and structure across different plant species is crucial for optimizing its extraction and transformation.

The global market for starch has grown rapidly, with significant contributions to food processing, adhesive and paper industries, and increasingly, bioethanol and bioplastic production. However, native starches often exhibit limitations, such as poor solubility, thermal instability, and retrogradation, which restrict their performance in many applications. Consequently, various modification techniques have been developed to improve or tailor starch's properties.

In parallel, advances in biotechnology, process engineering, and circular economy models are fostering more sustainable and efficient ways of extracting and utilizing starch from both conventional and unconventional sources. These innovations include enzymatic treatments, physical and chemical modifications, and even *in planta* genetic strategies to enhance starch functionality directly within crops.

This chapter explores the structure, sources, processing methods, and economic relevance of amylaceous biomasses. It discusses the mechanisms of starch biosynthesis, modern techniques for extraction and modification, and the role of starch-based biomasses in the context of global sustainability and industrial circularity.

3.2 Structure and Properties of Starch

Energy is essential to every living organism, which leads to the necessity of storing it in some way. Each life-form does so in its own way—while animals have glycogen, plants have starch.

Starch is a polysaccharide composed of two main structures, amylose and amylopectin, both of which are chains of glucose. This basal building block, glucose, is a monosaccharide with six carbon atoms and three hydroxyl groups at positions 2, 3, and 6. These hydroxyl groups allow the formation of both intermolecular and intramolecular hydrogen bonds, creating structures that will be further described in this chapter. They also enable interactions with water, which significantly affect starch's solubility and its capacity to absorb water during processes like gelatinization. As a whole, starch is mainly hydrophilic. Additionally, hydroxyl groups serve as reactive sites for chemical modifications of starch, such as esterification, etherification, and cross-linking. The glucose molecules in starch are connected by covalent linkages called glycosidic bonds, which are typically denoted by the carbon atoms they unite and their stereochemistry (α or β) (BeMiller and Whistler 2009).

Amylose is a linear polymer in which glucose subunits are united by α-1,4-glycosidic bonds, resulting in the formation of single or double helical coils. Although very rare, amylose can occasionally form α-1,6-glycosidic bonds, leading to branching. Its molecular weight typically ranges from 10^5 to 10^6 Da. Due to strong intermolecular hydrogen bonding and its linear structure, amylose is less soluble in water than amylopectin. The linear chains also allow the formation of strong and rigid gels upon cooling, further stabilized by hydrogen bonds between molecules. Amylose retrogrades quickly and is less digestible by enzymes, contributing to the formation of resistant starch, a type of starch that resists digestion in the small intestine of animals and can be fermented in the large intestine, feeding the gut microbiota (Bojarczuk). Amylose usually comprises about 20–30% of starch, though this varies with plant species, age, and conditions. For example, amylomaize can contain up to 70% amylose (BeMiller and Whistler 2009).

Amylopectin is a highly branched polymer, containing both α-1,4- and α-1,6-glycosidic bonds between glucose subunits—approximately one in every twenty glycosidic bonds is an α-1,6 branch. With a molecular weight ranging from 10^7 to 10^9 Da, amylopectin is one of the largest known biological molecules. The branching disrupts intense hydrogen bonding, making amylopectin more soluble in water than amylose. This also results in the formation of weaker gels, slower retrogradation, and greater digestibility by enzymes, making it more accessible to animals that consume plants. Amylopectin typically comprises about 70–80% of starch, though species that produce waxy starches can have up to 99% amylopectin content (BeMiller and Whistler 2009).

The glucose chains in amylopectin can be classified in two ways: by size and by type. Short chains have a degree of polymerization (DP) between 12 and 60, while long chains have a DP between 30 and 45, and very long chains have a DP greater than 60. When classified by type, C-chains are the "main" chains containing the single reducing end of the molecule, B-chains have one or more branches, and A-chains do not carry any branches (BeMiller and Whistler 2009).

The most scientifically accepted description of amylopectin's structure is the cluster model, which describes its branched organization as consisting of clusters of short chains arranged hierarchically. At the smallest level, branched glucose chains form clusters. Neighboring chains with a degree of polymerization of less than 10 create densely packed double helices, contributing to the crystalline regions of starch, while the α-1,6-glycosidic bonds form amorphous regions. These clusters are arranged in a larger, semicrystalline structure, with alternating crystalline and amorphous lamellae exhibiting a periodicity of around 10 nm. Other models, such as the Hizukuri model, provide additional insights into amylopectin's structure. Hizukuri demonstrated that most B-chains participate in more than one crystalline cluster and further divided them into B1, B2, B3, and B4 based on their involvement in multiple structures (Hizukuri 1985; Nakamura and Kainuma 2022).

For long-term energy storage, most plants form starch granules, densely packed structures that are insoluble in water but still bioavailable to necessary enzymes. These granules vary in size from 0.1 to 200 μm and can take on a variety of shapes, such as spheres, ellipsoids, and tubules. Their distinct shapes and sizes make them useful for identifying plant species via microscopy (BeMiller and Whistler 2009).

Starch granules are insoluble in water and can be classified as A-type, B-type, or C-type based on their diffraction patterns when analyzed with X-ray techniques such as X-ray diffraction and small-angle X-ray scattering. A-type starch, found primarily in cereals like maize, wheat, rice, and barley, has a densely packed, monoclinic crystalline structure. The amylopectin double helices form a hexagonal pattern, with water molecules located between the helices, resulting in less dense packing of the chains. This structure makes A-type starch less resistant to enzymatic digestion. A-type starch typically forms smaller granules and weaker gels, with a higher gelatinization temperature (BeMiller and Whistler 2009).

In contrast, B-type starch has a more open, hexagonal crystalline structure. Here, amylopectin double helices are arranged in a more spaced-out manner, with water molecules occupying larger channels. Despite this looser overall structure, the amylopectin chains are densely packed within clusters, providing greater resistance to enzymatic digestion and forming stronger gels. B-type starch granules are usually larger and have a lower gelatinization temperature. They are more commonly found in tubers and high-amylose cereals like amylomaize (BeMiller and Whistler 2009).

Finally, C-type starch is an intermediate between A- and B-types and is found in legumes, some roots, and seeds. It consists of a mixture of A- and B-type crystalline regions, and its properties vary depending on the proportion and arrangement of these regions (BeMiller and Whistler 2009).

In addition to amylose and amylopectin, starch grain can contain minor components, such as lipids, proteins, phosphates, and trace minerals. Lipids, which typically make up no more than 1% of starch, are found primarily as free fatty acids, lysophospholipids, and starch–lipid complexes. The hydrophobic tails of lipids can be trapped within the helical structure of amylose, contributing to the formation of resistant starch. Lipids can also increase the gelatinization temperature by stabilizing crystalline regions. Proteins, often remnants of biosynthesis and degradation processes, are also present in small amounts. In some plants, such as potatoes, glucose units in amylopectin can be esterified with phosphates, enhancing solubility and swelling while lowering the gelatinization temperature. Both phosphates and proteins can carry trace amounts of minerals into the starch. Additionally, some colored grains may have phenolic compounds attached to their starch. Although these components are present in small amounts, they can significantly affect starch's properties and must be carefully analyzed, especially in the food industry, where they can influence allergies and nutritional value (BeMiller and Whistler 2009).

The study and analysis of starch's structure and chemical properties are essential for developing methods to extract and utilize its intricate and intertwined glucose chains effectively.

3.3 Key Plant Species for Starch Accumulation: Diversity, Distribution, and Influencing Factors

Living beings require lots of energy to stay alive, which they obtain either by consuming it (heterotrophy) or producing it in specialized structures (autotrophy). This energy, typically in the form of carbohydrates, must be broken down into simpler forms like monosaccharides for cellular use. However, storing these simpler forms is less efficient and harder to mobilize. As a result, carbohydrate reserves are formed, with glycogen being the most common in animals and fungi and starch in plants and algae (BeMiller and Whistler 2009).

Starch enables plants to store glucose produced during photosynthesis and mobilize it when needed, such as during periods without sunlight or during processes like germination and growth. For starch formation and storage, plants utilize a specialized organelle called the amyloplast, which is derived from the same structure that forms chloroplasts—the protoplasts. This is why, in some plants, chloroplasts can transform into amyloplasts and vice versa. During photosynthesis, glucose is quickly converted into sucrose for transport via the phloem. Once inside the amyloplasts, the enzyme adenosine diphosphate (ADP)-glucose pyrophosphorylase converts glucose-1-phosphate and adenosine triphosphate (ATP) into adenosine 5-′-diphosphate-glucose (ADP-glucose). Starch synthases (SSs) then catalyze the formation of ADP-glucose chains by linking them through α-1,4-glycosidic bonds. Granule-bound starch synthases (GBSSs) are responsible for amylose synthesis, while soluble starch synthases synthesize amylopectin C-chains. Branching enzymes (BEs) introduce α-1,6-glycosidic bonds to amylopectin, with BE type I enzymes forming cluster structures and BE type II enzymes refining the fine structure. Finally, enzymes like isoamylase (ISA) and pullulanase (PUL) trim excessive or misplaced branches, perfecting the granule structure (Pfister and Zeeman 2016).

Starch accumulation is closely tied to the day–night cycle. During the day, sunlight drives photosynthesis, producing carbohydrates that are stored as starch, primarily in the leaves. At night, plants consume this stored starch at a regulated rate, ensuring it is nearly—but not completely—exhausted by dawn. Starch also plays a critical role in plant growth. In young plants, starch turnover is tightly regulated to provide energy for growth, as evidenced by the stunted development of mutants with impaired starch metabolism. Root elongation and leaf expansion follow diurnal patterns correlated with starch degradation rates, with reduced growth at night. During reproductive development, large amounts of starch are mobilized to support the formation of flowers, cones, and seeds. Under stress conditions like nutrient or water scarcity, plants adjust starch metabolism to optimize carbon usage, prioritizing energy storage over other metabolic pathways such as protein synthesis (Stitt and Zeeman 2012).

Starch is also integral to gravitropism, the plant's response to gravitational stimuli. In graviresponsive organs, amyloplasts act as "weights," accumulating more starch granules in areas that need to move downward and fewer in areas that need to move upward. This can be confirmed by observing mutant plants with fewer or no amyloplasts, which exhibit slower and less pronounced and slower gravitropism (Morita 2010).

Humans have cultivated and consumed starch-rich plants since the dawn of agriculture. Barley was domesticated in China as early as 10,000 years ago, while corn has been cultivated since 7000 years ago, with evidence of interbreeding with teosinte plants by 5000 B.C. Wheat, the most widely consumed grain today, has been consumed since 6700 B.C. Corn, or maize, is particularly notable for its high starch content (60–75% by dry weight), making it a staple in human diets and industries like biofuel production. Rice, another major starch source, contains 70–80% starch by dry weight, while wheat and sorghum grains contain 60–75% starch, with sorghum often used as a gluten-free alternative (BeMiller and Whistler 2009; Rashwan et al. 2024).

Although cereals are the most commonly associated with starch, tubers, legumes, and even fruits can also be starch-rich. Cassava roots, for example, contain 60–80% amylopectin-rich starch. Potatoes and sweet potatoes have 10–20% starch content, with higher levels in mature plants. Unripe plantains contain 70–80% starch, which decreases to 20–30% as they ripen, accompanied by an increase in sugar content. Legumes, known for their high protein content, also contain 30–50% starch, with 29–37% amylose, making it resistant starch that is beneficial for consumption. Cereal starches typically have A-type granules, tubers and high-amylose cereals have B-type granules, and legumes generally have C-type granules (BeMiller and Whistler 2009; Rashwan et al. 2024).

Table 3.1 summarizes the starch content of different plants, as well as their proportions of amylose and amylopectin.

Higher amylose content in starch can be achieved by increasing GBSS activity, reducing amylopectin synthesis through SS inhibition, or decreasing branching by inhibiting BEs. Cereals with high amylopectin content are often called “waxy” or “glutinous,” while high-amylose varieties are named with the prefix “amylo-” (Li et al. 2023). For example, waxy corn, or glutinous corn, contains up to 100% amylopectin due to a mutation in the *Waxy1* gene, which originated in China in the 1900s. This starch is easier to gelatinize and produces a clear, viscous paste. Conversely, amylomaize, with a modified amylose-extender (*ae*) gene, contains 50–94% amylose and is used for resistant starch and biodegradable materials (Yu and Moon 2021). Wheat also has waxy varieties, with null mutations in GBSS genes leading to partially or fully waxy wheats, which are used to extend the shelf life of baked goods and produce high-quality noodles (Graybosch 1998). Rice, particularly the japonica variety, also has waxy forms, commonly used in Asian cuisine (Patindol et al. 2015).

Environmental factors significantly influence the starch content and composition. For instance, rice grown at lower temperatures exhibits an increased amylose-to-amylopectin ratio due to enhanced GBSS activity and reduced BE and SS activity. Elevated CO_2 levels, a key substrate for photosynthesis, have inconsistent effects on starch content, with studies showing both increases and decreases in amylose content in rice. Light levels also impact starch properties; shaded rice plants show reduced carbohydrate content and SS and GBSS activity, while BE activity increases. Water

Table 3.1 Starch content and amylose-to-amylopectin ratio in different plants

	Plant	Starch content[a] (%)	Amylose (%)	Amylopectin (%)	References
Cereals	Corn (*Zea mays*)	65–75	~25	~75	Yu and Moon (2021)
	Wheat (*Triticum aestivum*)	60–70	25–30	70–75	Kim and Kim (2021)
	Rice (*Oryza sativa*)	~90	7–32	68–93	Coffman and Juliano (2015)
Tubers	Potato (*Solanum tuberosum*)	70–80	20–30	70–80	Rashwan et al. (2024)
	Cassava (*Manihot esculenta*)	80–91	18–24	76–82	Hoover (2001) and Rashwan et al. (2024)
	Yam (*Dioscorea esculenta*)	75–86	5–30	70–95	Hoover (2001) and Rashwan et al. (2024)
Legumes	Chickpea (*Cicer arietinum*)	30–32	20–24	76–80	Hoover and Ratnayake (2002)
	Smooth pea (*Pisum sativum*)	30–36	22–24	76–78	Hoover and Ratnayake (2002)
	Lentil (*Lens culinaris*)	34–38	22–25	75–78	Hoover and Ratnayake (2002)

[a]Content referring to the dry weight of each plant

availability during the reproductive phase of cereals affects starch accumulation, with low water availability reducing the starch content and increasing amylopectin, while excess water can decrease starch content but increase relative amylose. Soil nutrients like nitrogen and phosphorus can decrease amylose content, while zinc, boron, and potassium may increase it. High salinity generally reduces the starch content, even in tolerant species (Patindol et al. 2015; Rashwan et al. 2024).

In conclusion, starch content and composition can vary drastically between species, varieties, growth stages, and even hours of the day. It is important to analyze all these factors when choosing plants for consumption and industry applications.

3.4 Extraction, Purification, and Modification of Starch

3.4.1 Overview of Starch Isolation (Extraction and Purification) and Modification

Starch is not present alone in the plant. It is associated with and entrapped by other plant components that are often not useful for industrial applications. Therefore, there is a need to isolate starch from plant tissues, using different methods of extraction and purification, to make it suitable for industrial use (Rashwan et al. 2024; Li et al. 2022; Compart et al. 2023). For that reason, the isolation, extraction, and purification of starch will be discussed in this section.

Also, starch from different botanical sources exhibits varied properties, such as poor solubility in water, low shear resistance, thermal resistance, and a tendency toward retrogradation in its native form, which affects its suitability for different industrial applications. Therefore, it is important to modify the structure, reactivity, and functionality of native starch so that it can be used for a wider range of purposes. Changes in surface properties; polarity; linearity of molecular chains; degree of substitution; polymeric, granular, and crystalline structure; solubility; viscosity; and gelatinization, among others, can be explored by physical, chemical, or biological methods (Li et al. 2022; Nawaz et al. 2020; Wang et al. 2020). The objective of these methods is to enhance native starch's intrinsic attributes or confer specific properties to align with industrial needs (Rashwan et al. 2024). Thus, starch modification will be reviewed in this section as well.

3.4.2 Key Techniques for Starch Isolation

Starch extraction involves the isolation of starch granules from plant tissues. Efficiency is crucial for achieving maximum yield and purity, and the method chosen influences this, as well as molecular structure and functional properties (Rashwan et al. 2024). The intention is to isolate starch in the purest form achievable, with the least damage possible to its structure. Yet the process faces obstacles due to strong interactions with macronutrients (such as proteins and lipids) and cell wall materials (Li et al. 2022). Therefore, all protocols aim to homogenize the plant material to break down the cell walls and release the cell contents (Compart et al. 2023). The technique chosen for use, whether physical, chemical, or enzymatic, will depend on the cost of extraction, environmental regulations, and the desired degree of purity (Li et al. 2022).

The most common methods for starch separation are wet milling and dry grinding. Wet milling consists of soaking grains in an alkaline or acidic solution. The goal is to reduce grain component interactions and inhibit microbial growth. The grain is milled, and its components can be separated by different unit operations, such as screening, centrifuging, and washing. However, this technique leads to a significant starch loss in the protein and bran fractions but generally results in a higher-purity product. Dry milling results in lower purity because it does not completely separate starch from the matrix and also causes higher starch damage (Li et al. 2022; Rashwan et al. 2024). However, it achieves a higher recovery and is extensively used in biofuel applications, such as corn-based bioethanol production (Lee et al. 2021; Bothast and Schlicher 2005).

Alkaline extraction results in higher starch yields because of the breakdown of structural components. This method is associated with the production of starch with the highest degrees of crystallinity and whitest color due to efficient impurity removal. Furthermore, starches from hot water soaking and alkaline washing exhibit higher gelatinization temperatures, which may indicate better thermal stability. Enzymatic extraction often preserves the native structure, maintaining some properties (Rashwan et al. 2024). In comparison, alkaline extraction is more cost-effective than enzyme extraction but less environmentally friendly. Both alkaline and enzymatic methods show similar yields, crystallinity, and thermal behaviors (Rashwan et al. 2024; Li et al. 2022).

Efficient filtration of starch from liquids is a fundamental step for enhancing product quality in some industries. It can be done using the sedimentation method, filtration media, or membrane filtration. Sedimentation relies on density differences between starch and the liquid to achieve separation. Flocculants can be added to accelerate the process, and after that, the supernatant is removed, leaving the starch sediment alone at the bottom. The simplicity of the method can affect starch purity. Different filtration media can be used based on the size of starch particles and the required precision. It is important to note that they need to be replaced or cleaned to avoid clogging.

Membrane technology can be used to filter out specific constituents in the starch industry. Selectively permeable or semipermeable membranes are used to separate starch macromolecules from solvents and small molecules. This technology has been researched for more than two decades, and research has shown reduced energy cost while maintaining the same quality as conventional processes. One example of a large-scale application of membrane filtration in the starch-processing industry is in the clarification of corn syrups. Ultrafiltration and microfiltration membranes are widely used in starch filtration, and they provide a high-precision separation method, resulting in high-purity starch. It is important to control parameters such as pressure and temperature to prevent membrane fouling and damage.

Upon comparing different methods, significant differences in yield, purity, and functional properties were revealed. It is important to choose the technique based on the desired starch characteristics and source material. Moreover, progress in extraction technology not only leads to the optimization of yield, quality, and energy consumption but also supports industrial sustainability by improving the value of agricultural by-products, when starch is extracted from unconventional sources (Rashwan et al. 2024; Dorantes-Fuertes et al. 2024).

3.4.3 Starch Modification Methods

Modification can be defined as any alteration to the structure of a starch molecule caused by environmental, operational, and processing factors. The goal is to enhance the structural and functional quality of starch and achieve better results in formulations (Nawaz et al. 2020). The properties targeted include, for example, solubility, texture, adhesion, and thermal tolerance (Alcázar-Alay and Meireles 2015). Native starch can be modified by physical, chemical, enzymatic, and biotechnological methods (Nawaz et al. 2020).

3.4.3.1 Physical Modifications of Native Starch

Physical modifications are environmentally friendly, cost-effective techniques that alter starch properties without involving chemical reagents. These methods aim to improve starch functionality—such as solubility, swelling, viscosity, or thermal stability—through thermal, mechanical, or radiation-based treatments (Wang et al. 2020; Nawaz et al. 2020).

Thermal treatments include hydrothermal processing (gelatinization in water), heat–moisture treatment (HMT), and annealing. HMT, conducted at low moisture and elevated temperatures, reduces starch solubility and increases gelatinization temperature, while annealing promotes molecular reorganization, enhancing crystallinity and stability (Compart et al. 2023; Alcázar-Alay and Meireles 2015). Extrusion, another thermal method, uses high shear and pressure to disrupt starch granules, decreasing viscosity but improving digestibility and water solubility (Nawaz et al. 2020).

Radiation-based techniques, such as microwave and gamma irradiation, are also employed to alter starch structure. Microwaves increase water absorption and modify granule morphology, whereas gamma rays reduce crystallinity and induce minor surface changes (Wang et al. 2020). Mechanical processes like grinding and micronization reduce particle size and disrupt crystalline regions, affecting functional properties such as swelling and pasting.

Other physical methods include ultrasonication, high-pressure treatment, and freeze–thaw cycling. These techniques further diversify the range of structural changes and functionalities achievable through physical modification (Rashwan et al. 2024; Compart et al. 2023). Compared to chemical modifications, physical methods are simpler to implement, safer for food-grade applications, and more sustainable.

3.4.3.2 Chemical Modifications of Native Starch

Cost-effectiveness and simplicity make chemical methods of modification the most common at the industry level. However, these traditional methods produce a significant amount of effluents that are potentially harmful to the environment, and trace elements can be introduced into starch, limiting its application (Rashwan et al. 2024).

Chemical techniques are capable of allowing a wider range of functionalization and application of starch. Chemical methods use the hydroxyl groups (C2, C3, and C6 positions) of starches as active sites for the introduction of functional groups (Compart et al. 2023; Rashwan et al. 2024). The changes depend on the source of starch, reaction conditions, substituent types, degree of substitution, and distribution of substituents in the starch molecule (Wang et al. 2020).

Oxidation involves introducing carbonyl and carboxyl functional groups onto starch chains, followed by depolymerization of the molecules. Oxidized starch presents lower viscosity, better paste stability and film-forming ability, enhanced hydrophilicity, disruption of molecular linearity and crystallinity, retrogradation, and enthalpy reduction (Nawaz et al. 2020; Wang et al. 2020; Compart et al. 2023).

Etherification can introduce different groups into the starch molecule and, therefore, has different classifications, such as hydroxyethylation, hydroxypropylation, and carboxymethylation (Nawaz et al. 2020). The introduction of hydroxypropyl groups into starch chains, for example, destroys inter- and intramolecular hydrogen bonds, weakening the granular structure (Wang et al. 2020). Etherified starch presents improved properties concerning thermal stability, rheological behavior, and, in some cases, ionic activity (Compart et al. 2023).

Esterification, as well, can be classified into acetylation, fatty acetylation, phosphorylation, and succinylation (Nawaz et al. 2020). It involves the conversion of three available hydroxyl groups of glucosyl residues (C6, C2, and C3 hydroxyl groups in that order of reactivity) into alkyl or aryl derivatives. The result can be either organic or inorganic starch esters. In this kind of modification, the hydrogen-bonding ability of amylose and amylopectin is compromised, which leads to improved hydrophobicity, swelling capacity, and lower retrogradation rates (Wang et al. 2020; Compart et al. 2023). Acetylation is one of the most common methods. It results in structural reorganization due to steric obstruction, which promotes the penetration of water in amorphous regions, thereby increasing the swelling capacity. Amphiphilic polymers, such as starch esterified with octenyl succinic anhydride (OSA), are desirable because of their wide range of applications (Wang et al. 2020). Succinylation is used to reduce gelatinization temperature and retrogradation and increase the thickening power, viscosity, and water retention. Phosphorylation, although being a natural modification, can be used industrially to increase viscosity, pasting transparency, and gelatinization characteristics (Compart et al. 2023).

Cationization treats starch with diverse cationic molecules (amino, ammonium, imino, phosphonium, or sulfonium), giving it a positive ionic charge. The results are an increase in solubility, stability, dispersibility, and clarity. This technique has variations such as dry, wet, and semidry cationization, and each one has effects on starch properties (Nawaz et al. 2020).

Cross-linking is the most commonly used method for chemically changing starch properties. It is the covalent interconnection of linear and branched chains of a polymer and involves a molecular reaction between reactive hydroxyl groups on starch and multifunctional reagents, such as sodium trimetaphosphate (STMP), sodium tripolyphosphate (STPP), epichlorohydrin (ECH), and phosphoryl chloride (POCl3), leading to the formation of ether or ester bonds. The resulting starch shows a higher tolerance to acid, heat, and shear forces, in addition to improved characteristics, such as a decrease in retrogradation, higher viscosity, and better paste clarity. The source of starch granules, methods, and parameters for cross-linking have a fundamental influence on the final product's properties (Wang et al. 2020; Compart et al. 2023; Alcázar-Alay and Meireles 2015).

Acid hydrolysis is an ancient starch modification method. It consists of processing a starch paste in an inorganic acid (HCl or $H2SO_4$) solution. The temperature is below gelatinization temperature for a specified time, depending on the viscosity and the degree of conversion that is desired. This technique causes cleavage of starch chains randomly in both α-1,4 and α-1,6 bonds, considering that α-1,4 bonds and amorphous regions containing α-1,6 bonds are more accessible to acid penetration and hydrolysis. This modification increases the double helix content and crystallinity (Wang et al. 2020). It also enhances properties such as solubility, pasting viscosity, gelatinization enthalpy, and swelling power. It consists of the hydrolysis of the glycosidic linkage (Compart et al. 2023).

Grafting is the introduction of a polymer chain or monomers into the starch, causing copolymerization and combining features of both parts. As a result, modified starch exhibits higher biodegradability, thermal stability, and hydrodynamic radius and improved flocculation properties (Compart et al. 2023).

3.4.3.3 Enzymatic Modifications of Native Starch

Usually, enzymatic methods are based on the use of hydrolyzing enzymes, and their advantages include the production of fewer by-products, higher yield, better process control, greater homogeneity in products, mild reaction conditions, and environmental friendliness (Compart et al. 2023; Wang et al. 2020; Rashwan et al. 2024). Enzymatic modification can be divided into endo-amylase, exo-amylase, debranching enzymes, and transferases. Other potential types are also possible (Compart et al. 2023).

The starch-modifying enzymes originate from various microorganisms, including *Thermus* sp., *Bacillus stearothermophilus*, *Bacillus acidopullulyticus*, *Aspergillus niger*, *Bacillus subtilis*, *Bacillus megaterium*, *Neisseria polysaccharea*, *Deinococcus geothermalis*, *Rhodothermus obamensis*, *Aquifex aeolicus*, and *Deinococcus radiodurans*. Additionally, some enzymes are also derived from plants, such as ramie leaf (*Boehmeria nivea*) and barley (*Hordeum vulgare*) (Rashwan et al. 2024).

The results of endo-amylases (which cleave α-1,4-glycosidic bonds within amylose and amylopectin chains) are oligosaccharides of different sizes. Exo-amylases act on external glucose residues of amylose and amylopectin and, therefore, produce only glucose (for glucoamylase and alpha-glucosidase) or maltose and B-limit dextrin (for B-amylase). Debranching enzymes isoamylase and type I pullulanase exclusively hydrolyze α-1,6-glycosidic bonds. Transferases hydrolyze an α-1,4-glycosidic bond of a donor molecule and transfer the cleaved residue from the donor to a glycosidic acceptor with the formation of a new glycosidic bond. Each type of enzyme has its own particularities and cleavage preferences (Wang et al. 2020). Using specific enzymes such as pullulanase and transglucosidase results in a reduction in starch digestibility and an increase in amylose content and crystallinity, enhancing the starch's resistance to breakdown. These modifications boost the pasting properties and stability of starch, making it suitable for applications that need controlled glucose release, for example (Rashwan et al. 2024).

Enzyme modification has limitations in its applications because of the low thermostability and catalytic efficiency of some enzymes. Enzyme engineering is important to improve the properties of industrially relevant enzymes. Also, an enzyme by itself is often unable to substantially alter the properties of starch; thus, a blend of enzymes is necessary to achieve desired results (Compart et al. 2023).

Amyloglucosidase, alpha-glucosidase, and alpha-amylase are reported for use in biofuel production. The former and the latter are used, for example, to produce fermentable sugars for bioethanol production (Compart et al. 2023).

3.4.3.4 Genetic/In Planta Modification of Starch

Another relevant line of research to mention is the *in planta* modification approach, aimed at improving characteristics already during the plant's development through genetic manipulations of key traits. However, this strategy is not yet accepted in multiple countries (Apriyanto et al. 2022). Genetic modifications allow transforming starch-producing crops to produce starches with strategic properties straight from the plant. Amylose content, branch chain length, and resistance to retrogradation are some of the traits this strategy can aim (Rashwan et al. 2024). This perspective still remains under discussion in many countries when it comes to genetic tools. Additionally, the relationship between genetics and the regulation of each starch parameter is not completely straightforward since a gene often has effects on multiple properties (Compart et al. 2023).

As seen, undesirable characteristics of starch make modifications necessary to overcome these drawbacks and enhance its properties and functionalities for industrial applications across different sectors.

3.4.4 Methods of Separation for Starch Analysis

Size exclusion chromatography (SEC), field flow-fractionation (FFF), analytical ultracentrifugation, and hydrodynamic chromatography are some methods that are used for starch analysis and obtaining valuable information. The first two are widely used. SEC is mainly used to provide molecular weight distribution information and can be used, for example, to investigate starch-branching enzymes and the amylose content of enzymatically modified starches and study the effect of different processing conditions on starch molecular size distribution. FFF provides high-resolution separation of particulates ranging from 1 nm to 100 μm. It can help study starch amylolysis and the distribution of starch granules during the process (Rübsam et al. 2012; Ovando-Martínez et al. 2013).

3.4.5 *Role of Innovation in Optimizing Efficiency, Sustainability, and Scalability of Starch Processing, Considering a Circular Bioeconomy*

Eco-friendly methods for extracting and modifying plant-based starch are key areas of innovation, fueled by the growing demand for starch across various industries and the increasing emphasis on sustainability. There are drawbacks when it comes to traditional starch-processing methods, such as high energy consumption and chemical waste. Methods such as physical and enzymatic are favorable alternatives. The transition to sustainable starch extraction and modification techniques presents promising opportunities to minimize the environmental impact of this vital industry while ensuring its economic sustainability. The continuous advancement and implementation of these technologies play a key role in fostering eco-friendly practices in starch production and application (Dorantes-Fuertes et al. 2024).

It is also important to consider parts of plants that are usually discarded, such as roots, some tubers, pulses, pseudocereals, trees, herbs/shrubs, and fruits. New techniques capable of taking advantage of these types of waste are fundamental in a circular bioeconomy concept. New initiatives that provide unconventional starch sources and their extraction methods may reduce the current demand for resources used in starch extraction. Biofuels can be produced with this approach. More specifically, bioethanol was produced using starch extracted from anchovy pulp and waste. The study shows the potential of unconventional sources for this field, contributing to environmental and economic sustainability and applying the circular bioeconomy concept (Dorantes-Fuertes et al. 2024).

3.5 Economic Aspects and Global Production of Starch

3.5.1 *Introduction*

Starch, a polysaccharide primarily composed of amylose and amylopectin molecules, is one of the most abundant carbohydrates in nature. Its versatility has established it as a critical resource not only for human nutrition but also for biofuel production and various industrial applications, positioning it as a cornerstone of circular biotechnology (Vandenberghe et al. 2022). As a natural polymer, starch consists of amylose, a helical structure formed by α-1,4-glycosidic bonds, and amylopectin, a branched structure containing both α-1,4- and α-1,6-glycosidic linkages. This structural composition renders starch a vital dietary energy source, predominantly found in staple crops such as maize, wheat, potato, and cassava. However, its significance extends beyond the food sector, as it constitutes a strategic resource within the global bioeconomy, with a market valuation exceeding $ 126 billion in 2024 and an anticipated annual growth rate of 5.2% by 2030 (Bhuyar et al. 2022).

In the food industry, starch is valued not only for its fundamental nutritional role but also for its extensive application as a functional additive (Awodi et al. 2022). It is estimated that approximately 60% of the globally produced starch is allocated to food applications, ranging from staple products, such as bread and pasta, to processed foods, including sauces, desserts, and snacks (Mirzaei et al. 2024). The versatility of starch-based compounds has further positioned them as essential ingredients in gluten-free food formulations and in the development of innovative products, such as plant-based meat, aligning with the rising demand for sustainable and health-conscious alternatives (Liu et al. 2022).

Beyond its traditional function in nutrition, starch has emerged as a pivotal component in biofuel production, particularly in bioethanol manufacturing, reinforcing its role as a sustainable alternative to fossil fuels (Bhuyar et al. 2022; Jin et al. 2020; Ziaei-Rad et al. 2021). In 2022, global bioethanol production reached approximately 130 billion liters, with a substantial fraction derived from starch-rich feedstocks, including maize, rice, wheat, and cassava (Safarian and Unnthorsson 2018). This approach is particularly relevant in major bioenergy-producing countries such as the United States and Brazil, where maize and sugarcane (rich in both sucrose and starch) serve as primary raw materials for the biofuel industry (Nazli et al. 2024; Pleissner et al. 2016).

In the industrial sector, starch represents a fundamental input for the production of bioplastics, adhesives, textiles, and pharmaceutical products, owing to its biodegradable nature and adaptability to various transformation processes. Its ability to form biodegradable films has positioned starch as a promising alternative to petroleum-derived plastics, in alignment with global sustainability trends and circular economy principles (Ioannidou et al. 2020). Indeed, the starch-based bioplastics market is projected to expand at an annual growth rate of 10.5% between 2023 and 2030, driven by the increasing demand for eco-friendly materials (Tayyab et al. 2018).

3.5.2 Major Producers of Different Types of Starch

This versatile polysaccharide can be extracted from various plant sources, each possessing unique properties that determine its final applications (Rezania et al. 2020). Among the most common starch sources, corn stands out due to its high profitability and extensive use in the food industry and biofuel production, owing to its availability and versatility (Nitsos et al. 2018). Cassava starch has emerged as an economical and sustainable option, particularly valued in bioplastic production due to its low production costs and ability to generate biodegradable materials (Ancy et al. 2024). Potato starch, recognized for its high purity and excellent gelling properties, is widely employed in applications requiring specific textures. Meanwhile, wheat and rice starches are appreciated for their smoothness and biocompatibility, making them preferable for pharmaceutical and cosmetic industries, where skin tolerance is critical (Olivato 2024). In addition to traditional sources, emerging

Table 3.2 Global starch production (2024)

Crop	Main producing countries	2024 production (million tons)	Economic value (2024)	2030 projection (million tons)	References
Corn	United States, China, Brazil	365	$60 billion	400	Mirzaei et al. (2024), Vilpoux and Silveira Junior (2023)
Cassava/ manioc	Thailand, Nigeria, Brazil	32	$5 billion	35	Mirzaei et al. (2024)
Potato	Europe, China, India	26	$8 billion	28	Mirzaei et al. (2024)
Wheat	China, India, Russia	16	$4 billion	17	Vilpoux and Silveira Junior (2023), Liargovas and Papageorgiou (2024)
Rice	China, India, Bangladesh	10.5	$3 billion	11	Mirzaei et al. (2024)

alternatives are being investigated. Pea and quinoa starches not only offer sustainability advantages but are also allergen-free, making them promising options for specialized applications (Halim et al. 2024).

Corn starch is one of the most widely used starches in industrial applications due to its high carbohydrate content and low production cost. It comprises approximately 75–80% amylopectin and 20–25% amylose (Liu et al. 2022), along with small amounts of proteins (0.3–0.5%) and lipids (0.6–0.8%), which can influence its gelatinization properties. Its high amylopectin content makes it ideal for applications requiring high viscosity and stability, particularly in the food and adhesive industries (Oktay et al. 2024). Corn starch is the most produced starch worldwide, with the United States being one of the largest producers. Table 3.2 shows its production in 2024; the United States produced 365 million tons of corn, generating an economic value of approximately $62 billion; China achieved a production of 290 million tons, contributing $62 billion to its economy (Vilpoux and Silveira Junior 2023). Brazil, on the other hand, produced 125 million tons, generating $21 billion, reflecting its significance in the global agricultural landscape (Vandenberghe et al. 2022). However, the commercialization of corn starch is regulated by international standards such as the Codex Alimentarius, which establishes quality and safety guidelines. In the United States, the Farm Bill influences subsidies and corn production policies (Awodi et al. 2022). Recent studies, such as that by Liu et al. (2022), have demonstrated that gene editing techniques can enhance corn yields by up to 15%, not only optimizing production but also reinforcing its role as a key raw material in the bioplastics industry (Liu et al. 2022).

Cassava starch is characterized by its high purity and low protein and fat content. It consists of 70–75% amylopectin and 25–30% amylose (de Oliveira Schmidt et al. 2022), with less than 0.1% protein and less than 0.2% lipids, making it ideal for applications requiring transparency and a smooth texture, such as in bioplastics and processed foods (Ancy et al. 2024). Nigeria, the world's leading cassava producer, reached 62 million tons, although much of this production is allocated to local consumption. Brazil, producing 21 million tons, added $5.2 billion to its economy (Lee et al. 2021; Louw et al. 2023). The commercialization of cassava starch is subject to regulations set by organizations such as the World Trade Organization (WTO), which oversees export tariffs and quotas. De Oliveira Schmidt et al. (2022) demonstrated that the implementation of biofertilizers has proven to be an effective strategy, increasing cassava yields by 20%, thereby enhancing its global market competitiveness and reinforcing its position as a sustainable starch source for industrial applications (de Oliveira Schmidt et al. 2022).

Potato starch is known for its high purity and phosphate content, which confer unique gelling properties. It is composed of 75–80% amylopectin and 20–25% amylose, with traces of proteins (0.1–0.2%) and lipids (0.05–0.1%). Its high phosphate content makes it ideal for applications in the food industry, where creamy and stable textures are required (Vithu et al. 2020). The Netherlands, Germany, and Poland collectively produced 26 million tons in 2024, with an economic value of $8.3 billion. China and India also increased their production to 10.5 and 8.2 million tons, respectively (Das et al. 2021). The commercialization of potato starch within the European Union (EU) is regulated by the Common Agricultural Policy (CAP), which includes subsidies and quality controls. Vithu et al. (2020) demonstrated that optimizing irrigation and selecting specific potato varieties can increase starch production by 10% (Vithu et al. 2020).

Wheat and rice starches are composed of 70–75% amylopectin and 25–30% amylose (Kumar et al. 2022). They contain an additional 0.5% protein and 0.3% lipids, making them suitable for bakery and confectionery applications, where spongy textures and elasticity are required (Gronchi et al. 2019). China, India, and Russia are the leading producers, with a combined production of 16 million tons in 2024, generating $4.2 billion (Pradyawong et al. 2018). Rice starch, mainly produced in China, India, and Bangladesh, reached 10.5 million tons, with a market value of $4.2 billion. Both scenarios are subject to regulations such as those established by the US Food and Drug Administration (FDA), which sets limits on pesticide residues and heavy metals. Kumar et al. (2022) demonstrated that employing drying technologies during the extraction phase can enhance the efficiency of rice starch production (Kumar et al. 2022).

Finally, some emerging starch sources, such as pea and quinoa, are characterized by their sustainability. Although their production is still in its early stages, Watcharakitti et al. (2023) have demonstrated that these sources could become viable on a large scale in the near future due to the benefits they offer (Watcharakitti et al. 2023). Pea starch is a notable emerging source due to its high amylose content, making it ideal for applications requiring high heat resistance. It consists of 30–35% amylose and 65–70% amylopectin, along with approximately 1–2% protein and

0.5% lipids (Watcharakitti et al. 2023). Its composition makes it suitable for functional food and bioplastic production. Quinoa starch is valued for its high nutritional content and functional properties. It is composed of 60–65% amylopectin and 35–40% amylose, with approximately 1–1.5% protein and 0.2% lipids (Watcharakitti et al. 2023). This composition makes it ideal for applications in healthy food and cosmetics, where moisturizing properties, a smooth texture, and high purity are required.

3.5.3 *Economic Impact*

In the food industry, starch serves as a thickening, stabilizing, and gelling agent in sauces, dairy products, and snacks. Corn starch, due to its high amylopectin content, is ideal for enhancing viscosity and stability in processed foods. Conversely, potato starch, characterized by its high purity and phosphate content, is primarily used in applications requiring a creamy and transparent texture, such as soups and sauces (Aristizábal-Marulanda and Cardona Alzate 2019). Cassava starch, with its low protein and lipid content, is preferred for products requiring a smooth texture and neutral flavor profile. The global food starch market was valued at approximately $51.5 billion in 2024, with an annual growth rate of 3% (Mirzaei et al. 2024).

In the biofuels sector, starch is a fundamental raw material for bioethanol production. Corn starch, owing to its high availability and low cost, remains the predominant feedstock for bioethanol production. Its substantial carbohydrate content (75–80%) facilitates efficient conversion into fermentable sugars (Ziaei-Rad et al. 2021). In Brazil, corn starch is utilized for bioethanol production due to its high yield per hectare and cost-effectiveness (Nitsos et al. 2018). By 2024, starch demand in the biofuel industry accounted for 20% of the total market, with an estimated value of $15 billion (Vilpoux and Silveira Junior 2023).

In biodegradable materials, potato starch is widely used due to its high purity and gelling properties, making it ideal for producing transparent and durable films. Meanwhile, corn starch, valued for its viscosity and stability, is extensively employed in bioplastic production for packaging and textiles (Ancy et al. 2024). This market has exhibited an annual growth rate of 10%, reaching a valuation of $8.8 billion in 2024. Increasing environmental awareness and regulatory restrictions on petroleum-based plastics have driven this demand (Watcharakitti et al. 2023).

3.5.3.1 Trade Policies for Starch

The United States, as the largest producer of corn starch, implements agricultural subsidies under the Farm Bill, reducing production costs and promoting exports. Additionally, the United States imposes tariffs on cassava starch imports to protect domestic producers (Vilpoux and Silveira Junior 2023). China, on the other hand, has established export quotas for corn and rice starch to ensure domestic supply while investing in technologies to enhance starch extraction efficiency (Liu et al. 2022). The European Union enforces the Common Agricultural Policy (CAP), which includes subsidies for potato and wheat starch production, alongside stringent quality standards for commercialization. The EU further encourages starch utilization in bioplastics through tax incentives (Liargovas and Papageorgiou 2024). Thailand, as the leading global exporter of cassava starch, has strengthened its market position through trade agreements with Asian countries and the European Union. These agreements have not only reduced tariffs but have also facilitated export expansion, reinforcing Thailand's dominance in the sector (de Oliveira Schmidt et al. 2022). Tariffs and quotas may restrict access to key markets, whereas subsidies and trade agreements can stimulate production and exports.

3.5.3.2 Starch Production Costs

Corn starch is among the most widely produced starches globally; however, its production is also relatively costly due to high raw material expenses and energy-intensive processing requirements. The extraction cost of corn starch is approximately $280 per ton, accounting for labor and industrial processing costs, making it competitive for high-value applications such as biofuels and bioplastics (19). Cassava starch benefits from lower raw material costs and a less energy-intensive extraction process, resulting in a production cost of approximately $150 per ton. This cost advantage, along with lower labor and industrial processing expenses, makes cassava starch particularly suitable for bioplastics and food applications in tropical regions (de Oliveira Schmidt et al. 2022; Liargovas and Papageorgiou 2024). Potato starch is highly valued for its purity and functional properties; however, its production cost is significantly higher due to expensive raw materials. In 2023, the cost of potato starch was approximately $375 per ton, representing a major portion of the overall production cost. This positions potato starch for specialized applications in high-quality food products and bioplastics (Liargovas and Papageorgiou 2024; Sekulić et al. 2024). The total production cost of wheat starch is around $330 per ton, higher than that of other starches, making it more suitable for applications in the food and adhesive industries (Kumar et al. 2022; Liargovas and Papageorgiou 2024). Pea and quinoa starches are even more expensive due to smaller production scales and less efficient processing methods. The production costs for pea and quinoa starches are approximately $510 and $630 per ton, respectively, limiting their use to specialized applications such as functional foods and high-end cosmetics (Watcharakitti et al. 2023; Sekulić et al. 2024).

3.5.4 Growth in Global Demand

The increasing demand for starch is closely linked to the global need to reduce dependence on petroleum and its derivatives, as well as to mitigate their environmental impact. Consequently, the use of starch in nonfood applications has experienced significant growth in recent years. Bioplastic production accounts for approximately 15% of global starch consumption, primarily utilized in the synthesis of biopolymers such as polylactic acid (PLA), polyhydroxyalkanoates (PHAs), polyhydroxybutyrates (PHBs), and biodegradable films. This sector has exhibited an annual growth rate of 20% since 2020, driven by the rising demand for sustainable packaging and regulatory policies targeting single-use plastics (Watcharakitti et al. 2023). The biofuel industry consumes 10% of the total starch produced, particularly in countries such as the United States and Brazil, where corn and cassava serve as primary feedstocks (Vilpoux and Silveira Junior 2023; Louw et al. 2023; Suriyachai et al. 2024). Additionally, other industrial applications, including the production of adhesives, textiles, pharmaceuticals, and cosmetics, account for 15% of the global starch utilization. In these sectors, starch is valued for its functional properties, biocompatibility, and high purity (Kumar et al. 2022).

3.5.4.1 Benefits of Biodegradable Materials

Starch-derived biodegradable materials offer significant environmental and economic advantages, particularly in reducing pollution. Unlike conventional plastics, which can persist for centuries before degrading, starch-based bioplastics decompose within 3–6 months under appropriate conditions (Watcharakitti et al. 2023). Moreover, these materials have a substantially lower carbon footprint, as the production of bioplastics generates up to 60% fewer greenhouse gas emissions compared to petroleum-derived plastics (Vithu et al. 2020).

3.5.4.2 Preference for Plant-Based Ingredients

Plant-based ingredients, such as starch, are favored in the manufacturing of biodegradable materials due to their biocompatibility and low toxicity, making them ideal for applications involving food contact and medical products (Kumar et al. 2022) Additionally, their functional attributes derived from the molecular structure of amylose and amylopectin enable the development of materials with tunable mechanical and thermal properties (Liu et al. 2022). Finally, starch is considered a sustainable resource, as its primary sources, including corn and cassava, are renewable crops that can be cultivated on a large scale, ensuring a stable and sustainable supply chain (Lozowski 2012; Chen et al. 2021).

3.5.4.3 Leading Industries in Biodegradable Material Development

Several international industries have been at the forefront of developing and commercializing starch-based biodegradable materials. NatureWorks LLC, a pioneer in polylactic acid (PLA) production in the United States, utilizes corn starch for applications in food packaging, textiles, and disposable products (Watcharakitti et al. 2023). In Italy, Novamont manufactures bioplastics derived from potato and corn starch, primarily for compostable bags and agricultural films (Vithu et al. 2020). Tate & Lyle, a leader in the industrial production of modified starch in the United Kingdom, supplies starch-based solutions for bioplastics and biodegradable adhesives (Kumar et al. 2022), Meanwhile, Biopac, based in India, specializes in biodegradable packaging derived from cassava and corn starch, positioning itself as a key player in the disposable product market (de Oliveira Schmidt et al. 2022).

The global starch production in 2024, as shown in Table 3.3, reflects a clear shift toward sustainable applications, such as bioplastics and biofuels, driven by countries like the United States, Brazil, and Thailand. The United States, with 360 million tons of corn starch, leads the global production, dedicating a significant portion to biofuels and bioplastics (Liu et al. 2022), while China, producing 290 million tons, prioritizes the food industry and adhesives (Vilpoux and Silveira Junior 2023). Brazil, the third-largest producer with 130 million tons, combines food and energy uses, although its yield (6.5 t/ha) is lower compared to other leaders (Mirzaei et al. 2024). Meanwhile, Brazil and Thailand stand out in cassava starch production, with Thailand reaching 32 million tons and a yield of 10.3 t/ha, solidifying its role as a key player in bioplastics (de Oliveira Schmidt et al. 2022). Nigeria, the largest cassava producer, with 65 million tons, focuses its output on food, reflecting local food security priorities (Kumar et al. 2022). Europe excels in efficient potato starch production, with countries like the Netherlands and Germany surpassing 25 t/ha, leveraging its properties for food and bioplastics (Liargovas and Papageorgiou 2024). India and Russia, though with lower yields, are increasing their rice and wheat starch production, respectively, primarily for food and adhesives (Mirzaei et al. 2024). These regional differences highlight the need to balance food security with industrial opportunities, emphasizing that investment in agricultural technology and sustainable policies, as seen in Thailand and Europe, can significantly improve efficiency and competitiveness in the global starch market.

3.5.5 Challenges in Starch Production and Marketing

3.5.5.1 Price Standardization for Sustainability

Price volatility is one of the major challenges affecting both starch producers and consumers. The price of corn, a primary raw material for starch production, has fluctuated between $200 and $300 per ton in recent years, depending on factors such as climatic conditions, trade policies, and biofuel demand (Vilpoux and Silveira Junior 2023). This not only complicates long-term planning but also discourages investment in sustainable technologies. Proposed mechanisms for price stabilization include the implementation of long-term contracts between producers and buyers, as well as government policies to regulate agricultural markets and reduce fluctuations. According to a study by Mirzaei et al. (2024), price standardization could reduce uncertainty in the supply chain and promote the adoption of more sustainable agricultural practices (Mirzaei et al. 2024).

Table 3.3 Global starch production and cultivated area (2024)

Country	Type of starch	Production (million tons)	Cultivated area (million ha)	Yield (ton/ha)	Main use of starch	References
United States	Corn	360	36	10.0	Biofuels, bioplastics	Liu et al. (2022), Vilpoux and Silveira Junior (2023), Vithu et al. (2020), and Kumar et al. (2022)
	Potato	5.2	0.2	26.0	Food, bioplastics	
	Rice	9.8	1.1	8.9	Food, cosmetics	
China	Corn	290	31	9.4	Food, adhesives	Mirzaei et al. (2024), Vilpoux and Silveira Junior (2023), Liargovas and Papageorgiou (2024)
	Rice	210	30	7.0	Food, cosmetics	
	Potato	8.0	0.3	26.7	Food, bioplastics	
Brazil	Corn	130	20	6.5	Biofuels, food	Mirzaei et al. (2024), de Oliveira Schmidt et al. (2022)
	Cassava	22	2.6	8.5	Biofuels, bioplastics	
	Rice	11.5	1.8	6.4	Food	
Thailand	Cassava	32	3.1	10.3	Bioplastics, food	Mirzaei et al. (2024), de Oliveira Schmidt et al. (2022), Liargovas and Papageorgiou (2024)
	Rice	20	9.0	2.2	Food	
Netherlands	Potato	10.5	0.4	26.3	Food, bioplastics	Liargovas and Papageorgiou (2024)
	Wheat	1.2	0.2	6.0	Food, adhesives	
Germany	Potato	8.5	0.3	28.3	Food, adhesives	Mirzaei et al. (2024), Liargovas and Papageorgiou (2024)
	Wheat	22	3.0	7.3	Food, adhesives	
India	Rice	11	4.2	2.6	Food, cosmetics	Mirzaei et al. (2024), Vilpoux and Silveira Junior (2023), Kumar et al. (2022)
	Wheat	105	30	3.5	Food, adhesives	
	Corn	30	9.5	3.2	Food, biofuels	

(continued)

Table 3.3 (continued)

Country	Type of starch	Production (million tons)	Cultivated area (million ha)	Yield (ton/ha)	Main use of starch	References
Nigeria	Cassava	65	7.2	9.0	Food, biofuels	Mirzaei et al. (2024), Vilpoux and Silveira Junior (2023), de Oliveira Schmidt et al. (2022), and Kumar et al. (2022)
	Corn	12	5.0	2.4	Food	
	Rice	5.0	2.0	2.5	Food	
Russia	Wheat	16	10.5	1.5	Food, adhesives	Mirzaei et al. (2024) Vilpoux and Silveira Junior (2023), Kumar et al. (2022)
	Potato	30	1.2	25.0	Food	
	Corn	14	2.5	5.6	Food, biofuels	
France	Potato	7.5	0.3	25.0	Food, bioplastics	Vithu et al. (2020), Liargovas and Papageorgiou (2024)
	Wheat	35	5.0	7.0	Food, adhesives	

3.5.5.2 Supply Chain and Its Maintenance

The supply chain encompasses agricultural production, industrial processing, and final distribution. To maintain an efficient supply chain, optimizing logistics, strengthening relationships between key stakeholders, and investing in infrastructure are crucial (Wallerstein 2020). Reducing postharvest losses is essential for ensuring market stability. In Nigeria, it is estimated that up to 30% of cassava production is lost due to inadequate storage and transportation infrastructure, highlighting the need to invest in technological solutions that minimize these waste losses (de Oliveira Schmidt et al. 2022). These issues result in an economic loss of approximately $500 million annually. Linking farmers, processors, and distributors is essential to ensuring a steady flow of raw materials. Finally, improving processing and storage facilities, especially in developing regions, can increase efficiency and reduce costs. According to the study by Shukla et al. (2007), the implementation of digital technologies such as blockchain is necessary to improve traceability and transparency in the supply chain, reducing inefficiencies and costs by 15–20%, thereby enhancing profitability and contributing to the sustainability of the industry (Shukla et al. 2007).

3.5.5.3 Biotechnological Advances to Improve Efficiency

Biotechnology is revolutionizing the production and conversion of starch, making these processes more efficient and sustainable. Biotechnological approaches, such as crop modification and gene editing, have enabled the development of corn, cassava, and potato varieties with higher starch content and greater resistance to pests and diseases. According to Liu et al. (2022), the use of gene editing techniques can increase corn yield by 15%, raising global starch production to 50 million tons annually (Liu et al. 2022). In conversion, enzymes used in starch hydrolysis have been optimized to increase the efficiency of conversion to fermentable sugars. According to Vithu et al. (2020), the use of thermostable enzymes can reduce bioethanol production costs by 20%, representing a saving of approximately $10 per ton of processed starch (Vithu et al. 2020). Furthermore, biorefinery processes are gaining popularity by enabling the production of multiple products (energy, chemicals, and materials) from this biomass, which can simultaneously produce bioethanol, bioplastics, and food, maximizing economic value and minimizing waste (Watcharakitti et al. 2023).

3.6 Biotechnological and Industrial Applications of Starch-Based Biomasses

Starch serves as a key substrate for bio-based industries due to its abundance, versatility, renewability, and cost-effectiveness. It is utilized in various sectors, including food, pharmaceutical, packaging, and automotive industries. It offers a sustainable alternative to synthetic polymers and supports the development of eco-friendly products. These products help reduce the environmental impact of traditional industrial processes, promote the use of renewable resources, and contribute to the overall goal of a circular bioeconomy (van Loon et al. 2021). With ongoing research and technological advancements, the potential applications of starch in various industries continue to expand, making it an invaluable resource for a sustainable future.

3.6.1 Starch in Energy Production

3.6.1.1 Bioethanol Production

A well-established sector within the bio-based industry that employs starch as a primary substrate is the biorefinery. These facilities are designed to convert corn, cassava, wheat, and other grains into bioethanol and coproducts (Cherubini 2010). Bioethanol has emerged as the most widely utilized renewable fuel in the transportation sector (Kurambhatti et al. 2018). Ethanol production has surged due to its demand as a fuel additive, driven by environmental efforts and government mandates for cleaner energy (Rausch and Belyea 2006). As a result, bioethanol has become a key player in the transition toward sustainable energy solutions.

Corn is the predominant starch-based substrate for ethanol conversion on a global scale, with the United States leading as the foremost producer, followed by Brazil, which is emerging as a significant producer of corn-based ethanol (Zhou and Yan 2024; Eckert et al. 2018). It can be processed for ethanol production using wet milling or dry grinding. In wet milling, the corn kernel is separated into germ, fiber, and starch. The fiber and gluten are used in animal feed, while the starch is processed into corn starch or corn syrup or fermented into ethanol (Murthy et al. 2006). This technique is common in grain milling industries for food-grade products. Most ethanol comes from dry-grind technology, which involves grinding the whole corn kernel, using enzymes to break starch into sugars, and fermenting it to ethanol. Dry milling produces less ethanol per bushel of corn than wet milling but is less capital-intensive and offers various profitable coproducts (Rajagopalan et al. 2005). The appeal of corn biorefineries largely stems from these lucrative coproducts, including dried distillers' grains with solubles (DDGS) for animal feed (Boaitey et al. 2012) and extracted corn oil, utilized in oil refineries for biodiesel production, another important biofuel (Rausch and Belyea 2006).

While corn is a well-established feedstock for ethanol production, cassava stands out due to its high starch content and adaptability to marginal lands, making it a promising candidate for large-scale bioethanol production (Kurambhatti et al. 2018; Zhou and Yan 2024). Cassava starch exhibits ethanol production and fermentation profiles comparable to those of corn, making it an attractive alternative substrate. Additionally, the extraction of corn starch is complex, requiring steeping to loosen the protein matrices surrounding the starch. In contrast, cassava starch is easier to extract due to its lower protein content (Pradyawong et al. 2018).

3.6.1.2 Other Energy-Related Applications

Although bioethanol production reduces the environmental footprint compared to fossil fuels, it still generates plant waste. As production scales up, efficient waste management becomes crucial. Excess organic matter can be converted into valuable by-products like biogas through anaerobic digestion (AD) (Teixeira et al. 2009; Alrefai et al. 2020). AD uses microorganisms to convert waste into biogas, which can be used for electricity or heat or upgraded to bio-methane for vehicle fuel (Li et al. 2011; Tippayawong and Thanompongchart 2010). The AD process encompasses several stages: hydrolysis, acidogenesis, acetogenesis, and methanogenesis (Chen et al. 2008). The residual by-product, rich in nitrogen, serves as an effective bio-fertilizer (Li et al. 2011).

Carbohydrate-rich, nitrogen-deficient solid wastes from ethanol production can be utilized for hydrogen generation through bioprocess technologies. The process requires specific organisms, known as photoheterotrophic bacteria, as well as light and stringent control of environmental conditions (De Vrije et al. 2002). This approach offers cost-effective energy production while simultaneously addressing waste management (Kapdan and Kargi 2006).

Ethanol-to-jet (ETJ) fuel is produced by converting ethanol into jet fuel through a series of catalytic processes. This includes dehydration to produce ethylene, oligomerization to form longer-chain hydrocarbons, and hydrogenation to create paraffins. ETJ significantly reduces greenhouse gas emissions compared to traditional petroleum-based jet fuels and utilizes the existing corn ethanol infrastructure (Han et al. 2017).

Starch plays a crucial role in bioenergy production, particularly in generating bioethanol from corn and cassava (Pradyawong et al. 2018). Biorefineries not only produce renewable fuel but also valuable coproducts like animal feed and biodiesel precursors (Cherubini 2010). Efficient waste management techniques, such as anaerobic digestion, convert plant waste into biogas and bio-fertilizer, while carbohydrate-rich waste can be utilized for hydrogen generation (Kapdan and Kargi 2006). Innovations like ethanol-to-jet (ETJ) fuel highlight the environmental benefits by reducing emissions and leveraging the existing infrastructure (Han et al. 2017). These processes enhance energy sustainability and support a circular bioeconomy (van Loon et al. 2021).

3.6.2 Starch Biological Product Synthesis

Starch is one of the most prevalent biodegradable polymers in nature, extensively used in bioplastics, organic acids, and pharmaceutical production. Its popularity stems from its environmental friendliness, ease of processing, abundance, non-toxicity, and biodegradability (Amaraweera et al. 2021).

3.6.2.1 Organic Acid Production

The production of lactic acid from starch entails the enzymatic hydrolysis of starch into monosaccharides followed by fermentation by microorganisms (Khlestkin et al. 2018). Recent advancements, such as simultaneous saccharification and fermentation (SSF) by *Lactobacillus casei*, have significantly enhanced process efficiency (Nagarajan and Palaniraj 2012). The use of waste starch and genetically engineered microorganisms, including *Aspergillus oryzae* and *Corynebacterium glutamicum*, further augments productivity and reduces costs, thereby rendering the process more sustainable and economically viable (Tsuge et al. 2013; Wakai et al. 2014; Liaud et al. 2015).

Citric acid, widely used in food and pharmaceuticals, is produced through submerged fermentation using microorganisms like *A. niger* or *Candida* sp. This process uses carbohydrate sources such as molasses and starch-based media, often incorporating agro-industrial waste products (Lende et al. 2021). Starchy materials like corn starch, hydrolysate starch, and yam bean starch are common substrates.

Advances in treating and purifying these materials have improved citric acid yields, making production more efficient and cost-effective (Soccol et al. 2006; Sarangbin and Watanapokasin 1999).

3.6.2.2 Bioplastics

Starch is a promising raw material for producing biodegradable bioplastics due to its availability, renewability, and cost-effectiveness (Coppola et al. 2021; Balla et al. 2021). Composed of branched amylopectin and linear amylose, starch's structure allows it to withstand extreme heat and shear (Jiang et al. 2020). Bioplastics are primarily manufactured from polysaccharides, like starch; proteins; and alternative carbon sources, such as algae and wastewater by-products (Coppola et al. 2021). Thermoplastic starch (TPS) is a common bioplastic, produced through enzymatic saccharification and microbial fermentation or by modifying starch with hydrophilic plasticizers (Mojibayo et al. 2020). Various starch sources, including cassava, corn, potatoes, wheat, and tapioca, have been used to create bioplastics with different properties (Babalola and Olorunnisola 2019).

Starch, particularly corn starch processed with sulfuric acid, is crucial for producing polylactic acid (PLA), providing an economic pathway to lactic acid. PLA is often blended with starch to enhance biodegradability and reduce costs, though this can increase brittleness (Jem et al. 2009).

Another method involves using starch as a substrate for bacteria to synthesize biodegradable polymers like polyhydroxyalkanoates (PHAs). Starch is enzymatically or chemically hydrolyzed into substrates such as glucose or volatile fatty acids (VFAs). The biosynthesis of PHAs occurs in aqueous solutions with sustainable resources, making it an environmentally friendly alternative to petroleum-based methods (Choi and Lee 1999; Chen and Steinbüchel 2010; Koller et al. 2009).

Other starch-based bioplastics have been developed from potato processing wastewater, and activated sludge from wastewater treatment can produce polyhydroxybutyrates (PHBs) through thermal cracking (Liu et al. 2019). Starch can also be added to biopolymers like polyethylene (PE), PLA, and PSB to promote degradation, increase biodegradability, and reduce costs (Chen and Steinbüchel 2010).

3.6.2.3 Starch in Pharmaceutical Compounds

Starch is a widely used pharmaceutical excipient due to its nontoxic, odorless, inexpensive, and biocompatible nature. Both enzymatically and chemically modified starches are commonly employed in pharmaceuticals and drug delivery systems (Karaki et al. 2016). Enzymatic methods, although costly for polymer industries, are valuable in the pharmaceutical sector for their ability to release drugs at a constant rate (Amaraweera et al. 2021; Hong et al. 2016). Chemically modified starches, such as hydrophilic and hydrophobic carboxymethyl derivatives, are used to develop microparticles, nanoparticles, and hydrogels, enhancing drug delivery efficiency (Jem et al. 2009). Modified starches, achieved through physical or chemical processes, offer improved paste clarity, increased swelling capacity, and better stability under various conditions, expanding their versatility in pharmaceutical applications. These diverse applications make starch an essential component in drug delivery systems and other pharmaceutical products (Oyeyinka et al. 2023).

3.6.3 Conclusion

Starch-based biomasses hold immense potential for a wide range of biotechnological and industrial applications. As a renewable, versatile, and cost-effective resource, starch serves as a key substrate for bio-based industries, contributing to the development of eco-friendly products and sustainable solutions. From bioethanol production to bioplastics and pharmaceutical compounds, the diverse applications of starch underline its significance in promoting a circular bioeconomy. A summary of these applications can be seen in Table 3.4. Ongoing research and technological advancements continue to expand the potential uses of starch, reinforcing its role as a valuable resource for a sustainable future.

3.7 By-Products Generated by Starch Processing and Its Applications on a Circular Economy Approach

The extraction of starch from crops generates significant residues that modern circular economy practices now repurpose into valuable by-products. These materials originate from various stages of processing, beginning with the initial separation of plant components through to the purification of post-distillation stillage (Naik et al. 2010).

Lignocellulosic residues such as rice straw, corn stover, and potato peels contain polysaccharides in forms less accessible than starch. While their conversion requires more intensive processing through hydrolysis, fermentation, or gasification, they serve as important feedstocks for second-generation biofuels (Naik et al. 2010). Grain husks represent another valuable by-product, rich in both dietary fibers and amino acids. These components offer multiple health benefits, including improved gastrointestinal function and positive effects on intestinal microbiota. The carbohydrates extracted from husks through chemical processing show potential for diabetes management and immune regulation while also exhibiting antioxidant, antitumor, and blood pressure–lowering properties (Jiao et al. 2022).

The processing of starch-rich grains like corn and wheat typically begins with either dry or wet milling. Dry milling operations produce wet distillers' grains (WDG), a nutrient-rich by-product containing bran, germ, and unfermented

Table 3.4 The main industrial applications of starch-based biomasses

	Product	Feedstock	Production process	Applications	References
Biofuels	Bioethanol	Corn, cassava, wheat, other grains	Wet milling or dry grinding of corn, starch liquefaction, saccharification, and fermentation	Renewable fuel in the transportation sector	Cherubini (2010), Kurambhatti et al. (2018), Murthy et al. (2006), and Rajagopalan et al. (2005)
	Biogas	Excess organic matter can be from bioethanol production	Anaerobic digestion of organic matter	Electricity and heat generation, upgraded to bio-methane for vehicle fuel	Teixeira et al. (2009), Alrefai et al. (2020), and Li et al. (2011)
	Biodiesel	Extracted corn oil	Extraction of corn oil from biorefineries	Renewable fuel in the transportation sector	Rausch and Belyea (2006)
	Hydrogen	Residual starch from ethanol production and other grains	Conversion of sugars into hydrogen gas using photoheterotrophic bacteria	Clean energy source and industrial feedstock	de Vrije et al. (2002), Kapdan and Kargi (2006)
	Ethanol-to-jet fuel	Corn-based, cellulosic, or sugar-based ethanol	Conversion of ethanol into sustainable aviation fuel (SAF) using advanced processing technologies	Sustainable aviation fuel (SAF)	Han et al. (2017) and Khlestkin et al. (2018)
Organic acids	Lactic acid	Starch from various sources	Enzymatic breakdown of starch into monosaccharides, followed by microbial fermentation	Precursor for bioplastics and other chemicals, used in food and pharmaceutical industries	Khlestkin et al. (2018), Nagarajan and Palaniraj (2012)
	Citric acid	Corn starch, hydrolysate starch, yam bean starch	Submerged fermentation using microorganisms with starchy substrates	Used in the food and pharmaceutical industries	Lende et al. (2021), Soccol et al. (2006), Sarangbin and Watanapokasin (1999)
Bioplastics	Thermoplastic starch (TPS)	Cassava, corn, potatoes, wheat, tapioca	Enzymatic saccharification and microbial fermentation or modification of starch with hydrophilic plasticizers	Packaging materials, disposable cutlery, agricultural films	Mojibayo et al. (2020) and Babalola and Olorunnisola (2019)

(continued)

Table 3.4 (continued)

	Product	Feedstock	Production process	Applications	References
	Polylactic acid (PLA)	Corn starch processed with sulfuric acid	Produced from lactic acid derived from corn starch	Food packaging, disposable cutlery, medical implants, 3D printing	Jem et al. (2009)
	Polyhydroxyalkanoates (PHAs)	Starch hydrolyzed into glucose or volatile fatty acids	Biosynthesis by bacteria using hydrolyzed starch substrates in aqueous solutions	Biomedical applications, packaging materials	Choi and Lee (1999), Chen and Steinbüchel (2010), Koller et al. (2009)
	Polyhydroxybutyrates (PHBs)	Potato processing wastewater, activated sludge from wastewater treatment	Produced through thermal cracking and activated sludge	Food packaging, drug delivery systems, medical implants	Liu et al. (2019)
Pharmaceutical compounds	Enzymatically modified starches	Starch from various sources	Enzymatic modification to release drugs at a constant rate	Used in drug delivery systems for controlled and sustained drug release	Amaraweera et al. (2021), Karaki et al. (2016), Hong et al. (2016), and Oyeyinka et al. (2023)
	Chemically modified starches	Starch from various sources	Hydrophilic and hydrophobic modifications to develop microparticles, nanoparticles, and hydrogels	Used for enhanced drug delivery efficiency, improved paste clarity, increased swelling capacity, better stability under various conditions	Choi and Lee (1999), Karaki et al. (2016), and Oyeyinka et al. (2023)

endosperm components. This material is recovered through centrifugation of post-distillation stillage and can be further dried to create dried distillers' grains (DDG) for improved storage and transportation (Iram et al. 2020; Stock et al. 1999). Following WDG separation, the remaining liquid fraction, known as thin stillage, contains residual grain particles, fermentation microorganisms, and other nutrients. This can be concentrated through evaporation to produce condensed distillers' solubles (CDS), which may be marketed separately or combined with WDG to form wet distillers' grains with solubles (WDGS) (Stock et al. 1999).

Global production of these by-products has shown significant growth. According to the US Grains Council, the United States currently produces over 44 million metric tons of DDG annually, with exports doubling from 5 million tons in 2009 to more than 10 million tons by 2023 (US Grains Council 2025). Brazil, as the world's second-largest bioethanol producer, anticipates DDG production reaching 4 million tons by 2025 (Silva 2024).

The primary application for DDG remains animal feed due to its high content of fiber, protein, antioxidants, and fats. However, its low carbohydrate content makes it unsuitable as a sole dietary component, a limitation that can be mitigated by adding syrup to create DDGS. The high fiber content makes DDGS particularly suitable for ruminants, which can effectively digest and ferment these fibers; because of this, the main usage of DDGS is for beef cattle feed. For monogastric animals like poultry and aquaculture species, DDGS serves as a supplemental feed component due to their limited ability to process high-fiber materials. A significant concern in DDGS utilization is potential aflatoxin contamination, which has prompted research into alternative applications and methods to mitigate production excesses (Iram et al. 2020).

Besides animal nutrition, researchers have developed several alternative uses for DDG. Its application in cat litter production takes advantage of water-absorbing properties enhanced with clumping agents and odor neutralizers. While effective, this application faces economic challenges due to the need for hexane extraction to achieve optimal absorption (Vaughn et al. 2011). Additional studies have explored the use of foaming agents to inhibit urease activity and reduce ammonia production in such applications (Tsengas and Teuscher 2002). DDG is also being studied as a fertilizer, demonstrating its ability to improve soil quality, suppress weed growth, and enhance plant growth rates (Nelson et al. 2009).

Oils can also be extracted from both DDG and thin stillage, and they can be sold in their crude form for use in animal feed or be used for the production of biodiesel (Iram et al. 2020).

During starch extraction by wet milling, depending on the biomass used, whether it be cereals, tubers, roots, or fruits, one or two protein-rich by-products are generated, which are gluten and steep liquor, with the latter being more related to cereal processing (Kringel et al. 2020; Deepak and Jayadeep 2022). Both products are used in the feed industry for the nutrition of cattle, poultry, and swine. Corn gluten meal, obtained after starch separation at the end of the wet milling process, is a highly nutrient-dense co-product containing approximately 60% protein. In contrast, corn steep liquor is particularly rich in organic nitrogen, accounting for around 45% of its dry matter content. (Loy and Lundy 2019). The reasons why those products are used in the feed industry include an increase in the efficiency of absorption or their ability to provide nutrients that are not available in other feed resources. Therefore, many studies have been made to enhance the already natural properties of those products.

In terms of gluten meals derived from corn starch production, one of the value-enriching alternatives encompasses the modification of gluten in order to overcome the lack of essential amino acids and the hydrophobicity of the material (Kopparapu et al. 2022). The modifications can occur through physical, chemical, glycosylation, microbial fermentation, or enzymatic process. Regarding microbial fermentation, in one study, to enhance the nutritional value, a co-fermentation of *Bacillus subtilis* MZ01, *Candida utilis* CICC 31188, and keratinase was applied, resulting in a higher small peptide yield, allowing an increase in protein digestibility and amino acid transporter protein mRNA expression in experiments with piglets (Chi et al. 2024). The yield of peptides and the study of their properties, when they are obtained from enzymatic processes, are the focus of other studies (Yu et al. 2023; Singh et al. 2022). Still, one of the main objectives for the use of enzymes relates to the increase of the solubility of the material. One example would be a study conducted with wheat gluten protein, submitted to *Bacillus subtilis* MTCC 2423 protease hydrolysis, which resulted in significant improvements in solubility, which is a matter of great importance to the usage of gluten hydrolysates as additives for functional foods (Rajendhran et al. 2024). Something that highlights this application is the study of combinations of hydrolysate fractions of corn and soy, which, given in vitro studies, may provide the desired functional attributes due to a balance in the content of hydrophobic and hydrophilic amino acids (Mirzaee et al. 2024).

On the matter of steep liquor, also known as condensed fermented extractives, the most common usage encompasses the supplementation of dried feed resources, such as dry distillers' grain due to the presence of vitamins and trace elements, alongside the protein levels (Loy and Lundy 2019). However, other alternatives are still studied. Corn steep liquor may also be used to enhance the production of microbial lipids and enzymes, such as extracellular lipase and protease, in a *Yarrowia lipolytica* CBS 2075–growing medium.

Despite other materials used for starch production, given the usage of corn for either the feed or the fuel industry, it was used as the base model for the present applications. The extrapolation of the possibilities to other biomass by-products is not only possible but encouraged (Miranda et al. 2024).

Furthermore, throughout the extraction and processing, wastewater is produced, turning into a solution rich in organic matter, with a high biochemical oxygen demand (BOD) and chemical oxygen demand (COD), containing soluble carbohydrates, such as starch (Hernández-Carmona et al. 2017). This organic matter can be treated in order to reduce those demands or produce energy sources by coagulation–flocculation, adsorption, anaerobic digestion, microbial fermentation, and other techniques. Coagulation–flocculation is a well-known model for wastewater treatment using metal salt coagulants, synthetic organic coagulants, and more recently, biopolymer coagulants. Other alternatives may encompass the usage of clay minerals adjunct with biopolymers, such as a system of chitosan-modified bentonite as a coagulation–flocculation agent (Zhang et al. 2023). Those combinations can improve the treatment of wastewater due to the adsorption of the components that may be separated in specific conditions, giving also an alternative to the recovery of the soluble carbohydrates adhered in the system.

Still, one of the main alternatives for wastewater treatment and repurposing is anaerobic digestion, producing biogas as its final product. For biogas production, a multitude of possible process configurations are available, given pretreatments and reactor configurations. An alternative is the usage of anaerobic baffled reactors (ABRs), which are characterized by the sequential flow through multiple compartments. In this situation, one study determined that the microbiota from each compartment differed from others and that, with a pH decrease, each compartment would tend to specialize its production of volatile fatty acids and yield of hydrogen and methane, evidencing the importance of the conditions for anaerobic digestion (Zhang et al. 2022). Another alternative would be an anaerobic sequencing batch biofilm reactor (AnSBBR). In one study, in which the substrate was acidified cassava wastewater for biogas production, it was determined that the ratios of $NaHCO_3$ would directly impact digestion and the COD removal efficiency (Devens et al. 2024). However, overall, the content of methane in the biogas maintained stable values. Therefore, there is evidence that, like every process involving a series of microorganisms, the conditions to which they are exposed have a great impact on the yield of the process.

Other alternatives for the valorization of wastewater include products like biopesticides and single-cell proteins, obtained by fermentation (Ndao et al. 2021). Regarding single-cell protein production, their use is connected to the growing demand for animal and plant proteins in such a way as to offer a possible alternative to those products. In this matter, a study used potato starch–processing wastewater as the main medium for a coculture fermentation of the combinations of five yeast strains: *Candida utilis*, *Candida tropicalis*, *Saccharomycopsis fibuligera, Saccharomyces cerevisiae,* and *Geotrichum candidum.* After passing through an optimization process, the process with three of the strains had a single-cell protein yield of 3.06 g/L and a water-soluble protein content of 17.9% (Tian et al. 2023). Therefore, this study provides a unique and interesting alternative for the valorization of wastewater as a by-product.

Finally, it is clear that many alternatives are available for the repurposing of starch-based industries' by-products, and they can provide not only an advantage in economic terms but also in the sustainable usage of resources. At the pace the by-products are repurposed, a reduced quantity of waste is disposed in the environment, enabling a better natural cycling of nutrients and aiming for a more sustainable industry and, consequently, society. Furthermore, due to the fact that the by-products are rich in important compounds for the overall industrial system, the repurposing is also an efficient use of energy and material, demanding a reduced extraction of resources at the beginning of the process to produce the same diversity of products.

3.8 Conclusion

Amylaceous biomasses play a central role in the transition toward a sustainable and circular bioeconomy. As renewable, abundant, and versatile sources of starch, crops such as corn, cassava, potato, rice, and wheat offer strategic advantages for food systems, industrial processes, and bioenergy platforms. Understanding the molecular structure, biosynthetic pathways, and environmental influences on starch accumulation enables the development of more efficient extraction and valorization strategies.

The integration of physical, chemical, enzymatic, and genetic modification techniques has significantly expanded the functionality of starches, allowing their use in applications ranging from bioplastics and adhesives to bioethanol and biodegradable films. Furthermore, innovations in processing technologies, such as membrane filtration and high-pressure treatment, enhance sustainability and reduce waste, aligning with the principles of circularity.

Looking forward, emerging research on unconventional starch sources, gene-edited crops, and biorefineries that generate multiple high-value products will further elevate the importance of amylaceous biomasses. These approaches, combined with advances in biotechnology and green chemistry, have the potential to transform starch-based systems into key pillars of low-carbon development.

This chapter contributes to the scientific literature by providing a comprehensive, updated overview of starch science, from molecular properties to economic and technological applications, positioning starch-rich biomasses as drivers of innovation in bio-based industries.

References

Alcázar-Alay SC, Meireles MAA (2015) Physicochemical properties, modifications and applications of starches from different botanical sources. Food Sci Technol (Brazil) 35(2):215–236. https://doi.org/10.1590/1678-457X.6749

Alrefai AM, Alrefai R, Benyounis KY, Stokes J (2020) Impact of starch from cassava peel on biogas produced through the anaerobic digestion process. Energies (Basel) 13. https://doi.org/10.3390/EN13112713

Amaraweera SM, Gunathilake C, Gunawardene OHP (2021) Development of starch-based materials using current modification techniques and their applications: a review. Molecules 26(22): 68–80

Ancy A, Lazar M, Saritha Chandran A, Ushamani M (2024) Development of ecofriendly and sustainable bioplastics from cassava starch: tailoring the properties using nanoparticles. Sustain Chem Pharm 37:101377

Apriyanto A, Compart J, Fettke J (2022) A review of starch, a unique biopolymer – structure, metabolism and in planta modifications. Plant Sci. Elsevier Ireland Ltd. https://doi.org/10.1016/j.plantsci.2022.111223

Aristizábal-Marulanda V, Cardona Alzate CA (2019) Methods for designing and assessing biorefineries: review. Biofuels Bioprod Biorefin 13(3):789–808

Awodi PS, Ogbonna JC, Nwagu TN (2022) Bioconversion of mango (Mangifera indica) seed kernel starch into bioethanol using various fermentation techniques. Heliyon 8(6):e09707

Babalola OA, Olorunnisola AO (2019) Evaluation of coconut (Cocos nucifera) husk fibre as a potential reinforcing material for bioplastic production. In: By-products of palm trees and their applications. Materials Research Forum LLC, Millersville, pp 195–200

Balla E, Daniilidis V, Karlioti G (2021) Poly(lactic acid): a versatile biobased polymer for the future with multifunctional properties-from monomer synthesis, polymerization techniques and molecular weight increase to PLA applications. Polymers (Basel) 13(11):1822

BeMiller J, Whistler R (2009) Starch: chemistry and technology, 3rd edn. Academic Press, Cambridge

Bhuyar P, Trejo M, Mishra P, Unpaprom Y, Velu G, Ramaraj R (2022) Advancements of fermentable sugar yield by pretreatment and steam explosion during enzymatic saccharification of Amorphophallus sp. starchy tuber for bioethanol production. Fuel 323:124406

Boaitey A, Christensen C, Smyth S, Makkar H (2012) Biofuel co-products as livestock feed – opportunities and challenges. Food and Agriculture Organization of the United Nations, Rome

Bothast RJ, Schlicher MA (2005) Biotechnological processes for conversion of corn into ethanol. Appl Microbiol Biotechnol. https://doi.org/10.1007/s00253-004-1819-8

Chen GGQ, Steinbüchel A (2010) Plastics from bacteria: natural functions and applications, 1st edn. Springer, Berlin Heidelberg

Chen J, Zhang B, Luo L, Zhang F, Yi Y, Shan Y (2021) A review on recycling techniques for bioethanol production from lignocellulosic biomass. Renew Sustain Energy Rev 149:111370

Chen Y, Cheng JJ, Creamer KS (2008) Inhibition of anaerobic digestion process: a review. Bioresour Technol 99:4044–4064

Cherubini F (2010) The biorefinery concept: using biomass instead of oil for producing energy and chemicals. Energy Convers Manag 51:1412–1421. https://doi.org/10.1016/j.enconman.2010.01.015

Chi Z, Feng Y, Wei X (2024) A co-fermentation technology based on small peptide production and protein structure modification increases the edible value of corn gluten meal. LWT 198:116054. https://doi.org/10.1016/J.LWT.2024.116054

Choi J, Lee SY (1999) Factors affecting the economics of polyhydroxyalkanoate production by bacterial fermentation. Appl Microbiol Biotechnol 51:13–21. https://doi.org/10.1007/s002530051357

Coffman WR, Juliano BO (2015) Rice, pp 101–131

Compart J, Singh A, Fettke J, Apriyanto A (2023) Customizing starch properties: a review of starch modifications and their applications. Polymers. Multidisciplinary Digital Publishing Institute (MDPI). https://doi.org/10.3390/polym15163491

Coppola G, Gaudio MT, Lopresto CG (2021) Bioplastic from renewable biomass: a facile solution for a greener environment. Earth Syst Environ 5:231–251

Das N, Jena PK, Padhi D, Kumar Mohanty M, Sahoo G (2021) A comprehensive review of characterization, pretreatment and its applications on different lignocellulosic biomass for bioethanol production. Biomass Convers Biorefinery 13(2):1503–1527

DDGS – U.S. Grains Council (2025) Retrieved 28 Mar 2025, from https://grains.org/buying-selling/ddgs/

de Oliveira Schmidt VK, de Vasconscelos GMD, Vicente R, de Souza CJ, Della-Flora IK, Degang L (2022) Cassava wastewater valorization for the production of biosurfactants: surfactin, rhamnolipids, and mannosileritritol lipids. World J Microbiol Biotechnol 39(2):1–17

De Vrije T, De Haas GG, Tan GB (2002) Pretreatment of Miscanthus for hydrogen production by *Thermotoga elfii*. Int J Hydrog Energy 27(11–12):1381–1390. https://doi.org/10.1016/S0360-3199(02)00124-6

Deepak TS, Jayadeep PA (2022) Prospects of maize (corn) wet milling by-products as a source of functional food ingredients and nutraceuticals. Food Technol Biotechnol 60:109. https://doi.org/10.17113/FTB.60.01.22.7340

Devens KU, Ribeiro AR, Trevisan AP (2024) Optimization of alkalinity supplementation in the AnSBBR using acidified cassava starch wastewater. Biochem Eng J 202:109175. https://doi.org/10.1016/J.BEJ.2023.109175

Dorantes-Fuertes MG, López-Méndez MC, Martínez-Castellanos G, Meléndez-Armenta RA, Jiménez-Martínez HE (2024) Starch extraction methods in tubers and roots: a systematic review. Agronomy. Multidisciplinary Digital Publishing Institute (MDPI). https://doi.org/10.3390/agronomy14040865

Eckert CT, Frigo EP, Albrecht LP (2018) Maize ethanol production in Brazil: characteristics and perspectives. Renew Sust Energ Rev 82:3907–3912

Graybosch RA (1998) Waxy wheats: origin, properties, and prospects. Trends Food Sci Technol 9(4):135–142

Gronchi N, Favaro L, Cagnin L, Brojanigo S, Pizzocchero V, Basaglia M (2019) Novel yeast strains for the efficient saccharification and fermentation of starchy by-products to bioethanol. Energies 12(4):714

Halim A, Torley PJ, Farahnaky A, Majzoobi M (2024) Investigating the effects of acid hydrolysis on physicochemical properties of quinoa and faba bean starches as compared to cassava starch. Foods 13(23):3885

Han J, Tao L, Wang M (2017) Well-to-wake analysis of ethanol-to-jet and sugar-to-jet pathways. Biotechnol Biofuels 10. https://doi.org/10.1186/s13068-017-0698-z

Hernández-Carmona F, Morales-Matos Y, Lambis-Miranda H, Pasqualino J (2017) Starch extraction potential from plantain peel wastes. J Environ Chem Eng 5:4980–4985. https://doi.org/10.1016/J.JECE.2017.09.034

Hizukuri S (1985) Relationship between the distribution of the chain length of amylopectin and the crystalline structure of starch granules. Carbohydr Res 141(2):295–306

Hong Y, Liu G, Gu Z (2016) Recent advances of starch-based excipients used in extended-release tablets: a review. Drug Deliv 23:12–20

Hoover R (2001) Composition, molecular structure, and physicochemical properties of tuber and root starches: a review. Carbohydr Polym 45(3):253–267

Hoover R, Ratnayake WS (2002) Starch characteristics of black bean, chick pea, lentil, navy bean and pinto bean cultivars grown in Canada. Food Chem 78(4):489–498

Ioannidou SM, Pateraki C, Ladakis D, Papapostolou H, Tsakona M, Vlysidis A (2020) Sustainable production of bio-based chemicals and polymers via integrated biomass refining and bioprocessing in a circular bioeconomy context. Bioresour Technol 307:123093

Iram A, Cekmecelioglu D, Demirci A (2020) Distillers' dried grains with solubles (DDGS) and its potential as fermentation feedstock. Appl Microbiol Biotechnol 104(14):6115–6128. https://doi.org/10.1007/S00253-020-10682-0/TABLES/4

Jem KJ, Van Der Pol JF, De Vos S (2009) Microbial lactic acid, its polymer poly(lactic acid), and their industrial applications. Microbiology monographs, vol 14. https://doi.org/10.1007/978-3-642-03287_5_13

Jiang T, Duan Q, Zhu J (2020) Starch-based biodegradable materials: challenges and opportunities. Adv Ind Eng Polym Res 3:8–18

Jiao Y, Chen HD, Han H, Chang Y (2022) Development and utilization of corn processing by-products: a review. Foods 11(22):3709. https://doi.org/10.3390/FOODS11223709

Jin X, Ma J, Song J, Liu GQ (2020) Saccharification and detoxification of Na2CO3 pretreated rice straw with on-site manufactured enzymes secreted by Aspergillus fumigatus to enhance bioethanol yield. Renew Energy 166:117–124

Kapdan IK, Kargi F (2006) Bio-hydrogen production from waste materials. Enzym Microb Technol 38:569–582

Karaki N, Aljawish A, Humeau C (2016) Enzymatic modification of polysaccharides: mechanisms, properties, and potential applications: a review. Enzym Microb Technol 90:1–18

Khlestkin VK, Peltek SE, Kolchanov NA (2018) Review of direct chemical and biochemical transformations of starch. Carbohydr Polym 181:460–476

Kim KH, Kim JY (2021) Understanding wheat starch metabolism in properties, environmental stress condition, and molecular approaches for value-added utilization. Plants 10(11):2282

Koller M, Atlić A, Dias M (2009) Microbial PHA production from waste raw materials. Microbiology monographs, vol 14. https://doi.org/10.1007/978-3-642-03287-5_5

Kopparapu NK, Duan Y, Huang L, Katrolia P (2022) Review on utilisation of corn gluten meal, a by-product from corn starch industry for production of value-added products. Int J Food Sci Technol 57:5592–5599. https://doi.org/10.1111/IJFS.15541

Kringel D, El Halal SLM, da Rosa Zavareze E, Dias ÁRG (2020) Methods for the extraction of roots, tubers, pulses, pseudocereals, and other unconventional starches sources: a review. Starch – Stärke 72:1900234. https://doi.org/10.1002/STAR.201900234

Kumar A, Sahoo U, Lal MK, Tiwari RK, Lenka SK, Singh NR (2022) Biochemical markers for low glycemic index and approaches to alter starch digestibility in rice. J Cereal Sci 106:103501

Kurambhatti CV, Kumar D, Rausch KD (2018) Ethanol production from corn fiber separated after liquefaction in the dry grind process. Energies (Basel) 11. https://doi.org/10.3390/en11112921

Li C, Dhital S, Gidley MJ (2023) High amylose wheat foods: a new opportunity to improve human health. Trends Food Sci Technol 135:93–101

Lee U, Kwon H, Wu M, Wang M (2021) Retrospective analysis of the U.S. corn ethanol industry for 2005–2019: implications for greenhouse gas emission reductions. Biofuels Bioprod Biorefin 15(5):1318–1331. https://doi.org/10.1002/bbb.2225

Lende SV, Karemore H, Umekar MJ (2021) Review on production of citric acid by fermentation technology. GSC Biol Pharm Sci 17:085–093. https://doi.org/10.30574/gscbps.2021.17.3.0313

Li M, Tian Y, Dhital S (2022) Starch extraction. In: Starch and starchy food products. CRC Press, pp 17–40. https://doi.org/10.1201/9781003088929-2

Li Y, Park SY, Zhu J (2011) Solid-state anaerobic digestion for methane production from organic waste. Renew Sust Energ Rev 15:821–826

Liargovas P, Papageorgiou C (2024) The common agricultural policy: purpose, evolution and future projections. In: The European integration, vol 2, pp 197–251. https://doi.org/10.1007/978-3-031-47176-6_6

Liaud N, Rosso MN, Fabre N (2015) L-lactic acid production by Aspergillus brasiliensis overexpressing the heterologous ldha gene from *Rhizopus oryzae*. Microb Cell Fact 14. https://doi.org/10.1186/s12934-015-0249-x

Liu F, Li J, Zhang XL (2019) Bioplastic production from wastewater sludge and application. IOP Conf Ser Earth Environ Sci. Institute of Physics Publishing 344(1):012071

Liu X, Zhang S, Jiang Y, Yan T, Fang C, Hou Q (2022) Use of CRISPR/Cas9-based gene editing to simultaneously mutate multiple homologous genes required for pollen development and male fertility in maize. Cells 11(3):439

Louw J, Dogbe ES, Yang B, Görgens JF (2023) Prioritisation of biomass-derived products for biorefineries based on economic feasibility: a review on the comparability of techno-economic assessment results. Renew Sust Energ Rev 188:113840

Loy DD, Lundy EL (2019) Nutritional properties and feeding value of corn and its coproducts. In: Corn: chemistry and technology, 3rd edn, pp 633–659. https://doi.org/10.1016/B978-0-12-811971-6.00023-1

Lozowski D (2012) Gevo and Beta Renewables sign JDA to develop process for cellulosic isobutanol. Chem Eng 119(8):63–64

Miranda SM, Lopes M, Belo I (2024) Exploring the use of hexadecane by Yarrowia lipolytica: effect of dissolved oxygen and medium supplementation. J Biotechnol 380:29–37. https://doi.org/10.1016/J.JBIOTEC.2023.12.006

Mirzaee H, Ahmadi Gavlighi H, Nikoo M (2024) Improved antioxidant, antihypertensive, and antidiabetic activities and tailored emulsion stability and foaming properties of mixture of corn gluten and soy protein hydrolysates via enzymatic processing and fractionation. Food Sci Nutr 12:9749–9763. https://doi.org/10.1002/FSN3.4532

Mirzaei A, Azarm H, Noshad M (2024) Designing sustainability comprehensive indicator for the food supply chain under climate change: a systematic literature review. Ecol Indic 159:111722

Mojibayo I, Samson AO, Johnson OY (2020) A preliminary investigation of cassava starch potentials as natural polymer in bioplastic production. Am J Interdiscip Innov Res 02:31–39. https://doi.org/10.37547/tajiir/Volume02Issue09-05

Morita MT (2010) Directional gravity sensing in gravitropism. Annu Rev Plant Biol 61:705–720

Murthy GS, Singh V, Johnston DB (2006) Evaluation and strategies to improve fermentation characteristics of modified dry-grind corn processes. Cereal Chem 83:455–459. https://doi.org/10.1094/CC-83-0455

Nagarajan P, Palaniraj R (2012) Kinetic studies in production of lactic acid from waste potato starch using *Lactobacillus casei*. Int J ChemTech Res 4(4):1601–1614

Naik SN, Goud VV, Rout PK, Dalai AK (2010) Production of first and second generation biofuels: a comprehensive review. Renew Sust Energ Rev 14(2):578–597. https://doi.org/10.1016/J.RSER.2009.10.003

Nakamura Y, Kainuma K (2022) On the cluster structure of amylopectin. Plant Mol Biol 108(4–5):291–306

Nawaz H, Waheed R, Nawaz M, Shahwar D (2020) Physical and chemical modifications in starch structure and reactivity. www.intechopen.com

Nazli RI, Aslankaciran C, Yang D, Liaqat W, Polat M, Cavdar AS (2024) Assessment of sweet sorghum genotypes for bioethanol production potential and bagasse combustion characteristics in a semi-arid Mediterranean climate. Agronomy 14(11):2626

Ndao A, Sellamuthu B, Kumar LR (2021) Biopesticide production using *Bacillus thuringiensi*s kurstaki by valorization of starch industry wastewater and effluent from aerobic, anaerobic digestion. Syst Microbiol Biomanuf 1:494–504. https://doi.org/10.1007/S43393-021-00043-X/METRICS

Nelson KA, Motavalli PP, Smoot RL (2009) Utility of dried distillers grain as a fertilizer source for corn. J Agric Sci 1(1). https://doi.org/10.5539/jas.v1n1p3

Nitsos C, Rova U, Christakopoulos P (2018) Organosolv fractionation of softwood biomass for biofuel and biorefinery applications. Energies 11(1):50

Oktay S, Pizzi A, Köken N, Bengü B (2024) Chemical modification techniques of corn starch for synthesis wood adhesive. Int J Adhes Adhes 128:103545

Olivato JB (2024) Starch: a natural, safe, and multifunctional ingredient for cosmetic formulations. In: Starch industries: processes and innovative products in food and non-food uses. Academic Press, Cambridge, pp 255–269

Ovando-Martínez M, Whitney K, Simsek S (2013) Analysis of starch in food systems by high-performance size exclusion chromatography. J Food Sci 78(2). https://doi.org/10.1111/1750-3841.12037

Oyeyinka SA, Corrado M, Hazard B, Onarinde BA (2023) Fermentation of starch. In: Starch: advances in modifications, technologies and applications, pp 431–455. https://doi.org/10.1007/978-3-031-35843-2_18

Patindol JA, Siebenmorgen TJ, Wang YJ (2015) Impact of environmental factors on rice starch structure: a review. Starch/Staerke 67(1–2):42–54

Pfister B, Zeeman SC (2016) Formation of starch in plant cells. Cell Mol Life Sci 73(14):2781

Pleissner D, Qi Q, Gao C, Rivero CP, Webb C, Lin CSK (2016) Valorization of organic residues for the production of added value chemicals: a contribution to the bio-based economy. Biochem Eng J 116:3–16

Pradyawong S, Juneja A, Bilal Sadiq M, Noomhorm A, Singh V (2018) Comparison of cassava starch with corn as a feedstock for bioethanol production. Energies 11(12):3476. https://doi.org/10.3390/en11123476

Rajagopalan S, Ponnampalam E, Mccalla D, Stowers M (2005) Enhancing profitability of dry mill ethanol plants: process modeling and economics of conversion of degermed defibered corn to ethanol. Appl Biochem Biotechnol 120:37–50

Rajendhran HP, Vaidyanathan VK, Venkatraman S, Karthik P (2024) Optimization of enzymatic hydrolysis by protease produced from Bacillus subtilis MTCC 2423 to improve the functional properties of wheat gluten hydrolysates. Int J Food Sci 2024:5053510. https://doi.org/10.1155/2024/5053510

Rashwan AK et al (2024) Plant starch extraction, modification, and green applications: a review. Environ Chem Lett 22(5):2483–2530. https://doi.org/10.1007/s10311-024-01753-z

Rausch KD, Belyea RL (2006) The future of coproducts from corn processing. Appl Biochem Biotechnol 128(1):47–86

Rezania S, Oryani B, Cho J, Talaiekhozani A, Sabbagh F, Hashemi B (2020) Different pretreatment technologies of lignocellulosic biomass for bioethanol production: an overview. Energy 199: 117457

Rübsam H, Krottenthaler M, Gastl M, Becker T (2012) An overview of separation methods in starch analysis: the importance of size exclusion chromatography and field flow fractionation. Starch – Stärke 64(9):683–695. https://doi.org/10.1002/STAR.201100188

Safarian S, Unnthorsson R (2018) An assessment of the sustainability of lignocellulosic bioethanol production from wastes in Iceland. Energies 11(6):1493

Sarangbin S, Watanapokasin Y (1999) Yam bean starch: a novel substrate for citric acid production by the protease-negative mutant strain of Aspergillus Niger. Carbohydr Polym 38(3):219–224

Sekulić NM, Kovačević M, Jovičić R (2024) The EU state aid regime in agriculture – legal aspect. Ekonomika Poljopr 71(1):239–252

Shukla P, Garai D, Zafar M, Gupta K, Shrivastava S (2007) Process parameters optimization for lipase production by Rhizopus oryzae KG-10 under submerged fermentation using response surface methodology. J Appl Sci Environ Sanit 2(3):93–103

Silva E (2024) Produção de DDGS deve crescer 31,6% na safra 2024/25, estima Unem

Singh U, Kaur D, Mishra V, Krishania M (2022) Combinatorial approach to prepare antioxidative protein hydrolysate from corn gluten meal with dairy whey: preparation, kinetics, nutritional study and cost analysis. LWT 153:112437. https://doi.org/10.1016/J.LWT.2021.112437

Soccol CR, Vandenberghe LPS, Rodrigues C, Pandey A (2006) New perspectives for citric acid production and application. Food Technol Biotechnol 44(2):141–149

Stitt M, Zeeman SC (2012) Starch turnover: pathways, regulation and role in growth. Curr Opin Plant Biol 15(3):282–292

Suriyachai N, Imman S, Kreetachat T, Laosiripojana N, Khongchamnan P, Champreda V (2024) Enhanced bioethanol production from cassava waste: optimization of thermochemical pretreatment for maximum sugar recovery and characterisation of potential fractions. Int J Sustain Energy 43(1). https://doi.org/10.1080/14786451.2024.2411833

Tayyab M, Noman A, Islam W, Waheed S, Arafat Y, Ali F, Zaynab M, Lin S, Zhang H, Lin W (2018) Bioethanol production from lignocellulosic biomass by environment-friendly pretreatment methods: a review. Appl Ecol Environ Res 16(1):225–249

Teixeira E d M, Pasquini D, Curvelo AAS (2009) Cassava bagasse cellulose nanofibrils reinforced thermoplastic cassava starch. Carbohydr Polym 78:422–431. https://doi.org/10.1016/j.carbpol.2009.04.034

Tian Y, Li J, Meng J, Li J (2023) High-yield production of single-cell protein from starch processing wastewater using co-cultivation of yeasts. Bioresour Technol 370:128527. https://doi.org/10.1016/J.BIORTECH.2022.128527

Tippayawong N, Thanompongchart P (2010) Biogas quality upgrade by simultaneous removal of CO2 and H2S in a packed column reactor. Energy 35:4531–4535. https://doi.org/10.1016/j.energy.2010.04.014

Tsengas S, Teuscher JS (2002) Cat litter (Patent US6817315B1). United States

Tsuge Y, Tateno T, Sasaki K (2013) Direct production of organic acids from starch by cell surface-engineered *Corynebacterium glutamicum* in anaerobic conditions. AMB Expr 3:72. https://doi.org/10.1186/2191-0855-3-72

van Loon P, Diener D, Harris S (2021) Circular products and business models and environmental impact reductions: current knowledge and knowledge gaps. J Clean Prod 288. https://doi.org/10.1016/j.jclepro.2020.125627

Vandenberghe LPS, Valladares-Diestra KK, Bittencourt GA, Zevallos Torres LA, Vieira S, Karp SG (2022) Beyond sugar and ethanol: the future of sugarcane biorefineries in Brazil. Renew Sustain Energy Rev 167:112721

Vaughn SF, Berhow MA, Winkler-Moser JK, Lee E (2011) Formulation of a biodegradable, odor-reducing cat litter from solvent-extracted corn dried distillers grains. Ind Crop Prod 34(1): 999–1002. https://doi.org/10.1016/J.INDCROP.2011.03.005

Vilpoux OF, Silveira Junior JFS (2023) Global production and use of starch. In: Starchy crops morphology, extraction, properties and applications: vol 1: underground starchy crops of South American origin: production, processing, utilization and economic perspectives, vol 1, pp 43–66. https://doi.org/10.1016/B978-0-323-90058-4.00014-1

Vithu P, Dash SK, Rayaguru K, Panda MK, Nedunchezhiyan M (2020) Optimization of starch isolation process for sweet potato and characterization of the prepared starch. J Food Meas Charact 14(3):1520–1532

Wakai S, Yoshie T, Asai-Nakashima N (2014) L-lactic acid production from starch by simultaneous saccharification and fermentation in a genetically engineered Aspergillus oryzae pure culture. Bioresour Technol 173:376–383. https://doi.org/10.1016/j.biortech.2014.09.094

Wallerstein D (2020) Food-energy-water (FEW) nexus: rearchitecting the planet to accommodate 10 billion humans by 2050. Resour Conserv Recycl 155:104658

Wang S, Wang J, Liu Y, Liu X (2020) Starch modification and application. In: Starch structure, functionality and application in foods. Springer Singapore, pp 131–149. https://doi.org/10.1007/978-981-15-0622-2_8

Watcharakitti J, Win EE, Nimnuan J, Smith SM (2023) Modified starch-based adhesives: a review. Polymers 14(10):2023

Yu JK, Moon YS (2021) Corn starch: quality and quantity improvement for industrial uses. Plants 11(1):92

Yu X, Chen Y, Qi Z (2023) Preparation and identification of a novel peptide with high antioxidant activity from corn gluten meal. Food Chem 424:136389. https://doi.org/10.1016/J.FOODCHEM.2023.136389

Zhang L, Ban Q, Li J, Wang T (2022) Simultaneous production of hydrogen-methane and spatial community succession in an anaerobic baffled reactor treating corn starch processing wastewater. Chemosphere 300:134503. https://doi.org/10.1016/J.CHEMOSPHERE.2022.134503

Zhang Y, Li M, Zhang G (2023) Efficient treatment of the starch wastewater by enhanced flocculation–coagulation of environmentally benign materials. Sep Purif Technol 307:122788. https://doi.org/10.1016/J.SEPPUR.2022.122788

Zhou J, Yan S (2024) A comprehensive review of corn ethanol fuel production: from agricultural cultivation to energy application. J Energy Biosci 15:208–220. https://doi.org/10.5376/jeb.2024.15.0020

Ziaei-Rad Z, Fooladi J, Pazouki M, Gummadi SN (2021) Lignocellulosic biomass pre-treatment using low-cost ionic liquid for bioethanol production: an economically viable method for wheat straw fractionation. Biomass Bioenergy 151:106140

Chapter 4
Oilseed Biomasses as Feedstocks for Bioenergy and Value-Added Bioproducts

Neeraj Kumar Singh, Ayodeji Amobonye, Theresa Fadipe, Branly-Natalien Nguena-Dongue, Abdulameed Taiye Hassan, and Santhosh Pillai

Abstract The global demand for energy, driven by the growing human population and industrial advancements, coupled with depleting fossil fuel reserves, has increased the search for alternative, sustainable energy sources. Oilseed crops are cultivated worldwide due to their role in the human diet and the wide range of industrial applications of their seeds. As a result, the disposal of large volumes of oilseed biomass generated after processing and harvesting poses a significant challenge for waste management and environmental sustainability. On the contrary, this biomass also presents a potentially sustainable and renewable source for bioenergy production. Oilseed biomass residue can be transformed into bioenergy such as biogas, bioethanol, biohydrogen, and bio-oil through biological and thermochemical methods. The transformation of oilseed biomass requires pretreatment to disrupt its recalcitrant structure, thereby enhancing bioenergy yield. Techno-economic assessments have also indicated that the low cost of seed biomass improves its economic viability in the bioenergy sector. Therefore, this chapter explores compositional characteristics and the diverse spectrum of bioenergy derived from oilseed biomass, detailing its pretreatment methods, conversion technologies, and associated bioprocesses. This chapter also offers detailed insights into the environmental and economic aspects through life cycle assessment, along with its potential application

N. K. Singh · T. Fadipe · B.-N. Nguena-Dongue · S. Pillai (✉)
Department of Biotechnology and Food Science, Faculty of Applied Sciences, Durban University of Technology, Durban, South Africa
e-mail: santhoshk@dut.ac.za

A. Amobonye
Department of Polymer Chemistry and Technology, Kaunas University of Technology, Kaunas, Lithuania

A. T. Hassan
Department of Biochemistry, Kwara State University, Malete, Nigeria

Advanced Medical and Dental Institute, Universiti Sains Malaysia, Penang, Malaysia

A. L. Woiciechowski et al. (eds.), *Biomass for Bioenergy and Bioproducts*, Biofuel and Biorefinery Technologies 17, https://doi.org/10.1007/978-3-032-19898-3_4

in the circular bioeconomy. Furthermore, this chapter highlights the potential of oilseed biomass as a sustainable and renewable alternative energy source that contributes towards improving waste management.

4.1 Introduction

Oilseeds are plants grown primarily for the oil found in their seeds. These plant crops have played significant roles in human civilisation as their oily extracts have been used for centuries as key components of human nutrition, as a source of energy, and in countless other industrial applications. Oilseeds are major components of both human and animal diets due to their predominant lipid content and their significant amount of carbohydrates, fibre, proteins, vitamins, minerals, and a wide range of phytochemicals (Vichare and Morya 2024). It was previously surmised that soybeans, rapeseed, cotton, sunflower, peanuts, palm oil, and copra account for more than 90% of the global oilseed market, which comprises 25 major oilseed crops (Mirdoraghi et al. 2024). The production of oilseeds has been noted to have doubled in the past 25 years, mainly due to the increasing demand for vegetable oils, and the production has been driven by the development of high-yielding varieties and quality seeds, as well as climate-resilient technologies (Kastakova et al. 2023). In addition to balancing the demand and supply of these multifunctional commodities, managing the waste generated subsequent to oil extraction from the oilseeds has been identified as a daunting challenge.

Generally, the agricultural and food processing industries produce huge quantities of waste materials after the extraction of the products of interest from the raw material or feedstock. Such wastes require a huge amount of energy to be processed, leading to carbon dioxide (CO_2) generation during natural decomposition or artificial processing (Şahin and Elhussein 2018). Specifically, processing oilseeds through mechanical pressing or solvent extraction generates a huge amount of by-products, including peels, oilcake, and oil sludge. It has been reported that oilcake, which is the solid residual biomass remaining after a significant portion of oils has been extracted from oilseeds, is the most predominant agricultural residue in the vegetable oil industry (Thangaraju et al. 2021; Petraru and Amariei 2022). The oilseed cakes, which can be either edible (mahua, mustard, peanut, soybean, sunflower, and oilseed cakes) or inedible (castor, jatropha, neem, sesame, and simarouba oilseed cakes), are traditionally managed by being used as human food/animal feed or for other uses (energy, fertiliser, etc.) based on their palatability (Thangaraju et al. 2021). However, the compositional analysis of these waste biomasses has revealed that they contain copious amounts of desirable materials and compounds, which have led to attempts at further valorising them across various industries (Bárta et al. 2021; Abdelazim Mohdaly and Ramadan 2022).

For instance, flours derived from flax, hemp, milk thistle, poppy, pumpkin, rapeseed, safflower, and sunflower seeds were demonstrated to be highly rich in crude fibre, proteins, and phenolic compounds, suggesting further industrial applications (Bárta et al. 2021). The biomass of gingelly, sometimes referred to as Indian sesame, showed high promise for utilisation as feedstock in diverse industrial applications with its significant content of cellulose (38.07%), volatiles (75.19%), and calorific value (avg. 16.98 MJ/kg). It was specifically highlighted for its relatively high ethanol titre of 24.8 g/L and volumetric ethanol productivity of 1.03 g/L/h, compared to millet and pulse stem biomass (Nanda et al. 2024). Recently, soybean oil sludge and soybean oil sludge fatty acids were shown to improve the resistance to permanent deformation when utilised as recycling agents and binders in asphalt mixtures (Xu et al. 2024). Hence, it has been posited that by applying the biorefinery approach, which is an economical and integrated approach that aims to convert biomass into energy and high-value products, the agricultural waste from the vegetable oil industry can be valorised into energy, proteins, antioxidant compounds, phytochemicals, and dietary fibre, to mention a few (Sarkar et al. 2021; Salaria and Kaur 2024).

Hence, this chapter extensively discusses the significance of oilseed biomass in bioenergy and the production of various value-added commodities. In this regard, the biochemical composition and physico-chemical characteristics of various oilseed biomasses are presented. Furthermore, the different chemical, physical, and biological pretreatment technologies necessary for the efficient bioconversion of oilseed biomass are highlighted, while the bioenergy production process is also examined. Focus is also placed on other valorisation attempts at oilseed biomass, including extracting valuable chemicals and producing biofuels other than ethanol (e.g. biodiesel from residual oil, utilisation in materials production, and fertiliser formulation). This chapter also considers the environmental and economic sides of oilseed biomass valorisation by evaluating the life cycle assessment, carbon footprint, water and land use implications, economic feasibility, and circular bioeconomy potentials. Finally, some future trends and knowledge gaps in the utilisation of oilseed biomass are proposed.

4.2 Composition and Characteristics of Oilseed Biomass

Oilseed biomass includes press cakes, oil meals, straw, and various residues produced during oil extraction, and their chemical composition differs significantly depending on the crop, the species, growing conditions, and processing methods (Arrutia et al. 2020). These factors influence the nutritional, structural, and industrial applications of the by-products, and their variability has been noted to influence their potential valorisation in food production, animal feed, and bioenergy generation. For instance, press cakes and oil meals, which are the solid residues left after oil extraction, are rich in protein and contain other valuable nutrients, including carbohydrates, lipids, minerals, and fibres (Vichare and Morya 2024); hence, they have been found useful in various applications such as biofuel production, fertilisers, and alternative protein for human and animal consumption (Arrutia et al. 2020; Radu et al. 2024). The average chemical composition of oil meal, oilcake, and other

Table 4.1 Chemical composition of selected oilseed biomass

Oilseed type	Biomass type	Cellulose (%)	Hemicellulose (%)	Lignin (%)	Crude protein (%)	Crude fat (%)	Ash (%)	Reference
Soybean	Meal	5.8	25.0	ND	45.8	2.7	6.2	Wang et al. (2024a)
	Cake	16.2	15.5	2.2	50.4	ND	6.3	Sousa et al. (2023b)
	Straw	29.4	12.8	25.3	3.2	1.1	4.4	Koester et al. (2024)
Sunflower	Meal	13.8	11.9	12.2	33.7	0.2	7.4	Efthymiou et al. (2021)
	Cake	14.3	11.1	8.3	40.2	ND	8.4	Sousa et al. (2023b)
	Stem	53.0	14.6	10.7	5.3	ND	1.9	Malik et al. (2005)
Rapeseed	Meal	12.4	7.2	32.4	39.5	ND	1.3	Remón et al. (2018)
	Cake	15.5	13.6	7.7	39.8	ND	6.7	Sousa et al. (2023b)
	Straw	37.3	23.7	26.3	ND	ND	3.1	Halysh et al. (2022)
Oil palm	Kernel meal	34.9	19.6	10.2	13.7	6.9		Zhu et al. (2024)
	Kernel cake	36.7	29.5	8.9	17.0	0.3	4.7	Sathitkowitchai et al. (2022)
	Empty fruit bunch (EFB)	34.6	6.2	17.1	5.0	2.1	11.7	Rachmah et al. (2022)
Castor	Meal	12.2	4.6	30.1	44.8	2.7	12.4	Paulino et al. (2018)
	Cake	6.8	4.4	9.8	43.5	8.0	6.3	De Castro et al. (2016)
	Hull	43.4	25.2	2.9	5.7	0.6		Furtado et al. (2019)
Jatropha	Meal	12.8	13.8	7.7	38.8	8.7	9.4	Berihun et al. (2020)
	Cake	20.0	5.5	23.4	22.4	19.1	5.8	Gomes et al. (2022)
	Hull	32.7	2.1	14.5	ND	ND	8.9	Romuli et al. (2021)

ND Not determined

residues is presented in Table 4.1. The inherent genetic makeup of each oilseed species dictates fundamental differences in their biomass composition and most suitable applications. For instance, sunflower seeds yield oil meals with considerable oil content and protein, while rapeseed may have a slightly different balance (Gagour et al. 2022). Furthermore, genetic variation within species leads to further diversification in biomass composition, and breeding programmes exploit these variations to enhance specific traits. For example, high-oil rapeseed varieties yield oil meals with improved oil quality (Khalaf et al. 2021), while different sunflower varieties differ in hull thickness and lignin content, affecting their biofuel potential (Lindström et al. 2022). Varietal differences also influence the functional properties of oilseed biomass, such as the protein structure in sunflower meal, where 11S globulin and 2S albumin are dominant protein fractions (Li et al. 2024).

Environmental factors during the cultivation of oilseed crops significantly influence their biomass composition. Soil fertility influences the mineral profile, oil content, and protein accumulation in oil meals across various species (Rashmi et al. 2022; Szpunar-Krok and Wondołowska-Grabowska 2025), while drought conditions increase lignin deposition in straw and stems (Lee et al. 2023), reducing digestibility but improving biofuel potential. Climate variations affect fatty acid composition in the residual oil, with higher temperatures promoting oleic acid accumulation in oilseeds such as sunflower and rapeseed (Werteker et al. 2010). Additionally, nutrient availability during seed maturation determines cellulose, hemicellulose, and protein levels in oilseed biomass. At the same time, environmental stresses can influence phenolic compounds, impacting the nutritional properties of seeds, which is reflected in the residual biomass (Kefale et al. 2024). Soybean oil meal and oilcake are rich in protein, making them valuable for animal feed, while rapeseed oil meal is equally nutritious but contains glucosinolates, anti-nutritional compounds that interfere with thyroid function and reduce feed efficiency (Vichare and Morya 2024; Vujetić et al. 2025). Similarly, castor and jatropha oil meals/oilcakes are characterised by the presence of toxic compounds like ricin, curcin, and phorbol esters, which necessitate detoxification before use (De Barros et al. 2024; Cheikhyoussef and Cheikhyoussef 2023).

The composition of oilseed biomass is profoundly influenced by the extraction and processing techniques used. The method of oil extraction, whether solvent extraction or mechanical pressing (like cold-pressing), directly affects the residual oil content in oil meals and oilcakes across all species. Cold-pressing typically retains more residual oil in the cake, enhancing its calorific value, while solvent extraction yields a protein-concentrated but lower-fat meal (Rakita et al. 2023). Also, cold-pressed oilseed cakes retain higher levels of bioactive compounds, such as polyphenols and antioxidants (Rani et al. 2021). The pretreatment process, like dehulling, significantly increases the protein percentage but reduces the fibre content of the oil meal and oilcake (Kracht et al. 2004). Similarly, thermal treatments applied during processing can modify protein structures, reducing anti-nutritional factors in rapeseed meal and castor oilcake while also affecting protein digestibility in them (Salazar-Villanea et al. 2016; Melgar et al. 2024).

4.2.1 Lipids

Lipids, primarily triglycerides, are a key component of oilseeds, serving structural, protective, and energy storage functions. They act as a concentrated energy source within the seed, consisting of saturated fatty acids (SFAs), monounsaturated fatty acids (MUFAs), and polyunsaturated fatty acids (PUFAs) (Asokapandian et al. 2021). However, oil meal and oilcake, the by-products of oil extraction, generally contain low residual lipid content, which varies based on the seed type and extraction technique (Hallouch et al. 2024). For instance, mechanical pressing usually leaves 5%–22% lipids depending on the efficiency of the press, while mechanical pressing coupled with solvent extraction can yield a lipid content below 2% (Cravotto et al. 2023; Hallouch et al. 2024). The residual oil could also range from 0.2% in sunflower meals to 19.1% in jatropha cake (Table 4.1). Such residual oil contains essential fatty acids, similar to the original seed oil, and contributes to the nutritional value while dictating the suitability of the oil meal/cake for different industrial uses (Hallouch et al. 2024; Toomer et al. 2024). For example, defatted soybean cake typically has a low lipid content, often below 2% (Rani and Badwaik 2021). On the contrary, soybean residual oil is rich in polyunsaturated fatty acids like linoleic acid (C18:2, omega-6) and alpha-linolenic acid (C18:3, omega-3). They also contain monounsaturated oleic acid (C18:1), saturated palmitic acid (C16:0), and stearic acid (C18:0), which makes them a source of high-quality fats in animal feeds (Toomer et al. 2024). Residual oil in meals and cakes from PUFA-rich oilseeds, like sunflower, enhances flow properties in feed but is prone to oxidation, requiring stabilisation (Hallouch et al. 2024). In contrast, MUFA-rich sources like rapeseed offer better oxidative stability due to higher oleic acid content (El Idrissi et al. 2024; Stojanović et al. 2023). Saturated fatty acid (SFA)–dominant oils, like palm kernel oil, are less ideal for biodiesel due to viscosity but are valuable in cosmetics and biopolymer production (Hallouch et al. 2024; Appaiah et al. 2014). On the contrary, castor meal, which is dominated by ricinoleic acid, is limited by ricin toxicity, while detoxified jatropha meal has potential feed applications (Princess 2024). Unlike oil meal and cakes, straw, hull, and stem fractions of oilseeds contain minimal storage lipids but serve as valuable lignocellulosic feedstocks for biofuels, biocomposites, and fibre production.

4.2.2 Proteins and Amino Acids

Beyond its valued oil content, oilseed biomass is also a remarkable source of protein, a critical macronutrient that constitutes a substantial portion of its biomass and contributes significantly to its nutritional and industrial value. It has been observed that the protein content in oilcake or oil meal is always considerably higher than in the seed and straw because oil extraction removes a significant portion of the seed's lipids, concentrating the proteins and the other nutrients in the residual meal or cake

(Arrutia et al. 2020). Amino acids are the fundamental components of proteins, linked together via peptide bonds to form complex protein structures. Soybean meal and cake are particularly notable for their high protein composition, containing 40%–50% crude protein, and the plant is primarily cultivated as a source of protein-rich feed for animals (Wrigley et al. 2015). Additionally, soybean meal and cake offer a balanced array of essential amino acids (such as lysine, leucine, and isoleucine) necessary for human and animal nutrition; however, they contain lower amounts of sulfur-containing amino acids like methionine and cysteine (Ahmed et al. 2023; Wang et al. 2024a). Despite this, soybean meal and cake are the preferred choices in formulating food products for animals and humans due to their high palatability and digestibility.

Rapeseed meal and cake contain 30%–40% crude protein and have a balanced amino acid profile comparable to soybean meal, which is particularly rich in methionine and cystine (Sharma et al. 2024). However, it may also contain anti-nutritional compounds (glucosinolates, phytic acid, tannins, erucic acid, etc.), which reduce the digestibility of proteins and reduce nutrient availability (Axentii and Codină 2024). Similarly, sunflower meal and cake are valuable sources of proteins (30%–40%) and are notably rich in sulfur-containing amino acids, methionine, and cysteine (Hadidi et al. 2024). Additionally, they tend to be richer in alanine, arginine, and valine, with low lysine, which can reduce their nutritional and bioenergy application (Petraru et al. 2021). On the other hand, palm kernel cake and meal contain a moderate protein level of 15%–20%, lacking in some essential amino acids like lysine and methionine, and often require supplementation (Abd Rahim et al. 2022). However, it contains a significant amount of glutamic acid and aspartic acid, which are non-essential amino acids (Azizi et al. 2021). Jatropha meals, while offering substantial protein content (30%–40%), have amino acids such as lysine and valine, a low amount of tryptophan, and high glutamic and aspartic acid levels, supporting their use in fertilisers and bioproducts after detoxification (Zhang et al. 2024). Likewise, castor meal and cake have a relatively high crude protein content (30%–45%), which contains a high amount of toxic ricin, arginine, and glutamic acid but are low in lysine and require detoxification to remove ricin (Diarra and Seidavi 2020). In contrast, the straws and hulls of these oilseed crops generally have much lower protein and amino acid concentrations but are primarily composed of structural carbohydrates and lignin (Halysh et al. 2022; Cui et al. 2019).

4.2.3 Carbohydrates

The carbohydrate fraction of oilseed biomass exists in various forms, with dietary fibre being a major component, including cellulose, hemicellulose, and lignin (Aguirre et al. 2018). The polysaccharides (cellulose and hemicellulose) are the key structural components of plant cell walls, and interwoven within their matrix is lignin (Wang et al. 2024b). While lignin is a complex aromatic polymer, though not classified as a carbohydrate, it gives rigidity and hydrophobicity to the plant cell wall, making the extraction and conversion of the carbohydrate components more challenging (Ali et al. 2024). Despite its recalcitrance, lignin is often quantified alongside carbohydrates in biomass composition studies because it significantly affects the overall digestibility and valorisation potential of oilseed residues (Sikiru et al. 2024). The composition of carbohydrates in biomass depends on the plant tissue structure, processing techniques, and oil extraction methods. These factors collectively influence its suitability for various applications, ranging from animal feed to high-value biorefinery feedstocks. The highest cellulose content was recorded in the sunflower stem (53.0%), while the sunflower meal had the lowest (5.8%), indicating that the sunflower stem mainly contains cellulose and lignin. The hemicellulose content ranged from 2.1% in jatropha seed hull to 29.5% in oil palm kernel cake, while the lignin content varied from 2.2% in soybean cake to 32.4% in rapeseed meal (Efthymiou et al. 2021; Sousa et al. 2023b).

4.2.4 Ash and Mineral Content

Ash is the total amount of inorganic material that remains after combustion, which reflects the mineral content of biomass per gramme (Hallouch et al. 2024). While present in relatively modest proportions compared to the major constituents of biomass, ash and mineral content are nonetheless consequential components of oilseed biomass, contributing to its overall profile and utility. Furthermore, the composition of ash and minerals exhibits notable variability across different oilseed species and even within varieties and across different residues, reflecting genetic and environmental influences (Stojanović et al. 2023). Oilseed biomasses contain abundant macro- (potassium (K), phosphorus (P), calcium (Ca), and magnesium (Mg)) and micro- (iron, zinc, and manganese) minerals (Kolláthová et al. 2019). For instance, rapeseed press cakes are particularly high in calcium and magnesium, whereas sunflower press cakes are rich in potassium and phosphorus (Stojanović et al. 2023). The ash content ranges from 1.3% to 12.4%, with castor bean meal recording the highest (12.4%), indicating significant mineral content; however, due to the presence of toxic compounds like ricin, detoxification is necessary before further industrial use (Table 4.1). Similarly, oil palm empty fruit bunch has a high amount of ash (11.7%) and is rich in potassium, calcium, and magnesium, contributing to its utility in animal feed formulations. Rapeseed (canola) is rich in essential minerals but contains glucosinolates, compounds that can interfere with thyroid function and reduce the palatability of feed (Kumar et al. 2023a). Therefore, while not a dominant fraction by mass, the ash and mineral constituents are integral to a holistic understanding of oilseed biomass properties and applications.

4.3 Physical Properties of Oilseed Biomass

Oilseed biomasses exhibit a wide range of physical properties, and understanding them is essential for optimising the equipment design, handling, storage, and processing of oilseed biomass, thereby enhancing its viability as a renewable energy source. One of the critical functional properties is bulk density, which is the mass of seeds per unit volume, including both the solid particles and the void spaces between them (Clarke and Preto 2011). However, several factors, including particle size, shape, moisture content, and the degree of compaction, influence the bulk density of the biomass. Materials with smaller particle sizes or irregular shapes tend to have lower bulk densities due to increased void spaces (Demirel et al. 2020). In contrast, larger or more uniform particles can be packed more efficiently, resulting in higher bulk densities (Bojanowska 2023). Moisture content also plays a significant role; as biomass absorbs moisture, it can swell, affecting both its mass and volume, influencing the bulk density (Shojaeiarani et al. 2019). However, bulk density varies considerably across the different types of oilseed biomasses, ranging from 130 kg/m^3 to 1117.4 kg/m^3 (Table 4.2). Notably, sunflower husks exhibit the highest bulk density due to their compact structure and lower porosity, increasing energy density and suitability as a biofuel (Cui et al. 2019). In contrast, rapeseed straw has one of the lowest bulk densities, likely due to its fibrous and less dense composition (Stolarski et al. 2013).

Table 4.2 Physical properties of oilseed biomass

Oilseed type	Biomass type	Bulk (kg/m^3)	Moisture (%)	Particle size (mm)	Reference
Soybean	Meal	220.0	7.4	2.3	Singh and Krishnaswamy (2023)
	Cake	510.0	9.6	ND	Rani and Badwaik (2021)
	Stalk	242.3	8.2	0.4	Major (2012)
Sunflower	Meal	468.1	7.4	1.2	Vidosavljević et al. (2025)
	Cake	539.1	5.6	1.2	De Castro et al. (2016)
	Husk	1117.4	12	ND	Cui et al. (2019)
Rapeseed	Meal	538.0	10.0	0.1–3.0	Bojanowska (2023)
	Cake	383.0	9.8	ND	Stolarski et al. (2013)
	Straw	130.0	12.0	ND	Stolarski et al. (2013)
Oil palm	Kernel cake	623.0	7.9	2.0	Razuan et al. (2011)
	Empty fruit bunch (EFB)	536.8	6.2	0.5–1.0	Sang et al. (2025)
	Oil palm frond	177.0	7.5	$\leq$2	Kongto et al. (2022)
Castor	Cake	244.6	8.8	2.4	De Castro et al. (2016)
Jatropha	Cake	320.0	5.2	$\leq$0.0005	Waindim et al. (2022)
	Hull	720.0	10.2	ND	Doshi et al. (2014)

ND Not determined

Moisture content significantly influences various aspects of biomass utilisation, including its energy content, combustion efficiency, storage stability, and mechanical durability. For instance, higher moisture content can reduce the net heating value of biomass, as additional energy is required to evaporate the water during combustion (Shojaeiarani et al. 2019). However, the moisture content of biomass fuels is highly variable, depending on factors such as biomass type, season, and the time that has elapsed between harvesting and utilisation. The moisture content spanned from about 5.2% to 12% (Table 4.2). Sunflower husks, for example, have a moisture content of 12%, which could be due to their hygroscopic nature and environmental exposure (Cui et al. 2019). In contrast, jatropha cake has a lower moisture content of 5.2%, possibly resulting from processing methods that reduce moisture to enhance storage stability (Waindim et al. 2022). On the other hand, particle size distribution also varies among oilseed biomasses, with average sizes ranging from as small as 0.0005 mm to as large as 2.4 mm.

Particle size refers to the dimensions of individual particles within a material, as well as the flowability, compaction, and reactivity of the biomass during processing. In biomass materials, particle size is inherently variable due to the irregular shapes and sizes of the constituent particles (Chaloupková et al. 2019). Jatropha cake exhibits the smallest particle size, possibly due to fine-milling processes aimed at increasing the surface area for subsequent applications (Waindim et al. 2022). On the other hand, castor cake has larger particles, reflecting the inherent characteristics of the biomass (De Castro et al. 2016).

4.4 Pretreatment Technologies for Oilseed Biomass

Oilseed biomass residue is a potential feedstock for bioenergy production. It comprises plant cell wall components, which are primarily made up of cellulose (23.8–54.6 wt%), hemicellulose (11.9–29.8 wt%), and lignin (12.7–37.3 wt%) (Yong and Wu 2022). However, lignin, a recalcitrant structure associated with cellulose and hemicellulose through intramolecular and intermolecular hydrogen bonds, provides high resistance to biological degradation (Baruah et al. 2018), thus hindering the biological conversion of seed biomass residue cellulose into bioenergy. Hence, there is a need for suitable pretreatments to make cellulose more accessible to enzymes or microbial activity in the quest for bioenergy production. These pretreatment techniques should prevent the formation of inhibitors and enhance the digestibility of seed biomass cost-effectively by preserving sugars (Galbe and Zacchi 2012). The pretreatment of oilseed biomass is usually carried out via physical, chemical, and biological processes (Fig. 4.1).

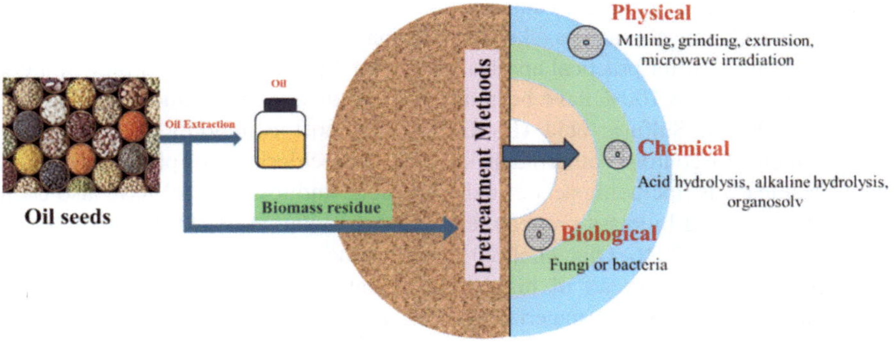

Fig. 4.1 Classification of various pretreatment methods for oilseed biomass. (Adapted and modified from Sudalai et al. 2024)

4.4.1 Physical Pretreatment of Oilseed Biomass

The main purpose of the physical pretreatment method is to break down the compact structure of the oilseed biomass for further applications. Methods such as milling, grinding, extrusion, and microwave irradiation disrupt the morphological structure and increase the surface pore size, making the cellulose more accessible to microbial activity for the production of biofuel and other value-added products (Sidana and Yadav 2022). These pretreatment processes effectively minimise the degree of polymerisation and the crystallinity of cellulose biomass. In addition to improving the accessibility to cellulose, mechanical pretreatment, such as milling and grinding, also improves the downstream process by increasing the interaction between the chemical solvent, fermentative microbes, and enzymes (Ab Rasid et al. 2021). It has also been found that physical pretreatment, such as chopping and grinding, improves the digestibility through acid pretreatment in oilseed biomass (Ranjan et al. 2024). Microwave irradiation breaks the internal structure of oilseed biomass, thereby enhancing the digestibility and fermentation efficiency (Amin et al. 2017). Microwave pretreatment of canola oilseeds improves the enzymatic saccharification at 713 W (Agu et al. 2017). However, cellulosic loss is a major drawback of physical pretreatment, particularly during milling and grinding (Khan et al. 2022).

4.4.2 Chemical Pretreatment of Oilseed Biomass

Chemical pretreatment methods involve using inorganic and organic compounds to break the polymeric bonds present in the oilseed biomass. Studies have shown that acidic, alkaline, and organic solvent pretreatment are the commonly used chemical pretreatment methods (Galbe and Wallberg 2019). Chemical pretreatment disrupts the complexity of oilseed biomass, leading to decreased crystallinity, depolymerisation of cellulose, and degradation of the lignin compounds (Dimos et al. 2019). Chemicals such as alkalis, acids, ionic liquids, and organic solvents are effectively utilised for practical application due to their low price, and pretreatment efficiencies of up to 100% have been achieved at lower concentrations (Galbe and Wallberg 2019). Sulfuric acid (H_2SO_4) is the most commonly used acid for pretreatment. In contrast, other acids such as hydrochloric acid, phosphoric acid, and organic acids (including acetic acid, citric acid, and formic acid) have also been studied for oilseed biomass pretreatment (Antonopoulou et al. 2016; Casoni et al. 2019; Tayeh et al. 2016; Qiao et al. 2019). Dilute acids in the 0.1%–10% range are typically employed instead of highly concentrated acids to minimise corrosive damage to the expensive material used in reactor construction (Baruah et al. 2018). However, acid pretreatment leads to inhibitors like furfural and 5-hydroxymethylfurfural (HMF). These inhibitors are generally formed under acidic conditions through the degradation of hemicellulose into hexose and pentose sugars, which dehydrate to form furfural and HMF by losing three molecules of water from pentose and hexose sugar, respectively (Lee and Wu 2021; Tripathi et al. 2023; Naresh Kumar et al. 2019). In contrast, inhibitors are not generally produced during the alkaline pretreatment, which is more effective for disrupting the bonds between lignin and other components (Valles et al. 2021). Organic acid pretreatment is a potential alternative to inorganic acid pretreatment due to its lower environmental toxicity and minimal production of inhibitors (Rattanaporn et al. 2018). A comparative study of different organic solvents, such as acetic acid, citric acid, and oxalic acid, in the pretreatment of oilseed biomass (palm) showed that the highest sugar recovery was in oxalic acid (Rattanaporn et al. 2018). This study also revealed that citric acid pretreatment resulted in minimal inhibitor production and the highest fermentation efficiency. Additionally, alkaline pretreatment partially solubilises hemicellulose and cellulose, thereby improving the digestibility and accessibility of cellulose (Amin et al. 2017). Sodium hydroxide (NaOH) is the most commonly used alkali for the pretreatment of oilseed biomass, followed by potassium hydroxide (KOH) (Oruganti et al. 2023) and calcium hydroxide ($Ca(OH)_2$) (Yoruklu et al. 2020; Sharmada et al. 2016).

4.4.3 Biological Pretreatment of Oilseed Biomass

Biological pretreatment of oilseed biomass involves the use of microorganisms or the enzymes they produce to break down lignin and hemicellulose (Hafid et al. 2021; Sudalai et al. 2024). Widely utilised microorganisms in the pretreatment process include actinomycetes, white rot fungi, brown soft rot fungi, and bacterial communities, which effectively disrupt the lignin and hemicellulose components with minimal damage to the cellulose (Sudalai et al. 2024). Pretreatment of sunflower seed cake using a combination of fungal strains (*Rhizopus oryzae* and *Aspergillus ibericus*) produces a maximum of cellulase activity of 135 U/g (Sousa et al. 2023a). Similarly, pretreatment of palm oilseed biomass using a locally isolated fungal species, *Pycnoporus sanguineus*, resulted in a cellulose content of 35.31%, much higher than the previously reported value (Hafid et al. 2021). Further, another study revealed that the pretreatment of castor seeds using *Irpex lacteus* improved the degradation percentage by nearly 31% (Testa et al. 2023). Biological pretreatment offers several advantages over other pretreatment methods, including low energy requirements, less environmental toxicity, and a high yield of desired products under ambient environmental conditions (Wang et al. 2020). However, biological pretreatment also faces some techno-economic issues concerning slow microbial growth rate, large area requirements, and the need to maintain optimal growth conditions to maximise yield efficiency.

4.4.4 Emerging Pretreatment Technologies for Oilseed Biomass

Emerging pretreatment methods of oilseed biomass have been identified to include steam explosions for rapeseeds (Wang et al. 2021), hydrothermal treatment of oilseed cake (Petrovič et al. 2024), and ultrasonic treatment of edible and non-edible oilseeds (Mohanty et al. 2022; Fernández-Cegrí et al. 2012). For instance, experiments performed at five different sonication times and specific energy levels led to a 53.8% increase in the average sugar yield relative to both the untreated oilcake and the deep eutectic solvent (DES) pretreatment methods (Fernández-Cegrí et al. 2012). Recently, steam explosion has been widely used to break down the lignin linkage through physical and chemical methods involving high pressure and temperature. Steam explosion pretreatment was applied to palm kernel cake before hydrolysis, and an 80.6% improvement in hydrolysis rate was recorded with this pretreatment (Li et al. 2018). The application of DES in oilseed biomass pretreatment has also been gaining interest owing to its low cost, biodegradable properties, and low toxicity (Parodi et al. 2023; Alhassan and Kumar 2016). Most studies have applied DES formulations containing chlorine chloride, organic acids, polyol, alkaline earth metal hexahydrate, urea, and organic salts to remove hemicellulose and lignin. It has been found that DES pretreatment of oil palm biomass can remove up to 36.42% of lignin (Ong et al. 2019).

4.5 Bioenergy Production from Oilseed Biomass

4.5.1 Anaerobic Digestion of Oilseed Biomass for Biogas Production

The generation of biogas from oilseed biomass residue is a potential key driver of the circular economy, as different oilseed biomasses have been used for the production of different biofuels (Table 4.1). Biogas is generally a combustible gas produced under anaerobic conditions through bacterial action on biomass (Jameel et al. 2024). This process is generally referred to as anaerobic digestion, or fermentation process, and typically involves municipal waste, sewage, agricultural waste, or green waste (Bernet et al. 2024). In addition to fermentation, biogas can also be produced through gasification, resulting in a mixture of gases, including methane, hydrogen, nitrogen (N), and carbon monoxide (Abanades et al. 2022). Among these biogases, methane and hydrogen have high calorific values, making them a suitable substitute for biofuels. Furthermore, biogases are capable of running heavy engines, boilers, and turbines. Biogas production from both edible and non-edible oilcakes has been reviewed by Mohanty et al. (2022). The studies found that palm oilcake and date seeds achieved a higher methane production efficiency at 37 °C using sludge inoculum in an anaerobic batch reactor (Kanchanasuta and Pisutpaisal 2016; Ourradi et al. 2022). In a more recent study, de-oiled jatropha cake mixed with cow dung slurry produced biogas of up to 0.216 m^3/kg of total solids (TS), with methane concentrations ranging from 62.33% to 69.16% under mesophilic conditions (Sharma et al. 2022). Researchers have observed that thermal and acidic pretreatment of *Jatropha curcas* oilcake can deactivate protease and phytate inhibitors, thereby enhancing methanogenic activity in anaerobic digesters and improving biogas production (Jabłoński et al. 2017). Additionally, *Madhuca indica* and hingan de-oiled seed cake have been utilised for the production of up to 198–233 L/kg of biogas when operated under laboratory conditions in the mesophilic temperature range (Deshpande et al. 2012). These processes also recover nitrogen and other secondary and tertiary nutrients, which can be used as fertilisers.

4.5.2 Fermentation of Oilseed Biomass for Bioethanol Production

Bioethanol is considered one of the most promising alternatives to fossil fuels (Cabrera-Jiménez et al. 2022), and oilseed biomass residue is a potential feedstock for bioethanol production due to the high carbohydrate content of more than 42% w/w (Muktham et al. 2016). Globally, selective oilseed biomasses such as palm seed cake, castor, and oats are commonly utilised for bioethanol production (Alavijeh and Yaghmaei 2016; Sudalai et al. 2024). After oil extraction, the residual seed biomass is pulverised and hydrolysed to release sugars, which are then utilised by fermentative microbes for bioethanol production (Sudalai et al. 2024). It was reported that bioethanol production from pretreated castor bean cake using *Saccharomyces cerevisiae* yielded approximately 0.465 g/g under fermentation conditions of pH 5.0, 100 rpm, and 30 °C (Abada et al. 2019). A study by Marisa Raita et al. investigated the use of palm kernel cake for bioethanol production and found an ethanol concentration of 9.9 g/L, equivalent to 92 ± 2% of the theoretical yield (Raita et al. 2016). Additionally, researchers achieved a bioethanol yield of 10.3%

from hydrolysed karanja oilseed residue, following pretreatment at 121 °C and 15 psi for 90 min, using commercial yeast under fermentation conditions at 32 °C (Doshi and Srivastava 2013).

4.5.3 Biohydrogen Production from Oilseed Biomass

Biohydrogen has gained global interest in the field of sustainable energy due to its eco-friendly nature (Bhandari and Adhikari 2024). In addition, it has a 2.5 times higher calorific value than other fossil fuels, making it highly applicable as a fuel source (Ong et al. 2021). Traditional hydrogen production processes, such as steam reforming, non-catalytic partial oxidation, and autothermal reforming, are less economical and less eco-friendly (Sharma et al. 2020). In contrast, the biological hydrogen production process is relatively more economical and eco-friendly, making it a more suitable alternative (Akram et al. 2024).

Biohydrogen is produced through dark fermentation, often in integrated systems using co-culturing of microorganisms, combining sequential processes such as photo-fermentation and dark fermentation to enhance overall hydrogen productivity (Bundhoo 2019). Dark fermentation is commonly carried out by microorganisms such as *Klebsiella*, *Clostridium*, *Thermoanaerobacterium thermosaccharolyticum*, and *Enterobacter aerogenes* (Dahiya et al. 2021). Several oilseed biomasses have been reported to be used as feedstock for biohydrogen production. For example, de-oiled palm kernel cake pretreated with 7% (v/v) H_2SO_4 produced 1575 mL/L of hydrogen using 70% of the pretreated substrate and *Clostridium acetobutylicum* YM1 as the fermenting microorganism (Azman et al. 2016). In another study, *Jatropha curcas* seed cake using *Enterobacter aerogenes* produced 68.2 mL of H_2/g volatile solids (VS) through dark fermentation without biomass pretreatment (Lopes et al. 2015). Furthermore, cotton seed biomass residue, obtained after biodiesel production, was pretreated using 10% NaOH (w/w dry matter). The residual sugars in the pretreated biomass yielded 84%–112% hydrogen using the thermophilic bacterium *Caldicellulosiruptor saccharolyticus* (Panagiotopoulos et al. 2013).

4.5.4 Thermochemical Conversion of Oilseed Biomass

Thermochemical conversion of biomass using heat and chemical reactions to convert biomass into energy, fuels, and chemicals contributes to global energy demand and supports rural development (Bisen et al. 2025). The main thermochemical conversion technologies include combustion, torrefaction, pyrolysis, and gasification, and studies have shown the processing of oilseed feedstock in oxygen-rich environments (Fig. 4.2). Direct combustion of oilseed biomass through conventional thermochemical processes has been employed to convert chemical energy into mechanical energy and heat for powering boilers, steam turbines, or turbo generators (Cruz

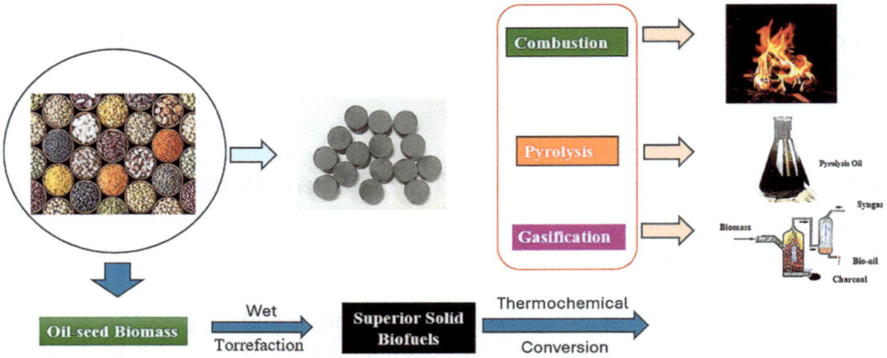

Fig. 4.2 Thermochemical conversion of oilseed biomass. (Adapted and modified (He et al. 2018))

et al. 2020). The combustion properties depend on several characteristics, such as the particle size, flow rate, and moisture content of oilseed biomass (Bates and Ghoniem 2014). It was found that olive residues decompose at approximately 200 °C under oxidising conditions, resulting in the production of a material with a high calorific value (Magalhaes et al. 2017). Olive seed combustion showed a high heating value of 33.3 MJ kg^{-1} (Pohlmann et al. 2017).

The transformation of oilseed biomass into a liquid biofuel through the pyrolysis process is another promising approach for bioenergy production (Qiu et al. 2022). The bio-oil produced from pyrolysis is believed to be easier to transport than raw biomass. In addition, the bio-oil can be blended with other heavy biofuels and emits less NO and SO_2 compared to heavy fuel oils (Hou et al. 2016). It can also be used as raw material for producing biodiesel and hydrogen (Sánchez-Borrego et al. 2021). During pyrolysis, the seed biomass is decomposed at 350–700 °C in the absence of oxygen (Jourabchi et al. 2014). Bio-oil and water are collected through a condensation process, which separates non-condensable gases such as syngas and biogas from the mixture (Miranda et al. 2021), while the residual material in the reactors is used as biochar for further industrial application (Krishna et al. 2016). Based on the operating conditions, oilseed biomass typically undergoes pyrolysis in three phases. In the first phase, light volatile compounds and moisture are removed from seed biomass between room temperature and 150 °C. In the second phase, cellulose, hemicellulose, and lignin fractions are decomposed at temperatures between 150 °C and 440 °C. In the third phase, lignin is completely degraded between 440 °C and 600 °C (Ferreira et al. 2020; Thangalazhy-Gopakumar et al. 2018).

Researchers have also investigated the use of the oilseed biomass residue from African star apple and silk cotton for bio-oil production at an optimum temperature of 400 °C; conversion rates of 72% and 68% were recorded for African star apple and silk cotton seed cake, respectively (Sokoto et al. 2020). The performance of pyrolysis can be improved using catalysts such as lanthanum (La), tin, titanium, phosphorus, and zinc. For instance, vacuum pyrolysis of rapeseed straw exhibited higher performance with 5 wt% lanthanum (La) as a catalyst (Karnjanakom et al. 2017). Furthermore, increasing temperature beyond a threshold temperature leads to decreased bio-oil yield and increased biogas production (Jiang et al. 2019). This is attributed to the secondary cracking of oil vapour, which produces non-condensable biogas (Ameh et al. 2024). Syngas is produced through gasification, which converts solid seed residue biomass to gas via a thermochemical process, including thermal cracking reactions (Fu et al. 2013).

The gasification process typically involves four steps: drying removes moisture, pyrolysis breaks down biomass, devolatilisation releases combustible gases, and reduction and combustion convert the remaining char into syngas through chemical reactions at high temperatures (Asadullah 2014). However, gasification faces challenges related to the formation of unwanted by-products, such as ash, impurities, and tar, which should be removed to evade downstream processing challenges (Xu et al. 2011). Notably, a gasifier operated with hazelnut shells as the feedstock produced more syngas compared to cotton stalks, attributed to the calorific value of hazelnut shells being 1.5 times higher than that of cotton stalk (Xu et al. 2011). A previous study also showed that palm kernel shells produce syngas with 80% higher hydrogen when using a fluidised bed gasification system (Khan et al. 2014). *Jatropha curcas* seed residue has also been used in decentralised gasification (combustion) plants to produce high-quality syngas (Pambudi et al. 2017). Another study found that groundnut shells gasified at an operating temperature of 894 °C, a pressure of 1 bar, and an equivalence ratio (ER) of 0.05 maximised the hydrogen yield in syngas, achieving a 0.64 molar fraction of hydrogen (Faraji and Saidi 2022).

Residual seed biomass can be converted into solid fuel with properties similar to coal through torrefaction (Gan et al. 2019). This thermochemical process involves heating the seed biomass to generate heat and power between 200 °C and 300 °C (Zhang et al. 2016). Torrefaction improves energy density and ignition properties while reducing moisture content and microbial activity (Zhang et al. 2017). Oilseeds such as sunflower, palm, olive, rapeseed, and jatropha have been used to produce high-energy fuel through the torrefaction process (Cruz et al. 2020). It has been found that the torrefaction of de-oiled jatropha seed kernel biomass increased the higher heating value by up to 1.243 after treatment at 300 °C for 60 min, with a particle size of 0.5–1.0 mm (Gan et al. 2019). Additionally, de-oiled marula seeds have been studied for torrefaction at temperatures between 275 °C and 300 °C to achieve coal-like properties (Pahla et al. 2018). Other reports also suggest that de-oiled jatropha seed kernels improve their energy value through torrefaction at temperatures between 200 °C and 300 °C, with holding times of 15–60 min (Hsu et al. 2018).

4.6 Environmental and Economic Considerations of Oilseed Biomass Valorisation

Increased demand for fossil energy has resulted in significant impacts, such as greenhouse gas (GHG) emissions, depletion of fossil resources, and risks to environmental, economic, and social sustainability (Wang and Azam 2024). However, the feasibility of more environmentally friendly approaches, such as the conversion of oilseed biomass residue to bioenergy, requires a thorough evaluation of their environmental impact, as well as other economic and technical considerations (Olofsson and Börjesson 2018). Bioenergy from biomass, such as oilseed biomass, shows promise in reducing reliance on fossil fuels. For instance, some studies have highlighted the high performance of bioenergy production systems based on oilseed compared to those based on biodiesel (Cocco 2011). Fiorentino et al. (2014) also evaluated a biorefinery from oilseed biomass of the non-food oil crop *Brassica carinata* relative to a system that only produces biodiesel and thermal energy. It was observed that despite the conversion of lignocellulosic residues into value-added products such as ethyl levulinate and formic acid, both systems rely on 90% non-renewable energy sources for the total energy use. The main environmental impacts of life cycle assessment attributed to the sustainable utilisation of this biomass have been identified as abiotic depletion and global warming, primarily by agricultural practices and extensive use of nitrogen fertilisers. Nevertheless, a deeper understanding of the life cycle and the techno-economic feasibility of commercialising oilseed bioenergy remains an area for exploration, including the demonstration of its benefits and the policy implications for sustainable production (Arias et al. 2024). Ultimately, future oilseed biomass valorisation holds promise; however, its success depends on the balance of strategies for economic return and ecological impact.

4.6.1 Greenhouse Gas Emissions and Carbon Footprint

Substituting conventional energy with bioenergy derived from crops is expected to reduce greenhouse gas (GHG) emissions, carbon footprint, and other emissions (Bamber et al. 2023). Studies have demonstrated that using oilseed crops and derivatives for various bioenergy applications can decrease greenhouse gas (GHG) emissions compared to conventional diesel and petroleum fuel. For instance, the life cycle analysis of the reduction of GHG emissions for biodiesel and renewable diesel production from soybean, canola, and carinata oils ranged from 40% to 69%, while using tallow, used cooking oil, and distillers' corn oil for biodiesel and renewable diesel could achieve higher GHG reductions of 79%–86% compared to petroleum diesel (Xu et al. 2022) The use of camelina oil for biodiesel production can reduce GHG emissions by 67%–92% compared to petroleum diesel (Li and Mupondwa 2014). Agricultural processes, particularly using nitrogen fertilisers, have been identified as one of the most important factors negatively affecting the life cycle assessment result of production systems involving oilseed crops for bioenergy production (Krohn and Fripp 2012). For instance, among the 66.5%–70.3% of GHG emissions attributed to camelina biodiesel, 52%–61% of CO_2 emissions are derived from the use of fertilisers, with nitrogen fertiliser alone contributing between 40% and 52% of the total, primarily due to direct and indirect N_2O emission expressed in CO_2 equivalent (Li and Mupondwa 2014). The growth stage of rapeseed has been identified as the highest GHG contributor during biodiesel production, with an environmental impact of above 68% due to intensive fertiliser doses and agricultural practices required (González-García et al. 2013). Despite the contribution to GHG emissions in some scenarios of the conversion of oilseed biomass to bioenergy, this process remains sustainable as the life cycle emission of biodiesel from *Camelina sativa*, soybean, spring camelina, and future *Camelina sativa* reduced the GHG emission by 37%–73%, 63%–85%, 57%–73%, and 55%–71%, respectively, when compared to fossil fuels like petroleum diesel that had GHG emissions higher than 90 g CO_2 eq. per MJ (Krohn and Fripp 2012).

4.6.2 Water and Land Use Implications

The production of energy crops, crucial raw materials for sustainable bioenergy feedstocks, may require more land and water resources to meet the growing demand for plant biomass bioenergy (Guzman et al. 2019). For instance, the competition between biofuel production and food crops for land considerably reduces the availability of land for food crops, leading to increased food costs and contributing to food insecurity (Ben Fradj et al. 2016). A study by Van Stappen et al. (2016) on a consequential life cycle assessment of installing a farm-scale biogas plant suggested that feedstock intended for animal feed should not be used because replacing them in animal diets has additional impacts, mainly in terms of extra agricultural land. The massive development of biofuel from oilseed crops, such as corn, soybean, or canola, has raised several concerns, including the food versus fuel debate and the diversion of resources like agricultural products and land from their original use (Searchinger et al. 2008). Another critical issue is the conversion of land with significant carbon storage, such as forests or peat bogs, into land with lower carbon storage, resulting in released carbon into the atmosphere and creating a carbon debt" that may take a long time for biofuel to compensate (Fargione et al. 2008). The careful management and cultivation of energy crops on unused or marginal land can help reduce CO_2 levels and energy consumption from other sources by sequestering and converting atmospheric carbon dioxide into biomass without disrupting regular agricultural production (Kumar et al. 2023b).

4.6.3 Economic Feasibility and Market Potential

Bioenergy and bioproducts derived from oilseed biomass not only offer environmental benefits but also present economic opportunities (Cruz et al. 2020), especially as bioenergy components are well-suited for the automotive sector due to their resemblance to gasoline and diesel (Othman et al. 2017). For instance, biodiesel produced from waste palm oil methyl esters (WPOMEs) and canola oil methyl esters (COMEs) exhibited heating values 8.9% and 8.2% lower than petroleum-based diesel fuel (PBDF), respectively. Additionally, the densities of WPOME and COME are 4.17% and 5.14% higher than PBDF, respectively (Ozsezen and Canakci 2011). Bioenergy from oilseed biomass can be more economically viable in regions with abundant biomass resources than fossil fuels, particularly in agricultural economies where feedstock costs are lower and readily available (Alazaiza et al. 2024). A study by Kumar et al. (2021) suggests that while producing biofuel may require substantial initial capital investment, some feedstocks, such as modified energy cane, have competitive production costs compared to conventional oilseed crops, like soybean, with lower capital investment ($0.91/L). Modular bio-oil plants, capable of processing various biomass feedstocks, provide financial security to investors due to their multiple product streams and cost-effectiveness at small scales (Hussin et al. 2023). Economic analysis of *Brassica carinata* oil crop biorefinery and bioenergy systems indicates a gross income of 4247 US $/ha and 1074 US $/ha, respectively (Fahd et al. 2012). The rates of energy gain per hectare for the biorefinery and energy-oriented patterns of the *B. carinata* oil crop are 25.5 GJ and 14.7 GJ, respectively (Fiorentino et al. 2014). However, the economic viability of these projects often relies on subsidies or favourable market conditions, such as high oil prices (Mungodla et al. 2019). Nevertheless, in the energy-oriented production chain, the product value is marginally higher than production costs, whereas the biorefinery option provides an income 3.7 times higher than the economic investment (Fiorentino et al. 2014).

4.6.4 Sustainability Aspects and Circular Economy Principles

The transformation of oilseed biomass into bioenergy and bioproducts offers several sustainable social, economic, and environmental benefits. It significantly reduces greenhouse gas emissions compared to fossil fuels, assisting in climate change mitigation (Mignogna et al. 2024) while also reducing waste and environmental pollution (Chavan et al. 2022). This process of transforming oilseed biomass into valuable products efficiently utilises resources by recycling and reusing biomass, limiting waste, and conserving natural resources (Leong et al. 2021). Chong et al. (2013) demonstrated that oil palm empty fruit bunch is an efficient substrate for biohydrogen production, with a yield of 1.98 $molH_2/mol_{xylose}$ corresponding to 690 mL H_2/L of the medium. Moreover, converting oilseed biomass into biofuels and bioproducts presents an economic opportunity by providing alternative revenue for farmers and reducing their dependence on fossil fuels (Mignogna et al. 2024). In addition to its significant bioenergy potential, characterised by low inputs, high yields, soil carbon sequestration, and potential ecosystem service benefits, the miscanthus crop presents an economic opportunity for rural regions, as farmers receive €75 Mg^{-1} dry matter (DM) at the farm gate, compared to a delivered pellet price of €329 Mg^{-1} DM (Styles et al. 2015). Hence, this concept is in line with various United Nations (UN) Sustainable Development Goals (SDGs), including Affordable and clean energy (SDG 7); Industry, innovation and infrastructure (SDG

9); Sustainable cities and communities (SDG 11); Responsible consumption and production (SDG 12); Climate action (SDG 13); Life below water (SDG 14); and Life on land (SDG 15). It is believed that aligning bioenergy and bioeconomy policies with international initiatives such as the Paris Agreement on Climate Change, the United Nations 2030 Sustainable Agenda (Sustainable Development Goals), or the Aichi Agreement can guide decision-makers towards the most beneficial chains in today's globalised world (Lago et al. 2019).

4.7 Bioproducts from Oilseed Biomass

Oilseed biomass is a valuable source of bioactive compounds, particularly phenolic compounds and flavonoids, which exhibit strong antioxidant, antimicrobial, and nutraceutical properties. These compounds are widely utilised in the pharmaceutical, cosmetic, and food industries due to their health benefits and bioactivity (Ancuta and Sonia 2020). In the food and nutraceutical industries, in particular, oilseed biomass has been identified as a valuable source of polyunsaturated/essential fatty acids. For instance, residues derived from thistle and sesame were recorded with a linoleic acid content of ~47%–55% (Mihai et al. 2022). Similarly, the linolenic contents of flax meals, grape seeds meals, and hemp flour were 64.65%, 33.62%, and 33.22%, respectively (Mihai et al. 2022). Purslane seeds and seed cakes were observed to have linoleic and α-linolenic acid contents of ~34% and 33%, respectively; it was also observed that the biomass and the amount of specific fatty acids varied significantly with the extraction method (Petropoulos et al. 2020). Many of the phenolic compounds from oilseed biomass have been identified with various pharmacological effects, thus highlighting their potential in the pharmaceutical industry. For instance, Zardo et al. (2020) examined phenolic extraction from canola seed cake, identifying 24 bioactive compounds, including sinapic acid derivatives, which are well known for their acetylcholinesterase inhibition, antimutagenicity, and antioxidant activities. Further studies on individual oilseeds have provided insight into their dominant phenolic components. Izzo et al. (2020) reported that quercetin-3-glucoside (28.59 mg/100 g) was the major phenolic compound in hemp extracts. For pumpkin, Kulczyński and Gramza-Michałowska (2019) identified rutin (46.9 ± 0.1 mg/100 dry weight) and kaempferol (36.24 ± 0.08 mg/100 dry weight) as the dominant flavonoids, whereas Švarc Gajić et al. (2022) reported higher levels of myricetin compared to rutin and kaempferol. Additionally, gallic acid was reported as a predominant phenolic in oilseed cakes, particularly in acid-modified extracts (Švarc Gajić et al. 2022). Quercetin, kaempferol, and rutin were identified in olive pomace and leaves, confirming their widespread occurrence in oilseed biomass residues (Cho et al. 2020; Markhali et al. 2020). Besides phenolics, other secondary metabolites have been sourced from oilseed biomass. For instance, the residual seed cake of *Camellia oleifera* was recorded as a rich source of triterpenoid saponins with significant anti-proliferative activity, and they were identified as camellia saponin B1 and three novel saponins, oleifera saponin C1–C3 (1–3) (Zong et al. 2015).

Triterpenoid saponins are known to be validated ingredients in cosmetics and personal care products, serving as both active and functional components (Kanlayavattanakul et al. 2024).

The different components of oilseed biomass have been found to be useful in material science. Proteins extracted from oilseed residues can be utilised to produce biodegradable films with superior mechanical and barrier properties, making them ideal for food packaging applications. A recent example of this is soy protein isolate (SPI), combined with plasticisers like glycerol, which has been successfully employed to improve film flexibility (Wang et al. 2022). Additionally, epoxidised soybean oil served as a bio-based plasticiser and coating material, offering a sustainable alternative to fossil-derived products (Calovi et al. 2024). Furthermore, research on polyhydroxyalkanoate (PHA) production using *Ralstonia eutropha* ATCC 17699 through continuous anaerobic treatment of palm oil mill effluent (POME) demonstrated an optimal PHA yield of 11.4 g/L (Hassan et al. 2019). Biocomposites are composed of natural fillers and polymers, offering an environmentally friendly and cost-effective alternative to conventional materials. The use of oilseed biomass, especially as fillers in biocomposite production, has the potential to facilitate sustainability while reducing material costs. Hassan et al. (2019) reinforced polypropylene (PP) with cellulose derived from oil palm empty fruit bunch (OPEFB), with results showing that the PP–cellulose composites exhibited improved tensile strength, impact resistance, and stiffness due to strong interfacial adhesion and compatibility with the PP matrix. Saba et al. (2017) utilised OPEFB-derived nanofillers for nanocomposite fabrication. The findings reveal that the nanocomposites increased in density and reached maximum thermomechanical properties due to improved interfacial bonding. Similarly, studies have shown that incorporating nanocellulose from oil palm mesocarp fibre (OPMF) into biocomposites enhances both the morphological and mechanical properties of the reinforced nanocomposites (Norrahim et al. 2018). Nanocellulose derived from sunflower seed husks has been integrated into composites, enhancing mechanical strength and biodegradability (Hubbe and Grigsby 2020). Similarly, rapeseed and flaxseed husk fibres have been utilised in biodegradable composites for sustainable construction applications (Hassan et al. 2019; Kumar and Leggate 2022). For instance, Hassan et al. (2019) developed a method to reinforce bioplastic-coated fertilisers (BpFs) with OPEFB using poly lactic acid (PLA) and Nitrogen (N) phosphorous (P) and Potassium (K) fertilisers. This approach increased the biodegradation rate of OPEFB-BpF by 12.8% while reducing thermal stability, ensuring a continuous and steady nutrient release. The waste from oilseed processing has also been found to have potential in organic farming in the form of fertilisers, as the ash and digestate are by-products from oilseed biomass, obtained via combustion and anaerobic digestion, which contain valuable nutrients that can be repurposed as bio-based fertilisers. These materials serve as eco-friendly alternatives to synthetic fertilisers, improving soil health while minimising environmental pollution. Ash-based fertilisers are rich in potassium (K), phosphorus (P), calcium (Ca), and magnesium (Mg) but are low in nitrogen (Kramar 2020). For instance, oilseed ash from sunflower husks has been successfully used as a potash substitute, enriching the soil with essential nutrients (Alauddin 2021). Meanwhile, digestate-based fertilisers are rich in nitrogen (N), phosphorus (P), potassium (K), and organic matter, which are beneficial for soil conditioning. However, digestate from rapeseed meal has been used in circular agriculture systems to reduce the need for chemical fertilisers while improving soil fertility (Pohl et al. 2019).

4.8 Conclusion

This chapter highlights the immense potential application of ample wastes and by-products generated from the vegetable oil industry for sustainable bioenergy production and other value-added bioproducts. It is believed that oilseed crops are second only to cereals in industrial importance; furthermore, their diversity, extensive cultivation, and substantial amounts of residues make them appealing options for a bio-based economy. In the coming years, waste biomass from oilseeds will continue to rise due to an increasing demand for vegetable oil; hence, it is essential to valorise this biomass in ways that will facilitate the development of low-cost and novel products while effectively managing waste disposal. However, exploring new opportunities through sustained innovation is necessary to fully realise the promise of oilseed biomass. This includes improving the genetics of oilseed crops to maximise biomass output and composition, developing more effective conversion techniques to increase energy and bioproduct yields, and creating effective and efficient pretreatment technologies to overcome biomass recalcitrance. Research into new bioproducts beyond conventional bioenergy applications is also crucial to fully maximise the benefits accruable from oilseed biomass. With the potential to play a bigger part in supplying the world's expanding need for sustainable materials and energy, oilseed biomass has a bright future. This will be propelled by ongoing research and innovations, as well as the integration of biorefinery concepts and circular economy principles.

References

Ab Rasid NS, Shamjuddin A, Rahman AZA, Amin NAS (2021) Recent advances in green pre-treatment methods of lignocellulosic biomass for enhanced biofuel production. J Clean Prod 321:129038

Abada E, Al-Fifi Z, Osman M (2019) Bioethanol production with carboxymethylcellulase of Pseudomonas poae using castor bean (Ricinus communis L.) cake. Saudi J Biol Sci 26:866–871

Abanades S, Abbaspour H, Ahmadi A, Das B, Ehyaei M, Esmaeilion F, El Haj Assad M, Hajilounezhad T, Jamali D, Hmida A (2022) A critical review of biogas production and usage with legislations framework across the globe. Int J Environ Sci Technol 19(4):3377–3400

Abd Rahim N, Luthfi AI, Abdul PM, Jahim JM, Bukhari NA (2022) Towards sustainable production of bio-based lactic acid via a bio-based technical route: recent developments and the use of palm kernel cakes in the bioconversion. Bioresources 17:3781

Abdelazim Mohdaly AA, Ramadan MF (2022) Characteristics, composition and functional properties of seeds, seed cake and seed oil from different Brassica carinata genotypes. Food Biosci 48:100752

Agu OS, Tabil LG, Dumonceaux T (2017) Microwave-assisted alkali pre-treatment, densification and enzymatic saccharification of canola straw and oat hull. Bioengineering 4:25

Aguirre M, Kiegle E, Leo G, Ezquer I (2018) Carbohydrate reserves and seed development: an overview. Plant Reprod 31:263–290

Ahmed HGM-D, Naeem M, Faisal A, Fatima N, Tariq S, Owais M (2023) Enriching the content of proteins and essential amino acids in legumes. In: Legumes biofortification. Springer

Akram F, Fatima T, Ibrar R, Ul Haq I (2024) Biohydrogen: production, promising progressions and challenges of a green carbon-free energy. Sustain Energy Technol Assess 69:103893

Alauddin M (2021) Integrated effects of NPK fertilizers and organic manures on the growth, yield and nutrient content of sunflower (Helianthus annus L.) (doctoral dissertation, University of Dhaka).Production. Cellulose 25(7):3853–3859

Alavijeh MK, Yaghmaei S (2016) Biochemical production of bioenergy from agricultural crops and residue in Iran. Waste Manag 52:375–394

Alazaiza MYD, Ahmad Z, Albahnasawi A, Nassani DE, Alenezi RA (2024) Biomass processing technologies for bioenergy production: factors for future global market. Int J Environ Sci Technol 21:2307–2324

Alhassan Y, Kumar N (2016) Single step biodiesel production from Pongamia pinnata (Karanja) seed oil using deep eutectic solvent (DESs) catalysts. Waste Biomass Valorization 7:1055–1065

Ali S, Rani A, Dar MA, Qaisrani MM, Noman M, Yoganathan K, Asad M, Berhanu A, Barwant M, Zhu D (2024) Recent advances in characterization and valorization of lignin and its value-added products: challenges and future perspectives. Biomass 4:947–977

Ameh VI, Ayeleru OO, Nomngongo PN, Ramatsa IM (2024) Bio-oil production from waste plant seeds biomass as pyrolytic lignocellulosic feedstock and its improvement for energy potential: a review. Waste Manag Bull 2:32

Amin FR, Khalid H, Zhang H, Rahman SU, Zhang R, Liu G, Chen C (2017) Pretreatment methods of lignocellulosic biomass for anaerobic digestion. AMB Express 7:1–12

Ancuța P, Sonia A (2020) Oil press-cakes and meals valorization through circular economy approaches: a review. Appl Sci 10(21):7432

Antonopoulou G, Vayenas D, Lyberatos G (2016) Ethanol and hydrogen production from sunflower straw: the effect of pretreatment on the whole slurry fermentation. Biochem Eng J 116: 65–74

Appaiah P, Sunil L, Prasanth Kumar P, Gopala Krishna A (2014) Composition of coconut testa, coconut kernel and its oil. J Am Oil Chem Soc 91:917–924

Arias A, Nika C-E, Vasilaki V, Feijoo G, Moreira MT, Katsou E (2024) Assessing the future prospects of emerging technologies for shipping and aviation biofuels: a critical review. Renew Sust Energ Rev 197:114427

Arrutia F, Binner E, Williams P, Waldron KW (2020) Oilseeds beyond oil: press cakes and meals supplying global protein requirements. Trends Food Sci Technol 100:88–102

Asadullah M (2014) Barriers of commercial power generation using biomass gasification gas: a review. Renew Sust Energ Rev 29:201–215

Asokapandian S, Sreelakshmi S, Rajamanickam G (2021) Lipids and oils: an overview. In: Food biopolymers: structural, functional and nutraceutical properties. Springer, Cham, pp 389–411

Axentii M, Codină GG (2024) Exploring the nutritional potential and functionality of hemp and rapeseed proteins: a review on unveiling anti-nutritional factors, bioactive compounds, and functional attributes. Plants 13:1195

Azizi MN, Loh TC, Foo HL, Teik Chung EL (2021) Is palm kernel cake a suitable alternative feed ingredient for poultry? Animals 11:338

Azman NF, Abdeshahian P, Kadier A, Shukor H, Al-Shorgani NKN, Hamid AA, Kalil MS (2016) Utilization of palm kernel cake as a renewable feedstock for fermentative hydrogen production. Renew Energy 93:700–708

Bamber N, Turner I, Dutta B, Heidari MD, Pelletier N (2023) Consequential life cycle assessment of grain and oilseed crops: review and recommendations. Sustainability 15:6201

Bárta J, Bártová V, Jarošová M, Švajner J, Smetana P, Kadlec J, Filip V, Kyselka J, Berčíková M, Zdráhal Z, Bjelková M, Kozak M (2021) Oilseed cake flour composition, functional properties and antioxidant potential as effects of sieving and species differences. Foods 10:2766

Baruah J, Nath BK, Sharma R, Kumar S, Deka RC, Baruah DC, Kalita E (2018) Recent trends in the pretreatment of lignocellulosic biomass for value-added products. Front Energy Res 6:141

Bates RB, Ghoniem AF (2014) Modeling kinetics-transport interactions during biomass torrefaction: the effects of temperature, particle size, and moisture content. Fuel 137:216–229

Ben Fradj N, Jayet PA, Aghajanzadeh-Darzi P (2016) Competition between food, feed, and (bio)-fuel: a supply-side model based assessment at the European scale. Land Use Policy 52:195–205

Berihun K, Negesse T, Tolera A (2020) Effects of different treatment methods on the composition of nutrients and Phorbol Ester levels of Jatropha (Jatropha curcas L.) seed meal produced in Ethiopia. J Agric Environm Sci 5:55–63

Bernet N, Escudié R, Steyer J-P (2024) Anaerobic digestion: application to the energy recovery of agricultural co-products and food industry waste. In: Green chemistry and agro-food industry: towards a sustainable bioeconomy. Springer

Bhandari R, Adhikari N (2024) A comprehensive review on the role of hydrogen in renewable energy systems. Int J Hydrog Energy 82:923–951

Bisen D, Chouhan APS, Pant M, Chakma S (2025) Advancement of thermochemical conversion and the potential of biomasses for production of clean energy: a review. Renew Sust Energ Rev 208:115016

Bojanowska M (2023) Handling and transportation properties of rapeseed biomass as affected by particle size. Zesz Nauk Akad Mors w Szczecinie 74:5–15

Bundhoo ZM (2019) Potential of bio-hydrogen production from dark fermentation of crop residues: a review. Int J Hydrog Energy 44:17346–17362

Cabrera-Jiménez R, Mateo-Sanz JM, Gavaldà J, Jiménez L, Pozo C (2022) Comparing biofuels through the lens of sustainability: a data envelopment analysis approach. Appl Energy 307: 118201

Calovi M, Zanardi A, Rossi S (2024) Recent advances in bio-based wood protective systems: a comprehensive review. Appl Sci 14(2):736

Casoni AI, Gutierrez VS, Volpe MA (2019) Conversion of sunflower seed hulls, waste from edible oil production, into valuable products. J Environ Chem Eng 7:102893

Chaloupková V, Ivanova T, Krepl V (2019) Particle size and shape characterization of feedstock material for biofuel production. Agron Res 17(5):1861–1873

Chavan S, Yadav B, Atmakuri A, Tyagi RD, Wong JWC, Drogui P (2022) Bioconversion of organic wastes into value-added products: a review. Bioresour Technol 344:126398

Cheikhyoussef N, Cheikhyoussef A (2023) Bioactive phytochemicals from castor (Ricinus Communis Linneo) seed oil processing by-products. In: Bioactive phytochemicals from vegetable oil and oilseed processing by-products. Springer

Cho HH, Strezov V (2020) A comparative review on the environmental impacts of combustion-based electricity generation technologies. Energy Fuel 34(9):10486–10502

Chong PS, Jahim JM, Harun S, Lim SS, Mutalib SA, Hassan O, Nor MTM (2013) Enhancement of batch biohydrogen production from prehydrolysate of acid treated oil palm empty fruit bunch. Int J Hydrog Energy 38:9592–9599

Clarke S, Preto F (2011) Biomass densification for energy production. Citeseer

Cocco D (2011) Life-cycle assessment of bioenergy production systems from oilseed rape crops. Proc Inst Mech Eng Part A 225:63–73

Cravotto C, Claux O, Bartier M, Fabiano-Tixier A-S, Tabasso S (2023) Leading edge technologies and perspectives in industrial oilseed extraction. Molecules 28:5973

Cruz G, Silva AV, Da Silva JB, De Nazare Caldeiras R, De Souza ME (2020) Biofuels from oilseed fruits using different thermochemical processes: opportunities and challenges. Biofuels Bioprod Biorefin 14:696–719

Cui X, Yang J, Shi X, Lei W, Huang T, Bai C (2019) Pelletization of sunflower seed husks: evaluating and optimizing energy consumption and physical properties by response surface methodology (Rsm). Processes 7:591

Dahiya S, Chatterjee S, Sarkar O, Mohan SV (2021) Renewable hydrogen production by dark-fermentation: current status, challenges and perspectives. Bioresour Technol 321:124354

De Barros CR, Ferreira LMM, Fraga I, Mourão JL, Rodrigues MAM (2024) Detoxification methods of Jatropha curcas seed cake and its potential utilization as animal feed. Fermentation 10:256

De Castro AM, Dos Reis Castilho L, Freire DMG (2016) Characterization of babassu, canola, castor seed and sunflower residual cakes for use as raw materials for fermentation processes. Ind Crop Prod 83:140–148

Demirel C, Gürdil GAK, Kabutey A, Herak D (2020) Effects of forces, particle sizes, and moisture contents on mechanical behaviour of densified briquettes from ground sunflower stalks and hazelnut husks. Energies 13:2542

Deshpande N, Kale N, Deshmukh S (2012) A study on biogas generation from Mahua (Madhuca Indica) and Hingan (Balanites Aegyaptiaca) oil seedcake. Energy Sustain Dev 16:363–367

Diarra SS, Seidavi A (2020) Advances in the utilisation of Castor (Ricinus Communis Linneo) seed meal as protein supplement in poultry diets. Agrofor Syst 94:1333–1341

Dimos K, Paschos T, Louloudi A, Kalogiannis KG, Lappas AA, Papayannakos N, Kekos D, Mamma D (2019) Effect of various pretreatment methods on bioethanol production from cotton stalks. Fermentation 5:5

Doshi P, Srivastava G (2013) Sustainable approach to produce bioethanol from Karanja (Pongamia Pinnata) oilseed residue. Turk J Agric For 37:781–788

Doshi P, Srivastava G, Pathak G, Dikshit M (2014) Physicochemical and thermal characterization of nonedible oilseed residual waste as sustainable solid biofuel. Waste Manag 34:1836–1846

Efthymiou M-N, Pateraki C, Papapostolou H, Lin CSK, Koutinas A (2021) Restructuring the sunflower-based biodiesel industry into a circular bio-economy business model converting sunflower meal and crude glycerol into succinic acid and value-added co-products. Biomass Bioenergy 155:106265

El Idrissi ZL, Oubihi A, Ibourki M, El Youssfi M, Gharby S, El Guezzane C, Ullah R, Iqbal Z, Goh KW, Gallo M (2024) Peanut (Arachis hypogaea L.) flour and oilcake flour: exploring the influence of roasting and varietal differences on proximal composition, elemental profiling, antimicrobial and antioxidant properties. Food Chem X 24:101791

Fahd S, Fiorentino G, Mellino S, Ulgiati S (2012) Cropping bioenergy and biomaterials in marginal land: the added value of the biorefinery concept. Energy 37:79–93

Faraji M, Saidi M (2022) Process simulation and optimization of groundnut shell biomass air gasification for hydrogen-enriched syngas production. Int J Hydrog Energy 47:13579–13591

Fargione J, Hill J, Tilman D, Polasky S, Hawthorne P (2008) Land clearing and the biofuel carbon debt. Science 319:1235–1238

Fernández-Cegrí V, De La Rubia M, Raposo F, Borja R (2012) Impact of ultrasonic pretreatment under different operational conditions on the mesophilic anaerobic digestion of sunflower oil cake in batch mode. Ultrason Sonochem 19:1003–1010

Ferreira M, Oliveira B, Pinheiro W, Correa N, França L, Ribeiro N (2020) Generation of biofuels by slow pyrolysis of palm empty fruit bunches: optimization of process variables and characterization of physical-chemical products. Biomass Bioenerg 140:105707

Fiorentino G, Ripa M, Mellino S, Fahd S, Ulgiati S (2014) Life cycle assessment of Brassica carinata biomass conversion to bioenergy and platform chemicals. J Clean Prod 66:174–187

Fu J, Tang C, Jin W, Thi LD, Huang Z, Zhang Y (2013) Study on laminar flame speed and flame structure of syngas with varied compositions using Oh-Plif and spectrograph. Int J Hydrog Energy 38:1636–1643

Furtado RN, Carneiro MSDS, Coutinho DN, Cândido MJD, Silva EBD (2019) Fermentative losses and chemical composition of elephant grass silage added with castor bean hull. Rev Ciênc Agron 50:140–147

Gagour J, Ahmed MN, Bouzid HA, Oubannin S, Bijla L, Ibourki M, Hajib A, Koubachi J, Harhar H, Gharby S (2022) Proximate composition, physicochemical, and lipids profiling and elemental profiling of rapeseed (Brassica napus L.) and sunflower (Helianthus annuus L.) grown in Morocco. Evid Based Complement Alternat Med 2022:3505943

Galbe M, Wallberg O (2019) Pretreatment for biorefineries: a review of common methods for efficient utilisation of lignocellulosic materials. Biotechnol Biofuels 12:294

Galbe M, Zacchi G (2012) Pretreatment: the key to efficient utilization of lignocellulosic materials. Biomass Bioenergy 46:70–78

Gan YY, Ong HC, Ling TC, Chen W-H, Chong CT (2019) Torrefaction of de-oiled Jatropha seed kernel biomass for solid fuel production. Energy 170:367–374

Gomes TG, Hadi SIIA, De Aquino Ribeiro JA, Segatto R, Mendes TD, Helm CV, Júnior AFC, Miller RNG, Mendonça S, De Siqueira FG (2022) Phorbol ester biodegradation in Jatropha curcas cake and potential as a substrate for enzyme and Pleurotus pulmonarius edible mushroom production. Biocatal Agric Biotechnol 45:102498

González-García S, García-Rey D, Hospido A (2013) Environmental life cycle assessment for rapeseed-derived biodiesel. Int J Life Cycle Assess 18:61–76

Guzman JG, Ussiri DAN, Lal R (2019) Soil physical properties following conversion of a reclaimed minesoil to bioenergy crop production. Catena 176:289–295

Hadidi M, Aghababaei F, Mcclements DJ (2024) Sunflower meal/cake as a sustainable protein source for global food demand: towards a zero-hunger world. Food Hydrocoll 147:109329

Hafid HS, Baharuddin AS, Mokhtar MN, Omar FN, Mohammed MAP, Wakisaka M (2021) Enhanced laccase production for oil palm biomass delignification using biological pretreatment and its estimation at biorefinary scale. Biomass Bioenergy 144:105904

Hallouch O, Ibourki M, Devkota KP, Gharby S (2024) Proximate composition, antioxidant activity, lipids and elemental profiling of argan, almond, sesame, nigella, soybean and sunflower oil press cakes reveal a great potential of valorization. https://doi.org/10.21203/rs.3.rs-3944130/v1

Halysh V, Skiba M, Nesterenko A, Kulik T, Palianytsia B (2022) Structural characterization of by-product Lignins from Organosolv rapeseed straw pulping and their application as biosorbents. J Polym Res 29:510

Hassan MA, Ahmad Farid MA, Shirai Y, Ariffin H, Othman MR, Samsudin MH, Hasan MY (2019) Oil palm biomass biorefinery for sustainable production of renewable materials. Biotechnol J 14 (6):1800394

He C, Tang C, Li C, Yuan J, Tran K-Q, Bach Q-V, Qiu R, Yang Y (2018) Wet torrefaction of biomass for high quality solid fuel production: a review. Renew Sust Energ Rev 91:259–271

Hou S-S, Huang W-C, Rizal FM, Lin T-H (2016) Co-firing of fast pyrolysis bio-oil and heavy fuel oil in a 300-Kwth furnace. Appl Sci 6:326

Hsu T-C, Chang C-C, Yuan M-H, Chang C-Y, Chen Y-H, Lin C-F, Ji D-R, Shie J-L, Wu C-H, Chiang S-W (2018) Upgrading of Jatropha-seed residue after mechanical extraction of oil via torrefaction. Energy 142:773–781

Hubbe MA, Grigsby W (2020) From Nanocellulose to wood particles: a review of particle size vs. the properties of plastic composites reinforced with cellulose-based entities. Bioresources 15:1

Hussin F, Hazani NN, Khalil M, Aroua MK (2023) Environmental life cycle assessment of biomass conversion using hydrothermal technology: a review. Fuel Process Technol 246:107747

Jabłoński SJ, Kułażyński M, Sikora I, Łukaszewicz M (2017) The influence of different pretreatment methods on biogas production from Jatropha curcas oil cake. J Environ Manag 203:714–719

Jameel MK, Mustafa MA, Ahmed HS, Jassim Mohammed A, Ghazy H, Shakir MN, Lawas AM, Khudhur Mohammed S, Idan AH, Mahmoud ZH (2024) Biogas: production, properties, applications, economic and challenges: a review. Results Chem 7:101549

Jiang L, Wang Y, Dai L, Yu Z, Yang Q, Yang S, Jiang D, Ma Z, Wu Q, Zhang B (2019) Co-pyrolysis of biomass and Soapstock in a downdraft reactor using a novel Zsm-5/sic composite catalyst. Bioresour Technol 279:202–208

Jourabchi SA, Gan S, Ng HK (2014) Pyrolysis of Jatropha curcas pressed cake for bio-oil production in a fixed-bed system. Energy Convers Manag 78:518–526

Kanchanasuta S, Pisutpaisal N (2016) Waste utilization of palm oil decanter cake on biogas fermentation. Int J Hydrog Energy 41:15661–15666

Kanlayavattanakul M, Mersni D, Lourith N (2024) Plant-derived Saponins and their prospective for cosmetic and personal care products. Bot Stud 65:32

Karnjanakom S, Suriya-Umporn T, Bayu A, Kongparakul S, Samart C, Fushimi C, Abudula A, Guan G (2017) High selectivity and stability of Mg-doped Al-Mcm-41 for in-situ catalytic upgrading fast pyrolysis bio-oil. Energy Convers Manag 142:272–285

Kastakova E, Bato V, Dubenova D (2023) Competitiveness on the global oilseeds market. J Econ Res Bus Adm 144:3–13

Kefale H, You J, Zhang Y, Getahun S, Berhe M, Abbas AA, Ojiewo CO, Wang L (2024) Metabolomic insights into the multiple stress responses of metabolites in major oilseed crops. Physiol Plant 176:E14596

Khalaf AE, Abd Al-Aziz SA, Ali SM, Mohdaly AA, Rady MM, Majrashi A, Ali EF, Yassein AA (2021) Morphological formation, fatty acid profile, and molecular identification of some landraces of Ethiopian brassica as a promising crop to support breeding programs. Plants 10: 1431

Khan Z, Yusup S, Ahmad MM, Rashidi NA (2014) Integrated catalytic adsorption (Ica) steam gasification system for enhanced hydrogen production using palm kernel shell. Int J Hydrog Energy 39:3286–3293

Khan MU, Usman M, Ashraf MA, Dutta N, Luo G, Zhang S (2022) A review of recent advancements in pretreatment techniques of lignocellulosic materials for biogas production: opportunities and limitations. Chem Eng J Adv 10:100263

Koester DL, Gomes LM, Dresch AP, Matsuo TT, Dos Santos DF, Bender JP, Deon VG, De Amorin SG, Quast LB, Pinto VZ (2024) Biodegradable starch foams reinforced by food-chain side streams. Int J Biol Macromol 275:133386

Kolláthová R, Varga B, Ivanišová E, Bíro D, Rolinec M, Juráček M, Šimko M, Gálik B (2019) Mineral profile analysis of oilseeds and their by-products as feeding sources for animal nutrition. Slovak J Anim Sci 52:9–15

Kongto P, Palamanit A, Ninduangdee P, Singh Y, Chanakaewsomboon I, Hayat A, Wae-Hayee M (2022) Intensive exploration of the fuel characteristics of biomass and biochar from oil palm trunk and oil palm fronds for supporting increasing demand of solid biofuels in Thailand. Energy Rep 8:5640–5652

Kracht W, Dänicke S, Kluge H, Keller K, Matzke W, Hennig U, Schumann W (2004) Effect of dehulling of rapeseed on feed value and nutrient digestibility of rape products in pigs. Arch Anim Nutr 58:389–404

Kramar V (2020) Possible utilization directions of ash from biomass combustion. Biomass ash as fertilizer in agriculture. UABIO Bioenergy Association Ukraine 27:1–71

Krishna BB, Biswas B, Kumar J, Singh R, Bhaskar T (2016) Role of reaction temperature on pyrolysis of cotton residue. Waste Biomass Valorization 7:71–78

Krohn BJ, Fripp M (2012) A life cycle assessment of biodiesel derived from the "niche filling" energy crop Camelina in the USA. Appl Energy 92:92–98

Kulczyński B, Gramza-Michałowska A (2019) The profile of secondary metabolites and other bioactive compounds in Cucurbita pepo L. and Cucurbita moschata pumpkin cultivars. Molecules 24(16):2945

Kumar D, Long SP, Arora A, Singh V (2021) Techno-economic feasibility analysis of Engineered energycane-based biorefinery co-producing biodiesel and ethanol. GCB Bioenergy 13:1498–1514

Kumar C, Leggate W (2022) An overview of bio-adhesives for engineered wood products. Int J Adhes Adhes 118:103187

Kumar MS, Mawlong I, Kumar A, Singh K, Gurung B, Rani R, Rai P (2023a) Evaluation of major anti-nutritional factors in oilseed brassica. Vegetos 36:591–598

Kumar V, Singh P, Sharma J, Saini S, Sharma P, Sharma A (2023b) Rice straw management through biofuel, biochar, mushroom cultivation, and paper production to overcome environmental pollution in North India. Waste Dispos Sustain. Energy 5:483–510

Lago C, Herrera I, Caldés N, Lechón Y (2019) Chapter 1 - Nexus bioenergy–bioeconomy. In: Lago C, Caldés N, Lechón Y (eds) The role of bioenergy in the bioeconomy. Academic Press

Lee CBTL, Wu TY (2021) A review on solvent systems for furfural production from lignocellulosic biomass. Renew Sust Energ Rev 137:110172

Lee BR, Park SH, Muchlas M, La VH, Al Mamun M, Bae DW, Kim TH (2023) Differential response of Phenylpropanoid pathway as linked to hormonal change in two brassica Napus cultivars contrasting drought tolerance. Physiol Plant 175:E14115

Leong HY, Chang C-K, Khoo KS, Chew KW, Chia SR, Lim JW, Chang J-S, Show PL (2021) Waste biorefinery towards a sustainable circular bioeconomy: a solution to global issues. Biotechnol Biofuels 14:87

Li X, Mupondwa E (2014) Life cycle assessment of Camelina oil derived biodiesel and jet fuel in the Canadian prairies. Sci Total Environ 481:17–26

Li Y-X, Yi P, Liu J, Yan Q-J, Jiang Z-Q (2018) High-level expression of an engineered B-Mannanase (Mrmman5a) in Pichia Pastoris for Manno-oligosaccharide production using steam explosion pretreated palm kernel cake. Bioresour Technol 256:30–37

Li Z, Xiang F, Huang X, Liang M, Ma S, Gafurov K, Gu F, Guo Q, Wang Q (2024) Properties and characterization of sunflower seeds from different varieties of edible and oil sunflower seeds. Foods 13:1188

Lindström LI, Franchini MC, Nolasco SM (2022) Sunflower fruit hullability and structure as affected by genotype, environment and canopy shading. Ann Appl Biol 180:338–347

Lopes SL, Fragoso R, Duarte E, Marques PA (2015) Bioconversion of Jatropha curcas seed cake to hydrogen by a strain of Enterobacter Aerogenes. Fuel 139:715–719

Izzo L, Castaldo L, Narváez A, Graziani G, Gaspari A, Rodríguez-Carrasco Y, Ritieni A (2020) Analysis of phenolic compounds in commercial Cannabis sativa L. inflorescences using UHPLC-Q-Orbitrap HRMS. Molecules 25(3):631

Magalhaes D, Kazanç F, Riaza J, Erensoy S, Kabaklı Ö, Chalmers H (2017) Combustion of Turkish Lignites and olive residue: experiments and kinetic modelling. Fuel 203:868–876

Major P (2012) Availability and physical properties of residues from major agricultural crops for energy conversion through thermochemical processes. Am J Agric Biol Sci 7:312–321

Malik UR, Hasany SM, Subhani MS (2005) Sorptive potential of sunflower stem for Cr (III) ions from aqueous solutions and its kinetic and thermodynamic profile. Talanta 66:166–173

Markhali FS, Teixeira JA, Rocha CM (2020) Olive tree leaves—a source of valuable active compounds. PRO 8(9):1177

Melgar ORA, Silva RR, Da Silva FF, Santos LV, Lima ACR, Machado SLM, Dueñez WYS, Da Conceição Santos M, Devia DCC, Paixão TR (2024) Finishing of grazing crossbred steers supplemented with detoxified Castor bean meal (Ricinus Communis L.) in the rainy-dry transition period. Trop Anim Health Prod 56:111

Mignogna D, Szabó M, Ceci P, Avino P (2024) Biomass energy and biofuels: perspective, potentials, and challenges in the energy transition. Sustainability 16:7036

Mihai AL, Negoiță M, Horneț G-A, Belc N (2022) Valorization potential of oil industry by-products as sources of essential fatty acids. Processes 10:2373

Miranda NT, Motta IL, Maciel Filho R, Maciel MRW (2021) Sugarcane bagasse pyrolysis: a review of operating conditions and products properties. Renew Sustain Energy Rev 149:111394

Mirdoraghi M, Farahani SM, Rezazadeh A (2024) Oilseeds in intercropping systems: strategies to increase oil quality and fatty acid profile, a review. J Agric Food Res 17:101229

Mohanty A, Rout PR, Dubey B, Meena SS, Pal P, Goel M (2022) A critical review on biogas production from edible and non-edible oil cakes. Biomass Convers Biorefinery 12(3):949–966

Muktham R, Ball AS, Bhargava SK, Bankupalli S (2016) Bioethanol production from non-edible de-oiled Pongamia pinnata seed residue-optimization of acid hydrolysis followed by fermentation. Ind Crop Prod 94:490–497

Mungodla SG, Linganiso LZ, Mlambo S, Motaung T (2019) Economic and technical feasibility studies: technologies for second generation biofuels. J Eng, Des Technol 17:670–704
Nanda S, Mishra A, Priyadarsini A, Barpanda T, Baral AK, Jena S, Jena PK, Mallick B, Dash M, Swain N (2024) A comparative thermo-chemical characterization of oilseed, millet and pulse stem biomass for bioethanol production. Heliyon 10:e36946
Naresh Kumar M, Ravikumar R, Thenmozhi S, Ranjith Kumar M, Kirupa Shankar M (2019) Choice of pretreatment technology for sustainable production of bioethanol from lignocellulosic biomass: bottle necks and recommendations. Waste Biomass Valorization 10:1693–1709
Norrahim MNF, Ariffin H, Yasim-Anuar TAT, Ghaemi F, Hassan MA, Ibrahim NA, Yunus WMZW (2018) Superheated steam pretreatment of cellulose affects its electrospinnability for microfibrillated cellulose production. Cellulose 25(7):3853–3859
Olofsson J, Börjesson P (2018) Residual biomass as resource—life-cycle environmental impact of wastes in circular resource systems. J Clean Prod 196:997–1006
Ong VZ, Wu TY, Lee CBTL, Cheong NWR, Shak KPY (2019) Sequential ultrasonication and deep eutectic solvent pretreatment to remove lignin and recover xylose from oil palm fronds. Ultrason Sonochem 58:104598
Ong ES, Rabbani AH, Habashy MM, Abdeldayem OM, Al-Sakkari EG, Rene ER (2021) Palm oil industrial wastes as a promising feedstock for biohydrogen production: a comprehensive review. Environ Pollut 291:118160
Oruganti RK, Sunar SL, Panda TK, Shee D, Bhattacharyya D (2023) Kraft lignin recovery from de-oiled Jatropha curcas seed by potassium hydroxide pretreatment and optimization using response surface methodology. Bioresour Technol Rep 23:101572
Othman MF, Adam A, Najafi G, Mamat R (2017) Green fuel as alternative fuel for diesel engine: a review. Renew Sust Energ Rev 80:694–709
Ourradi H, Lahboubi N, Habchi S, Hanine H (2022) Methane production from date seed cake (Phoenix Dactylifera L.) using mesophilic fed-batch anaerobic digestion. Clean Waste Syst 2: 100009
Ozsezen AN, Canakci M (2011) Determination of performance and combustion characteristics of a diesel engine fueled with canola and waste palm oil methyl esters. Energy Convers Manag 52: 108–116
Pahla G, Mamvura T, Muzenda E (2018) Torrefaction of waste biomass for application in energy production in South Africa. S Afr J Chem Eng 25:1–12
Pambudi NA, Laukkanen T, Syamsiro M, Gandidi IM (2017) Simulation of Jatropha curcas shell in gasifier for synthesis gas and hydrogen production. J Energy Inst 90:672–679
Panagiotopoulos I, Pasias S, Bakker R, De Vrije T, Papayannakos N, Claassen P, Koukios E (2013) Biodiesel and biohydrogen production from cotton-seed cake in a biorefinery concept. Bioresour Technol 136:78–86
Parodi E, La Nasa J, Ribechini E, Petri A, Piccolo O (2023) Extraction of proteins and residual oil from flax (Linum usitatissimum), Camelina (Camelina sativa), and sunflower (Helianthus annuus) oilseed press cakes. Biomass Convers Biorefinery 13:1915–1926
Paulino ADS, De Almeida VV, Oliveira AC, Oliveira HC, Garcia R, Silva RR, Santos PD, Silva YAD, Bispo SB, De Lima-Júnior DM (2018) Influence of increased doses of detoxified Castor bean meal on chemical composition and characteristics of sugarcane silage. Chil J Agr Res 78: 503–510
Petraru A, Amariei S (2022) Sunflower oilcake as a potential source for the development of edible membranes. Membranes 12:789
Petraru A, Ursachi F, Amariei S (2021) Nutritional characteristics assessment of sunflower seeds, oil and cake. Perspective of using sunflower oilcakes as a functional ingredient. Plants 10:2487
Petropoulos SA, Fernandes Â, Arampatzis DA, Tsiropoulos NG, Petrović J, Soković M, Barros L, Ferreira IC (2020) Seed oil and seed oil byproducts of common purslane (Portulaca Oleracea L.): a new insight to plant-based sources rich in Omega-3 fatty acids. LWT 123:109099

Petrovič A, Predikaka TC, Vohl S, Hostnik G, Finšgar M, Čuček L (2024) Hydrothermal conversion of oilseed cakes into valuable products: influence of operating conditions and whey as an alternative process liquid on product properties and their utilization. Energy Convers Manage 313:118640

Pohl M, Sánchez-Sánchez M, Mumme J (2019) Anaerobic digestion of wheat straw and rape oil cake in a two-stage solid-state system. Renew Energy 141:359–367

Pohlmann JG, Osório E, Vilela AC, Diez MA, Borrego AG (2017) Pulverized combustion under conventional (O2/N2) and oxy-fuel (O2/Co2) conditions of biomasses treated at different temperatures. Fuel Process Technol 155:174–182

Princess AP (2024) Comparative analysis of the chemical and physical properties of seeds and oil from two variants of castor plant (Ricinus Communis L.). Int J Agric Res Innov Technol. https://cambridgeresearchpub.com/ijaat/article/view/132

Qiao H, Cui J, Ouyang S, Shi J, Ouyang J (2019) Comparison of dilute organic acid pretreatment and a comprehensive exploration of citric acid pretreatment on corn cob. J Renew Mater 7: 1197–1207

Qiu B, Tao X, Wang J, Liu Y, Li S, Chu H (2022) Research progress in the preparation of high-quality liquid fuels and chemicals by catalytic pyrolysis of biomass: a review. Energy Convers Manag 261:115647

Rachmah ANL, Fatmawati A, Widjaja A (2022) Impact of surfactant-aided subcritical water pretreatment process conditions on the reducing sugar production from oil palm empty fruit bunch. IOP Conf Ser: Earth Environ Sci 963:012005. Iop Publishing

Radu O, Covaliov BE, Capcanari T (2024) Technological properties and functional food potential of oilseed cakes. Ukr Food J 13:287–302

Raita M, Ibenegbu C, Champreda V, Leak DJ (2016) Production of ethanol by thermophilic oligosaccharide utilising Geobacillus Thermoglucosidasius Tm242 using palm kernel cake as a renewable feedstock. Biomass Bioenergy 95:45–54

Rakita S, Kokić B, Manoni M, Mazzoleni S, Lin P, Luciano A, Ottoboni M, Cheli F, Pinotti L (2023) Cold-pressed oilseed cakes as alternative and sustainable feed ingredients: a review. Foods 12:432

Rani R, Badwaik LS (2021) Functional properties of oilseed cakes and defatted meals of mustard, soybean and flaxseed. Waste Biomass Valoriz 12(10):5639–5647

Rani H, Sharma S, Bala M (2021) Technologies for extraction of oil from oilseeds and other plant sources in retrospect and prospects: a review. J Food Process Eng 44:E13851

Ranjan A, Welz P, Mthethwa T (2024) Investigation of an effective acid pre-treatment method for the valorisation of canola fines. Biomass Convers Biorefinery 14:12013–12026

Rashmi I, Kumawat A, Munawery A, Karthika KS, Sharma GK, Kala S, Pal R (2022) Soil amendments: an ecofriendly approach for soil health improvement and sustainable oilseed production. In: Oilseed crops-uses, biology and production. Intechopen

Rattanaporn K, Tantayotai P, Phusantisampan T, Pornwongthong P, Sriariyanun M (2018) Organic acid pretreatment of oil palm trunk: effect on enzymatic Saccharification and ethanol production. Bioprocess Biosyst Eng 41:467–477

Razuan R, Finney K, Chen Q, Sharifi V, Swithenbank J (2011) Pelletised fuel production from palm kernel cake. Fuel Process Technol 92:609–615

Remón J, Matharu AS, Clark JH (2018) Simultaneous production of lignin and polysaccharide rich aqueous solutions by microwave-assisted hydrothermal treatment of rapeseed meal. Energy Convers Manag 165:634–648

Romuli S, Karaj S, Correa CR, Kruse A, Müller J (2021) Physico-mechanical properties and thermal decomposition characteristics of pellets from Jatropha curcas L. residues as affected by water addition. Biofuels 12:1149–1156

Saba N, Jawaid M, Paridah MT, Alothman O (2017) Physical, structural and thermomechanical properties of nano oil palm empty fruit bunch filler-based epoxy nanocomposites. Ind Crop Prod 108:840–843

Şahin S, Elhussein EAA (2018) Valorization of a biomass: phytochemicals in oilseed by-products. Phytochem Rev 17:657–668

Salaria A, Kaur N (2024) A detailed review on valorizing oilseed cakes: extraction of bioactive compounds from agro-industrial waste. Bioscene 21:254–278

Salazar-Villanea S, Bruininx EM, Gruppen H, Hendriks WH, Carré P, Quinsac A, Van Der Poel AF (2016) Physical and chemical changes of rapeseed meal proteins during toasting and their effects on in vitro digestibility. J Anim Sci Biotechnol 7:1–11

Sánchez-Borrego FJ, Alvarez-Mateos P, Garcia-Martin JF (2021) Biodiesel and other value-added products from bio-oil obtained from agrifood waste. Processes 9:797

Sang CS, Samad NAFA, Saleh SB (2025) Enhancing biofuel pellet quality using torrefaction and co-pelletization of palm kernel shell and empty fruit bunch. Environ Sci Pollut Res 32(32): 19390–19401

Sarkar N, Chakraborty D, Dutta R, Agrahari P, Bharathi SD, Singh AA, Jacob S (2021) A comprehensive review on oilseed cakes and their potential as a feedstock for integrated biorefinery. J Adv Biotechnol Exp Ther 4:376–387

Sathitkowitchai W, Ayimbila F, Nitisinprasert S, Keawsompong S (2022) Optimization of Mannanase hydrolysis of steamed palm kernel cake to improve nutritional quality. Biocatal Agric Biotechnol 43:102433

Švarc-Gajić J, Rodrigues F, Moreira MM, Delerue-Matos C, Morais S, Dorosh O, Silva AM, Bassani A, Dzedik V, Spigno G (2022) Chemical composition and bioactivity of oilseed cake extracts obtained by subcritical and modified subcritical water. Bioresour Bioprocess 9(1):114

Searchinger T, Heimlich R, Houghton RA, Dong F, Elobeid A, Fabiosa J, Tokgoz S, Hayes D, Yu T-H (2008) Use of U.S. croplands for biofuels increases greenhouse gases through emissions from land-use change. Science 319:1238–1240

Sharma S, Basu S, Shetti NP, Aminabhavi TM (2020) Waste-to-energy nexus for circular economy and environmental protection: recent trends in hydrogen energy. Sci Total Environ 713:136633

Sharma AK, Sahoo PK, Mukherjee M, Patel A (2022) Assessment of sustainable biogas production from co-digestion of Jatropha de-oiled cake and cattle dung using floating drum type digester under psychrophilic and mesophilic conditions. Clean Technol 4:529–541

Sharma S, Lindquist JC, Hwang D-C (2024) Canola/rapeseed as a potential source of alternative protein. Food Rev Intl 40:2306–2320

Sharmada N, Punja A, Shetty SS, Shet VB, Goveas LC, Rao CV (2016) Optimization of pre-treatment of de-oiled oil seed cake for release of reducing sugars by response surface methodology. Bioethanol 2. https://doi.org/10.1515/bioeth-2016-0006

Shojaeiarani J, Bajwa DS, Bajwa SG (2019) Properties of densified solid biofuels in relation to chemical composition, moisture content, and bulk density of the biomass. Bioresources 14: 4996–5015

Sidana A, Yadav SK (2022) Recent developments in lignocellulosic biomass pretreatment with a focus on eco-friendly, non-conventional methods. J Clean Prod 335:130286

Sikiru S, Abioye KJ, Adedayo HB, Adebukola SY, Soleimani H, Anar M (2024) Technology projection in biofuel production using agricultural waste materials as a source of energy sustainability: a comprehensive review. Renew Sust Energ Rev 200:114535

Singh P, Krishnaswamy K (2023) Non-Gmo-high oleic soybean meal value addition and studying the functional and reconstitution behavior. Int J Food Prop 26:708–728

Sokoto MA, Biswas B, Kumar J, Bhaskar T (2020) Slow pyrolysis of defatted seeds cakes of African star apple and silk cotton for production of bio-oil. Renew Energy 146:1710–1716

Sousa D, Salgado JM, Cambra-López M, Dias A, Belo I (2023a) Bioprocessing of oilseed cakes by fungi consortia: impact of enzymes produced on antioxidants release. J Biotechnol 364:5–12

Sousa D, Salgado JM, Cambra-López M, Dias A, Belo I (2023b) Biotechnological valorization of oilseed cakes: substrate optimization by simplex centroid mixture design and scale-up to tray bioreactor. Biofuels Bioprod Biorefin 17:121–134

Stojanović ZS, Uletilović DD, Kravić SŽ, Kevrešan ŽS, Grahovac NL, Lončarević IS, Đurović AD, Marjanović Jeromela AM (2023) Comparative study of the nutritional and chemical composition of new oil rape, safflower and mustard seed varieties developed and grown in Serbia. Plants 12:2160

Stolarski MJ, Szczukowski S, Tworkowski J, Krzyżaniak M, Gulczyński P, Mleczek M (2013) Comparison of quality and production cost of briquettes made from agricultural and Forest origin biomass. Renew Energy 57:20–26

Styles D, Gibbons J, Williams AP, Dauber J, Stichnothe H, Urban B, Chadwick DR, Jones DL (2015) Consequential life cycle assessment of biogas, biofuel and biomass energy options within an arable crop rotation. GCB Bioenergy 7:1305–1320

Sudalai S, Prabakaran S, Varalakksmi V, Kireeti IS, Upasana B, Yuvasri A, Arumugam A (2024) A review on oilcake biomass waste into biofuels: current conversion techniques, sustainable applications, and challenges: waste to energy approach (Wte). Energy Convers Manage 314: 118724

Szpunar-Krok E, Wondołowska-Grabowska A (2025) Effect of fertilization with ash from biomass combustion on the fatty acid composition of winter rapeseed oil. Agronomy 15:231

Tayeh HA, Levy-Shalev O, Azaizeh H, Dosoretz CG (2016) Subcritical hydrothermal pretreatment of olive mill solid waste for biofuel production. Bioresour Technol 199:164–172

Testa G, Corinzia SA, Calcagno S, Cosentino SL, Piccitto A (2023) A biorefinery model for the production of oil and biomethane using Castor plants. GCB Bioenergy 15:1339–1354

Thangalazhy-Gopakumar S, Lee CW, Gan S, Ng HK, Lee LY (2018) Comparison of bio-oil properties from non-catalytic and in-situ catalytic fast pyrolysis of palm empty fruit bunch. Mater Today Proc 5:23456–23465

Thangaraju S, Pulivarthi MK, Natarajan V (2021) Waste from oil-seed industry: a sustainable approach. In: Sustainable food waste management: concepts and innovations. Springer Singapore, Singapore, pp 177–190

Toomer OT, Oviedo-Rondón EO, Ali M, Joseph M, Vu T, Fallen B, Mian R (2024) Full-fat soybean meals as an alternative poultry feed ingredient—feed processing methods and utilization—review and perspective. Animals 14:2366

Tripathi M, Sharma M, Bala S, Connell J, Newbold JR, Rees RM, Aminabhavi TM, Thakur VK, Gupta VK (2023) Conversion technologies for valorization of hemp lignocellulosic biomass for potential biorefinery applications. Sep Purif Technol 320:124018

Valles A, Capilla M, Álvarez-Hornos F, García-Puchol M, San-Valero P, Gabaldón C (2021) Optimization of alkali pretreatment to enhance rice straw conversion to butanol. Biomass Bioenergy 150:106131

Van Stappen F, Mathot M, Decruyenaere V, Loriers A, Delcour A, Planchon V, Goffart JP, Stilmant D (2016) Consequential environmental life cycle assessment of a farm-scale biogas plant. J Environ Manag 175:20–32

Vichare SA, Morya S (2024) Exploring waste utilization potential: nutritional, functional and medicinal properties of oilseed cakes. Front Food Sci Technol 4:1441029

Vidosavljević S, Bojanić N, Dragojlović D, Stojkov V, Sedlar T, Banjac V, Fišteš A (2025) Application of optimized dry fractionation process for nutritional enhancement of different sunflower meals. Processes 13:255

Vujetić J, Spasevski N, Dragojlović D (2025) Processing tehniques of removing antinutrients from oilseed cake by products use in animal feed. Food Feed Res 52(1):37–51

Waindim Y, Gundu D, Ashwe A, Kuhe A (2022) Performance evaluation of black soap (toilet soap) blended with de-oiled jatropha cake additive as solid lubricant in direct cold extrusion of aluminium alloys. Int J Eng Sci Tech 6(4):66–83

Wang J, Azam W (2024) Natural resource scarcity, fossil fuel energy consumption, and total greenhouse gas emissions in top emitting countries. Geosci Front 15:101757

Wang C, Zhang J, Hu F, Zhang S, Lu J, Liu S (2020) Bio-pretreatment promote hydrolysis and acidification of oilseed rape straw: roles of fermentation broth and micro-oxygen. Bioresour Technol 308:123272

Wang W, Yang B, Li W, Zhou Q, Liu C, Zheng C (2021) Effects of steam explosion pretreatment on the bioactive components and characteristics of rapeseed and rapeseed products. Lwt 143: 111172

Wang C, An X, Lu Y, Li Z, Gao Z, Tian S (2022) Biodegradable active packaging material containing grape seed ethanol extract and corn starch/κ-carrageenan composite film. Polymers 14(22):4857

Wang C, Zhang Y, Wei S, Fu L, Wang Y, Jin M (2024a) Bioconversion of soybean meal into gut microbiota-targeting polysaccharides via fermentation by bacillus subtilis. J Clean Prod 464: 142787

Wang Y, Xu T, Liu K, Zhang M, Cai XM, Si C (2024b) Biomass-based materials for advanced supercapacitor: principles, progress, and perspectives. Aggregate 5:E428

Werteker M, Lorenz A, Johannes H, Berghofer E, Findlay C (2010) Environmental and varietal influences on the fatty acid composition of rapeseed, soybeans and sunflowers. J Agron Crop Sci 196:20–27

Wrigley CW, Corke H, Seetharaman K, Faubion J (2015) Encyclopedia of food grains. Academic Press

Xu D, Tree DR, Lewis RS (2011) The effects of syngas impurities on syngas fermentation to liquid fuels. Biomass Bioenergy 35:2690–2696

Xu H, Ou L, Li Y, Hawkins TR, Wang M (2022) Life cycle greenhouse gas emissions of biodiesel and renewable diesel production in the United States. Environ Sci Technol 56:7512–7521

Xu H, Hungwe D, Yang P, Yu M, Cheng S, Yoshikawa K, Takahashi F (2024) Oil sludge addition enables prediction of biomass pyrolysis product profiles by synergistic behaviors between biomass components and oil sludge. Appl Energy 362:123036

Yong KJ, Wu TY (2022) Second-generation bioenergy from oilseed crop residues: recent technologies, techno-economic assessments and policies. Energy Convers Manag 267:115869

Yoruklu HC, Koroglu EO, Ozdemir OK, Demir A, Ozkaya B (2020) Bioenergy production from cotton straws using different pretreatment methods. Int J Hydrog Energy 45:34720–34729

Zardo I, Rodrigues NP, Sarkis JR, Marczak LD (2020) Extraction and identification by mass spectrometry of phenolic compounds from canola seed cake. J Sci Food Agric 100(2):578–586

Zhang Y, Yao A, Song K (2016) Torrefaction of cultivation residue of Auricularia Auricula-Judae to obtain biochar with enhanced fuel properties. Bioresour Technol 206:211–216

Zhang F, Tian X-F, Fang Z, Shah M, Wang Y-T, Jiang W, Yao M (2017) Catalytic production of Jatropha biodiesel and hydrogen with magnetic carbonaceous acid and base synthesized from Jatropha hulls. Energy Convers Manag 142:107–116

Zhang Z, Zhao H, Chen X, Tian G, Liu G, Cai J, Jia G (2024) Enhancing pig growth and gut health with fermented Jatropha curcas cake: impacts on microbiota, metabolites, and neurotransmitters. J Anim Physiol Anim Nutr 108:1243–1257

Zhu X, Deng Z, Wang Q, Hao S, Liu P, He S, Li X (2024) Improvement in palm kernel meal quality by solid-sate fermentation with Bacillus velezensis, Saccharomyces cerevisiae and Lactobacillus paracasei. Fermentation 10:655

Zong J, Wang R, Bao G, Ling T, Zhang L, Zhang X, Hou R (2015) Novel triterpenoid Saponins from residual seed cake of camellia Oleifera Abel. Show anti-proliferative activity against tumor cells. Fitoterapia 104:7–13

Chapter 5
Lignocellulosic Biomass, Sustainable Biorefinery Applications, and Circular Economy

Khyati Joshi, Lachi Wankhede, Gaurav Bhardwaj, Sara Magdouli, and Satinder Kaur Brar

Abstract Lignocellulosic biomass (LCB), including agricultural residues, forestry by-products, and food processing waste, is a renewable and abundant resource with high potential for sustainable valorization. Rich in cellulose, hemicellulose, and lignin, it can be converted into a variety of value-added products, including bioethanol, biomethane, organic acids, enzymes, bioplastics, antioxidants, and bio-based materials. Technological advances in pretreatment, enzymatic hydrolysis, and microbial conversion have enhanced the efficiency of these processes, helping overcome challenges, such as biomass recalcitrance and low product yields. Integrated biorefineries and circular economy approaches further support the full utilization of lignocellulosic waste, reducing environmental impact while generating economic value. This chapter discusses recent progress in lignocellulosic biomass valorization and its role in sustainable biorefineries and circular bioeconomy systems.

5.1 Introduction

Lignocellulosic biomass (LCB) represents one of the most abundant renewable resources on Earth, originating from diverse sources, including agricultural residues, forestry by-products, and industrial wastes (Mujtaba et al. 2023). Composed primarily of cellulose, hemicellulose, and lignin, these complex polymers provide the raw material for an expanding array of bioproducts. While traditional valorization has largely focused on bioethanol production, recent advances have significantly

K. Joshi · L. Wankhede · G. Bhardwaj · S. K. Brar (✉)
Department of Civil Engineering, Lassonde School of Engineering, York University, Toronto, ON, Canada
e-mail: satinder.brar@lassonde.yorku.ca

S. Magdouli
Department of Civil Engineering, Faculty of Engineering, University of Ottawa, Ottawa, ON, Canada

A. L. Woiciechowski et al. (eds.), *Biomass for Bioenergy and Bioproducts*, Biofuel and Biorefinery Technologies 17, https://doi.org/10.1007/978-3-032-19898-3_5

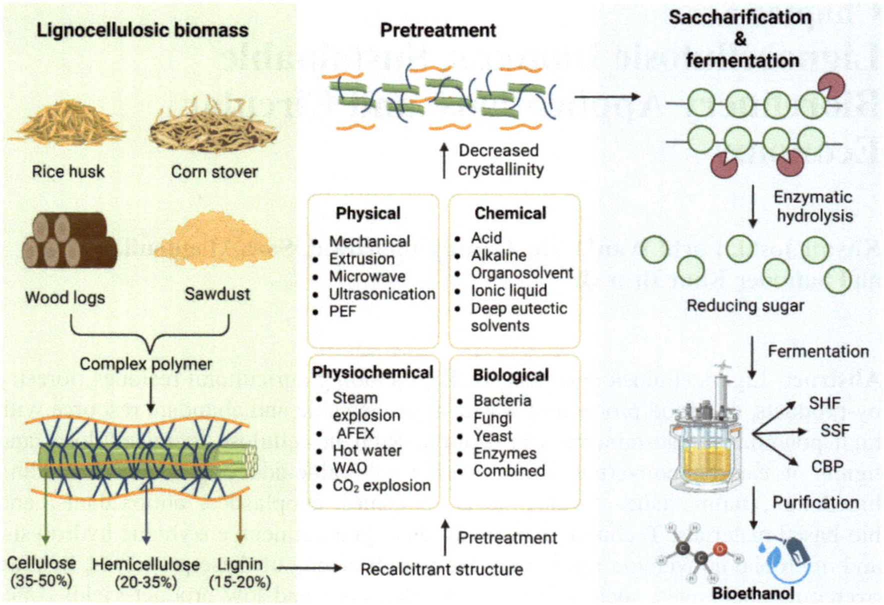

Fig. 5.1 Overview of the biofuel production process using lignocellulosic biomass as the feedstock

broadened the scope of applications to include biogas, furfural, organic acids, bioplastics, and other value-added compounds (Fig. 5.1) (Mujtaba et al. 2023). These products are derived through integrated biorefinery approaches that aim to maximize resource efficiency and economic viability (Mujtaba et al. 2023).

Globally, the volume of lignocellulosic residues generated from crop production, forest harvesting, and industrial processing reaches hundreds of millions of tons annually (Limayem and Ricke 2012). The strategic utilization of these residues is essential not only for waste reduction but also for fostering sustainable bioeconomies (Rajesh Banu et al. 2021). Biorefineries that process lignocellulosic feedstocks are increasingly adopting multiproduct frameworks that enhance flexibility and resilience, enabling the simultaneous generation of fuels, chemicals, and materials. This diversified valorization helps meet growing demands for renewable energy and sustainable materials while reducing environmental footprints (Rajesh Banu et al. 2021).

Countries worldwide have recognized the potential of LCB valorization and invested in research, technology development, and industrial implementation. From large-scale ethanol plants in Brazil to emerging bioplastic and biochemical production in Europe and Asia, these global successes demonstrate diverse pathways for converting lignocellulosic feedstocks into commercial products (Lopes et al. 2016; Mogany et al. 2024; Liang et al. 2025). Policy support, innovation in pretreatment and conversion technologies, and optimized supply chains are the key drivers behind these advancements (Lopes et al. 2016; Mogany et al. 2024; Liang et al. 2025).

This chapter provides an in-depth discussion on the valorization of lignocellulosic biomass within the context of sustainable biorefineries and circular economy principles. It highlights the productive chains linking biomass availability to bioproducts, illustrates successful examples of residue reuse, and addresses the challenges and opportunities in biorefinery development, process optimization, and circular economy. By integrating scientific insights and practical case studies, this chapter aims to offer a comprehensive perspective on the evolving role of lignocellulosic biomass in global sustainable development.

5.2 Sources of LCB and Their Composition

LCB is the most abundant renewable feedstock on Earth, offering a sustainable and cost-effective source for biorefineries. The major sources of LCB include agricultural residues, forestry resources, and other plant-based waste materials (Chowdhury et al. 2019). Agricultural residues produced during farming activities include rice husk, wheat straw, sugarcane bagasse (SCB), and corn stover, while forestry resources include wood chips, sawdust, bark, and logging residues. These sources are particularly advantageous due to their widespread availability, lignocellulosic content, and potential for large-scale bioconversion (Chowdhury et al. 2019; Antolini and Reviews 2021).

Structurally, LCB is primarily composed of cellulose (35%–50%), hemicellulose (20%–35%), and lignin (15%–20%), with minor amounts of inorganic compounds such as silica, calcium, potassium, and magnesium (Fig. 5.2) (Chowdhury et al. 2019). Cellulose microfibrils, held together by extensive hydrogen bonding, provide rigidity and mechanical strength to plant cell walls (Chowdhury et al. 2019). Hemicellulose, a heterogeneously branched polysaccharide, forms a matrix around cellulose fibers, interacting through hydrogen and van der Waals forces (Spiridon and Popa 2008). Lignin, a highly branched aromatic polymer, is intricately interwoven within the cell wall, covalently linked to hemicellulose through ether and ester bonds (Shah et al. 2022). This structural complexity contributes to the resistance of LCB to enzymatic hydrolysis, necessitating efficient pretreatment strategies to enhance its conversion into biofuels. In this section, the structure of cellulose, hemicellulose, and lignin will be discussed in detail.

5.2.1 Cellulose

Cellulose is widely recognized as the most abundant organic polymer on Earth. It is a linear polymer composed of β-D-glucose units connected by β-1,4-glycosidic bonds. These glucose units are rotated 180° relative to one another, allowing the chains to

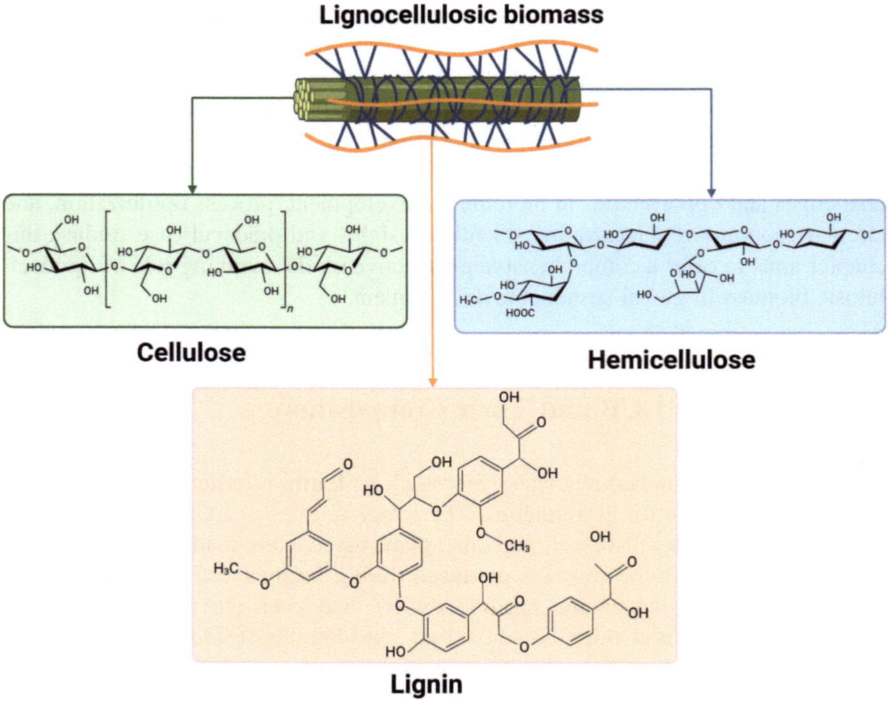

Fig. 5.2 Components of lignocellulosic biomass and their structure

align into extended microfibrils that contain both crystalline and amorphous regions (Banik 2008). The crystalline zones, stabilized by extensive intra- and intermolecular hydrogen bonding, contribute to its mechanical strength and chemical resistance, with crystallinity levels often ranging between 60% and 70% (Doelker et al. 1987). The degree of polymerization can vary widely—from a few hundred to over 10,000 glucose units—depending on the source and processing conditions (Balan 2014). The conversion of cellulose into fermentable sugars involves pretreatment to disrupt the compact structure, followed by enzymatic hydrolysis using cellulases (including endoglucanases, exoglucanases, and β-glucosidases) to break the polymer into glucose monomers, which can then be fermented into fuels such as ethanol (Soccol et al. 2019).

5.2.2 *Hemicellulose*

Hemicellulose is a heterogeneous group of branched polysaccharides that is less ordered than cellulose. Unlike the uniform structure of cellulose, hemicellulose is composed of a variety of sugar monomers, including xylose, mannose, glucose, galactose, and arabinose, and features diverse glycosidic linkages (both β- and α-configurations) and extensive branching (Soccol et al. 2019). This amorphous structure results in a lower degree of polymerization (typically between 50 and 200), which makes hemicellulose more amenable to hydrolysis under milder conditions (Liu et al. 2014). The composition can vary with the biomass type: Hardwoods mainly contain O-acetyl-4-O-methylglucuronoxylan (10%–35%) with some glucomannan (3%–5%), whereas softwoods contain O-acetyl-galactoglucomannan (10%–15%), along with some arabino-4-*O*-methylglucuronoxylan (5%–10%) (Asmadi et al. 2017). The conversion of hemicellulose into fermentable sugars involves releasing a mix of pentose and hexose sugars through pretreatment and the action of hemicellulases (such as xylanases and mannanases), which are then fermented into biofuels (Houfani et al. 2020).

5.2.3 Lignin

Lignin is a complex, amorphous aromatic polymer that functions as the "glue" binding cellulose and hemicellulose together within the plant cell wall (Sivagurunathan et al. 2021). It is synthesized from three primary monolignols—coniferyl alcohol, p-coumaryl alcohol, and sinapyl alcohol—which form guaiacyl (G), p-hydroxyphenyl (H), and syringyl (S) units, respectively. The polymer's structure is highly irregular and cross-linked, with the predominant linkage being the β-O-4 aryl ether bond (making up about 50%–60% of the bonds), along with various other linkages such as β–β, β–5, and 5–5 (Sivagurunathan et al. 2021). Lignin imparts rigidity, hydrophobicity, and resistance to microbial degradation. Although its recalcitrant nature poses challenges for the enzymatic or chemical conversion of the polysaccharides during biofuel production, lignin can be valorized by burning it for process heat or further processing it into high-value aromatic chemicals through depolymerization techniques (Tian et al. 2023).

5.3 LCB: Global Production and Biorefinery Applications

LCB is produced globally in massive quantities, with annual yields reaching billions of tons from agricultural, forestry, and industrial sources. Major producers include countries with large agricultural and forest sectors, such as the United States, China, Brazil, India, and Canada, highlighting their strategic importance for sustainable bioeconomies worldwide (Bhat et al. 2024).

The concept of a biorefinery represents a sustainable industrial system that integrates biomass conversion processes to produce fuels, power, heat, and value-added biochemicals from biomass (Fig. 5.3) (Bhat et al. 2024). This model is analogous to petroleum refineries but utilizes renewable resources, particularly LCB, as the primary feedstock. Biorefineries aim to achieve maximum valorization

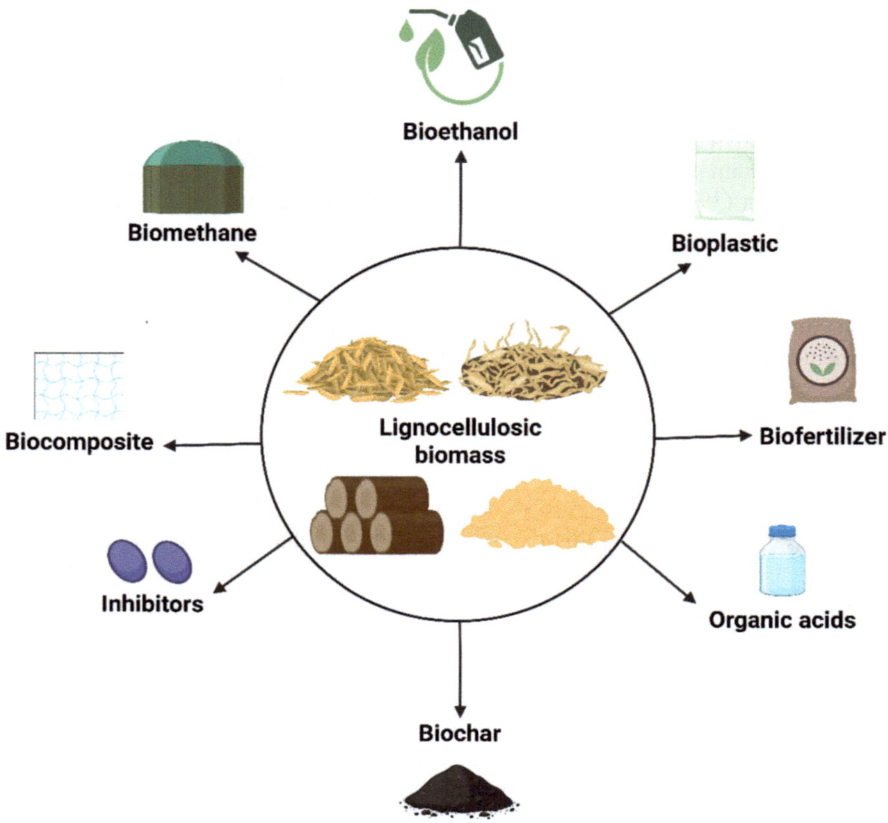

Fig. 5.3 Applications of lignocellulosic biomass in the production of value-added chemicals

of all biomass components, including cellulose, hemicellulose, and lignin, through integrated, multiproduct systems that align with the principles of the circular economy (Uzochukwu et al. 2025).

Lignocellulosic residues are categorized into agricultural residues, forestry residues, and industrial by-products from primary biomass processing (Table 5.1). These residues are largely underutilized and often treated as waste, yet they contain high energy potential and valuable biochemical precursors (Uzochukwu et al. 2025). The valorization of such residues contributes significantly to reducing greenhouse gas emissions, minimizing dependence on fossil fuels, and supporting rural economies (Uzochukwu et al. 2025). In this section, the composition, world production, and importance of these LCB residues will be discussed.

Table 5.1 Examples of LCB and their composition

LCB type	Example	Cellulose (%)	Hemicellulose (%)	Lignin (%)	Reference
Agricultural residues	Corn stover	38	23	20	Wan and Li (2010)
	Rice husk	37.1	29.4	24.1	Kalita et al. (2015)
	Sugarcane bagasse	40–45	30–35	20–30	Cardona et al. (2010)
	Wheat straw	33–40	20–25	15–20	Talebnia et al. (2010)
	Peanut shells	44.8	5.6	36.1	Pączkowski et al. (2021)
	Coconut husk	20–30	15–30	50	Anuchi et al. (2022)
	Banana pseudostem	30.08	27.79	6.08	Li et al. (2016)
	Cotton stalks	32–46	20–28	15–30	Cai et al. (2024)
	Soybean straw	34	16	22	Wan et al. (2011)
Forestry resources	Birch sawdust	36.6	21.8	22	Monção et al. (2021)
	Bamboo	45	24	20	Li et al. (2015)
	Pine	42	21	30	Sannigrahi et al. (2010)
	Fir	45	22	30	Di Blasi et al. (2010)
	Switchgrass	31	24	18	Lee et al. (2009)
	Elephant grass	36	24	28	Scholl et al. (2015)
	Eucalyptus bark	51	26	23	Arteaga-Pérez et al. (2015)
	Willow bark	41.7	16.7	29.3	Taherzadeh et al. (1997)
	Cedar shavings	39.3	12	28.8	Fayoud et al. (2016)

5.3.1 Agricultural Residues

5.3.1.1 Corn Stover

Corn stover, the nonedible part of the maize plant (leaves, stalks, cobs, and husks), represents a substantial biomass source (Aghaei et al. 2022). Corn is the most widely produced cereal crop worldwide, cultivated in many countries, with the United States and China accounting for about half of the global market (Tanklevska et al. 2020). The United States alone generates approximately 250 million tons of corn stover annually, contributing to global maize production that exceeds 1.16 billion metric tons (da Rosa et al. 2025; Ssemugenze et al. 2025). Its composition, comprising cellulose (38%), hemicellulose (23%), and lignin (20%), makes it suitable for second-generation bioethanol, biogas, and bioplastic production (Wan and Li 2010). This potential has been demonstrated in several studies. For example, in a study by

Lau and Dale, ethanol production from ammonia fiber expansion (AFEX)–pretreated corn stover using *Saccharomyces cerevisiae* reached 191.5 g/kg, with an ethanol concentration of 40.0 g/L (Lau and Dale 2009).

Corn stover is also utilized for the production of organic acids. For instance, He et al. reported lactic acid yields of 79.6 wt% from the hemicellulosic fraction of fractionated corn stover (He et al. 2016). For sugar alcohol production, Rodrigues and Jeffries achieved a xylitol yield of 0.61 g/g xylose using a mutant *Pichia stipitis* strain and corn stover hydrolysate (Rodrigues et al. 2011). Other examples, such as those by Xue et al., Li et al., and Yu et al., have demonstrated the efficient conversion of corn stover into microbial lipids, bioplastics, and platform chemicals, further supporting its versatility as a sustainable biorefinery feedstock (Xue et al. 2015; Yu et al. 2020; Li et al. 2023).

5.3.1.2 Rice Husk

Rice husk, a by-product that accounts for approximately 20%–22% of the weight of rice grains, is generated in vast quantities worldwide, exceeding 100 million tons annually (Ishida et al. 2025). Major producers include China, India, Indonesia, Bangladesh, and Vietnam (El-Sayed 2025). Composed of approximately up to 50% cellulose, 25%–30% lignin, and 15%–20% silica, rice husk is widely explored for applications in bio-silica recovery, activated carbon production, and composite materials (Nasir et al. 2022).

Its high silica content, in particular, has drawn interest for material applications. For example, Villota-Enríquez and Rodríguez-Páez extracted bio-silica from rice husk ash with a surface area of ~202 m^2/g, demonstrating its potential for dye removal (Villota-Enríquez and Rodríguez-Páez 2023). Rice husk has also been successfully used in developing composite materials. In a study by Choi et al., combining rice husk with waste polystyrene resulted in strong, water-resistant building composites, with optimal mechanical properties achieved at a 30% binder content (Choi et al. 2006).

Rice husk is also widely utilized for bioethanol production. In a study by Pabón et al., ethanol yields of up to 43.9% using *Saccharomyces cerevisiae* SC 1507 have been reported (Arismendy Pabón et al. 2020). These diverse applications highlight the value of rice husk as a versatile resource in biorefinery pathways.

5.3.1.3 Sugarcane Bagasse

Sugarcane bagasse (SCB), the fibrous residue left after juice extraction from sugarcane, is generated in substantial quantities, with over 540 million metric tons produced globally from more than 1.6 billion metric tons of sugarcane (Ajala et al. 2021; Ungureanu et al. 2022). Brazil, India, Thailand, and China are the largest contributors of sugarcane bagasse. Rich in cellulose (40%–45%), hemicellulose (30%–35%), and lignin (20%–30%), SCB is widely used for energy generation in sugar mills and is gaining attention in emerging biorefineries for the production of biobutanol, bioplastics, furfural, and levulinic acid (LA) (Victor et al. 2014; Ajala et al. 2021; Ungureanu et al. 2022; Yong et al. 2024).

Its suitability for microbial fermentation has made SCB a preferred substrate in biofuel production. For example, Su et al. reported 6.4 g/L butanol and 11.9 g/L total acetone–butanol–ethanol (ABE) using *Clostridium beijerinckii* under a sequential pretreatment and fermentation approach (Su et al. 2015). In another study, Kong et al. enhanced sugar yield to 248.41 g/L using gamma-valerolactone-assisted hydrolysis, resulting in 14.26 g/L ABE with 9.3 g/L butanol (Kong et al. 2016).

SCB is also a promising substrate for bioplastic production. In a study by Yong et al., enzymatically hydrolyzed SCB was used by *Bacillus cereus* to produce polyhydroxyalkanoates (PHAs), achieving a yield of 90.39 g/L and a PHA accumulation of up to 62.1% (w/w) under optimized fermentation conditions (Yong et al. 2024). In addition, SCB is explored for integrated bioproduct generation. Ntimbani et al. evaluated ethanol and furfural coproduction using furfural residues from SCB, achieving ethanol yields ranging from 77% to 95% depending on the strain and process parameters. Optimal conditions (170 °C and 0.25 wt.% H_2SO_4) resulted in the generation of 10.91 g ethanol and 7.64 g furfural per 100 g of dry bagasse (Ntimbani et al. 2022).

Other studies have demonstrated the potential of SCB in biogas, xylitol, and levulinic acid production, emphasizing its role as a versatile and abundant lignocellulosic feedstock in biorefinery applications (Rao et al. 2006; Victor et al. 2014; Agarwal et al. 2022).

5.3.1.4 Peanut Shells

Peanut shells, accounting for 20%–30% of the total peanut biomass and produced in large quantities globally, especially in China, India, Nigeria, and the United States, have gained considerable attention as a valuable lignocellulosic feedstock in sustainable biorefineries (Hauser 2018). Their composition, with approximately ~45% cellulose, ~6% hemicellulose, and ~36% lignin, makes them suitable for various bioconversion processes (Pączkowski et al. 2021). For instance, Nazir et al. demonstrated the production of biochar from peanut shells via pyrolysis at different temperatures, highlighting improved biochar properties and soil amendment potential at higher temperatures (Ganguly and Das 2024). Ganguly and Das reported efficient bioethanol production from pretreated peanut shells, achieving ethanol yields of up to 1.96 g/g using *Saccharomyces cerevisiae* under anaerobic fermentation while also valorizing the fermentation residue as an effective biochar adsorbent for wastewater treatment (Ganguly and Das 2024). Furthermore, Nisar et al. explored catalytic pyrolysis of peanut shells, which resulted in enhanced bio-oil production with a diverse range of hydrocarbons (C4–C31), indicating the feasibility of peanut shells as a feedstock for biofuel generation (Nisar et al. 2022). These examples collectively signify the multifaceted applications of peanut shells in biorefineries, ranging from bioenergy and biofuels to soil improvement and environmental remediation, supporting their role in circular economy frameworks.

5.3.1.5 Banana Pseudostem

Banana pseudostem, a major agricultural residue, especially abundant in India, the Philippines, Indonesia, and Ecuador, is gaining recognition as a valuable LCB for sustainable biorefinery applications (Emmanuel et al. 2025). With approximately 30% cellulose, 28% hemicellulose, and a low lignin content (6%), it is well-suited for bioethanol and biogas production, as well as for fibers used in textiles, biocomposites, and handicrafts, supporting local circular economies (Li et al. 2016; Sangamithirai and Vasugi 2020). Gupta et al. explored ethanol production from concentrated banana sap obtained from pseudostems, achieving a maximum ethanol yield of 2.5 g/L by supplementing the banana sap with industrial by-products and fermenting with *Saccharomyces cerevisiae* strains (Gupta et al. 2019). Ingale et al. demonstrated that fungal pretreatment of banana pseudostem, followed by fermentation with *S. cerevisiae* NCIM 3570, yielded up to 17.1 g/L ethanol, with 84% conversion efficiency, highlighting its potential as an economical substrate for bioethanol (Ingale et al. 2014). Additionally, Li et al. investigated methane production from fresh banana pseudostem through anaerobic digestion, reporting methane yields of up to 347 mL/g volatile solids (VS) under optimized conditions, confirming its suitability for biogas generation (Li et al. 2016). These studies exemplify the diverse valorization pathways of banana pseudostem in biorefineries, contributing to renewable energy production and sustainable waste management in banana-producing regions.

5.3.1.6 Wheat Straw

Wheat straw is a major agricultural residue generated worldwide, with global wheat production exceeding 760 million tons and approximately 350 million tons of straw produced annually (Tufail et al. 2021; Taghizadeh-Alisaraei et al. 2023). The largest producers include China, India, and Russia (Taghizadeh-Alisaraei et al. 2023). Wheat straw is composed mainly of cellulose (33%–40%), hemicellulose (20%–25%), and lignin (15%–20%) and is used as a feedstock for biofuels such as bioethanol, biogas, and biohydrogen, as well as for biopolymers, biofertilizers, and biochar applied in soil remediation and carbon sequestration (Talebnia et al. 2010; Abbas et al. 2020).

Kaparaju et al. investigated bioethanol, biohydrogen, and biogas production from wheat straw in a biorefinery approach, reporting ethanol yields of 0.41 g ethanol/g glucose from enzymatic hydrolysis and fermentation, hydrogen production of 178 mL H_2/g sugars via dark fermentation, and methane yields of 0.324 and 0.381 m^3 CH_4/kg volatile solids from process effluents (Kaparaju et al. 2009). Risberg et al. evaluated co-digestion of steam-exploded wheat straw with cattle manure and found stable but relatively low methane yields ranging from 0.13 to 0.21 NL CH_4/kg volatile solids (VS), with little effect from pretreatment or temperature variation (Risberg et al. 2013). Mancini et al. showed that chemical pretreatments improved biomethane yields by 11%–15%, with cumulative methane production reaching 274 mL CH_4/g VS (Mancini et al. 2018). Han et al. applied ammonia percolation pretreatment on wheat straw and achieved high delignification (>70%), enzymatic digestibility (>90%), and an ethanol concentration of 24.15 g/L after fermentation with *Saccharomyces cerevisiae* (Han et al. 2009). Last, Chen and Sun developed a novel solid-state fermentation process with periodic air-forced pressure oscillation (SSF-PAPO) to convert steam-exploded wheat straw into bio-organic fertilizer, achieving high microbial biomass (2×10^{11} cfu/g) and substantial degradation of cellulose (48.6%) and hemicellulose (84.8%) (Chen and Sun 2007).

These studies collectively demonstrate the versatile potential of wheat straw as a sustainable feedstock for multiple bio-based products through various pretreatment and bioconversion strategies.

5.3.2 Forestry Residues

5.3.2.1 Bamboo

Bamboo forests span over 30 million hectares globally, predominantly in Asia and Africa, with China, India, and Indonesia being the leading producers (Dutta et al. 2025). Bamboo residues, including leaves and culm shavings, contain approximately 45% cellulose, 24% hemicellulose, and 20% lignin (Li et al. 2015). These residues are increasingly used for producing bioethanol, biocomposites, nanocellulose, and paper pulp. Bamboo biorefineries are being developed in countries like India and China for integrated biomass valorization (Kuttiraja et al. 2013; Li et al. 2015; Norrrahim et al. 2025).

Kuttiraja et al. investigated the enzymatic saccharification and fermentation of bamboo processing waste and reported a bioethanol yield of 143 L per dry ton, with 43% overall process efficiency after dilute alkaline pretreatment (Kuttiraja et al. 2013). Deshmukh and Pathan tested several pretreatment methods and highlighted the efficiency of NaOH with ultrahigh-pressure explosion in enhancing enzymatic hydrolysis and ethanol conversion (Deshmukh and Pathan 2024). In a separate study, Pivsa-art demonstrated the potential of bamboo fibers in reinforcing biodegradable poly(butylene succinate) (PBS)-based biocomposites, improving mechanical properties, and supporting sustainable alternatives to petroleum-derived materials (Pivsa-Art and Pivsa-Art 2021).

In conclusion, bamboo offers significant potential as a fast-growing and renewable lignocellulosic feedstock for sustainable fuel and material production.

5.3.2.2 Softwood Residues (Pine and Fir)

The current softwood harvest in the Southeastern United States alone amounts to 150 million cubic meters annually, generating substantial residues in the form of bark, sawdust, and trimmings (OSDP 2025). Pine, a major softwood species, contains approximately 42% cellulose, 21% hemicellulose, and 30% lignin, making it suitable for applications such as bioadhesives, phenolic resins, biochar, and lignin-derived polymers (Gaskin et al. 2010; Sannigrahi et al. 2010; Saražin et al. 2020). Fir residues are similarly used, with biorefinery processes increasingly targeting high-value compounds like furfural and lignosulfonates (Qin et al. 2016; Ren et al. 2019).

Pine sawdust conversion to levulinic acid (LA) and furfural was studied by Lappalainen and Dong, who used sulfuric acid–catalyzed mechanical pretreatment and microwave-assisted heating in a biphasic water–toluene system. Their approach significantly enhanced furfural yield (up to 85%) and improved LA production efficiency (up to 38%) under optimized conditions (Lappalainen and Dong 2019).

In another study, pine needle waste biomass from the Himalayan region was utilized by Dwiwedi et al. for bioethanol production. Following NaOH and microwave pretreatment, enzymatic hydrolysis, and co-fermentation, the study achieved an ethanol yield of 17.65 g/L, with over 91% fermentation efficiency, highlighting its potential to mitigate forest fire risks while generating biofuel (Dwivedi et al. 2024).

In conclusion, softwood residues like pine and fir offer significant potential for conversion into high-value bio-based chemicals and fuels, aligning with both environmental and industrial sustainability goals.

5.3.2.3 Hardwood Residues (Eucalyptus and Poplar)

Eucalyptus and poplar plantations are prevalent in Brazil, Australia, China, and parts of Europe, with Brazil alone covering over 6 million hectares of eucalyptus (Bellink and Verburg 2023). Hardwood residues from these plantations contain approximately 38%–51% cellulose, 17%–38% hemicellulose, and 21%–31% lignin (Tarasov et al. 2018). While they have been traditionally used in the pulp and paper industry, there is growing interest in their application within biorefineries for the production of platform chemicals like hydroxymethylfurfural (HMF), levulinic acid, and acetic acid (Patil 2016; Kłosowski et al. 2019).

Mikulski and Kłosowski demonstrated efficient ethanol production from beechwood, achieving 91.78 g/L glucose and 49.07 g/L ethanol after high-pressure microwave-assisted pretreatment using 1% NaOH (Mikulski and Kłosowski 2023). In another study, Dedes et al. optimized organosolv-pretreated beechwood for fructose and HMF production, attaining 104.5 g/L fructose and an HMF yield of 44.6% using formic acid (Dedes et al. 2021). Kłosowski et al. reported levulinic acid yields of up to 16.5% from cherry wood chips through microwave-assisted acid catalysis under controlled pressure and acid concentration (Kłosowski et al. 2019).

These studies highlight the potential of hardwood residues as valuable lignocellulosic resources for producing both biofuels and high-value biochemicals using integrated biorefinery approaches.

5.3.2.4 Elephant Grass

Elephant grass (*Pennisetum purpureum*) is a fast-growing, high-yield energy crop cultivated widely in tropical regions such as Brazil and Kenya. It can produce over 35–41 tons of dry biomass per hectare annually (Negawo et al. 2017). Its composition, cellulose (36%), hemicellulose (24%), and lignin (28%), varies with cultivar and maturity, making it suitable for a range of bioenergy applications, including bioethanol, biogas, pellets, and fiber-reinforced composites (Scholl et al. 2015; Johannes et al. 2024).

Iyyappan et al. demonstrated a dual biological strategy for converting elephant grass into bioethanol. Biodelignification was achieved using a recombinant *Trichoderma reesei* strain expressing *Lip8H* and *MnP1* genes with NiO nanoparticles, followed by saccharification and fermentation using *Kluyveromyces marxianus*. The approach yielded 84.52 g/L of reducing sugars and 14.65 g/L of ethanol, highlighting the crop's potential for second-generation biofuel production (Iyyappan et al. 2023).

For biogas generation, Haryanto et al. evaluated the anaerobic co-digestion of elephant grass and cow dung in batch digesters. Among the treatments, the 25:25:100 (cow dung:grass:water) ratio produced the highest biogas volume (524.3 L), though the methane content remained relatively low (~31.37%). The study indicated that mechanical pretreatment could improve methane yields by enhancing substrate breakdown (Haryanto et al. 2018).

These studies highlight elephant grass's versatility as a renewable lignocellulosic biomass for both bioethanol and biogas production, especially when supported by integrated pretreatment strategies.

5.3.3 Industrial Residues: Valorization in Biorefineries

5.3.3.1 Pulp and Paper Sludge

Paper sludge, a fibrous waste generated in large quantities by the pulp and paper industry, contains residual cellulose, lignin, and high levels of inorganic fillers such as $CaCO_3$. In the European Union (EU) alone, over 11 million tons are produced annually, with additional volumes globally (Furszyfer Del Rio et al. 2022). Due to its composition, paper sludge is increasingly being valorized into biofuels and bioproducts. For example, Mendes et al. achieved 88% sugar conversion after acid pretreatment, enabling an ethanol production of up to 10.5 g/L using *Pichia stipites* (Mendes et al. 2014). Hay et al. reported biohydrogen yields of 9.62 mL/mL from paper mill effluent using *Rhodobacter sphaeroides* with ultrasound-enhanced photofermentation. Anaerobic digestion of paper sludge has also shown promising results for biomethane production (Hay et al. 2016). Priadi et al. reported methane yields of 14.7 mL/g VS for untreated sludge, which increased to 269 mL/g VS upon co-digestion with cow manure (Priadi et al. 2014).

These results indicate that paper sludge can serve as a valuable feedstock for ethanol, hydrogen, and methane generation, supporting integrated biorefinery approaches for industrial waste management.

5.3.3.2 Sawdust and Wood Chips

Sawdust and wood chips, by-products of sawmilling and forest operations, are rich in cellulose (~36%) and lignin (~28%), making them suitable feedstocks for biorefineries (Wang et al. 2024a). Tulashie et al. reported bioethanol yields of up to 206 mL per 100 g sawdust waste following acid hydrolysis and fermentation with *Saccharomyces cerevisiae* (Tulashie et al. 2023). Similarly, Cotana et al. achieved 10.6 g ethanol per 100 g pine wood chips after steam explosion and enzymatic hydrolysis. Fast pyrolysis is also being explored for bio-oil generation (Cotana et al. 2014). Guida et al. optimized process parameters to yield 39.5 wt% bio-oil from sawdust at 500 °C, while char yield decreased with rising temperature, dropping from 54.61 to 29.47 wt% (Guida et al. 2020). These findings highlight the versatility of lignocellulosic wood residues in generating biofuels (ethanol and bio-oil) and high-value products (biochar and green chemicals), supporting their integration into circular bioeconomy models.

5.3.3.3 Fruit Peels and Agro-Industrial Waste

Fruit wastes such as citrus peel, pomegranate husk, and apple pomace are rich in fermentable sugars, pectin, and phenolic compounds, making them valuable feedstocks for bioproducts (Lucarini et al. 2021). The food processing industry generates millions of tons of such residues annually, especially in countries like the United States, Brazil, and China (Russo et al. 2021). These wastes have been utilized to produce pectin-based films, organic acids, antioxidants, and bioethanol (Russo et al. 2021). Casabar et al. demonstrated bioethanol production from pineapple peel, achieving a yield of 5.98 g/L after 48 h of fermentation following alkaline pretreatment and microbial hydrolysis (Casabar et al. 2019). Choi et al. reported enzymatic saccharification efficiencies of approximately 90% for various fruit wastes, including citrus peels and apple pomace, and developed a d-limonene removal system to alleviate fermentation inhibition, resulting in ethanol yields as high as 29.5 g/L, with over 90% conversion efficiency (Choi et al. 2015). Ricci et al. achieved lactic acid concentrations of 209.65 g/kg via fermentation of orange peel with *Lactobacillus casei*, with yields of up to 0.88 g/g, highlighting the importance of microbial strain selection (Ricci et al. 2019). Collectively, these studies illustrate the significant potential of fruit waste streams for sustainable bioconversion into high-value chemicals and fuels.

5.4 Global Successes in LCB Valorization

5.4.1 Brazil: Sugarcane Bagasse Biorefineries

Brazil represents a global leader in the industrial utilization of lignocellulosic biomass, predominantly through its well-established sugarcane industry. Sugarcane bagasse, a by-product of sugar and ethanol production, is generated in enormous quantities, estimated at over 757 million tons annually (Ungureanu et al. 2022). The country has developed a robust infrastructure for cogeneration using bagasse, where the fibrous residue fuels boilers to produce steam and electricity, often supplying surplus energy to the national grid (Ungureanu et al. 2022). Beyond traditional energy applications, Brazilian biorefineries have evolved to adopt integrated approaches, exploiting bagasse for second-generation bioethanol production through enzymatic hydrolysis of cellulose and hemicellulose (Chandel et al. 2021). Notably, companies like GranBio have pioneered commercial-scale cellulosic ethanol plants, incorporating advanced fermentation technologies that demonstrate the economic viability and environmental benefits of biorefineries (Chandel et al. 2021). This industrial model not only adds value to residual biomass but also reduces fossil fuel dependence and greenhouse gas emissions, aligning with Brazil's commitments to sustainable development (Chandel et al. 2021).

5.4.2 United States: Corn Stover–Based Bioethanol Plants

In the United States, corn stover constitutes one of the most abundant agricultural residues, with annual availability exceeding 250 million tons (da Rosa et al. 2025). Several commercial ventures, including those by companies such as POET-DSM and DuPont (now part of Verbio), have advanced the production of cellulosic ethanol using corn stover (Malik 2023). While large-scale production at these facilities has since been discontinued, their efforts provided critical insights into process optimization and demonstrated the viability of cellulosic bioethanol as a sustainable alternative to fossil-derived fuels (Malik 2023). These biorefineries utilize state-of-the-art pretreatment and enzymatic hydrolysis processes to convert cellulose and hemicellulose fractions into fermentable sugars. The integration of corn stover into bioethanol production serves dual purposes: enhancing feedstock security by diversifying beyond corn kernels and increasing the overall energy yield per hectare (Malik 2023). In addition to biofuels, stover-derived lignin is being developed for use in bioplastics and as a renewable precursor for carbon fiber production (Li et al. 2024b). The US Department of Energy has played a pivotal role in supporting pilot and demonstration-scale projects, highlighting the country's commitment to establishing a sustainable bioeconomy that utilizes lignocellulosic residues (Davis and Bartling 2023).

5.4.3 Europe: Forest Residue Valorization in the Nordic Countries

The Nordic countries, including Sweden, Finland, and Norway, have extensive forest resources and a long-standing tradition of the forest industry. These nations generate significant volumes of forestry residues, such as sawdust, bark, and branches, which have become central feedstocks for biorefineries (Fredman et al. 2021). Industrial operations focus on converting these residues into bioenergy, biochemicals, and biomaterials. For instance, in Sweden, large pulp and paper mills have integrated biorefinery platforms that valorize black liquor lignin and extract hemicellulose sugars to produce a variety of platform chemicals (Agnihotri and Horváth 2024). Finland has invested heavily in producing lignin-based adhesives, promoting a circular use of wood residues (Good News from Finland 2025). The European Union's Horizon 2020 program has supported multiple projects aiming to scale-up forest residue biorefineries, demonstrating the commercial viability of converting forestry waste streams into biofuels and green chemicals. This Nordic model exemplifies a highly integrated forest biorefinery approach, which balances industrial efficiency with sustainable forest management principles (Velghe et al. 2021).

5.4.4 China: Bamboo and Agricultural Residue Integration

China's vast bamboo forests contribute an estimated 4.6 million hectares of biomass suitable for biorefinery applications (Li et al. 2024a). Industrial utilization of bamboo residues, such as culm shavings and leaves, has been expanding, focusing on bioethanol, nanocellulose, and biocomposite production (Deshmukh and Pathan 2024; Verma et al. 2024). Several studies in China have demonstrated the feasibility of producing bio-based materials from bamboo residues, considering the rapid renewability of bamboo (Zhao et al. 2024). Additionally, agricultural residues, including rice husk, wheat straw, and corn stover, are increasingly incorporated into integrated biorefineries that supply bioenergy and biochemicals to local industries (Wang et al. 2024b; Ji et al. 2025). The government's policy incentives and subsidies have fostered public–private partnerships aimed at scaling up lignocellulosic biomass valorization (Ogwu et al. 2025). China's approach highlights the strategic importance of utilizing a diverse portfolio of biomass residues to reduce pollution from open-field burning and promote rural development through bio-based industries.

5.4.5 *India: Banana Pseudostem and Agricultural Waste Biorefineries*

India generates massive quantities of lignocellulosic residues from both its agricultural and forestry sectors, with banana alone accounting for approximately 26.20 million tons annually, producing large quantities of banana pseudostem (Prasad and Das 2024). The country has witnessed growing interest in valorizing such residues within biorefinery frameworks to produce bioethanol, biogas, and bioplastics. Various state-run and private initiatives focus on converting banana pseudostem fibers into biocomposites and textiles, while the cellulose and hemicellulose fractions are processed for fermentable sugars (The Better India 2025; The Hindu 2025; Gupta et al. 2024). Agricultural residues such as sugarcane bagasse, wheat straw, and rice husk are also utilized in cogeneration and biogas plants, particularly in the sugar and rice milling industries (Patil et al. 2022). Indian biorefineries emphasize the dual objective of reducing rural pollution (notably crop residue burning) and promoting sustainable energy solutions, thereby contributing to national targets on renewable energy and carbon emission reductions (Sharma et al. 2015). Research institutions and start-ups are actively exploring innovative bioproducts, including enzyme production and organic acids, reinforcing the potential for a circular bioeconomy in the country (Thatoi et al. 2025).

5.5 Challenges in LCB Valorization

LCB represents one of the most abundant and renewable carbon sources on Earth, composed primarily of cellulose, hemicellulose, and lignin (Mujtaba et al. 2023). Despite its potential as a sustainable feedstock for the production of biofuels, biochemicals, and biomaterials, its valorization remains a major technological and scientific challenge. The intrinsic recalcitrance of LCB is a primary obstacle; the tightly bound and crystalline cellulose fibers are embedded in a matrix of hemicellulose and lignin, forming a highly heterogeneous and rigid structure that resists enzymatic and chemical breakdown (Joshi and Manjare 2024). This complexity necessitates energy-intensive and often chemically harsh pretreatment methods, such as steam explosion, dilute acid, ammonia fiber expansion (AFEX), and organosolv processes, to disrupt the lignin–carbohydrate complex (LCC) and improve the accessibility of polysaccharides (Joshi and Manjare 2024). However, these pretreatments can lead to the formation of fermentation inhibitors, such as furfural, hydroxymethylfurfural (HMF), acetic acid, and phenolic compounds, which hinder microbial metabolism and reduce the overall process efficiency (Alawad and Ibrahim 2022).

Another significant challenge lies in the variability of LCB composition, which is influenced by plant species, growth conditions, maturity, and postharvest storage (Chaudhary et al. 2023). Such variability affects process reproducibility and scalability, especially in large-scale biorefinery operations. Enzymatic hydrolysis, a key step in valorization, suffers from issues related to enzyme inhibition by residual lignin and the nonproductive binding of cellulases (Agrawal et al. 2021). The high cost of enzyme cocktails, compounded by the need for high dosages due to the inaccessibility of substrate, poses a barrier to economic feasibility (Mohanram et al. 2013). Moreover, lignin valorization itself remains underdeveloped; although lignin accounts for 15%–30% of the biomass by weight and nearly 40% of its energy content, it is largely underutilized and often burned for process heat despite its potential to be converted into high-value aromatic compounds, resins, and advanced materials (Sethupathy et al. 2022).

Furthermore, the downstream recovery and purification of target compounds such as ethanol, organic acids (e.g., levulinic acid and succinic acid), and platform chemicals (e.g., HMF and furfural) are technologically challenging (Singhvi and Gokhale 2019). These processes typically involve complex separation steps, including distillation, solvent extraction, membrane filtration, and adsorption, which add to the cost and energy burden of the overall process (Singhvi and Gokhale 2019). The microbial strains commonly used for fermentation (e.g., *Saccharomyces cerevisiae* and *Zymomonas mobilis*) cannot often metabolize pentose sugars (like xylose and arabinose) present in hemicellulose hydrolysates, further reducing the sugar utilization efficiency (Xia et al. 2019). Metabolic engineering of robust and inhibitor-tolerant microorganisms is progressing, but strain development and regulatory hurdles still impede commercial deployment.

Environmental considerations also play a role, as many pretreatment processes and catalysts involve toxic solvents or generate waste streams requiring additional treatment (Prasad et al. 2022). Achieving a closed-loop, environmentally sustainable, and economically viable LCB valorization process requires holistic approaches, including integrated biorefinery concepts, the development of green pretreatment technologies, adaptive biocatalysts, and process intensification strategies (Hassan et al. 2018). Overcoming these multifaceted challenges is critical for transitioning lignocellulosic biomass from a promising feedstock to a cornerstone of the global bioeconomy.

5.6 Conclusion

Valorization of lignocellulosic biomass is pivotal in the development of sustainable biorefineries and the realization of circular economy frameworks. By efficiently converting abundant and renewable lignocellulosic feedstocks, including agricultural residues, forestry by-products, industrial wastes, and food processing leftovers, into a diverse portfolio of value-added products, this approach addresses critical environmental and economic challenges. These feedstocks, rich in cellulose, hemicellulose, lignin, and other bioactive compounds, serve as raw materials for the production of biofuels such as bioethanol, biomethane, and biohydrogen, as well as biochemicals, including organic acids (e.g., lactic and citric acids), enzymes, bioplastics, antioxidants, and bio-based composites.

The transformation processes utilize advanced pretreatment methods, enzymatic hydrolysis, and fermentation technologies to overcome biomass recalcitrance and maximize the release of fermentable sugars and bioactive components. Innovations such as integrated biorefineries and coprocessing strategies enhance material and energy efficiency, improving economic feasibility and reducing waste generation. Furthermore, valorization pathways contribute significantly to waste management by diverting lignocellulosic residues from landfills and mitigating associated environmental pollution.

However, several technical and economic challenges remain, including the complexity of feedstock heterogeneity, inhibitory by-product formation during processing, enzyme cost and efficiency, and scale-up constraints. Addressing these issues through research and development in genetic engineering, process intensification, and catalyst development is essential to unlock the full potential of lignocellulosic biomass.

The broad spectrum of products derived from LCB supports diverse industrial sectors, ranging from energy and materials to pharmaceuticals and agriculture, fostering rural development and job creation. Such diversification strengthens the bioeconomy by minimizing dependence on fossil resources and enhancing the sustainable use of renewable biomass.

Ultimately, the valorization of LCB embodies a sustainable pathway that integrates environmental stewardship, economic growth, and social benefits. It is a key enabler for achieving global sustainability goals by advancing renewable energy production, waste reduction, and resource circularity, thereby contributing to a more sustainable and low-carbon future.

References

Abbas A, Naveed M, Azeem M et al (2020) Efficiency of wheat straw biochar in combination with compost and biogas slurry for enhancing nutritional status and productivity of soil and plant. Plants 9:1516. https://doi.org/10.3390/PLANTS9111516

Agarwal NK, Kumar M, Ghosh P et al (2022) Anaerobic digestion of sugarcane bagasse for biogas production and digestate valorization. Chemosphere 295:133893. https://doi.org/10.1016/J.CHEMOSPHERE.2022.133893

Aghaei S, Karimi Alavijeh M, Shafiei M, Karimi K (2022) A comprehensive review on bioethanol production from corn stover: worldwide potential, environmental importance, and perspectives. Biomass Bioenergy 161:106447. https://doi.org/10.1016/J.BIOMBIOE.2022.106447

Agnihotri S, Horváth IS (2024) Integrated products biorefinery options within the Swedish pulp and paper industry: current status. Sustain Chem Environ 7:100128. https://doi.org/10.1016/J.SCENV.2024.100128

Agrawal R, Verma A, Singhania RR et al (2021) Current understanding of the inhibition factors and their mechanism of action for the lignocellulosic biomass hydrolysis. Bioresour Technol 332:125042. https://doi.org/10.1016/J.BIORTECH.2021.125042

Ajala EO, Ighalo JO, Ajala MA et al (2021) Sugarcane bagasse: a biomass sufficiently applied for improving global energy, environment and economic sustainability. Bioresour Bioprocess 8(1):1–25. https://doi.org/10.1186/S40643-021-00440-Z

Alawad I, Ibrahim H (2022) Pretreatment of agricultural lignocellulosic biomass for fermentable sugar: opportunities, challenges, and future trends. Biomass Convers Biorefinery 14(5): 6155–6183. https://doi.org/10.1007/S13399-022-02981-5

Antolini E, Reviews C (2021) Lignocellulose, cellulose and lignin as renewable alternative fuels for direct biomass fuel cells. ChemSusChem 14:189–207. https://doi.org/10.1002/CSSC.202001807

Anuchi SO, Campbell KLS, Hallett JP (2022) Effective pretreatment of lignin-rich coconut wastes using a low-cost ionic liquid. Sci Rep 12(1):1–11. https://doi.org/10.1038/s41598-022-09629-4

Arismendy Pabón AM, Felissia FE, Mendieta CM et al (2020) Improvement of bioethanol production from rice husks. Cellul Chem Technol 54:689–698. https://doi.org/10.35812/CELLULOSECHEMTECHNOL.2020.54.68

Arteaga-Pérez LE, Segura C, Bustamante-García V et al (2015) Torrefaction of wood and bark from *Eucalyptus globulus* and *Eucalyptus nitens*: focus on volatile evolution vs feasible temperatures. Energy 93:1731–1741. https://doi.org/10.1016/J.ENERGY.2015.10.007

Asmadi M, Kawamoto H, Saka S (2017) Characteristics of softwood and hardwood pyrolysis in an ampoule reactor. J Anal Appl Pyrolysis 124:523–535. https://doi.org/10.1016/J.JAAP.2017.01.029

Balan V (2014) Current challenges in commercially producing biofuels from lignocellulosic biomass. ISRN Biotechnol 2014:463074. https://doi.org/10.1155/2014/463074

Banik SP (2008) Biomolecules (introduction, structure and functions)—carbohydrates. Indian Institute of Chemical Biology, Kolkata

Bellink M, Verburg RW (2023) A system lock-in blocks the uptake of mixed sustainable Eucalyptus plantations in Brazil. Land Use Policy 134:106882. https://doi.org/10.1016/J.LANDUSEPOL.2023.106882

Bhat MA, Bhat MA, Jan S et al (2024) Lignocellulosic biomass in circular economy: a techno-transition in carbon neutrality towards sustainable energy production. Biomass Bioenergy 189: 107349. https://doi.org/10.1016/J.BIOMBIOE.2024.107349

Cai C, Wang Z, Ma L et al (2024) Cotton stalk valorization towards bio-based materials, chemicals, and biofuels: a review. Renew Sust Energ Rev 202:114651. https://doi.org/10.1016/J.RSER.2024.114651

Cardona CA, Quintero JA, Paz IC (2010) Production of bioethanol from sugarcane bagasse: status and perspectives. Bioresour Technol 101:4754–4766. https://doi.org/10.1016/J.BIORTECH.2009.10.097

Casabar JT, Unpaprom Y, Ramaraj R (2019) Fermentation of pineapple fruit peel wastes for bioethanol production. Biomass Convers Biorefinery 9:761–765. https://doi.org/10.1007/S13399-019-00436-Y/TABLES/3

Chandel AK, Forte MBS, Gonçalves IS et al (2021) Brazilian biorefineries from second generation biomass: critical insights from industry and future perspectives. Biofuels Bioprod Biorefin 15: 1190–1208. https://doi.org/10.1002/BBB.2234;WGROUP:STRING:PUBLICATION

Chaudhary G, Chaudhary N, Saini S et al (2023) Assessment of pretreatment strategies for valorization of lignocellulosic biomass: path forwarding towards lignocellulosic biorefinery. Waste Biomass Valorization 15(1):1–36. https://doi.org/10.1007/S12649-023-02219-Z

Chen H, Sun F (2007) Novel bioconversion of wheat straw to bio-organic fertilizer in a solid-state bioreactor. Bioprocess Biosyst Eng 30:99–105. https://doi.org/10.1007/S00449-006-0105-9/FIGURES/3

Choi NW, Mori I, Ohama Y (2006) Development of rice husks–plastics composites for building materials. Waste Manag 26:189–194. https://doi.org/10.1016/J.WASMAN.2005.05.008

Choi IS, Lee YG, Khanal SK et al (2015) A low-energy, cost-effective approach to fruit and citrus peel waste processing for bioethanol production. Appl Energy 140:65–74. https://doi.org/10.1016/J.APENERGY.2014.11.070

Chowdhury R, Ghosh S, Manna D et al (2019) Hybridization of sugar-carboxylate-syngas platforms for the production of bio-alcohols from lignocellulosic biomass (LCB)—a state-of-the-art review and recommendations. Energy Convers Manag 200:112111. https://doi.org/10.1016/J.ENCONMAN.2019.112111

Cotana F, Cavalaglio G, Gelosia M et al (2014) Production of bioethanol in a second generation prototype from pine wood chips. Energy Procedia 45:42–51. https://doi.org/10.1016/J.EGYPRO.2014.01.006

da Rosa RG, Ferreira VC, Barroso TLCT et al (2025) Corn biomass as a feedstock for biorefinery: a bibliometric analysis of research on bioenergy, biofuel and value-added products. Biofuels Bioprod Biorefin 19:206–230. https://doi.org/10.1002/BBB.2693

Davis R, Bartling A (2023) Biochemical conversion of lignocellulosic biomass to hydrocarbon fuels and products: 2022 State of Technology and Future Research. National Renewable Energy Laboratory (NREL), Golden. https://doi.org/10.2172/1962809

Dedes G, Karnaouri A, Marianou AA et al (2021) Conversion of organosolv pretreated hardwood biomass into 5-hydroxymethylfurfural (HMF) by combining enzymatic hydrolysis and isomerization with homogeneous catalysis. Biotechnol Biofuels 14:1–11. https://doi.org/10.1186/S13068-021-02022-9/FIGURES/4

Deshmukh M, Pathan A (2024) Transformations of bamboo into bioethanol through biorefinery. Environ Sci Pollut Res Int 31:3343–3360. https://doi.org/10.1007/S11356-023-31510-8/TABLES/2

Di Blasi C, Branca C, Galgano A (2010) Biomass screening for the production of furfural via thermal decomposition. Ind Eng Chem Res 49:2658–2671. https://doi.org/10.1021/IE901731U

Doelker E, Gurny R, Schurz J et al (1987) Degrees of crystallinity and polymerization of modified cellulose powders for direct tableting. Powder Technol 52:207–213. https://doi.org/10.1016/0032-5910(87)80106-7

Dutta S, Gorain S, Roy J et al (2025) Bamboo for global sustainability: a systematic review of its environmental and ecological implications, climate action, and biodiversity contributions. Environ Rev 33:1–26. https://doi.org/10.1139/ER-2025-0032

Dwivedi D, Rathour RK, Sharma V et al (2024) Co-fermentation of forest pine needle waste biomass hydrolysate into bioethanol. Biomass Convers Biorefinery 14:8829–8841. https://doi.org/10.1007/S13399-022-02896-1/FIGURES/6

El-Sayed SA (2025) Chemical products yielded from different pyrolysis processes of rice waste residues: a comprehensive review. Biomass Convers Biorefinery 2025:1–41. https://doi.org/10.1007/S13399-025-06624-3

Emmanuel JK, Mtashobya LA, Mgeni ST (2025) Potential contributions of Banana fruits and residues to multiple applications: an overview. Nat Prod Commun 20:1934578X251320151. https://doi.org/10.1177/1934578X251320151

Fayoud N, Tahiri S, Alami Younssi S et al (2016) Kinetic, isotherm and thermodynamic studies of the adsorption of methylene blue dye onto agro-based cellulosic materials. Desalin Water Treat 57:16611–16625. https://doi.org/10.1080/19443994.2015.1079249

Fredman P, Haukeland JV, Tyrväinen L et al (2021) Nature-based tourism in a Nordic context. In: Nordic perspectives on nature-based tourism: from place-based resources to value-added experiences. Edward Elgar Publishing, pp 2–15. https://doi.org/10.4337/9781789904031.00008

Furszyfer Del Rio DD, Sovacool BK, Griffiths S et al (2022) Decarbonizing the pulp and paper industry: a critical and systematic review of sociotechnical developments and policy options. Renew Sust Energ Rev 167:112706. https://doi.org/10.1016/J.RSER.2022.112706

Ganguly P, Das P (2024) Integral approach for second-generation bio-ethanol production and wastewater treatment using peanut shell waste: yield, removal, and ANN studies. Biomass Convers Biorefinery 14:689–699. https://doi.org/10.1007/S13399-021-02277-0/TABLES/1

Gaskin JW, Speir RA, Harris K et al (2010) Effect of peanut hull and pine chip biochar on soil nutrients, corn nutrient status, and yield. Agron J 102:623–633. https://doi.org/10.2134/AGRONJ2009.0083

Good News from Finland (2025) Finnish researchers find constructive solution that sticks. https://www.goodnewsfinland.com/articles/breaking-news/2022/finnish-researchers-find-constructive-solution-that-sticks/. Accessed 8 June 2025

Guida MY, Lanaya SE, Laghchioua FE et al (2020) Production of bio-oil and bio-char from pyrolysis of sawdust wood waste (SWW). Prog Agric Eng Sci 16:61–80. https://doi.org/10.1556/446.2020.00012

Gupta G, Baranwal M, Saxena S, Reddy MS (2019) Utilization of banana stem juice as a feedstock material for bioethanol production. Clean (Weinh) 47:1900047. https://doi.org/10.1002/CLEN.201900047

Gupta PK, Basu S, Rana V et al (2024) Utilization of non-concentrated banana pseudostem sap waste for converting to bioethanol: in vitro and in silico evidence. Waste Manag Bull 2:109–119. https://doi.org/10.1016/J.WMB.2024.07.002

Han M, Moon SK, Kim Y et al (2009) Bioethanol production from ammonia percolated wheat straw. Biotechnol Bioprocess Eng 14:606–611. https://doi.org/10.1007/S12257-008-0320-0/METRICS

Haryanto A, Hasanudin U, Afrian C, Zulkarnaen I (2018) Biogas production from anaerobic codigestion of cowdung and elephant grass (*Pennisetum purpureum*) using batch digester. IOP Conf Ser Earth Environ Sci 141:012011. https://doi.org/10.1088/1755-1315/141/1/012011

Hassan SS, Williams GA, Jaiswal AK (2018) Emerging technologies for the pretreatment of lignocellulosic biomass. Bioresour Technol 262:310–318. https://doi.org/10.1016/J.BIORTECH.2018.04.099

Hauser A (2018) Peanuts. J Agric Food Inf 19:195–202. https://doi.org/10.1080/10496505.2018.1483677

Hay JXW, Wu TY, Ng BJ et al (2016) Reusing pulp and paper mill effluent as a bioresource to produce biohydrogen through ultrasonicated *Rhodobacter sphaeroides*. Energy Convers Manag 113:273–280. https://doi.org/10.1016/J.ENCONMAN.2015.12.041

He T, Jiang Z, Wu P et al (2016) Fractionation for further conversion: from raw corn stover to lactic acid. Sci Rep 6:1–11. https://doi.org/10.1038/SREP38623

Houfani AA, Anders N, Spiess AC et al (2020) Insights from enzymatic degradation of cellulose and hemicellulose to fermentable sugars—a review. Biomass Bioenergy 134:105481. https://doi.org/10.1016/J.BIOMBIOE.2020.105481

Ingale S, Joshi SJ, Gupte A (2014) Production of bioethanol using agricultural waste: banana pseudo stem. Braz J Microbiol 45:885–892. https://doi.org/10.1590/S1517-83822014000300018

Ishida S, Kudo S, Asano S, Hayashi JI (2025) Multi-step pre-treatment of rice husk for fractionation of components including silica. Front Chem 13:1538797. https://doi.org/10.3389/FCHEM.2025.1538797/BIBTEX

Iyyappan J, Pravin R, Al-Ghanim KA et al (2023) Dual strategy for bioconversion of elephant grass biomass into fermentable sugars using *Trichoderma reesei* towards bioethanol production. Bioresour Technol 374:128804. https://doi.org/10.1016/J.BIORTECH.2023.128804

Ji K, Pei J, Huang W (2025) Quantitative assessment of carbon reduction potential in regional agricultural systems based on bioenergy: a case study of Sichuan Province, China. Bioenergy Res 18:1–17. https://doi.org/10.1007/S12155-025-10833-Z/FIGURES/3

Johannes LP, Minh TTN, Xuan TD (2024) Elephant grass (*Pennisetum purpureum*): a bioenergy resource overview. Biomass 4:625–646. https://doi.org/10.3390/BIOMASS4030034

Joshi M, Manjare S (2024) Chemical approaches for the biomass valorisation: a comprehensive review of pretreatment strategies. Environ Sci Pollut Res 31(36):48928–48954. https://doi.org/10.1007/S11356-024-34473-6

Kalita E, Nath BK, Deb P et al (2015) High quality fluorescent cellulose nanofibers from endemic rice husk: isolation and characterization. Carbohydr Polym 122:308–313. https://doi.org/10.1016/J.CARBPOL.2014.12.075

Kaparaju P, Serrano M, Thomsen AB et al (2009) Bioethanol, biohydrogen and biogas production from wheat straw in a biorefinery concept. Bioresour Technol 100:2562–2568. https://doi.org/10.1016/J.BIORTECH.2008.11.011

Kłosowski G, Mikulski D, Menka A (2019) Microwave-assisted one-step conversion of wood wastes into levulinic acid. Catalysts 9:753. https://doi.org/10.3390/CATAL9090753

Kong X, Xu H, Wu H et al (2016) Biobutanol production from sugarcane bagasse hydrolysate generated with the assistance of gamma-valerolactone. Process Biochem 51:1538–1543. https://doi.org/10.1016/J.PROCBIO.2016.06.013

Kuttiraja M, Sindhu R, Varghese PE et al (2013) Bioethanol production from bamboo (Dendrocalamus sp.) process waste. Biomass Bioenergy 59:142–150. https://doi.org/10.1016/J.BIOMBIOE.2013.10.015

Lappalainen K, Dong Y (2019) Simultaneous production of furfural and levulinic acid from pine sawdust via acid-catalysed mechanical depolymerization and microwave irradiation. Biomass Bioenergy 123:159–165. https://doi.org/10.1016/J.BIOMBIOE.2019.02.017

Lau MW, Dale BE (2009) Cellulosic ethanol production from AFEX-treated corn stover using *Saccharomyces cerevisiae* 424A(LNH-ST). Proc Natl Acad Sci USA 106:1368–1373. https://doi.org/10.1073/PNAS.0812364106/SUPPL_FILE/0812364106SI.PDF

Lee JM, Shi J, Venditti RA, Jameel H (2009) Autohydrolysis pretreatment of Coastal Bermuda grass for increased enzyme hydrolysis. Bioresour Technol 100:6434–6441. https://doi.org/10.1016/J.BIORTECH.2008.12.068

Li X, Sun C, Zhou B, He Y (2015) Determination of hemicellulose, cellulose and lignin in moso bamboo by near infrared spectroscopy. Sci Rep 5(1):1–11. https://doi.org/10.1038/srep17210

Li C, Liu G, Nges IA et al (2016) Fresh banana pseudo-stems as a tropical lignocellulosic feedstock for methane production. Energy Sustain Soc 6:1–9. https://doi.org/10.1186/S13705-016-0093-9/FIGURES/4

Li R, Zhu X, Peng F, Lu F (2023) Biodegradable, colorless, and odorless PLA/PBAT bioplastics incorporated with corn Stover. ACS Sustain Chem Eng 11:8870–8883. https://doi.org/10.1021/ACSSUSCHEMENG.3C00691/ASSET/IMAGES/LARGE/SC3C00691_0010.JPEG

Li D, Wei J, Wu J et al (2024a) The expansion of Moso bamboo (*Phyllostachys edulis*) forests into diverse types of forests in China from 2010 to 2020. Forests 15:1418. https://doi.org/10.3390/F15081418

Li J, Wang Y, Xu C et al (2024b) Bioplastic derived from corn Stover: life cycle assessment and artificial intelligence-based analysis of uncertainty and variability. Sci Total Environ 946: 174349. https://doi.org/10.1016/J.SCITOTENV.2024.174349

Liang J, Sun Y, Yang Y et al (2025) Agricultural biotechnology in China: product development, commercialization, and perspectives. aBIOTECH 6(2):1–27. https://doi.org/10.1007/S42994-025-00209-4

Limayem A, Ricke SC (2012) Lignocellulosic biomass for bioethanol production: current perspectives, potential issues and future prospects. Prog Energy Combust Sci 38:449–467. https://doi.org/10.1016/J.PECS.2012.03.002

Liu C, Wang H, Karim AM et al (2014) Catalytic fast pyrolysis of lignocellulosic biomass. Chem Soc Rev 43:7594–7623. https://doi.org/10.1039/C3CS60414D

Lopes ML, Paulillo SCDL, Godoy A et al (2016) Ethanol production in Brazil: a bridge between science and industry. Braz J Microbiol 47:64–76. https://doi.org/10.1016/J.BJM.2016.10.003

Lucarini M, Durazzo A, Bernini R et al (2021) Fruit wastes as a valuable source of value-added compounds: a collaborative perspective. Molecules 26:6338. https://doi.org/10.3390/MOLECULES26216338

Malik VS (2023) The bioethanol production for blending into petrol. J Plant Biochem Biotechnol 32:1–7. https://doi.org/10.1007/S13562-023-00826-4/METRICS

Mancini G, Papirio S, Lens PNL, Esposito G (2018) Increased biogas production from wheat straw by chemical pretreatments. Renew Energy 119:608–614. https://doi.org/10.1016/J.RENENE.2017.12.045

Mendes CVT, Rocha JMS, Carvalho MGVS (2014) Valorization of residual streams from pulp and paper mills: Pretreatment and bioconversion of primary sludge to bioethanol. Ind Eng Chem Res 53:19398–19404. https://doi.org/10.1021/IE503021Y/ASSET/IMAGES/LARGE/IE-2014-03021Y_0003.JPEG

Mikulski D, Kłosowski G (2023) High-pressure microwave-assisted pretreatment of softwood, hardwood and non-wood biomass using different solvents in the production of cellulosic ethanol. Biotechnol Biofuels Bioprod 16:1–19. https://doi.org/10.1186/S13068-023-02272-9/TABLES/1

Mogany T, Bhola V, Bux F (2024) Algal-based bioplastics: global trends in applied research, technologies, and commercialization. Environ Sci Pollut Res 31:38022–38044. https://doi.org/10.1007/S11356-024-33644-9/METRICS

Mohanram S, Amat D, Choudhary J et al (2013) Novel perspectives for evolving enzyme cocktails for lignocellulose hydrolysis in biorefineries. Sustainable Chem Processes 1:1–12. https://doi.org/10.1186/2043-7129-1-15

Monção M, Hrůzová K, Rova U et al (2021) Organosolv fractionation of birch sawdust: establishing a lignin-first biorefinery. Molecules 26:6754. https://doi.org/10.3390/MOLECULES26216754

Mujtaba M, Fernandes Fraceto L, Fazeli M et al (2023) Lignocellulosic biomass from agricultural waste to the circular economy: a review with focus on biofuels, biocomposites and bioplastics. J Clean Prod 402:136815. https://doi.org/10.1016/J.JCLEPRO.2023.136815

Nasir A, Adrus N, Bohari SPM (2022) Rice husk as potential resource for biomedical applications: a review. Cellul Chem Technol 56:911–928. https://doi.org/10.35812/CELLULOSECHEMTECHNOL.2022.56.81

Negawo AT, Teshome A, Kumar A et al (2017) Opportunities for Napier grass (*Pennisetum purpureum*) improvement using molecular genetics. Agronomy 7:28. https://doi.org/10.3390/AGRONOMY7020028

Nisar J, Ahmad A, Ali G et al (2022) Enhanced bio-oil yield from thermal decomposition of peanut shells using termite hill as the catalyst. Energies 15:1891. https://doi.org/10.3390/EN15051891

Norrahim MNF, Norizan MN, Knight VF et al (2025) Bamboo-based nanocomposites: preparation, processes and its emerging applications. In: Bamboo-based polymer composites. Woodhead Publishing, pp 205–226. https://doi.org/10.1016/B978-0-443-33445-0.00008-0

Ntimbani RN, Farzad S, Görgens JF (2022) Furfural production from sugarcane bagasse along with co-production of ethanol from furfural residues. Biomass Convers Biorefinery 12:5257–5267. https://doi.org/10.1007/S13399-021-01313-3/TABLES/2

Ogwu MC, Kosoe EA, Osawaru ME (2025) Raw materials for sustainable bioeconomy development in the global south. In: Sustainable bioeconomy development in the global south. Springer Nature Singapore, pp 37–63. https://doi.org/10.1007/978-981-96-0640-5_2

OSDP (2025) A market study of the Southeast United States—Open Science and Data Platform. https://osdp-psdo.canada.ca/dp/en/search/metadata/NRCAN-CFS-1-4182. Accessed 7 June 2025

Pączkowski P, Puszka A, Gawdzik B (2021) Effect of eco-friendly Peanut Shell powder on the chemical resistance, physical, thermal, and thermomechanical properties of unsaturated polyester resin composites. Polymers 13:3690. https://doi.org/10.3390/POLYM13213690

Patil RA (2016) Production of acetic acid in Kraft pulp mill biorefinery using bi-polar membrane electrodialysis. The University of Maine

Patil S, Konde K, Behera S (2022) Bio-circular economy: an opportunity for diversification for sugar industries in compressed biogas (CBG) and organic fertilizer production. Sugar Tech 24: 1079–1092. https://doi.org/10.1007/S12355-022-01130-6/FIGURES/9

Pivsa-Art S, Pivsa-Art W (2021) Eco-friendly bamboo fiber-reinforced poly(butylene succinate) biocomposites. Polym Compos 42:1752–1759. https://doi.org/10.1002/PC.25930;JOURNAL:JOURNAL:15480569;PAGE:STRING:ARTICLE/CHAPTER

Prasad SS, Das U (2024) Banana waste as a nutraceuticals product. In: Nutraceuticals from fruit and vegetable waste. Wiley, pp 175–194. https://doi.org/10.1002/9781119803980.CH6;CTYPE:STRING:BOOK

Prasad BR, Padhi RK, Ghosh G (2022) A review on key pretreatment approaches for lignocellulosic biomass to produce biofuel and value-added products. Int J Environ Sci Technol 20(6): 6929–6944. https://doi.org/10.1007/S13762-022-04252-2

Priadi C, Wulandari D, Rahmatika I, Moersidik SS (2014) Biogas production in the anaerobic digestion of paper sludge. APCBEE Procedia 9:65–69. https://doi.org/10.1016/J.APCBEE.2014.01.012

Qin Y, Yu L, Wu R et al (2016) Biorefinery lignosulfonates from sulfite-pretreated softwoods as dispersant for graphite. ACS Sustain Chem Eng 4:2200–2205. https://doi.org/10.1021/ACSSUSCHEMENG.5B01664/ASSET/IMAGES/LARGE/SC-2015-016648_0001.JPEG

Rajesh Banu J, Preethi KS et al (2021) Lignocellulosic biomass based biorefinery: a successful platform towards circular bioeconomy. Fuel 302:121086. https://doi.org/10.1016/J.FUEL.2021.121086

Rao RS, Jyothi CP, Prakasham RS et al (2006) Xylitol production from corn fiber and sugarcane bagasse hydrolysates by *Candida tropicalis*. Bioresour Technol 97:1974–1978. https://doi.org/10.1016/J.BIORTECH.2005.08.015

Ren S, Lei H, Zhang Y et al (2019) Furfural production from microwave catalytic torrefaction of Douglas fir sawdust. J Anal Appl Pyrolysis 138:188–195. https://doi.org/10.1016/J.JAAP.2018.12.023

Ricci A, Diaz AB, Caro I et al (2019) Orange peels: from by-product to resource through lactic acid fermentation. J Sci Food Agric 99:6761–6767. https://doi.org/10.1002/JSFA.9958

Risberg K, Sun L, Levén L et al (2013) Biogas production from wheat straw and manure—impact of pretreatment and process operating parameters. Bioresour Technol 149:232–237. https://doi.org/10.1016/J.BIORTECH.2013.09.054

Rodrigues RCLB, Kenealy WR, Jeffries TW (2011) Xylitol production from DEO hydrolysate of corn stover by *Pichia stipitis* YS-30. J Ind Microbiol Biotechnol 38:1649–1655. https://doi.org/10.1007/S10295-011-0953-4

Russo C, Maugeri A, Lombardo GE et al (2021) The second life of citrus fruit waste: a valuable source of bioactive compounds. Molecules 26:5991. https://doi.org/10.3390/MOLECULES26195991

Sangamithirai K, Vasugi N (2020) Banana fibre—a potential source of sustainable textiles. J Appl Hortic 22:133–136. https://doi.org/10.37855/JAH.2020.V22I02.24

Sannigrahi P, Miller SJ, Ragauskas AJ (2010) Effects of organosolv pretreatment and enzymatic hydrolysis on cellulose structure and crystallinity in Loblolly pine. Carbohydr Res 345:965–970. https://doi.org/10.1016/J.CARRES.2010.02.010

Saražin J, Schmiedl D, Pizzi A, Šernek M (2020) Bio-based adhesive mixtures of pine tannin and different types of lignins. Bioresources 15:9401–9412. https://doi.org/10.15376/BIORES.15.4.9401-9412

Scholl AL, Menegol D, Pitarelo AP et al (2015) Elephant grass pretreated by steam explosion for inducing secretion of cellulases and xylanases by *Penicillium echinulatum* S1M29 solid-state cultivation. Ind Crop Prod 77:97–107. https://doi.org/10.1016/J.INDCROP.2015.08.051

Sethupathy S, Murillo Morales G, Gao L et al (2022) Lignin valorization: status, challenges and opportunities. Bioresour Technol 347:126696. https://doi.org/10.1016/J.BIORTECH.2022.126696

Shah TA, Zhihe L, Zhiyu L et al (2022) Composition and role of lignin in biochemicals. In: Lignin—chemistry, structure, and application. IntechOpen. https://doi.org/10.5772/INTECHOPEN.106527

Sharma N, Nainwal S, Jain S, Jain S (2015) Emerging biorefinery technologies for Indian forest industry to reduce GHG emissions. Ecotoxicol Environ Saf 121:105–109. https://doi.org/10.1016/J.ECOENV.2015.04.050

Singhvi MS, Gokhale DV (2019) Lignocellulosic biomass: hurdles and challenges in its valorization. Appl Microbiol Biotechnol 103:9305–9320. https://doi.org/10.1007/S00253-019-10212-7/FIGURES/3

Sivagurunathan P, Raj T, Mohanta CS et al (2021) 2G waste lignin to fuel and high value-added chemicals: approaches, challenges and future outlook for sustainable development. Chemosphere 268:129326. https://doi.org/10.1016/J.CHEMOSPHERE.2020.129326

Soccol CR, Faraco V, Karp SG et al (2019) Lignocellulosic bioethanol: current status and future perspectives. In: Biomass, biofuels, biochemicals: biofuels: alternative feedstocks and conversion processes for the production of liquid and gaseous biofuels. Elsevier, pp 331–354. https://doi.org/10.1016/B978-0-12-816856-1.00014-2

Spiridon I, Popa VI (2008) Hemicelluloses: major sources, properties and applications. In: Monomers, polymers and composites from renewable resources. Elsevier, pp 289–304. https://doi.org/10.1016/B978-0-08-045316-3.00013-2

Ssemugenze B, Ocwa A, Kuunya R et al (2025) Enhancing maize production through timely nutrient supply: the role of foliar fertiliser application. Agronomy 15:176. https://doi.org/10.3390/AGRONOMY15010176

Su H, Liu G, He M, Tan F (2015) A biorefining process: sequential, combinational lignocellulose pretreatment procedure for improving biobutanol production from sugarcane bagasse. Bioresour Technol 187:149–160. https://doi.org/10.1016/J.BIORTECH.2015.03.107

Taghizadeh-Alisaraei A, Tatari A, Khanali M, Keshavarzi M (2023) Potential of biofuels production from wheat straw biomass, current achievements and perspectives: a review. Biofuels 14: 79–92. https://doi.org/10.1080/17597269.2022.2118779;REQUESTEDJOURNAL:JOURNAL:TBFU20;JOURNAL:JOURNAL:TBFU20;WGROUP:STRING:PUBLICATION

Taherzadeh MJ, Eklund R, Gustafsson L et al (1997) Characterization and fermentation of dilute-acid hydrolyzates from wood. Ind Eng Chem Res 36:4659–4665. https://doi.org/10.1021/IE9700831

Talebnia F, Karakashev D, Angelidaki I (2010) Production of bioethanol from wheat straw: an overview on pretreatment, hydrolysis and fermentation. Bioresour Technol 101:4744–4753. https://doi.org/10.1016/J.BIORTECH.2009.11.080

Tanklevska N, Petrenko V, Karnaushenko A, Melnykova K (2020) World corn market: analysis, trends and prospects of its deep processing. Agric Resour Econ 6:96–111. https://doi.org/10.22004/AG.ECON.305555

Tarasov D, Leitch M, Fatehi P (2018) Lignin–carbohydrate complexes: properties, applications, analyses, and methods of extraction: a review. Biotechnol Biofuels 11(1):1–28. https://doi.org/10.1186/S13068-018-1262-1

Thatoi HN, Chattaraj S, Mishra RR et al (2025) Contributions of biotechnology industries of India to global bioeconomy: an overview. 3 Biotech 15:1–21. https://doi.org/10.1007/S13205-025-04215-7/FIGURES/8

The Better India (2025) Taruwar Agro: zero-waste startup turns banana fibre into eco-friendly products. https://thebetterindia.com/367817/bihar-college-friends-turn-banana-tree-waste-into-useful-products-with-startup-taruwar-agro/. Accessed 8 June 2025

The Hindu (2025) NRCB announces initiative to utilise banana waste—The Hindu. https://www.thehindu.com/news/cities/Tiruchirapalli/nrcb-announces-initiative-to-utilise-banana-waste/article33812393.ece. Accessed 8 June 2025

Tian Q, Xu P, Huang D et al (2023) The driving force of biomass value-addition: selective catalytic depolymerization of lignin to high-value chemicals. J Environ Chem Eng 11:109719. https://doi.org/10.1016/J.JECE.2023.109719

Tufail T, Saeed F, Afzaal M et al (2021) Wheat straw: a natural remedy against different maladies. Food Sci Nutr 9:2335–2344. https://doi.org/10.1002/FSN3.2030

Tulashie SK, Akpari EEA, Appiah G et al (2023) Acid hydrolysis of sawdust waste into bioethanol. Biomass Convers Biorefinery 13:5743–5756. https://doi.org/10.1007/S13399-021-01725-1/FIGURES/8

Ungureanu N, Vlăduț V, Biriș SȘ (2022) Sustainable valorization of waste and by-products from sugarcane processing. Sustainability 14:11089. https://doi.org/10.3390/SU141711089

Uzochukwu JS, Onu CE, Ezechukwu C et al (2025) Multi-product biorefineries for biofuels and value-added products: advances and future perspectives. SSRN Electron J. https://doi.org/10.2139/SSRN.5202547

Velghe F, De Wilde F, Snellinx S et al (2021) Volatile fatty acid platform—a cornerstone for the circular bioeconomy. FEMS Microbiol Lett 368:56. https://doi.org/10.1093/FEMSLE/FNAB056

Verma YK, Singh AK, Paswan MK, Gurmaita PK (2024) Preparation and characterization of bamboo based nanocellulose by ball milling and used as a filler for preparation of nanocomposite. Polymer 308:127396. https://doi.org/10.1016/J.POLYMER.2024.127396

Victor A, Pulidindi IN, Gedanken A (2014) Levulinic acid production from *Cicer arietinum*, cotton, *Pinus radiata* and sugarcane bagasse. RSC Adv 4:44706–44711. https://doi.org/10.1039/C4RA06246A

Villota-Enríquez MD, Rodríguez-Páez JE (2023) Bio-silica production from rice husk for environmental remediation: removal of methylene blue from aqueous solutions. Mater Chem Phys 301:127671. https://doi.org/10.1016/J.MATCHEMPHYS.2023.127671

Wan C, Li Y (2010) Microbial pretreatment of corn stover with *Ceriporiopsis subvermispora* for enzymatic hydrolysis and ethanol production. Bioresour Technol 101:6398–6403. https://doi.org/10.1016/J.BIORTECH.2010.03.070

Wan C, Zhou Y, Li Y (2011) Liquid hot water and alkaline pretreatment of soybean straw for improving cellulose digestibility. Bioresour Technol 102:6254–6259. https://doi.org/10.1016/J.BIORTECH.2011.02.075

Wang K, Hu X, Zhang Q et al (2024a) Effect of temperature on sawdust and sawdust pellets during dry and wet torrefaction. Green Mater 13:428–439. https://doi.org/10.1680/JGRMA.24.00045/ASSET/IMAGES/SMALL/JGRMA.24.00045-F6.GIF

Wang X, Liu Q, Yang L et al (2024b) Perceptions of biomass energy sustainability in policy scenarios of China. Heliyon 10:e37180. https://doi.org/10.1016/j.heliyon.2024.e37180

Xia J, Yang Y, Liu CG et al (2019) Engineering *Zymomonas mobilis* for robust cellulosic ethanol production. Trends Biotechnol 37:960–972. https://doi.org/10.1016/J.TIBTECH.2019.02.002/ASSET/818169D9-67D0-455D-BDEC-F1878BF377D6/MAIN.ASSETS/GR4.JPG

Xue YP, Jin M, Orjuela A et al (2015) Microbial lipid production from AFEX™ pretreated corn Stover. RSC Adv 5:28725–28734. https://doi.org/10.1039/C5RA01134E

Yong TK, Musa MNS, Abdulla R et al (2024) Synthesis and characterization of bioplastics, polyhydroxyalkanoates produced from sugarcane bagasse by using *Bacillus cereus*. Biocatal Agric Biotechnol 58:103181. https://doi.org/10.1016/J.BCAB.2024.103181

Yu H, Xu Y, Hou J et al (2020) Fractionation of corn stover for efficient enzymatic hydrolysis and producing platform chemical using p-toluenesulfonic acid/water pretreatment. Ind Crop Prod 145:111961. https://doi.org/10.1016/J.INDCROP.2019.111961

Zhao B, Yu Y, Xu Y, Guo H (2024) Bamboo as a sustainable construction material for residential buildings in the cold and severe cold regions of China. Archit Eng Des Manag 20:1511–1528. https://doi.org/10.1080/17452007.2023.2195614

Chapter 6
Algal Biomass

Juliana Corrêa Cardoso, Carlos Ricardo Soccol, Hissashi Iwamoto, Denisse Tatiana Molina-Aulestia, Andrele Lauth Oeckslér, Luca Gavio, and Júlio Cesar de Carvalho

Abstract The biotechnological potential of microalgae and macroalgae across various industrial sectors is widely recognized and extensively studied. As photosynthetic organisms capable of producing high concentrations of O_2, algae offer significant advantages, such as lower production costs and reduced space requirements for cultivation, contributing to the overall cost-effectiveness of bioproducts. The bioactive compounds derived from algae have diverse applications worldwide, including nutraceuticals, human and animal nutrition, pigments, biofuels, and other industries. The growing demand for sustainable production alternatives with minimal environmental impact and reliance on clean energy has intensified research efforts into algae-based products. As a result, the use of algal biomass has been widely explored to fully leverage its properties, reduce production costs, and develop high-quality nutritional products, such as food supplements for humans and animals. Additionally, algae-derived products with high added value, such as pigments, oils, and thickeners, along with coproducts like bio-oil, biodiesel, bioethanol, and biofertilizers, have gained significant attention. The market for algae-based products is valued between US$ 12 and 30 billion and is expected to grow even further in the coming years. In this context, optimizing cultivation conditions and minimizing process costs are key research areas aimed at enhancing the production of high-quality functional bioproducts, particularly in the biofuels industry.

6.1 Introduction

Microalgae and macroalgae comprise a diverse group of photosynthetic organisms found in various aquatic environments. Algae phyla can be classified into the kingdoms Eubacteria, Plantae, Chromista, or Protozoa (Fig. 6.1). Microalgae are

J. C. Cardoso (✉) · C. R. Soccol · H. Iwamoto · D. T. Molina-Aulestia · A. L. Oecksler
L. Gavio · J. C. de Carvalho (✉)
Bioprocess Engineering and Biotechnology Department, Federal University of Paraná, Curitiba, Paraná, Brazil
e-mail: julianacardoso@ufpr.br; jccarvalho@ufpr.br

A. L. Woiciechowski et al. (eds.), *Biomass for Bioenergy and Bioproducts*, Biofuel and Biorefinery Technologies 17, https://doi.org/10.1007/978-3-032-19898-3_6

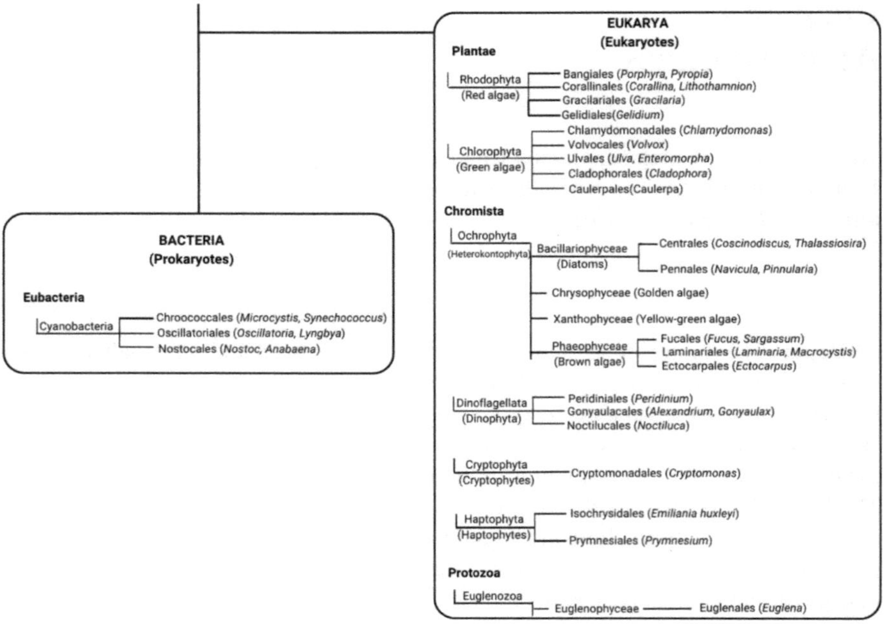

Fig. 6.1 Phylogenetic classification of algae into domains, kingdoms, and phyla, with some groups further divided into classes and then orders

unicellular, microscopic organisms found in fresh or brackish waters and are the main component of marine phytoplankton. They can be eukaryotic and are classified into the phyla Chlorophyta, Rhodophyta, Cryptophyta, Dinophyta, Haptophyta, Ochrophyta, or Euglenozoa. They also include cyanobacteria, which are prokaryotic but photosynthetic and are therefore often referred to as microalgae (Guiry 2024). Macroalgae, on the other hand, are multicellular, eukaryotic organisms that can reach several meters in length. Also known as seaweeds, they are found in various aquatic environments, including deep-sea habitats. Macroalgae can be classified into Chlorophyta (green algae), Rhodophyta (red algae), and Ochrophyta (Phaeophyceae—brown algae) (Babich et al. 2024; Guiry 2024; Lu et al. 2025).

Algae are essential organisms in the aquatic ecosystem. They are part of the marine benthos and plankton and play a crucial role in primary production within these environments, serving as sources of high concentrations of proteins, lipids, and carbohydrates. They also have a high potential for CO_2 biofixation, converting it into O_2 and energy-rich compounds through photosynthesis while producing high-value bioproducts (Cesário et al. 2018; de Morais et al. 2019). In addition, certain microalgae play a significant role in the food chain by providing carotenoids, which contribute to the pink coloration of flamingos that consume them. The bioaccumulation effect is crucial for the bioremediation of polluted environments, particularly in the removal of nitrogen, phosphorus, and heavy metals, as well as in the biosorption of contaminants such as phenols and organic pollutants. Thus, they can also serve as environmental bioindicators (Thoré et al. 2023; Babich et al. 2024; Singh Mehra et al. 2025).

Under eutrophic conditions, microalgae can proliferate uncontrollably, causing severe harm to marine life. Additionally, some cyanobacteria can produce cyanotoxins that affect animals that consume them, while invasive macroalgae can cause harm by changing whole ecosystems. Overall, both micro- and macroalgae are highly beneficial to the marine ecosystem and present significant biotechnological potential for various industries. In many cases, they also have lower production costs compared to the production of bioactive compounds from land plants because they do not need irrigation, farmland, and, in some cases, fertilizers (Arias et al. 2023; Thoré et al. 2023; Amponsah et al. 2024).

6.1.1 A Bit of History and the Current Production

There is historical evidence that seaweeds have been consumed since the Neolithic period by coastal dwellers, who collected and incorporated these organisms into their diets (Buschmann and Camus 2019). The use of *Spirulina* by the Aztecs in Lake Texcoco, Mexico, and by the Kanembu in Lake Chad, Africa, has been documented. They consumed it as a blue-green cake collected from the lakes, and this form of consumption likely took place between the fourteenth and sixteenth centuries. Research on microalgae began in 1890 with the isolation of *Chlorella* by Dutch researchers. From the 1940s to the 1960s, studies on microalgae were conducted, primarily focusing on their potential as a source of lipids and carotenoids. In the 1970s, research into microalgae as a protein source began, leading to the commercial production of *Chlorella* and *Spirulina* in Japan, Taiwan, and Mexico. Subsequently, in the 1980s and 1990s, β-carotene extracted from *Dunaliella salina* and astaxanthin from *Haematococcus pluvialis* began to be produced and marketed, respectively (Borowitzka 2013; Thoré et al. 2023).

In 1989, Mexico produced 4375 tons of wild *Spirulina*. By 2003, *Spirulina* production in China reached 16,483 tons, increasing to 56,208 tons in 2019. *Spirulina* accounted for 99.6% of the total microalgae cultivated in China in 2019. Chile was the second-largest producer in 2019, with a total production of 903 tons, of which *Spirulina* accounted for 861 tons. *Haematococcus pluvialis* was the second most cultivated microalga in 2019 in both China and Chile, reaching 200 and 42 tons, respectively (Cai et al. 2021).

Macroalgae were first consumed as human food in the fourth and sixth centuries in Japan and China, and to this day, these countries, along with the Republic of Korea, remain the largest consumers of these organisms. The gelling properties of red algae hydrocolloids, extracted with hot water, were already being explored in Japan as early as 1658. In the nineteenth century, carrageenan extraction from the red alga Irish moss became widespread. From 1930 onward, the thickener alginate, extracted from brown algae, began to be commercially marketed. The production and sale of seaweed extracts as thickeners and gelatin substitutes were consolidated after World War II, driven by an increased demand for agar, alginate, and carrageenan (McHugh 2003; Buschmann and Camus 2019).

Global commercial production of macroalgae is significantly higher than that of microalgae. Currently, China produces approximately 5 million tons of wet edible seaweed biomass per year, with the largest production coming from the brown alga *Laminaria japonica*. In the Republic of Korea, more than 50% of seaweed production consists of the brown alga *Undaria pinnatifida*, with an annual production of 800,000 tons of wet macroalgae. In Japan, macroalgae production currently stands at 600,000 tons of wet algae, 75% of which is dedicated to *Porphyra* cultivation (McHugh 2003; Buschmann and Camus 2019).

6.1.2 Algae Culture

Micro- and macroalgae can be cultivated under autotrophic or mixotrophic conditions, while certain microalgae can also be grown under heterotrophic conditions. Cultivation can occur either in controlled environments with regulated temperature and lighting or under natural conditions. Culture media containing essential nutrients such as carbon, nitrogen, and phosphorus can be used to promote the accumulation of high concentrations of proteins, carbohydrates, and lipids in the biomass. The use of wastewater-based culture media has been extensively studied, as it facilitates the reuse of municipal and industrial effluents for the production of various bioproducts, contributing to the development of a circular economy (Singh Mehra et al. 2025; Martínez-Herrera et al. 2025).

For microalgae cultivation, large-scale production commonly employs raceways or photobioreactors, and they can be produced in either open or closed systems depending on the type of product. In both cases, agitation and forced aeration systems may enhance the CO_2 supply and maintain system homogeneity. In macroalgae farming, artificial aquaculture ponds, built on land, can be used, where controlled artificial lighting and water circulation via pumps can be implemented. However, most commonly, the cultivation occurs in semi-confined areas within the marine environment, in longline or raft cultivation methods. Factors such as culture contamination and water evaporation in open systems must be assessed and monitored, as they directly influence the biomass quality. Adjustments in cultivation parameters, including light intensity, temperature, pH, agitation, and nutrient availability, can either enhance or hinder the production of target bioactive compounds (Thoré et al. 2023; Amponsah et al. 2024; Narayanan 2024; Martínez-Herrera et al. 2025).

6.1.3 Algae Biomass Composition and Products

Like most organisms, algae contain macrocomponents that serve structural or storage functions, as well as smaller but essential components that can be valuable as products. Macrocomponents include carbohydrates, such as structural or cell wall polysaccharides, like cellulose or alginate, and storage polysaccharides, like starch or laminarin. Proteins and lipids with various functions are also present. Minor components include nucleic acids and photosynthetic or photoprotective pigments. It is important to note that algae generally lack lignin, a natural polymer responsible for the rigidity of terrestrial plants. This absence somewhat facilitates cell rupture and access to the bioactive compounds within the cell (Martone et al. 2009; Cesário et al. 2018). Regarding the nutritional composition of algae, variations in lipid, protein, and carbohydrate concentrations are directly influenced by the organisms' growth conditions, primarily by factors such as light intensity, temperature, salinity, and the concentrations of carbon, nitrogen, and phosphorus in the medium, as well as cultivation duration and harvest timing. Additionally, some species have a greater tendency to produce and store higher amounts of these nutrients (Cesário et al. 2018; Øverland et al. 2019).

The protein content of microalgae is typically high, ranging from 50% to 75% of dry mass in certain species. It is one of the main components of microalgae and serves as a crucial source of protein and essential amino acids for the human and animal food industries. Macroalgae, in contrast, have lower protein concentrations than microalgae, generally ranging from 7% to 50% of dry mass (Chew et al. 2017; Cesário et al. 2018; de Carvalho et al. 2020; Arias et al. 2023).

The carbohydrate content of microalgae can vary between 6% and 57% of dry biomass, depending on the species and cultivation conditions. Microalgae can produce mono-, oligo-, and polysaccharides with both structural and energy storage functions. The carbohydrates that make up the cell wall can be linked to proteins and lipids, forming glycoproteins and glycolipids. Storage polysaccharides generally include glycogen, produced by cyanobacteria; starch and amylopectin, synthesized by chlorophytes; floridean starch, produced by rhodophytes; and chrysolaminarin, synthesized by diatoms (Chew et al. 2017; Barkia et al. 2019; de Carvalho et al. 2020). Additionally, exopolysaccharides, which have a more complex structure and can contain between 9 and 12 monosaccharides, have been the focus of studies for industrial applications as hydrocolloids. These polysaccharides are primarily excreted by cyanobacteria, which use mucilage to protect their cells from adverse environmental factors (Delattre et al. 2016).

Macroalgae contain polysaccharides at concentrations ranging from 4% to 76% of their dry mass, depending on the species. The highest carbohydrate levels are found in species of the genera *Ascophyllum*, *Porphyra*, *Palmaria*, and *Ulva* (Kraan 2012; Carrasqueira et al. 2025). Energy-storing polysaccharides include starch, found in red algae, and laminarin, found in brown algae. Structural polysaccharides, which are components of the cell wall, include cellulose, present in both green and brown algae, and agar, present in red algae. Additionally, macroalgae are potential producers of high-value complex polysaccharides, such as ulvan from green algae, fucoidan and alginate from brown algae, and agar and carrageenan from red algae. These polysaccharides are widely used as thickeners and gelling agents, among other applications. Moreover, some of these polysaccharides are being studied for their various biological properties, including antioxidant, anti-inflammatory, and anticancer activities, mainly attributed to fucoidan, agar, alginate, and laminarin (Chew et al. 2017; Cesário et al. 2018; Carrasqueira et al. 2025; Lu et al. 2025).

The lipids synthesized by algae can be classified into two groups: structural lipids (polar) and storage lipids (nonpolar). The latter, known as triacylglycerols, function as energy reserves and can be recovered through solvent extraction, serving as a key raw material for biodiesel production. On average, microalgae can produce approximately 17% lipids (by dry mass). Certain species, such as *Chlorella* sp., *Scenedesmus* sp., *Dunaliella* sp., *Schizochytrium* sp., *Nannochloropsis* sp., and *Botryococcus braunii*, are considered major lipid producers, accumulating over 20–30% by dry mass (Finkel et al. 2016; Cesário et al. 2018; de Carvalho et al. 2020, 2023). Macroalgae, on the other hand, produce fewer lipids than microalgae, typically ranging from 0.8% to 8% of dry mass, depending on the species and growth conditions. Although the yield is low, macroalgae can produce omega-3 and long-chain polyunsaturated fatty acids (PUFAs), which are high-value-added products (Øverland et al. 2019; Arias et al. 2023).

Due to the nutritional content and bioactivity of micro- and macroalgae, a wide range of bioproducts can be developed from algal biomass. After cultivation, harvesting, and extraction or processing, these products can be commercialized in the food, pharmaceutical, cosmetic, biofuel, fertilizer, and biorefinery industries, serving as natural alternatives to certain synthetic products with high environmental impacts. As food, some macroalgae are consumed directly by humans, especially in countries such as Japan, China, and Korea. *Porphyra* sp. is a genus of red macroalgae, best known for its use as nori seaweed in sushi. It is an example of an organism approved by the Food and Drug Administration (FDA) as Generally Recognized as Safe (GRAS) for human consumption (FAO 2009b; Chew et al. 2017; de Carvalho et al. 2020). In animal feed, macroalgae of the genus *Ulva* have shown promise as an alternative protein source (Øverland et al. 2019). Table 6.1 presents micro- and macroalgae species that produce commercially valuable bioproducts and some of their characteristics (Shah et al. 2016; Cai et al. 2021; Ma et al. 2023).

Despite being promising raw materials for the development of various bioproducts and possibly more economical than some land crops due to their rapid growth, lower nutrient consumption, and reduced space requirements, the production of single products, such as biofuels alone, can be unsustainable due to the high processing costs and low yields. This is where the concept of biorefineries comes in, allowing for the full utilization of algal biomass fractions to generate high-value-added coproducts and biofuels (Amponsah et al. 2024; Narayanan 2024).

Table 6.1 Top ten microalgae and macroalgae used as sources of commercial bioproducts

Algae	Bioactive compound produced	Optimal growth conditions	Growth rate	Cell composition (DW)	Major producers	References
Haematococcus pluvialis (microalga)	Astaxanthin (antioxidant)	19–28 °C, freshwater, high light intensity	0.3–0.5 day^{-1}	Proteins: 17–45%[a] Lipids: 20–37%[a] Carbohydrates:15–40%[a]	China and Chile	Shah et al. (2016), Cai et al. (2021), and Ma et al. (2023)
Dunaliella salina (microalga)	Beta-carotene (provitamin A)	25–35 °C, hypersaline conditions, strong light	0.2–0.5 day^{-1}	Proteins: 37% Lipids: 7.8% Carbohydrates: 33%	Spain	Tertychnaya et al. (2020) and Cai et al. (2021)
Spirulina sp. (*Arthrospira*) (microalga)	Phycocyanin (pigment and antioxidant) Food and feed Nutraceutical	30–35 °C, alkaline medium, moderate-to-high light	0.4–0.6 day^{-1}	Proteins: 53–70% Lipids: 6–20% Carbohydrates:12–24%	China, Chile, and France	Cai et al. (2021) and Kartik et al. (2021)
Chlorella vulgaris (microalga)	Lipids (biofuel) Nutraceutical Food and feed	20–28 °C, freshwater, moderate light	0.5–0.7 day^{-1}	Proteins: 19–65% Lipids: 4–26% Carbohydrates:12–42%	France and Taiwan	Kumar et al. (2013), Cai et al. (2021), and Fang et al. (2024)
Nannochloropsis sp. (microalga)	Omega-3 fatty acids (eicosapentaenoic acid (EPA))	20–24 °C, marine environment, high light intensity	0.1–0.2 day^{-1}	Proteins: 31.7% Lipids: 15% Carbohydrates: 9%	The United States and Japan	Kartik et al. (2021) and Ma et al. (2023)
Sargassum sp. (brown algae)	Food and feed, fucoxanthin	22–31 °C, marine water	–	Proteins: 13.36% Lipids: 0.39% Carbohydrates: 38.75%	China and the Republic of Korea	Bakker et al. (2023) and Cai et al. (2021)
Laminaria digitata (brown algae)	Food and feed, fucoxanthin	20 °C, marine water	–	Proteins: 6.6% Carbohydrates: 51.9%	China, the Republic of Korea, and Japan	Bakker et al. (2023), FAO (2009a), and Cai et al. (2021)
Porphyra sp. (red algae)	Biomass (food and feed)	20–28 °C, marine water	–	Proteins: 21.14% Lipids: 0.19% Carbohydrates: 20.59%	China, the Republic of Korea, and Japan	Bakker et al. (2023), FAO (2009b), and Cai et al. (2021)

(continued)

Table 6.1 (continued)

Algae	Bioactive compound produced	Optimal growth conditions	Growth rate	Cell composition (DW)	Major producers	References
Ulva sp. (green algae)	Ulvan (polysaccharide)	20–30 °C marine water	–	Proteins: 18.35% Lipids: 0.59% Carbohydrates: 32.65%	China and South Africa	Bakker et al. (2023), Cai et al. (2021), and Hiraoka (2021)
Kappaphycus alvarezii (red algae)	Carrageenan (polysaccharide)	28–36 °C in an alkaline environment	–	Proteins: 4.86% Lipids: 0.57% Carbohydrates: 41.15%	Indonesia, the Philippines, and Malaysia	Bakker et al. (2023) and Tee et al. (2015)

[a]Variation in nutritional composition is influenced by the vegetative or flagellate phase of the microalgae

6.2 Productive Chain (from Cultivation to Product)

In general, the development of algae-based products involves steps such as algae cultivation and growth, followed by harvesting or recovery. Drying processes can be applied to further preserve the biomass, after which cell rupture, extraction, and other treatments are carried out to obtain various bioproducts from the algae biomass (Chew et al. 2017; Cesário et al. 2018).

6.2.1 Algae Growth

Macroalgae are macroscopic organisms that reproduce asexually and/or sexually (with alternating generations), and their cultivation conditions differ from those of microalgae. Algae of the genus *Porphyra*, for example, have a life cycle phase as spores, which are cultivated in closed tanks under controlled lighting and temperature conditions for gamete production. During the stalk formation phase, macroalgae are cultivated in the open sea using floating, semifloating, fixed-net, or freezer-net systems for 40–50 days. These systems can be used in various climatic and tidal conditions, supported by buoys, piles, or anchors to secure the structure, and can be arranged either horizontally or vertically in the water column (FAO 2009b; Iwamoto et al. 2024a, Iwamoto et al. 2024b).

Microalgae are microorganisms that primarily reproduce asexually through mitosis, often generating two identical cells, though some species may produce multiple offspring per division. Under optimal growth conditions, their population can expand rapidly. In general, they require light, which is used to convert CO_2 and water into carbohydrates through photosynthesis; an optimal growth temperature, which can typically range between 16 and 35 °C depending on the strain; water; nutrients; aeration; and agitation to prevent sedimentation and ensure that all cells receive light and nutrients (Fang et al. 2024). Microalgal growth typically occurs in five distinct phases. The first is the lag phase, during which growth is minimal as the strains acclimate to the cultivation conditions. This is followed by the exponential phase, characterized by a rapid increase in cell division. The third phase is the deceleration phase, where the rate of cell division slows down due to limitations in nutrients or light availability. Subsequently, the stationary phase is reached, marked by a cessation of cell division because of resource depletion. The final phase is the death phase, wherein the number of dead cells surpasses the number of viable cells. For industrial-scale microalgal cultivation aiming at high productivity, it is desirable to minimize the duration of the lag phase, indicating that the strain can quickly adapt to new conditions. Furthermore, biomass should ideally be harvested prior to the stationary phase to prevent the accumulation of dead cells, thereby ensuring higher biomass quality. Depending on the species, cultivation conditions, and inoculation rate, microalgae cultures reach the stationary phase in media between 7 and 20 days (Ohse et al. 2008; Farag 2013; Delrue et al. 2017).

6.2.2 Biomass Recovery

As they are microscopic cells suspended in a culture medium, microalgae separation must be carried out after cultivation using recovery methods such as filtration, centrifugation, or flocculation, depending on the strain. Some microalgae species, such as *Nostoc* sp., have the ability to sediment without the need for a flocculant, while *Chlorella* sp. can settle and separate from the medium with the use of flocculants. *Spirulina* sp. is more difficult to sediment, so filtration and centrifugation are more commonly used for its separation (Ponnusamy et al. 2014; Cesário et al. 2018; de Carvalho et al. 2020). Macroalgae are harvested mechanically or manually. Manual methods have been the most commonly used, with tools such as scythes and forks to remove naturally grown algae and, more recently, on shallow-water or raft algae cultivation. Mechanical harvesting is more common in larger cultivation, such as for *Porphyra*, and involves dredging cutters, rotating blades, or suction devices, typically operated by boats or ships. A washing and sorting process to remove possible dirt and fouling organisms can be carried out at this stage, and then the algae are cut, dried, and prepared for consumption or further processing (FAO 2009b; Iwamoto et al. 2024a, Iwamoto et al. 2024b).

Depending on the product of interest, fermentation processes or strain treatments can be applied after cultivation. However, the algae are usually harvested and dried through methods such as drying, freeze-drying, or spray-drying to ensure longer preservation. In some cases, the biomass can also be used fresh. The harvesting and drying processes are the most energy-intensive stages in downstream algae production. Some industries use the dried biomass for animal and human supplementation, as well as for nutraceutical applications. Wet biomass can also be used for cell disruption processes and other treatments to obtain specific products (Ponnusamy et al. 2014; de Carvalho et al. 2020; Babich et al. 2024).

6.2.3 Bioproduct Extraction and Biomass Treatment

The nutritional composition of algae is directly associated with the genus or species, as well as the cultivation and growth conditions to which they are subjected, particularly regarding lighting, temperature, salinity, agitation, aeration, and nutrient availability. Commercially valuable components present in microalgae and macroalgae, such as proteins, lipids, carbohydrates, and pigments, are found inside the cell. Proteins are typically more dispersed throughout cellular structures, such as ribosomes, free enzymes, and protein complexes, while carbohydrates are present in higher concentrations in the cell wall. Lipids, on the other hand, can be found in cellular storage organelles, along with some polysaccharides (Gordalina et al. 2021; Martins et al. 2025).

Most algae possess a rigid cell wall (though not as rigid as that of plants), which plays a crucial role in algal protection and defense. The composition of the cell wall varies between microalgae and macroalgae species but generally consists of layers of different biopolymers, mainly cellulose and other fibrous polysaccharides, as well as glycoproteins. Since the main components of interest are intracellular, cell disruption is a crucial step in bioprocessing, with the choice of method depending on the species and the composition of the cell wall. After disruption, extraction techniques are applied to isolate these molecules, tailored to each type of product. To fully valorize the biomass, fractionation methodologies have been implemented, optimizing the use of all commercially relevant components efficiently and reducing production costs (de Carvalho et al. 2020; Gordalina et al. 2021; Martins et al. 2025).

Disruption techniques can be classified as mechanical, where force is applied to lyse the cell, for example, high-pressure homogenization, ball milling, and ultrasound. On the other hand, nonmechanical methods rely on chemical, physical, or enzymatic processes to lyse the cell or reduce membrane permeability. Most nonmechanical methods are easier to scale up and tend to have lower impacts, minimizing the release of undesirable cellular contents and preserving the purity of the target molecules (de Carvalho et al. 2020; Iwamoto et al. 2024a).

For example, in the extraction of carotenoids from macro- and microalgae, a combination of cell disruption and extraction using nonpolar solvents, such as hexane and ethyl acetate, can be employed, while polar solvents, such as alcohols and acetone, are used for the extraction of xanthophylls (Iwamoto et al. 2024b). For the extraction of water-soluble protein pigments, such as phycobiliproteins from cyanobacteria, cell disruption is typically followed by the use of aqueous solvents, such as buffers (Devi et al. 2020). After extracting these bioproducts, separation and purification processes such as dialysis, precipitation, and chromatography can be applied. To be marketed, these pigments are usually spray-dried and sold in powder form or encapsulated (Generalić Mekinić et al. 2023; Iwamoto et al. 2024a).

Algal polysaccharides, particularly the high-value ones from macroalgae, such as agar, alginate, and carrageenan, can be extracted with or without cell disruption. However, cell disruption using ultrasound, enzymes, or microwaves has been proven to improve efficiency. The extraction process is followed by using aqueous solvents that dissolve the polysaccharides from the algae, after which they can be purified. Acidic and alkaline solvents can also be used, depending on the polarity of the polysaccharides (Lu et al. 2025).

After the extraction of these bioproducts, algal biomass may still contain significant amounts of carbohydrates, proteins, and lipids, allowing for the production of additional coproducts, particularly in the bioenergy sector, thereby reducing process costs. Various processes can be applied to this residual biomass, as well as to fresh or dried biomass. Thermochemical conversion, for example, is based on the thermal decomposition of organic matter through processes such as hydrothermal liquefaction, pyrolysis, gasification, and direct combustion. These methods can produce bio-oil, synthesis gas, and electricity. Biochemical conversion includes fermentation processes, often involving consortia with other organisms, mainly fungi and bacteria. Alcoholic fermentation can produce bioethanol, while anaerobic digestion generates biogas. Transesterification of the fatty acids from algal biomass with methanol can produce biodiesel (Fig. 6.2) (Chew et al. 2017; de Carvalho et al. 2020; Babich et al. 2024; Narayanan 2024).

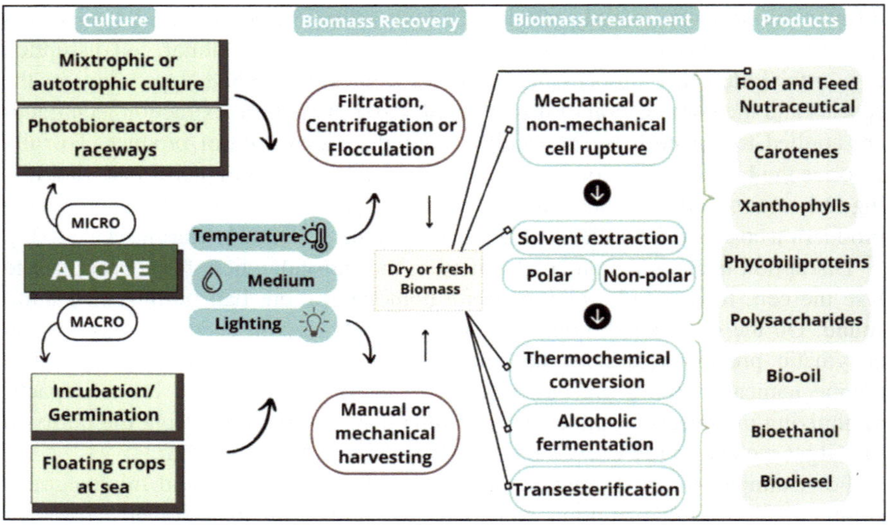

Fig. 6.2 Flowchart of the main stages in the production of bioproducts from micro- and macroalgae. It provides a brief overview of the cultivation processes and conditions, recovery methods, and possible biomass treatments for obtaining commercial products

6.3 Main Applications

Algae have high productivity and are considered solutions to various challenges, ranging from biofuel production and carbon fixation to food generation (Jacob-Lopes et al. 2015; Cardoso et al. 2020; Coutinho Rodrigues et al. 2021). However, when it comes to commercial products, the most established applications are primarily in nutraceuticals derived from microalgae, such as *Spirulina*, *Chlorella*, and *Haematococcus*, which can also serve as bio-inputs in animal feed, and in biofertilizers extracted from macroalgae. Most commercially available biofertilizers consist of pressed extracts from *Ascophyllum* sp., with more recent options based on *Kappaphycus* sp. The relatively limited diversity of these products is partly because large-scale production requires robust strains that grow quickly and are resistant to contamination. Additionally, converting algae into bioproducts involves complex processes, and many solutions already proven on a small scale have yet to be sufficiently studied for industrial-scale feasibility (de Carvalho et al. 2019, 2020; Scapini et al. 2024).

6.3.1 *Algae as Fuel*

Due to their rapid growth and high lipid and carbohydrate content, microalgae may be ideal raw materials for biofuel production. They are widely investigated in the production of biodiesel and bioethanol, particularly the species *Nannochloropsis* sp.,

Botryococcus braunii, *Chlorella* sp., *Scenedesmus* sp., and *Chlamydomonas reinhardtii*. On the other hand, macroalgae are more extensively studied for bioethanol and biogas production, as they have a high carbohydrate content. Some of the most studied species include *Kappaphycus alvarezii*, *Chondrus crispus*, *Ulva* sp., *Saccharina* sp., and *Ascophyllum nodosum* (Jalilian et al. 2020).

Oleaginous algae, when cultivated under stress conditions (nutrient deficiency, for example), can accumulate lipids in their biomass, making them a suitable resource for energy generation. Nutrient deprivation is a strategy applied to increase lipid concentration in microalgae. However, these environmental stresses can also significantly compromise biomass growth (Xue et al. 2021). The application of salinity and light stress in *Chlorella vulgaris* cultures significantly improved lipid content and fatty acid profile, making the biomass more viable for biodiesel production (Song et al. 2022).

After algal cultivation, biomass recovery and disruption are required for carbohydrate and lipid extraction. The composition of algal biodiesel depends on the fatty acid profile, which varies with the species and cultivation conditions, and the oil content typically ranges from 20% to 70%. It consists mainly of fatty acid methyl esters (FAMEs), and biodiesel is produced through transesterification, where the lipids react with alcohol and a catalyst. Although conventional transesterification is the primary method, supercritical and enzymatic approaches are also explored. Moreover, nanocatalysts like Fe_3O_4/ZnMg (Al)O have been shown to improve production efficiency, achieving a 94% yield at 65 °C with a 12:1 alcohol-to-oil ratio over 3 hours (Chen et al. 2018; Hassan Pranta and Muk Cho 2025).

Algal bioethanol, composed of ethanol (C_2H_5OH), is derived from the conversion of microalgal carbohydrates. In this case, hydrolysis can be performed using acidic, enzymatic, or physicochemical methods, followed by fermentation with microorganisms such as *Saccharomyces cerevisiae* and *Zymomonas mobilis*. In particular, simultaneous saccharification and fermentation (SSF) is more efficient than separate hydrolysis and fermentation (SHF). Ethanol separation is facilitated by yeast flocculation, and hydrolysate neutralization is required before fermentation. Yields vary depending on the species and pretreatment, with *Dunaliella* sp. (1% H_2SO_4 and SSF) achieving 0.21 g g^{-1} dcw, and *Scenedesmus dimorphus* reaching 0.15 g g^{-1} dcw (Maity and Mallick 2022).

Algal biogas primarily consists of methane (CH_4) and carbon dioxide (CO_2), along with traces of nitrogen (N_2), hydrogen (H_2), water vapor, and hydrogen sulfide (H_2S). It is produced through anaerobic digestion (AD), which converts carbohydrates, lipids, and proteins via hydrolysis, acidogenesis, acetogenesis, and methanogenesis. Thus, various pretreatment methods such as grinding, ultrasound, microwave, thermal, enzymatic, and catalytic treatments enhance biomass solubility and degradation. Furthermore, biogas quality can be improved by removing CO_2 and H_2S, with studies demonstrating over 99% H_2S removal by microalgae. Yields vary depending on species and treatment methods, with the highest recorded at 740 mL g^{-1} volatile solids (VS, representing the organic fraction of the total solids that can be biodegraded) from de-oiled Taihu algae (thermochemical pretreatment at 70 °C for 3 hours) and 615 ± 7 mL g^{-1} VS from *Laminaria* sp. (mechanical pretreatment at 38 ± 1 °C for 15 minutes) (Abusweireh et al. 2023).

In summary, biofuels derived from macro- and microalgae offer several advantages over conventional sources. Their high biomass productivity throughout the year stands out, as the cultivation of these organisms is not dependent on seasonality, unlike terrestrial oil crops. Additionally, water demand is reduced since they can be cultivated using wastewater, simultaneously contributing to effluent treatment by absorbing nutrients such as nitrogen and phosphorus. Likewise, another significant environmental advantage is the ability of algae to capture large amounts of carbon dioxide during growth, making them a promising alternative for mitigating industrial greenhouse gas emissions (Wang et al. 2022; Sing et al. 2024).

However, challenges such as high production costs, particularly in cultivation, recovery, and compound extraction, still limit the competitiveness of this class of biofuels compared to fossil fuels and other renewable sources (Sing et al. 2024). Therefore, from an economic perspective, ensuring process viability is essential by adopting strategies such as integrating biofuel production into biorefinery pathways for other value-added products, reusing nitrogen-rich residual components, and utilizing wastewater. The choice of appropriate cultivation and extraction methods plays a fundamental role in determining the overall cost (Wang et al. 2022).

6.3.2 Algae as Food and Feed

The need to feed millions of people while reducing environmental impact has driven a movement for new and more sustainable food sources. Micro- and macroalgae have been cultivated for thousands of years as food sources. They are a viable source of nutritious biomass due to their richness in essential nutrients, such as proteins, lipids, carbohydrates, vitamins, pigments, polyunsaturated fatty acids (PUFAs), and minerals (Bhatnagar et al. 2024). They are recognized for their nutritional value and versatility, integrating them into various food products, reflecting the growing consumer awareness and demand for healthier and environmentally friendly options.

Macroalgae species, renowned for their high protein content and beneficial amino acid composition, have long been valued as important sources of protein. Both macroalgae and microalgae offer substantial protein levels, ranging from 5% to 70% on a dry weight basis. Examples of notable species include *Porphyra tenera* (nori), with 47% protein, and *Arthrospira platensis*, with up to 70% protein. While *Chlorella* can contain around 40% carbohydrates and *Laminaria* more than 50%, these algae are important for human nutrition, as well as for the production of gums and thickeners of high industrial interest, which are already well-accepted and marketed as additives and gelatin substitutes (Table 6.1) (Matos et al. 2022; de Carvalho et al. 2023; Lu et al. 2025).

The presence of fiber is also an important nutritional factor, as it aids digestion and promotes satiety. Fibers are classified as soluble or insoluble. Soluble fibers tend to form gels when in contact with water, while insoluble fibers do not form gels but can retain water. The fibers found in algae consist of a mixture of compounds, including carbohydrates and polymers. Red algae, such as *Porphyra*, can produce around 15–22% of soluble fiber relative to their dry weight, while brown algae, such as *Fucus*, can contain 27–40% of insoluble fiber (Peñalver et al. 2020).

Both microalgae and macroalgae find diverse applications in food products, whether consumed whole or fractionated for specific components. For example, macroalgae, such as *Caulerpa* sp., are consumed fresh and raw, while microalgae, such as *Spirulina*, are dried and incorporated into various foods without further processing. Macroalgae that are commonly marketed are fresh, dried, or powdered, and drying is the most commonly used preservation method. The latest trends highlight the use of algae powders and extracts to enrich food products (Bakker et al. 2023; Matos et al. 2022). In addition, their biomass is processed to extract specific fractions or components that improve the nutritional profile, color, texture, and other characteristics of foods while serving as thickening, stabilizing, and gelling additives in food applications (Matos et al. 2022).

Despite the high nutritional quality of algae-based products, in some cases, supplementation with algal biomass can negatively affect consumer interest, mainly due to its color and smell. Techniques to mitigate these effects have been widely studied in an effort to develop healthy and appealing food products for consumers (Siladji et al. 2024).

The FAO (2014) estimated that 38% of the 23.8 million tons of seaweeds from the 2012 global harvest was consumed by humans in forms recognizable as seaweed (e.g., kelps and nori/laver), excluding additional consumption of hydrocolloids (e.g., agars, alginates, and carrageenan) used as thickening agents in foods and beverages. Human consumption of algae-based foods varies by nation, with Japanese diets accounting for a recent (2010–2014) annual per capita consumption ranging from 9.6 (2014) to 11.0 (2010) grams of macroalgae per day (Wells et al. 2017).

Despite challenges, such as high production costs, algae products applied in high-end supplements are cost-effective substitutes for food and feed. Overall, the trend toward increasing nutritional demand for algae-based products globally stems from increased attention to health and wider use of food additives. In addition to their nutritional value, algae are increasingly marketed as functional foods or nutraceuticals. These terms have no legal status in many nations but describe foods that contain bioactive compounds, or phytochemicals, that may benefit health beyond the role of basic nutrition (e.g., anti-inflammatory and disease prevention) (Naik et al. 2024).

6.3.3 *Algae as Nutraceuticals*

Nutraceutical foods or supplements are products designed to promote medical and health benefits, primarily in the prevention and treatment of certain diseases. This market includes both human and animal foods (Borowitzka 2013). Antioxidant substances such as pigments, polysaccharides, amino acids, polyunsaturated fatty acids, phytosterols, and phenolic compounds are produced by macroalgae and microalgae. However, plants are still considered the primary sources of exogenous antioxidants for human nutrition supplementation. Due to economic advantages in production, as well as greater resilience and resistance to adverse conditions compared to higher plants, the use of algal bioactives as nutraceuticals has shown promise. Additionally, antioxidant molecules such as phycobiliproteins and mycosporine-like amino acids are synthesized only by microalgae and not by plants (Esim et al. 2024; Lu et al. 2025).

Micro- and macroalgae are also widely used in the food industry, both as whole biomass for human and animal supplementation and as specific biomass fractions, such as pigments, lipids, and proteins, among others, as shown in Table 6.1. As nutraceuticals, microalgae, particularly those of the genera *Spirulina* and *Chlorella*, and macroalgae such as *Laminaria*, *Fucus*, and *Porphyra* are commercially available worldwide. They are also approved by the FDA as Generally Recognized as Safe (GRAS) for human consumption and exhibit immunomodulatory and antioxidant activities, in addition to serving as a protein supplement (Borowitzka 2013; de Carvalho et al. 2023; Singh Mehra et al. 2025).

Chrysolaminarin, a polysaccharide extracted from the microalga Odontella aurita, has been reported as a potential antioxidant and proposed for use as a natural antioxidant in aquaculture, food, and pharmaceuticals. Additionally, this molecule has demonstrated hepatoprotective activity by preventing lipid accumulation. Other polysaccharide molecules and pigments obtained from different algae have shown not only antioxidant properties but also anti-hemolytic, anti-inflammatory, lipid peroxidation inhibition, and free metal ion chelation activities. Mycosporine-like amino acids play a key role in absorbing UV light and exhibit biological activities such as antioxidant, anti-inflammatory, protein glycation inhibition, and collagenase activity inhibition. For these reasons, they are widely used in the sunscreen industry and in antiaging products (Esim et al. 2024; Lu et al. 2025).

Some studies have identified that fucoxanthin and β-carotene extracted from the brown algae *Padina tetrastromatica* exhibit activity against oxidative stress in rats. Additionally, the fucoxanthin present in *Laminaria japonica* has shown hepatoprotective effects under oxidative stress. Docosahexaenoic acid (DHA) is an essential omega-3 fatty acid, meaning it is not produced by the human body and must be obtained through food or supplementation. It primarily accumulates in the eyes and brain tissue. *Spirulina platensis* can be an excellent source of DHA, producing approximately 9.1% of the cell's total fatty acid concentration. Algae-derived DHA is already available on the market as a nutraceutical in the form of oil capsules (Generalić Mekinić et al. 2023; Narayanan 2024).

6.3.4 Algae for Value-Added Products

Both microalgae and macroalgae are potential pigment producers, playing crucial roles in light absorption for photosynthesis and providing protection against cell oxidation and other light-induced damage. Most macro- and microalgae produce carotenoids and xanthophylls, fat-soluble pigments typically synthesized by higher plants or chemically synthesized for industrial purposes, which involve high costs and environmental impacts. The synthesis of these pigments by algae has garnered significant interest in their large-scale production, extraction, and commercialization, as this approach would be more cost-effective and require less energy and space (Generalić Mekinić et al. 2023; Babich et al. 2024).

In general, microalgae such as *Haematococcus* produce high concentrations of astaxanthin, while *Dunaliella* produces β-carotene, and *Chlorella* is known for its lutein production. Green macroalgae, such as those of the genus *Ulva*, have been reported as carotenoid producers, mainly lutein. Brown algae, such as *Laminaria* and *Sargassum*, predominantly produce fucoxanthin. Red algae, in addition to producing carotenoids, such as α- and β-carotene, zeaxanthin, and lutein, can also synthesize protein-based, water-soluble pigments such as phycocyanin and phycoerythrin, similar to cyanobacteria (de Carvalho et al. 2020; Generalić Mekinić et al. 2023). Factors such as the strain's place of origin, time of year, salinity, temperature, lighting, pH, CO_2 availability, and other nutrients can directly affect the production of lipids, carbohydrates, and proteins (Narayanan 2024).

Spirulina, a filamentous blue-green microalga, is rich in proteins, vitamins, essential amino acids, and polyunsaturated fatty acids. It contains phycocyanin, a blue pigment with antioxidant and anti-inflammatory properties. *Spirulina* is widely used to reduce cholesterol, enhance immune function, and as an antiviral agent due to its sulfated polysaccharides. Since 1975, it has been commercially available in Japan and is used in health foods, like protein bars, and cosmetics, such as face masks (Mobin and Alam 2017).

Similarly, Chlorella is a unicellular green microalga that contains chlorophyll a and b, β-1,3-glucan, lutein, and various antioxidant compounds (Mobin and Alam 2017). It has been shown to boost immune response, lower blood lipids, and act as an antitumor agent. Furthermore, it supports detoxification and wound healing and exhibits hepatoprotective properties. Its biomass and carbohydrates are commonly applied in the food industry, animal nutrition, and dietary supplements (Iwamoto et al. 2024b).

Dunaliella salina is a flagellated extremophile green microalga renowned for its high β-carotene content, up to 14 mg/g of dry biomass. It has strong antioxidant properties and is used in animal feed, fish feed, and pharmaceuticals (β-carotene supplements). Its carotenoids and xanthophylls have been associated with anticancer, antihypertensive, broncholytic, and analgesic effects (Mobin and Alam 2017).

Last, *Haematococcus pluvialis* is a unicellular microalga recognized as the primary source of astaxanthin, a potent antioxidant that is widely used in aquaculture for fish pigmentation (salmon meat), pharmaceuticals (astaxanthin supplements), and nutraceuticals due to its anti-inflammatory, immunomodulatory, and macular degeneration–preventing effects (Ambati et al. 2014).

In general terms, macroalgae are a rich source of bioactive compounds with industrial applications. Protein production varies among different groups, with brown algae being the lowest producers at around 15%, green algae at about 26%, and red algae being the highest, reaching a value of 47%. Species of the genus *Porphyra* sp. can reach protein concentrations comparable to those found in soybean meal (Øverland et al. 2019; Arias et al. 2023).

Macroalgae also contain glycolipids and phospholipids with anti-inflammatory properties. Red and brown algae are rich in eicosapentaenoic acid and arachidonic acid, while green algae contain oleic, palmitic acids, and polyunsaturated fatty acids (PUFAs), very important molecules for human health, associated with reducing the risk of chronic diseases, like linoleic acid. These oils are crucial in nutraceuticals and pharmaceuticals, but despite their health benefits, the extraction efficiency and cost-effectiveness are still a problem at the industrial scale (Biris-Dorhoi et al. 2020).

Different methods can be used to extract macroalgal lipids. Traditional methods involve the use of solvents such as chloroform, but due to its toxicity, alternative approaches have been considered, including supercritical CO_2 extraction and pressurized liquid extraction, which can achieve yields between 35% and 57%. Lipid extraction trials with the green alga *Utricularia rigida* involved freezing and grinding the biomass, followed by immersion in methanol and $CHCl_3$, with phase separation by centrifugation. The lipid phase was then recovered and dried to a powder-like consistency (Arias et al. 2023).

Otherwise, the carbohydrate content includes agar, alginates, carrageenan, fucoidan, and laminarin. They are known for antioxidant, anti-inflammatory, and neuroprotective effects, in addition to industrial gelling applications, and for this, they are widely used in food, pharmaceuticals, and cosmetics. Pigments such as beta-carotene, astaxanthin, chlorophyll, and fucoxanthin offer antioxidant and chemopreventive benefits that drive commercialization, though maintaining stability remains a problem. Additionally, phenolic compounds, especially phlorotannins, exhibit antioxidant, antimicrobial, and neuroprotective properties (Biris-Dorhoi et al. 2020).

Species such as *Gracilaria manilaensis* and *Kappaphycus alvarezii* are considered great producers of carrageenan. One of the main methods for extracting this polysaccharide from these macroalgae involves using potassium and sodium hydroxide through alkaline treatment, followed by washing, soaking, and grinding of the macroalgae. A viscous filtrate is obtained after extraction and filtration, followed by purification, neutralization with HCl, heating, coagulation with KCl, precipitation in alcohol, filtration, and drying. The yield after all these steps can reach approximately 44%. However, to obtain a less pure but more cost-effective polysaccharide, a single precipitation step using ethanol can recover up to 95% of the gum (Arias et al. 2023).

While demand grows in all the fields mentioned, the challenges also grow, mainly in terms of standardization, sustainable extraction methods, seasonal variation in algae composition, and optimization of large-scale cultivation. The high costs of algae bioproducts are also an obstacle that makes algae-based products less economically competitive compared to conventional alternatives. Cultivation requires controlled environments, nutrients, and significant water and energy inputs, while harvesting is labor-intensive due to the small size and dilute concentration of algae in water. Also, the extraction of valuable compounds, such as biofuels or proteins, involves costly technologies like centrifugation or chemical solvents. Despite the industrial-scale obstacles, research into the use of macroalgae in the food, pharmaceutical, and cosmetic industries continues to expand. This growth is driven by their rich bioactive content, rapid growth rate, and minimal environmental requirements, making them a promising sustainable alternative (Gupta et al. 2024).

6.3.5 Other Products

Micro- and macroalgae can also be used as biofertilizers due to the presence of cytokinins, auxins, abscisic acid, and gibberellins, hormones that act as biostimulants for plant growth. Extracts from brown macroalgae such as *Ascophyllum nodosum* and *Sargassum vulgare*, as well as the red alga *Kappaphycus alvarezii*, are already being marketed as plant biostimulants (Arias et al. 2023).

Recent research has identified a high potential for the production of polyhydroxybutyrate (PHB) biopolymers, which are key components in bioplastic production, by various species of microalgae. The use of algal consortia and specific cultivation conditions—such as nitrogen or phosphorus deprivation, supplementation with an organic carbon source, and adjustments in temperature, light intensity, and CO_2 addition—can enhance the production of these biopolymers. The microalga *Scenedesmus* sp. and the cyanobacterium *Spirulina platensis*, cultivated in a mixotrophic medium, produced approximately 30% and 19% PHB, respectively (Martínez-Herrera et al. 2025).

6.3.6 Algae-Based Biorefineries

While integral biomass products use the whole of the biomass harvested, the fractionated products leave residual biomasses, presenting an opportunity for increasing the number of products toward the total use of biomass. Some single fractions are valuable enough to sustain the bioprocess (e.g., astaxanthin production); however, it is the diversification of products that can lead to the feasibility of producing biofuels, which must have a competitive price to be used as drop-in biofuels (de Carvalho et al. 2024).

The challenge in fractionating algal biomass is that product diversification requires more steps and inputs. Therefore, an economic balance is crucial, and algae biorefineries must be developed following techno-economic analysis based on lab-scale data. Possible routes of processing are presented in Table 6.2.

The initial steps in algal biomass processing involve removing the bulk of water. For microalgae, this is done by filtration or centrifugation. The resulting biomass is often dried by spray or tray drying or, in some cases, processed as a wet paste. For macroalgae, the process consists of collection, often manual or partially automated, and draining. The algae can be used as fresh or air-dried biomass, further processed into sheets or strips (for Nori or Kombu), or pressed to remove part of the water, and

Table 6.2 Examples of approaches for biomass valorization in biorefineries

Alga	Use	Minimal processing	Basic fractionation	High-level fractionation	Examples
Spirulina	Biomass, rich in protein	Concentration, dewatering, drying	Phycocyanin extraction	Phycocyanin, protein, cell wall, carotenoids, bioactive peptides	Obeid et al. (2025)
Chlorella	Biomass, rich in protein	Concentration, dewatering, drying	Lipid extraction	Lipids, protein, cell wall, carotenoids, phytohormones, antioxidants	Chovancek et al. (2025)
Haematococcus	Biomass, rich in astaxanthin	Concentration, dewatering, drying	Astaxanthin SFE extraction	Astaxanthin, protein meal, soluble proteins	Ba et al. (2016) and Jiang et al. (2019)
Dunaliella	Biomass, rich in beta-carotene	Concentration, dewatering, drying	Beta-carotene extraction	Beta-carotene, lipids, protein meal, or bioenergy	Francavilla et al. (2015) and de Souza Félix et al. (2024)
Gracilaria	Agar production	–	Polysaccharide extraction	Agar, agarose, nutritional fiber, or hydrolysates for biofuels	Kumar et al. (2013) and Martins et al. (2023)
Kappaphycus	Carrageenan production	–	Pressing, polysaccharide extraction	Agar, agarose, nutritional fiber, biofertilizers, pigments, or biogas	Ingle et al. (2018) and Rudke et al. (2020)

then oven-dried (Banerjee et al. 2024). Polysaccharides can be extracted with wet or dry biomass, requiring commercial-grade alkali and acids for pH adjustment.

Biorefinery integration strategies can start with low-complexity approaches, such as using residual algal biomass for biofuel production. Biofuels typically have a low added value, so extracting other fractions from the biomass without significantly increasing the biofuel production costs is crucial. These fractions may help subsidize biofuel production. For example, algae like *Scenedesmus* or *Chlorella* have high productivity but relatively low carotenoid or lipid content. However, in an economy of scale, extraction may be done before large amounts of biomass are processed into biofuels. This nonpolar fraction can then produce valuable products such as nutritional products. The specific approach depends on the microalgal composition, as different algae are optimized for different end products, whether biofuels or high-value nutritional lipids like DHA. If the biomass is of a higher quality, as for SFE-extracted biomass, it can be used for higher-value applications (de Carvalho et al. 2023).

As there is a multitude of possibilities in biorefineries, it is wise to evaluate the processes in silico first, starting with material balances of possible routes, then checking the yields in a laboratory setting, and then progressing to a techno-economic analysis, where prices for raw materials, energy, and equipment are balanced against potential income (Korsah et al. 2025). Despite the complexity, there is usually no need to start from scratch: Valuable fractions and possible routes of valorization can be inspired by existing processes.

Another important aspect of algal biorefineries is using residues of other processes as culture media, especially for microalgae production. Microalgae are produced on fertilizer-based media on a small scale, but on a large scale, the use of residues is one of the strategies for lowering costs (de Carvalho et al. 2024); macroalgae, conversely, are typically produced at sea, and the labor costs tend to be proportionally higher.

6.4 Market

It is evident that the global production and commercialization of algae have increased over the years, driven both by advancements in research and new discoveries regarding algal products. This growth is also attributed to the relatively low production costs compared to other animal and plant-based raw materials (Borowitzka 2013). The global seaweed market has observed significant growth in recent years, with a valuation of US$ 717.14 million in 2018 and a projected US$ 1365.8 million by 2027, reflecting a compound annual growth rate (CAGR) of 5.35% between 2019 and 2027 (Wells et al. 2017).

In 2017, the market for microalgae-based products was valued at approximately US$32 billion. Projections indicate that this market will exceed US$53 billion by 2026 (Rafa et al. 2021). Figure 6.3 shows some of the main algae-based products on the market and their selling prices.

Fig. 6.3 Market values of selected micro- and macroalgae-based products worldwide. The graph compares the average prices of these products, highlighting significant differences among high-value-added items. (Data collected through online searches in the US market)

The production of *Chlorella* microalgae, primarily by companies in Taiwan, ranges from 130 to 450 tons per year, with an average market value of US$44 per kilogram (Iwamoto et al. 2024a, Iwamoto et al. 2024b). On the other hand, products derived from *Spirulina* biomass, marketed in the form of powder, tablets, or capsules, generate approximately US$100 million in annual revenue, with prices ranging from US$100 to US$400 per kilogram (Costa et al. 2019).

The production and commercialization of carotenoids extracted from algae have experienced significant growth, particularly in the nutraceutical, food, feed, pharmaceutical, and cosmetic industries. In 2019, the global carotenoid market was valued at approximately US$1.5 billion, with projections indicating that it could reach US$2 billion in revenue by 2026 (Generalić Mekinić et al. 2023).

Among the most widely produced carotenoids is astaxanthin, extracted from *Haematococcus*, which is typically sold at an average price ranging from US$25 to US$700 per kilogram. Production is primarily concentrated in the United States and Japan, reaching approximately 300 tons per year (Shah et al. 2016). The production of β-carotene has been driven by the microalga *Dunaliella salina*, which contains a higher concentration of β-carotene than any other organism. This compound is considered a high-value product due to its quality and bioactivity, in addition to being marketed as a natural, non-synthetic alternative. Depending on its formulation and purity, β-carotene derived from *Dunaliella salina* can be sold for between 300 and 1500 US$ kg^{-1}. As of 2013, the global β-carotene market was valued at approximately US$ 270 million, with natural-source β-carotene accounting for 20–30% of total sales. Lutein, a carotenoid extracted from microalgae such as *Chlorella, Muriellopsis, and Scenedesmus*, generated revenues of US$135 million in 2015 and is expected to reach up to US$370 million between 2020 and 2027, with projections of approximately US$490 million by 2029 (Borowitzka 2013; Iwamoto et al. 2024b).

Phycobiliproteins (phycocyanin, allophycocyanin, and phycoerythrin) are protein pigments exclusive to cyanobacteria and are primarily derived from *Spirulina*. The average price of these pigments ranges from US$160 to 500 per kilogram, depending on the product's purity, which varies from 0.7 (food grade) to 4 or higher (analytical grade) (Borowitzka 2013; Costa et al. 2019).

Between 2019 and 2024, global macroalgae production generated around US $12.5 billion to US$14.7 billion in revenue, with a total output of approximately 35 million tons. Algae of the *Laminaria* genus accounted for close to US$4.6 billion, followed by *Porphyra* with roughly US$2.7 billion in revenue. *Kappaphycus* generated about US$2.4 billion, while *Gracilaria* and *Undaria* each reached around US $2 billion in 2019 (Cai et al. 2021; Narayanan 2024). Macroalgae hydrocolloids represent a significant portion of the revenue generated from products derived from these organisms. In 2009, the total revenue from agar, alginates, and carrageenan reached approximately US$1 billion, with average prices of around US$18 per kg for agar, US$12 per kg for alginate, and US$10.5 per kg for carrageenan. By 2019, this revenue had increased to approximately US$1.7 billion, with China being the world's largest producer of hydrocolloids, accounting for 30% of the global production (Bixler and Porse 2011).

Until 2010, the market costs of omega-3 derived from algae were still higher than those produced from fish, reaching an average price of US$140 per kg, possibly due to production costs. In the same way, the production of biofuels from microalgal oils requires an assessment of production costs in comparison to other fuels. The average production cost of diesel is approximately US$0.49 per kg, while ethanol reaches US $2.6 per kg. In optimistic projections, using algae cultivated in wastewater, the average production cost can reach approximately US$0.73 per kg of dry biomass. These costs can be reduced to around US$0.60 per kg when associated with coproducts, such as the use of residual biomass for animal feed or high-value-added products (Borowitzka 2013; Rafa et al. 2021). Despite these challenges, the biofuel market derived from microalgae in Japan reached a revenue of US$1.17 billion in 2017 and has continued to receive investments in research, improvements, and innovation in this field (Loke Show 2022).

6.5 Conclusion and Future Perspectives

The use of microalgae biomass has seen promising advances in scientific research worldwide. This is due to the diverse properties, active compounds, and products that can be derived from these organisms. The main bioproducts obtained from algae, which have been widely marketed globally, include food products such as dried biomass for dietary supplementation, food composition, nutraceuticals, and even animal feed. These primarily come from microalgae and cyanobacteria, such as *Spirulina*, *Chlorella*, and *Haematococcus*, as well as macroalgae like *Porphyra*, *Laminaria*, and *Kappaphycus*. Pigments derived from these organisms are used as food colorings and dietary supplements and in human nutrition, depending on their activities and functions. Additionally, macroalgal polysaccharides are sold as gelling agents or thickeners for various industries. Oils extracted from algae also serve as food supplements and are utilized in biofuel production.

The costs of production, recovery, and treatment of biomass and the extraction, recovery, and purification of these products have been a challenge for industries and researchers investing in this type of production. That is why new studies and innovations around microalgae bioproducts, especially with regard to reducing production and downstream costs, are of great interest to industries and the market for these products. The idea of biorefineries can help solve some of these problems, so that the same algal biomass generates two or more products, including high-value-added pigments or proteins and oils, and finally a total extraction of the biomass properties is carried out, using this fraction for the production of biofuels or other products.

References

Abusweireh RS, Rajamohan N, Sonne C, Vasseghian Y (2023) Algae biogas production focusing on operating conditions and conversion mechanisms – a review. Heliyon 9:e17757. https://doi.org/10.1016/j.heliyon.2023.e17757

Ambati R, Phang S-M, Ravi S, Aswathanarayana R (2014) Astaxanthin: sources, extraction, stability, biological activities and its commercial applications—a review. Mar Drugs 12:128–152. https://doi.org/10.3390/md12010128

Amponsah L, Chuck C, Parsons S (2024) Life cycle assessment of a marine biorefinery producing protein, bioactives and polymeric packaging material. Int J Life Cycle Assess 29:174–191. https://doi.org/10.1007/s11367-023-02239-w

Arias A, Feijoo G, Moreira MT (2023) Macroalgae biorefineries as a sustainable resource in the extraction of value-added compounds. Algal Res 69:102954. https://doi.org/10.1016/j.algal.2022.102954

Ba F, Ursu AV, Laroche C, Djelveh G (2016) Haematococcus pluvialis soluble proteins: extraction, characterization, concentration/fractionation and emulsifying properties. Bioresour Technol 200:147–152. https://doi.org/10.1016/j.biortech.2015.10.012

Babich O, Ivanova S, Michaud P et al (2024) Fermentation of micro- and macroalgae as a way to produce value-added products. Biotechnol Rep 41:e00827. https://doi.org/10.1016/j.btre.2023.e00827

Bakker A, Khan S, Hasan M (2023) Seaweed: a prominent source of protein and other nutrients. https://doi.org/10.5281/zenodo.8302372

Banerjee S, Mandari V, Shalini M et al (2024) Cost-effective downstream processing of algal biomass for industrial-scale biofuels production. In: Recent trends and developments in algal biofuels and biorefinery. Springer Nature Switzerland, pp 239–262

Barkia I, Saari N, Manning SR (2019) Microalgae for high-value products towards human health and nutrition. Mar Drugs 17:304. https://doi.org/10.3390/md17050304

Bhatnagar P, Gururani P, Parveen A et al (2024) Algae: a promising and sustainable protein-rich food ingredient for bakery and dairy products. Food Chem 441:138322. https://doi.org/10.1016/j.foodchem.2023.138322

Biris-Dorhoi E-S, Michiu D, Pop CR et al (2020) Macroalgae—a sustainable source of chemical compounds with biological activities. Nutrients 12:3085. https://doi.org/10.3390/nu12103085

Bixler HJ, Porse H (2011) A decade of change in the seaweed hydrocolloids industry. J Appl Phycol 23:321–335. https://doi.org/10.1007/s10811-010-9529-3

Borowitzka MA (2013) High-value products from microalgae—their development and commercialisation. J Appl Phycol 25:743–756. https://doi.org/10.1007/s10811-013-9983-9

Buschmann AH, Camus C (2019) An introduction to farming and biomass utilisation of marine macroalgae. Phycologia 58:443–445. https://doi.org/10.1080/00318884.2019.1638149

Cai J, Lovatelli A, Garrido Gamarro E et al (2021) Seaweeds and microalgae: an overview for unlocking their potential in global aquaculture development. FAO Fisheries and Aquaculture Circular NFIA/C1229 (En)

Cardoso LG, Duarte JH, Andrade BB et al (2020) Spirulina sp. LEB 18 cultivation in outdoor pilot scale using aquaculture wastewater: high biomass, carotenoid, lipid and carbohydrate production. Aquaculture 525:735272. https://doi.org/10.1016/j.aquaculture.2020.735272

Carrasqueira J, Bernardino S, Bernardino R, Afonso C (2025) Marine-derived polysaccharides and their potential health benefits in nutraceutical applications. Mar Drugs 23:60. https://doi.org/10.3390/md23020060

Cesário MT, da Fonseca MMR, Marques MM, de Almeida MCMD (2018) Marine algal carbohydrates as carbon sources for the production of biochemicals and biomaterials. Biotechnol Adv 36:798–817. https://doi.org/10.1016/j.biotechadv.2018.02.006

Chen Y, Liu T, He H, Liang H (2018) Fe_3O_4/ZnMg(Al)O magnetic nanoparticles for efficient biodiesel production. Applied Organometallic Chemistry 32(5). https://doi.org/10.1002/aoc.4330

Chew KW, Yap JY, Show PL et al (2017) Microalgae biorefinery: high value products perspectives. Bioresour Technol 229:53–62. https://doi.org/10.1016/j.biortech.2017.01.006

Chovancek E, Poque S, Bayram E et al (2025) Stepwise processing of Chlorella sorokiniana confers plant biostimulant that reduces mineral fertilizer requirements. Bioresour Technol 418:131923. https://doi.org/10.1016/j.biortech.2024.131923

Costa JAV, Freitas BCB, Rosa GM et al (2019) Operational and economic aspects of Spirulina-based biorefinery. Bioresour Technol 292:121946. https://doi.org/10.1016/j.biortech.2019.121946

Coutinho Rodrigues OH, Itokazu AG, Rörig L et al (2021) Evaluation of astaxanthin biosynthesis by Haematococcus pluvialis grown in culture medium added of cassava wastewater. Int Biodeterior Biodegrad 163:105269. https://doi.org/10.1016/j.ibiod.2021.105269

de Carvalho JC, Sydney EB, Assú Tessari LF, Soccol CR (2019) Culture media for mass production of microalgae, 2nd edn. Elsevier B.V.

de Carvalho JC, Magalhães AI, de Melo Pereira GV et al (2020) Microalgal biomass pretreatment for integrated processing into biofuels, food, and feed. Bioresour Technol 300:122719. https://doi.org/10.1016/j.biortech.2019.122719

de Carvalho JC, Aulestia DTM, Cardoso J et al (2023) Chapter 15: Process integration opportunities applied to microalgae specialty chemicals production. In: Microalgae-based systems. De Gruyter, pp 299–324

de Carvalho JC, Sydney EB, Kirnev PCS et al (2024) Microalgal products from agro-industry wastes. In: Microalgal bioengineering. Springer International Publishing, Cham, pp 127–150

de Morais MG, de Morais EG, Duarte JH et al (2019) Biological CO2 mitigation by microalgae: technological trends, future prospects and challenges. World J Microbiol Biotechnol 35:78. https://doi.org/10.1007/s11274-019-2650-9

de Souza Félix FCC, Hidalgo VB, de Carvalho AKF et al (2024) Assessing the application of marine microalgae *Dunaliella salina* in a biorefinery context: production of value-added biobased products. Biofuels Bioprod Biorefin 18:439–452. https://doi.org/10.1002/bbb.2577

Delattre C, Pierre G, Laroche C, Michaud P (2016) Production, extraction and characterization of microalgal and cyanobacterial exopolysaccharides. Biotechnol Adv 34:1159–1179. https://doi.org/10.1016/j.biotechadv.2016.08.001

Delrue F, Alaux E, Moudjaoui L et al (2017) Optimization of Arthrospira platensis (Spirulina) growth: from laboratory scale to pilot scale. Fermentation 3:59. https://doi.org/10.3390/fermentation3040059

Devi AC, Tavanandi HA, Govindaraju K, Raghavarao KSMS (2020) An effective method for extraction of high purity phycocyanins (C-PC and A-PC) from dry biomass of Arthrospira maxima. J Appl Phycol 32:1141–1151. https://doi.org/10.1007/s10811-019-02033-y

Esim N, Dawar P, Arslan NP et al (2024) Natural metabolites with antioxidant activity from micro- and macro-algae. Food Biosci 62:105089. https://doi.org/10.1016/j.fbio.2024.105089

Fang Y, Cai Y, Zhang Q et al (2024) Research status and prospects for bioactive compounds of Chlorella species: composition, extraction, production, and biosynthesis pathways. Process Saf Environ Prot 191:345–359. https://doi.org/10.1016/j.psep.2024.08.114

FAO. (2009a). Laminaria japonica. In Cultured aquatic species fact sheets. Text by Chen, J. Edited and compiled by Valerio Crespi and Michael New. https://www.fao.org/fishery/docs/CDrom/aquaculture/I1129m/file/en/en_japanesekelp.htm

FAO. (2009b). Porphyra spp. In Cultured aquatic species fact sheets. Text by Chen, J. & Pu Xu. Edited and compiled by Valerio Crespi and Michael New. https://www.fao.org/fishery/docs/CDrom/aquaculture/I1129m/file/en/en_nori.htm

FAO (2014). The State of World Fisheries and Aquaculture 2014. Rome. 223 pp.

Farag I (2013) Resources conservation in microalgae biodiesel production. Int J Eng Tech Res 1: 49–56

Finkel ZV, Follows MJ, Liefer JD et al (2016) Phylogenetic diversity in the macromolecular composition of microalgae. PLoS One 11:e0155977. https://doi.org/10.1371/journal.pone.0155977

Francavilla M, Kamaterou P, Intini S et al (2015) Cascading microalgae biorefinery: fast pyrolysis of Dunaliella tertiolecta lipid extracted-residue. Algal Res 11:184–193. https://doi.org/10.1016/j.algal.2015.06.017

Generalić Mekinić I, Šimat V, Rathod NB et al (2023) Algal carotenoids: chemistry, sources, and application. Foods 12:2768. https://doi.org/10.3390/foods12142768

Gordalina M, Pinheiro HM, Mateus M et al (2021) Macroalgae as protein sources—a review on protein bioactivity, extraction, purification and characterization. Appl Sci 11:7969. https://doi.org/10.3390/app11177969

Guiry MD (2024) How many species of algae are there? A reprise. Four kingdoms, 14 phyla, 63 classes and still growing. J Phycol 60:214–228. https://doi.org/10.1111/jpy.13431

Gupta A, Kang K, Pathania R et al (2024) Harnessing genetic engineering to drive economic bioproduct production in algae. Front Bioeng Biotechnol 12. https://doi.org/10.3389/fbioe.2024.1350722

Hassan Pranta M, Muk Cho H (2025) A comprehensive review of the evolution of biodiesel production technologies. Energy Convers Manage 328:119623. https://doi.org/10.1016/j.enconman.2025.119623

Hiraoka M (2021) Massive Ulva green tides caused by inhibition of biomass allocation to sporulation. Plants 10:2482. https://doi.org/10.3390/plants10112482

Ingle K, Vitkin E, Robin A et al (2018) Macroalgae biorefinery from Kappaphycus alvarezii: conversion modeling and performance prediction for India and Philippines as examples. Bioenergy Res 11:22–32. https://doi.org/10.1007/s12155-017-9874-z

Iwamoto H, Soccol CR, Aulestia DTM et al (2024a) Algae as ingredients for food production: opportunities to be explored. In: Trending topics on fermented foods. Springer Nature Switzerland, Cham, pp 303–324

Iwamoto H, Soccol CR, Molina-Aulestia DT et al (2024b) Lutein from microalgae: an industrial perspective of its production, downstream processing, and market. Fermentation 10:106. https://doi.org/10.3390/fermentation10020106

Jacob-Lopes E, Mérida LGR, Queiroz MI, Zepka LQ (2015) Microalgal biorefineries. In: Biomass production and uses. InTech

Jalilian N, Najafpour GD, Khajouei M (2020) Macro and micro algae in pollution control and biofuel production – a review. ChemBioEng Rev 7:18–33. https://doi.org/10.1002/cben.201900014

Jiang M, Zhao HH, Zai SW et al (2019) A defatted microalgae meal (Haematococcus pluvialis) as a partial protein source to replace fishmeal for feeding juvenile yellow perch Perca flavescens. J Appl Phycol 31:1197–1205. https://doi.org/10.1007/s10811-018-1610-3

Kartik A, Akhil D, Lakshmi D et al (2021) A critical review on production of biopolymers from algae biomass and their applications. Bioresour Technol 329:124868. https://doi.org/10.1016/j.biortech.2021.124868

Korsah MA, Gyau T, Lyon D, Danquah MK (2025) Techno-economic analysis of downstream processing of microalgae. In: Algal bioreactors. Elsevier, pp 309–321

Kraan S (2012) Algal polysaccharides, novel applications and outlook. In: Carbohydrates – comprehensive studies on glycobiology and glycotechnology. InTech

Kumar S, Gupta R, Kumar G et al (2013) Bioethanol production from Gracilaria verrucosa, a red alga, in a biorefinery approach. Bioresour Technol 135:150–156. https://doi.org/10.1016/j.biortech.2012.10.120

Loke Show P (2022) Global market and economic analysis of microalgae technology: status and perspectives. Bioresour Technol 357:127329. https://doi.org/10.1016/j.biortech.2022.127329

Lu S-Y, Zhou T, Shabbir I et al (2025) Marine algal polysaccharides: multifunctional bioactive ingredients for cosmetic formulations. Carbohydr Polym 353:123276. https://doi.org/10.1016/j.carbpol.2025.123276

Ma G, Li L, Hong Y, Yu M (2023) Cultivation of Haematococcus pluvialis in source-separated urine for biomass production and astaxanthin accumulation. Algal Res 69:102945. https://doi.org/10.1016/j.algal.2022.102945

Maity S, Mallick N (2022) Trends and advances in sustainable bioethanol production by marine microalgae: a critical review. J Clean Prod 345:131153. https://doi.org/10.1016/j.jclepro.2022.131153

Martínez-Herrera RE, González-Meza GM, Meléndez-Sánchez ER (2025) Toward sustainable bioplastics: the potential of algal biomass in PHA production and biocomposites fabrication. Process Biochem 150:276–287. https://doi.org/10.1016/j.procbio.2025.01.019

Martins A, Pinto FR, Barroso S et al (2023) Valorization of the Red Seaweed Gracilaria gracilis through a biorefinery approach. J Vis Exp. https://doi.org/10.3791/65923

Martins PL, Duarte LC, Pereira H et al (2025) Evaluation of different fractionation methods for the simultaneous protein and carbohydrate extraction from microalgae. Biomass Convers Biorefinery 15:999–1011. https://doi.org/10.1007/s13399-024-05279-w

Martone PT, Estevez JM, Lu F et al (2009) Discovery of lignin in seaweed reveals convergent evolution of cell-wall architecture. Curr Biol 19:169–175. https://doi.org/10.1016/j.cub.2008.12.031

Matos ÂP, Novelli E, Tribuzi G (2022) Use of algae as food ingredient: sensory acceptance and commercial products. Front Food Sci Technol 2. https://doi.org/10.3389/frfst.2022.989801

McHugh DJ (2003) A guide to the seaweed industry. FAO Fisheries Technical Paper 441

Mobin S, Alam F (2017) Some promising microalgal species for commercial applications: a review. Energy Procedia 110:510–517. https://doi.org/10.1016/j.egypro.2017.03.177

Naik B, Mishra R, Kumar V et al (2024) Micro-algae: revolutionizing food production for a healthy and sustainable future. J Agric Food Res 15:100939. https://doi.org/10.1016/j.jafr.2023.100939

Narayanan M (2024) Marine algae biomass: a viable and renewable resource for biofuel production: a review. Algal Res 82:103687. https://doi.org/10.1016/j.algal.2024.103687

Obeid S, Rida H, Peydecastaing J et al (2025) Coupling ultrasound and membrane filtration for the fractionation of Spirulina platensis sp. and the recovery of phycocyanin and pigment-free proteins. Biotechnol Lett 47:8. https://doi.org/10.1007/s10529-024-03541-9

Ohse S, Bianchini Derner R, Ávila Ozório R et al (2008) Crescimento de microalgas em sistema autotrófi co estacionário. Biotemas 21(2):7–18

Øverland M, Mydland LT, Skrede A (2019) Marine macroalgae as sources of protein and bioactive compounds in feed for monogastric animals. J Sci Food Agric 99:13–24. https://doi.org/10.1002/jsfa.9143

Peñalver R, Lorenzo JM, Ros G et al (2020) Seaweeds as a functional ingredient for a healthy diet. Mar Drugs 18:301. https://doi.org/10.3390/md18060301

Ponnusamy S, Reddy HK, Muppaneni T et al (2014) Life cycle assessment of biodiesel production from algal bio-crude oils extracted under subcritical water conditions. Bioresour Technol 170:454–461. https://doi.org/10.1016/j.biortech.2014.07.072

Rafa N, Ahmed SF, Badruddin IA et al (2021) Strategies to produce cost-effective third-generation biofuel from microalgae. Front Energy Res 9. https://doi.org/10.3389/fenrg.2021.749968

Rudke AR, de Andrade CJ, Ferreira SRS (2020) Kappaphycus alvarezii macroalgae: an unexplored and valuable biomass for green biorefinery conversion. Trends Food Sci Technol 103:214–224. https://doi.org/10.1016/j.tifs.2020.07.018

Scapini T, Woiciechowski AL, Manzoki MC et al (2024) Microalgae-mediated biofixation as an innovative technology for flue gases towards carbon neutrality: a comprehensive review. J Environ Manag 363:121329. https://doi.org/10.1016/j.jenvman.2024.121329

Shah MMR, Liang Y, Cheng JJ, Daroch M (2016) Astaxanthin-producing green microalga Haematococcus pluvialis: from single cell to high value commercial products. Front Plant Sci 7. https://doi.org/10.3389/fpls.2016.00531

Siladji C, Djordjevic V, Milijasevic JB et al (2024) Micro- and macroalgae in meat products. Foods 13:826. https://doi.org/10.3390/foods13060826

Sing N, Mahali K, Roy S (2024) An overview on biofuels: advantages and disadvantages. In: Recent developments in chemistry and biochemistry research, vol 7. BP International, pp 98–118

Singh Mehra K, Abrar I, Kant Bhatia R, Goel V (2025) A comprehensive review of algae consortium for wastewater bioremediation and biodiesel production. Energy Convers Manag 325:119428. https://doi.org/10.1016/j.enconman.2024.119428

Song X, Liu B-F, Kong F et al (2022) Overview on stress-induced strategies for enhanced microalgae lipid production: application, mechanisms and challenges. Resour Conserv Recycl 183:106355. https://doi.org/10.1016/j.resconrec.2022.106355

Tee MZ, Yong YS, Rodrigues KF, Yong WTL (2015) Growth rate analysis and protein identification of Kappaphycus alvarezii (Rhodophyta, Gigartinales) under pH induced stress culture. Aquac Rep 2:112–116. https://doi.org/10.1016/j.aqrep.2015.09.001

Tertychnaya TN, Manzhesov VI, Andrianov EA, Yakovleva SF (2020) New aspects of application of microalgae Dunaliella Salina in the formula of enriched bread. In: IOP conference series: earth and environmental science. Institute of Physics Publishing

Thoré ESJ, Muylaert K, Bertram MG, Brodin T (2023) Microalgae. Curr Biol 33:R91–R95. https://doi.org/10.1016/j.cub.2022.12.032

Wang S, Mukhambet Y, Esakkimuthu S, Abomohra AE-F (2022) Integrated microalgal biorefinery – routes, energy, economic and environmental perspectives. J Clean Prod 348: 131245. https://doi.org/10.1016/j.jclepro.2022.131245

Wells ML, Potin P, Craigie JS et al (2017) Algae as nutritional and functional food sources: revisiting our understanding. J Appl Phycol 29:949–982. https://doi.org/10.1007/s10811-016-0974-5

Xue J, Balamurugan S, Li T et al (2021) Biotechnological approaches to enhance biofuel producing potential of microalgae. Fuel 302:121169. https://doi.org/10.1016/j.fuel.2021.121169

Chapter 7
Cellulose

Ariane Fátima Murawski de Mello, Luciana Porto de Souza Vandenberghe, Gustavo Amaro Bittencourt, Lucia Ramos Neyra, Clara Matte Borges Machado, Gabriel Dudeque, and Carlos Ricardo Soccol

Abstract The awareness about the impact of human action on the environment is constantly rising. With population growth, demands for energy and materials will only increase. However, the dependence on nonrenewable resources will not be able to meet these demands. Therefore, alternatives based on biomass are essential. Cellulose is the main polysaccharide found in the environment, primarily in plants but also in other organisms. Cellulose's properties make it suitable for applications in several industries, such as medical, pharmaceutical, food, and electronics. Besides, the characteristics of this polysaccharide can be tailored depending on the manufacturing method or further modifications. This chapter will deal with the different aspects of cellulose, from its obtainment and natural properties to its applications and commercial products.

7.1 Introduction

Cellulose, a polymer composed of glucose monomers linked by β-1,4-glycosidic bonds (Jenkins et al. 2012), is the most abundant polysaccharide in the world (Heinze 2015). It is found mainly in plants (e.g., wood, crops, and agro-industrial residues) (Fig. 7.1) but also in other organisms such as aquatic animals, marine algae, fungi, and bacteria (Heinze 2015). Cellulose's wide availability, along with its fascinating characteristics of renewability, biodegradability, and biocompatibility, makes it an attractive material for different applications. There are reports of cellulose being used since antiquity in China for papermaking (de Avila et al. 2021). Nowadays, the growing interest in alternative materials makes cellulose a promising product.

A. F. M. de Mello · L. P. de Souza Vandenberghe (✉) · G. A. Bittencourt · L. R. Neyra · C. M. B. Machado · G. Dudeque · C. R. Soccol
Department of Bioprocess Engineering and Biotechnology, Federal University of Paraná, Curitiba, Paraná, Brazil
e-mail: lvandenberghe@ufpr.br

A. L. Woiciechowski et al. (eds.), *Biomass for Bioenergy and Bioproducts*, Biofuel and Biorefinery Technologies 17, https://doi.org/10.1007/978-3-032-19898-3_7

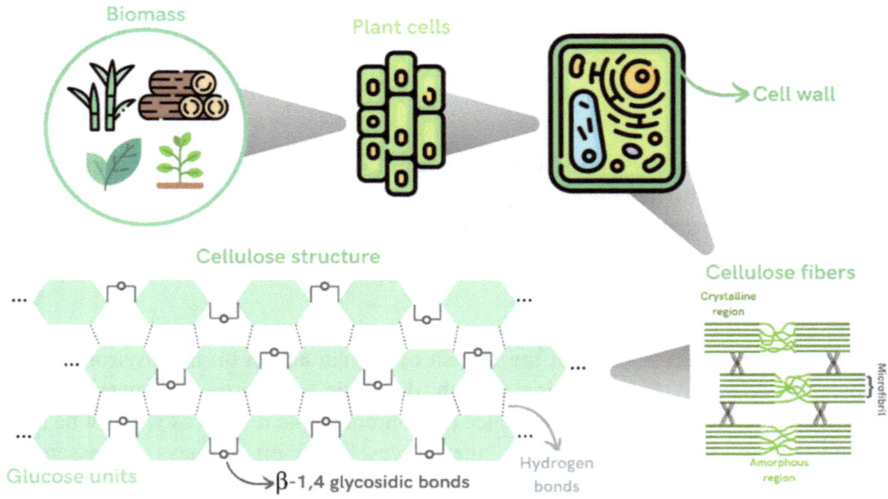

Fig. 7.1 Cellulose source and structure

With population growth, demand for energy and materials is constantly rising (Wang et al. 2021a). However, there is still a high dependence on nonrenewable and fossil sources for obtaining energy and product. Along with this, the environmental impacts of petroleum exploitation (e.g., damage to ecosystems, oil spills, and water use, among others) are constantly being questioned (U.S. Energy Information and Admnistration 2022). Therefore, renewable and environmentally friendly alternatives are essential for a greener future. Cellulose can replace some petroleum-based materials in different applications, having the required properties for industries. Besides, the possibility of tailoring some of cellulose's characteristics by employing different manufacturing methods and/or by molecular modification can also greatly expand the possibilities of applications (Heinze 2015).

This chapter will deal with cellulose's different sources and properties, along with the modes of manufacturing this polysaccharide. Furthermore, common and innovative applications will be discussed, along with the available commercial products based on this polymer.

7.2 Cellulose Properties

7.2.1 Sources of Cellulose

Cellulose is a ubiquitous biopolymer, occurring in a variety of natural environments and across many kingdoms of life. It is estimated that 10^{11}–10^{12} tons of cellulose is produced annually, primarily coming from photosynthesis in plants (Motaung and Linganiso 2018). In plants, almost 40% of their carbon content occurs in the form of cellulose. To illustrate, cellulose corresponds to 45–55% of bamboo and 40–50% of wood weight. It is also possible to find higher contents of cellulose in plants, such as in cotton (more than 90% wt) and ramie (70–75% wt) (Heinze 2015). Nevertheless, organisms other than plants can also synthesize cellulose. Marine algae are a promising source of cellulose due to their abundance and renewability (Machado et al. 2024). In green seaweed, the percentage of cellulose can reach up to 34% on dry weight, which is remarkable compared to the percentages found in brown seaweed (up to 10% dry weight) and red seaweed (up to 18% dry weight) (Baghel et al. 2021). Even though marine algae present a considerably lower content of cellulose compared to plants, their availability makes them a relevant source.

Cellulose can also be found in marine animals, such as ascidians, and microorganisms, such as bacteria and fungi. In particular, bacterial cellulose has received great attention due to its highly ordered and crystalline structure, which can reach up to 90% of crystallinity, allowing its use in different fields. The main genera of bacteria to produce cellulose are *Gluconacetobacter*, *Agrobacterium*, *Pseudomonas*, *Rhizobium*, and *Sarcina* (Girard et al. 2024).

7.2.2 Cellulose Structure and Morphology

Even though cellulose can be obtained from different sources, its molecular structure consists of D-glucopyranose ring units in a $^4C_1$-chair configuration, which is the most stable conformation (Rao et al. 1967). This stability comes from an intramolecular hydrogen bond network, spanning from the O(3′)-H hydroxyl to the O (5) ring oxygen in adjacent glucose units and also connecting the O(2)-H hydroxyl to the O(6′) hydroxyl in successive residues (Fig. 7.2a). The units are linked by β-1,4-glycosydic bonds, resulting in alternate turning of 180°, which creates a twisted, unbranched cellulose chain (Habibi et al. 2010). Each glucose monomer, also known as an anhydroglucose unit (AGU), presents three reactive hydroxyl groups (Heinze 2015). A primary hydroxyl group is found at C6, and two secondary hydroxyl groups are found at C2 and C3. The hydroxyl group in cellulose is positioned in the equatorial plane, while the hydrogen atoms are placed in the axial plane (Habibi et al. 2010).

Every cellulose chain presents directional asymmetry in its axis due to the polycondensation reactions that take place to link glucose units. On one end, cellulose possesses a reducing terminal, with a hemiacetal function in C1, whereas the nonreducing end at the opposing extremity shows a pendant hydroxyl group (Habibi et al. 2010). Cellulose chains can appear in nature with a variable degree of polymerization, which varies between 1,000 and 30,000, corresponding to 500–15,000 nm of chain length (Heinze 2015). When cellulose chains are packed together, they can form crystalline and amorphous regions. The latter is marked by cellulose chains that are randomly organized, with turns between adjacent polymers, resulting in less dense regions that are prone to react with other molecules and suffer enzymatic attack. In contrast, crystalline cellulose is composed of closely packed

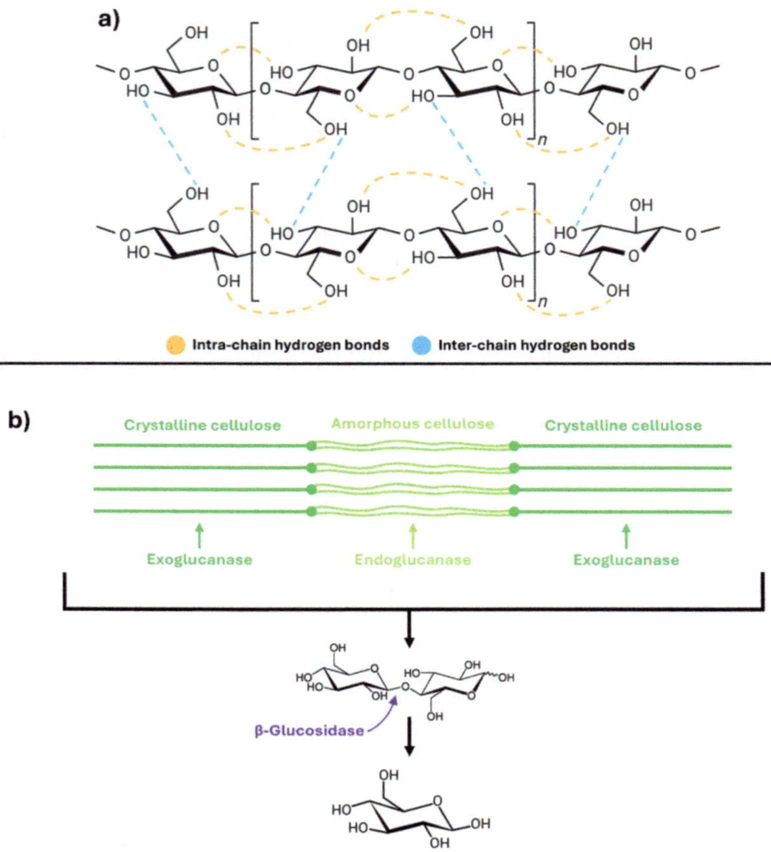

Fig. 7.2 (**a**) Structure of cellulose and (**b**) the synergic action of exocellulases, endocellulases, and β-glucosidase on cellulose structure

chains arranged in a defined direction, leading to highly ordered and dense regions (Heinze 2015). However, the direction of cellulose chains in crystalline cellulose can vary, giving rise to polymorphs. Currently, there are six identified cellulose polymorphs: I, II, III_I, III_{II}, IV_I, and IV_{II}.

Native cellulose was initially thought to have a single crystal structure, named cellulose I. However, evidence for two types of cellulose I was obtained when infrared spectroscopy was applied to marine algae and bacterial cellulose. These findings revealed that these sources of cellulose produced a polymer with different orientations of fibers in crystalline regions compared to plant cellulose, resulting in the classification of cellulose I as I_α and I_β (Marrinan and Mann 1956; Mann and Marrinan 1958). Solid-state nuclear magnetic resonance spectroscopy also helped elucidate cellulose polymorphism, demonstrating that cellulose I regions presented both cellulose I_α and I_β in different proportions (Atalla and Vanderhart 1984; Vanderhart and Atalla 1984). Cellulose I_α is formed by single-chained triclinic units predominant in marine algae and bacteria, while cellulose I_β is predominant in plants and ascidians and is formed by double-chained monoclinic units. The more thermodynamically stable polymorph I_β can be converted into I_α polymorph and vice versa by physicochemical treatment, as demonstrated by Sugiyama et al. (1990). A detailed account of how native cellulose polymorphs are structured in a cellulose fiber was given by Horikawa (2022).

Cellulose II polymorph is rarely observed in nature but is commonly produced by chemical treatment. This polymorph can be produced by chemical regeneration, where cellulose I is dissolved in solutions of heavy metals, such as amine or diamine complexes, and then precipitated in water. Mercerization is also used to produce cellulose II. In this case, cellulose I is swollen in a concentrated swelling agent solution, becoming cellulose II after the removal of the swelling agent. Cellulose II polymorph is monoclinic and presents a more stable structure than native cellulose. Because of the chemical treatment, the hydrogen bonding network changes in cellulose II, yielding an antiparallel arrangement of fibers that confers higher resistance (Habibi et al. 2010).

Cellulose III can be produced by exposing cellulose I or II to ammonia or amines and subsequently removing the swelling agent, resulting in polymorphs III_I or III_{II}, respectively. Each cellulose III subpolymorph structure resembles its parent polymorph. Cellulose III_I presents chains in a parallel configuration in the same manner as cellulose I, while cellulose III_{II} presents chains in an antiparallel orientation (Spedding 1962). Moreover, cellulose III can be converted into its parent polymorph by hydrothermal treatment with or without chemical agents (Wu et al. 2020). Polymorphs IV_I and IV_{II} can be reversibly prepared by heating cellulose III_I or III_{II} up to 260 °C in glycerol or in ionic liquids (ILs) (Gardiner and Sarko' 1985; Tokumasu et al. 2024).

Native cellulose has a hierarchical structure. Inside plant cells, glucose units are linked by glycosidic bonds, forming a glucan chain, named elementary fibril. These fibrils compose the basic unit of cellulose and have been reported to measure 100 nm in length and 1.5–3.5 nm in width. Secreted elementary fibrils start bundling together, giving rise to microfibrils of cellulose, with their typical width ranging from 10 to 30 nm and a length of hundreds of nanometers (Krassig 1993). The packing of the glucan chains into the microstructure of cellulose relies partially on a complex hydrogen bond network, also depending on other kinds of interactions, such as hydrophobic interaction and solvation energy (Altaner et al. 2014). In plant cell walls, the microfibrils are organized as macrostructures, called cellulose fibers. These structures are surrounded by amorphous cellulose, which is covered by hemicellulose and lignin, creating lignocellulosic biomass. The morphology of cellulose fibers changes with the source. For instance, cellulose fibers in cotton are twisted, whereas cellulose fibers in wood are untwisted. Furthermore, unlike native cellulose, where fibers are organized in a parallel fashion, cellulose polymorphs II, III, and IV are randomly distributed in the macrostructure. This is due to the random precipitation of cellulose fibers after the removal of the swelling agent (Heinze 2015). The enzymatic hydrolysis of cellulose occurs with the synergistic action of three enzymes (Fig. 7.2b): endoglucanases (which act internally on cellulose, exposing nonreducing terminals); cellobiohydrolases, or exoglucanases (release cellobiose and other oligosaccharides); and β-glucosidases (generate glucose monomers) (Pino et al. 2018; Anoop Kumar et al. 2019).

7.2.3 Cellulose Biosynthesis

Cellulose is synthesized in plants as a constituent of the cell walls. Plant cells produce two types of walls: the primary cell wall (PCW) and the secondary cell wall (SCW). The PCW is ubiquitous in plant cells, while the SCW is exclusive of some mature cells specialized in water transport or mechanical support. The PCW is produced during cell growth, consisting of a thin and flexible layer composed of polysaccharides and proteins. The main function of PCW is to provide shape to the protoplast and maintain it during cell enlargement. When plant cells grow, their cell walls suffer an irreversible extension, called polymer creep, in which cellulose microfibrils and other polysaccharides slide within the walls. This process can increase the surface area of cells impressively. Such an enlargement is caused by the higher water uptake in cells, augmenting the vacuole size and pressing the walls outward, while new polymers are integrated into the expanded cell wall (Cosgrove 2005).

Cellulose biosynthesis occurs on the membrane of plant cells and is conducted by large enzyme complexes, called cellulose synthase complexes (CSCs) (Kimura et al. 1999). The CSC is formed by different cellulose synthase (CESA) units, which were initially identified in the model plant *Arabidopsis thaliana* and are involved in the synthesis of cellulose with different functions. For example, CESA 1, CESA 3, and CESA 6 constitute the CSC for PCW synthesis, whereas CESA 4, CESA7, and CESA8 are involved in SCW synthesis (Arioli et al. 1998; Taylor et al. 2003; Kumar et al. 2016). CSC presents a hexameric structure, often called a rosette, which is believed to come from the dimerization of CESA units by two zinc fingers in the N-terminal regions, which then form trimers that create the complex (Kurek et al. 2002; Atanassov et al. 2009). CESA units catalyze the formation of β-1,4-glycosidic bonds between two UDP–glucose moieties and extrude glucan chains from the membrane. A cellulose chain is formed when 18 glucan chains bundle together in the outside of the cell, possibly aided by proteins anchored on the outer side of the plasma membrane (Cosgrove 2014).

Other proteins involved in cellulose biosynthesis are Korrigan (KOR) endoglucanase, companion of cellulose synthase (CC), and cellulose synthase interacting (CSI) proteins (Kumar et al. 2016). The exact function of these proteins in cellulose synthesis is currently unclear, but some studies suggest that they play a role in microfibril crystallization and CSC stabilization on the membrane. On one hand, genetic studies demonstrated that the co-expression of KOR endoglucanase is crucial for the development of crystalline cellulose (Bhandari et al. 2006). This is in line with findings that demonstrate KOR endoglucanases processing cellulose chains as they are excreted, releasing them from the CSC and contributing to the formation of microfibrils (Vain et al. 2014). On the other hand, CC and CSI are believed to work together to align CSC with microtubules. The association of CSC with microtubules is needed for not only the movement of the rosette along the membrane but also its delivery to the correct sites on the membrane and stability during adverse conditions. CC1 protein was recently demonstrated to sustain the CSC and microtubule association during salt stress (Kesten et al. 2019). CSI1's role as a scaffold between CSC and microtubules was also demonstrated, showing its participation in CSC delivery on the membrane, movement across it, and stability (Bringmann et al. 2012; Zhu et al. 2018).

7.2.4 Cellulose Properties

Cellulose is an odorless, tasteless, chiral, and biodegradable polymer. Cellulose is composed of β-D-glucopyranoside units, which impart chirality to the molecule. Moreover, it presents high thermal resistance, requiring 320 °C and a pressure of 25 MPa to become amorphous in water and producing volatile compounds due to thermal degradation between 400 and 550 °C (Krumm et al. 2016). Cellulose also presents notable tensile strength, which can vary from 7.5 to 7.7 GPa in elementary fibers to 124–157 MPa in cellulose composites (Liu et al. 2019). Because of the abundance of reactive hydroxyl groups on its surface, cellulose chains can undergo intense chemical modification. In certain organic solvents, hydroxyl groups in cellulose can be substituted by chemical radicals, suffering sulfation, acetylation, and methylation reactions, for example. These modifications can introduce novel properties to the polymer, such as explosiveness, solubility in water, and bioactivity (Heinze 2015).

Cellulose is insoluble in water. This behavior has been traditionally associated with the complex hydrogen bond network within cellulose microfibrils (Kondo 1997). The reactive hydroxyl groups of AGUs interact with each other, creating intra- and intermolecular hydrogen bonds and giving rise to a myriad of 3D shapes. The intramolecular hydrogen bonds are formed by pairs of the hydroxyl group at C3 and the adjacent ether oxygen and the hydroxyl groups at C6 and C2 (Heinze 2015). When combined with β-glycosidic linkages, the intramolecular hydrogen bond network is responsible for the cellulose polymer's high viscosity and rigidity, leading to crystalline structures (Krassig 1993). Furthermore, the intermolecular hydrogen bonds are produced by the interaction of oxygen atoms at C3 and hydroxyl groups at C6 in adjacent cellulose molecules (Gardner and Blackwell 1974). However, this intricate hydrogen bond network alone is not responsible for the strong interaction between cellulose chains. Recently, the influence of hydrophobic and weak interactions on not only intermolecular cellulose interactions but also other properties has been emphasized (Wohlert et al. 2022). Hydrophobic interactions emerge in cellulose chains by the approximation of carbon atoms, while weak interactions result from the interplay of carbon, hydrogen, and oxygen atoms. Some studies suggest that these interactions play a more fundamental role in cellulose arrangement than hydrogen bonds.

7.3 Cellulose Production

Due to its structural characteristics and abundant presence in the biosphere, as previously mentioned, cellulose has historically been a highly important biopolymer for human societies since antiquity (Hon 1994; Alén 2018). Over time, cellulose has evolved from a naturally occurring component to a cornerstone of modern industry, playing an essential role in communication, textiles, chemicals, and, more recently, sustainable solutions for the future (Carolin et al. 2023).

7.3.1 Wood-Pulping Process

Pulping refers to different processes used to extract cellulose from feedstocks, transforming them into a mass of liberated fibers by dissolving the lignin that binds components together, in a process known as delignification (Schieppati et al. 2023). Despite achieving lower yields of 45–55% of cellulose recoveries, kraft-based processes account for around 75% of the worldwide applied processes for papermaking due to the excellent strength properties of the resulting pulp and their well-established system for the efficient recovery of cooking chemicals, energy, and by-products (Das and Houtman 2004).

The conventional kraft process takes place at high temperatures (160–170 °C) and pressures (7 bar) under strongly alkaline conditions, using aqueous solutions of sodium hydroxide and sodium sulfide (Sewsynker-Sukai et al. 2020). Due to the lack of selectivity in kraft pulping, around 50% of the raw wood degrades and dissolves into black liquor. As a result, besides degraded lignin fragments, it contains a substantial amount of carbohydrate-derived materials (Fu et al. 2022). Unlike the alkaline kraft method, the sulfite process operates across a broad pH range, from acidic (pH 1–2) to alkaline (pH 9–13; alkaline sulfite–anthraquinone cooking), leading to significant variations in pulp yield and quality (Sewsynker-Sukai et al. 2020; Fu et al. 2022). The active sulfur compounds also change depending on the pH. Similar to kraft pulping, acid sulfite and bisulfite cooking degrade approximately 50% of wood into spent liquor, which contains lignosulfonates and sugars (Alén 2018; Sewsynker-Sukai et al. 2020).

7.3.2 Advanced Methods for Cellulose Production

With the increasing awareness of environmental degradation and the urgent need for sustainable solutions, the production of cellulose has significantly evolved. The development of more efficient and environmentally friendly methods is aimed at the maximization and diversification of renewable biomass resource exploration.

The development of advanced pretreatment strategies has significantly improved cellulose extraction, enhancing process efficiency, reducing environmental impact, and increasing cellulose purity (Abolore et al. 2024).

7.3.2.1 Pretreatment

Lignocellulosic biomasses (wood and non-wood) are complex and abundant materials composed of three major components: cellulose, hemicellulose, and lignin (Abolore et al. 2024). These components are structurally and chemically intertwined, making the extraction of cellulose challenging. The pretreatment process is essential to breaking down its rigid structure, making cellulose more accessible for further processing (Bhatia et al. 2020). There are numerous pretreatment strategies, each with its own set of advantages, depending on the specific biomass and the intended use of the resulting cellulose. There is an abundance of reviews that address in detail and depth the various pretreatment methods for lignocellulosic biomass (Bhatia et al. 2020; Kumar et al. 2020; Mankar et al. 2021; Liu et al. 2022; Liu and Cheng 2024; Lv et al. 2024).

In the context of this chapter, various pretreatment methods are used to selectively target and break down the lignin and hemicellulose components, leaving the cellulose fraction largely intact. Chemical pretreatments, particularly those involving alkaline or acid treatments, are widely applied to dissolve lignin and solubilize hemicellulose, allowing the preferential isolation of cellulose (Woiciechowski et al. 2020). Acid pretreatment is commonly employed as an effective process to hydrolyze hemicellulose, while alkaline pretreatments are prevalent to remove lignin. Additionally, the organosolv process, which utilizes organic solvents such as ethanol or acetone, offers a selective method to dissolve lignin, enabling the extraction of cellulose (Zhang et al. 2016; Luo et al. 2023). Recent developments have also focused on more integrated processes, combining different methods, trying to improve the global efficiency and cost-effectiveness of cellulose extraction.

7.3.2.2 Purification

Once the biomass undergoes pretreatment and the cellulose is partially separated from the other components, purification processes are necessary to remove the remaining impurities. Bleaching is a widely applied method for purification, in which oxidative agents, such as hydrogen peroxide (H_2O_2) or sodium chlorite ($NaClO_2$), are employed to remove the residual lignin and improve cellulose characteristics (Wang and Zhao 2021). H_2O_2 bleaching can increase the cellulose content, crystallinity, and brightness in various biomass sources (Gil-Guillén et al. 2024), while $NaClO_2$ can achieve up to 92.6% delignification, also resulting in high-purity hemicellulose (86.1%) and xylo-oligosaccharides (96.3%) with enhanced antioxidant activity (Wang et al. 2021b). This process is typically followed by additional washing steps to ensure that all unwanted impurities are thoroughly removed.

Another promising approach for purification is solvent-based extraction. Ionic liquids (ILs) and deep eutectic solvents (DESs) are used to selectively dissolve and remove noncellulosic components. ILs and DESs represent emerging alternative reaction media with comparable physicochemical characteristics. DESs are particularly useful for processing polysaccharides, enabling selective solubilization, delignification, and nanofibrillation of cellulose (Zdanowicz et al. 2018). These processes enable the precipitation of pure cellulose, which can then be further processed for various applications.

7.3.2.3 Nanocellulose Production

With the advancement in understanding the structures and properties of cellulose fibers, nanocellulose has emerged as a material of significant interest for production. Its unique characteristics, such as high surface area, versatility, exceptional mechanical strength, and biocompatibility, make it a promising biomaterial for various high-performance applications, ranging from biomedical uses to advanced composites (Ahmed et al. 2025; Zhang et al. 2025). Due to their hydroxyl groups and strong hydrogen bonding, cellulose fibers aggregate themselves within an entangled composite between highly ordered crystalline regions, which provide stiffness and strength, and disordered amorphous regions, which enhance flexibility (Dufresne 2012). Nanocellulose is the natural fiber extracted from cellulose, containing at least one nanoscale dimension and 100% of cellulose chemical composition, typically having a high aspect ratio with lengths up to several micrometers (Phanthong et al. 2018).

The attributes of nanocellulose, including morphology, dimensions, and surface characteristics, are determined by the origin of the cellulose, the method of isolation, and the processing techniques applied (Phanthong et al. 2018). Cellulose nanofibrils, also referred to as cellulose nanofibers (CNFs), nanofibrillated cellulose, or microfibrillated cellulose, consist of interwoven bundles of elongated and flexible nanofibrils 1–100 nm in diameter and 500–2000 nm in length, containing both crystalline and amorphous regions (Wang et al. 2024). These are obtained mainly through mechanical treatments such as homogenization, microfluidization, grinding, cryo-crushing, or high-intensity ultrasonication, which apply a force to fragment cellulose fibrils along their longitudinal axis (Phanthong et al. 2018). Cellulose nanocrystals, also recognized as nanocrystalline cellulose, are rod-like or whisker-shaped crystals with a high degree of crystallinity (54–88%), also composed entirely of cellulose (Wang et al. 2024). Cellulose nanocrystals are distinguished by their elevated tensile strength (7500 MPa), rigid configuration, large specific surface area (250–500 m^2/g), optical attributes, chemical purity, and high aspect ratio (~10–70) (Zhang et al. 2025).

Acid hydrolysis with sulfuric acid is one of the most commonly employed methods for nanocellulose manufacturing due to its ability to selectively hydrolyze amorphous cellulose regions while preserving crystalline domains. The major drawback is the generation of acidic wastewater, necessitating extensive washing cycles.

Maiti et al. (2013) applied sulfuric acid concentrations around 47%, requiring neutralization steps with deionized water and subsequent treatment with 0.5 N sodium hydroxide (NaOH). An alternative to acid hydrolysis is 2,2,6,6-tetramethylpiperidine-1-oxyl radical (TEMPO)–mediated oxidation, which converts hydroxyl groups into carboxylates, enhancing dispersion and stability (Tang et al. 2024). The reaction employs TEMPO/NaBr/NaClO under alkaline conditions or TEMPO/NaClO/$NaClO_2$ under neutral-to–mildly acidic conditions (Phanthong et al. 2018; Wang et al. 2024). The resulting oxidized nanocellulose fibers exhibit uniform diameters of 3–4 nm, with high aspect ratios. Due to the presence of functional carboxyl groups, TEMPO oxidization–based cellulose nanofibrils have been studied in different fields, including biomedicine, wastewater treatment, bioelectronics, and others (Tang et al. 2024).

Among mechanical processes, high-pressure homogenization involves forcing cellulose slurry through a chamber with high pressure and high velocity, generating shear forces that can reduce the fibril size to 10–20 nm, though at the cost of reduced crystallinity (Li et al. 2012; Phanthong et al. 2018). Ultrasonication relies on acoustic cavitation, inducing bubble collapse and fiber fragmentation; for instance, processing pure wood pulp at 70 °C for 5 h has yielded nanocellulose with widths of 10–100 nm and an 85.38% yield (Tang et al. 2013). The primary disadvantage of the mechanical process is its high energy consumption. As a result, it is typically combined with other pretreatment methods to reduce energy usage. Hybrid chemo-mechanical approaches improve efficiency by integrating chemical pretreatments with mechanical disintegration. For instance, atmospheric pressure liquefaction combined with TEMPO oxidation and acid hydrolysis has yielded nanocellulose crystals from banana pseudo-stem with 95.6% crystallinity (Meng et al. 2019).

Despite their industrial adoption, conventional methods face economic barriers, including multistep processes, high chemical costs, wastewater treatment challenges, and high energy consumption in the mechanical processes. Consequently, future scaling efforts must prioritize single-step approaches capable of directly extracting nanocellulose from biomass residues while reducing processing costs.

7.3.2.4 Bacterial Cellulose Production

Among the advanced processes recently being researched is the microbial cellulose, produced by certain bacteria, which offers a promising alternative to plant-derived cellulose, possessing superior properties such as high purity, water-holding capacity, and unique structural characteristics (Lin et al. 2020). Additionally, bacterial cellulose nanofibrils can be converted into macrofibers with exceptional strength, potentially replacing synthetic fibers derived from fossil fuels (Raghavendran et al. 2020). Bacteria from the genus *Komagataeibacter* sp. are the most studied species for bacterial cellulose production due to the higher purity of cellulose (Raghavendran et al. 2020).

The production of bacterial cellulose involves the fermentation of simple sugars, such as glucose, or glycerol by bacteria. These bacteria secrete cellulose in the form of nanofibers, which can then be harvested and processed. The post-processing steps, including purification and drying, are crucial to obtaining the final cellulose product in the desired form, such as films, membranes, or hydrogels. Recent research has focused on optimizing bacterial cellulose production using low-cost, sustainable media derived from agricultural, industrial, and food processing wastes (El-Gendi et al. 2022). According to the review of El-Gendi et al. (2022), final concentrations of bacterial cellulose reached using sugarcane straw, citrus pulp water, and vinasse were 4.51, 8.77, and 7.02 g/L, respectively, by *Komagataeibacter xylinus* fermentation. The large-scale production of bacterial cellulose faces key challenges, including the need for the reduction of fermentation time through the optimization of cultivation conditions and the limited integration of modern fermentation strategies and molecular biology tools that hinders advancements in production efficiency (Rahman et al. 2021; El-Gendi et al. 2022).

7.4 Cellulose Applications

Cellulose has been applied in different fields for a long time, with reports of its use in paper production as early as 105 A.D. in China (de Avila et al. 2021). Ever since, its use has grown in different fields, such as the food, packaging, cosmetic, pharmaceutical, biomedical, and, more recently, electronics industries (Table 7.1). In fact, a very common household item is made of regenerated cellulose—the transparent film cellophane. This material has several applications in packaging due to its durability, transparency, and elasticity (Shokri and Adibki 2013). However, its production is not eco-friendly, as harsh chemicals are required. Even so, cellophane is biodegradable (El Bourakadi et al. 2024).

Other than packaging films, cellulosic materials can also be used in mulch films. Chen et al. (2025) prepared an agricultural film of polyvinylidene fluoride (PVDF) and cellulose acetate (CA) with a heat dissipation function, which promoted higher germination and growth of vegetables when compared to traditional polyethylene films. Also, cellulose fibers can be used in edible films and coatings in the food industry to protect fruits and vegetables from moisture loss, extending their shelf life (El Bourakadi et al. 2024). In a study conducted by Três et al. (2024), cellulose nanofibers (CNFs) were added as reinforcing agents in biodegradable films, enhancing their mechanical properties and biodegradability, showing potential to be used in food products.

Cellulose derivatives and nanocellulose can be used in the cosmetic field since they are able form a gel-like material, establishing a protective barrier and delivering bioactive compounds (de Avila et al. 2021). For example, it is reported that carboxymethyl cellulose (CMC) is capable of stimulating collagen formation, showing potential as a bioactive and structural ingredient in skin care formulations (Costa et al. 2022). Additionally, Kim et al. (2024) analyzed cellulose microbeads that showed biocompatibility, solubility in different solvents, mechanical robustness, and pH stability, indicating their potential to be used in cosmetic products.

Table 7.1 Application of cellulose in different fields, based on its type and respective characteristics

Field	Application	Cellulose type	Properties and advantages	Reference
Food industry	Stabilizer	MC, HPMC	Formation of gel-like structure when cooling, maintaining consistent texture and stability	El Bourakadi et al. (2024)
	Emulsifier	CMC	Prevention of the separation of oil and water phases	El Bourakadi et al. (2024)
	Functional foods	CF	Increase in dietary fiber content	El Bourakadi et al. (2024)
	Edible films	CA, CMC	Reduced moisture loss, promoting extended shelf life and preventing spoilage	El Bourakadi et al. (2024)
	Gluten-free products	HPC	Improvement in texture and structure by mimicking gluten properties	El Bourakadi et al. (2024)
Biomedical	Tissue engineering	NCC	Presence of a natural fiber network of fibrils or nanorods, facilitating cell attachment and proliferation	Seddiqi et al. (2021)
		CMC	Film formation capacity, rheological control– and viscosity-increasing agent, binder and stabilizer	Seddiqi et al. (2021)
	Wound healing	NFC	High surface area–to-volume ratio, high water-holding ability, and high porosity	Seddiqi et al. (2021)
	Drug delivery	NFC	Flexibility, conducive elasticity, low density, low toxicity, and a reactive surface that can be used for grafting specific chemical groups	Seddiqi et al. (2021)
		CANF	Biocompatibility, biodegradability, high shear strength and shear modulus, regenerative properties, high affinity with other substances, and suitable flexural and tensile strength	Seddiqi et al. (2021)

(continued)

Table 7.1 (continued)

Field	Application	Cellulose type	Properties and advantages	Reference
Packaging	Films	MC	Enhanced mechanical and antibacterial characteristics	El Bourakadi et al. (2024)
Electronics	Power sources, energy storage devices, supercapacitors, and rechargeable lithium batteries	NC	Good optical properties, flexibility, lightweight, and eco-friendly characteristics	Motaung (2021)

Abbreviations: *MC* methylcellulose, *HPMC* hydroxypropyl methylcellulose, *CMC* carboxymethylcellulose, *CF* cellulose fibers, *CA* cellulose acetate, *HPC* hydroxypropyl cellulose, *NC* nanocellulose, *NCC* nanocrystalline cellulose, *NFC* nanofibrillar cellulose, *CANF* cellulose nanofiber acetate

In the pharmaceutical industry, cellulose-based materials are used in drug delivery systems, as excipients for tablets and capsules and in the formulation of ointments, creams, and syrups while also being used in wound healing and dressing (El Bourakadi et al. 2024). The use of such materials can enhance the properties of already existing polymeric matrices. For example, Palmer et al. (2011) noticed a slower drug release after incorporating CMC into a hydrophilic matrix, when compared to a single-polymer matrix. Similarly, Godakanda et al. (2019) formulated a fiber mat loaded with naproxen by blending ethyl cellulose (EC) and polyvinyl pyrrolidone (PVP). The final product showed great potential to be used as wound dressings for pain and inflammation.

Cellulosic products can also be used in the medical field as carriers for bioactive compounds and in the development of artificial blood vessels and bioresorbable surgical sutures (El Bourakadi et al. 2024). It is reported that the use of methyl cellulose (MC) in support hydrogel ink could enhance the scaffold's properties, being attractive for 3D plotting in the biomedical field (Ahlfeld et al. 2018). Also, the addition of cellulose nanocrystals (CNCs) in nanocomposite films can enhance the final composite's biocompatibility, indicating its potential toward tissue engineering, especially skin tissue regeneration (Nasalapure et al. 2025).

Furthermore, cellulose derivatives are suitable for energy applications due to their flexibility, durability, wettability, and thermal stability, in addition to being biocompatible and biodegradable (Verma et al. 2024). These materials can be used as flexible and wearable electronics, triboelectric and piezoelectric generators, thermoelectric and solar energy conversion devices, and storage devices like electrodes, polymer separators, polymer electrolytes, and rechargeable aqueous batteries (Teng et al. 2023). One example is the work developed by Abdalkarim et al. (2021), where a composite reinforced with magnetic CNC was developed to be used as an energy storage material. The authors noticed that the final composite showed excellent thermal stability and high efficiency for thermal energy conversion and storage. Another example is the development of a cellulose-based energy-harvesting and interactive device by Zhang et al. (2021). The material fabricated by the authors could drive commercial electronics and control an electronic piano, in addition to being completely degraded after 8 h of hydrolysis with cellulolytic enzymes.

7.5 Advancements and Innovation in Cellulose Applications

Other than being applied in the food, biomedical, packaging, and electronics industries, cellulose-based materials have been innovatively used in magnetic products, bio-ink, wastewater treatment, and seawater desalination (Fig. 7.3). The addition of magnetic nanoparticles to cellulosic derivatives enables the creation of a new class of smart materials that can adapt to changing external magnetic fields, making them attractive to be used in optical and medical equipment, electrical fabrics, loudspeakers, magnetic shielding, separation of biomolecules, and environmental accumulations (Verma et al. 2024). For instance, Galland et al. (2013) fabricated CNFs decorated with magnetic ferrite nanoparticles, applying the final product in a

Fig. 7.3 Possible applications for cellulose-based materials, namely food, biomedical, packaging, and electronic industries; magnetic products; bio-ink; wastewater treatment; and seawater desalination

loudspeaker prototype. The obtained cellulosic product enabled music generation by a single static coil, which was suitable for application in small loudspeakers, hearing aids, and noise reduction technologies.

Additionally, cellulose materials can be used as bio-ink, being the most cutting-edge in wound dressing technologies, while also showing potential to be applied in nanostructures for functionalized molecule stabilization (Verma et al. 2024). As an example, Mietner et al. (2021) developed CNF-based hydrogel inks combined with metallic cations (Fe^{3+}, Al^{3+}, Ca^{2+}, and Mg^{2+}). The gels with incorporated Ca^{2+} and Mg^{2+} were suitable for 3D printing, showing potential in tissue engineering and wound dressing applications. In an innovation developed by Yaopeng et al. (2020, CN106267370B), the authors disclose obtaining a silk fibroin/cellulose 3D printing ink, with high viscosity, high forming and curing speed, mild printing conditions, high precision, excellent mechanical properties, and good biocompatibility. According to the inventors, the obtained cellulosic bio-ink can be used to repair and reconstruct damaged tissues and organs in the human body.

Furthermore, cellulose can be used in water treatment applications due to its innate hydrophilicity and ready interaction with water molecules, being suitable for the removal of heavy metals, oil, microorganisms, and dyes (Sharma et al. 2025). It is reported that a cellulose–graphene oxide composite can efficiently treat wastewater and remove rhodamine B (an organic dye) while also being easily regenerated and suitable for use in further cycles (Tian et al. 2018). Cellulose can also be used in the form of Carbon Nanotubes (CNTs) to improve ultrafiltration membranes' anti-fouling properties, making the membranes suitable to remove water contaminants (Han et al. 2021). In this sense, Xiuping (2021, CN110270317B) disclosed a method to prepare a cellulose-based hydrogel suitable for wastewater treatment.

Cellulosic materials also have great advantages in seawater desalination due to their natural origin, low cost, and porous hydrophilicity, showing excellent water transport capabilities, high interfacial evaporation efficiency, and environmentally friendly solutions (Zhu et al. 2025). In fact, Li et al. (2023) fabricated a CNT–CNF composite film that exhibited light-induced desalination capability, reaching an evaporation efficiency of 84.5% and a reduction of Na^{+}, Ca^{2+}, K^{+}, and Mg^{2+} by four orders of magnitude. The authors concluded that this was a scalable green technology with potential in seawater desalination.

7.6 Cellulose Market and Commercial Products

The cellulose sector has a significant role in the global economy due to its high revenue generation, substantial investment, and impact on various economic sectors (Vieira et al. 2024). In 2023, the global cellulose market was valued at USD 44.48 billion, with an estimated annual growth rate of 9.67% from 2023 to 2031 (Data Bridge 2024). Cellulose contains abundant hydroxyl and polar groups, which can be chemically modified to introduce various active functional groups. This results in a wide range of cellulose materials with diverse properties. Commercial cellulose derivatives include methyl cellulose (MC), ethyl cellulose (EC), carboxymethyl cellulose (CMC), hydroxypropyl cellulose (HPC), hydroxypropyl methyl cellulose (HPMC), and cellulose acetate (CA), among others (Zhou et al. 2022). Most companies that commercialize these products are based in Asia. Table 7.2 summarizes some of the key companies that distribute commercial cellulose-based products.

Cellulose-based products are widely used for water and aqueous fluid absorption. The most common absorbent products include paper towels and tissue paper. Additionally, cellulose-based materials can serve as superabsorbents in hygiene products, offering a sustainable alternative to hydrogels, which are typically derived from synthetic polymers. Cellulose hydrogels can absorb and retain large amounts of a liquid relative to their own mass (Bashari et al. 2018). Another application involves cellulose nanofibers (CNFs). For instance, Nippon Paper Crecia commercializes highly effective antibacterial and deodorant sheets using CNFs modified via the TEMPO catalytic oxidation method. The company introduced *Hada Care Acty*, the first adult diapers made from functional CNF sheets, and later launched *Poise*, a mild incontinence pad utilizing the same CNF technology (Nippon Paper 2025).

Cellulose esters, compared with nitrocellulose, are non-yellowing and offer excellent clarity and stability, making them ideal for the nail care industry as film formers. Films formed with Eastman™ cellulose acetate propionate (CAP-482-0.5) provide an excellent resin for nail lacquer topcoats. Additionally, Eastman cellulose acetate butyrate (CAB) and Eastman cellulose acetate propionate (CAP) are used in skin care formulations, contributing to oil stabilization, high solubility, and compatibility with other ingredients (Eastman 2024). Another key application of cellulose is in construction, coatings, and ceramics, where its water retention, thickening, and adhesion properties are highly beneficial. For example, Celopro® offers tailored solutions for building materials by enhancing workability, setting, and durability. These properties improve the performance of cement-based mortars, tile adhesives, and joint compounds, ensuring excellent stability, even under varying environmental conditions (Celotech 2025).

7.7 Conclusions and Future Perspectives

Cellulose is an intriguing polymer. Its high availability in the environment, along with its characteristics of biodegradability and biocompatibility, makes it a promising material for several industries. Different methodologies can be applied for obtaining this polysaccharide from plants, with the kraft process being the most common. Besides, acid hydrolysis and TEMPO can be applied for obtaining nanocellulose. The addition of functional groups can ensure the final material's specific properties for different applications. Future research is focused on new and green methodologies for efficiently manufacturing this polysaccharide, along with finding new applications for such a valuable molecule.

Table 7.2 Companies that distribute cellulose-based products

Company	Cellulose-based product	Application	Reference
Anomera	DextraCel®	Bioplastic production, coating, and adhesives	Anomera (2025)
	CellAno™	Coatings	
Ashland	Blanose™	Thickener and stabilizer, adding texture and enhancing mouthfeel, in the food and beverage industry	Ashland (2025)
Celanese	Clarifoil®	Packaging of products from baked goods to perfume, clothing, and confectionery	Celanese (2025)
	Cellulose acetate flake	Packaging film, textiles, fiber, filtration, and wood coating	
	Cellulose acetate nonwoven	Textiles, fiber, and filtration	
	Cellulose acetate tow	Textiles, fiber, cigarette filtration	
CelluForce	CelluRods®	Stabilization of emulsions, reinforcing agent in plastics, rubbers, composites, coatings, and adhesives	CelluForce (2025)
Cellulose Lab	Cellulose nanocrystals (CNCs)	Oil spill cleaning, engineering, pharmaceuticals, and vessel substitutes to biodegradable food packaging	Cellulose Lab Inc. (2025)
	Nanofibrillated cellulose (NFC) and microfibrillated cellulose (MFC)	Food packaging, medical, and pharmaceutical applications	
Celotech	Celopro®: hydroxypropyl methyl cellulose (HPMC)	Wall putty, gypsum plaster, tile adhesive	Celotech (2025)
Daicel	ACETY®	Eyeglass frames and tools (screwdriver handles)	Daicel Corporation (2025)
	BELLOCEA®	Foundations and other cosmetic products	
	CAFBLO®	Cutlery, packaging containers, clothing, clear file holders, and toothbrushes	
	Cigatow	Cigarette filters	
	CMC	Sanitary products, batteries, foods, and cosmetics	
	MOLSEP®	Water treatment, food processing, ultrapure water production, production of membrane modules for pharmaceuticals, and biotechnology applications	
Dow	Cellosize™	Thickener of cosmetic products	Dow (2025)

(continued)

Table 7.2 (continued)

Company	Cellulose-based product	Application	Reference
Hebei Hao Shuo	Hydroxyethyl methyl cellulose (MHEC)	Synthesis of latex, oil drilling, construction, cosmetics, detergents, and macromolecular polymerization	Hebei Hao Shuo Chemical Co. (2025)
	Hydroxyethyl cellulose (HEC)		
	HPMC	Construction (production of building materials and additives)	
Kehao Shijiazhuang	HPMC	Construction, pharmaceuticals, chemicals, and cosmetics	Kehao Shijiazhuang (2025)
MelaColl	MelaColl™ HPMC	Tile adhesive and wall putty/skim coat	MelaColl (2025)
	MelaColl™ HEMC		
	MelaColl™ HEC	Latex paint, oil drilling, and daily chemicals	
Nippon Paper	Hada Care Acty and Poise	Absorbent sheets	Nippon Paper (2025)
Nouryon	Akucell® and Cekol®	Improved mouthfeel and protein stabilization	Nouryon (2025)
	AkuPure™	Used in batteries and battery storage	
	Celect® and Finnfix®	Mineral flotation and depressants	
Shanghai Honest	CMC	Oil drilling, building and ceramics, paper coating, toothpaste, textile, and battery	Shanghai Honest Chem. Co. (2025)
	Ethyl cellulose	Paints for wooden products, metal surfaces, paper, rubber, thermotropic products, etc.	
	HPMC	Construction, painting industry, and synthetic resins	

References

Abdalkarim SYH, Ouyang Z, Yu HY et al (2021) Magnetic cellulose nanocrystals hybrids reinforced phase change fiber composites with highly thermal energy storage efficiencies. Carbohydr Polym 254:117481. https://doi.org/10.1016/j.carbpol.2020.117481

Abolore RS, Jaiswal S, Jaiswal AK (2024) Green and sustainable pretreatment methods for cellulose extraction from lignocellulosic biomass and its applications: a review. Carbohydr Polym Technol Appl 7:100396

Ahlfeld T, Köhler T, Czichy C et al (2018) A methylcellulose hydrogel as support for 3d plotting of complex shaped calcium phosphate scaffolds. Gels 4. https://doi.org/10.3390/gels4030068

Ahmed S, Islam MS, Antu UB et al (2025) Nanocellulose: a novel pathway to sustainable agriculture, environmental protection, and circular bioeconomy. Int J Biol Macromol 285: 137979

Alén R (2018) Manufacturing cellulosic fibres for making paper: a historical perspective. In: Technological transformation in the global pulp and paper industry 1800–2018: comparative perspectives. Springer International Publishing, Cham, pp 13–34
Altaner CM, Thomas LH, Fernandes AN, Jarvis MC (2014) How cellulose stretches; synergism between covalent and hydrogen bonding. Biomacromolecules 15:791–798
Anomera (2025) Anomera. https://anomera.ca/anomera-products/. Accessed 12 Feb 2025
Anoop Kumar V, Suresh Chandra Kurup R, Snishamol C, Nagendra Prabhu G (2019) Role of cellulases in food, feed, and beverage industries. In: Green bio-processes. Springer Singapore, pp 323–343
Arioli T, Peng L, Betzner AS et al (1998) Molecular analysis of cellulose biosynthesis in arabidopsis. Science 279(5351):717–720. https://doi.org/10.1126/science.279.5351.717
Ashland (2025) Blanose. https://www.ashland.com/industries/food-and-beverage/dairy/blanose-cellulose-gum. Accessed 13 Feb 2025
Atalla RH, Vanderhart DL (1984) Native cellulose: a composite of two distinct crystalline forms. Science 223:283–285
Atanassov II, Pittman JK, Turner SR (2009) Elucidating the mechanisms of assembly and subunit interaction of the cellulose synthase complex of Arabidopsis secondary cell walls. J Biol Chem 284:3833–3841
Baghel RS, Reddy CRK, Singh RP (2021) Seaweed-based cellulose: applications, and future perspectives. Carbohydr Polym 267:118241
Bashari A, Rouhani Shirvan A, Shakeri M (2018) Cellulose-based hydrogels for personal care products. Polym Adv Technol 29:2853–2867
Bhandari S, Fujino T, Thammanagowda S et al (2006) Xylem-specific and tension stress-responsive coexpression of KORRIGAN endoglucanase and three secondary wall-associated cellulose synthase genes in aspen trees. Planta 224:828–837. https://doi.org/10.1007/s00425-006-0269-1
Bhatia SK, Jagtap SS, Bedekar AA et al (2020) Recent developments in pretreatment technologies on lignocellulosic biomass: effect of key parameters, technological improvements, and challenges. Bioresour Technol 300:122724. https://doi.org/10.1016/j.biortech.2019.122724
Bringmann M, Li E, Sampathkumar A et al (2012) POM-POM2/CELLULOSE SYNTHASE INTERACTING1 is essential for the functional Association of Cellulose Synthase and Microtubules in *Arabidopsis*. Plant Cell 24:163–177. https://doi.org/10.1105/tpc.111.093575
Carolin F, Kamalesh T, Kumar PS et al (2023) A critical review on sustainable cellulose materials and its multifaceted applications. Ind Crop Prod 203. https://doi.org/10.1016/j.indcrop.2023.117221
Celanese (2025) Celanese. https://www.celanese.com/en/products. Accessed 12 Feb 2025
CelluForce (2025) CelluRods. https://celluforce.com/products/. Accessed 12 Feb 2025
Cellulose Lab Inc (2025) Cellulose Lab. https://www.celluloselab.com/
Celotech (2025) Celotech. https://www.celotech.com/products-category/hydroxypropyl-methyl-cellulose-hpmc/?gad_source=1&gclid=CjwKCAiAqrG9BhAVEiwAaPu5ziVmMS3Tro30Wv_jOlUiaKjbsn7Bj2wP7puNp9GlLQnevDPgkjmyYRoCbQAQAvD_BwE. Accessed 12 Feb 2025
Chen Y, Dai Y, Zhu Y et al (2025) Transparent cellulose acetate/polyvinylidene difluoride films with heat dissipation for agricultural mulch application. Ind Crop Prod 224:120302. https://doi.org/10.1016/j.indcrop.2024.120302
Cosgrove DJ (2005) Growth of the plant cell wall. Nat Rev Mol Cell Biol 6:850–861
Cosgrove DJ (2014) Re-constructing our models of cellulose and primary cell wall assembly. Curr Opin Plant Biol 22:122–131
Costa EM, Pereira CF, Ribeiro AA et al (2022) Characterization and evaluation of commercial carboxymethyl cellulose potential as an active ingredient for cosmetics. Appl Sci (Switz) 12. https://doi.org/10.3390/app12136560
Daicel Corporation (2025) Daicel's cellulose products. https://www.daicel.com/en/cellulose/products/. Accessed 12 Feb 2025

Das TK, Houtman C (2004) Evaluating chemical-, mechanical-, and bio-pulping processes and their sustainability characterization using life-cycle assessment. Environ Prog 23:347–357. https://doi.org/10.1002/ep.10054

Data Bridge (2024) Global cellulose market - industry trends and forecast to 2031. https://www.databridgemarketresearch.com/reports/global-cellulose-market. Accessed 13 Feb 2025

de Avila Delucis R, de Cademartori PHG, Fajardo AR, Amico SC (2021) Cellulose and its derivatives: properties and applications. In: Polysaccharides, pp 221–252. https://doi.org/10.1002/9781119711414.ch11

Dow (2025) Dow. https://www.dow.com/en-us/market/mkt-beauty-personal-care/cellosize-hec-thickeners.html. Accessed 12 Feb 2025

Dufresne A (2012) Chapter 1. Nanocellulose: potential reinforcement in composites. pp. 1–32. https://doi.org/10.1039/9781849735315-00001

Eastman (2024) Eastman. https://www.eastman.com/Markets/Personal-Care-Cosmetics-Ingredients/Product_Portfolio/Pages/Film_Formers_-_CE.aspx. Accessed 12 Feb 2025

El Bourakadi K, Semlali FZ, Hammi M, El Achaby M (2024) A review on natural cellulose fiber applications: empowering industry with sustainable solutions. Int J Biol Macromol 281:135773. https://doi.org/10.1016/j.ijbiomac.2024.135773

El-Gendi H, Taha TH, Ray JB, Saleh AK (2022) Recent advances in bacterial cellulose: a low-cost effective production media, optimization strategies and applications. Cellulose 29:7495–7533. https://doi.org/10.1007/s10570-022-04697-1

Fu Q, Shofner ML, Sinquefield SA, Nair S (2022) Recovery and enrichment of organic acids from Kraft black liquor by an adsorption-based process. ACS Sustain Chem Eng 10:11165–11175. https://doi.org/10.1021/acssuschemeng.2c02544

Galland S, Andersson RL, Salajková M et al (2013) Cellulose nanofibers decorated with magnetic nanoparticles-synthesis, structure and use in magnetized high toughness membranes for a prototype loudspeaker. J Mater Chem C Mater 1:7963–7972. https://doi.org/10.1039/c3tc31748j

Gardner KH, Blackwell J (1974) The structure of native cellulose Abstract Biopolymers 13(10): 1975–2001. 10.1002/bip.36.v13:10 10.1002/bip.1974.360131005

Gardiner ES, Sarko' A (1985) Packing analysis of carbohydrates and polysaccharides. 16. The crystal structures of celluloses IV, and IVll. Can J Chem 63(1):173–180

Gil-Guillén I, Freitas PAV, González-Martínez C, Chiralt A (2024) Obtaining cellulose fibers from almond shell by combining subcritical water extraction and bleaching with hydrogen peroxide. Molecules 29:3284. https://doi.org/10.3390/molecules29143284

Girard VD, Chaussé J, Vermette P (2024) Bacterial cellulose: a comprehensive review. J Appl Polym Sci 141:e55163

Godakanda VU, Li H, Alquezar L et al (2019) Tunable drug release from blend poly(vinyl pyrrolidone)-ethyl cellulose nanofibers. Int J Pharm 562:172–179. https://doi.org/10.1016/j.ijpharm.2019.03.035

Habibi Y, Lucia LA, Rojas OJ (2010) Cellulose nanocrystals: chemistry, self-assembly, and applications. Chem Rev 110:3479–3500. https://doi.org/10.1021/cr900339w

Han JC, Zhu YK, Wang LF et al (2021) Modification of regenerated cellulose ultrafiltration membranes with multi-walled carbon nanotubes for enhanced antifouling ability: field test and mechanism study. Sci Total Environ 780:146657. https://doi.org/10.1016/j.scitotenv.2021.146657

Hebei Hao Shuo Chemical Co (2025) Hebei Hao Shuo. https://www.hpmcfactory.com/. Accessed 12 Feb 2025

Heinze T (2015) Cellulose: structure and properties. Adv Polym Sci 271:1–52. https://doi.org/10.1007/12_2015_319

Hon DN (1994) Cellulose: a random walk along its historical path. Cellulose 1:1–25

Horikawa Y (2022) Structural diversity of natural cellulose and related applications using delignified wood. J Wood Sci 68:54

Jenkins DJA, Augustin LSA, Malick A et al (2012) Glucose: chemistry and dietary sources. Enc Hum Nutr 2–4:372–380. https://doi.org/10.1016/B978-0-12-375083-9.00133-1

Kehao Shijiazhuang (2025) Kehao Shijiazhuang. https://xhhpmc.com/product-hydroxypropyl-methylcellulose-hpmc/?gad_source=1&gclid=CjwKCAiAqrG9BhAVEiwAaPu5zs_F2gRAEENB_jnt__1ia8QZ2jDTnpZr4DPOsb-AGn_FFMUqEhT5rxoCVg4QAvD_BwE. Accessed 12 Feb 2025

Kesten C, Wallmann A, Schneider R et al (2019) The companion of cellulose synthase 1 confers salt tolerance through a Tau-like mechanism in plants. Nat Commun 10:857. https://doi.org/10.1038/s41467-019-08780-3

Kim EH, Lee J, Cho KY et al (2024) Physicochemical investigation of cellulose microbeads produced through cross-flow membrane emulsification for cosmetic applications. Cellulose 32:467–481. https://doi.org/10.1007/s10570-024-06314-9

Kimura S, Laosinchai W, Itoh T et al (1999) Immunogold labeling of rosette terminal cellulose-synthesizing complexes in the vascular plant vigna angularis. Plant Cell 11:2075–2086

Kondo T (1997) The relationship between intramolecular hydrogen bonds and certain physical properties of regioselectively substituted cellulose derivatives. Journal of Polymer Science Part B: Polymer Physics 35(4):717–723. https://doi.org/10.1002/(SICI)1099-0488(199703)35:43.0.CO;2-J

Krassig HA (1993) Cellulose: structure, acessibility and reactivity. Gordon and Breach Science

Krumm C, Pfaendtner J, Dauenhauer PJ (2016) Millisecond pulsed films unify the mechanisms of cellulose fragmentation. Chemistry of Materials 28(9):3108–3114. https://doi.org/10.1021/acs.chemmater.6b00580

Kumar M, Campbell L, Turner S (2016) Secondary cell walls: biosynthesis and manipulation. J Exp Bot 67:515–531

Kumar B, Bhardwaj N, Agrawal K et al (2020) Current perspective on pretreatment technologies using lignocellulosic biomass: an emerging biorefinery concept. Fuel Process Technol 199. https://doi.org/10.1016/j.fuproc.2019.106244

Kurek I, Kawagow Y, Jacob-Wilk D et al (2002) Dimerization of cotton fiber cellulose synthase catalytic subunits occurs via oxidation of the zinc-binding domains. Proc Natl Acad Sci U S A 99:11109–11114

Li J, Wei X, Wang Q et al (2012) Homogeneous isolation of nanocellulose from sugarcane bagasse by high pressure homogenization. Carbohydr Polym 90:1609–1613. https://doi.org/10.1016/j.carbpol.2012.07.038

Li Y, Wang J, Fu C et al (2023) Designing flexible CNT/CNF films with highly light-absorbing for solar energy harvesting: seawater desalination, photothermal power generation and light- driven actuators. Energy Convers Manag 289:117160. https://doi.org/10.1016/j.enconman.2023.117160

Lin D, Liu Z, Shen R et al (2020) Bacterial cellulose in food industry: current research and future prospects. Int J Biol Macromol 158:1007–1019. https://doi.org/10.1016/j.ijbiomac.2020.04.230

Liu Y, Li B, Mao W et al (2019) Strong cellulose-based materials by coupling sodium hydroxide–anthraquinone (NaOH–AQ) pulping with hot pressing from wood. ACS Omega 4(4):7861–7865. https://doi.org/10.1021/acsomega.9b00411

Liu S, Cheng G (2024) Developments and perspectives on lignin-first biomass pretreatment for efficient enzymatic hydrolysis and isolation of lignin with minimized degradation. Ind Crops Prod 208:117926. https://doi.org/10.1016/j.indcrop.2023.117926

Liu B, Liu L, Deng B et al (2022) Application and prospect of organic acid pretreatment in lignocellulosic biomass separation: a review. Int J Biol Macromol 222:1400–1413

Luo H, Zhou T, Cao J et al (2023) Utilization of lignocellulosic biomass by glycerol organosolv pretreatment for biobutanol production integrated with bioconversion of residual glycerol into value-added products. Bioresour Technol 387. https://doi.org/10.1016/j.biortech.2023.129661

Lv Y, Zhang Y, Xu Y (2024) Understanding and technological approach of acid hydrolysis processing for lignocellulose biorefinery: panorama and perspectives. Biomass Bioenergy 183:107133

Machado B, Costa SM, Costa I et al (2024) The potential of algae as a source of cellulose and its derivatives for biomedical applications. Cellulose 31:3353–3376

Maiti S, Jayaramudu J, Das K et al (2013) Preparation and characterization of nano-cellulose with new shape from different precursor. Carbohydr Polym 98:562–567. https://doi.org/10.1016/j.carbpol.2013.06.029

Mankar AR, Pandey A, Modak A, Pant KK (2021) Pretreatment of lignocellulosic biomass: a review on recent advances. Bioresour Technol 334. https://doi.org/10.1016/j.biortech.2021.125235

Mann J, Marrinan HJ (1958) Crystalline modifications of cellulose. Part II. A study with plane-polarized infrared radiation. J Polym Sci 32:357–370. https://doi.org/10.1002/pol.1958.1203212507

Marrinan HJ, Mann J (1956) Infrared spectra of the crystalline modifications of cellulose. J Polym Sci 21:301–311. https://doi.org/10.1002/pol.1956.120219812

MelaColl (2025) MelaColl. https://melacoll.com/products/. Accessed 12 Feb 2025

Meng F, Wang G, Du X et al (2019) Extraction and characterization of cellulose nanofibers and nanocrystals from liquefied banana pseudo-stem residue. Compos B Eng 160:341–347. https://doi.org/10.1016/j.compositesb.2018.08.048

Mietner JB, Jiang X, Edlund U et al (2021) 3D printing of a bio-based ink made of cross-linked cellulose nanofibrils with various metal cations. Sci Rep 11:1–9. https://doi.org/10.1038/s41598-021-85865-4

Motaung TE (2021) Recent applications and innovations of cellulose based materials: a critical review. Cellul Chem Technol 55:1–12. https://doi.org/10.35812/CelluloseChemTechnol.2021.55.01

Motaung TE, Linganiso LZ (2018) Critical review on agrowaste cellulose applications for biopolymers. Int J Plast Technol 22:185–216

Nasalapure AV, Pattadkal S, Ghatti V et al (2025) Design of cellulose nanocrystal-integrated poly (vinyl alcohol)/polyethylene glycol electrospun nanofiber scaffolds for biomedical applications. J Appl Polym Sci 142(6):e56463. https://doi.org/10.1002/app.56463

Nippon Paper (2025) Nippon paper. https://www.nipponpapergroup.com/english/research/organize/cnf.html. Accessed 12 Feb 2025

Nouryon (2025) Nouryon. https://www.nouryon.com/products/cellulose-ethers/cmc/. Accessed 12 Feb 2025

Palmer D, Levina M, Nokhodchi A et al (2011) The influence of sodium carboxymethylcellulose on drug release from polyethylene oxide extended release matrices. AAPS PharmSciTech 12:862–871. https://doi.org/10.1208/s12249-011-9648-4

Phanthong P, Reubroycharoen P, Hao X et al (2018) Nanocellulose: extraction and application. Carbon Resour Convers 1:32–43

Pino MS, Rodríguez-Jasso RM, Michelin M et al (2018) Bioreactor design for enzymatic hydrolysis of biomass under the biorefinery concept. Chem Eng J 347:119–136. https://doi.org/10.1016/j.cej.2018.04.057

Raghavendran V, Asare E, Roy I (2020) Bacterial cellulose: biosynthesis, production, and applications. In: Advances in microbial physiology, vol 77. Academic Press, pp 89–138

Rahman SSA, Vaishnavi T, Vidyasri GS et al (2021) Production of bacterial cellulose using *Gluconacetobacter kombuchae* immobilized on *Luffa aegyptiaca* support. Sci Rep 11:2912. https://doi.org/10.1038/s41598-021-82596-4

Rao VS, Sundararajan PR, Ramakrishnan C, Ramachandran GN (1967) Conformational studies of amylose. In: Conformation of biopolymers. Academic Press, pp 721–737

Schieppati D, Patience NA, Galli F et al (2023) Chemical and biological delignification of biomass: a review. Ind Eng Chem Res 62:12757–12794

Seddiqi H, Oliaei E, Honarkar H et al (2021) Cellulose and its derivatives: towards biomedical applications. Springer Netherlands

Sewsynker-Sukai Y, Naomi David A, Gueguim Kana EB (2020) Recent developments in the application of kraft pulping alkaline chemicals for lignocellulosic pretreatment: potential beneficiation of green liquor dregs waste. Bioresour Technol 306:123225

Shangai Honest Chem. Co (2025) Honest. http://www.honestsh.net/english/index.asp. Accessed 12 Feb 2025

Sharma K, Choudhary P, Majeed A et al (2025) Cellulose based membranes, hydrogels and aerogels for water treatment application. Ind Crops Prod 225:120474. https://doi.org/10.1016/j.indcrop.2025.120474

Shokri J, Adibki K (2013) Application of cellulose and cellulose derivatives in pharmaceutical industries. In: Cellulose-medical, pharmaceutical and electronic applications. IntechOpen. https://doi.org/10.5772/55178

Spedding H (1962) The action of ethylamine on cellulose part III—the formation of cellulose III and its infra-red spectrum between 1500 and 650 cm^{-1}. Polymer (Guildf) 3:211–214

Sugiyama J, Okano T, Yamamoto H et al (1990) Transformation of Valonia cellulose crystals by an alkaline hydrothermal treatment. Macromolecules 23(12):3196–3198. https://doi.org/10.1021/ma00214a029

Tang L, Huang B, Lu Q et al (2013) Ultrasonication-assisted manufacture of cellulose nanocrystals esterified with acetic acid. Bioresour Technol 127:100–105. https://doi.org/10.1016/j.biortech.2012.09.133

Tang Z, Lin X, Yu M et al (2024) Recent advances in TEMPO-oxidized cellulose nanofibers: oxidation mechanism, characterization, properties and applications. Int J Biol Macromol 259:129081

Taylor NG, Howells RM, Huttly AK et al (2003) Interactions among three distinct CesA proteins essential for cellulose synthesis. Proc Natl Acad Sci U S A 100(3):1450–1455. https://doi.org/10.1073/pnas.0337628100

Teng CP, Tan MY, Pei J et al (2023) Advances in cellulose-based composites for energy applications. Materials 16:3856

Tian S-Y, Guo J-H, Zhao C et al (2018) Preparation of cellulose/graphene oxide composite membranes and their application in removing organic contaminants in wastewater. J Nanosci Nanotechnol 19:2147–2153. https://doi.org/10.1166/jnn.2019.15808

Tokumasu N, Nakano T, Yoshida S, Kimura Y, Endo T (2024) Facile preparation of cellulose IV II using ionic liquids. CrystEngComm 26:5777–5784

Três M, Pagno CH, de Moraes CT et al (2024) Production of cellulose nanofibers from pineapple peel (*Ananas comosus*) and application in biodegradable films. Packag Technol Sci 38:117–129. https://doi.org/10.1002/pts.2860

U.S. Energy Information and Admnistration (2022) Oil and petroleum products explained. https://www.eia.gov/energyexplained/oil-and-petroleum-products/oil-and-the-environment.php. Accessed 15 Mar 2025

Vain T, Crowell EF, Timpano H et al (2014) The cellulase KORRIGAN is part of the cellulose synthase complex. Plant Physiol 165:1521–1532. https://doi.org/10.1104/pp.114.241216

Vanderhart DL, Atalla RH (1984) Studies of microstructure in native celluloses using solid-state 13C NMR. Macromolecules 17:8. https://doi.org/10.1021/ma00138a009

Verma J, Petru M, Goel S (2024) Cellulose based materials to accelerate the transition towards sustainability. Ind Crops Prod 210:118078. https://doi.org/10.1016/j.indcrop.2024.118078

Vieira SR, da Silva JBA, Pessôa LC et al (2024) Cellulose processing using ionic liquids: an analysis of patents and technological trends. Heliyon 10. https://doi.org/10.1016/j.heliyon.2024.e39590

Wang T, Zhao Y (2021) Optimization of bleaching process for cellulose extraction from apple and kale pomace and evaluation of their potentials as film forming materials. Carbohydr Polym 253:117225. https://doi.org/10.1016/j.carbpol.2020.117225

Wang J, Wang L, Gardner DJ et al (2021a) Towards a cellulose-based society: opportunities and challenges. Cellulose 28:4511–4543. https://doi.org/10.1007/s10570-021-03771-4

Wang Y, Yang Y, Qu Y, Zhang J (2021b) Selective removal of lignin with sodium chlorite to improve the quality and antioxidant activity of xylo-oligosaccharides from lignocellulosic biomass. Bioresour Technol 337:125506. https://doi.org/10.1016/j.biortech.2021.125506

Wang Y, Wang Z, Lin Y et al (2024) Nanocellulose from agro-industrial wastes: a review on sources, production, applications, and current challenges. Food Res Int 192:114741

Woiciechowski AL, Dalmas Neto CJ, de Souza Vandenberghe LP et al (2020) Lignocellulosic biomass: acid and alkaline pretreatments and their effects on biomass recalcitrance – conventional processing and recent advances. Bioresour Technol 304:122848. https://doi.org/10.1016/j.biortech.2020.122848

Wohlert M, Benselfelt T, Wågberg L et al (2022) Cellulose and the role of hydrogen bonds: not in charge of everything. Abstract Cellulose 29(1):1–23. https://doi.org/10.1007/s10570-021-04325-4

Wu Q, Xu J, Zhu S et al (2020) Crystalline stability of cellulose III nanocrystals in the hydrothermal treatment and NaOH solution. Carbohydr Polym 249. https://doi.org/10.1016/j.carbpol.2020.116827

Xiuping S (2021) Preparation method of cellulose-based hydrogel adsorbent for heavy metal wastewater treatment. CN110270317B

Yaopeng Z, Li H, Yufang Z, et al (2020) Silk fibroin/cellulose 3D printing ink. CN106267370B

Zdanowicz M, Wilpiszewska K, Spychaj T (2018) Deep eutectic solvents for polysaccharides processing. A review. Carbohydr Polym 200:361–380. https://doi.org/10.1016/j.carbpol.2018.07.078

Zhang Z, Harrison MD, Rackemann DW et al (2016) Organosolv pretreatment of plant biomass for enhanced enzymatic saccharification. Green Chem 18:360–381. https://doi.org/10.1039/c5gc02034d

Zhang J, Hu S, Shi Z et al (2021) Eco-friendly and recyclable all cellulose triboelectric nanogenerator and self-powered interactive interface. Nano Energy 89:106354. https://doi.org/10.1016/j.nanoen.2021.106354

Zhang L, Ren N, Zhang J et al (2025) Cellulose nanocrystals: sustainable production and emerging fruit coating applications. Chem Eng J 509:161190. https://doi.org/10.1016/j.cej.2025.161190

Zhou J, Wang H, Du C et al (2022) Cellulose for sustainable triboelectric nanogenerators. Adv Energy Sustain Res 3. https://doi.org/10.1002/aesr.202100161

Zhu X, Li S, Pan S et al (2018) CSI1, PATROL1, and exocyst complex cooperate in delivery of cellulose synthase complexes to the plasma membrane. Proc Natl Acad Sci U S A 115. https://doi.org/10.1073/pnas.1800182115

Zhu M, Zhang R, Yuan S, Liu P (2025) Advances in cellulose materials for interfacial water evaporation and their applications in seawater desalination. Desalination 601:118559. https://doi.org/10.1016/j.desal.2025.118559

Chapter 8
Hemicellulose: Structures, Properties, and Its Biorefinery Applications

Swarnima Agnihotri

Abstract Hemicellulose, a major component of lignocellulosic biomass, plays a pivotal role in the transition toward a sustainable bioeconomy. As the second most abundant polysaccharide in nature, hemicellulose is integral to biomass valorization due to its structural diversity, biodegradability, and functional versatility. It is found in significant quantities in agricultural residues, forestry by-products, and industrial side streams, making it a globally available and underutilized resource. Recent trends in biorefinery research emphasize the importance of hemicellulose in producing biofuels, biochemicals, and biodegradable materials. Its amorphous structure, composed of various sugar monomers, such as xylose, mannose, and arabinose, imparts unique physicochemical properties that are advantageous for multiple applications. This chapter explores the global production and availability of hemicellulose, its structural and functional characteristics, and its emerging role in biorefinery platforms. Emphasis is placed on sustainable extraction methods, value-added applications, and the challenges that must be addressed to fully harness its potential in a circular bioeconomy.

8.1 Introduction

Hemicellulose, a structurally diverse group of heteropolysaccharides, is a key component of lignocellulosic biomass and plays a central role in the development of sustainable biorefinery systems. Unlike cellulose, which is a linear polymer of glucose, hemicellulose is composed of branched chains of various sugar monomers, such as xylose, arabinose, mannose, galactose, and glucose, along with uronic acids. This amorphous and heterogeneous structure accounts for 20–35% of plant biomass, depending on the botanical source, and contributes significantly to the mechanical integrity and porosity of plant cell walls (Chen 2015; Rao et al. 2023). As the second most abundant polysaccharide in nature, hemicellulose represents a vast, renewable,

S. Agnihotri (✉)
Swedish Centre for Resource Recovery, University of Borås, Borås, Sweden
e-mail: swarnima.agnihotri@hb.se

A. L. Woiciechowski et al. (eds.), *Biomass for Bioenergy and Bioproducts*, Biofuel and Biorefinery Technologies 17, https://doi.org/10.1007/978-3-032-19898-3_8

and underutilized resource. Its biodegradability, chemical modifiability, and functional versatility make it an attractive candidate for the production of bio-based materials, chemicals, and fuels. Recent advances have highlighted its potential in applications ranging from biodegradable packaging and pharmaceuticals to prebiotics and antioxidants (Mathura et al. 2024). In particular, hemicellulose-derived oligosaccharides are gaining attention for their health-promoting properties and functional roles in food systems. In the context of the circular bioeconomy, hemicellulose valorization is increasingly recognized as a strategic pathway to reduce reliance on fossil resources and mitigate environmental impacts. Biorefineries are evolving to incorporate hemicellulose as a key intermediate for the production of high-value compounds such as furfural, xylitol, and organic acids (Rao et al. 2023; Schmidt et al. 2023). However, challenges remain in achieving efficient extraction, selective fractionation, and cost-effective conversion processes at industrial scales.

This chapter provides a comprehensive overview of hemicellulose, focusing on its global availability, structural and physicochemical characteristics, and its emerging role in integrated biorefinery platforms. Emphasis is placed on sustainable extraction and processing methods, value-added applications, and the technological and economic challenges that must be addressed to fully harness its potential in a circular and sustainable bioeconomy.

8.2 Global Production and Availability

Hemicellulose is a major component of lignocellulosic biomass, widely distributed across agricultural residues, forestry by-products, and industrial waste streams. These biomass sources are not only renewable and abundant but also geographically diverse, making them attractive for sustainable biorefinery applications. Agricultural residues such as corn stover, wheat straw, rice husks, and sugarcane bagasse remain primary sources, with sugarcane bagasse containing up to 35% hemicellulose, predominantly xylans and arabinans (Mujtaba et al. 2023). Forestry residues, including hardwood and softwood by-products like sawdust and bark, are also rich in hemicellulose—hardwoods typically contain glucuronoxylans, while softwoods are abundant in galactoglucomannans (Badgujar and Bhanage 2015). Additionally, industrial waste streams from the pulp and paper industry, such as spent liquors and sludge, offer underutilized yet hemicellulose-rich feedstocks (Stoklosa and Hodge 2014). Globally, the theoretical availability of lignocellulosic biomass exceeds 1800 billion tons annually, with hemicellulose comprising approximately 20–35% of this mass. This equates to a potential production of 360–630 billion tons of hemicellulose per year. However, actual industrial-scale extraction remains limited due to technological, economic, and logistical constraints. Regionally, Asia dominates in agricultural residue generation, particularly from rice and sugarcane cultivation. North America and Europe contribute significantly through forestry residues and industrial waste, while Latin America, especially Brazil, is a key supplier of sugarcane bagasse. The global hemicellulose market was valued at approximately USD 1.5 billion in 2023 and is projected to grow at a compound annual growth rate (CAGR) of 4.5–7% through 2030, driven by increasing demand in the food, pharmaceutical, and paper industries (Sharma 2024; Research 2025).

Despite its abundance, several challenges hinder the full-scale utilization of hemicellulose. Its complex association with cellulose and lignin necessitates energy-intensive pretreatment processes. Current extraction methods often yield impure fractions, limiting its applicability in high-value products. Moreover, the economic viability of hemicellulose valorization is challenged by the need to compete with fossil-based alternatives. Variability in biomass composition across regions and seasons further complicates process standardization and optimization. To address these issues, recent research emphasizes integrated biorefinery models, green extraction technologies, and AI-driven process optimization to enhance yield, purity, and cost-efficiency (Research 2024; Sharma 2024).

8.3 Structure and Composition of Hemicellulose

Hemicellulose is a structurally diverse group of branched polysaccharides embedded within the plant cell wall matrix, where it interacts with cellulose microfibrils and lignin to provide mechanical strength and flexibility. Unlike cellulose—a linear, crystalline polymer composed solely of glucose with a high degree of polymerization (10,000+ units)—hemicellulose is amorphous, shorter (typically 100–200 sugar units), and composed of various sugar monomers. These include pentoses (xylose and arabinose), hexoses (mannose, glucose, and galactose), and uronic acids (glucuronic acid and 4-O-methylglucuronic acid), linked primarily through β-(1 → 4) or β-(1 → 3) glycosidic bonds (Paz-Cedeno et al. 2022), as shown in Fig. 8.1.

The presence of side chains and branching patterns varies by plant source, influencing hemicellulose's solubility, reactivity, and functional properties. Its amorphous nature and diverse functional groups make it more water-soluble, less thermally stable, and more chemically reactive than cellulose. Additionally, its glycosidic linkages are more accessible to microbial enzymes, enhancing biodegradability (Gautam et al. 2025). Hemicelluloses are classified based on their dominant sugar residues and structural motifs:

Xylans: Abundant in hardwoods and cereal grains, with a β-(1 → 4)-linked xylose backbone and substitutions such as arabinose, glucuronic acid, or acetyl groups.

Mannans: Found in softwoods, composed of β-(1 → 4)-linked mannose units, often with galactose (galactomannans) or glucose (glucomannans).

Arabinogalactans: Present in larch wood and some dicotyledonous plants, featuring a galactose backbone with arabinose side chains.

Glucuronoxylans: Common in hardwoods, substituted with 4-O-methylglucuronic acid and acetyl groups, affecting solubility and reactivity.

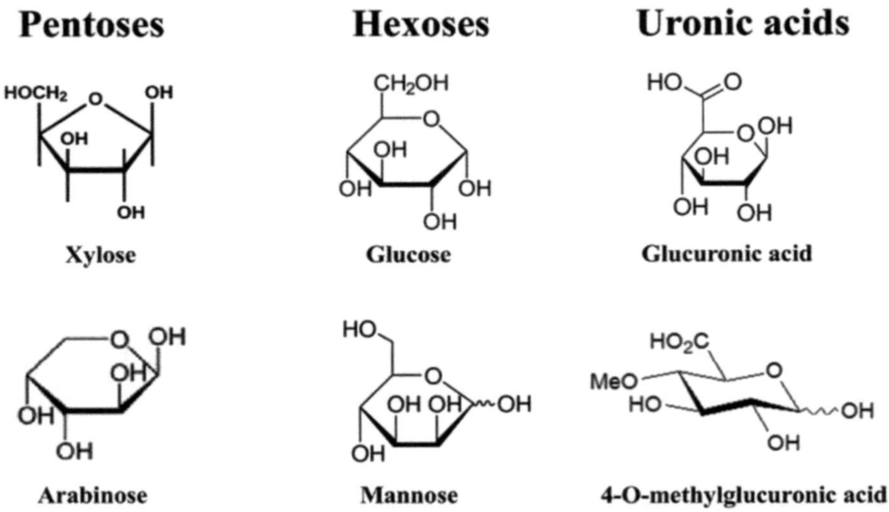

Fig. 8.1 Chemical structures of the main sugar monomers found in hemicellulose

Two notable hemicellulose types with industrial relevance are *xyloglucans* and *arabinoxylans*. Xyloglucans have a cellulose-like β-(1 → 4)-linked glucose backbone with xylose branches, while arabinoxylans feature a β-(1 → 4)-linked xylan backbone substituted with arabinose. These polysaccharides can be extracted using strong alkaline solutions, and their structural features influence their processing and suitability for applications (Gautam et al. 2025).

The structural heterogeneity of hemicellulose enables its use in a wide range of applications, including biofuels, biodegradable packaging, and functional food ingredients. Recent advances in extraction and modification techniques—such as alkaline, acid, and microwave-assisted hydrothermal methods—have further expanded its utility in sustainable technologies (Mathura et al. 2024). Figure 8.2 illustrates the general chemical structure of hemicellulose, highlighting its repeating sugar units, glycosidic linkages, and functional groups, such as hydroxyl, methylene, ester, and carboxylate groups:

8.4 Properties of Hemicelluloses

Hemicellulose exhibits a wide range of physicochemical and functional properties that are highly dependent on its botanical origin, structural complexity, and extraction or modification methods. These properties are critical in determining its behavior in biological systems and its suitability for industrial applications.

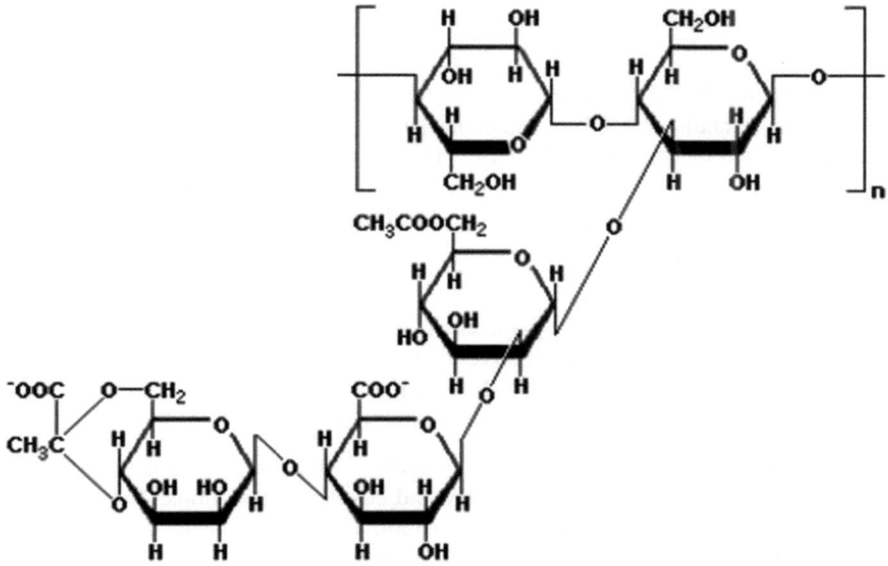

Fig. 8.2 Structure of hemicellulose. (Kulkarni et al. 2012)

8.4.1 Physicochemical Properties

Solubility Hemicellulose is generally soluble in water and alkaline solutions due to its amorphous structure and the presence of hydrophilic functional groups, such as hydroxyl and carboxyl groups. The solubility is influenced by the degree of branching, acetylation, and the type of sugar residues present. For instance, steam explosion treatment has been shown to enhance the solubility and functional performance of hemicellulose-rich dietary fibers by disrupting the cell wall matrix and increasing the availability of soluble fractions (Tang et al. 2023).

Thermal Stability Hemicellulose decomposes at lower temperatures (typically 200–260 °C) compared to cellulose, which is attributed to its lower molecular weight and less ordered structure. This property is particularly relevant in thermochemical conversion processes such as pyrolysis and gasification, where hemicellulose contributes to the early stages of thermal degradation (Soni and Soni 2025).

Reactivity The presence of multiple functional groups—especially hydroxyl, carboxyl, and acetyl groups—makes hemicellulose highly reactive and suitable for chemical modifications. These modifications include esterification, etherification, and cross-linking, which can enhance its mechanical strength, hydrophobicity, or compatibility with other polymers. Such reactivity is essential for tailoring hemicellulose-based materials for specific applications like coatings, hydrogels, and packaging films (Huang et al. 2024).

8.4.2 Functional Properties

Film-Forming Ability Hemicellulose can form flexible, transparent films, making it a promising candidate for biodegradable packaging. Its film-forming capacity is enhanced when blended with plasticizers (e.g., glycerol) or other biopolymers such as starch or chitosan. These films exhibit good oxygen barrier properties and moderate mechanical strength, suitable for food packaging applications (Huang et al. 2024).

Biodegradability As a natural polysaccharide, hemicellulose is readily degraded by microbial enzymes in soil and aquatic environments. This biodegradability makes it an environmentally friendly alternative to synthetic polymers, especially in single-use applications. Its degradation products are nontoxic and can be assimilated into natural biogeochemical cycles (Soni and Soni 2025).

Polymer Interactions Hemicellulose interacts with cellulose through hydrogen bonding and with lignin via hydrophobic and covalent interactions. These interactions are crucial in maintaining the structural integrity of plant cell walls and influence the mechanical, thermal, and barrier properties of lignocellulosic composites. In engineered materials, these interactions can be exploited to design bio-based composites with tailored performance (Tang et al. 2023).

8.5 Extraction and Purification Techniques for Hemicelluloses

The extraction and purification of hemicellulose from lignocellulosic biomass are critical steps in its valorization for industrial applications. Due to the complex and heterogeneous nature of plant cell walls, a variety of pretreatment and separation techniques have been developed to efficiently isolate hemicellulose while preserving its structural integrity.

8.5.1 Extraction

The extraction of hemicellulose from lignocellulosic biomass involves a range of pretreatment methods designed to break down the complex plant cell wall matrix and release hemicellulosic polymers. Among these, *acid hydrolysis* is a widely used technique that employs dilute or concentrated acids—typically sulfuric or hydrochloric acid—to cleave the glycosidic bonds in hemicellulose (Alam et al. 2024). While this method is effective in releasing monomeric sugars, it often leads to the degradation of sugars and the formation of inhibitory by-products such as furfural and hydroxymethylfurfural (HMF), especially under harsh conditions. To mitigate these effects, acid hydrolysis is frequently combined with detoxification steps or conducted under milder conditions to preserve sugar integrity (Jincy and Femina 2023). *Alkaline extraction* is another common and efficient method, particularly for recovering xylan-rich hemicellulose. This technique typically uses sodium hydroxide or potassium hydroxide to disrupt ester linkages between hemicellulose and lignin, thereby enhancing solubilization while preserving the sugar backbone. Alkaline extraction is scalable and yields high-purity hemicellulose suitable for further chemical modification or functional applications (Lu et al. 2021; Abolore et al. 2024). *Steam explosion* is a physical pretreatment method that involves exposing biomass to high-pressure steam, followed by rapid decompression. This sudden pressure release disrupts the lignocellulosic matrix, increasing the accessibility of hemicellulose and other polysaccharides. Steam explosion is considered a green and energy-efficient method, especially when integrated with enzymatic hydrolysis for downstream processing (Jincy and Femina 2023). *The organosolv* method is an effective and increasingly studied technique for extracting hemicellulose from lignocellulosic biomass using organic solvents such as ethanol, methanol, or acetone, often in combination with acid catalysts. This process disrupts the lignin–carbohydrate complex, enabling the selective solubilization and recovery of hemicellulose while preserving its structural integrity. Organosolv extraction is particularly valued for producing high-purity hemicellulose fractions with minimal contamination from lignin or cellulose. Recent advancements have demonstrated its scalability and potential for integration into biorefinery platforms, although challenges such as solvent recovery and energy input remain (Tofani et al. 2023; Thoresen et al. 2024). *Enzymatic hydrolysis* utilizes specific enzymes such as xylanases and mannanases to selectively cleave hemicellulose chains under mild, environmentally friendly conditions. This method preserves the functional integrity of the released oligosaccharides and avoids the formation of toxic by-products. However, it requires careful optimization of enzyme mixtures, reaction time, and pH to achieve high yields (Shi et al. 2025). In addition to these conventional methods, advanced techniques such as *microwave-assisted hydrothermal extraction* and *ionic liquid pretreatment* are gaining attention. These methods offer enhanced selectivity, reduced processing times, and minimal degradation of hemicellulose. They are particularly promising for producing high-quality hemicellulose fractions with tailored functional properties (Jincy and Femina 2023).

8.5.2 Purification

Following extraction, purification of hemicellulose is essential to remove residual lignin, cellulose, and other impurities. Common purification strategies include *antisolvent precipitation,* where non-solvents like ethanol or acetone are added to precipitate hemicellulose from the solution (Haimer et al. 2010). This method is simple, scalable, and effective for obtaining high-purity hemicellulose. *Ultrafiltration*, a membrane-based technique, is used to separate hemicellulose based on molecular weight, allowing for the concentration and fractionation of oligosaccharides. For analytical-scale purification, chromatographic techniques such as *ion-exchange* and *size-exclusion chromatography* offer high resolution and are useful for characterizing hemicellulose fractions (Arkell 2015). *Deep eutectic solvents (DESs)* facilitate the selective dissolution of lignin and partial hydrolysis of hemicellulose, thereby improving the accessibility and recovery of hemicellulosic sugars (Alam et al. 2024). They are particularly effective in cleaving β-O-4 linkages in lignin while preserving the structural integrity of hemicellulose. This selective fractionation enhances the purity of hemicellulose and reduces the need for harsh chemicals or energy-intensive steps (Yao et al. 2024). Recent studies have demonstrated that DES pretreatment not only improves sugar yields but also maintains the functional properties of hemicellulose, making it suitable for downstream applications such as biofilms, hydrogels, and bioplastics (Tong et al. 2023). Moreover, DESs can be recycled and reused, further enhancing the environmental and economic sustainability of the process (Alsoy Altinkaya 2024).

The choice of extraction and purification method significantly influences the yield, purity, molecular weight distribution, and functional properties of the final hemicellulose product. Moreover, the environmental impact and economic feasibility of these processes are key considerations in the development of sustainable biorefinery platforms.

8.6 Applications of Hemicelluloses in Biorefineries

Hemicellulose, as a major component of lignocellulosic biomass, plays a pivotal role in the development of sustainable biorefineries. Its structural diversity and functional versatility enable its transformation into a wide array of value-added products across multiple sectors.

8.6.1 Biofuels and Biochemicals

Hemicellulosic sugars, primarily xylose and arabinose, can be fermented by engineered microbial strains into bioethanol, biobutanol, and other advanced biofuels. These processes are central to second-generation biofuel production, which utilizes nonfood biomass and reduces competition with food crops. In addition to fermentation, chemical conversion of hemicellulose yields platform chemicals such as furfural, xylitol, lactic acid, and acetic acid, which serve as precursors for bioplastics, solvents, and pharmaceuticals. Emerging technologies are also exploring the use of hemicellulose hydrolysates for biohydrogen and biogas production through anaerobic digestion and dark fermentation, offering promising routes for renewable energy generation (Huang et al. 2021; Malik et al. 2022; Kazmi et al. 2025).

8.6.2 Biodegradable Polymers and Materials

Due to its film-forming ability and compatibility with other biopolymers, hemicellulose is increasingly used in the development of biodegradable films, coatings, and composites. When blended with materials like starch, polylactic acid (PLA), or polyhydroxyalkanoates (PHAs), hemicellulose enhances mechanical strength, barrier properties, and biodegradability. These composites are particularly valuable in food packaging, mulch films, and agricultural applications, where environmental sustainability is a priority. The hydrophilic nature of hemicellulose can be modified through chemical treatments to improve water resistance and durability (Cheng et al. 2024; Mathura et al. 2024; Siddiqui et al. 2024).

8.6.3 Functional Food and Feed Applications

Hemicellulose-derived oligosaccharides, such as xylooligosaccharides (XOSs) and arabinoxylans, exhibit prebiotic activity, promoting the growth of beneficial gut microbiota like *Bifidobacteria* and *Lactobacilli.* These compounds are being incorporated into functional foods, nutraceuticals, and animal feed formulations to enhance digestion, nutrient absorption, and immune function. Their resistance to digestion in the upper gastrointestinal tract and selective fermentation in the colon make them ideal candidates for gut health applications (Gullón et al. 2014; Jana et al. 2021; Ali et al. 2025).

8.6.4 Pharmaceuticals and Cosmetics

Hemicellulose and its derivatives possess antioxidant, antimicrobial, and film-forming properties, making them suitable for use in pharmaceutical formulations and cosmetic products. In drug delivery, hemicellulose-based hydrogels and films are being developed as controlled-release matrices for oral and topical applications. In cosmetics, hemicellulose is used in moisturizers, emulsifiers, and skin care formulations due to its biocompatibility and ability to retain moisture. These applications align with the growing demand for natural, biodegradable, and nontoxic ingredients in personal care products (Xu et al. 2022; Solihat et al. 2024; Salehi and Rashidinejad 2025).

8.7 Challenges and Future Prospects

Despite the growing interest in hemicellulose as a renewable feedstock for bio-based products, several technical, economic, and biological challenges continue to hinder its full-scale industrial deployment. One of the primary obstacles is the complexity of extraction and purification. Hemicellulose is intricately bound to cellulose and lignin within the plant cell wall, requiring energy-intensive and often multistep processes for its isolation. Conventional methods such as acid or alkaline hydrolysis can lead to sugar degradation or require harsh conditions, while greener alternatives like enzymatic or deep eutectic solvent–based extractions are still under optimization for scalability and cost-effectiveness (Mathura et al. 2024).

Another significant barrier is the relatively low market value of some hemicellulose-derived products, such as xylitol or furfural, which limits the economic viability of hemicellulose valorization unless integrated into a broader biorefinery framework. Additionally, the lack of robust microbial and enzymatic systems capable of efficiently converting pentose-rich hydrolysates (like xylose and arabinose) into high-value chemicals remains a bottleneck. Many industrial microbes are naturally inefficient at fermenting these sugars, and inhibitors generated during pretreatment further complicate bioconversion (Arzami et al. 2022; Mujtaba et al. 2023).

Looking ahead, future research and innovation must focus on several key areas. First, the development of highly efficient and selective biocatalysts—both enzymes and microbial strains—is essential. Advances in synthetic biology and metabolic engineering offer promising avenues for tailoring microbes to convert hemicellulosic sugars into targeted biofuels, bioplastics, and pharmaceuticals with high yield and tolerance to inhibitors. Second, there is a pressing need to engineer microbial consortia or chassis organisms that can co-utilize mixed sugar streams and operate under industrial conditions. This includes improving tolerance to acidic or oxidative environments and enhancing substrate uptake and product secretion pathways. Finally, the integration of hemicellulose valorization into circular biorefinery models is critical. This involves designing closed-loop systems where hemicellulose is not treated as a by-product but as a central component of biomass valorization. Coupling hemicellulose processing with lignin and cellulose streams and aligning with principles of the circular bioeconomy can significantly improve resource efficiency, reduce waste, and enhance the overall sustainability of biorefineries (Mathura et al. 2024).

In conclusion, while challenges remain, the future of hemicellulose lies in interdisciplinary innovation—combining advances in green chemistry, biotechnology, and systems engineering to unlock its full potential as a cornerstone of the next-generation bioeconomy.

8.8 Conclusions

Hemicellulose stands at the forefront of sustainable biorefinery innovation, offering a renewable, multifunctional platform for the production of biofuels, biochemicals, biodegradable materials, and health-promoting compounds. Its structural heterogeneity, once considered a challenge, has become a key asset, enabling tailored applications across diverse sectors, including energy, packaging, agriculture, food, pharmaceuticals, and cosmetics. The physicochemical and functional properties of hemicellulose, such as its solubility, reactivity, film-forming ability, and biodegradability, make it a compelling alternative to petroleum-based polymers. Its compatibility with other biopolymers and its ability to be chemically or enzymatically modified further expand its utility in creating high-performance, eco-friendly materials. Moreover, hemicellulose-derived oligosaccharides and derivatives are gaining traction in functional foods and biomedical applications due to their prebiotic, antioxidant, and antimicrobial properties. However, the path to commercial realization is not without obstacles. The complexity of extraction and purification, the low market value of some derivatives, and the limited efficiency of current microbial and enzymatic systems remain significant barriers. Addressing these challenges requires a multidisciplinary approach that combines advances in green chemistry, process engineering, synthetic biology, and systems integration. Looking ahead, the future of hemicellulose lies in the development of robust biocatalysts, engineered microbial platforms, and integrated biorefinery models that valorize all biomass components in a circular and sustainable manner. Innovations such as deep eutectic solvents, organosolv fractionation, and microbial consortia engineering are already paving the way for more efficient and environmentally friendly processing routes.

In essence, hemicellulose is not merely a by-product of lignocellulosic biomass; it is a strategic biopolymer with the potential to drive the transition toward a circular bioeconomy. With continued research, policy support, and industrial collaboration, hemicellulose-based technologies can play a transformative role in building a more sustainable and resilient future.

References

Abolore RS, Jaiswal S, Jaiswal AK (2024) Green and sustainable pretreatment methods for cellulose extraction from lignocellulosic biomass and its applications: a review. Carbohydrate Polymer Technologies and Applications 7:100396

Alam MM, Greco A, Rajabimashhadi Z, Esposito Corcione C (2024) Efficient and environmentally friendly techniques for extracting lignin from lignocellulose biomass and subsequent uses: a review. Cleaner Materials 13:100253

Ali K, Niaz N, Waseem M, Ashraf W, Hussain M, Khalid MU, Tahir AB, Raza A, Khan IM (2025) Xylooligosaccharides: a comprehensive review of production, purification, characterization, and quantification. Food Res Int 201:115631

Alsoy Altinkaya S (2024) A perspective on cellulose dissolution with deep eutectic solvents. Front Membrane Sci Technol 3:1382054

Arkell A (2015) Membrane filtration for fractionation of biomass compounds – process design for the separation of hemicelluloses and lignin from alkaline solutions

Arzami AN, Ho TM, Mikkonen KS (2022) Valorization of cereal by-product hemicelluloses: fractionation and purity considerations. Food Res Int 151:110818

Badgujar KC, Bhanage BM (2015) Factors governing dissolution process of lignocellulosic biomass in ionic liquid: current status, overview and challenges. Bioresour Technol 178:2–18

Chen H (2015) 3 – lignocellulose biorefinery feedstock engineering. In: Chen H (ed) Lignocellulose biorefinery engineering. Woodhead Publishing, pp 37–86

Cheng J, Gao R, Zhu Y, Lin Q (2024) Applications of biodegradable materials in food packaging: a review. Alexandria Eng J 91:70–83

Gautam D, Rana V, Sharma S, Walia YK, Kumar K, Umar A, Ibrahim AA, Baskoutas S (2025) Hemicelluloses: a review on extraction and modification for various applications. ChemistrySelect 10(24):e06050

Gullón P, Gullón B, Cardelle-Cobas A, Alonso JL, Pintado M, Gomes AM (2014) Effects of hemicellulose-derived saccharides on behavior of lactobacilli under simulated gastrointestinal conditions. Food Res Int 64:880–888

Haimer E, Wendland M, Potthast A, Henniges U, Rosenau T, Liebner F (2010) Controlled precipitation and purification of hemicellulose from DMSO and DMSO/water mixtures by carbon dioxide as anti-solvent. J Supercrit Fluids 53:121–130

Huang L-Z, Ma M-G, Ji X-X, Choi S-E, Si C (2021) Recent developments and applications of hemicellulose from wheat straw: a review. Front Bioeng Biotechnol 9:690773

Huang Y, Li C, Zheng S, Fu X, Huang Q, Liu G, Chen Q (2024) Influence of three modification methods on the structure, physicochemical, and functional properties of insoluble dietary fiber from Rosa roxburghii Tratt pomace. Molecules 29(9):2111

Jana UK, Kango N, Pletschke B (2021) Hemicellulose-derived oligosaccharides: emerging prebiotics in disease alleviation. Front Nutr 8:670817

Jincy EM, Femina KS (2023) Heteropolymer in biomass: hemicellulose extraction and modifications. In: Thomas S, Hosur M, Pasquini D, Jose Chirayil C (eds) Handbook of biomass. Springer Nature Singapore, Singapore, pp 1–32

Kazmi A, Sultana T, Ali A, Nijabat A, Li G, Hou H (2025) Innovations in bioethanol production: a comprehensive review of feedstock generations and technology advances. Energy Strategy Rev 57:101634

Kulkarni V, Butte K, Rathod S (2012) Natural polymers – a comprehensive review. Int J Res Pharmaceut Biomed Sci 3:1597–1613

Lu Y, He Q, Fan G, Cheng Q, Song G (2021) Extraction and modification of hemicellulose from lignocellulosic biomass: A review. Green Process Synth 10(1):779–804

Malik K, Sharma P, Yang Y, Zhang P, Zhang L, Xing X, Yue J, Song Z, Nan L, Yujun S, El-Dalatony MM, Salama E-S, Li X (2022) Lignocellulosic biomass for bioethanol: insight into the advanced pretreatment and fermentation approaches. Ind Crop Prod 188:115569

Mathura SR, Landázuri AC, Mathura F, Andrade Sosa AG, Orejuela-Escobar LM (2024) Hemicelluloses from bioresidues and their applications in the food industry—towards an advanced bioeconomy and a sustainable global value chain of chemicals and materials. Sustain Food Technol 2(5):1183–1205

Mujtaba M, Fernandes Fraceto L, Fazeli M, Mukherjee S, Savassa SM, Araujo de Medeiros G, do Espírito Santo Pereira A, Mancini SD, Lipponen J, Vilaplana F (2023) Lignocellulosic biomass from agricultural waste to the circular economy: a review with focus on biofuels, biocomposites and bioplastics. J Clean Prod 402:136815

Paz-Cedeno FR, Solorzano-Chavez EG, Dias LM, Otaviano CA, Bustamante LJA, Monti R, Martínez-Galán JP, Masarin F (2022) Composition and chemical structure of hemicelluloses and polysaccharides with capability of gel formation. In: Brienzo M (ed) Hemicellulose biorefinery: a sustainable solution for value addition to bio-based products and bioenergy. Springer Nature Singapore, Singapore, pp 111–137

Rao J, Lv Z, Chen G, Peng F (2023) Hemicellulose: structure, chemical modification, and application. Prog Polym Sci 140:101675

Research GI (2024) Global Hemicellulose Market: Supply, Demand and Key Producers, 2023–2029. Retrieved July 17, 2025, from https://www.globalinforesearch.com/reports/1405092/hemicellulose

Research CM (2025) Hemicellulose Market Report. Retrieved July 17, 2025, from https://www.cognitivemarketresearch.com/hemicellulose-market-report

Salehi M, Rashidinejad A (2025) Multifaceted roles of plant-derived bioactive polysaccharides: a review of their biological functions, delivery, bioavailability, and applications within the food and pharmaceutical sectors. Int J Biol Macromol 290:138855

Schmidt AR, Dresch AP, Alves Junior SL, Bender JP, Treichel H (2023) Applications of brewer's spent grain hemicelluloses in biorefineries: extraction and value-added product Obtention. Catalysts 13(4):755

Sharma R (2024) Hemicellulose market report: Global Forecast from 2025 to 2033. Retrieved July 17, 2025, from https://dataintelo.com/report/global-hemicellulose-market

Shi R, Zhang Z, Zhang J, Chen C, Li W, Lin Y, Shi X, Zhao P, Zhang T, Yan Q, Cheng X (2025) A comparative study on enhanced enzymatic hydrolysis of diverse herbaceous and woody wastes by promising dilute acid and alkaline pretreatments. Bioresour Bioprocess 12(1):36

Siddiqui SA, Yang X, Deshmukh RK, Gaikwad KK, Bahmid NA, Castro-Muñoz R (2024) Recent advances in reinforced bioplastics for food packaging—a critical review. Int J Biol Macromol 263:130399

Solihat NN, Hidayat AF, Ilyas RA, Thiagamani SMK, Azeele NIW, Sari FP, Ismayati M, Bakshi MI, Garba ZN, Hussin MH, Restu WK, Syafii W, Ariyanta HA, Fatriasari W (2024) Recent antibacterial agents from biomass derivatives: characteristics and applications. J Bioresour Bioprod 9(3):283–309

Soni SK, Soni R (2025) Biodegradable waste composition and characterization: understanding the building blocks of green biorefinery solutions. In: Soni SK, Soni R (eds) Green biorefinery solutions: transforming biodegradable waste into resources. Springer Nature Singapore, Singapore, pp 25–56

Stoklosa R, Hodge D (2014) Integration of (hemi)-cellulosic biofuels technologies with chemical pulp production, pp 73–100

Tang X, Wang Z, Zheng J, Kan J, Chen G, Du M (2023) Physicochemical, structure properties and in vitro hypoglycemic activity of soluble dietary fiber from adlay (Coix lachryma-jobi L. var. ma-yuen Stapf) bran treated by steam explosion. Front Nutr 10:1124012

Thoresen PP, Fahrni J, Patel A, Enman J, Gustafsson T, Rova U, Christakopoulos P, Matsakas L (2024) Organosolv-derived lipids from hemicellulose and cellulose, and pre-extracted tannins as additives upon hydrothermal liquefaction (HTL) of spruce bark lignins to bio-oil. BMC Biotechnol 24(1):96

Tofani G, Jasiukaitytė E, Grilc M, Likozar B (2023) Organosolv biorefinery: resource-based process optimisation, pilot technology scale-up and economics. Green Chem 26:189–201

Tong J, Hu W, Qin Y, Liu Y (2023) Deep eutectic solvent pretreatment for green preparation of nanocellulose. Cellulose 30(8):4773–4792

Xu Y, Liu K, Yang Y, Kim MS, Lee CH, Zhang R, Xu T, Choi SE, Si C (2022) Hemicellulose-based hydrogels for advanced applications. Front Bioeng Biotechnol 10:1110004

Yao Z, Chong G, Guo H (2024) Deep eutectic solvent pretreatment and green separation of lignocellulose. Appl Sci 14. https://doi.org/10.3390/app14177662

Chapter 9
Lignin

Alvaro Luiz Mathias

Abstract Lignin is the most abundant aromatic biopolymer on Earth and serves as an essential structural component in vascular plants. It is produced on a large scale—over 50 million tons annually—as a by-product of the pulp and paper industry, although only a small fraction is commercially utilized. This chapter provides a comprehensive overview of lignin's ecological functions, availability, chemical structure, physicochemical properties, and its degradation and valorization pathways. It highlights conventional industrial applications such as resins and dispersants, as well as advanced uses in bioplastics, nanomaterials, energy storage, and green chemistry. High-value lignin-derived products include carbon fibers, UV-protective films, smart materials, and biomedical hydrogels. Emerging technologies such as enzymatic processing, catalytic conversion, and additive manufacturing are discussed, along with digital tools that facilitate the design of lignin-based materials. Overall, lignin is positioned as a strategic renewable resource with growing relevance across diverse industries and scientific domains.

9.1 Introduction

Historically, lignin has primarily been used as a fuel in pulp production processes. However, its potential for higher-value commercial applications has gradually gained attention. Lignosulfonate products began to be used in leather tanning and as industrial dispersants, while kraft lignins found applications in paints, ceramics, and rubber formulations. Additionally, lignin has been directly or indirectly associated with the production of vanillin, ethanol, yeasts, and adhesives (McCarthy and Islam 2000).

The global transition toward sustainable materials and green technologies has never been more pressing. Lignin, a traditionally undervalued by-product of the pulp

A. L. Mathias (✉)
Postgraduate Program in Urban and Industrial Environment (PPGMAUI), Federal University of Paraná, Curitiba, Brazil
e-mail: mathias@ufpr.br

A. L. Woiciechowski et al. (eds.), *Biomass for Bioenergy and Bioproducts*, Biofuel and Biorefinery Technologies 17, https://doi.org/10.1007/978-3-032-19898-3_9

industry, is increasingly recognized as a strategic renewable resource (Prasad et al. 2007). This chapter invites readers to explore the multifaceted roles of lignin—from plant structural integrity to smart materials and nanotechnological innovations. More than just a chemical feedstock, lignin represents a bridge between circular economy principles and the United Nations Sustainable Development Goals (SDGs). Its valorization supports clean energy (SDG 7), water treatment (SDG 6), environmentally responsible production (SDG 12), and climate resilience (SDG 13) (United Nations, General Assembly 2015).

With its expanding applications in energy, second-generation ethanol, healthcare, electronics, and agriculture, lignin is emerging as a central component of the next-generation biorefineries. This chapter offers both foundational insights and advanced perspectives for scientists, engineers, and innovators interested in reimagining lignin not as industrial waste but as a cornerstone of a more intelligent and sustainable future.

9.2 Importance of Lignin

9.2.1 Ecological and Industrial Significance

Lignin is a three-dimensional aromatic macromolecule that plays a fundamental role in the architecture and functionality of higher plants. It is a major component of lignocellulosic biomass—ranking just after cellulose and hemicellulose—and is the most abundant aromatic biopolymer in nature (Azadi et al. 2013; Demirbaş 2001). Its importance can be viewed along two axes: its ecological functions and its economic and technological relevance (Zakzeski et al. 2010).

9.2.1.1 Ecological Role

Lignin serves several physiological functions in vascular plants. Primarily, it acts as a structural reinforcement agent, conferring rigidity to secondary cell walls. This enables plants to grow tall and resist mechanical forces like wind and gravity. For instance, lignification of xylem cells is essential for mechanical support and efficient transport of water and nutrients throughout the stem and roots (Wang et al. 2018; Vanholme et al. 2019).

Additionally, lignin acts as a natural barrier against pathogenic microorganisms, particularly fungi (Vanholme et al. 2019). This resistance is linked to its highly recalcitrant structure, characterized by strong carbon–carbon and ether bonds that resist enzymatic degradation. Thus, lignin contributes to plant defense and longevity.

Lignin also plays a pivotal role in the global carbon cycle. It is estimated that between 67.5 and 157.5 Gt of carbon are stored in the form of lignin, representing a significant fraction of the terrestrial biomass carbon pool (Bar-On et al. 2018). Given that lignocellulosic tissues—wood, bark, and leaves—constitute the bulk of this biomass, and that lignin represents 15–35% of it (Azadi et al. 2013), lignin stands among the largest organic carbon reservoirs on Earth, surpassing the total biomass of all animals, estimated at only 2 Gt (Bar-On et al. 2018). Furthermore, over 220 billion tons of residual biomass are generated annually, providing an abundant source of lignin along with associated cellulose and hemicellulose (Ariyanta et al. 2023).

9.2.1.2 Economic Relevance

Despite being the second most abundant biopolymer on Earth, lignin remains largely underutilized and is burned chiefly for energy in pulp and paper mills. However, it holds great potential as a feedstock for high-value products. Global lignin production is estimated to reach 100 million tons annually, with projections for growth driven by the bioeconomy and energy transition. The main commercial types include kraft lignin, lignosulfonates, and organosolv lignin, the latter being sulfur-free and more suitable for advanced industrial applications (Bajwa et al. 2019).

Various extraction methods have been explored for agricultural residues: kraft ($NaOH/Na_2S$, 150–170 °C), soda (NaOH, 160–180 °C), organosolv (organic solvents, 120–200 °C), steam explosion (160–240 °C), biological extraction (30–50 °C using enzymes or microbes), and ionic liquids (80–120 °C). These technologies aim to valorize lignin, particularly from agricultural residues. Applications have been proposed in multiple domains (Laurichesse and Avérous 2014; Raj et al. 2025). For example, Alcell™ organosolv lignin has a higher phenolic group content and lower carboxyl content, highlighting significant structural differences relevant to distinct applications (Gosselink et al. 2004):

- Environmental: Adsorbents for heavy metals and dyes.
- Energy: Binders in lithium batteries and renewable energy sources.
- Functional Materials: Superabsorbents, antioxidants, antibacterial and antifungal agents, UV blockers, and controlled-release systems.
- Industrial: Resins, adhesives, and composites.
- Cosmetics and Packaging: Protective and sustainable formulations.

Recent advances in biorefineries and conversion technologies have enabled the transformation of lignin into carbon fibers (Gbenebor et al. 2023), bio-oils, phenolic resins, BTX, vanillin, polyurethanes, and materials for biomedical, environmental, and energy applications. Other potential products include flame retardants, hydrogels, and nanostructured materials. Life cycle and techno-economic analyses show that lignin-derived products can reduce greenhouse gas emissions and improve industrial sustainability. Overcoming challenges related to lignin's structural heterogeneity and its purification is key to broader industrial adoption (Bajwa et al. 2019).

9.2.1.3 Challenges

Despite its abundance, less than 2% of the lignin produced annually is effectively utilized in industrial applications. The vast majority continues to be incinerated, primarily due to the structural complexity of lignin, processing difficulties, and the absence of well-established value chains for lignin-derived products. Overcoming these challenges requires advancements in separation technologies, chemical characterization, lignin modification, and integration into existing production chains (Haq et al. 2020; Baghel 2025).

Therefore, lignin represents a key component in the development of a low-carbon bioeconomy. A paradigm shift is needed—one that views lignin not as a mere by-product of the pulp and paper industry but as a strategic, bio-based feedstock (Prasad et al. 2007). An analogy can be drawn between the petroleum refineries of the last century and the emerging biorefineries that are shaping a new, sustainable model. However, although they share principles of feedstock fractionation and conversion, biorefineries face distinct challenges—such as the need for hydrophilic catalysts, in contrast to the hydrophobic catalysts traditionally designed for petroleum-derived substrates (Zakzeski et al. 2010; Laurichesse and Avérous 2014).

9.2.2 Global Reserves and Availability

9.2.2.1 Primary Sources of Lignin

Lignin is present in all vascular plants, although its concentration varies according to the biomass (Table 9.1). Softwoods have a high lignin content (about 32.5%), hardwoods around 21.9%, and grasses like corn stover and wheat straw range from 17.6% to 21%. Agricultural residues such as olive pomace may reach up to a 55.29% lignin content. These variations influence the selection of lignocellulosic feedstocks for lignin-based industrial applications (Demirbaş 2001).

Plant residues represent a valuable lignin source. Sugarcane yields up to 95% recovery via organosolv, corn 90% (alkaline acid), wheat 97% (alkaline/

Table 9.1 Average chemical composition of different classes of lignocellulosic biomass[a]

Class	Cellulose (%)	Hemicellulose (%)	Lignin (%)	Energy[b] (kJ/g)
Softwood (conifers)	42.68	24.82	32.5	19.55
Hardwood (leafy)	45.85	32.26	21.89	18.97
Poaceae[c] (corn straw)	51.53	30.88	17.59	18.27
Poaceae[c] (wheat straw)	33.82	45.2	20.98	18.55
Other (olive cake)	23.08	21.63	55.29	21.53

Adapted from Demirbaş (2001)
[a]Demirbaş (2001)
[b]Higher calorific value
[c]Gramineae

organosolv), rice up to 93% (organosolv), banana 78.86%, grape 100% (autohydrolysis), apple 59.4%, triticale 70%, and cassava 90% (ionic liquids). The yield depends on the biomass and the extraction technique—alkaline, acidic, organosolv, autohydrolysis, or ionic liquids—highlighting the importance of process optimization (Ariyanta et al. 2023).

Lignocellulosic cell walls are primarily composed of cellulose, hemicellulose, and protolignin. Cellulose consists of linear microfibrils formed by D-glucose chains joined by β-(1 → 4)-glycosidic bonds, conferring rigidity. Hemicellulose is a branched heteropolysaccharide associated with cellulose fibrils, enhancing structural cohesion. Protolignin, in turn, is a three-dimensional matrix made of phenolic units such as coniferyl, p-coumaryl, and sinapyl alcohols. It fills the gaps between cellulose and hemicellulose, providing additional rigidity and resistance to biological and chemical degradation (Higuchi 1980).

9.2.2.2 Global and Regional Production

Industrial lignin production is concentrated in countries with large pulp and paper operations, such as the United States, Canada, Scandinavian countries, Brazil, China, and Indonesia. The United States leads in kraft lignin production, while Europe has a strong presence in lignosulfonate manufacturing (Baghel 2025).

Black liquor, a by-product of the kraft pulping process, represents the largest source of industrial lignin (Baghel 2025). This liquor contains not only lignin but also dissolved hemicelluloses and other compounds and is commonly used for cogeneration of energy in pulp mills. However, commercial technologies such as the LignoBoost process now enable the recovery of lignin from black liquor, enabling its conversion into higher-value products (Goliszek et al. 2024).

In addition to wood, urban and agro-industrial residues represent an underexplored but promising lignin source. Materials such as rice husks, fruit waste, vinasse, brewery spent grains, and nutshells contain lignin that can be extracted using various methods (soda, organosolv, deep eutectic solvents). These residues are abundant in tropical and subtropical regions and offer logistical advantages when integrated into local biorefineries (Prasad et al. 2007; Ariyanta et al. 2023).

9.2.2.3 Logistics and Feasibility

The economic feasibility of lignin valorization depends on several factors: proximity to biomass sources, transportation and separation costs, scale of operations, and integration with existing production lines (e.g., lignocellulosic ethanol, green chemicals) (Bajwa et al. 2019).

Integrated biorefineries, which simultaneously utilize cellulose, hemicellulose, and lignin, are emerging as the most promising model for making lignin use economically viable. These facilities combine pretreatment technologies, enzymatic catalysis, second-generation ethanol fermentation (Prasad et al. 2007), and chemical synthesis to produce a wide range of products from lignocellulosic residues. Although still in the early stages, modern biorefineries currently produce limited outputs such as ethanol and bio-oils and operate with low levels of process integration between chemical and energy systems (Zakzeski et al. 2010).

9.2.3 Chemical Structure and Molecular Characteristics

9.2.3.1 Monomeric Units

Lignin is composed of three primary monomeric units (Fig. 9.1) derived from the shikimic acid pathway and phenylpropanoid metabolism (Higuchi 1980):

- Coniferyl Alcohol (G-unit): Predominant in softwoods
- Sinapyl Alcohol (S-unit): Dominant in hardwoods
- p-Coumaryl Alcohol (H-unit): Common in grasses and herbs

Typical lignin is formed by radical-mediated polymerization catalyzed by peroxidases and laccases (Zakzeski et al. 2010), resulting in a complex, irregular three-dimensional structure. However, catechyl lignin found in the milled wood lignin (MWL) of Camellia oleifera bark exhibits a more linear and homogeneous architecture, consisting of caffeyl alcohol units linked via benzodioxane structures. This unique structure offers high thermal stability and selective pyrolysis, yielding catechols (~18.76%), thereby supporting simpler and more efficient biorefinery pathways (Gao et al. 2024).

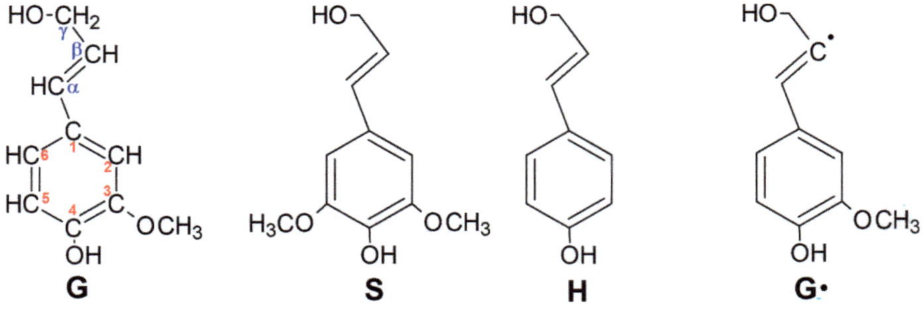

Fig. 9.1 Protolignin precursors: coniferyl alcohol (G), sinapyl alcohol (S), and p-coumaryl alcohol (or p-Hydroxycinnamyl alcohol, H). Example of a radical: guaiacyl (G·)

9.2.3.2 Linkages and Architecture

Biologically, protolignin is synthesized via enzymatic oxidation of three primary phenolic monomers: coniferyl, sinapyl, and p-coumaryl alcohols. Enzymes such as peroxidases and laccases catalyze the formation of free radicals, which undergo nonenzymatic polymerization (Fig. 9.2), leading to a heterogeneous macromolecule (Higuchi 1980). The most common interunit linkage is the β-O-4 ether bond, which contributes to the partial linearity of lignin's structure.

Other notable linkages include (Zakzeski et al. 2010):

- β-5 (phenylcoumaran)
- 5-5 (biphenyl)
- 4-O-5 (diaryl ether)
- β-1 (1,2-diarylpropane)
- β-β (resinol)

The resulting protolignin (Fig. 9.3) is a polyphenolic network functionally integrated into the plant cell wall (Wu et al. 2024). Its native structure is substantially altered during conventional extraction processes (Wang et al. 2018).

The type and abundance of these linkages influence lignin's solubility and reactivity. For instance, hardwood lignins, which contain higher proportions of β-O-4 linkages (60–62%), tend to be more linear and more straightforward to depolymerize. In contrast, softwood lignins are more condensed, with greater frequencies of 5-5 (19–27%) and β-5 (9–12%) linkages, rendering them more chemically resistant. These structural attributes are crucial for determining the lignin's behavior in catalytic and chemical conversion processes (Zakzeski et al. 2010).

9.2.3.3 Protolignin Versus Technical Lignin

Protolignin refers to the native form of lignin present in the plant cell wall before undergoing any chemical modification. It retains its original, highly complex, and cross-linked structure. In contrast, technical lignin is the form extracted industrially using various chemical treatments, which alter its structure and properties significantly (McCarthy and Islam 2000).

The characteristics of lignin vary significantly depending on the extraction method employed. Kraft, soda, organosolv, and lignosulfonate lignins are the primary sources of technical lignin (Laurichesse and Avérous 2014). For instance, kraft lignin typically exhibits a high degree of recondensation and a reduced number of free functional groups due to the harsh alkaline-sulfide conditions used during its isolation (Yan et al. 2024). In contrast, organosolv lignin better preserves the native lignin structure. It maintains higher chemical reactivity, making it more suitable for advanced applications such as the production of vanillin, phenolic alcohols, and other high-value aromatic compounds (Zakzeski et al. 2010).

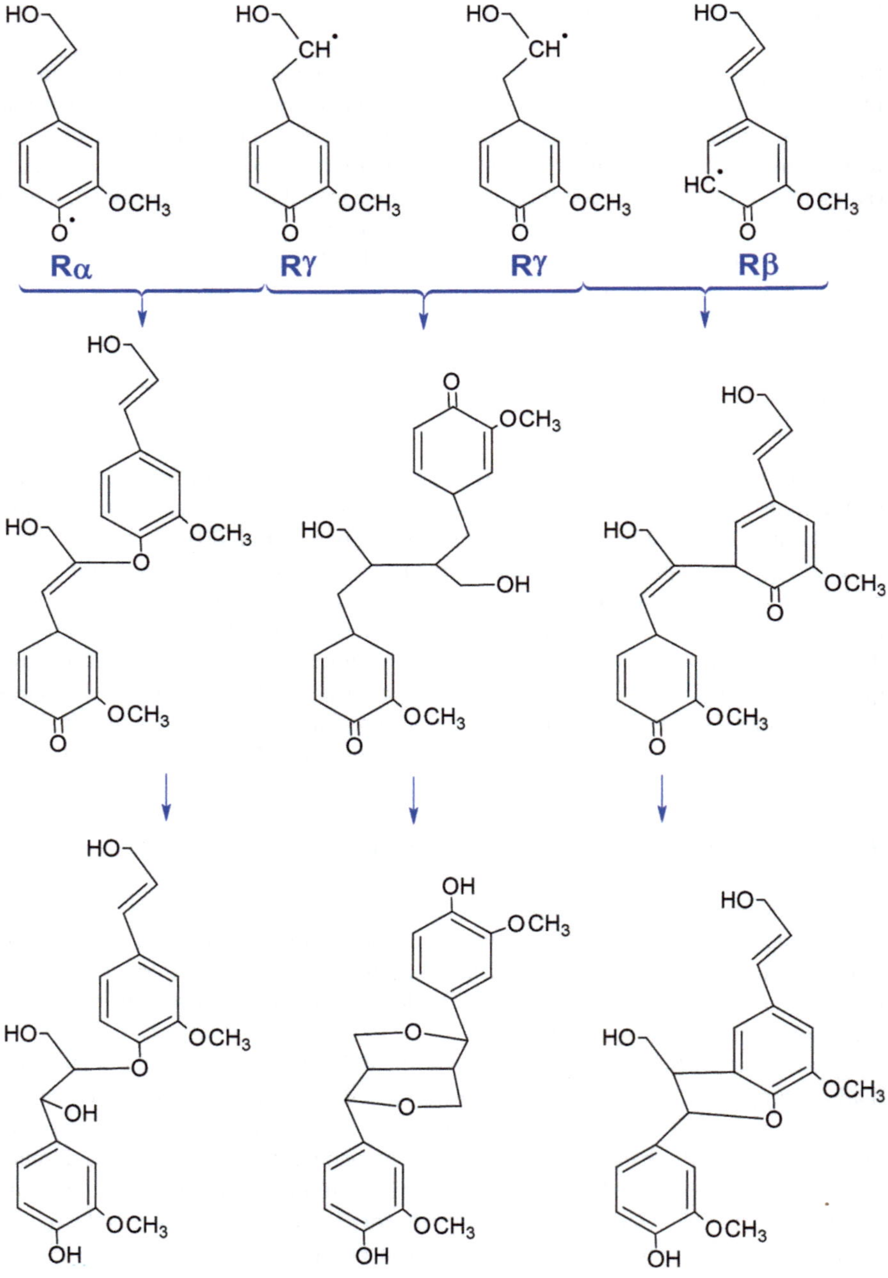

Fig. 9.2 Termination reactions of phenylpropanoid radicals: formation of Guaiacylglycerol-β-Coniferyl Ether (Rα + Rδ yields β-O-4 aryl ether, the most common), DL-Pinoresinol (Rδ + Rδ), and Dehydrodiconiferyl Alcohol (Rβ + Rδ)

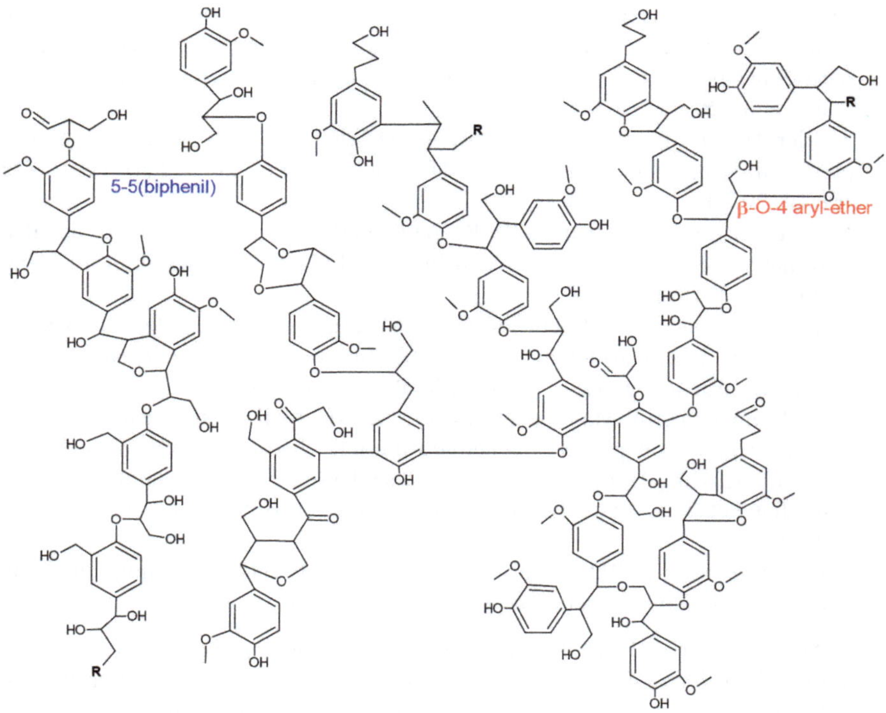

Fig. 9.3 Molecule proposed for conifer protolignin by Sakakibara (McCarthy and Islam 2000)

High-value products derived from lignin often rely on chemical modification techniques such as esterification, epoxidation, alkylation, acylation, oxidation, and ring-opening polymerization. These strategies are tailored to exploit the presence of reactive functional groups, including phenolic hydroxyls, alcohols, and carboxylic acids. Modified lignins exhibit improved properties such as enhanced solubility, reactivity, and compatibility with both synthetic and natural polymers (Laurichesse and Avérous 2014).

Applications of these functionalized lignins range from adhesives, foams, films, and thermosetting and thermoplastic resins to nanocomposites and biodegradable polymers (Laurichesse and Avérous 2014). Thus, the extraction technique plays a pivotal role in determining the physicochemical characteristics of lignin and directly influences its suitability for industrial formulations and the development of bio-based products.

9.2.3.4 Analytical Techniques

A range of advanced analytical methods is used to investigate the structural features and functional properties of lignin and its transformation products:

- Elemental Analysis (CHNS/O): Determines the carbon, hydrogen, nitrogen, sulfur, and oxygen content in lignin, providing insights into its composition, purity, and potential for energy and material applications (Ramos et al. 1999; Sahoo et al. 2011).
- Fourier-Transform Infrared Spectroscopy (FTIR): Identifies functional groups present in lignin (Gosselink et al. 2004; Gbenebor et al. 2023; Faix 1991; Vaz Jr et al. 2025).
- Nuclear Magnetic Resonance (NMR), particularly 2D HSQC: Provides detailed information on monomeric units and interunit linkages (Gosselink et al. 2004; Jiang et al. 2018; Hao et al. 2025).
- High-Performance Size Exclusion Chromatography (HPSEC): Determines the molecular weight distribution, offering insights into polydispersity and macromolecular structure (Jiang et al. 2018; Mathias et al. 1995).
- Gas Chromatography-Mass Spectrometry (GC-MS): Allows the identification of volatile degradation products generated through thermal or enzymatic breakdown, along with silanized monosaccharides released from lignocellulosic hydrolysis (Jiang et al. 2018; Mathias et al. 1995; Mathias and Rodrigues 1995; Majeke et al. 2021).
- X-ray Diffraction (XRD): Analyzes the crystallinity and interactions of lignin with cellulose and hemicellulose (Gbenebor et al. 2023).

These and other tools are fundamental for understanding lignin's structure and supporting its valorization through the design of materials with targeted properties. The choice of technique, however, may vary depending on the intended application, as shown in Table 9.2.

9.2.4 Physicochemical Properties

Lignin exhibits unique physicochemical properties stemming from its highly aromatic and functionalized structure. These properties directly influence its thermal and chemical behavior as well as its interaction with other materials. Understanding these traits is fundamental for their use in industrial formulations and the development of novel materials (Azadi et al. 2013; Raj et al. 2025; Mennani et al. 2025; Thammasang et al. 2025).

9.2.4.1 Thermal Resistance and Energy Potential

Lignin demonstrates remarkable thermal stability, enabling it to resist degradation at temperatures above 200 °C (Demirbaş 2001). This resilience stems from its complex structure, characterized by a dense network of aromatic rings and robust carbon–carbon and ether linkages between phenolic monomers. Thermogravimetric analysis (TGA) typically reveals that lignin begins to undergo significant weight loss between

Table 9.2 Potential information obtained from analytical characterization of lignin and related molecules

Method	What it analyzes	What it tells us
TAPPI (e.g., T222, T249)	Lignin, sugars, N, ash, silicates	Overall composition; impurities from biomass and extraction
CHNS/O	Elemental content	Methoxylation, purity, and energy potential
FTIR	Functional groups	Aromatic rings, hydroxyls, carbonyls
^{1}H NMR	Proton types	Detects –OCH_3, aromatic, and vinyl protons
^{13}C NMR	Carbon types	Aliphatic, aromatic, carboxylic carbons; supports LCC structure mapping
HSQC NMR	H–C correlations	Interunit linkages (β-O-4, β-5, β-β); G-, H-, S-units
^{31}P NMR	Hydroxyl groups (after tagging)	Differentiates phenolic, aliphatic, carboxylic –OH; reactivity insight
Non-aqueous titration	Phenolic and carboxylic –OH	Accurate quantification using organic solvents (e.g., DMF)
Aqueous titration	Total acidity	Less selective; estimates carboxyl content
HPSEC	Molar mass distribution (Mn, Mw)	Estimates size; affected by solubility and adsorption
HPLC (after acetylation)	Total hydroxyls	Quantifies –OH groups; possible underestimation
VPO	Absolute Mn	Accurate for low-mass lignins
XRD	Crystallinity and structure	Shows interaction with cellulose/hemicellulose

Note: *TAPPI* Technical Association of the Pulp and Paper Industry (e.g., T222, T249), *CHNS/O* Elemental Analysis of Carbon, Hydrogen, Nitrogen, Sulfur, and Oxygen, *FTIR* Fourier-Transform Infrared Spectroscopy, *^{1}H NMR* Proton Nuclear Magnetic Resonance, *^{13}C NMR* Carbon-13 Nuclear Magnetic Resonance, *HSQC NMR* Heteronuclear Single Quantum Coherence NMR (^{1}H–^{13}C), *^{31}P NMR* Phosphorus-31 Nuclear Magnetic Resonance, *HPSEC* High-Performance Size Exclusion Chromatography, *HPLC* High-Performance Liquid Chromatography, *VPO* Vapor Pressure Osmometry, *XRD* X-ray Diffraction, *Mn, Mw* Number-average and Weight-average Molar Mass, respectively, *LCC* Lignin–Carbohydrate Complexes

220 and 250 °C, with maximum degradation rates occurring between 300 and 400 °C, depending on its botanical origin and extraction method (Zakzeski et al. 2010; Gbenebor et al. 2023; Sahoo et al. 2011; Tong et al. 2024).

This high thermal stability contributes directly to lignin's superior higher heating value (HHV) compared to cellulose and hemicellulose, making it an effective energy source. As noted by Demirbaş (2001), the lignin content of biomass is a key determinant of its energy potential. Accordingly, lignin is extensively used in cogeneration systems, particularly in the pulp and paper industry, where its combustion helps meet energy demands while facilitating the recovery of chemicals such as NaOH. This dual functionality aligns with broader environmental goals: it supports Sustainable Development Goal (SDG) 12 by enabling chemical recycling and SDG 13 by promoting the use of renewable, nonfossil carbon sources (Li et al. 2024a; Pang et al. 2025).

9.2.4.2 Solubility and Reactivity

A lignin's solubility varies according to its type (Zakzeski et al. 2010) and the solvent system used (Wang et al. 2018). Kraft lignin, for instance, is partially soluble in alkaline solutions (pH >10) but insoluble in neutral water. In contrast, lignosulfonates are highly water-soluble due to the presence of sulfonate groups (Zakzeski et al. 2010).

Key functional groups in lignin include (Laurichesse and Avérous 2014):

- Phenolic Hydroxyls (–OH): Enable etherification, esterification, and oxidation reactions.
- Methoxyl Groups (–OCH_3): Influence thermal stability and electronic properties.
- Carboxylic Acids (–COOH): Enhance solubility and react with amines and alcohols.
- Carbonyl Groups (–C=O): Typically formed via partial oxidation.

This functional diversity gives lignin great chemical versatility as a reagent, modifier, or functional filler in various formulations (McCarthy and Islam 2000; Zakzeski et al. 2010; Ariyanta et al. 2023; Raj et al. 2025; Baghel 2025).

9.2.4.3 Influence of Extraction Method

The method employed to extract lignin plays a pivotal role in determining its molecular architecture, functional group availability, and physicochemical properties (Zakzeski et al. 2010; Ariyanta et al. 2023; Laurichesse and Avérous 2014; Raj et al. 2025; Yan et al. 2024). Each extraction technique—whether alkaline, acidic, or organic solvent-based—results in distinct lignin fractions with varying degrees of condensation, molecular weights, and functionalization levels. For instance, kraft lignin, obtained through a sulfur-containing alkaline pulping process, tends to exhibit a high molecular weight and a highly condensed aromatic structure. These characteristics lead to reduced chemical reactivity but enhanced thermal stability, making it more suitable for energy recovery and structural applications (Yan et al. 2024). In contrast, sulfite lignin (lignosulfonates) is rich in sulfonate groups, which impart high water solubility and polyelectrolyte behavior, favoring its use in dispersants, emulsifiers, and colloidal systems (Ariyanta et al. 2023).

Organosolv lignin, derived from organic solvent-based processes, retains much of the native lignin structure and typically displays lower molecular weight and greater chemical purity, making it highly attractive for applications in fine chemical synthesis, functional additives, and polymer composites (Laurichesse and Avérous 2014). Meanwhile, soda lignin, produced under sulfur-free alkaline conditions and often sourced from non-wood biomass, offers lower sulfur content and improved environmental compatibility, facilitating its incorporation into sustainable materials and green chemistry platforms (Raj et al. 2025).

- Kraft Lignin: Exhibits high molecular weight and a more condensed structure, resulting in lower chemical reactivity.
- Sulfite Lignin (lignosulfonates): Highly functionalized, water-soluble, and suitable for colloidal applications.
- Organosolv Lignin: Lower molecular weight, less structural degradation, and ideal for fine chemical synthesis and composite materials.
- Soda Lignin: Lower sulfur content, making it more appropriate for environmentally sustainable applications.

Therefore, the selection of the lignin extraction method must be carefully matched to the intended end-use, considering essential parameters such as solubility, thermal behavior, chemical reactivity, and environmental impact. This alignment is critical for the efficient valorization of lignin in advanced material systems and circular bioeconomy strategies.

9.2.4.4 Polymer Compatibility

Lignin has been employed in various polymer formulations, both as a filler and as a functional component (Bajwa et al. 2019; Laurichesse and Avérous 2014; Goliszek et al. 2024; Sahoo et al. 2011; Adjaoud et al. 2024; Han et al. 2024; Iswanto et al. 2024; Nurcahyani et al. 2024; Ye et al. 2024; Dai et al. 2025; Shafiq et al. 2024).

- PVA/Lignin: Biodegradable films with enhanced UV resistance.
- PU/Lignin: Partial substitution of polyol in flexible and rigid polyurethane foams.
- Epoxy/Lignin: Replacement of bisphenol A, resulting in improved mechanical properties.

This represents an area of high demand for further research focused on the development of advanced materials derived from lignin. Despite widespread recognition of its aromatic content and potential for sustainable polymer applications, lignin remains underutilized, with only a small fraction being converted into high-value products. Recent studies aim to overcome challenges such as the structural heterogeneity of technical lignin, proposing chemical functionalization approaches to enhance mechanical, thermal, and antioxidant properties in composites, adhesives, and coatings.

In this context, there is growing interest in technologies that modify and compatibilize lignin with biodegradable polymers, targeting applications in packaging, biomedical devices, structural components, and 3D printing. Therefore, developing integrated strategies to optimize the extraction, purification, and application of lignin in engineering materials is a critical frontier for advancing its integration into high-performance, sustainable production chains.

Nevertheless, challenges remain in terms of compatibility, particularly regarding lignin dispersion in polymer matrices and its low miscibility with nonpolar polymers. To address these limitations, strategies such as chemical modification (e.g., acetylation, alkylation) and the use of compatibilizers have been adopted.

9.2.4.5 Optical and Antioxidant Properties

Lignin exhibits strong absorption in the ultraviolet (UV) region, primarily due to the presence of conjugated aromatic rings and functional groups such as phenols and quinones. This characteristic has spurred its use as a natural UV blocker in cosmetics, smart coatings, and packaging films, serving as an alternative to synthetic additives like titanium dioxide. Moreover, the abundance of phenolic groups and conjugated structures endows lignin with antioxidant properties, which are crucial for inhibiting oxidative degradation in food, pharmaceutical, and polymer formulations. Antioxidant activity is typically assessed using methods like DPPH (Vaz Jr et al. 2025; Li et al. 2024b) and tends to be more pronounced in lignins with lower molecular weight and higher concentrations of free phenolic hydroxyl groups.

The presence of guaiacyl (G) and syringyl (S) units is directly associated with a greater electron-donating capacity, while a high content of free phenolic groups favors the neutralization of free radicals such as DPPH·. Methylene groups at the α-position of the side chain and electron-donating substituents, such as hydroxyl and methoxyl groups, enhance the stability of the resulting phenoxyl radicals. On the other hand, a high content of carbonyl groups or the presence of extended π-conjugated systems tends to reduce antioxidant activity. Based on these structural features, the descending order of antioxidant activity among the evaluated technical lignins is as follows: birch lignins (organosolv and soda), grass lignins, softwood kraft lignins, hardwood kraft lignins, and finally, lignosulfonate lignins. The latter exhibits the lowest activity due to extensive sulfonation and reduced phenolic content. These structural correlations underscore the potential of lignin chemical modification for targeted antioxidant applications (Ponomarenko et al. 2025).

Recent studies further support these findings, demonstrating that lignin-based materials can be engineered for multifunctional performance. For example, Gbenebor et al. (2023) highlighted the structural attributes of lignin that influence UV-shielding and oxidative stability. Han et al. (2024) and Li et al. (2024c) reported the development of lignin-containing films and composites with improved UV protection and radical scavenging activity. Additionally, Raj et al. (2025) explored the potential of lignin-based hydrogels in wastewater treatment, emphasizing their optical responsiveness. Surface-functionalized lignin composites, as investigated by Thammasang et al. (2025), have shown enhanced hydrophobicity and UV durability. Moreover, the catalytic functionalization of lignin, as proposed by Mennani et al. (2025), opens new avenues for designing advanced materials with synergistic antioxidant and optoelectronic properties. Collectively, these findings highlight lignin as a versatile, renewable feedstock and support its growing relevance for high-value materials and sustainable chemical production.

9.2.4.6 Rheological Properties and Surface Activity

Lignin demonstrates intrinsic surfactant behavior, particularly in its sulfonated form—lignosulfonates—which possess functional groups that enable them to act as dispersants, emulsifiers, and plasticizers in aqueous systems (McCarthy and Islam

2000; Zakzeski et al. 2010; Baghel 2025; Lora 2008; Xu et al. 2024). These amphiphilic properties are widely utilized in a variety of commercial applications, such as self-compacting concrete, agricultural pesticides, livestock feed additives, and industrial paints, where lignin enhances formulation stability and flow behavior. Additionally, lignin can contribute intrinsic antioxidant and antimicrobial properties, along with favorable surface characteristics for the development of functional biomaterials and bio-based formulations—as demonstrated for lignin extracted from bamboo biomass through mandelic acid hydrolysis followed by alkaline extraction (Liao et al. 2024).

From a rheological perspective, lignin-containing solutions exhibit non-Newtonian flow behavior, with viscosity influenced by parameters such as concentration, pH, and ionic strength. These characteristics reflect the polyelectrolyte nature of certain lignin derivatives and allow for tuning the flow properties of suspensions and emulsions. Moreover, chemical modification of lignin has enabled the development of hydrogels, foams, and thin films with adjustable mechanical strength and water absorption capacity—key attributes for biomedical, agricultural, and packaging applications.

Overall, lignin presents a complex and multifunctional profile, combining high thermal stability and calorific value—which support its use in energy systems—with variable solubility and reactivity depending on the extraction process. These parameters directly affect their compatibility with polymers and their functionalization potential. In addition, lignin's optical, antioxidant, and surfactant properties broaden its relevance in high-performance materials. Its rheological versatility further enhances its applicability in aqueous formulations, positioning lignin as a pivotal bio-based component in sustainable industrial technologies. The careful selection of lignin type—tailored to the target application—is thus essential for unlocking its full potential.

9.2.4.7 Applicability of Technical Lignins: Key Properties and Commercial Potential

Based on the physicochemical properties previously discussed, it becomes clear that different types of technical lignin exhibit distinct structural and functional characteristics that determine their compatibility with various industrial applications. Table 9.3 provides a comparative overview of the main commercial lignin types—lignosulfonates, kraft lignin, and soda lignin—highlighting their sources, functional properties, and established commercial uses (Lora 2008).

Lignosulfonates, derived from the sulfite pulping process, are water-soluble polyelectrolytes due to the presence of sulfonate groups ($–SO_3^-$). These features make them suitable as dispersants in concrete, paints, and pesticides, as well as dust suppressants and mineral binders. In contrast, Kraft lignin has a highly condensed structure with thiol groups (–SH), is insoluble in water, and offers superior thermal stability. It is widely employed as a phenol substitute in phenolic resins, as a conductive additive, and as a precursor for technical materials.

Table 9.3 Comparative overview of major commercial lignins: sources, key properties, and applications

Lignin type	Process/source	Key characteristics	Commercial applications
Sulfonated (lignosulfonates)	Sulfite pulping process (acidic or neutral); softwood and hardwood	Water-soluble; sulfonated structure (–SO_3^- groups); high molecular weight; polyelectrolyte nature	Dispersant (concrete, paints, pesticides), binder for mineral pellets, asphalt emulsifier, vanillin precursor, food additives, dust control agent
Kraft (thiols-containing lignin)	Kraft process (NaOH and Na_2S); wood-based	High purity; highly condensed structure; water-insoluble; presence of thiol groups (–SH)	Emulsifier, paint additive, dye dispersant, phenol substitute in phenolic resins, conductive polymers, prepolymer for technical materials
Soda lignin	Sulfur-free alkaline process (soda); nonwoody fibers (e.g., straw, bagasse, flax, hemp)	Sulfur-free; simpler and more homogeneous structure; higher chemical reactivity	Phenol replacement in resins, industrial dispersants, animal feed additives, biodegradable polymer precursor, eco-friendly emulsifiers

Adapted from Lora (2008)

Soda lignin, on the other hand, originates from sulfur-free alkaline processes applied to nonwoody feedstocks such as sugarcane bagasse, flax, and straw. It features a more homogeneous structure, higher chemical reactivity, and lower environmental impact, making it a promising candidate for biodegradable resins, green polymers, and animal feed additives.

This functional diversity allows technical lignin to be explored as a key component in the development of advanced materials, ranging from adhesives and coatings to structural and bio-based composites. By considering the source, extraction method, and chemical composition of each lignin type, their application can be strategically directed toward sustainable industrial solutions aligned with the principles of the circular bioeconomy.

9.2.5 Degradation and Conversion Processes

9.2.5.1 Chemical Processes

Chemical routes generally involve the use of catalysts under moderate to high temperature and pressure conditions:

9.2.5.1.1 Oxidation

Oxidation is a widely studied route for the valorization of lignin, enabling its transformation into valuable phenolic compounds such as vanillin, acetovanillone, syringaldehyde, and ferulic acid. This process typically employs oxidizing agents such as molecular oxygen or hydrogen peroxide, often in the presence of copper or manganese catalysts, under alkaline conditions. Studies by Mathias and Rodrigues (1995) and Fargues et al. (1996a, b) demonstrated the selective production of vanillin from kraft pine lignin using oxygen and an alkaline medium, with acetovanillone as a secondary product; however, vanillin formed can undergo degradation through consecutive side reactions within the reaction medium. Zakzeski et al. (2010) further highlighted the relevance of oxidative catalysis for lignin valorization, emphasizing the use of environmentally friendly oxidants and transition-metal catalysts. Recent developments, such as the FeNi-catalyzed oxidative systems proposed by Mennani et al. (2025), aim to enhance selectivity and sustainability, reinforcing the potential of oxidation as a key pathway in lignin biorefinery schemes.

9.2.5.1.2 Reduction and Reductive Hydrolysis

Reduction and reductive hydrolysis are promising pathways for lignin valorization, enabling the production of valuable phenolic monomers such as guaiacol, syringol, catechol, and other partially hydrogenated aromatic compounds. These processes typically employ molecular hydrogen in the presence of noble metal catalysts such as Ni, Pd, Ru, or Mo, often supported on materials like carbon, alumina, or silica (Zakzeski et al. 2010; Tong et al. 2024; Li et al. 2024a). Reductive hydrolysis involves the selective cleavage of ether bonds—particularly β-O-4 linkages—while preserving the aromatic ring through mild hydrogenation.

The conversion efficiency is significantly influenced by the origin (Zakzeski et al. 2010) and structural complexity of lignin (Laurichesse and Avérous 2014). Less-condensed lignins, such as those obtained through organosolv extraction or steam explosion, exhibit higher reactivity and better monomer yields (Wang et al. 2023a; Zheng et al. 2024). In contrast, highly cross-linked lignins tend to resist depolymerization and often require pretreatment to enhance their reactivity. However, when structurally less-condensed lignins—such as those rich in syringyl units and methoxyl groups—are used, they can undergo efficient pyrolysis at 900 °C without chemical activation, yielding carbon materials with abundant micropores. This structural transformation leads to a high surface area and improved electrochemical properties, making such lignins ideal precursors for porous carbon or coke-like materials in energy storage applications (Yi et al. 2025). Lignin is a key biorefinery feedstock for producing fuels and chemicals. Co-pyrolysis with coal (e.g., straw + lignite) enhances liquid yield and char reactivity, especially at 600 °C and 73% biomass, where biomass donates hydrogen for better conversion (Zhang et al. 2007). Catalytic depolymerization targets β-O-4 bond cleavage, yielding phenolics like vanillin. Metal catalysts (Ni, Ru, Pd) in alcohol or water improve efficiency. Hardwood lignins, richer in S-units, are more reactive than softwood or kraft types, which often need pretreatment. Organosolv or steam-exploded lignins, with higher β-O-4 content, offer better conversion. These approaches support selective, scalable lignin valorization in green chemical production (Zakzeski et al. 2010; Zhang et al. 2007).

9.2.5.1.3 Pyrolysis

Pyrolysis has emerged as a key thermochemical strategy for lignin valorization, enabling its transformation into high-value aromatic compounds, bio-oil, char, and syngas. Studies have shown that the type of catalyst used—ranging from metal oxides to industrial waste-derived materials—significantly influences product selectivity and quality. Wang et al. (2023a) highlight how catalytic pyrolysis and tandem systems can enhance deoxygenation and aromatic yields, aligning with industrial symbiosis and circular economy models. Sarve et al. (2022) demonstrated the production of bio-oil with high calorific value and low oxygen content using boric acid, while activated char derived from KOH treatment exhibited remarkable environmental remediation potential. Tong et al. (2024) emphasize optimal pyrolysis temperatures between 450 and 550 °C to maximize phenol-rich fractions, while Branca and Di Blasi (2024) contributed kinetic models to guide reactor design. Innovative approaches, such as Flash Joule Heating, have also been proposed for efficient lignin-to-carbon conversion (Guan et al. 2025), opening new avenues for producing graphitic materials. Together, these advancements position pyrolysis as a crucial platform for sustainable lignin-based fuel and chemical production.

9.2.5.1.4 Catalytic Cracking

Catalytic cracking has emerged as a promising pathway for the valorization of lignin into valuable aromatic compounds. The use of solid acid catalysts such as zeolites (e.g., HZSM-5), metal oxides, and hybrid materials enables the selective depolymerization of lignin into monocyclic aromatics like phenols, benzene, toluene, and xylene. Zakzeski et al. (2010) emphasized the potential of catalytic systems to enhance product yield and selectivity by breaking the complex aromatic polymer structure of lignin. Recent studies have advanced this concept further: Wang et al. (2023a) reviewed catalytic pyrolysis as an integrated strategy in biorefinery models, while Zheng et al. (2024) demonstrated microwave-assisted co-pyrolysis of industrial lignin and polyethylene using bimetallic zeolite-based catalysts for the efficient generation of petrochemical substitutes.

Overall, the range of chemical processes available for lignin valorization—including oxidation, reductive depolymerization, pyrolysis, and catalytic cracking—offers a versatile toolbox for converting this abundant aromatic biopolymer into value-added chemicals, fuels, and functional materials. Each route presents unique advantages: oxidation enables the selective formation of phenolic compounds like vanillin and syringaldehyde using green oxidants and metal catalysts; reductive hydrolysis yields stable monomers under mild hydrogenation conditions and aligns with petrochemical substitution; pyrolysis produces a diverse mixture of phenolic-rich bio-oils, syngas, and char, preserving aromatic cores and supporting integration into biorefineries; and catalytic cracking facilitates depolymerization into lighter aromatic hydrocarbons through zeolite- or metal oxide-based systems.

These pathways complement each other and can be tailored to different lignin types and end-use targets, maximizing resource efficiency. In this context, lignin chemistry stands at the crossroads of green chemistry, industrial decarbonization, and circular bioeconomy. Advances in catalysis, process design, and lignin pretreatment will continue to play a crucial role in unlocking the full potential of lignin as a renewable feedstock for sustainable chemical production.

9.2.5.2 Biological Processes

Biological conversion involves microorganisms or enzymes capable of degrading lignin:

9.2.5.2.1 Enzymes

Several studies have investigated the enzymatic depolymerization of lignin as a selective and sustainable approach for its valorization. Enzymes such as lignin peroxidase (LiP) and quinone reductase (QR) have shown promising results in breaking down technical lignins from soda-anthraquinone and sulfite pulping processes, significantly reducing their molecular weight by up to 52% and enabling the production of aromatic compounds like vanillin, 4-vinylguaiacol, and 4-vinylphenol (Majeke et al. 2021). Kraft lignin from *Pinus pinaster* was oxidized using chemical, enzymatic, and sequential strategies. Chemical oxidation with molecular oxygen at 120–138 °C achieved up to 1.2 g/L of vanillin, demonstrating superior efficiency due to its ability to disrupt the highly recalcitrant lignin structure, which naturally resists microbial and enzymatic degradation. In contrast, enzymatic oxidation with the ligninase from *Acinetobacter anitratus*—though effective in depolymerizing lignin into low-molecular-weight aromatics such as veratraldehyde—did not produce detectable vanillin and, when used after chemical pretreatment, even degraded it. While less effective for vanillin production, enzymatic oxidation operates under mild, eco-friendly conditions and highlights the potential of bacterial ligninases to convert technical lignins into fine aromatic chemicals. These findings support the use of chemical oxidation for efficient vanillin recovery and the broader integration of enzymatic processes into biorefineries for sustainable lignin valorization (Ramos et al. 1999; Mathias et al. 1995; Lopretti et al. 1993).

Moreover, the use of bacterial ligninases, such as those from *Acinetobacter anitratus*, offers advantages in selectivity and substrate specificity, potentially minimizing unwanted side reactions. Their ability to target specific linkages within the lignin macromolecule—particularly β-O-4 bonds—enhances the yield of desirable monomers like vanillin and veratraldehyde (Lopretti et al. 1993). Still, maintaining enzyme activity over long reaction periods, scaling up the process, and recovering products efficiently remain bottlenecks for industrial applications. Advances in enzyme engineering, immobilization techniques, and integration with downstream catalytic steps may help overcome these issues. Overall, enzymatic routes remain a promising complement to thermochemical and catalytic strategies in lignin valorization for fine chemicals.

9.2.5.2.2 Microorganisms

Lignin-degrading microorganisms play a pivotal role in the biological depolymerization of lignin, offering environmentally friendly pathways for its conversion into high-value bioproducts. White-rot fungi, particularly basidiomycetes, are among the most effective degraders due to their secretion of extracellular enzymes such as lignin peroxidase (LiP), manganese peroxidase (MnP), and laccase, which oxidatively cleave the complex aromatic structure of lignin (Ariyanta et al. 2023).

In addition to fungi, several bacterial strains, including species of Pseudomonas and Rhodococcus, have shown the ability to metabolize lignin and transform it into valuable compounds such as vanillin, aromatic acids, lipids, and polyhydroxyalkanoates (PHAs). Oleaginous bacteria have also been investigated for their capacity to convert lignin-derived substrates into biodiesel, expanding the role of microbes in lignin-based biofuel production (Zhang et al. 2025). Furthermore, microbial platforms have been developed for the synthesis of cis, cis-muconic acid, a precursor for bioplastics and nylon, demonstrating the synthetic versatility of lignin bioconversion (Ariyanta et al. 2023).

Recent findings highlight another critical microbial-driven transformation of lignin: its role in the formation of humic acids (HAs) during composting. Using plant residues with distinct lignin types—such as rice straw (G-S-H), twigs (G-S), and pine needles (G)—it was shown that microbial degradation of β-O-4, β-5, and β-β linkages facilitates the release of functional groups. These contribute significantly to HA formation, particularly through C–O bonds that reorganize into aromatic, aliphatic, and hydroxyl structures. Units rich in guaiacyl (G) and syringyl (S) lignin, for instance, accounted for up to 48.4% of aromatic core formation in HAs. FTIR, 2D-COS, and 2D-HSQC-NMR confirmed the gradual transformation of lignin into humic components, with Firmicutes, Actinobacteria, and Proteobacteria playing active roles in the process (Wang et al. 2024a).

Beyond composting, advanced analytical tools are enhancing our understanding of lignin's structural diversity across plant tissues. A recent study applied a rapid and high-resolution method—nonuniform sampling (NUS) combined with 2D HSQC NMR spectroscopy—to characterize lignin extracted from various parts of tobacco plants using deep eutectic solvents (DES). The ChCl/OA/GVL formulation was the most effective at preserving native β-O-4 linkages. This structural integrity was crucial for assessing reactivity and valorization potential. The optimized NUS/2D HSQC approach reduced acquisition time from 7.5 to 1.42 hours, maintaining resolution with <2% relative error. Tobacco lignin displayed SGH-type units, with guaiacyl (G) and β-O-4 bonds dominating. Composition varied between roots, stalks, stems, and leaves, influencing thermal stability and chemical reactivity. Roots and stalks were richer in S-units and β-O-4 linkages, suggesting enhanced processability for high-value applications, while leaves and stems had higher contents of G, H, and ferulic acid (FA), associated with greater oxidative reactivity. This technique marks a significant advancement for the structural elucidation of lignin from agricultural residues, aiding biorefinery integration and the development of lignin-derived functional materials (Hao et al. 2025).

Together, these advances underscore the biotechnological potential of microbial lignin degradation, particularly when coupled with genetic engineering and enzyme immobilization strategies that enhance selectivity, stability, and recyclability. Microbial valorization of lignin thus stands as a promising pillar of the emerging bioeconomy, enabling applications ranging from renewable fuels to bioactive compounds and functional biomaterials.

Enzymatic biorefineries for extracted lignin: These systems combine ligninolytic enzymes with other conversion routes (e.g., fermentation, biosynthesis) to produce bioaromatics and organic acids (Taghizadeh-Alisaraei et al. 2019; Ahmad et al. 2020). Enzymatic studies span the entire matrix of the plant cell wall, involving not only the genetic modification potential of plants (Monties 1998) but also microbial and enzymatic processes to release the structural monomers of cellulose and hemicellulose. Lignin is bound to hemicellulose monosaccharides—especially xylans and mannans—through covalent bonds such as ester, benzyl ether, and glycosidic linkages (Jiang et al. 2018), and once separated, it can interfere with enzymatic actions (Ramos et al. 1999). One of the major challenges is reducing lignin's interference with hydrolytic enzyme activity (Prasad et al. 2007), such as endoglucanases, exoglucanases, and β-glucosidases (for cellulose), and xylanases, β-xylosidases, mannanases, arabinofuranosidases, galactosidases, and acetylxylan esterases (for hemicelluloses) (Zeng et al. 2012).

9.2.5.2.3 Enzymatic Biorefineries for Lignocellulosic Material

Enzymatic valorization of lignin has gained increasing relevance in biorefinery contexts, where lignin-degrading enzymes such as peroxidases and laccases play a central role in depolymerizing the lignin matrix to release phenolic intermediates. These systems, referred to as enzymatic biorefineries, integrate enzymatic delignification with downstream processes such as fermentation and biosynthesis, enabling the production of bioaromatics and organic acids (Taghizadeh-Alisaraei et al. 2019; Ahmad et al. 2020). The enzymatic approach encompasses the entire lignocellulosic matrix, from plant genetic modification (Monties 1998) to the use of microbial and enzymatic systems capable of breaking down cellulose and hemicellulose into fermentable sugars (Prasad et al. 2007). However, a major bottleneck is the inhibitory effect of lignin on hydrolytic enzymes—such as endoglucanases, exoglucanases, and β-glucosidases for cellulose; and xylanases, β-xylosidases, mannanases, arabinofuranosidases, galactosidases, and acetylxylan esterases for hemicellulose—limiting saccharification efficiency (Zeng et al. 2012). To address this, strategies include optimizing lignin removal or modifying its structure to reduce nonproductive binding and enhance accessibility to polysaccharide substrates.

9.2.5.3 Technological Challenges

Selectivity is one of the greatest challenges in lignin conversion processes. During degradation, secondary repolymerization reactions may occur, leading to the formation of insoluble residues (char) and reducing the yield of valuable products.

Other challenges include:

- Catalyst inhibition by intermediate products (Zakzeski et al. 2010).
- Complexity in product separation and purification (Prasad et al. 2007; Tong et al. 2024).
- Need for scalable continuous processes to ensure economic viability and industrial feasibility (Azadi et al. 2013; Zakzeski et al. 2010; Yi et al. 2025).

9.2.5.4 Solutions in Development

To overcome the challenges, some approaches have stood out:

- Bifunctional Catalysis: Combines acidic and metallic sites to promote selective breakdown and simultaneous hydrogenation (Zakzeski et al. 2010; Ariyanta et al. 2023; Sarve et al. 2022).
- Continuous Flow and Microreactors: Increase the efficiency and selectivity of reactions (Zakzeski et al. 2010; Ariyanta et al. 2023).
- Kinetic and Computational Modeling: Modeling tools are essential for optimizing fractionation and lignin purification, guiding process choices, and more efficient technological routes (Bécsy-Jakab et al. 2024). Simulations based on density functional theory (DFT), ab initio molecular dynamics (AIMD), and KMC have been applied to predict reaction products, evaluate the stability of chemical bonds, and investigate the interaction between lignin and specific compounds (Tong et al. 2024). Furthermore, modeling the kinetics of product formation and degradation is essential for the design and operation of industrial reactors (Fargues et al. 1996a, b; Branca and Di Blasi 2024; Wang et al. 2024a).

These advances contribute to transforming lignin into a viable raw material for the renewable-based industry.

9.2.5.5 Conversion into Base Products

The main target products resulting from lignin conversion include:

- Vanillin and Syringaldehyde: Widely used in the fragrance and food industries (Zakzeski et al. 2010; Vanholme et al. 2019; Mathias et al. 1995).

- Phenolic Acids (Ferulic, Gallic, P-coumaric): Known for their antioxidant and antimicrobial applications (Zhang et al. 2025).
- Basic Aromatics (Benzene, Toluene, Xylene): With the potential to replace fossil-derived compounds in the petrochemical industry (Wang et al. 2023a; Zheng et al. 2024).
- Lignin-Based Polyols: Employed in the production of polyurethane foams, lignin-derived vitrimers (CAO or Covalently Adaptable Polymers), and recyclable materials (Bajwa et al. 2019; Sahoo et al. 2011; Adjaoud et al. 2024; Iswanto et al. 2024; Nurcahyani et al. 2024; Dai et al. 2025; Shafiq et al. 2024; Feldman et al. 1986).

Optimizing the yields of these compounds is a central goal of lignin conversion strategies.

9.2.5.6 Environmental and Life Cycle Assessment

The valorization of lignin must also consider its environmental benefits. Life cycle analysis (LCA) studies indicate that replacing petroleum derivatives with lignin can reduce CO_2 emissions. In addition, the use of lignin avoids the emission of gases and toxic compounds generated during its combustion, such as NOx, SOx, and fine particles. The conversion of lignin into chemical compounds and advanced materials adds value and reduces the ecological footprint of biomass-based production chains (Bajwa et al. 2019; Guan et al. 2025).

9.2.6 Consolidated Industrial Applications

Lignin is already used industrially in several applications, although most of them are still limited in nature or involve low-value-added products. In this section, we address the main consolidated applications by sector.

9.2.6.1 Pulp and Paper

In the paper industry, lignin is removed from pulp during the bleaching process. Residual lignin is burned to generate energy, contributing to the energy self-sufficiency of the mills. However, processes such as LignoBoost allow the selective extraction of lignin from black liquor, producing dry lignin for use as a substitute for synthetic additives in high-performance coatings (Goliszek et al. 2024).

9.2.6.2 Lignosulfonates

Lignosulfonates are the most commercialized lignin derivatives. Obtained by lignin sulfitation in acidic processes, they are soluble in water and function as:

- Plasticizers for concrete and mortar (Baghel 2025; Lora 2008; Wang et al. 2023b).
- Dispersants in pesticides and paints (McCarthy and Islam 2000; Lora 2008; Wang et al. 2023b).
- Binders in biomass pellets and animal feed (Baghel 2025; Lora 2008).
- Soil and dust stabilizers on roads (Ariyanta et al. 2023).

It is estimated that only 1–2% of lignin is utilized in material applications, such as the production of additives, resins, composites, and other high-value-added products (Zakzeski et al. 2010; Baghel 2025).

9.2.6.3 Resins and Adhesives

Lignin can partially replace phenol in phenolic resins, used in wood panels, plywood, and laminates (Iswanto et al. 2024; Feldman et al. 1986). This substitution reduces the environmental impact and the cost of formulation. In polyurethane foams, lignin is used as a partial polyol, resulting in products with better thermal resistance and dimensional stability (Iswanto et al. 2024; Dai et al. 2025; Feldman et al. 1986; Wang et al. 2025a).

9.2.6.4 Activated Carbon and Carbon

Controlled carbonization of lignin results in activated carbon with high porosity, used in air and water filters. Lignin is also used as a precursor of graphitized carbon, with applications in batteries and supercapacitors (Raj et al. 2025; Han et al. 2024; Li et al. 2024c; Yi et al. 2025; Jiang et al. 2020; Duarah et al. 2020; Wang et al. 2024b; Guerroudj et al. 2024; Lu et al. 2025).

9.2.6.5 Bioplastics and Composites

Lignin can be incorporated into polymers such as PLA, PVA, PVC, and HDPE, forming lignin composites with greater resistance to UV radiation, mechanical rigidity, and lower cost. Despite dispersion challenges, advances in compatibility have driven their adoption in packaging, electronics, and automotive materials (Han et al. 2024; Duarah et al. 2020; Zhang et al. 2024a; Wang et al. 2024c).

9.2.6.6 Transparent Wood Composites

Transparent wood is made by removing or modifying the lignin to reduce light absorption, then filling the structure with polymers of similar refractive index. Eco-friendly methods (e.g., $NaClO_2$, H_2O_2, DES) allow lignin decolorization while preserving wood structure. Bio-based resins like chitosan and PG resin replace petrochemical polymers. The material offers high light transmittance (up to 91%), strength (up to 258 MPa), and low thermal conductivity (~0.2 W/mK). It is ideal for windows, structural panels, and smart coatings with UV-shielding and thermal storage functions (Wu et al. 2024).

9.2.7 Commercial Importance and Global Market

9.2.7.1 Market Overview

According to Baghel (2025), the global lignin market was valued at approximately US$881 billion in 2024 and is expected to reach US$1.81 trillion by 2037, reflecting a compound annual growth rate (CAGR) of 5.7%. The increased demand for sustainable materials and the replacement of petrochemicals have driven this growth.

9.2.7.2 Key Segments

The main segments that use lignin are (Zakzeski et al. 2010; Baghel 2025; Lora 2008; Duarah et al. 2020):

- Precursor: In resins, activated carbon, polymers, and composite materials.
- Civil Construction: Concrete additives represent 63% of global consumption.
- Agrochemicals: As dispersants in pesticide formulations.
- Animal Feed: As binders and digestibility enhancers.
- Paints and Coatings: Used as thickeners and dispersants.
- Cosmetics: A natural antioxidant in creams and sunscreens.

9.2.7.3 Commercial Barriers and Challenges

Despite the potential, the commercial valorization of lignin faces barriers such as (Prasad et al. 2007; Zakzeski et al. 2010; Ariyanta et al. 2023):

- Processing Cost: Separation and purification are still expensive.
- Competition with fossil derivatives, such as phenol and bisphenol A.
- Quality Standardization: lignin variability according to biomass and process.
- Lack of Market Infrastructure: Few consolidated production chains.

9.2.7.4 Strategic Initiatives

Government and private initiatives have been encouraging the use of lignin (European Commission 2021; United States – DOE 2023):

- European Union: Supports biorefineries through Horizon Europe.
- United States: The US Department of Energy invests in lignin technologies for bioenergy.
- Brazil: Partnerships between universities (e.g., Federal University of Paraná (UFPR), Federal University of Rio Grande do Sul (UFRGS), Federal University of São Carlos (UFSCar), Federal University of Viçosa (UFV), University of Campinas (UNICAMP), University of São Paulo (USP), etc.), Embrapa, and the sugar and alcohol sector.

Strengthening these initiatives can accelerate the insertion of lignin as a green commodity in the global market.

9.2.8 Generation of High-Value-Added Products

The conversion of lignin into high-value-added products has been one of the main areas of research in recent years. Its abundance, aromatic richness, and functional diversity make lignin a promising raw material to replace synthetic compounds derived from petroleum in sectors such as health, electronics, energy, cosmetics, and smart packaging.

9.2.8.1 Nanotechnology and Health Applications

Lignin has been used in the production of nanoparticles with antioxidant (Gbenebor et al. 2023), antimicrobial (Raj et al. 2025), and anti-inflammatory properties. Lignin nanoparticles (LNPs) are biocompatible (Gbenebor et al. 2023; Guo et al. 2025; Chaudhary and Sinha 2023), biodegradable, and can be loaded with drugs, genes, or active ingredients, being applied in controlled-release systems (Duarah et al. 2020; Wu et al. 2024; Chaudhary and Sinha 2023).

These particles have been studied for the delivery of anticancer agents, wound treatment, and dermatological formulations (Gbenebor et al. 2023; Guo et al. 2025). C-lignin can also be produced for high-performance adhesives and natural antioxidants. C-lignin, due to its structural regularity, is especially suitable for this application due to its ease of size control and surface functionalization (Gao et al. 2024).

In addition, lignin can be used in the formulation of smart hydrogels with properties that respond to stimuli such as pH, temperature, or light, used in smart dressings and tissue regeneration systems (Yi et al. 2025). The presence of phenolic groups also confers antiseptic properties to these materials (Mennani et al. 2025).

9.2.8.2 Energy: Supercapacitors and Carbon Fibers

In the energy sector, lignin has stood out as a sustainable alternative to produce carbon fibers, used in high-performance lightweight composites for the aerospace

(Wang et al. 2025a; Guerroudj et al. 2024; Prashak et al. 2023), automotive (Azadi et al. 2013; Shafiq et al. 2024; Wang et al. 2025a), and sports (Azadi et al. 2013) sectors. Lignin is thermoplastic (Sahoo et al. 2011) and can be molded and stabilized for subsequent carbonization, yielding fibers with carbon content exceeding 90% (Guan et al. 2025).

Furthermore, lignin has been used in the production of battery electrodes, serving as a precursor to porous carbon or as a conductive binder. The antioxidant and conductive properties of lignin favor its use in lithium-ion, zinc-air, and sodium-ion batteries (Han et al. 2024; Yi et al. 2025) and supercapacitors (Raj et al. 2025; Li et al. 2024c; Yi et al. 2025; Jiang et al. 2020; Duarah et al. 2020; Wang et al. 2024b; Guerroudj et al. 2024; Lu et al. 2025).

9.2.8.3 Green Chemistry and Smart Materials

Lignin is a natural candidate for replacing toxic substances such as bisphenol A (BPA), widely used in plastics and epoxy resins (McCarthy and Islam 2000; Gbenebor et al. 2023; Monties 1998; Goldstein 1981).

Other examples include:

- Formaldehyde-free expandable foams and adhesives, using lignin as a phenolic reagent (Bécsy-Jakab et al. 2024; Zhu et al. 2024).
- Smart films for packaging and fabrics, sensitive to humidity, pH, or temperature, useful in food preservation (Han et al. 2024; Dai et al. 2025; Li et al. 2024b; Liao et al. 2024; Wang et al. 2023b; Ye et al. 2024; Priyadarshi et al. 2024).
- Microcapsules for controlled fertilizers that release nutrients gradually in response to environmental stimuli (Zhang et al. 2024a).
- Sustainable cosmetics and personal products (Xu et al. 2024; Acurio-Cerda et al. 2025; Amieva et al. 2025).
- Lignin can serve as a support for metal catalysts, as its functional groups (–OH, –COOH, –OCH_3) facilitate metal immobilization through coordination interactions or in situ reduction (Pang et al. 2025).

Lignin has found increasing applications in the cosmetics industry, mainly as a photoprotective agent (McCarthy and Islam 2000; Gosselink et al. 2004; Goliszek et al. 2024; Gao et al. 2024; Mennani et al. 2025; Han et al. 2024; Liao et al. 2024; Duarah et al. 2020; Wang et al. 2024c; Chaudhary and Sinha 2023). Its structure efficiently absorbs UV radiation, offering a green alternative to titanium dioxide and zinc oxide, which have been questioned due to their environmental impacts.

Furthermore, lignin can be incorporated into lotions, creams, shampoos, and makeup such as:

- Emulsion stabilizer (Baghel 2025; Xu et al. 2024; Guo et al. 2025; Acurio-Cerda et al. 2025)
- Antimicrobial agent (Thammasang et al. 2025; Dai et al. 2025; Zhang et al. 2021)

- Fragrance ingredient (Zakzeski et al. 2010) (vanillin derived from lignin), which has been increasing its production at the Borregaard Ingredients Corporation from softwood lignin (https://www.borregaard.com/company/news-archive/borregaard-increases-production-of-bio-based-vanillin-1)

Large cosmetics companies are already exploring specific lignins with high purity and controlled color, generated by processes such as organosolv and sulfur-free soda.

9.2.8.4 Innovative Examples

Recent research highlights pioneering applications of lignin in:

- Photochromic lenses, which change color under sunlight (Wang et al. 2025b).
- Self-cleaning coatings with functionalized lignin (Thammasang et al. 2025).
- Tissue engineering materials with slow release of bioactive compounds (Wang et al. 2024c; Chaudhary and Sinha 2023).
- Anti-COVID materials with antiviral properties based on lignin nanoparticles (Chaudhary and Sinha 2023).

These innovations position lignin as a key material for high-value and technology-intensive industries.

9.2.9 Advanced Technical and Technological Applications

With the advancement of chemical modification and characterization technologies, lignin has ceased to be considered a waste by-product and has become an advanced functional material. Technical and technological applications explore their structural complexity, selective functionalization, and optical, thermal, and electrical properties.

9.2.9.1 Smart Materials

Lignin has been used in stimuli-responsive materials, such as:

- Shape memory films: Developed by Yue et al. (2023), these lignin films can return to their original shape after thermal deformation, being useful in biodegradable sensors and actuators.
- Coatings with photosensitive properties: Lignins functionalized with natural chromophores are applied to surfaces that change color under light or heat, and are used in decoration, packaging, and security (Goliszek et al. 2024; Wang et al. 2024b).

In addition, lignin can be combined with conductive polymers (e.g., polyaniline, PEDOT: PSS or Poly(3,4-ethylenedioxythiophene):Poly(styrenesulfonate) to create electroactive materials, applicable in sensors, flexible electronics, and wearable energy systems (Wang et al. 2024b).

9.2.9.2 Desalination and Water Treatment

Lignin foams with high porosity and thermal resistance have been successfully applied in solar desalination systems, in which solar energy evaporates water and lignin acts as a photothermal surface. These materials achieve efficiencies above 80% in the conversion of sunlight into vapor (Liang et al. 2024).

Lignin has also been used as:

- Magnetic adsorbent for heavy metals: When functionalized with –COOH groups and iron oxide nanoparticles, lignin can remove ions such as Pb^{2+}, Cd^{2+}, and Hg^{2+} from contaminated water (Raj et al. 2025).
- Barrier against emerging contaminants: Pharmaceuticals and pesticides in treatment plants (Wang et al. 2023b).

These applications are aligned with the United Nations Sustainable Development Goals (SDGs), particularly SDG 6 (Clean Water and Sanitation).

9.2.9.3 Storage Systems and Renewable Energy

Lignin is used in:

- Membranes for fuel cells, offering greater chemical stability and low cost (Yi et al. 2025).
- Anodes and cathodes in batteries, replacing graphite in lithium-ion and sodium-ion systems (Raj et al. 2025; Yi et al. 2025).
- High-energy-density supercapacitors, with lignin activated by KOH or CO_2, generating surfaces with areas >1000 m^2/g (Fang et al. 2024).

These systems are key for electric mobility, intermittent energy storage (solar and wind), and portable electronics.

9.2.9.4 3D Printing and Additive Manufacturing

Lignin can be used as a filler or matrix in 3D printing filaments, enabling the creation of objects with high UV resistance, optimized thermal properties, and biodegradability. Sustainable architecture projects, industrial design, and functional prototyping have benefited from the incorporation of lignin (Prashak et al. 2023; Zhang et al. 2024b).

With dual-extrusion printers, it is possible to print graded-stiffness structures, combining lignin and PLA for biomedical applications and custom devices.

9.2.9.5 Artificial Intelligence and Computational Modeling

Computational models have been used to predict:

- The chemical reactivity of lignin under different conditions (Fargues et al. 1996a, b; Branca and Di Blasi 2024; Wang et al. 2024a).
- Interactions with catalysts and solvents (Tong et al. 2024).
- Mechanical and electrical properties of lignin-based composites (Bécsy-Jakab et al. 2024).

Techniques such as density functional theory (DFT) and ab initio molecular dynamics (AIMD) support the rational design of valorization pathways and new materials.

9.2.9.6 Circular Economy and Industry 5.0

Lignin plays a central role in the development of integrated biorefineries, which utilize the entire biomass to produce energy, materials, and chemicals. This contributes to:

- Reducing the carbon footprint (Prasad et al. 2007; Zakzeski et al. 2010; Li et al. 2024a; Rong et al. 2013)
- Decarbonizing industry (Rong et al. 2013)
- Minimizing lignocellulosic waste from agricultural, industrial, and urban sources (Prasad et al. 2007; Raj et al. 2025; Gbenebor et al. 2023; Gao et al. 2024; Hao et al. 2025; Branca and Di Blasi 2024; Zhang et al. 2025)

In addition, lignin aligns with the principles of Industry 5.0, which emphasizes sustainability, personalization, and human–machine integration. Applications such as wearable sensors, active packaging, and personalized medicine have lignin as a key enabling element.

Despite the broad spectrum of high-value applications already identified, the technological potential of lignin—particularly that derived from agricultural and agro-industrial residues—remains underexplored. Greater attention must be given to the diversity of lignin types, as their origin and structure significantly influence performance in the fabrication of the aforementioned products and those yet to be conceived.

9.3 Conclusion

Traditionally regarded as a by-product of the pulp and paper industry, lignin has emerged over the past decades as one of the most promising raw materials to drive a sustainable bioeconomy. Its abundance, complex aromatic structure, and diverse functionalization make it not only a renewable resource but also a versatile one, capable of yielding everything from basic products like vanillin and activated carbon to high-performance materials such as carbon fibers, smart sensors, battery electrodes, and biodegradable films with photoprotective properties.

This chapter has shown that lignin plays a key ecological role, acting as a structural support in plants, a barrier against pathogens, and a global carbon reservoir. Its industrial potential, though still underutilized, is immense: it is estimated that only 1–2% of global technical lignin is currently valorized in applications beyond energy combustion. This situation must change.

The efficient conversion of lignin into value-added chemicals and materials faces technological challenges—such as selective bond cleavage and material standardization—but progress in catalysis, continuous processing, targeted fermentation, and computational modeling is paving the way for lignin's full integration into industrial value chains.

Beyond technical and economic opportunities, lignin valorization is deeply aligned with several United Nations Sustainable Development Goals (SDGs):

- SDG 7 (Affordable and Clean Energy): Lignin-based batteries, supercapacitors, and biofuels.
- SDG 9 (Industry, Innovation, and Infrastructure): Development of smart materials and integrated biofactories.
- SDG 12 (Responsible Consumption and Production): Cascade use of lignocellulosic waste.
- SDG 13 (Climate Action): Replacing fossil-based inputs with renewable lignin.
- SDG 6 (Clean Water and Sanitation): Lignin-based adsorbents and foams for water treatment.

Therefore, lignin is not just a technological challenge but a strategic opportunity to reinvent industrial processes, replace toxic and fossil-derived compounds, and align material production with the principles of sustainability and circularity. Investing in research, development, and public policies focused on lignin is essential to transform this biomolecule into a key player in a green and innovative future.

Acknowledgments The author would like to thank the Postgraduate Program in Urban and Industrial Environment at the Federal University of Paraná, Brazil.

References

Acurio-Cerda K, Keloth R, Obewhere OA, Dishari SK (2025) Lignin-based membranes for health, food safety, environmental, and energy applications: current trends and future directions. Curr Opin Chem Eng 47:101098. https://doi.org/10.1016/j.coche.2025.101098

Adjaoud A, Brosse N, Verge P (2024) Lignin-derived thermosets and vitrimers: bridging the gap between renewable and recyclable materials. Mater Today Chem 42. https://doi.org/10.1016/j.mtchem.2024.102342

Ahmad B, Yadav V, Yadav A, Rahman MU, Zhong Yuan W, Li Z, Wang X (2020) Integrated biorefinery approach to valorize winery waste: a review from waste to energy perspectives. Sci Total Environ 719:137315. https://doi.org/10.1016/j.scitotenv.2020.137315

Amieva E, Pola L, Collado S, Oulego P, Calvo PÁ, Díaz M (2025) Understanding the relationship between acid precipitation and lignin odor in Kraft Black liquor. Cleaner Chem Eng 11. https://doi.org/10.1016/j.clce.2024.100143

Ariyanta HA, Sari FP, Sohail A, Restu WK, Septiyanti M, Aryana N, Fatriasari W, Kumar A (2023) Current roles of lignin for the agroindustry: applications, challenges, and opportunities. Int J Biol Macromol 240:124523. https://doi.org/10.1016/j.ijbiomac.2023.124523

Azadi P, Inderwildi OR, Farnood R, King DA (2013) Liquid fuels, hydrogen and chemicals from lignin: a critical review. Renew Sustain Energy Rev 21:506–523. https://doi.org/10.1016/j.rser.2012.12.0223

Baghel R (2025) Lignin waste market size & share, by product (lignosulfonate, kraft lignin); application (concrete admixtures, animal feed binders, dyestuff) – global supply & demand analysis, growth forecasts, statistics report 2025–2037. Research Nester. https://www.researchnester.com/reports/lignin-waste-market/6166

Bajwa DS, Pourhashem G, Ullah AH, Bajwa SG (2019) A concise review of current lignin production, applications, products and their environmental impact. Ind Crop Prod 139:111526. https://doi.org/10.1016/j.indcrop.2019.111526

Bar-On YM, Phillips R, Milo R (2018) The biomass distribution on Earth. Proc Natl Acad Sci USA 115(25):6506–6511. https://doi.org/10.1073/pnas.1711842115

Bécsy-Jakab VE, Savoy A, Saulnier BK, Singh SK, Hodge DB (2024) Extraction, recovery, and characterization of lignin from industrial corn stover lignin cake. Bioresour Technol 399:130610. https://doi.org/10.1016/j.biortech.2024.130610

Branca C, Di Blasi C (2024) Multi-step devolatization kinetics of Klason lignin isolated from beech wood and agro-industrial wastes. Fuel 374. https://doi.org/10.1016/j.fuel.2024.132469

Chaudhary M, Sinha VR (2023) Lignin-based carriers for drug delivery applications: from an industrial waste to a pharmaceutical aid. Eur Polym J 195:112206. https://doi.org/10.1016/j.eurpolymj.2023.112206

Dai S, Zang X, Zhang X, Ren M, Qiu H, Chen K (2025) Highly elastic lignin-based waterborne polyurethane with light color and its application in functional fabric coatings. Int J Biol Macromol 302. https://doi.org/10.1016/j.ijbiomac.2025.140521

Demirbaş A (2001) Relationships between lignin contents and heating values of biomass. Energy Convers Manage 42:183–188. https://doi.org/10.1016/S0196-8904(00)00050-9

Duarah P, Haldar D, Purkait MK (2020) Technological advancement in the synthesis and applications of lignin-based nanoparticles derived from agro-industrial waste residues: a review. Int J Biol Macromol 163:1828–1843. https://doi.org/10.1016/j.ijbiomac.2020.09.076

European Commission (2021) FLEXIBLE and modular FRACTionation of various lignocellulosic feedstocks, a stepping stone for a true circular bioeconomy (FRACTION). CORDIS – EU Research Results (Project no. 101023202). https://cordis.europa.eu/project/id/101023202

Faix O (1991) Classification of lignins from different botanical origins by FT-IR spectroscopy. Holzforschung 45(Suppl 1):21–27. https://doi.org/10.1515/hfsg.1991.45.s1.21

Fang W, Wu H, Ma K, Zuo B, She D, Geng Z, Liang H (2024) Efficient selective adsorption of Cr (VI) by S-doped porous carbon prepared from industrial lignin: waste increment and wastewater treatment. Int J Biol Macromol 278:134765. https://doi.org/10.1016/j.ijbiomac.2024.134765

Fargues C, Mathias A, Rodrigues AE (1996a) Kinetics of vanillin production from Kraft lignin oxidation. Ind Eng Chem Res 35(1):28–36. https://doi.org/10.1021/ie950267k

Fargues C, Mathias AL, Silva J, Rodrigues AE (1996b) Kinetics of vanillin oxidation. Chem Eng Technol 19(2):127–136. https://doi.org/10.1002/ceat.270190206

Feldman D, Lacasse M, Beznaczuk LM (1986) Lignin-polymer systems and some applications. Prog Polym Sci 12(4):271–299. https://doi.org/10.1016/0079-6700(86)90002-X

Gao W, Tu Q, Tang X, Duan S, Zeng J, Wang B, Xu J (2024) Characteristics of lignin isolated from an agricultural by-product: Camellia oleifera shell. Ind Crop Prod 219:118973. https://doi.org/10.1016/j.indcrop.2024.118973

Gbenebor OP, Olanrewaju OA, Usman MA, Adeosun SO (2023) Lignin from brewers' spent grain: structural and thermal evaluations. Polymers 15:2346. https://doi.org/10.3390/polym15102346

Goldstein IS (1981) Organic chemicals from biomass. CRC Press, Boca Raton

Goliszek M, Hosseinaei O, Matwijczuk A, Podkościelna B (2024) Comparing industrial and analytical kraft lignins as biobased UV-protective additives in coatings. Ind Crop Prod 214(118536):2024. https://doi.org/10.1016/j.indcrop.2024.118536

Gosselink RJA, De Jong E, Gebhardt T, Müller G, Gunnewijk L, Haan J, De Vries HJ, German CJ, Li J, Bösch P, Van der Putten JC (2004) Analytical protocols for characterisation of sulphur-free lignin. Ind Crops Prod 19:271–281. https://doi.org/10.1016/j.indcrop.2003.10.008

Guan W, Dong Z, Jiang H, Chen L, Yang H, Li T, Yang S, Hua D, Shao J, Yu J (2025) Flash Joule heating-driven lignin conversion: pyrolysis mechanisms and applications of graphitic carbon. Chem Eng J 504. https://doi.org/10.1016/j.cej.2024.158813

Guerroudj F, Fliri L, Bengtsson J, Gomes LC, Gazzola T, Hummel M, Bernin D (2024) Applications of NMR based methodologies investigating the behavior of lignin and cellulose towards bio-based carbon fibers production. Solid State Nucl Magn Reson 134:101977. https://doi.org/10.1016/j.ssnmr.2024.101977

Guo F, Hu X, Zhang X, Chen Y, Hou J (2025) Study on soil heavy metal contamination and its remediation using lignin-based adsorbents: a review. Environ Technol Innov 37. https://doi.org/10.1016/j.eti.2024.103958

Han M, Zhang Y, Zhang Y, Ye Q, Sillanpää M, Zhu X, Yang W (2024) A mini-review on polyvinyl alcohol/lignin (nano)composites: preparation, applications and perspectives. Sustain Chem Pharm 42:101861. https://doi.org/10.1016/j.scp.2024.101861

Hao J-H, Huang H-F, Zhang J-P, Liu X-C, Lan H-Q, Zhou A-L, Ge S-L, Xuan S-H, Liu Z-C, Yang J (2025) A novel approach for characterization of lignin structure in tobacco via fast 2D HSQC NMR spectroscopy with non-uniform sampling. Ind Crop Prod 223:120236. https://doi.org/10.1016/j.indcrop.2024.120236

Haq I, Mazumder P, Kalamdhad AS (2020) Recent advances in removal of lignin from paper industry wastewater and its industrial applications – a review. Bioresour Technol 312:123636. https://doi.org/10.1016/j.biortech.2020.123636

Higuchi T (1980) Lignin structure and morphological distribution in plant cell wall. In: Kirk TK, Higuchi T, Chang H-M (eds) Lignin biodegradation: microbiology, chemistry, and potential applications, vol 1. CRC Press, Boca Raton, pp 1–20

Iswanto AH, Lee SH, Hussin MH, Hamidon TS, Hajibeygi M, Manurung H, Solihat NN, Nurcahyani PR, Lubis MAR, Antov P, Savov V, Kristak L, Kawalerczyk J, Osvaldová LM, Farid S, Selvasembian R, Fatriasari W (2024) A comprehensive review of lignin-reinforced lignocellulosic composites: enhancing fire resistance and reducing formaldehyde emission. Int J Biol Macromol 283:137714. https://doi.org/10.1016/j.ijbiomac.2024.137714

Jiang B, Zhang Y, Guo T, Zhao H, Jin Y (2018) Structural characterization of lignin and lignin-carbohydrate complex (LCC) from ginkgo shells (Ginkgo biloba L.) by comprehensive NMR spectroscopy. Polymers 10(7):736. https://doi.org/10.3390/polym10070736

Jiang C, Wang J, Liu L, Shi J, Zhang Y, Zhang J, Qin C, Dong H, Zhang Y (2020) A sustainable reduction route of graphene oxide by industrial waste lignin for versatile applications in energy and environment. J Clean Prod 268:122019. https://doi.org/10.1016/j.jclepro.2020.122019

Laurichesse S, Avérous L (2014) Chemical modification of lignins: towards biobased polymers. Prog Polym Sci 39(7):1266–1290. https://doi.org/10.1016/j.progpolymsci.2013.11.004

Li C, Li W, Liu H, Li Y, Dai L, Li C, Si C (2024a) State of the art on the production and application of lignin-derived carbon nanoparticles. Ind Crop Prod 222. https://doi.org/10.1016/j.indcrop.2024.120123

Li J, Guan D, Xia S, Fan Y, Zhao K, Zhao Z, Zheng A (2024b) Recent advances, challenges, and opportunities in lignin valorization for value-added chemicals, biofuels, and polymeric materials. Energy Convers Manage 322. https://doi.org/10.1016/j.enconman.2024a.119123

Li K, Xu H, Liu Y, Zhong W, Jin Y, Wu W (2024c) Exploring the relationship between lignin structure and antioxidant property using lignin model compounds. Int J Biol Macromol 282:136786. https://doi.org/10.1016/j.ijbiomac.2024.136786

Liang C, Xia H, Yin L, Du C, Wu X, Wang J, Li S, Xu J, Zhang X, Wang Y, Qu W (2024) Carbon foam directly synthesized from industrial lignin powder as featured material for high efficiency solar evaporation. Chem Eng J 481:148375. https://doi.org/10.1016/j.cej.2023.148375

Liao H, Feng B, Song X, Zhang J, Zhang Z (2024) Unlocking full potential of bamboo waster: efficient co-production of xylooligosaccharides, lignin, and glucose through low-dosage mandelic acid hydrolysis with alkaline processing. Int J Biol Macromol 282. https://doi.org/10.1016/j.ijbiomac.2024.137165

Lopretti MI, Mathias AL, Rodrigues AE (1993) Activity of ligninase peroxidase from Acinetobacter anitratus and the degradation of Pinus pinaster lignin. Process Biochem 28: 543–547. https://doi.org/10.1016/0032-9592(93)85016-9

Lora J (2008) Industrial commercial lignins: sources, properties and applications. In: Belgacem MN, Gandini A (eds) Monomers, polymers and composites from renewable resources. Elsevier, Amsterdam, pp 225–259. https://doi.org/10.1016/B978-0-08-045316-3.00010-7

Lu F, Xu J, Li Z, Wang X, Zhou J, Wang Y, Tan F (2025) Self-synthesized carbon nanotubes exhibiting temperature-responsive effects to "modify" honeycomb lignin-based carbon aerogels for supercapacitor applications. J Energy Storage 110. https://doi.org/10.1016/j.est.2025.115348

Majeke BM, Collard F-X, Tyhoda L, Görgens JF (2021) The synergistic application of quinone reductase and lignin peroxidase for the deconstruction of industrial lignins. Bioresour Technol 319:124152. https://doi.org/10.1016/j.biortech.2020.124152

Mathias AL, Rodrigues AE (1995) Vanillin from Kraft Lignin. Holzforschung 49(3):273–278. https://doi.org/10.1515/hfsg.1995.49.3.273

Mathias AL, Lopretti MI, Rodrigues AE (1995) Chemical and biological oxidation of Pinus pinaster lignin for the production of vanillin. J Chem Technol Biotechnol 64(3):225–234. https://doi.org/10.1002/jctb.280640303

McCarthy JL, Islam A (2000) Lignin chemistry, technology, and utilization: a brief history. In: Glasser WG, Northey RA, Schuertz TF (eds) Lignin: historical, biological, and materials perspectives. ACS symposium series, vol 742. American Chemical Society, Washington, DC, pp 1–27

Mennani M, Abdellaoui Y, Ait Benhamou A, Lopez-Maldonado EA, Kasbaji M, El Achaby M, Moubarik A, Kassab Z (2025) FeNi bimetallic functionalized lignin catalyst for sustainable oxidation processes. Sustain Mater Technol 43:e01267. https://doi.org/10.1016/j.susmat.2025.e01267

Monties B (1998) Novel structures and properties of lignins in relation to their natural and induced variability in ecotypes, mutants and transgenic plants. Polym Degrad Stab 59:53–64. https://doi.org/10.1016/S0141-3910(97)00166-3

Nurcahyani PR, Sohail A, Antov P, Lee SH, Zhang J, Suhartana S, Iswanto AH, Bakshi MI, Fatria Sari W (2024) The influence of washing of rice straw lignin isolation on polyurethane synthesis for coating in biomedical application. Results Eng 24. https://doi.org/10.1016/j.rineng.2024.103160

Pang T, Wang G, Sui W, Xu T, Wang D, Si C (2025) Lignin-based support for metal catalysts: synthetic strategies, performance boost, and application advances. Coord Chem Rev 528. https://doi.org/10.1016/j.ccr.2024.216426

Ponomarenko J, Dzhbite T, Lauberts M, Volperts A, Dobele G, Telysheva G (2025) Analytical pyrolysis – a tool for revealing of lignin structure-antioxidant activity relationship. J Anal Appl Pyrolysis 113:360–369. https://doi.org/10.1016/j.jaap.2015.02.027

Prasad S, Singh A, Joshi HC (2007) Ethanol as an alternative fuel from agricultural, industrial and urban residues. Resour Conserv Recycl 50:1–39. https://doi.org/10.1016/j.resconrec.2006.05.007

Prashak VRA, Bourchak M, Alshahrani H, Juhany KA (2023) Development of cashew nut shell lignin-acrylonitrile butadiene styrene 3D printed core and industrial hemp/aluminized glass fiber epoxy biocomposite for morphing wing and unmanned aerial vehicle applications. Int J Biol Macromol 253:127068. https://doi.org/10.1016/j.ijbiomac.2023.127068

Priyadarshi R, Ghosh T, Purohit SD, Prasannavenkadesan V, Rhim J-W (2024) Lignin as a sustainable and functional material for active food packaging applications: a review. J Clean Prod 469. https://doi.org/10.1016/j.jclepro.2024.143151

Raj K, Vora T, Padmapriya G, Lal B, Devi A, Sharma RSK, Chahar M, Sudhakar L, Suman RJ, Nagraik R (2025) A comprehensive review of sustainable hydrogels from lignin for advanced wastewater solutions. Int J Biol Macromol 301:139963. https://doi.org/10.1016/j.ijbiomac.2025.139963

Ramos LP, Mathias AL, Silva FT, Cotrim AR, Ferraz AL, Chen C-L (1999) Characterization of residual lignin after SO_2-catalyzed steam explosion and enzymatic hydrolysis of Eucalyptus viminalis wood chips. J Agric Food Chem 47(6):2295–2302. https://doi.org/10.1021/jf9810665

Rong H, Gao B, Zhao Y, Sun S, Yang Z, Wang Y, Yue Q, Li Q (2013) Advanced lignin-acrylamide water treatment agent by pulp and paper industrial sludge: synthesis, properties and application. J Environ Sci 25(12):2367–2377. https://doi.org/10.1016/S1001-0742(12)60326-X

Sahoo S, Seydibeyoğlu MÖ, Mohanty AK, Misra M (2011) Characterization of industrial lignins for their utilization in future value added applications. Biomass Bioenergy 35:4230–4237. https://doi.org/10.1016/j.biombioe.2011.07.009

Sarve DT, Dutta R, Rastogi A, Ekhe JD (2022) Valorization of industrial waste lignin via pyrolysis in the presence of additives: formation, characterization, and application of fuel valued bio-oil and activated char. J Indian Chem Soc 99:100398. https://doi.org/10.1016/j.jics.2022.100398

Shafiq A, Bhatti IA, Amjed N, Zeshan M, Zaheer A, Kamal A, Naz S, Rasheed T (2024) Lignin derived polyurethanes: current advances and future prospects in synthesis and applications. Eur Polym J 209. https://doi.org/10.1016/j.eurpolymj.2024.112899

Taghizadeh-Alisaraei A, Motevali A, Ghobadian B (2019) Ethanol production from date wastes: adapted technologies, challenges, and global potential. Renew Energy 143:1094–1110. https://doi.org/10.1016/j.renene.2019.05.048

Thammasang S, Pimsawat N, Kasemsiri P, Knijnenburg JTN, Theerakulpisut S, Kamwilaisak K, Chindaprasirt P (2025) Surface modification of lignin-based activated carbon for enhanced hydrophobicity in oil-in-water emulsion separation, self-cleaning and antibacterial applications. Surf Interfaces 60:106044. https://doi.org/10.1016/j.surfin.2025.106044

Tong Y, Yang T, Wang J, Li B, Zhai Y, Li R (2024) A review on the overall process of lignin to phenolic compounds for chemicals and fuels: from separation and extraction of lignin to transformation. J Anal Appl Pyrolysis 181:106663. https://doi.org/10.1016/j.jaap.2024.106663

United Nations, General Assembly (2015) Transforming our world: the 2030 agenda for sustainable development. Resolution A/RES/70/1. United Nations, New York. https://www.un.org/en/development/desa/population/migration/generalassembly/docs/globalcompact/A_RES_70_1_E.pdf

United States – DOE, Department of Energy (2023) Biochemical conversion and lignin utilization. Bioenergy Technologies Office – BETO. DOE, Washington, DC. https://www.energy.gov/eere/bio

Vanholme R, De Meester B, Ralph J, Boerjan W (2019) Lignin biosynthesis and its integration into metabolism. Curr Opin Biotechnol 56:230–239. https://doi.org/10.1016/j.copbio.2019.02.018

Vaz S Jr, Barreto AE, Sampaio BL (2025) Catalytic degradation of kraft lignin to determine its antioxidant potential for industrial purposes. Heliyon 11:e42783. https://doi.org/10.1016/j.heliyon.2025.e42783

Wang W, Tan X, Yu Q, Wang Q, Qi W, Zhuang X, Wang Z, Yuan Z (2018) Effect of stepwise lignin removal on the enzymatic hydrolysis and cellulase adsorption. Ind Crop Prod 122:16–22. https://doi.org/10.1016/j.indcrop.2018.05.053

Wang S, Wan Z, Han Y, Jiao Y, Li Z, Fu P, Li N, Zhang A, Yi W (2023a) A review on lignin waste valorization by catalytic pyrolysis: catalyst, reaction system, and industrial symbiosis mode. J Environ Chem Eng 11:109113. https://doi.org/10.1016/j.jece.2022.109113

Wang T, Li H, Diao X, Lu X, Ma D, Ji N (2023b) Lignin to dispersants, adsorbents, flocculants and adhesives: a critical review on industrial applications of lignin. Ind Crop Prod 199:116715. https://doi.org/10.1016/j.indcrop.2023.116715

Wang B, Xu J, Lu F, Zhou J, Wang X, Tan F (2024a) Efficient utilization of lignin heterogeneity: employing a simple approach to modulate the structure-effect relationship of lignin microspheres for its high-quality applications in supercapacitors. Ind Crop Prod 221:119360. https://doi.org/10.1016/j.indcrop.2024.119360

Wang W-Y, Guo B-X, Wang R, Liu H-M, Qin Z (2024b) Revealing the structural changes of lignin in Chinese quince (Chaenomeles sinensis) fruit as it matures. Int J Biol Macromol 264:130718. https://doi.org/10.1016/j.ijbiomac.2024.130718

Wang Y, Tang S, Jiang L, Yuan Z, Zhang Y (2024c) A review of lignin application in hydrogel dressing. Int J Biol Macromol 281:135786. https://doi.org/10.1016/j.ijbiomac.2024a.135786

Wang J, Seidi F, Shi X, Li C, Huang Y, Xiao H (2025a) Unveiling the potential of dual-extrinsic/intrinsic self-healing lignin-based coatings for anticorrosion applications. Int J Biol Macromol 285:138073. https://doi.org/10.1016/j.ijbiomac.2024.138073

Wang L, Guo Z, Guo D, Li J, Janaswamy S, Yang G, Ji X, Lyu G (2025b) Constructing industrial lignin into robust and multifunctional superhydrophobic coating for effective anti-icing. Int J Biol Macromol 289. https://doi.org/10.1016/j.ijbiomac.2024.138891

Wu X, Lian H, Xia C, Deng J, Li X, Zhang C (2024) Mechanistic insights and applications of lignin-based ultraviolet shielding composites: a comprehensive review. Int J Biol Macromol 280:10.1016/j.ijbiomac.2024.135477

Xu J, Zhou J, Du B, Li X, Huang Y, Cao Q, Xu S, Wang X (2024) Research progress on the preparation and application of lignin-based Pickering emulsions: a review. Ind Crop Prod 222: 119723. https://doi.org/10.1016/j.indcrop.2024.119723

Yan B, Lin X, Xiao J, Fu Y, Zhang S (2024) Comparative study for the separation and depolymerization behavior of lignin from different separation methods. Fuel 358:130145. https://doi.org/10.1016/j.fuel.2023.130145

Ye H, You T, Nawaz H, Xu F (2024) A comprehensive review on polylactic acid/lignin composites — structure, synthesis, performance, compatibilization, and applications. Int J Biol Macromol 280:135886. https://doi.org/10.1016/j.ijbiomac.2024.135886

Yi Y, Hu S, Ma Y, Tang T, Liu C, Yan Y, Lei L, Hou Y (2025) The structure impact of lignin in pulping material on the energy storage performance of black liquor derived carbon cathodes for zinc ion hybrid capacitors. J Colloid Interface Sci 683:55–67. https://doi.org/10.1016/j.jcis.2024.12.037

Yue P, Hu Y, Yang Z, Peng F, Yang L (2023) Renewable and functional composite film from epoxidized Eucommia ulmoides gum and industrial lignin. Ind Crop Prod 194:116381. https://doi.org/10.1016/j.indcrop.2023.116381

Zakzeski J, Bruijnincx PCA, Jongerius AL, Weckhuysen BM (2010) The catalytic valorization of lignin for the production of renewable chemicals. Chem Rev 110(6):3552–3599. https://doi.org/10.1021/cr900354u

Zeng J, Helms GL, Gao D, Chen S (2012) Quantification of individual phenolic compounds in lignin samples and determination of their inhibitory effects on cellulase. Biotechnol Bioeng 109(2):346–356. https://doi.org/10.1002/bit.23242

Zhang L, Xu S, Zhao W, Liu S (2007) Co-pyrolysis of biomass and coal in a free fall reactor. Fuel 86:353–359. https://doi.org/10.1016/j.fuel.2006.07.004

Zhang L-X, Li C-X, Kakar MU, Khan MS, Wu P-F, Amir RM, Dai D-F, Naveed M, Li Q-Y, Saeed M, Shen J-Q, Rajput SA, Li J-H (2021) Resveratrol (RV): a pharmacological review and call for further research. Biomed Pharmacother 143:112164. https://doi.org/10.1016/j.biopha.2021.112164

Zhang B, Wang G, Huang X, Fang Y, Sui W, Parvez AM, Si C, Jia H (2024a) Formulation and application assessment of lignin-based biodegradable composite mulching film with emphasis on lignin enhancement. Ind Crop Prod 215. https://doi.org/10.1016/j.indcrop.2024.118634

Zhang S, Ji A, Meng X, Bhagia S, Yoo CG, Harper DP, Zhao X, Ragauskas AJ (2024b) Structure-property relationship between lignin structures and properties of 3D-printed lignin composites. Compos Sci Technol 249:110487. https://doi.org/10.1016/j.compscitech.2024.110487

Zhang C, Feng T, Deng X, Mohamed TA, Wu J (2025) Analyze the impact of lignin depolymerization process and its products on humic substance formation. Int J Biol Macromol 295: 139476. https://doi.org/10.1016/j.ijbiomac.2025.139476

Zheng X, Zhong Z, Zhang B, Du H, Wang W, Li Q, Yang Y, Qi R, Ye Q, Li Z (2024) Preparation of monocyclic aromatic hydrocarbons from industrial lignin residue and polyethylene co-pyrolysis by microwave-assisted in fluidized bed based on bimetal-loaded HZSM-5/MCM-41 core-shell catalyst. Fuel 364. https://doi.org/10.1016/j.fuel.2024.131100

Zhu Y, Zhang F, Bian R, Zeng G, Li J, Lyu Y, Li J (2024) Formaldehyde-free biomass adhesive based on industrial alkali lignin with high strength and toughness. Ind Crop Prod 222:119525. https://doi.org/10.1016/j.indcrop.2024.119525

Chapter 10
Biomass Pretreatment: Hydrolysate Products, Characterization, and Detoxification Strategies

Arion Zandoná Filho, Satinder Kaur Brar, Thiago Alessandre, Victor Brisk, and Khyati Joshi

Abstract This chapter outlines the procedures for processing lignocellulosic biomass to produce higher-value-added products and by-products compared to raw biomass, and the importance of the biomass in the present fuel world scenario. The increasing demand for renewable and sustainable energy sources has driven this research, leading to the development of analytical methods and process technologies for converting biomass, particularly lignocellulosic biomass, which is abundant and low cost, serving as an important alternative to fossil fuels and synthetic products. As a result, structures containing renewable carbon are considered accessible resources with significant potential for advancing sustainable development through innovative products and commodities. Its complex and resistant structure, which accounts for a large percentage of the total mass, is mainly composed of cellulose, hemicellulose, and lignin. This composition makes it challenging to access the minor components that are valuable for bioproduct production. Lignocellulosic biomass (LCB) pretreatment methods are discussed and categorized into physical, physicochemical, chemical, biological, and combined processes. The selection of the most suitable pretreatment method is multifaceted, depending on the specific raw material, desired products, and the economic and environmental viability of the process. Pretreatment methods are essential because they help break down the lignocellulosic matrix, making cellulose and hemicellulose more accessible for subsequent conversion processes, such as enzymatic hydrolysis and fermentation. The efficient conversion to produce bioproducts typically involves hydrolysis and fermentation, which break down complex polymers into simpler compounds. The chemical speciation of processes is assessed using selective analytical methods and techniques. A key aspect of this process is identifying potential inhibitors that can significantly impact

A. Z. Filho (✉) · T. Alessandre · V. Brisk
Department of Chemical Engineering, Federal University of Paraná, Curitiba, Paraná, Brazil
e-mail: a.zandona@ufpr.br

S. K. Brar · K. Joshi
Department of Civil Engineering, Lassonde School of Engineering, York University, Toronto, ON, Canada

A. L. Woiciechowski et al. (eds.), *Biomass for Bioenergy and Bioproducts*, Biofuel and Biorefinery Technologies 17, https://doi.org/10.1007/978-3-032-19898-3_10

fermentation efficiency and, therefore, the yield of bioproducts in downstream steps. Understanding the chemical composition and structural changes after hydrolysis is crucial for optimizing these processes and improving product yields. Analytical techniques are vital for characterizing hydrolyzed biomass. They help clarify the chemical and biochemical pathways, with spectroscopy providing insights into the molecular structure, functional groups, and both qualitative and quantitative composition. Chromatographic methods complement this analysis by enabling precise measurement of the compounds produced during hydrolysis. Collectively, these techniques are essential for a comprehensive assessment of biomass and the development of effective bioconversion strategies.

Graphical Abstract

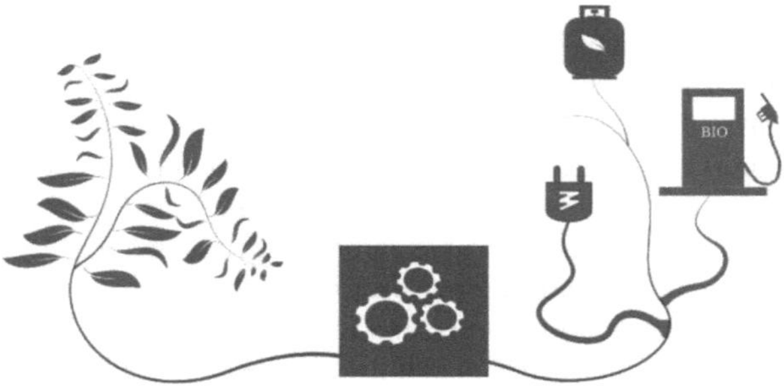

10.1 Introduction

Biomass is a renewable energy source that has emerged as a sustainable alternative to fossil fuels. It consists of organic matter from a plant or animal origin that can be converted into energy through various processes, such as combustion, anaerobic digestion, and thermochemical conversion. Global biomass resources are substantial, encompassing an estimated 1.8 trillion tons of terrestrial and 4 billion tons of aquatic biomass. Significant quantities of biomass are also generated from waste streams, including agricultural and forestry residues, animal by-products, aquatic and marine debris, algae, sewage, and municipal solid waste (Tursi 2019).

Figure 10.1 shows data from over 60 years of oil extraction and its use by modern society, clearly illustrating the gap between production capacity and consumption. The International Energy Agency (IEA) predicts that oil demand will peak in the next few years and then decline slightly, while the Organization of the Petroleum Exporting Countries (OPEC) has different projections, estimating that global oil demand will keep growing in the long term. This disagreement underscores the importance of renewable energy sources in the system, which encompasses consumption, the production of goods, and the development of new technologies.

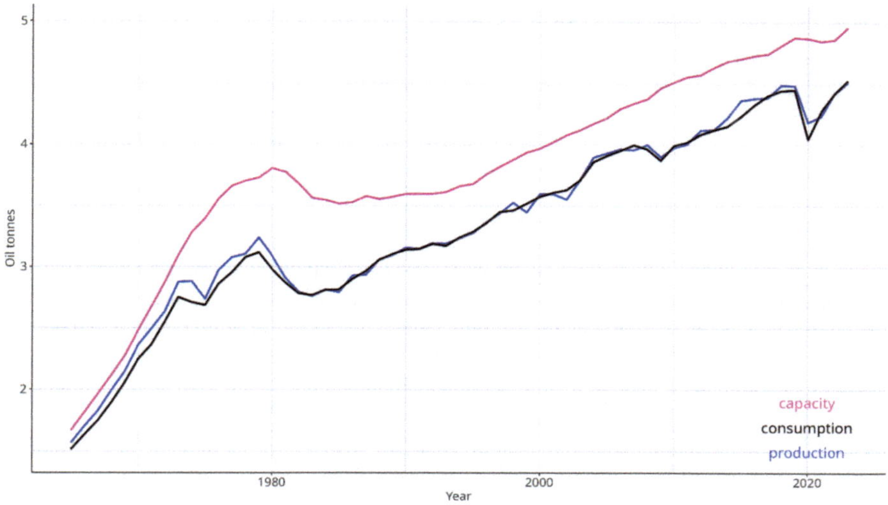

Fig. 10.1 Oil billions of tonnage production over 60 years ago

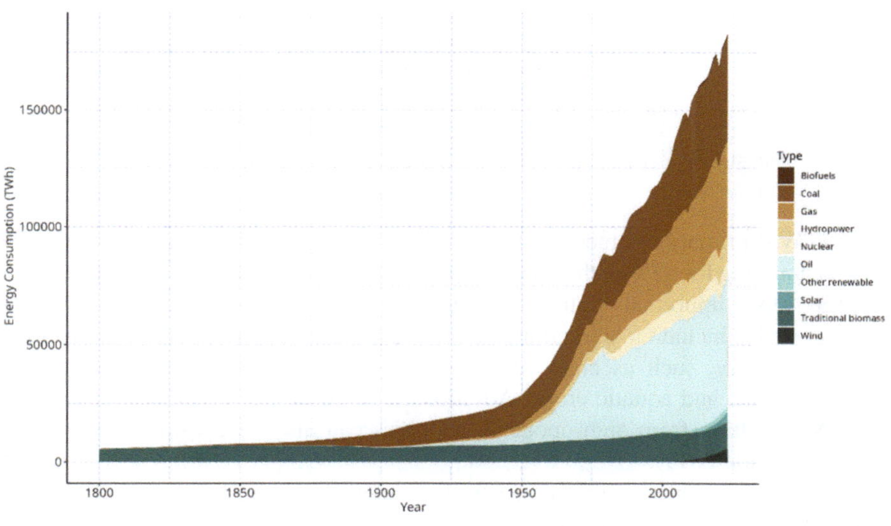

Fig. 10.2 Global energy consumption (TWh) by type of renewable energy

This topic aims to explore the main types of biomass used to produce fuels and energy, highlighting their characteristics, applications, and relevance in today's energy landscape. It is important to note that data on traditional biomass consumption is often estimated and may not be as precise as data for other energy sources. Science can represent energy demands in different ways. Below, in Fig. 10.2, you can see evidence of the consumption of various energy types depending on the fuel used, expressed in TWh on a single chart. The global energy consumption by fuel

Table 10.1 Classification and types of biomasses

Main categories	Types of biomasses
Herbaceous and agricultural biomass	Flowering plants and grasses: Bamboo, brassica, timothy, alfalfa, ryegrass, switchgrass, sugarcane, and others Straws derived from crops: Sunflower, mint, bean, barley, flax, oat, sesame, wheat, corn, rice, rape, rye, and similar species Miscellaneous residues: Husks, fruits, grains, vegetables, coir, pips, cakes, bagasse, fodder, pits, hulls, pulps, seeds, shells, stalks, cobs, and various food processing by-products
Wood-based biomass	Forestry residues: Branches, stems, and foliage Processing residues: Chips, sawdust, bark, pellets, briquettes, and lumps, as well as additional materials generated at sawmills from various wood species
Animal and human waste biomass	Animal-derived and related organic residues can include materials such as bones, poultry litter, meat-and-bone meal, assorted animal manures, and other similar wastes
Aquatic biomass	Freshwater or marine algae, lake and seaweed, microalgae, kelp and others
Contaminated biomass and industrial biomass wastes	General waste: Municipal solid waste, refuse-derived fuel, hospital waste Wood waste: Demolition wood, chipboard, fiber board, plywood, wood pallets and boxes, railway sleepers Paper waste: Waste papers, paperboard waste, paper-pulp sludge Other: Sewage sludge, tannery waste

Source: Adapted from Vassilev et al. (2012)

type (TWh) primarily comes from fossil fuels, which include coal, oil, and natural gas, accounting for 82% of the total energy mix, according to Earth.Org Ltd. in 2023.

The most important sources of biomass can be classified by their types (Table 10.1), including agricultural and forestry residues (by-products from the timber industry, such as chips and sawdust), residues from livestock activities, sewage, algae, and aquatic crops. Additionally, a portion of municipal solid waste (MSW) resulting from anthropogenic activities can also be considered a form of biomass (Tursi 2019).

Bioenergy production relies on various sources, including agricultural residues that serve as vital feedstock for many applications. As demand for renewable energy increases, the importance of these residues in achieving climate and energy goals is becoming more widely recognized (Sivabalan et al. 2021). Global energy crop production reached significant levels between 2019 and 2024, as shown in Table 10.2.

Wheat production has experienced modest growth due to improved yields, but its use in biofuels has declined and is unlikely to return to pre-pandemic levels. Corn, primarily used for ethanol production in major producing countries such as the United States and China, has seen increased production and consumption without a significant rise in harvested areas, indicating greater efficiency. Coarse grains

Table 10.2 Overview of global production and biofuel use of selected biomass

	Year 2019		Year 2024	
Biomass	Production (Mt)	Biofuel use (Mt)	Production (Mt)	Biofuel use (Mt)
Wheat	766	9	798	9
Corn	1152	182	1262	193
Coarse grains	299	10	308	6
Vegetable oils	211	30	234	41
Sugarcane	1731	377	1857	412

Source: Adapted from World Bioenergy Association (2024)

(barley, rye, sorghum, etc.) play a minor role in biofuel production, especially in the EU, and their biofuel use has decreased, although a slight increase is forecasted (Banerjee et al. 2009). Vegetable oils (palm, soybean, rapeseed) are crucial in this sector, with their use expanding substantially. Sugarcane remains a key feedstock for ethanol, particularly in Brazil, the top producer, and India, which has greatly boosted its ethanol output using sugarcane and other crops. Despite sugarcane's importance for biofuels, its primary purpose continues to be food and feed, and the increase in biofuel production has not significantly affected its harvested area (World Bioenergy Association 2024). Agricultural waste, or crop residues, is a major by-product of farming activities, originating from the initial stages of harvesting. Untreated agricultural waste left in fields can contribute to soil erosion and offer limited nutrient replenishment, which may affect soil fertility. Additionally, the practice of open field burning, sometimes used by farmers in developing countries, results in energy loss. If this energy were captured, it could be transformed into a valuable resource for consumers. For example, within the cereal group, rice straw—a common agricultural by-product—has a high heating value (HHV) of 15.54 MJ/kg. Oilseeds, another major category, include mustard and rapeseed shoots, with an HHV of 17 MJ/kg. Pulses, a key protein source, produce residues like tur arhar (Pigeonpea) shoots, with an HHV of 18.58 MJ/kg. Sugarcane bagasse, a by-product of sugarcane processing, has an HHV of 20 MJ/kg. Horticultural crop residues such as banana peel have an HHV of 17.4 MJ/kg (Sivabalan et al. 2021). Even more valuable than directly harnessing energy from these biomasses is the opportunity to repurpose them into more valuable forms. For example, thermochemical conversion processes can transform them into biocrude—a substance with a higher calorific value, ranging from 30.8 to 42.2 MJ/kg (Yang et al. 2020). Wood biomass, mainly composed of cellulose, hemicellulose, and lignin, serves as a versatile and renewable energy source. Significant sources include by-products from the wood and furniture industries, as well as residual tree parts such as roots, branches, bark, and foliage. Notably, about half of the biomass collected from forest operations consists of residues (Braghiroli and Passarini 2020). According to the FAO, global wood chip and particle production reached approximately 252 million units in 2023, while wood residue production was estimated at around 250 million cubic meters in the same year (Food and Agriculture Organization 2025). Regionally, Mansuy et al. (2017) estimated Canada's annual biomass availability from harvest residues at 14 ± 2 million tons (oven-dry mass). These sources—including wood chips, particles, residues, and others—can be used to generate energy through various thermal processes, such as combustion, gasification, pyrolysis, and hydrothermal liquefaction. Additional biomass sources often used for energy include industrial wood scraps and noncommercial timber (Tursi 2019). In 2022, wood biomass production continued to play a vital role in the global energy landscape. Over 1.9 billion m^3 of wood fuel was produced worldwide, with Africa and Asia accounting for 37% and 35%, respectively, followed by the Americas (18%) and Europe (9%). Wood charcoal, a solid energy carrier made from the carbonization of woody biomass, exceeded 57 million tonnes in production, with Africa leading at 65% of the global total, mainly driven by Ethiopia and Nigeria. Meanwhile, wood pellet production reached 48 million tonnes, with Europe accounting for 52% (approximately 25 million tonnes), and Germany emerged as the leading producer. The Americas produced 32% of the global total, with the United States and Canada dominating the region, together responsible for over 90% of pellet output. In the European Union, wood pellets remain a key part of the solid biofuel market, especially in residential and commercial sectors, which accounted for 56% of total consumption in 2022 (World Bioenergy Association 2024).

In the United States, industry is the largest consumer of wood and wood waste fuels, particularly in the wood product and paper manufacturing sectors (Zevallos Torres et al. 2020). These industries often repurpose mill by-products to generate steam and electricity, lowering their dependence on other energy sources. In 2023, industrial users consumed about 64% of the nation's wood fuels—representing 8.5% of their end-use and 7.2% of total energy consumption. The residential sector is the next major user, mainly for heating. In 2023, wood energy contributed 4.0% to residential end-use and 2.4% to overall residential energy. By 2020, approximately 11 million US households (8.9%) used wood for heating, with around 2.2 million relying on it as their primary fuel (EIA 2024). Manure, traditionally used as fertilizer, a soil amendment, and an energy source, plays a key role in the biomass field. As a renewable cellulosic biomass resource, manure is rich in recyclable components such as nutrients, organic matter, solids, energy, and fiber. It also varies depending on the animal species and their diets, which influence its suitability for different biomass applications (Hoyos-Sebá et al. 2024). Globally, Berendes et al. (2018) estimated that the worldwide recoverable livestock manure production for 2024 is approximately 3.12×10^{12} kg annually. As a result, manure is increasingly being integrated into biomass management systems, where it can be converted into bioenergy and other value-added products, helping to promote sustainable waste management and renewable energy generation. Given its role in bioenergy, understanding the energy content of manure is crucial. The high heating value (HHV) of manure differs among livestock species. On a dry basis, cow and horse manure hold the highest energy content, at 17.51 MJ/kg and 17.50 MJ/kg, respectively (Da Lio et al. 2021; Jiang et al. 2020), while sheep manure has an HHV of 16.02 MJ/kg (Miller & Miller, 2002). Conversely, poultry and pig manure have lower energy potentials, at 14.87 MJ/kg and 13.66 MJ/kg, respectively (Miller and Miller 2002; Wu et al. 2020). These HHV differences highlight the varied potential of manure as a biomass energy source, which is key to optimizing energy recovery and waste management. In summary, the diverse composition and significant energy potential of manure make it a valuable biomass resource for various technological applications. Processes such as combustion, pyrolysis, gasification, composting, and anaerobic digestion can efficiently convert manure into energy, biofuels, and nutrient-rich soil amendments, underscoring its role in sustainable waste management and renewable energy production (Hoyos-Sebá et al. 2024). Aquatic biomass, encompassing macroalgae, microalgae, and emerging plants, represents a versatile and promising resource with a broad range of applications. In nature, approximately 55,000 species and over 100,000 strains of brackish water, freshwater, and terrestrial algae convert sunlight, water, and CO_2 into a diverse array of metabolites and chemicals that accumulate as algal biomass (Vassilev and Vassileva 2016). Although these aquatic plants are compositionally similar to their terrestrial counterparts, their lower lignin content allows for milder pretreatment during processing. Rich in proteins, starches, sugars, and fats, aquatic biomass serves as an excellent feedstock for biorefineries, leading to the production of valuable commodities such as protein supplements, fish feed, soil compost, biofertilizers, and biofuels (Alam et al. 2021). Their high photosynthetic efficiency, fast growth, and substantial biomass production further enhance their appeal. At the same time, since they do not require arable land, they are ideal for integration into tertiary wastewater treatment systems, where extensive root structures help in pollution cleanup. The global demand for aquatic plants is rising, as evidenced by harvest data that increased from approximately 10.5 million tonnes in 2000 to nearly 34 million tonnes in 2018, with forecasts indicating a further 20% growth by 2030 compared to the previous levels in 2016. Major contributors to this harvest include China, Russia, Norway, and Japan (Arefin et al. 2021). However, the productivity of aquatic biomass depends on several factors, such as light intensity, CO_2 and O_2 levels, light/dark cycles, temperature, salinity, pH, and nutrient availability, which are crucial for maximizing yield (Vassilev and Vassileva 2016). In addition to their potential as biorefinery feedstocks, aquatic plants are emerging as a key resource for next-generation biofuels. Certain species can directly produce biofuel through processes that utilize the oil created in subsequent transformations, such as pyrolysis and transesterification. Beyond energy production, these plants provide valuable compounds, including pigments and vitamins, with lipids serving as the main component for biofuel conversion. Extracted lipids can be converted into biodiesel via hydro-treatment or directly into fatty acid methyl esters (FAME) through esterification. Additionally, hydrothermal processing can produce bio-oil or syngas, which can be further refined into ethanol, methanol, jet fuel, and renewable diesel. Other conversion methods, such as anaerobic digestion for biomethane and fermentation for ethanol, further highlight the versatility of aquatic biomass (Arefin et al. 2021; Fig. 10.3).

Municipal solid waste (MSW) refers to trash produced by households, offices, small institutions, and businesses in urban areas. Its makeup varies widely across the world, consisting of both biodegradable organic materials and non-biodegradable inorganic parts. Globally, about 2 billion tons of MSW are generated each year, with projections suggesting this will rise to 3.4 billion tons by 2050. Typically, roughly

Fig. 10.3 Biogas plant in a particular farm

70% of this waste is disposed of in landfills or dumpsites, 19% is recycled, and the remaining 11% is used for energy recovery (Kaza et al. 2018). Besides creating environmental issues, MSW also serves as a renewable and economically viable resource because of its potential for energy recovery and resource extraction through waste-to-energy (WtE) and other valorization methods. A variety of technologies, from thermochemical to biological processes, can convert MSW into solid, liquid, and gaseous fuels, helping to meet increasing energy demands. Incineration, the most common WtE method, involves burning waste to generate electricity, steam, and combined heat and power, while also reducing waste volume (Nanda and Berruti 2021). The effectiveness of WtE technologies is clear worldwide. For instance, the energy produced from MSW grew from 0.74 exajoules (EJ) in 2000 to 1.50 EJ in 2021. In Europe alone, nearly 500 WtE plants were operational by 2021, processing about 103 million tons of household and similar waste. This energy production not only supplies electricity to around 21 million Europeans but also provides heat to roughly 17 million residents (World Bioenergy Association 2024).

10.2 Biomass Pretreatment Process

To enable access and efficient use of the various biomasses available around the world to convert biomasses into bioproducts, chemicals, and fuels, it is necessary to disrupt the lignocellulosic structure of the vegetal cell, separate and recover the three main fractions of the lignocellulosic complex: cellulose, hemicellulose, and lignin. So, it is necessary to submit the biomass to a pretreatment method, even physical, chemical, biological, or combined, as follows:

10.2.1 Extrusion (Physical Method)

Extrusion is a thermomechanical process used to disrupt, mix, compact, pelletize, or physically shape various materials. Its primary function is to apply a high shear force to the material. Depending on the desired outcome, different equipment setups are employed. Typically, an extruder includes a central screw thread, with the number ranging from 1 to 4, and various groove shapes that affect the shearing process, along with a supply system, heating, and cooling blocks (Konan et al. 2024). With an automated feeding system, the treatment process can be performed in batch, fed batch, or continuous modes. This improves practicality in industrial uses and makes the process more cost-effective. Besides acting as a pretreatment for biomass, if needed, it can also be used to pelletize biomass for other commercial purposes (Xin et al. 2020). Compared to different physical methods, such as milling, extrusion uses less electrical energy per kg of biomass processed. It also does not produce inhibitors and requires low water consumption. When it comes to pretreatment time, extrusion is one of the fastest methods, typically taking 2–3 min. In contrast, biological or chemical processes can take hours or even days to treat the material properly. There are no restrictions on the raw material, as variations in the composition of different biomasses do not affect the process (Kapoor et al. 2018).

Modifications to the extrusion process involve adding reagents, alkali, or enzymes. These chemical additives aim to enhance the mechanical properties of certain biomasses, making them easier to pass through the extrusion screw. Using alkali, such as NaOH, mainly aims to partially dissolve lignin by breaking the bonds between lignin and hemicellulose. This helps improve movement through the equipment and allows for the recovery of biomass components. Some processes also add enzymes, such as amylases, glucoamylases, and pectinases, to facilitate biomass displacement and separate specific components. When using enzymes, it is essential to monitor temperature to prevent denaturation and loss of activity. To cut costs, recirculating the enzyme-containing extruded liquid is recommended, as enzymes are more expensive to produce (Senturk-Ozer et al. 2011). In a study on enhanced biomass utilization, sugarcane bagasse was pretreated with a twin-screw extruder operating at 100 rpm, 130 °C, and 10% moisture content. These conditions clearly improved the material's susceptibility to enzymatic hydrolysis. A notable result was a 330% increase in glucose yield.

Furthermore, the treatment induced a substantial structural modification, evidenced by a 78% reduction in cellulose crystallinity, which directly contributed to the enhanced efficiency of subsequent enzymatic hydrolysis (Fasheun et al. 2022). Another study by Yoo et al. (2012) examined the pretreatment of soybean hulls using a twin-screw extruder to enhance enzymatic hydrolysis. The study tested different speeds (280, 350, and 420 rpm) while adjusting the barrel temperature, which ranged from 40 to 80 °C. According to the authors, this method resulted in a significant increase in enzymatic hydrolysis, with a 155% rise in glucose yield compared to untreated biomass. In terms of biomethane production, a study by Souza et al. (2022) assessed the valorization of grass for renewable energy, demonstrating the effectiveness of twin-screw extrusion as a pretreatment method. This mechanical and thermal process significantly improved the material's digestibility for anaerobic digestion. Specifically, the biomethane potential (BMP) increased by 11% for fresh grass and by a more substantial 18% for ensiled grass.

10.2.2 Alkaline Extraction (Thermochemical Method)

Alkaline extraction is the most common chemical method for biomass pretreatment. It is known to solubilize lignin without harming the carbohydrate fractions, with hemicellulose being the most vulnerable to small solubilization or degradation. Pretreatments with alkali typically use 1–8% NaOH at temperatures ranging from 90 to 220 °C and durations of 5 min to several hours. Adjusting these factors is necessary to optimize the process for a specific type of biomass. Besides solubilizing lignin, cellulose may undergo mercerization, where its crystalline structure is disrupted, resulting in a more porous material that enhances enzyme access during the enzymatic hydrolysis stage (Xu and Cheng 2011). The study by Harun and Geok (2016) examined NaOH pretreatment of rice straw at 55 °C, using 2–12% (w/v) NaOH and a 1:20 (w/v) straw-to-solution ratio for 1 or 3 h. Pretreatment with 12% NaOH for 1 h significantly increased the glucan content by 85.6% and achieved the highest delignification rate of 79.6%.

In contrast, 3 h with 2% NaOH yielded the highest total carbohydrate content (79.2%), demonstrating the potential of optimized NaOH conditions to enhance the suitability of rice straw for sustainable processes. The investigation into the valorization of corn stover through alkaline pretreatment showed a significant increase in its biomethane potential. Pretreatment with 2.5% potassium hydroxide (KOH), applied for 1 day, produced a methane gas generation capacity of 295.0 mL/gVS. This marked a substantial 95.6% increase in methane yield compared to untreated corn stover, indicating the effectiveness of this method in enhancing anaerobic digestibility (Li et al. 2015).

10.2.3 Acid Treatment (Thermochemical Method)

Using different acids in biomass submitted to temperature helps to break down certain bonds, mainly in hemicellulose. Depending on the conditions, they are hydrolyzed to oligomers and even monosaccharides. The most common acids for this are hydrochloric, sulfuric, acetic, and phosphoric. Developing treatment depends on the amount of biomass, temperature, time, and acid load. The primary benefit of acid treatment, along with its associated cost, is the ability to produce a high glucose yield at the end. However, disadvantages include the need for acid-resistant reactors and acid recovery systems. It is also important to remember that harsher conditions can produce inhibitors, such as furanic and phenolic compounds, which can negatively affect the subsequent fermentation stage (Rezania et al. 2020).

10.2.4 Organosolv (Thermochemical Method)

The Organosolv treatment method utilizes organic solvents, such as ethanol, acetone, and ethylene glycol, for delignification and the solubilization of hemicellulose. This process is advantageous for industries because solvents can be recovered and reused. Typical temperatures for this pretreatment range from 100 to 200 °C, and some processes use acids to catalyze the reactions. Recent studies using a mixture of methanol and dioxane, combined with microwave technology, have produced a material with over 99% conversion of glucans from poplar wood. Organosolv pretreatment is currently the leading example of a biorefinery process where, on an industrial scale, biomass components are separated with minimal waste (Park et al. 2010).

10.2.5 Steam Explosion (STEX) (Thermophysicochemical Method)

Steam explosion pretreatment is a widely used method for producing material that can be readily hydrolyzed. The process mainly involves breaking bonds in the most amorphous regions of biomass, such as lignin and especially hemicellulose. It is also reported that changes in the crystalline structure of cellulose may occur (Kumari and Singh 2018). The method utilizes saturated steam at temperatures exceeding 160 °C and pressures exceeding 5 bar, with variations in residence time in the reactor ranging from seconds to minutes. Afterwards, the material undergoes rapid decompression and is then transferred to a collection system (Cantarella et al. 2004). In this way, the recalcitrant structure of the biomass is broken down, chemically by the action of the released acetic acid itself and physically by shear at the reactor outlet (Moreno et al. 2019). The characteristics of the biomass, including moisture, particle size, and plant material, influence the conditions required for effective treatment. The main advantages include the ability to avoid adding chemical compounds, the reuse of steam outlets, and relatively low energy costs. However, the main disadvantages are that the product does not contain immediately fermentable sugars and, primarily, it generates fermentation inhibitors (Overend and Chornet 1987).

10.2.6 Hydrothermal (Hot Water) (Thermochemical Method)

The hot water method is quite like the steam explosion method, except that it uses liquid water instead of steam. The liquid water is heated to temperatures ranging from 160 to 240 °C under pressure. Typically, an ideal pH range is between 4 and 7, which facilitates better action on the structure of hemicellulose, enabling its dissolution. Unlike with a steam explosion, there is no depressurization and ejection of material, which makes the physical operation of the installation easier (Cheah et al. 2020).

10.2.7 Wet Oxidation (Thermochemical Method)

The method called wet oxidation utilizes water at temperatures above 120 °C, with a pressure ranging from 0.5 to 2 MPa, for a period of up to 30 min. The primary parameters affecting this reaction are temperature, duration, and the ratio of oxygen to air in the system. Water at high temperatures and pressures generates more H^+ ions, lowering the pH and affecting the structure of the biomass, particularly the hemicelluloses, which can be solubilized (Rajak and Banerjee 2018).

10.2.8 Ammonia Fiber Explosion (AFEX) (Tphermohysicochemical Method)

The AFEX method, from an operational perspective, closely resembles STEX, with the main differences being in the treatment conditions. In AFEX, lower temperatures are typically used, ranging from 50 to 110 °C, and the process duration can vary from minutes to hours. Since ammonia is volatile, it diffuses more easily into the fibers, causing swelling and rupture of the constituents, which reduces the crystallinity of the cellulose and increases the material's porosity. Materials with high lignin content are not recommended for this treatment because ammonia does not diffuse as effectively into them. One of the main advantages of this method is the ease of recovering ammonia, which can be reused (Rezania et al. 2020).

10.2.9 Biological Method

Biological methods are known for being environmentally friendly, non-toxic because they do not produce inhibitors, low cost with minimal energy consumption. Several strains of bacteria and fungi can produce enzymes that act on lignin (ligninases), leaving carbohydrates available for hydrolysis and fermentation processes. White-rot fungi are particularly noteworthy because they specifically target lignin. The enzymes involved are primarily oxidases from the extracellular environment, such as laccases, manganese peroxidase, and lignin peroxidases. Factors affecting the efficiency of this pretreatment relate mainly to two elements: (a) biomass (particle size and moisture content), and (b) microorganism (aeration, incubation time, and cultivation medium) (Shankar et al. 2022).

10.2.10 Green Methods (GM)

Green methodologies are environmentally sustainable approaches for the degradation of biomass, offering sustainable and efficacious means of converting plant-derived materials into sugars suitable for fermentation processes.

10.2.10.1 Supercritical Fluids (GM)

A method that uses supercritical fluids leverages the best properties of both liquids and gases. The diffusion and viscosity properties resemble those of gases, while the density is similar to that of liquids. By definition, a fluid is considered supercritical when its temperature and pressure go beyond the critical point. The most common fluids used in industry are ammonia, water, methanol, and CO_2. CO_2 has the advantage of having the lowest critical temperature (31 °C) and pressure (73.8 bar). Like AFEX and STEX, supercritical CO_2 can also increase porosity and reduce the crystallinity of cellulose, making subsequent steps easier without forming inhibitors (Escobar et al. 2020).

10.2.10.2 Microwave (GM)

Microwaves are non-ionizing radiation in the range of 300–300,000 MHz that can selectively transfer heat to different substances. Microwaves were initially used in conventional chemical reactions at a laboratory scale; however, in recent years, their use on a pilot scale has become increasingly common as an alternative method for pretreating biomass (Li et al. 2016). There are two main groups of microwave and biomass treatment: solvolysis at moderate temperatures (below 200 °C), which aims to depolymerize biomass into compounds of commercial interest; and lignin pyrolysis (above 400 °C) to produce bio-oil or biogas. Overall, the main advantages of this process include rapid heat transfer, extremely short reaction times, selective energy transfer, ease of operation, and the absence of by-products. However, biomass in its physical form is not highly efficient at absorbing microwave energy and converting it to heat. Hence, it is necessary to add another material that can facilitate this process (Chen et al. 2017).

10.2.10.3 Ultrasound (GM)

Ultrasound is a form of non-ionizing radiation that ranges from 20 kHz to 1 MHz. At lower frequencies, it is used in chemical processes, while higher frequencies are applied in diagnostic medicine. Its effects on biomass promote the formation of oxidizing radicals on fiber surfaces. It can also break the α-O-4 and β-O-4 bonds in lignin, resulting in the fragmentation of biomass into carbohydrates and the formation of lignin cavitation bubbles. Combining ultrasound with other pretreatment methods enhances carbohydrate yields and lowers operational costs by allowing gentler conditions. For example, ultrasound has been combined with microwaves, AFEX, and alkaline extraction (Luo et al. 2013).

10.2.10.4 Gamma Rays (GM)

Biomass can be treated with ionizing radiation using gamma rays, which are typically produced by cobalt-60 or cesium-137. Gamma rays easily penetrate biomass, altering the structure of lignin and decreasing the crystallinity of cellulose. The amount of radiation needed to produce a material that enzymes can easily access varies depending on the type of biomass. Therefore, a specific study is required to determine the optimal pretreatment conditions (Hyun Hong et al. 2014).

10.2.10.5 Electron Beam Irradiation—EBI (GM)

The foundation of EBI treatment involves ionizing radiation generated by linear electron accelerators, aimed at overcoming the recalcitrance of biomass. This process can be accomplished through the reduction of cellulose crystallinity or by cleaving bonds between lignin and hemicellulose. Ultimately, biomass components are subjected to partial depolymerization, and when combined with other treatments, the overall effectiveness is enhanced (Mante et al. 2014).

10.2.10.6 Pulsed Electric Field (PEF)—(GM)

In PEF, biomass is placed in a chamber with electrodes on opposite sides. For a very short time interval (10^{-4} to 10^{-2} s), a current of 0.1 to 80 kV/cm is applied to the material. This causes the rupture of the cell membrane and facilitates substance exchange, with cellular components moving to the external environment, while enzymes enter. Although the exact mechanism of PEF is not fully understood, literature shows that enzymatic hydrolysis is enhanced after this treatment (Barba et al. 2015).

10.2.10.7 High-Pressure Homogenization—HAP (GM)

PAH is a widely used physical method for cell disruption and component extraction. The equipment can generate a material with a very uniform particle distribution from solids suspended in a liquid. High pressure is necessary to force this suspension through specific valves, ensuring the standardization of the material. As a result, the material produced is small in size. It has a high surface area, which improves contact between the sample and enzymes, thereby boosting the efficiency of the hydrolysis stage (Oliveira et al. 2012).

10.3 Lignocellulose Biomass Composition and Derivatives and Methods for Analysis

The industrial process requires numerous critical controls at each step. It needs specific characterization because it plays a crucial role in advancing our understanding of biomass, particularly after hydrolysis, a vital step in converting biomass into biofuels, biochemicals, and biomaterials. The National Renewable Energy Laboratory (NREL) and the U.S. Department of Energy are key supporters in applying official analytical methods for all types of biomass. Instrumental techniques such as nuclear magnetic resonance (NMR) spectroscopy, Raman spectroscopy, Fourier-transform infrared (FT-IR) spectroscopy, ultraviolet-visible (UV-Vis) spectroscopy, and mass spectrometry (MS) each offer unique capabilities for revealing the complex chemical makeup of biomass and the structural changes during the hydrolysis of natural biopolymers like cellulose, hemicellulose, and lignin.

NMR is an expensive method but provides unparalleled structural details and quantitative insights into biomass components. Raman and FT-IR spectroscopies allow rapid, non-destructive analysis of functional groups and structural modifications, with Raman especially useful for in situ monitoring in aqueous environments. UV-Vis spectroscopy is a cost-effective technique for measuring specific chromophores such as lignin and degradation products. Mass Spectrometry, particularly when combined with chromatography, offers high sensitivity and specificity for identifying and quantifying a wide range of compounds in complex hydrolysates. These techniques are vital for addressing real-world challenges in biomass conversion, including quantifying sugars, monitoring fermentation, characterizing structural changes, and identifying bottlenecks in hydrolysis. The field is continuously evolving, with emerging trends focused on advanced hyphenated techniques, real-time and in situ monitoring, sophisticated chemometrics, high-throughput analysis, and the characterization of innovative feedstocks and products. These advancements are expected to deepen our understanding of biomass chemistry, leading to more efficient, sustainable, and economically viable biorefinery processes. By harnessing spectrometric characterization, researchers and industries can accelerate the development of a bio-based economy, supporting energy security and environmental sustainability. The exploration of new biomass feedstocks (e.g., algae, agricultural residues, municipal solid waste) and the development of new biofuels and biochemicals will require tailored instrumental methods and approaches. This includes characterizing unconventional components, complex mixtures, and trace impurities that may impact process efficiency or product quality. Advanced methods and techniques will be essential for understanding the unique chemical properties of these new materials and optimizing their conversion pathways.

UV-Vis spectroscopy is primarily used in biomass characterization to measure the acid-soluble lignin content after hydrolysis. Lignin, with its aromatic structure, strongly absorbs UV light, especially around 280 nm. It can also detect and quantify various sugar degradation products, such as furfural (from pentoses) and hydroxymethylfurfural (HMF, from hexoses), which form during intense hydrolysis and can inhibit subsequent fermentation. UV-Vis spectroscopy was then used to determine lignin concentrations in aqueous solutions, with measurements taken at multiple wavelengths (212, 225, 237, 270, 280, and 287 nm). Calibration curves were created using precipitated lignin standards and then applied to analyze actual substrates produced from a one-step steam process on reed canary grass.

The FT-IR spectroscopy method is commonly used for analyzing the composition and structure of lignocellulosic biomass. It can identify structural changes and detect the presence of carbohydrates, proteins, and other components. It is particularly useful for comparing different biomasses and products that have been subjected to physical or chemical treatments, offering insights into the transformations that occur during hydrolysis, pretreatment, and other conversion processes. FT-IR can also be employed for the quantitative analysis of specific components within the biomass. Water's strong infrared absorption bands can interfere with the analysis of aqueous samples or those with high moisture content (FT-IR Spectroscopy for Biomass Characterization (n.d.), https://www.sciencedirect.com/science/article/abs/pii/S0960854213001089). This problem can be solved by drying samples or using ATR accessories. Raman spectroscopy is widely used to analyze lignocellulosic biomass, its derivatives, and downstream products. It evaluates chemical properties such as cellulose, lignin, or lignin monomer content, along with sugars released during acid or enzymatic hydrolysis. This is essential for assessing plant recalcitrance and developing effective biomass conversion strategies. Raman also tracks changes in biomass composition and structure during pretreatment and hydrolysis. Fluorescence from natural chromophores can cause issues by masking weaker Raman signals; however, using longer excitation wavelengths or fluorescence quenching methods can help mitigate this problem. NMR spectroscopy provides detailed information on the molecular structure and composition of lignocellulosic biomass, both in its native state and after hydrolysis. It can differentiate components such as cellulose, hemicellulose, and lignin, providing insights into their structural changes. Solid-state NMR is crucial for studying intact lignocellulosic biomass without prior dissolution, enabling the analysis of insoluble components, such as cellulose and lignin, in their native state. It provides valuable insights into crystallinity, molecular dynamics, and interactions among the components of biomass. This method provides precise measurements of residual xylan and cellulose in wood samples, as well as xylose and glucose in hydrolysates (Kenny et al. 2023). NMR has demonstrated good repeatability, with carbohydrate compositional results comparable to those obtained using published methods for sugar maple and aspen woods. Quantitative sugar analysis in wood hydrolysates can be performed using high-performance anion-exchange chromatography coupled with mass spectrometry (HPAEC-MS), which enables the analysis of oligosaccharides in lignocellulosic biomass hydrolysates (Mittal et al. 2009). Ultimately, the key technique for understanding which biomolecules are present is MS, essential in biomass conversion, producing a wide array of industrially valuable materials. The complexity and heterogeneity of biomass and its transformation products present significant analytical challenges, which advanced MS techniques help to address. MS can identify and quantify various components in biomass hydrolysates, including sugars, oligosaccharides, lignin fragments, and degradation products (Lupoi et al. 2015).

Sample preparation may be extensive, especially for complex biomass hydrolysates, to remove interfering substances. The analysis of oligosaccharides in lignocellulosic biomass hydrolysates utilizes high-performance anion-exchange chromatography coupled with mass spectrometry (HPAEC-MS), which has also been validated for linearity, reproducibility, and response factors. No single example is a collection of 21 different biomass hydrolysates, produced from various biomass types and multiple pretreatment-hydrolysis methods. HPAEC-MS effectively separated and detected numerous oligosaccharides in a single run based on retention time and mass. The technique could categorize peaks according to their degree of polymerization (DP) and building blocks (pentoses, hexoses, and aldonic acids). Disaccharides were identified as the primary remaining sugar forms in all 21 biomass hydrolysates, indicating a bottleneck in the hydrolysis process. Commonly used MS techniques for biomass analysis include the following:

- Gas Chromatography–mass spectrometry (GC-MS): Widely used for analyzing volatile and semi-volatile compounds produced during biomass pyrolysis and gasification. It can identify hydrocarbons, aldehydes, ketones, phenols, and furans. It is also helpful in assessing adsorbed organic compounds and residual tars in activated carbon analysis.
- Liquid Chromatography–mass spectrometry (LC-MS): Ideal for analyzing non-volatile and thermally sensitive components, such as bio-oils and liquefaction products. It helps identify complex mixtures of high-molecular-weight compounds like lignin-derived oligomers, organic acids, and bioactive compounds.
- Inductively Coupled Plasma-mass spectrometry (ICP-MS): Essential for detecting and quantifying trace elements (heavy metals and inorganic impurities) in biomass and its derivatives. These impurities can impact catalytic efficiency and product quality.
- Matrix-Assisted Laser Desorption/Ionization-mass spectrometry (MALDI-MS): This technique is especially useful for analyzing large biomolecules like lignin and its derivatives in biomass-based products. It ionizes high-molecular-weight species without causing fragmentation, making it ideal for examining the

molecular structure of lignocellulosic materials. Often, the analysis of oligosaccharides in lignocellulosic biomass hydrolysates is conducted using high-performance anion-exchange chromatography combined with mass spectrometry (HPAEC-MS). The use of integrated analytical approaches, often called hyphenated techniques, represents significant progress in the field.

Combining chromatographic methods such as GC, LC, and HPAEC with the high sensitivity and selectivity of mass spectrometry (e.g., GC-MS, LC-MS, HPAEC-MS) enables detailed analysis of complex biomass hydrolysates. Future innovations are expected to include increasingly advanced hyphenated systems, possibly integrating NMR or other optical spectroscopy techniques, to provide highly detailed, multidimensional insights into biomass composition and structural features.

The established procedures for quantifying structural carbohydrates and lignin in biomass involve acid hydrolysis followed by measuring both acid-soluble and acid-insoluble components. Sample preparation includes a two-step sulfuric acid hydrolysis. After hydrolysis, solids are separated from liquids, and each is analyzed using high-performance liquid chromatography (HPLC). The values from acid-soluble and acid-insoluble lignin are combined to determine total lignin content. For carbohydrate analysis, it is essential to fully convert polysaccharides into their monomeric forms; any remaining carbohydrate oligomers suggest incomplete hydrolysis and are not included in the quantification. During hydrolysis, polymer chains break down into monomers, each gaining a hydrogen and a hydroxyl group. An anhydro correction is then applied to accurately convert monomer measurements into structural polymer equivalents (Wolfrum et al. 2009). Table 10.3 shows the standard method for the determination and techniques for biomass speciation:

10.4 Pretreatment Chemical Changes and Hydrolysis Subproducts

The recalcitrant structure of lignocellulosic biomass (LCB) is a characteristic that must be overcome to enable its various uses. An effective pretreatment breaks the interactions between the components of LCB, allowing for fractionation or easing digestion (Worku et al. 2024). Plant-based (lignocellulosic) biomass mainly consists of cellulose, hemicellulose, and lignin, which together make up about 90% of its mass. These biopolymers form the structure of the plant cell wall: cellulose creates strong fibers, hemicellulose provides flexibility as a branched matrix, and lignin reinforces and defends against microbes and degradation. Their composition and arrangement determine how easily biomass can be broken down and processed, affecting the efficiency of hydrolysis, fermentation, and thermochemical methods. Understanding these interactions is crucial for improving biorefinery processes. The molecular structure of the main components of biomass is:

Table 10.3 Summary of biomass conversion methods and uses

Description	Techniques	Reference method
1. Summative mass closure	Concept	(1)
2. Determination of structural carbohydrates and lignin D- (+) glucose D- (+) xylose D- (+) galactose L- (+) arabinose D- (+) mannose Acetylation of lignin	UV–spectroscopy Spectrophotometer–UV HPLC–ionic acid NMR	NREL/TP-510-42618
3. Determination of extractives	Gravimetry UV–spectroscopy HPLC–MS GC–MS	NREL/TP-510-4619 TAPPI test method T204
4. Preparation of samples for compositional analysis	Gravimetry Lyophilization	NREL/TP-510-42620
5. Determination of total solids in biomass and total dissolved solids in liquid process samples	Gravimetry	NREL/TP-510-42621
6. Measurement of ash	Gravimetry UV–spectroscopy HPLC–MS	NREL/TP-510-42622
7. Determination of sugars, by-products, and degradation products in liquid fraction process samples: D- cellobiose D- (+) glucose D- (+) xylose D- (+) galactose L- (+) arabinose D- (+) mannose Xylitol Succinic acid L-lactic acid Glycerol, Acetic acid Ethanol 5-hydroxy-2-furaldehyde (HMF) Furfural(F)	HPLC–ionic acid HPLC–MS GC–MS NMR	NREL/TP-510-42623
8. Determination of protein content	Kjeldahl UV–spectroscopy	NREL/TP-510-4625
9. Determination of insoluble solids in pretreated biomass material	Gravimetry	NREL/TP-510-42627
10. Measurement of cellulase activities	Enzymatic activity UV–spectroscopy	NREL/TP-510-42628 ASTM D5298-16
11. Determination of total organic, and inorganic carbon in biological cultures and liquid fraction process samples	UV–spectroscopy	NREL/PR-2700-79268

(continued)

Table 10.3 (continued)

Description	Techniques	Reference method
12. Determination of carbohydrates in biomass by high-performance liquid chromatography	HPLC–ionic acid	ASTME1758-24
13. Determination insoluble lignin–klason	Gravimetry UV–spectroscopy HPLC–ionic acid NMR	ASTM D1166-84
14. Quantification of phenolic hydroxyl in lignin	UV–spectroscopy HPLC–ionic acid NMR	(2)
15. Determination of cis,cis- and cis, transmuconic acid from biological conversion	UV–spectroscopy HPLC–ionic acid	NREL/TP-5100-74473
16. Measurement of filter paper activities	Enzymatic activity UV–Spectroscopy	ASTM D5298 (3)
17. Measurement of cellulase activities	Enzymatic activity UV–spectroscopy	(4)
18. Measurement metals in ash	Gravimetry UV–spectroscopy Atomic absorption	NREL/TP-5100-42622
19. Measurement of aromatic and phenols	UV–spectroscopy HPLC–organic phase HPLC–MS GC–MS	(5)
20. Measurement organic acids	UV–Spectroscopy HPLC–Ionic acid HPLC-MS	(5)
21. Measurement hemicellulose (a) Determination hemicellulose-alpha (b) Determination hemicellulose-Beta	UV–spectroscopy Spectrophotometer–UV HPLC–ionic acid RX NMR	(6)

(1) https://www.nrel.gov/docs/gen/fy11/48087.pdf
(2) https://doi.org/10.1021/acssuschemeng.3c00115
(3)Yu et al. (2016) https://doi.org/10.1016/j.sjbs.2015.06.018
(4) https://doi.org/10.1351/pac198759020257
(5) https://doi.org/10.1021/acs.energyfuels.2c00769
(6) https://doi.org/10.1371/journal.pone.0256559

Cellulose	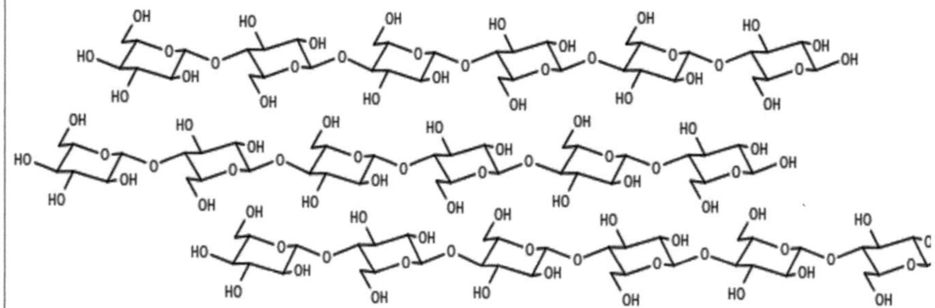
Hemicellulose	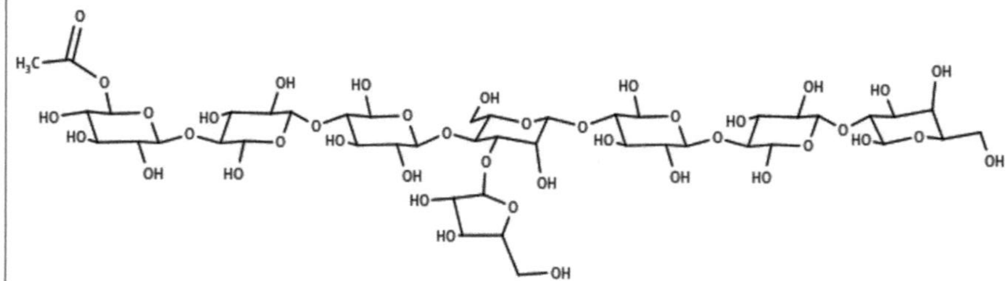

(continued)

Lignin	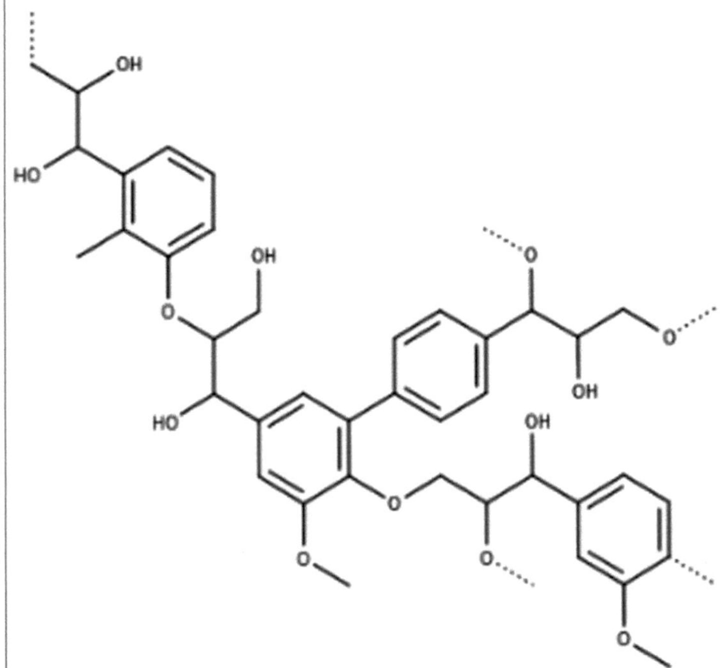

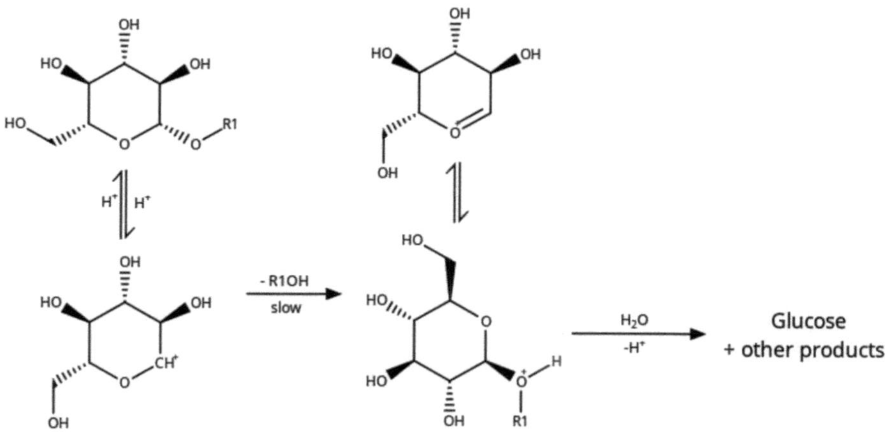

Fig. 10.4 Glycosidic bonds broken by acids

The harsh conditions used during pretreatment, which are intended to break down the complex lignocellulosic structure, often lead to the formation of various inhibitory compounds. These compounds, if not removed or diminished, can substantially reduce the efficiency of downstream enzymatic hydrolysis and microbial fermentation processes, thereby impacting the economic viability of lignocellulose-based biorefineries (Sluiter 2021). Understanding the types and formation of these inhibitors is essential for developing effective detoxification strategies to enhance the overall efficiency of lignocellulose bioconversion. Glycosidic bonds can be broken by acids, as shown in Fig. 10.4. In the initial stage, the oxygen in the glycosidic bond is protonated, and then the conjugate acid decomposes. Generally, the β anomer reacts faster compared to the α equivalent. Also, the shape of the monosaccharide ring is influential, since the furanosidic form has greater molecular tension compared to the pyranosidic form; thus, the rupture under greater tension in the furanosidic form is easier (Milessi et al. 2021).

Although monosaccharides and oligomers are relatively stable under mildly acidic conditions, carbohydrates can undergo reactions such as dehydration, fragmentation, and condensation. Pentoses in an acidic medium experience the removal of three water molecules, forming a compound called 5-hydroxy-methyl-furfural. This compound can also break down into levulinic and formic acid in an aqueous medium (Fig. 10.5) (Girisuta et al. 2008).

Similarly, pentoses, after dehydration produce furfural. Being a highly reactive compound, in an acidic medium, it forms new bonds that lead to the creation of oligomers and traditional aromatic structures, as shown in the figure. These larger, new structures can precipitate in the medium and are known as pseudolignins (Zeitsch 2000). One of the main changes that can happen with lignin is the breaking of the α and β ether bonds. The first step involves protonating the benzylic oxygen, followed by α-ether elimination of a phenolic group or alcohol, creating an intermediate that can follow three main pathways (Fig. 10.6).

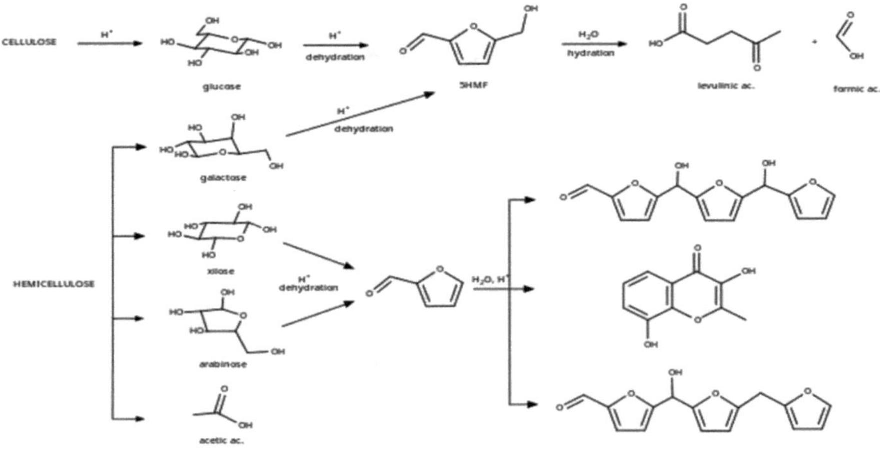

Fig. 10.5 Cellulose and hemicellulose under mildly acidic conditions and their reaction defines dehydration, fragmentation, and condensation

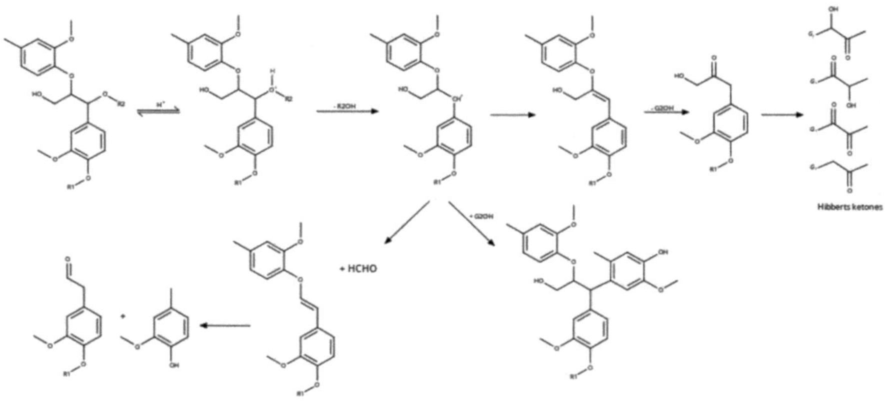

Fig. 10.6 Main changes with lignin breaking of the α and β ether bonds

In the first pathway, the intermediate forms a C_6C_3-enol ether, which is quickly hydrolyzed in an acidic medium to produce an α-β-ketol and then Hibbert's ketones (Cao et al. 2012).

The second possibility is that the intermediate shows a C–C break between the β and γ positions. The products of this breakdown are formaldehyde and C6C2-enol-ether, which slowly degrade to vanillin. In the third possibility, an intramolecular electrophilic addition occurs, resulting in α-6-diphenylmethane and α-5-condensed structures. Because of the presence of formaldehyde and phenolic aromatic compounds, new structures can form through condensation of this diphenyl type (Samuel et al. 2010). Understanding these properties of biomass in the presence of acids allows for the planning of pretreatment strategies with specific objectives to target one class or another of the constituents. Additionally, the type of product obtained can be aimed for, and processes can be optimized.

10.5 Processes to Detoxify Hydrolysates Before Fermentation

All the pretreatment methods, to a greater or lesser extent, besides the lignocellulosic structure disruption and recovering of the cellulose, hemicellulose, and lignin, some toxic compounds are produced due to the thermic degradation, chemical reaction or physical effect on the structural components of the lignocellulosic complex. So, some strategies must be followed to detoxify the obtained hydrolysate for future fermentation. The main categories of inhibitors from lignocellulosic biomass include the following:

(a) Aldehydes like furfural and HMF, when treated with reducing agents, can be converted into less reactive alcohol groups (e.g., furfuryl alcohol, HMF alcohol), reducing cytotoxicity. Detoxifying phenolics is more complex, but reducing agents may alter their redox state or promote polymerization into less harmful forms. Furan derivatives are formed from sugar dehydration during pretreatment, especially under acidic and high-temperature conditions. Furfural, produced from the dehydration of pentoses like xylose and arabinose in hemicellulose, is toxic to many microorganisms and impacts their growth, metabolism, and enzyme activity. Hydroxymethylfurfural (HMF), produced from the dehydration of hexoses like glucose and mannose in cellulose, can also inhibit microbial growth and fermentation.

(b) Weak acids, mainly organic acids formed from hemicellulose and lignin degradation or from the deacetylation of hemicellulose, include acetic acid. It is released during the deacetylation process and can lower the pH of hydrolysates, thereby affecting enzyme activity and microbial metabolism. Additionally, it acts as a proton uncoupler, disrupting cellular energy production. Formic acid and levulinic acid, derived from sugar breakdown, add to the overall toxicity of the hydrolysate.

(c) Phenolic compounds, resulting from lignin breakdown during pretreatment, are highly diverse and can strongly inhibit microorganisms. Examples include vanillin, syringaldehyde, coumaric acid, ferulic acid, and other lignin-derived monomers and oligomers. Their inhibition mechanisms involve phenolic compounds disrupting cell membrane integrity, inhibiting enzymes, and causing oxidative stress in microorganisms. The toxicity of these compounds varies depending on their chemical structure.

Detoxifying lignocellulosic hydrolysates is crucial before fermentation to remove inhibitory compounds that can hinder microbial growth and enzymatic activity (Basak et al. 2020). These inhibitors often include phenolic compounds, furan derivatives (such as furfural and HMF), and weak acids (like acetic and formic acid). Removing fermentation inhibitors, while essential for enhancing fermentability, often involves a trade-off with sugar yield. Although various technologies have been developed to remove these inhibitors, specific amounts of sugars are also inevitably lost during the treatment process. A systematic investigation of pretreatment parameters on both inhibitor formation and sugar recovery is necessary, highlighting a delicate balance. The formation of furfural and HMF (furan derivatives) is often linked to sugar degradation during pretreatment. Sulfuric acid pretreatment and inhibitors from sugarcane bagasse impact the production of fermentable sugars and ethanol; conditions that lead to inhibitor formation can also cause sugar loss. Therefore, while methods like laccase treatment and activated charcoal adsorption can effectively reduce furans and phenolics, they may also decrease sugar yield through co-removal or degradation. Fine-tuning these processes is crucial to maximizing fermentable sugar recovery while minimizing inhibitory compounds. The main detoxification methods are briefly described as follows.

10.5.1 Overliming

Overliming is a detoxification method mainly used in processing lignocellulosic hydrolysates, which are often produced during biofuel manufacturing. These hydrolysates contain various inhibitory compounds that can block further fermentation. The main idea of overliming involves adding calcium hydroxide, $Ca(OH)_2$, to significantly raise the hydrolysate's pH, usually to between 9 and 12. This pH increase helps remove or alter inhibitory substances, making the hydrolysate easier to ferment. It is a technique used in many industrial processes, particularly in sugar refining, to purify juices by raising the pH with lime (calcium hydroxide). This causes impurities to settle out, allowing them to be filtered away. The process removes some inhibitors and chemically changes others. It effectively reduces furans and phenolics but may also decrease sugar yields.

The precipitation of inhibitors, a keyway overliming functions, involves removing certain inhibitory compounds from the solution. When calcium hydroxide is added, it reacts with various substances in the hydrolysate, forming insoluble precipitates that can then be removed. A typical example is the formation of calcium sulfate (gypsum) when sulfuric acid is used during the lignocellulose pretreatment process. This process helps reduce the concentration of soluble salts, which benefits the fermentation process. Besides precipitating inhibitors, overliming also induces chemical changes in these compounds. The high pH environment created by adding calcium hydroxide initiates reactions that convert toxic compounds into less inhibitory or non-inhibitory forms. For instance, studies indicate that overliming can significantly decrease levels of furfural and hydroxymethylfurfural (HMF), two common furan derivatives known to inhibit fermentation strongly. Although the precise pathways of these chemical modifications are still under investigation, alkaline conditions encourage reactions that modify the chemical structure of these inhibitors. This process is commonly used in the bioconversion of lignocellulosic biomass to produce ethanol and other biochemicals. It is viewed as a promising and often cost-effective method for detoxifying dilute-acid hydrolysates, making them more suitable for fermentation by microorganisms such as *Saccharomyces cerevisiae.*

10.5.2 Adsorption

This method uses materials such as activated carbon, polymeric resins, or nanomaterials to bind and remove inhibitors. The advantage is that they do not alter the chemical structure of the inhibitors, enabling potential recovery and reuse. Nanomaterials are becoming highly effective adsorbents because of their large surface area and customizable properties. The removal of furan and phenolic compounds from simulated biomass hydrolysates through batch adsorption and continuous fixed-bed column adsorption highlights the use of activated charcoal for selectively removing these compounds. The mechanism depends on the hydrophobic nature of activated charcoal, which has a higher affinity for hydrophobic inhibitors (e.g., vanillic acid, 4-hydroxybenzoic acid, furfural) than for other components like xylose, sulfuric acid, and acetic acid. This approach is effective at achieving high removal efficiency for furans and phenolics.

10.5.3 Ion Exchange

Ion exchange resins are a highly effective and selective method for detoxifying lignocellulose hydrolysates by removing selectively charged inhibitory compounds. They can be customized to target specific inhibitors but may be costly and require regeneration, including understanding the types of inhibitory compounds in lignocellulose biotechnology, their specific applications for removal, and the main cost factors and regeneration steps involved. Recognizing these factors is crucial for optimizing the lignocellulose bioconversion process and maximizing its potential as a sustainable alternative to fossil fuels. Ion exchange resins are considered among the most effective methods for detoxifying lignocellulose hydrolysates prior to fermentation. The choice of resin (anion or cation exchange) depends on the specific inhibitory compounds that need removal. Anion-exchange resins are effective at removing negatively charged inhibitory compounds, such as weak acids like acetic acid, formic acid, and levulinic acid, which are commonly found in lignocellulose hydrolysates. Many studies have documented high levels of these acids. Although furfural and HMF are generally neutral, their removal by anion-exchange resins can occur through adsorption at certain pH levels where they may exhibit some anionic characteristics or interact with the resin matrix. Phenolic compounds derived from lignin are acidic and negatively charged, making them suitable for removal by anion-exchange resins. Cation exchange resins are designed to target positively charged compounds. While less often emphasized for directly removing inhibitors in lignocellulose hydrolysates, they can be used to eliminate heavy metal ions that may be present in the biomass or introduced during pretreatment; these metals can inhibit enzymes and microorganisms. Basic compounds, although less common as inhibitors in lignocellulose conversion, would also be targeted if present.

10.5.4 Membrane Filtration

Techniques like nanofiltration or reverse osmosis can separate inhibitors based on molecule size and charge, making them effective for concentrating sugars while removing small inhibitory molecules. Nanofiltration operates at lower pressures, effectively removing inhibitors while concentrating sugars, but fouling is a challenge. Reverse osmosis offers higher sugar retention but requires more energy and higher pressure.

Membrane processes like nanofiltration and reverse osmosis effectively remove inhibitors such as furfural, acetic acid, and phenolics while concentrating sugars. Nanofiltration provides high inhibitor rejection at lower pressures, which helps reduce fouling costs; reverse osmosis offers higher sugar retention but requires higher pressure and energy. Fouling is a major challenge, affecting operational costs and membrane lifespan. Ongoing research aims to optimize membrane selection, cleaning methods, and energy efficiency to enhance sustainability and economic viability in lignocellulose biorefineries. Cost factors include the membrane price, energy consumption, and fouling management. These techniques can be combined with ion exchange for optimal detoxification, balancing performance and cost. Pressure-driven membrane processes efficiently separate sugars from inhibitors. Nanofiltration operates at lower pressures, while reverse osmosis achieves higher sugar retention but at increased energy costs. Fouling remains a primary challenge, managed through pretreatment. Cost analysis indicates that membrane expenses, energy use, and fouling control are key considerations.

10.5.5 Biological Detoxification

This technique involves the use of microorganisms or enzymes to metabolize or transform inhibitors into less toxic forms. Examples include engineered yeast strains or bacteria capable of degrading furans and phenolics. In situ detoxification during fermentation is also being explored to simplify the bioprocess. For example, phenolic inhibitors from lignocellulose hydrolysates are laccases to produce fuels and chemicals are highlighted, as laccases are enzymes capable of oxidizing lignin and various phenolic compounds. This oxidation process detoxifies lignocellulose hydrolysates, thus enhancing fermentability. The mechanism involves environmentally friendly oxidation of these compounds. Further research into the specific mechanisms of furan and phenolic reduction by different methods is necessary. The other biological strategy to face the use of the hydrolysate is to select a microorganism that is tolerant to toxic compounds present and has the desirable ability for process. In this case, a specific microorganism selection step is necessary.

10.5.6 Chemical Reduction

Reducing agents, such as sulfite, dithionite, or dithiothreitol, can chemically transform inhibitors into less toxic substances. They are often used with other methods. A reducing agent donates electrons and becomes oxidized, which is essential for detoxifying inhibitors in bioprocesses. Common examples include sulfite, dithionite, and DTT, which are known for their ability to donate electrons and their stability. They specifically neutralize inhibitors like furfural or phenolic compounds during lignocellulose bioconversion. Their main role is to convert inhibitors such as furfural and phenolics into less harmful compounds during lignocellulose bioconversion. These agents donate electrons, reducing other compounds as they are oxidized. In lignocellulose bioprocesses, typical inhibitors like furfural, phenolics, and acids form during pretreatment and hinder fermentation. Reducing agents transform these inhibitors into less toxic compounds through electron transfer, neutralizing their toxicity. For example, sulfite acts as a mild reducing agent that detoxifies inhibitors like furfural and phenolics by converting aldehydes into less toxic alcohols. Using sulfite in bioprocessing, especially for detoxifying lignocellulosic hydrolysates, improves fermentation efficiency. This chemical conversion by reducing agents is a common detox strategy in bioprocesses, used alone or with other methods like biological detox or adsorption. The effectiveness depends on the reducing agent's reduction potential, selectivity, and compatibility with the overall process.

10.5.7 Evaporation and Extraction

Volatile inhibitors like furfural can be removed by evaporation, and liquid-liquid extraction can selectively eliminate specific inhibitors but may require solvent recovery. Evaporation might not be effective for nonvolatile inhibitors or may require high temperatures, which may degrade other valuable components in the mixture. This method involves using a solvent to extract inhibitors from the liquid mixture selectively. The solvent should be chosen carefully, with a high affinity for the inhibitor and a low affinity for other components. After extraction, the solvent containing the inhibitor must be separated from the original liquid, often through solvent recovery steps. Examples of solvents used for furfural removal include oleic acid, 1-butanol, isobutyl acetate, and MIBK. Deep eutectic solvents (DES) are also being explored as potential solvents for furfural extraction.

10.6 Final Remarks

The transition to cleaner and more sustainable energy sources is a global challenge that requires diverse and integrated solutions. In this context, biomass stands out as a renewable resource capable of significantly reducing dependence on fossil fuels or, at least, helping to decrease their use and exploitation. Understanding its structure in detail and optimizing conversion and treatment processes expands its potential to produce high-value biofuels and inputs, bringing environmental, economic, and social benefits. Therefore, it is crucial to identify the best processes to transform lignocellulosic biomass into higher-value products, minimizing the formation of biomolecules that inhibit fermentation and enabling biorefineries to meet the growing demand for renewable energy and high-value bioproducts. Biomass, especially lignocellulosic biomass, is an abundant and low-cost source, making it an essential alternative to fossil fuels. Its complex composition, mainly consisting of cellulose, hemicellulose, and lignin, presents obstacles to accessing its most valuable components. To overcome these obstacles, various pretreatment methods—such as physical, physicochemical, chemical, biological, or combinations thereof—are employed to break down the matrix, facilitating enzymatic hydrolysis and fermentation. The process involves hydrolysis and fermentation steps to generate bioproducts, with analytical techniques that identify inhibitors and monitor structural changes. Methods like spectroscopy and chromatography provide detailed information about chemical composition and transformation pathways, which are essential for improving bioconversion. The importance of the analytical methods discussed in this chapter is vital to ensure efficiency, safety, and quality at all stages, from biomass processing to final product monitoring. This integration of science, technology, and innovation provides a solid foundation for the sector's progress and social acceptance. Despite technical, logistical, and regulatory challenges, continuous technological advancements and the implementation of sustainable management strategies indicate a future where biomass will be a central piece in the energy matrix. Thus, investing in research, infrastructure, and appropriate public policies is not only desirable but essential to turn its potential into a concrete reality.

References

Alam SN, Singh B, Guldhe A (2021) Aquatic weed as a biorefinery resource for biofuels and value-added products: Challenges and recent advancements. Cleaner Engineering and Technology 4:100235. https://doi.org/10.1016/j.clet.2021.100235

Arefin MA, Rashid F, Islam A (2021) A review of biofuel production from floating aquatic plants: an emerging source of bio-renewable energy. Biofuels Bioprod Biorefin 15(2):574–591. https://doi.org/10.1002/bbb.2180

Banerjee S, Sen R, Pandey RA, Chakrabarti T, Satpute D, Giri BS, Mudliar S (2009) Evaluation of wet air oxidation as a pretreatment strategy for bioethanol production from rice husk and process optimization. Biomass Bioenergy 33:1680–1686

Barba FJ, Parniakov O, Pereira SA, Wiktor A, Grimi N, Boussetta N, Saraiva JA, Raso J, Martin-Belloso O, Witrowa-Rajchert D, Lebovka N, Vorobiev E (2015) Current applications and new opportunities for the use of pulsed electric fields in food science and industry. Food Res Int 77: 773–798

Basak B, Jeon B-H, Kim TH, Lee J-C, Chatterjee PK, Lim H (2020) Dark fermentative hydrogen production from pretreated lignocellulosic biomass: effects of inhibitory byproducts and recent trends in mitigation strategies. Renew Sust Energ Rev 133:110338. https://doi.org/10.1016/j.rser.2020.110338

Berendes DM, Yang PJ, Lai A, Hu D, Brown J (2018) Estimation of global recoverable human and animal faecal biomass. Nat Sustainability 1(11):679–685. https://doi.org/10.1038/s41893-018-0167-0

Braghiroli FL, Passarini L (2020) Valorization of biomass residues from forest operations and wood manufacturing presents a wide range of sustainable and innovative possibilities. Curr For Rep 6(2). https://doi.org/10.1007/s40725-020-00112-9

Cantarella M, Cantarella L, Gallifuoco A, Spera A, Alfani F (2004) Comparison of different detoxification methods for steam-exploded poplar wood as a substrate for the bioproduction of ethanol in SHF and SSF. Process Biochem 39(11):1533–1542. https://doi.org/10.1016/S0032-9592(03)00285-1

Cao S, Pu Y, Studer M, Wyman C, Ragauskas AJ (2012) Chemical transformations of *Populus trichocarpa* during dilute acid pretreatment. RSC Adv 2(29):10925–10936

Cheah WY, Sankaran R, Show PL, Ibrahim T, Baizura TN, Chew KW (2020) Pretreatment methods for lignocellulosic biofuels production: current advances, challenges and future prospects. Biofuel Res J 7:1115–1127

Chen W, Chen Y, Yang H, Xia M, Li K, Chen X, Chen H (2017) Co-pyrolysis of lignocellulosic biomass and microalgae: products characteristics and interaction effect. Bioresour Technol 245A:860–868

Da Lio L, Castello P, Gianfelice G, Cavalli R, Canu P (2021) Effective energy exploitation from horse manure combustion. Waste Manag 128:243–250. https://doi.org/10.1016/j.wasman.2021.04.035

EIA (2024) Biomass explained: wood and wood waste

Escobar ELN, Da Silva TA, Pirich CL, Corazza ML, Pereira Ramos L (2020) Supercritical fluids: a promising technique for biomass pretreatment and fractionation. Front Bioeng Biotechnol 8:1–18

Fasheun DO, de Oliveira RA, Bon EPS, da Silva AS, Teixeira RSS, Ferreira-Leitão VS (2022) Dry extrusion pretreatment of cassava starch aided by sugarcane bagasse for improved starch saccharification. Carbohydr Polym 285:119256. https://doi.org/10.1016/j.carbpol.2022.119256

Food and Agriculture Organization (2025) FAOSTAT – Forestry Production and Trade

FT-IR Spectroscopy for Biomass Characterization (n.d.). IntechOpen. https://www.intechopen.com/chapters/84973

Girisuta B, Danon B, Manurung R, Janssen LPBM, Heeres HJ (2008) Experimental and kinetic modelling studies on the acid-catalysed hydrolysis of the water hyacinth plant to levulinic acid. Bioresour Technol 99:8367–8375

Harun S, Geok SK (2016) Effect of sodium hydroxide pretreatment on rice straw composition. Indian J Sci Technol 9(21):1–9. https://doi.org/10.17485/ijst/2016/v9i21/95245

Hoyos-Sebá JJ, Arias NP, Salcedo-Mendoza J, Aristizábal-Marulanda V (2024) Animal manure in the context of renewable energy and value-added products: a review. Chem Eng Process Process Intensif 196:109660. https://doi.org/10.1016/j.cep.2023.109660

https://www.sciencedirect.com/science/article/abs/pii/S0960854213001089

Hyun Hong S, Taek Lee J, Lee S, Gon Wi S, Ju Cho E, Singh S, Sik Lee S, Yeoup Chung B (2014) Improved enzymatic hydrolysis of wheat straw by combined use of gamma ray and dilute acid for bioethanol production. Radiat Phys Chem 94:231–235

Jiang C, Lin Q, Wang C, Jiang X, Bi H, Bao L (2020) Experimental study of the ignition and combustion characteristics of cattle manure under different environmental conditions. Energy 197:117143. https://doi.org/10.1016/j.energy.2020.117143

Kapoor M, Semwal S, Gaur R, Kumar R, Gupta RP, Puri SK (2018) The pretreatment technologies for deconstruction of lignocellulosic biomass. Springer, Singapore, pp 395–421

Kaza S, Yao LC, Bhada-Tata P, Van Woerden F (2018) What a waste 2.0: a global snapshot of solid waste management to 2050. World Bank, Washington, DC. https://doi.org/10.1596/978-1-4648-1329-0

Kenny JK, Will Medlin J, Beckham GT (2023) Quantification of phenolic hydroxyl groups in lignin via 19F NMR. Spectroscopy. ACS sustainable. Chem Eng 11:5644–5655

Konan D, Rodrigue D, Koffi E, Elkoun S, Ndao A, Adjallé K (2024) Combination of technologies for biomass pretreatment: a focus on extrusion. Waste Biomass Valoriz 15(8):4519–4540

Kumari D, Singh R (2018) Pretreatment of lignocellulosic wastes for biofuel production: a critical review. Renew Sust Energ Rev 90:877–891

Li L, Chen C, Zhang R, He Y, Wang W, Liu G (2015) Pretreatment of corn Stover for methane production with the combination of potassium hydroxide and calcium hydroxide. Energy Fuel 29(9):5841–5846. https://doi.org/10.1021/acs.energyfuels.5b01170

Li H, Qu Y, Yang Y, Chang S, Xu J (2016) Microwave irradiation - a green and efficient way to pretreat biomass. Bioresour Technol 199:34–41

Luo J, Fang Z, Smith RL (2013) Ultrasound-enhanced conversion of biomass to biofuels. Prog Energy Combust Sci 41:56–93

Lupoi JS, Gjersing E, Davis MF (2015) Evaluating lignocellulosic biomass, its derivatives, and downstream products with Raman Spectroscopy. Front Bioeng Biotechnol 3:50

Mansuy N, Paré D, Thiffault E, Bernier PY, Cyr G, Manka F, Lafleur B, Guindon L (2017) Estimating the spatial distribution and locating hotspots of forest biomass from harvest residues and fire-damaged stands in Canada's managed forests. Biomass Bioenergy 97:90–99. https://doi.org/10.1016/j.biombioe.2016.12.014

Mante OD, Amidon TE, Stipanovic A, Babu SP (2014) Integration of biomass pretreatment with fast pyrolysis: an evaluation of electron beam (EB) irradiation and hot-water extraction (HWE). J Anal Appl Pyrolysis 110:44–54

Milessi TS, Corradini FA, Marçal JV, Baldez TO, Kopp W, Giordano RC, Giordano RL (2021) Xylooligosaccharides production chain in sugarcane biorefineries: from the selection of pretreatment conditions to the evaluation of nutritional properties. Ind Crop Prod 172:114056

Miller SF, Miller BG (2002) The occurrence of inorganic elements in various biofuels and its effect on the formation of melt phases during combustion. In: 2002 International joint power generation conference. ASME, pp 873–880. https://doi.org/10.1115/IJPGC2002-26177

Mittal A, Scott GM, Amidon TE, Kiemle DJ, Stipanovic AJ (2009) Quantitative analysis of sugars in wood hydrolyzates with ^{1}H NMR during the autohydrolysis of hardwoods. Bioresour Technol 100:6398–6406

Moreno AD, Tomás-Pejó E, Ballesteros M, Negro MJ (2019) Pretreatment technologies for lignocellulosic biomass deconstruction within a biorefinery perspective. In: Biofuels: alternative feedstocks and conversion processes for the production of liquid and gaseous biofuels. Academic Press, pp 379–399

Nanda S, Berruti F (2021) Municipal solid waste management and landfilling technologies: a review. Environ Chem Lett 19(2):1433–1456. https://doi.org/10.1007/s10311-020-01100-y

Oliveira SCT, Figueiredo AB, Evtuguin DV, Saraiva JA (2012) High pressure treatment as a tool for engineering of enzymatic reactions in cellulosic fibres. Bioresour Technol 107:530–534

Overend RP, Chornet E (1987) Fractionation of lignocellulosics by steam-aqueous pretreatments. Philos Trans R Soc Lond Ser A 321(1561):523–536

Park N, Kim H-Y, Koo B-W, Yeo H, Choi I-G (2010) Organosolv pretreatment with various catalysts for enhancing enzymatic hydrolysis of pitch pine (*Pinus rigida*). Bioresour Technol 101:7046–7053

Rajak RC, Banerjee R (2018) An eco-friendly process integration for second generation bioethanol production from laccase delignified Kans grass. Energy Convers Manag 157:364–371. https://doi.org/10.1016/j.enconman.2017.11.060

Rezania S, Oryani B, Cho J, Talaiekhozani A, Sabbagh F, Hashemi B, Rupani PF, Mohammadi AA (2020) Different pretreatment technologies of lignocellulosic biomass for bioethanol production: an overview. Energy 199:117457

Samuel R, Pu Y, Raman B, Ragauskas A (2010) Structural characterization and comparison of switchgrass ballmilled lignin before and after dilute acid pretreatment. Appl Biochem Biotechnol 162:62–74
Senturk-Ozer S, Gevgilili H, Kalyon DM (2011) Biomass pretreatment strategies via control of rheological behavior of biomass suspensions and reactive twin screw extrusion processing. Bioresour Technol 102:9068–9075
Shankar A, Saini S, Sharma KK (2022) Fungal-integrated second-generation lignocellulosic biorefinery: utilization of agricultural biomass for co-production of lignocellulolytic enzymes, mushroom, fungal polysaccharides, and bioethanol. Biomass Convers Biorefinery 14:1–15
Sivabalan K, Hassan S, Ya H, Pasupuleti J (2021) A review on the characteristic of biomass and classification of bioenergy through direct combustion and gasification as an alternative power supply. J Phys Conf Ser 1831(1). https://doi.org/10.1088/1742-6596/1831/1/012033
Sluiter J (2021) Biochemical conversion and lignin utilization. National Renewable Energy Laboratory
Souza MF, Devriendt N, Willems B, Guisson R, Biswas JK, Meers E (2022) Techno-economic feasibility of extrusion as a Pretreatment step for biogas production from grass. Bioenergy Res 15(2):1232–1239. https://doi.org/10.1007/s12155-021-10287-z
Tursi A (2019) A review on biomass: importance, chemistry, classification, and conversion. Biofuel Res J 6(2). https://doi.org/10.18331/BRJ2019.6.2.3
UV-Vis spectrophotometer for biomass analysis. Celignis. https://www.celignis.com/archive/equip.php?value=3/
Vassilev SV, Baxter D, Andersen LK, Vassileva CG, Morgan TJ (2012) An overview of the organic and inorganic phase composition of biomass. Fuel 94:1–33. https://doi.org/10.1016/j.fuel.2011.09.030
Vassilev SV, Vassileva CG (2016) Composition, properties and challenges of algae biomass for biofuel application: An overview. Fuel 181:1–33. https://doi.org/10.1016/j.fuel.2016.04.106
Wolfrum E, Lorenz A, de Leon N (2009) Correlating detergent fiber analysis and dietary fiber analysis data for corn Stover collected by NIRS. Cellulose 16(4):577–585
Worku LA, Bachheti RK, Bachheti A, Milessi TS, Chandel AK (2024) Understanding the biochemical changes at molecular level during biomass pretreatment: a comprehensive analysis. Cellulose 31(12):7281–7312
World Bioenergy Association (2024) Global bioenergy statistics report
Wu Q, Wang H, Zheng X, Liu F, Wang A, Zou D, Yuan J, Xiao Z (2020) Thermochemical liquefaction of pig manure: factors influencing on oil. Fuel 264:116884. https://doi.org/10.1016/j.fuel.2019.116884
Xin Y, Zhu X, Wang T (2020) Mixing characteristics of polymer melt in triple screw extruders with combined screws using finite element method. IOP Conf Ser Mater Sci Eng 768:022068
Xu J, Cheng JJ (2011) Pretreatment of switchgrass for sugar production with the combination of sodium hydroxide and lime. Bioresour Technol 102:3861–3868
Yang C, Wang S, Yang J, Xu D, Li Y, Li J, Zhang Y (2020) Hydrothermal liquefaction and gasification of biomass and model compounds: a review. Green Chem 22(23):8210–8232. https://doi.org/10.1039/D0GC02802A
Yoo J, Alavi S, Vadlani P, Behnke KC (2012) Soybean hulls Pretreated using thermo-mechanical extrusion—hydrolysis efficiency, fermentation inhibitors, and ethanol yield. Appl Biochem Biotechnol 166(3):576–589. https://doi.org/10.1007/s12010-011-9449-y
Yu X, Liu Y, Cui Y, Cheng Q, Zhang Z, Jia Hui L, Meng Q, Teng L, Ren X (2016) Measurement of filter paper activities of cellulase with microplate-based assay. Saudi J Biol Sci 23(1):S93–S98. https://doi.org/10.1016/j.sjbs.2015.06.018
Zeitsch KJ (2000) The chemistry and technology of furfural and its many by-products, 1st edn. Elsevier, Amsterdam, The Netherlands
Zevallos Torres LA, Woiciechowski AL, Tanobe VOA, Karp SG, Guimarães Lorenci LC, Faulds C, Soccol CR (2020) Lignin as a potential source of high-added value compounds: a review. J Clean Prod 263:121499. https://doi.org/10.1016/j.jclepro.2020.121499

Chapter 11
Biomass Pretreatment Technologies for Lignocellulosic Feedstocks Valorization in Sustainable Biorefineries

Khyati Joshi, Swarnima Agnihotri, Lachi Wankhede, Gaurav Bhardwaj, and Satinder Kaur Brar

Abstract Lignocellulosic biomass, primarily composed of cellulose, hemicellulose, and lignin, is the most abundant renewable organic resource on Earth and holds tremendous promise as a feedstock for biofuels, biochemicals, and biomaterials. Its utilization not only reduces reliance on fossil fuels but also contributes to climate change mitigation and rural economic development. Despite the potential, its structural complexity and recalcitrance present significant challenges to efficient bioconversion. The rigid lignin–carbohydrate matrix limits enzyme accessibility and often requires energy-intensive processing. Pretreatment is therefore a vital step in biorefinery workflows, aimed at disrupting the biomass structure, enhancing enzymatic hydrolysis, and facilitating downstream processing. Various pretreatment strategies, including physical, chemical, physicochemical, and biological methods, offer different advantages and limitations in terms of efficiency, cost, environmental impact, and scalability. Recent innovations focus on integrating pretreatment into sustainable biorefinery frameworks, optimizing inhibitor formation, and adapting processes for diverse feedstocks. This chapter focuses on the principles, mechanisms, and recent developments in lignocellulosic biomass pretreatment technologies to aid in the selection and optimization of strategies for specific biomass types and processing goals.

K. Joshi · L. Wankhede · G. Bhardwaj · S. K. Brar (✉)
Department of Civil Engineering, Lassonde School of Engineering, York University, Toronto, ON, Canada
e-mail: satinder.brar@lassonde.yorku.ca

S. Agnihotri
Swedish Centre for Resource Recovery, University of Borås, Borås, Sweden

A. L. Woiciechowski et al. (eds.), *Biomass for Bioenergy and Bioproducts*, Biofuel and Biorefinery Technologies 17, https://doi.org/10.1007/978-3-032-19898-3_11

11.1 Introduction

Lignocellulosic biomass (LCB), composed primarily of cellulose, hemicellulose, and lignin, represents the most abundant renewable organic resource on Earth. Its valorization into biofuels, biochemicals, and biomaterials is central to the development of sustainable biorefineries (Haq et al. 2021). In addition to reducing dependence on fossil fuels, LCB utilization contributes to climate change mitigation by enabling carbon-neutral or even carbon-negative production cycles (Nahak et al. 2022; Sharma et al. 2025). Moreover, it offers a promising route for rural economic development through the creation of decentralized bioprocessing facilities (Suota et al. 2021; Qin et al. 2023). However, the inherent recalcitrance of LCB, stemming from the complex and rigid matrix formed by lignin–carbohydrate complexes and crystalline cellulose, poses a significant barrier to efficient bioconversion (Chaudhary et al. 2023).

This structural resistance is not merely a physical impediment but also a biochemical challenge, as lignin acts as a protective barrier that limits enzyme access to polysaccharides while also binding nonproductively to hydrolytic enzymes (dos Santos et al. 2019). Additionally, the variability in biomass composition based on plant species, growth conditions, and harvest timing further complicates uniform processing (Velvizhi et al. 2022). For example, softwoods typically contain more lignin compared to hardwoods or agricultural residues, necessitating more severe pretreatment conditions (Suota et al. 2021). Hence, there is a growing interest in developing tailored pretreatment solutions that accommodate these compositional differences. An ideal pretreatment approach should strike a balance between effectiveness, cost-efficiency, environmental sustainability, and the minimal generation of fermentation inhibitors (Mukherjee et al. 2024).

Pretreatment is a crucial upstream step designed to overcome this recalcitrance by disrupting the structural integrity of lignocellulosic biomass. This process facilitates the selective deconstruction of the biomass matrix by enhancing the accessibility of cellulose to hydrolytic enzymes, solubilizing hemicellulose, and modifying or removing lignin (Yuan et al. 2021; Martín et al. 2022) (Fig. 11.1). In addition to physical and chemical disruption, pretreatment may also generate by-products such as furfural, hydroxymethylfurfural (HMF), and organic acids, which can inhibit microbial fermentation in later stages (Chen et al. 2024; Woźniak et al. 2025). Therefore, optimizing pretreatment conditions to minimize the formation of these inhibitory compounds is equally important for improving downstream process performance.

A wide array of pretreatment strategies has been developed, including physical (e.g., milling, extrusion), chemical (e.g., acid, alkali, organosolv), physicochemical (e.g., steam explosion, ammonia fiber expansion), and biological methods (e.g., fungal delignification) (Baksi et al. 2023). Each method offers distinct advantages and limitations in terms of efficiency, cost, environmental impact, and scalability.

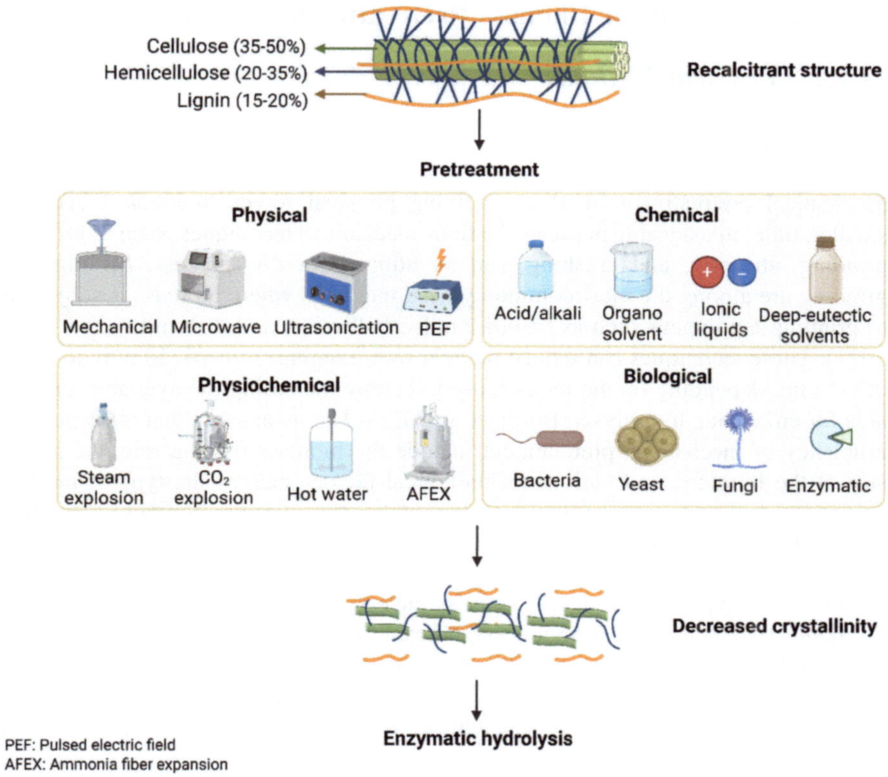

Fig. 11.1 Overview of the LCB pretreatment methods

Recent advances have focused on integrating pretreatment into holistic biorefinery frameworks, emphasizing not only sugar recovery but also the valorization of lignin and other coproducts. Innovative approaches such as deep eutectic solvents, ionic liquids, and hybrid pretreatment systems are gaining traction for their potential to fractionate biomass with a minimal environmental footprint selectively (Verdía Barbará et al. 2025). Moreover, the development of pretreatment technologies that are compatible with diverse biomass feedstocks and adaptable to industrial-scale operations remains a key research priority. This chapter provides a comprehensive overview of the principles, mechanisms, and recent developments in pretreatment technologies. It aims to guide researchers and practitioners in selecting and optimizing pretreatment strategies tailored to specific biomass types and downstream processing goals.

11.2 Lignocellulosic Biomass Pretreatment Methods

11.2.1 Physical Methods

11.2.1.1 Mechanical

Mechanical pretreatment involves applying physical forces to break LCB into smaller, more manageable particles. Various mechanical techniques, such as milling, grinding, shearing, and crushing, can be utilized for this process. Milling and grinding are among the most commonly used methods, where biomass is subjected to grinding equipment such as hammer mills, ball mills, and disk mills (Das et al. 2021). These techniques can reduce particle sizes ranging from 10–30 mm down to 0.2–2 mm, depending on the method used, thereby increasing the available surface area for enzymatic hydrolysis (Broda et al. 2022). The parameters that influence the efficiency of mechanical pretreatment include the biomass feeding rate, the initial size of the biomass, and various technological factors such as the type of milling equipment and operational settings. Additionally, the moisture content of the biomass plays a significant role in the milling process, as it affects the toughness and friability of the biomass. The time of milling is another important factor, with longer milling times typically producing finer particle sizes (Editors et al. 2023). While mechanical pretreatment can achieve significant biomass size reduction, it requires a considerable amount of energy. This high energy input often makes mechanical pretreatment economically infeasible on a large scale, especially when considering equipment depreciation, scale-up challenges, and capital and operating costs (Shukla et al. 2023).

11.2.1.2 Extrusion

Extrusion involves high shearing, moderate heating, and rapid mixing of LCB, resulting in the disruption of its complex structure. This process is highly efficient, combining mechanical, thermal, and chemical techniques simultaneously (Shukla et al. 2023). The extruder's rotating screw generates a significant amount of mechanical shearing forces between the biomass, the screw, and the barrel. This action leads to both temperature and pressure increase within the barrel, which softens the biomass and reduces its degree of crystallinity (Shukla et al. 2023). As the biomass undergoes this shearing and heating process, the lignin and hemicellulose components are partially depolymerized, improving the accessibility of cellulose to the biomass for hydrolysis. These changes ultimately result in a higher yield of fermentable sugars from the biomass.

Extrusion offers several advantages, including its ability to work effectively, even with high solids loading, making it suitable for large-scale operations. Additionally, the process does not produce furfural or 5-hydroxymethylfurfural (HMF), which are common by-products in other pretreatment methods that can inhibit enzymatic activity (Broda et al. 2022). As a result, extrusion is considered a clean and effective method for enhancing biomass conversion. However, the high energy requirement of extrusion, due to the substantial power needed to generate the necessary shear forces and heat, remains a significant drawback (Broda et al. 2022).

11.2.1.3 Microwave Irradiation

Microwave irradiation induces dielectric polarization, generating molecular collisions that weaken the interactions between cellulose, hemicellulose, and lignin. This leads to rapid heating at frequencies up to a million times per second, significantly reducing pretreatment time (Shukla et al. 2023). The deep penetration of microwave energy facilitates lignin degradation and alters the polysaccharide structure, enhancing their solubility and enzymatic digestibility. This method is highly effective in pretreating various feedstocks, such as switchgrass and Miscanthus, where it removes lignin effectively and increases reducing sugar yields ranging from 40% to 60% (Editors et al. 2023). While microwave irradiation offers several advantages, such as reduced processing time and energy input compared to conventional heating, the need for large microwave irradiators, which are energy-intensive and costly, hinders its scalability for industrial applications. Additionally, uneven heating can lead to the formation of inhibitors, reducing expected yields and increasing operational costs (Editors et al. 2023).

11.2.1.4 Ultrasonication

Ultrasonication is an advanced and eco-friendly technique used to enhance the digestibility of LCB. This method alters both the physical and chemical properties of the biomass, making it more amenable to enzymatic hydrolysis. The process works by generating ultrasound waves that produce rapid pressure changes within the biomass, creating microbubbles. These bubbles collapse with great force, causing disruption in the recalcitrant structure of LCB, thereby increasing its porosity and surface area. This leads to the breakage of complex lignocellulosic structures into simpler components like reducing sugars, which are more easily converted into biofuels (Shukla et al. 2023).

The application of ultrasound at slightly increased temperatures (around 50 °C) and the addition of an alkali can further improve the efficiency (Broda et al. 2022). Ultrasound has been found to enhance enzymatic hydrolysis, leading to a significant increase in sugar yields by up to 200%, compared to untreated biomass (Editors et al. 2023). However, the exact mechanisms underlying this enhancement are still not fully understood, but it is believed that the energy supplied by the ultrasound waves may break the H- bonds in the cellulose crystalline structure. The advantages of ultrasonication include its short processing time and the use of lower operating temperatures, which makes it a more energy-efficient and sustainable option compared to traditional methods. Additionally, it requires minimal chemical usage, which makes it an eco-friendly choice. However, the efficiency of the ultrasound method can vary depending on the ultrasonic frequency, the solvent used, and the design of the reactor (Editors et al. 2023).

11.2.1.5 Pulsed Electric Field

Pulsed electric field (PEF) treatment is a nonthermal, environmentally friendly method for pretreating LCB. The process involves applying short bursts of high-voltage electric pulses to the biomass, which induce electric fields that permeate the biomass cell walls (Das et al. 2021). This formation of pores in the cell membranes facilitates the release of intracellular contents and makes hemicellulose and cellulose more accessible for subsequent enzymatic hydrolysis.

The efficiency of PEF treatment can be influenced by several factors, including the electric field strength, pulse duration, and the frequency of the electric pulses. By optimizing these parameters, PEF can achieve significant improvements in biomass digestibility and sugar recovery. The method is particularly effective in pretreating various feedstocks, such as switchgrass and wood chips, where it facilitates higher sugar yields compared to untreated biomass (Kumar et al. 2011). Despite its benefits, PEF treatment does have limitations. One challenge is the need for high electric field intensities, which may require more energy input and specialized equipment. Additionally, PEF treatment is generally more effective when combined with other pretreatment methods, such as chemical or thermal treatments, to further enhance the breakdown of lignocellulosic structures and optimize overall biomass conversion (Bocker and Silva 2024).

11.2.2 Chemical Methods

11.2.2.1 Acid Pretreatment

Acid pretreatment is a widely used method to break down LCB and enhance its enzymatic digestibility for biofuel production. This process primarily targets the disruption of the lignin–carbohydrate matrix, solubilizing hemicellulose, and increasing cellulose accessibility for enzymatic hydrolysis. Acid pretreatment works by breaking van der Waals forces, hydrogen bonds, and covalent linkages that hold biomass components together, thereby improving the release of fermentable sugars (Amin et al. 2017).

The efficiency of acid pretreatment on biomass conversion is highly dependent on several operational parameters, including the type and concentration of acid, temperature, pressure, and reaction time. Strong acids such as sulfuric acid (H_2SO_4) and hydrochloric acid (HCl) are commonly used for pretreatment, either in concentrated or dilute forms. Alternatively, weaker organic acids like acetic acid (CH_3COOH) and formic acid (HCOOH) offer milder pretreatment conditions with potentially lower environmental impact (Editors et al. 2023). Dilute acid pretreatment, typically using acid concentrations ranging from 0.5% to 8% (w/w) and a solid-to-liquid (S:L) ratio between 1:5 and 1:20 (w/v), is widely used, as it enables high reaction rates and effective hemicellulose solubilization at elevated temperatures (100–200 °C) within 5–300 min (Oriez et al. 2019). This method enhances enzymatic digestibility by increasing cellulose accessibility while minimizing excessive sugar degradation. However, a major drawback of acid pretreatment is the formation of inhibitory compounds such as furfural and hydroxymethylfurfural (HMF), which can hinder microbial fermentation (Iwaki et al. 2013). Moreover, acid pretreatment presents challenges such as equipment corrosion and the need for downstream acid neutralization (Jönsson and Martín 2016). Therefore, optimizing pretreatment conditions and integrating detoxification strategies are essential to improve the sustainability, cost-effectiveness, and industrial feasibility of this method for biofuel production.

11.2.2.2 Alkaline Pretreatment

Alkaline pretreatment is a widely adopted chemical method for enhancing the enzymatic digestibility of LCB, primarily through selectively removing lignin and increasing cellulose accessibility. Unlike acid pretreatment, which primarily targets hemicellulose hydrolysis, alkaline pretreatment focuses on lignin solubilization by cleaving ester bonds and ether linkages, particularly the cross-links between lignin and hemicellulose (Abolore et al. 2024). This structural disruption increases the porosity of the biomass and improves the accessibility of cellulose to hydrolytic enzymes, thereby enhancing the overall efficiency of saccharification and subsequent bioethanol production. The effectiveness of alkaline pretreatment is influenced by several key parameters, including the type and concentration of alkali, reaction temperature, residence time, and the intrinsic composition of the biomass (Amin et al. 2017).

Strong alkalis such as sodium hydroxide (NaOH) and potassium hydroxide (KOH) are commonly employed due to their high delignification efficiency. Among these, NaOH is particularly effective and widely studied. Typical NaOH pretreatment conditions involve concentrations ranging from 5% to 20% (w/v), temperatures between 50 and 130 °C, and reaction times from 1 to 24 h (Iberahim et al. 2013; Modenbach and Nokes 2014).

During NaOH pretreatment, saponification of ester linkages in lignin and hemicellulose occurs, leading to lignin solubilization and partial hemicellulose removal. Additionally, cellulose fibers undergo swelling, which increases their surface area and enhances enzyme accessibility. This method is especially suitable for lignin-rich feedstocks such as agricultural residues (e.g., corn stover, wheat straw) and hardwoods (Modenbach and Nokes 2014; Kululo et al. 2025).

Despite its effectiveness, alkaline pretreatment has some limitations. The relatively long reaction times required for significant delignification can reduce process throughput. Moreover, the use of strong alkalis necessitates downstream neutralization and recovery steps, which can increase operational complexity and cost.

Nonetheless, its ability to significantly improve enzymatic hydrolysis yields makes it a valuable option in the pretreatment toolbox, particularly when integrated with efficient chemical recovery systems (Broda et al. 2022).

11.2.2.3 Organosolv

Organosolv pretreatment is a promising and increasingly refined method for the fractionation of LCB, aimed at enhancing enzymatic hydrolysis and facilitating the production of bioethanol and other bio-based products. This method employs organic solvents—often in combination with acid or base catalysts—to selectively solubilize lignin, while preserving the structural integrity of cellulose and minimizing hemicellulose degradation. By disrupting the lignin–carbohydrate complex (LCC), organosolv pretreatment significantly improves cellulose accessibility for enzymatic hydrolysis and subsequent fermentation (Zhang et al. 2016). The core mechanism of organosolv pretreatment involves the cleavage of ether and ester linkages within the lignin polymer and between lignin and hemicellulose. This disruption enhances the porosity and surface area of the biomass, facilitating enzyme penetration and action (Nair et al. 2023). The process is typically conducted at temperatures ranging from 120 to 200 °C for durations of 30 min to several hours, under moderate pressure to maintain solvent stability and prevent volatilization (Agnihotri et al. 2015; Parchami et al. 2022). Various organic solvents are used either alone or in combination with catalysts such as acids or bases to enhance lignin removal efficiency. The effectiveness of this pretreatment depends on the solvent type, concentration, reaction temperature, and time (Editors et al. 2023).

Ethanol and methanol are among the most commonly used, often combined with dilute acids such as sulfuric acid (H_2SO_4) or hydrochloric acid (HCl) to enhance hemicellulose solubilization and lignin removal (Wildschut et al. 2013; Agnihotri et al. 2015; Parchami et al. 2022). Acetone and tetrahydrofuran have also been explored for lignin removal, with tetrahydrofuran offering high selectivity in disrupting lignin–carbohydrate linkages. The use of ethylene glycol and glycerol, often referred to as glycol-based pretreatment, has gained attention due to their lower volatility and reduced environmental impact (Zhang et al. 2016).

One of the major advantages of this method is its high efficiency in lignin removal, which improves enzymatic hydrolysis and fermentation yields. Additionally, the extracted high-purity lignin can be recovered and valorized into value-added products, such as adhesives, resins, carbon fibers, and aromatic platform chemicals, reducing process waste (Xu and Ferdosian 2017; Wang et al. 2019). Despite its advantages, organosolv pretreatment faces several challenges. The high cost of organic solvents, the energy demands of solvent recovery and recycling, and the potential toxicity of residual solvents to fermentative microbes are significant barriers to large-scale implementation (Broda et al. 2022). However, recent advances in process intensification—such as continuous-flow organosolv systems, hybrid solvent formulations, and solvent recovery technologies—are addressing these limitations. Moreover, techno-economic analyses and life cycle assessments increasingly support the viability of organosolv pretreatment when integrated with lignin valorization and solvent recycling strategies (Long et al. 2025). In conclusion, organosolv pretreatment represents a technically robust and versatile approach to biomass deconstruction. Its ability to selectively remove lignin, preserve fermentable sugars, and integrate with downstream valorization pathways makes it a strong candidate for future biorefinery platforms. Continued research into solvent systems, catalyst optimization, and process integration will be essential to fully realize its potential in sustainable biofuel and biochemical production.

11.2.2.4 Ionic Liquids

Ionic liquid (IL) pretreatment is a novel and efficient method for breaking down LCB to enhance enzymatic hydrolysis and bioethanol production. Ionic liquids are salts composed of organic cations and inorganic or organic anions that are liquid at room temperature. They possess unique physicochemical properties such as low volatility, high thermal stability, and excellent solvation ability, making them particularly effective in dissolving lignin and disrupting the crystalline structure of cellulose (Coimbra et al. 2025).

Imidazolium salts are the most widely used ILs for the pretreatment of LCB. These ILs effectively dissolve lignin and disrupt hydrogen bonding in cellulose, leading to significant improvements in biomass digestibility (Das et al. 2021). The process is typically conducted at moderate temperatures (80–160 °C) for a few hours, depending on the ionic liquid type and biomass composition (Alayoubi et al. 2020).

One of the major advantages of IL pretreatment is its ability to efficiently fractionate lignocellulose into its three major components—cellulose, hemicellulose, and lignin, without significant degradation of fermentable sugars. This selective fractionation facilitates downstream processing and allows the recovery of high-purity lignin for value-added applications (Hou et al. 2017). Additionally, ILs can be recycled and reused, reducing overall process costs. However, several challenges hinder large-scale implementation, including the high cost of ILs, the need for effective IL recovery systems, and potential toxicity to microbial fermentation organisms. Furthermore, some ILs require extensive washing steps to remove residual solvents, which can lead to water-intensive processing (Broda et al. 2022).

11.2.2.5 Deep Eutectic Solvents

Deep eutectic solvents (DESs) have gained significant attention as an eco-friendly and efficient method for the pretreatment of LCB in bioethanol production. DESs are a class of green solvents formed by mixing a hydrogen bond acceptor (HBA) and a hydrogen bond donor (HBD), resulting in a eutectic mixture with a lower melting point than its components (Das et al. 2021). This unique property allows DESs to effectively break down the lignocellulosic structure by solubilizing lignin and hemicellulose while preserving cellulose for enzymatic hydrolysis. Their biodegradable nature, low toxicity, and ease of preparation make them a promising substitute for conventional pretreatment methods that rely on harsh chemicals (Broda et al. 2022).

The composition of DESs plays a crucial role in their efficiency for lignocellulose pretreatment. Choline chloride (ChCl) is the most commonly used HBA, often combined with HBDs such as urea, glycerol, lactic acid, oxalic acid, or ethylene glycol (R Filho et al. 2023). Acidic DESs, typically formed with organic acids like lactic or oxalic acid, effectively break down hemicellulose and weaken the lignin–carbohydrate complex, improving biomass digestibility. Alkaline DESs, such as ChCl–urea, mainly target lignin solubilization, disrupting the aromatic structure and increasing enzyme accessibility to cellulose (Yu et al. 2022). Compared to ILs, DESs offer a more cost-effective and sustainable approach, as they are derived from natural and biodegradable components (Broda et al. 2022).

As a subcategory of deep eutectic solvents (DES), natural deep eutectic solvents (NADES) have emerged as a sustainable and biocompatible alternative for the pretreatment of lignocellulosic biomass. While conventional DESs are typically composed of synthetic hydrogen bond donors and acceptors, NADES are formulated from naturally occurring, biodegradable components such as sugars, organic acids, amino acids, and choline derivatives. This composition not only enhances their environmental compatibility but also reduces toxicity, making NADES particularly suitable for integrated biorefinery applications (Meraj et al. 2024). Moreover, NADES exhibit favorable properties such as low volatility, high thermal stability, and recyclability. Their compatibility with enzymatic hydrolysis and microbial fermentation further supports their use in downstream processing without extensive detoxification. However, challenges such as high viscosity, limited solubility for certain biomass types, and the need for efficient recovery systems remain areas of active research (Usmani et al. 2023). Overall, NADES represent a promising green solvent system for lignocellulosic biomass pretreatment. Their natural origin, tunable properties, and compatibility with biological systems position them as a key component in the development of environmentally sustainable and economically viable biorefinery platforms.

11.2.3 Physicochemical Methods

11.2.3.1 Steam Explosion

Steam explosion (SE) is a widely accepted physicochemical pretreatment method used to break down the recalcitrant structure of LCB. Recognized for its environmental compatibility and cost-effectiveness, SE offers a viable alternative to conventional chemical pretreatments, as it operates without the need for added chemicals or catalysts. In the SE process, the biomass is subjected to high-pressure saturated steam (0.7–4.8 MPa) at temperatures ranging from 160 to 260 °C for a short residence time, typically from a few seconds to several minutes. This is followed by a rapid depressurization, causing the steam within the biomass structure to expand explosively. The resulting mechanical disruption breaks down the cell wall matrix, facilitating fiber separation and partial solubilization of hemicellulose and lignin components. This method is particularly advantageous for processing large-sized biomass, often eliminating the need for energy-intensive size reduction steps (Broda et al. 2022).

One of the key advantages of SE is its low energy demand compared to mechanical or chemical pretreatment methods. Moreover, it minimizes environmental impact by avoiding the generation of toxic chemical effluents and reducing the need for chemical recovery systems. SE also significantly enhances the enzymatic digestibility of cellulose by increasing substrate accessibility, thereby improving sugar yields and boosting the efficiency of downstream bioethanol production (Editors et al. 2023).

Despite these advantages, SE has certain limitations. One major drawback is the incomplete removal of lignin, which can impede enzymatic hydrolysis. Additionally, the process may generate fermentation inhibitors such as acetic acid, furfural, and hydroxymethylfurfural (HMF), necessitating detoxification steps that can increase overall processing costs and complexity(Abolore et al. 2024).

11.2.3.2 Ammonia Fiber Expansion

Ammonia fiber expansion (AFEX) is an efficient physicochemical pretreatment method that uses liquid ammonia to disrupt the complex structure of LCB. The process involves treating biomass with liquid or gaseous ammonia under moderate temperature conditions (90–100 °C) and high pressure for a short duration (30–60 min), followed by a rapid pressure release. This sudden expansion causes fiber swelling and an increase in the surface area, providing increased accessibility to cellulose for enzymatic hydrolysis (Editors et al. 2023).

AFEX facilitates selective delignification, partial hemicellulose depolymerization, and cellulose decrystallization, all of which enhance glucose yields in subsequent enzymatic saccharification (Das et al. 2021). Unlike acid-based pretreatment methods, AFEX does not require extensive washing steps before enzymatic hydrolysis (Broda et al. 2022). Additionally, ammonia is noncorrosive, nonpolluting, and easy to recover and reuse, reducing process costs and environmental impact (Das et al. 2021). The residual ammonia in the biomass serves as a nitrogen source for microbes during bioethanol fermentation, further improving process efficiency (Mohammadi et al. 2024).

Despite its advantages, AFEX is less effective for biomass with high lignin content, such as woody feedstocks, as it does not achieve complete lignin removal. The method is more suitable for agricultural residues and herbaceous biomass. Other ammonia-based pretreatment techniques, such as soaking in aqueous ammonia (SAA) and ammonia recycle percolation (ARP), have been developed to improve delignification efficiency. ARP, in particular, has demonstrated delignification rates of up to 85% when biomass is treated at 170 °C with circulating ammonia. Similarly, SAA, which operates at lower temperatures (40–90 °C), helps preserve glucan and xylan content while modifying the lignin–carbohydrate complex (Das et al. 2021).

11.2.3.3 Liquid Hot Water Pretreatment

Liquid hot water (LHW) pretreatment is a hydrothermal process that employs pressurized hot water to modify the structure of LCB. Operating at temperatures between 150 and 230 °C and pressures up to 5 MPa, this method ensures water remains in a liquid state, preventing excessive biomass degradation while enhancing hemicellulose solubilization and cellulose accessibility. Unlike SE, LHW does not involve rapid decompression, making it a gentler yet effective approach for biomass pretreatment (Li et al. 2017; Paul et al. 2023).

LHW pretreatment can be carried out using different approaches depending on how water interacts with biomass. In the cocurrent method, biomass slurry and water are heated together to a specific temperature and maintained for an optimized residence time before cooling. In the counter-current method, hemicellulose hydrolysis and sugar recovery are enhanced by allowing biomass and liquid hot water to flow in opposite directions (Chen et al. 2022).

During the LHW pretreatment process, water ionizes under high temperatures, forming hydronium ions that hydrolyze hemicellulose and facilitate the release of acetic acid from biomass. This reaction improves pentose sugar yield in the liquid phase, increases lignin removal in the solid phase, and enhances cellulose digestibility in subsequent enzymatic hydrolysis. One of the primary advantages of LHW is the minimal formation of inhibitory compounds, reducing the need for detoxification before fermentation. The method is also considered environmentally friendly and cost-effective since it does not require harsh catalysts or chemicals (Imman et al. 2015).

11.2.3.4 Wet Air Oxidation

Wet air oxidation (WAO) utilizes oxygen or air at high temperatures and pressures to break down LCB. Typically performed at temperatures above 120 °C and pressures of 0.5–2 MPa, this process is effective for biomass with high lignin content (Broda et al. 2022). The presence of oxygen as the primary chemical agent, along with the formation of hydrogen ions, facilitates oxidation and hydrolytic reactions, which leads to biomass degradation (Devi et al. 2022). The reaction is autocatalytic due to the formation of organic acids, such as acetic acid, which further enhance the breakdown of hemicellulose and lignin. Additionally, the process becomes exothermic at temperatures above 170 °C, allowing for self-sustained heat generation (Editors et al. 2023).

This method primarily affects hemicellulose and lignin, but the cellulose fraction remains largely intact, ensuring a higher yield of fermentable sugars after enzymatic hydrolysis. However, its industrial feasibility depends on factors such as oxygen supply requirements and process energy demands (Devi et al. 2022). Despite these challenges, its effectiveness in processing lignin-rich biomass makes it a valuable technique for biofuel production and other biorefinery applications.

11.2.3.5 CO_2 Explosion

CO_2 explosion is a biomass pretreatment method that shares similarities with steam explosion. In this method, CO_2 is introduced into a high-pressure chamber containing biomass, accompanied by agitation at an elevated temperature of around 200 °C (Senila et al. 2022). The CO_2 dissolves in the water present in the biomass, forming carbonic acid, which facilitates hemicellulose hydrolysis. As the pressure is rapidly released, the structural integrity of the biomass is disrupted, increasing the accessibility of cellulose for subsequent enzymatic hydrolysis. This technique is considered environmentally friendly due to CO_2's nontoxic and recyclable nature (Editors et al. 2023).

The CO_2 required for this process can be sourced from ethanol fermentation, allowing for continuous recycling without contributing to atmospheric CO_2 emissions. This method enhances glucose yield, facilitates delignification, and enables selective extraction of biomass components while preserving sugar stability due to lower operating temperatures compared to other pretreatment methods (Broda et al. 2022). This method is particularly advantageous due to its low critical temperature (31.1 °C) and pressure (73.8 bar), making it an energy-efficient option. Unlike ammonia-based pretreatments, this method generates fewer toxic by-products and is more cost-effective (Editors et al. 2023). However, the requirement for high-pressure reactors remains a significant challenge, limiting their scalability and industrial application.

11.2.4 Biological Methods

Biological pretreatment employs ligninolytic microorganisms, primarily fungi and bacteria, to reduce the recalcitrance of lignocellulosic biomass (LCB), thereby enhancing its digestibility for subsequent enzymatic hydrolysis and ethanol production (Sharma et al. 2022). The efficiency of biological pretreatment is influenced by several factors, including the choice of microorganisms, biomass particle size, and key process parameters, such as pH, temperature, and incubation duration (Mishra et al. 2018).

Among microbial agents, white-rot fungi are considered most efficient due to their high lignin degradation capabilities. These fungi produce a suite of ligninolytic enzymes, primarily lignin peroxidase (LiP), manganese peroxidase (MnP), and laccase, which work synergistically to break down the complex lignin polymer (Kumar and Chandra 2020). Commonly used species include *Phanerochaete chrysosporium*, known for its high LiP and MnP activity; *Ceriporiopsis subvermispora*, which selectively degrades lignin with minimal cellulose loss; *Trametes versicolor*, a prolific producer of laccase; and *Pleurotus ostreatus*, which secretes all three major ligninolytic enzymes(Wan and Li 2012; Kucharska et al. 2018). In addition, bacterial strains such as *Clostridium* spp., *Cellulomonas* spp., *Bacillus* spp., *Thermomonospora* spp., and *Streptomyces* spp. also contribute significantly to lignin and hemicellulose degradation during biological pretreatment (Abolore et al. 2024).

Biological pretreatment can be carried out either directly, by cultivating microorganisms on LCB, or indirectly, through enzyme extracts derived from these organisms. The selection of an appropriate microbial consortium or enzyme mix depends on the specific composition of the LCB. Optimizing incubation conditions, particularly temperature and pH, is critical for maximizing lignin and hemicellulose removal efficiency(Wu et al. 2022).

One major advantage of this method is its low energy requirement, as it does not involve harsh chemicals or expensive recycling processes. It also generates fewer inhibitory by-products, thereby reducing the need for detoxification and lowering downstream processing costs. Furthermore, the availability of a wide range of naturally occurring microbial species allows for flexible adaptation to different biomass types (Wu et al. 2022).

Despite its environmental benefits, the large-scale application of biological pretreatment faces several challenges. The process is relatively slow compared to other pretreatment techniques, requiring extended incubation periods to achieve significant delignification (Wu et al. 2022). Maintaining a sterile environment to prevent contamination can be costly and technically demanding. Additionally, some microorganisms may metabolize fermentable sugars during treatment, leading to reduced overall yields and increased processing costs (Zuliani et al. 2021).

11.3 Selection Criteria: A Perspective on Tailored Pretreatment Strategies

As the global push toward sustainable energy and a circular bioeconomy intensifies, the efficient utilization of LCB has become a cornerstone of next-generation biorefineries. As highlighted in the discussion, pretreatment remains a critical bottleneck in this process, as it dictates the accessibility of cellulose and hemicellulose for enzymatic hydrolysis and influences the overall yield, cost, and environmental footprint of biomass conversion. In this context, the selection of an appropriate pretreatment method must be both strategic and adaptive, i.e., guided by feedstock characteristics, process integration, and sustainability goals (Table 11.1). Feedstock-specific tailoring is increasingly recognized as essential.

For agricultural and herbaceous residues, which are typically lower in lignin and structurally less recalcitrant, milder methods such as Ammonia Fiber Expansion

Table 11.1 Different physical, chemical, and biological methods of LCB pretreatment

Pretreatment method	Suitable biomass	Mechanism	Key effect on biomass	Advantages	Disadvantages	References
Mechanical	All types	Milling, grinding	Reduces particle size, increases surface area	Simple and scalable, does not require chemicals, enhances enzymatic accessibility	High energy consumption, does not significantly alter lignin structure	Taherzadeh and Karimi (2008)
Extrusion	Agricultural residues	Shear and thermal forces	Disrupt LCB structure, increases porosity	Continuous process, effective for various biomass types	High energy input, limited delignification	Duque et al. (2017)
Microwave irradiation	Agricultural residues	Dielectric heating	Cell wall disruption, increased porosity	Rapid heating, enhances enzymatic hydrolysis	Requires integration with other methods, potential equipment cost	Shukla et al. (2023)
Ultrasonication	Agricultural residues	Acoustic cavitation	Cell wall disruption, increased porosity	Reduces particle size, enhances enzymatic digestibility	Limited delignification, requires high energy, potential equipment cost	Shukla et al. (2023)
Pulsed electric field (PEF)	Herbaceous biomass	Electroporation	Cell permeabilization	Nonthermal, preserves carbohydrate integrity	Not effective alone; needs combination with other treatments	Kovačić et al. (2021)
Acid pretreatment	Agricultural residues	Hydrolysis of hemicellulose and partial lignin removal using dilute/strong acids (e.g., H_2SO_4, HCl)	Breaks down hemicellulose to sugars	Effective hemicellulose removal, increases cellulose accessibility	Corrosive, formation of inhibitory by-products	Brodeur et al. (2011)
Alkali pretreatment	Herbaceous woody biomass	Saponification using alkalis (e.g., NaOH, KOH, ammonia)	Delignification, hemicellulose solubilization	Efficient lignin removal, improves cellulose digestibility	Long reaction time, requires a washing step	Brodeur et al. (2011)

(continued)

Table 11.1 (continued)

Pretreatment method	Suitable biomass	Mechanism	Key effect on biomass	Advantages	Disadvantages	References
Organoslv	Woody biomass	Lignin dissolution using organic solvents (e.g., ethanol, acetone)	Lignin removal, hemicellulose solubilization	High delignification efficiency, high quality lignin recovery, Fewer fermentation inhibitors, improved cellulose accessibility and sugar yield	Expensive solvents; solvent recovery required, residual solvent if not removed may inhibit Enzymes in saccharification step	Zhang et al. (2016)
Ionic liquids (ILs)	Mixed biomass	Disrupt hydrogen bonding	Dissolves cellulose and lignin	High selectivity, mild conditions	Expensive, toxicity concerns	Hou et al. (2017) and Broda et al. (2022)
Deep eutectic solvents (DES)	Hardwood, softwood	Lignin solubilization via hydrogen bond donor-acceptor interaction	Selective lignin solubilization	Low toxicity; biodegradable	Slow reaction rate; some DES can be viscous	Broda et al. (2022)
Steam explosion	Woody biomass, agri-waste	High-pressure steam + explosive reach	Fiber separation, partial hemicellulose/lignin solubilization	Low energy, chemical-free, scalable	Incomplete lignin removal; production of inhibitors	Broda et al. (2022) and Abolore et al. (2024)
Ammonia fiber explosion (AFEX)	Herbaceous residues	Uses ammonia under high pressure to swell and disrupt biomass	Delignification, cellulose decrystallization	Recyclable ammonia, no washing, improved saccharification	Not suitable for high-lignin feedstock	Brodeur et al. (2011)
Liquid hot water (LHW)	Agricultural and softwood biomass	Uses hot pressurized water	Hemicellulose hydrolysis, lignin redistribution	No chemical residues, minimal inhibitors, environmentally friendly	Needs a high-pressure setup	Imman et al. (2015), Li et al. (2017), and Paul et al. (2023)

Wet air oxidation	Woody and municipal biomass	Oxidation with air/O_2 at high T/P	Hemicellulose & lignin breakdown	Suitable for lignin-rich biomass, cellulose preserved	Energy-intensive, oxygen supply needed	Devi et al. (2022) and Broda et al. (2022)
CO_2 explosion	Mixed feedstocks	Carbonic acid formation + rapid depressurization	Hemicellulose hydrolysis, structure disruption	Nontoxic, mild conditions, recyclable CO_2, fewer inhibitors	High pressure is needed, limited delignification	Shukla et al. (2023)
Biological methods	All types (preprocessed preferred)	Microorganisms (fungi, bacteria) degrade lignin enzymaticaly	Selective lignin and hemicellulose removal	Environmentally friendly, low energy input, selective lignin and hemicellulose removal	Slow process, requires specific microbial strains, sterile conditions required	Wu et al. (2022)

(AFEX), Liquid Hot Water (LHW), and biological pretreatment offer effective hemicellulose solubilization with minimal inhibitor formation (Long et al. 2025). These methods are particularly attractive for decentralized biorefineries in agricultural regions, where low-cost and low-impact processing is prioritized.

Conversely, woody biomass, with its dense structure and high lignin content, demands more intensive pretreatment. Techniques such as alkaline pretreatment, steam explosion (SE), wet air oxidation, and organosolv pretreatment have demonstrated superior delignification and structural disruption capabilities (Mersmann et al. 2025). Among these, organosolv stands out for its ability to fractionate biomass into high-purity cellulose, hemicellulose, and lignin streams, facilitating integrated valorization and supporting the economic viability of lignin-derived products (Agnihotri et al. 2015; Parchami et al. 2022; Soldatos et al. 2024).

Inhibitor formation remains a key challenge, particularly for acid- and heat-intensive methods. To mitigate this, pretreatments like LHW, CO_2 explosion, and biological approaches are gaining traction for their ability to preserve sugar integrity and reduce the generation of fermentation inhibitors such as furfural and HMF (Larsson et al. 1999; Zhuang et al. 2016; Sun et al. 2022). Organosolv, when optimized for solvent recovery and reaction control, also offers a balance between efficiency and inhibitor suppression (Nair et al. 2023).

When the goal is maximizing sugar recovery, acid pretreatment and steam explosion remain among the most effective, particularly for hemicellulose hydrolysis and cellulose accessibility. However, these methods often require detoxification steps to ensure compatibility with microbial fermentation (Long et al. 2025). Organosolv again emerges as a versatile option, offering high sugar yields with the added benefit of lignin recovery for value-added applications.

From a sustainability and green chemistry perspective, the future of pretreatment lies in methods that minimize environmental impact while maximizing resource efficiency. Biological pretreatment, NADES, LHW, CO_2 explosion, AFEX, and organosolv (with solvent recycling) align well with these principles. These methods support the transition toward low-carbon, fossil-free biomanufacturing systems and are increasingly favored in techno-economic and life cycle assessments (Ferreira et al. 2022; Mohammadi et al. 2024; Long et al. 2025).

In summary, the pretreatment landscape is evolving from a one-size-fits-all paradigm to a feedstock- and product-driven strategy. The integration of pretreatment with downstream processing, coupled with advances in solvent recovery, inhibitor mitigation, and lignin valorization, will be pivotal in realizing the full potential of lignocellulosic biorefineries. As highlighted in recent reviews, the convergence of systems biology, process engineering, and green chemistry is shaping a new generation of pretreatment technologies that are not only efficient but also economically and environmentally sustainable (Narron et al. 2016; Dharmaraja et al. 2023).

11.4 Conclusion

Pretreatment of LCB is a critical and indispensable step toward unlocking its full potential as a sustainable feedstock for biofuels, biochemicals, and biomaterials. Overcoming the inherent recalcitrance caused by the complex lignin–carbohydrate matrix is essential to improve enzymatic hydrolysis and downstream processing efficiencies. While numerous pretreatment methods exist, ranging from physical and chemical to biological and novel solvent-based approaches, each carries unique advantages and challenges related to cost, environmental impact, scalability, and feedstock compatibility. The ongoing development of tailored and integrated pretreatment strategies that minimize inhibitor formation and maximize biomass fractionation is key to advancing the economic viability and sustainability of biorefineries.

Moreover, the successful implementation of pretreatment technologies must consider the variability of biomass composition influenced by species, growth conditions, and harvest time. This necessitates flexible and adaptable pretreatment solutions capable of handling diverse feedstocks without compromising process efficiency. Integrating pretreatment with downstream valorization pathways, such as lignin recovery and coproduct generation, will further enhance the overall sustainability and profitability of biorefineries. Continued innovation, supported by interdisciplinary research and pilot-scale demonstrations, is vital to translate promising pretreatment methods into industrial practice and to achieve a truly circular and carbon-neutral bioeconomy.

Acknowledgments The authors thank the Natural Sciences and Engineering Research Council (NSERC) of Canada Collaborative Research and Training Experience Program (CREATE, CREAT/543238–2020) Training in Applied Biotechnology for Environmental Sustainability (TABES), Discovery Grants (RGPIN-2020–06067; RGPIN-2020–06140), and Alliance Mission (ALLRP 570473–2021), as well as the James and Joanne Love Chair in Environmental Engineering for the financial support of this work.

References

Abolore RS, Jaiswal S, Jaiswal AK (2024) Green and sustainable pretreatment methods for cellulose extraction from lignocellulosic biomass and its applications: a review. Carbohydr Polym Technol Appl 7:100396. https://doi.org/10.1016/J.CARPTA.2023.100396

Agnihotri S, Johnsen IA, Bøe MS et al (2015) Ethanol organosolv pretreatment of softwood (Picea abies) and sugarcane bagasse for biofuel and biorefinery applications. Wood Sci Technol 49: 881–896. https://doi.org/10.1007/S00226-015-0738-4/FIGURES/9

Alayoubi R, Mehmood N, Husson E et al (2020) Low temperature ionic liquid pretreatment of lignocellulosic biomass to enhance bioethanol yield. Renew Energy 145:1808–1816. https://doi.org/10.1016/J.RENENE.2019.07.091

Amin FR, Khalid H, Zhang H et al (2017) Pretreatment methods of lignocellulosic biomass for anaerobic digestion. AMB Express 7:72. https://doi.org/10.1186/S13568-017-0375-4

Baksi S, Saha D, Saha S et al (2023) Pre-treatment of lignocellulosic biomass: review of various physico-chemical and biological methods influencing the extent of biomass depolymerization. Int J Environ Sci Technol 20:13895–13922. https://doi.org/10.1007/S13762-023-04838-4

Bocker R, Silva EK (2024) Pulsed electric field technology as a promising pre-treatment for enhancing orange agro-industrial waste biorefinery. RSC Adv 14:2116. https://doi.org/10.1039/D3RA07848E

Broda M, Yelle DJ, Serwańska K (2022) Bioethanol production from lignocellulosic biomass—challenges and solutions. Molecules 27:8717. https://doi.org/10.3390/MOLECULES27248717

Brodeur G, Yau E, Badal K et al (2011) Chemical and physicochemical pretreatment of lignocellulosic biomass: a review. Enzyme Res 2011:787532. https://doi.org/10.4061/2011/787532

Chaudhary G, Chaudhary N, Saini S et al (2023) Assessment of pretreatment strategies for valorization of lignocellulosic biomass: path forwarding towards lignocellulosic biorefinery. Waste Biomass Valoriz 15:1–36. https://doi.org/10.1007/S12649-023-02219-Z

Chen WH, Nižetić S, Sirohi R et al (2022) Liquid hot water as sustainable biomass pretreatment technique for bioenergy production: a review. Bioresour Technol 344:126207. https://doi.org/10.1016/J.BIORTECH.2021.126207

Chen J, Ma X, Liang M et al (2024) Physical–chemical–biological pretreatment for biomass degradation and industrial applications: a review. Waste 2:451–473. https://doi.org/10.3390/WASTE2040024

Coimbra P, Dias AMA, de Sousa HC (2025) Encapsulated ionic liquids: a comprehensive review of production methods and potential applications. Chem Eng J 504:159039. https://doi.org/10.1016/J.CEJ.2024.159039

Das N, Jena PK, Padhi D et al (2021) A comprehensive review of characterization, pretreatment and its applications on different lignocellulosic biomass for bioethanol production. Biomass Convers Biorefin 13(2):1503–1527. https://doi.org/10.1007/S13399-021-01294-3

Devi A, Bajar S, Kour H et al (2022) Lignocellulosic biomass valorization for bioethanol production: a circular bioeconomy approach. BioEnergy Res 15(4):1820–1841. https://doi.org/10.1007/S12155-022-10401-9

Dharmaraja J, Shobana S, Arvindnarayan S et al (2023) Lignocellulosic biomass conversion via greener pretreatment methods towards biorefinery applications. Bioresour Technol 369:128328. https://doi.org/10.1016/J.BIORTECH.2022.128328

dos Santos AC, Ximenes E, Kim Y, Ladisch MR (2019) Lignin–enzyme interactions in the hydrolysis of lignocellulosic biomass. Trends Biotechnol 37:518–531. https://doi.org/10.1016/J.TIBTECH.2018.10.010

Duque A, Manzanares P, Ballesteros M (2017) Extrusion as a pretreatment for lignocellulosic biomass: fundamentals and applications. Renew Energy 114:1427–1441. https://doi.org/10.1016/J.RENENE.2017.06.050

Editors A, Kumar Patel A, Carlos Magalhães Pires J et al (2023) The production of bioethanol from lignocellulosic biomass: pretreatment methods, fermentation, and downstream processing. Energies 16:7003. https://doi.org/10.3390/EN16197003

Ferreira SS, Brito TA, Santana APR et al (2022) Greenness of procedures using NADES in the preparation of vegetal samples: comparison of five green metrics. Talanta Open 6. https://doi.org/10.1016/J.TALO.2022.100131

Haq IU, Qaisar K, Nawaz A et al (2021) Advances in valorization of lignocellulosic biomass towards energy generation. Catalysts 11:309. https://doi.org/10.3390/CATAL11030309

Hou Q, Ju M, Li W et al (2017) Pretreatment of lignocellulosic biomass with ionic liquids and ionic liquid-based solvent systems. Molecules 22:490. https://doi.org/10.3390/MOLECULES22030490

Iberahim NI, Jahim JM, Harun S et al (2013) Sodium hydroxide pretreatment and enzymatic hydrolysis of oil palm mesocarp fiber. Int J Chem Eng Appl:101–105. https://doi.org/10.7763/IJCEA.2013.V4.272

Imman S, Arnthong J, Burapatana V et al (2015) Influence of alkaline catalyst addition on compressed liquid hot water pretreatment of rice straw. Chem Eng J 278:85–91. https://doi.org/10.1016/J.CEJ.2014.12.032

Iwaki A, Kawai T, Yamamoto Y, Izawa S (2013) Biomass conversion inhibitors furfural and 5-hydroxymethylfurfural induce formation of messenger RNP granules and attenuate translation activity in Saccharomyces cerevisiae. Appl Environ Microbiol 79:1661. https://doi.org/10.1128/AEM.02797-12

Jönsson LJ, Martín C (2016) Pretreatment of lignocellulose: formation of inhibitory by-products and strategies for minimizing their effects. Bioresour Technol 199:103–112. https://doi.org/10.1016/J.BIORTECH.2015.10.009

Kovačić Đ, Rupčić S, Kralik D et al (2021) Pulsed electric field: an emerging pretreatment technology in a biogas production. Waste Manag 120:467–483. https://doi.org/10.1016/J.WASMAN.2020.10.009

Kucharska K, Rybarczyk P, Hołowacz I et al (2018) Pretreatment of lignocellulosic materials as substrates for fermentation processes. Molecules 23. https://doi.org/10.3390/MOLECULES23112937

Kululo WW, Habtu NG, Abera MK et al (2025) Advances in various pretreatment strategies of lignocellulosic substrates for the production of bioethanol: a comprehensive review. Discover Appl Sci 7. https://doi.org/10.1007/S42452-025-06748-1

Kumar A, Chandra R (2020) Ligninolytic enzymes and its mechanisms for degradation of lignocellulosic waste in environment. Heliyon 6:e03170. https://doi.org/10.1016/J.HELIYON.2020.E03170

Kumar P, Barrett DM, Delwiche MJ, Stroeve P (2011) Pulsed electric field pretreatment of switchgrass and wood chip species for biofuel production. Ind Eng Chem Res 50:10996–11001. https://doi.org/10.1021/IE200555U

Larsson S, Palmqvist E, Hahn-Hägerdal B et al (1999) The generation of fermentation inhibitors during dilute acid hydrolysis of softwood. Enzyme Microb Technol 24:151–159. https://doi.org/10.1016/S0141-0229(98)00101-X

Li M, Cao S, Meng X et al (2017) The effect of liquid hot water pretreatment on the chemical-structural alteration and the reduced recalcitrance in poplar. Biotechnol Biofuels 10:1–13. https://doi.org/10.1186/S13068-017-0926-6/SCHEMES/1

Long B, Zhang F, Dai SY et al (2025) Engineering strategies to optimize lignocellulosic biorefineries. Nat Rev Bioeng 3:230–244. https://doi.org/10.1038/S44222-024-00247-5

Martín C, Dixit P, Momayez F, Jönsson LJ (2022) Hydrothermal pretreatment of lignocellulosic feedstocks to facilitate biochemical conversion. Front Bioeng Biotechnol 10:846592. https://doi.org/10.3389/FBIOE.2022.846592/BIBTEX

Meraj A, Jawaid M, Singh SP et al (2024) Isolation and characterisation of lignin using natural deep eutectic solvents pretreated kenaf fibre biomass. Sci Rep 14:1–16. https://doi.org/10.1038/s41598-024-59200-6

Mersmann L, Dias C, Pires J, Fernando AL (2025) The green shift in lignin recovery: sustainability and impact of pretreatment methods, pp 192–209. https://doi.org/10.1007/978-3-031-94484-0_16

Mishra S, Singh PK, Dash S, Pattnaik R (2018) Microbial pretreatment of lignocellulosic biomass for enhanced biomethanation and waste management. 3 Biotech 8:458. https://doi.org/10.1007/S13205-018-1480-Z

Modenbach AA, Nokes SE (2014) Effects of sodium hydroxide pretreatment on structural components of biomass. Trans ASABE 57:1187–1198. https://doi.org/10.13031/TRANS.56.10046

Mohammadi M, Alian M, Dale B et al (2024) Multifaced application of AFEX-pretreated biomass in producing second-generation biofuels, ruminant animal feed, and value-added bioproducts. Biotechnol Adv 72:108341. https://doi.org/10.1016/J.BIOTECHADV.2024.108341

Mukherjee P, Pal S, Sivaprakasam S (2024) Process parameter controls for efficient enzymatic hydrolysis of cellulosic biomass. In: Handbook of biorefinery research and technology, pp 1–29. https://doi.org/10.1007/978-94-007-6724-9_77-1

Nahak BK, Preetam S, Sharma D et al (2022) Advancements in net-zero pertinency of lignocellulosic biomass for climate neutral energy production. Renew Sustain Energy Rev 161:112393. https://doi.org/10.1016/J.RSER.2022.112393
Nair LG, Agrawal K, Verma P (2023) Organosolv pretreatment: an in-depth purview of mechanics of the system. Bioresour Bioprocess 10(1):1–29. https://doi.org/10.1186/S40643-023-00673-0
Narron RH, Kim H, Chang HM et al (2016) Biomass pretreatments capable of enabling lignin valorization in a biorefinery process. Curr Opin Biotechnol 38:39–46. https://doi.org/10.1016/J.COPBIO.2015.12.018
Oriez V, Peydecastaing J, Pontalier PY (2019) Lignocellulosic biomass fractionation by mineral acids and resulting extract purification processes: conditions, yields, and purities. Molecules 24:4273. https://doi.org/10.3390/MOLECULES24234273
Parchami M, Agnihotri S, Taherzadeh MJ (2022) Aqueous ethanol organosolv process for the valorization of Brewer's spent grain (BSG). Bioresour Technol 362:127764. https://doi.org/10.1016/J.BIORTECH.2022.127764
Paul M, Pandey NK, Banerjee A et al (2023) An insight into omics analysis and metabolic pathway engineering of lignin-degrading enzymes for enhanced lignin valorization. Bioresour Technol 379:129045. https://doi.org/10.1016/J.BIORTECH.2023.129045
Qin S, Liu K, Ren D et al (2023) All-natural and high-performance structural material based on lignin-reinforced cellulose. Mater Today Commun 36:106559. https://doi.org/10.1016/J.MTCOMM.2023.106559
R Filho GN, Emmerson Ferreira da Costa C, Wawoczny A et al (2023) Enzymatic valorization of lignocellulosic biomass—the influence of deep eutectic solvents and ionic liquids on the activity of cellulolytic enzymes. Sustainability 15(24):16726. https://doi.org/10.3390/SU152416726
Senila L, Scurtu DA, Kovacs E et al (2022) High-pressure supercritical CO2 pretreatment of apple orchard waste for carbohydrates production using response surface methodology and method uncertainty evaluation. Molecules 27:7783. https://doi.org/10.3390/MOLECULES27227783
Sharma S, Tsai ML, Sharma V et al (2022) Environment friendly pretreatment approaches for the bioconversion of lignocellulosic biomass into biofuels and value-added products. Environments 10:6. https://doi.org/10.3390/ENVIRONMENTS10010006
Sharma T, Chauhan PS, Patel M et al (2025) Carbon negative biofuels: a step ahead of carbon neutrality. Biofuels. https://doi.org/10.1080/17597269.2025.2450156;JOURNAL:JOURNAL:TBFU20;WGROUP:STRING:PUBLICATION
Shukla A, Kumar D, Girdhar M et al (2023) Strategies of pretreatment of feedstocks for optimized bioethanol production: distinct and integrated approaches. Biotechnol Biofuels Bioprod 16(1):1–33. https://doi.org/10.1186/S13068-023-02295-2
Soldatos P, Margellou A, Pappa C et al (2024) Conversion of beechwood organosolv lignin via fast pyrolysis and in situ catalytic upgrading towards aromatic and phenolic-rich bio-oil. Sustain Chem Environ 6:100107. https://doi.org/10.1016/J.SCENV.2024.100107
Sun J, Ding R, Yin J (2022) Pretreatment corn ingredient biomass with high pressure CO2 for conversion to fermentable sugars via enzymatic hydrolysis of cellulose. Ind Crop Prod 177. https://doi.org/10.1016/J.INDCROP.2022.114518
Suota MJ, da Silva TA, Zawadzki SF et al (2021) Chemical and structural characterization of hardwood and softwood LignoForce™ lignins. Ind Crops Prod 173:114138. https://doi.org/10.1016/J.INDCROP.2021.114138
Taherzadeh MJ, Karimi K (2008) Pretreatment of lignocellulosic wastes to improve ethanol and biogas production: a review. Int J Mol Sci 9:1621–1651. https://doi.org/10.3390/IJMS9091621
Usmani Z, Sharma M, Tripathi M et al (2023) Biobased natural deep eutectic system as versatile solvents: structure, interaction and advanced applications. Sci Total Environ 881:163002. https://doi.org/10.1016/J.SCITOTENV.2023.163002
Velvizhi G, Goswami C, Shetti NP et al (2022) Valorisation of lignocellulosic biomass to value-added products: paving the pathway towards low-carbon footprint. Fuel 313:122678. https://doi.org/10.1016/J.FUEL.2021.122678

Verdía Barbará P, Choudhary H, Nakasu PS et al (2025) Recent advances in the use of ionic liquids and deep eutectic solvents for lignocellulosic biorefineries and biobased chemical and material production. Chem Rev. https://doi.org/10.1021/ACS.CHEMREV.4C00754/ASSET/IMAGES/LARGE/CR4C00754_0035.JPEG

Wan C, Li Y (2012) Fungal pretreatment of lignocellulosic biomass. Biotechnol Adv 30:1447–1457. https://doi.org/10.1016/J.BIOTECHADV.2012.03.003

Wang H, Pu Y, Ragauskas A, Yang B (2019) From lignin to valuable products–strategies, challenges, and prospects. Bioresour Technol 271:449–461. https://doi.org/10.1016/J.BIORTECH.2018.09.072

Wildschut J, Smit AT, Reith JH, Huijgen WJJ (2013) Ethanol-based organosolv fractionation of wheat straw for the production of lignin and enzymatically digestible cellulose. Bioresour Technol 135:58–66. https://doi.org/10.1016/J.BIORTECH.2012.10.050

Woźniak A, Kuligowski K, Świerczek L, Cenian A (2025) Review of lignocellulosic biomass pretreatment using physical, thermal and chemical methods for higher yields in bioethanol production. Sustainability 17:287. https://doi.org/10.3390/SU17010287

Wu Z, Peng K, Zhang Y et al (2022) Lignocellulose dissociation with biological pretreatment towards the biochemical platform: a review. Mater Today Bio 16:100445. https://doi.org/10.1016/J.MTBIO.2022.100445

Xu C, Ferdosian F (2017) Conversion of lignin into bio-based chemicals and materials. https://doi.org/10.1007/978-3-662-54959-9

Yu H, Xue Z, Shi R et al (2022) Lignin dissolution and lignocellulose pretreatment by carboxylic acid based deep eutectic solvents. Ind Crops Prod 184. https://doi.org/10.1016/J.INDCROP.2022.115049

Yuan Y, Jiang B, Chen H et al (2021) Recent advances in understanding the effects of lignin structural characteristics on enzymatic hydrolysis. Biotechnol Biofuels 14(1):1–20. https://doi.org/10.1186/S13068-021-02054-1

Zhang K, Pei Z, Wang D (2016) Organic solvent pretreatment of lignocellulosic biomass for biofuels and biochemicals: a review. Bioresour Technol 199:21–33. https://doi.org/10.1016/J.BIORTECH.2015.08.102

Zhuang X, Wang W, Yu Q et al (2016) Liquid hot water pretreatment of lignocellulosic biomass for bioethanol production accompanying with high valuable products. Bioresour Technol 199:68–75. https://doi.org/10.1016/J.BIORTECH.2015.08.051

Zuliani L, Serpico A, De Simone M et al (2021) Biorefinery gets hot: thermophilic enzymes and microorganisms for second-generation bioethanol production. Processes 9:1583. https://doi.org/10.3390/PR9091583

Chapter 12
First-Generation Biofuels

Susan Grace Karp, Mariana Zanlorenzi Weber, Rafael Novaes Barros, Maísa Emanuelle Colodel, Araceli Shissel Hong Cabrera, Felipe Marcondes Stival, Pedro Henrique Martins Mussi Corrêa, Laura Sebben Galarce, Adenise Lorenci Woiciechowski, and Carlos Ricardo Soccol

Abstract Biofuels can be classified into four different categories or "generations", depending on the feedstock used or the technology employed in the production process. First-generation biofuels are the conventional fuels produced from food crops, while second-generation biofuels are derived from nonfood biomass, specifically lignocellulose-rich agricultural residues or energy crops. Third-generation biofuels, in turn, are produced essentially from algal biomass, and fourth-generation fuels focus on carbon capture technologies and engineered crops or microorganisms, being potentially carbon negative. First-generation biofuels hold the largest market share, benefiting from decades of production that have led to process optimization and a well-established industrial infrastructure. This chapter will describe the main aspects of first-generation liquid biofuels, including bioethanol, biodiesel, and bio-jet fuels.

12.1 Introduction

According to reports from the International Energy Agency (IEA), liquid biofuels are estimated to account for 30% of the global bioenergy demand by 2050. These fuels will be essential for the decarbonization of sectors that are challenging to electrify, such as road transport, shipping, and aviation (IEA Bioenergy 2022; Pelkmans et al. 2024).

Depending on the feedstock used or the technology employed in the production process, biofuels can be classified into four different generations, with first-generation biofuels being produced from food crops (Padder et al. 2024). First-generation biofuels hold the largest market share, benefiting from decades of

S. G. Karp (✉) · M. Z. Weber · R. N. Barros · M. E. Colodel · A. S. H. Cabrera · F. M. Stival · P. H. M. M. Corrêa · L. S. Galarce · A. L. Woiciechowski · C. R. Soccol
Department of Bioprocess Engineering and Biotechnology, Federal University of Paraná, Curitiba, Paraná, Brazil
e-mail: susan.karp@ufpr.br

A. L. Woiciechowski et al. (eds.), *Biomass for Bioenergy and Bioproducts*, Biofuel and Biorefinery Technologies 17, https://doi.org/10.1007/978-3-032-19898-3_12

production that have led to process optimization and a well-established industrial infrastructure (Padder et al. 2024). Among first-generation biofuels, bioethanol is the most widely produced worldwide, reaching approximately 113.3 million m^3 in 2023. The United States leads production with 52% of the total, followed by Brazil (28%) and India (5%) (Renewable Fuels Association 2025). First generation bioethanol is produced through the fermentation of sugar- and starch-rich materials, primarily sugarcane and corn, as well as other sources such as potatoes, cassava, beetroot, and various crops, followed by the distillation of the fermented mash (Abdul Kareem Joyia et al. 2024).

First-generation biodiesel stands out as the second most widely produced biofuel globally, with an estimated production of 76.3 million tons in 2024. The United States, Brazil, and Indonesia together account for 60% of the global production. Indonesia, the world's leading crude palm oil (CPO) producer with 48 million tons in 2021, uses CPO as its main biodiesel feedstock (Wirawan et al. 2024). In 2021, biodiesel production in Indonesia consumed 7.34 million tons of CPO, nearly 15% of the total production (Halimatussadiah et al. 2021). In the United States, biodiesel production is primarily based on soybean oil, followed by waste oils, fats and greases, corn oil, canola oil, and recycled residues (EIA 2024). In Brazil, soybean oil was the most widely used feedstock for biodiesel production in 2023, accounting for approximately 69.15% of the total production (5.2 billion liters), followed by other materials such as animal fats, beef tallow, and palm oil (EPE 2023).

In addition to these, first-generation aviation biofuels, or sustainable aviation fuels (SAF), recorded a production of 600 million liters in 2023, accounting for only 0.53% of the global aviation fuel demand (Jones 2024). This technology is still in its growth phase and can potentially be integrated into current industrial infrastructure. For example, existing corn dry-grind facilities may be used to convert sugars into 1,4-dimethylcyclooctane (DMCO), a renewable jet-A fuel blend. The conversion process begins with milling, followed by saccharification, similar to the process used for converting corn into ethanol (Batten et al. 2023).

Although first-generation biofuels offer a feasible pathway to strengthen domestic energy security, concerns about the feedstock supply and competition with food production for land have stimulated the pursuit of new biofuel sources (Padder et al. 2024). Second-generation biofuels are derived from nonfood biomass, specifically energy crops or lignocellulose-rich agricultural residues, with typical feedstocks including straw, sugarcane bagasse, and forestry residues (Patel and Shah 2021). In the case of bioethanol production, lignocellulosic biomass undergoes pretreatment and enzymatic hydrolysis to convert polysaccharides (mainly cellulose and hemicellulose) into fermentable sugars (Patel and Shah 2021). Second-generation biofuels offer significant benefits, avoiding competition with food resources, and reducing environmental impacts compared to first-generation biofuels. Furthermore, they allow for integrated production by leveraging the existing first-generation plant infrastructure with the necessary 2G technology (Carpio and de Souza 2017).

Biodiesel derived from nonedible feedstocks, such as jatropha, waste cooking oil, forestry residues, biomass sources, and algae, is classified as a second-generation biofuel. The significance of using these feedstocks lies in their ability to reduce reliance on edible oils, promoting a more sustainable and resource-efficient biofuel production (Bhuiya et al. 2016; Varghese et al. 2025).

Third-generation biofuels, in turn, are produced essentially from algal biomass, including microalgae and macroalgae (Abbasi et al. 2021). In recent years, the potential of microalgae as a biofuel feedstock has attracted global interest, driven by their relatively fast growth, high oil yield, effective CO_2 capture, the capacity to grow in non-potable water, suitability for cultivation on nonarable land, and the absence of competition with food production (Abbasi et al. 2021). Additionally, third-generation biofuels stand out for providing a higher biomass yield per unit of area compared to terrestrial crops (Correa et al. 2017; Padder et al. 2024). Microalgae, the primary source of third-generation biofuels, contain 20–70% lipid and require various extraction methods for subsequent conversion into biodiesel (Chaos-Hernández et al. 2023). Third-generation bioethanol is another advanced biofuel that can be produced from macro- or microalgae, through cultivation, harvesting, cell rupture, saccharification, and fermentation (Müller et al. 2023).

To address the limitations of first-, second-, and third-generation biofuels, fourth-generation biofuels are under development, focusing on carbon capture technologies and the use of engineered crops or microorganisms, being often classified as carbon negative fuels. Fourth-generation biofuels are produced through advanced technologies, like genetic engineering, CRISPR, and genome editing (Ashokkumar et al. 2024). Genetically modified microorganisms used for fourth-generation biofuels may include microalgae, yeasts, fungi, and cyanobacteria (Azizi et al. 2018). In contrast to the microbial-mediated methods, thermochemical techniques are also being employed for biomass conversion to biofuels, which include pyrolysis, gasification, and hydrothermal liquefaction (HTL) (Ashokkumar et al. 2024).

12.2 First-Generation Bioethanol

By definition, first-generation bioethanol comprises ethanol produced from renewable biomass materials with great quantities of carbohydrates, such as starches and sugars (de Almeida and Colombo 2021), represented by the conventional food crops. Those biomass types are related to agricultural production worldwide, and their access depends on the geographical region in which they are produced and on the fluctuation in the production yield, due to climatological and seasonal reasons (Abdul Kareem Joyia et al. 2024). In the USA, the main crop used is corn grain, accounting for up to 94% of its ethanol production (U.S. Department of Energy Alternative Fuels Data Center 2025). On the other hand, in Brazil, the most common material is sugarcane, followed by corn (Ramos et al. 2024; Grandis et al. 2024). In the European Union, the main materials are more diverse, comprising wheat, maize, sugar beet, barley, and rye (Bioethanol – European Biomass Industry Association 2025). In China, whose total capacity is still not at a maximum, refineries can be found using corn, wheat, cassava, and sweet sorghum as their basic raw material (Wu et al. 2021). Given the different compositions of feedstocks, there are some differences regarding the ethanol production process, prior to fermentation and distillation.

Sugar beet crops contain monosaccharides as their main component, which represent around 75% of their dry matter and are present in their roots, making them suitable for ethanol production (Marzo et al. 2019). When preparing sugar beet for ethanol fermentation, the first step is to remove solid components, such as dust, leaves, and mud, by washing the root bulbs with pressurized water (Mall et al. 2021). After that, the roots are chopped into thin strips of about 5–10 cm in length (cossettes) and dipped into hot water, making the raw juice available (Marzo et al. 2019). The pulp generated in this process is pressed, and part of the retained juice may also be made available. After that, the whole juice is supplemented, if necessary, with nutrients and goes through the step of pH correction (Pavlečić et al. 2021). The following stage includes sterilization before the juice is complete and ready for fermentation.

A similar processing method can be applied to sweet sorghum. In this, the saccharides are present in the stalks of the plant in a way similar to that of sugarcane, containing sucrose, glucose, and fructose (Ray et al. 2019). First, the stalks are crushed by a roller mill, and the bagasse is collected. Then, the juice passes through a clarification process in order to separate any solid debris by decantation (Kasegn et al. 2024). After that, the system undergoes a filtration process that may be followed by a concentration step to ensure that the sugar concentration is adequate for the subsequent fermentation.

Another variety of sorghum, namely amylaceous, contains approximately 70–80% of carbohydrates in its grains, primarily starch, and therefore undergoes a process similar to that of corn grains (Tanwar et al. 2023). First, the sorghum grains are dried and milled (Stamenković et al. 2020). Then, they are subjected to liquefaction and hydrolysis to convert the starch into available sugars for fermentation. There are different possible routes for the hydrolysis step, including acid, alkaline and enzymatic hydrolysis (Audilakshmi and Swarnalatha 2019). Ethanol production from wheat and barley is closely similar to that of sorghum grain. This is because they are grains with relatively high starch content, therefore needing to be processed by milling, which can be of a dry or wet nature, liquefaction, hydrolysis, and finally fermentation (Li et al. 2022; Mohanty and Swain 2019).

12.2.1 Sugarcane Ethanol

The Brazilian sugar and ethanol sector is fundamental to the economy, serving as a key pillar of agribusiness and generating billions of dollars in annual revenue. In the 2023–2024 harvest, Brazil produced 35.9 billion liters of ethanol (UNICA Data 2025). The sector operates in 1200 municipalities and cultivates approximately 10 million hectares, generating employment and contributing to regional development (UNICA data 2025).

Sugar and ethanol exports generated US$19.8 billion between April 2023 and March 2024, positioning the sector as the third-largest exporter in the country. The sugarcane energy chain generates over US$100 billion, with a sectoral GDP of US $40 billion, representing 2% of the national GDP (UNICA Data 2025).

Sugarcane leads Brazil's renewable energy matrix, playing a strategic role in the country's energy landscape by contributing 15.4% of the national energy supply and 33% of the domestic renewable energy (EPE 2023). Bioelectricity generated from sugarcane biomass supplied 20,973 GWh in 2023 (4% of the electricity demand), with the potential to expand from 15% to 55% by 2030, driven by RenovaBio (UNICA 2025; UNICA Data 2025).

The production of sugar and ethanol undergoes a series of physical, chemical, and biological transformations. The initial step of the process involves the harvesting and preparation of the sugarcane, which comprises removing tips and leaves to reduce impurities and retain only the sucrose-rich stem (Karimi and Chisti 2017). In sugar and ethanol processing facilities, sugarcane is washed, crushed, and pressed in order to extract its juice. Approximately 70% of the mass is converted into juice, while the remaining 30% forms bagasse, which is used as fuel for energy generation (NOVACANA 2024; Propeq 2025).

The extracted juice undergoes a purification process to remove solid impurities. The clarification of the extracted juice requires precise temperature control, maintained within the range of 30–70 °C, and the careful addition of calcium oxide (CaO). This stage, which is essential for the subsequent processing steps, induces the precipitation of impurities, which are subsequently removed by filtration. The clarified broth can be then sterilized at 105 °C to eliminate microorganisms and evaporate residual water (Amorim et al. 2011; Basso et al. 2011; Raízen 2022).

After the upstream processing steps, the broth is transferred to fermentation tanks, where yeasts, particularly *Saccharomyces cerevisiae*, convert sugars into ethanol and CO_2 within 6–12 h, resulting in a fermented mash with approximately 7–10% ethanol (Karimi and Chisti 2017; Raízen 2022). After fermentation, the mash is processed via centrifugation to remove yeast biomass and then distilled to achieve an ethanol concentration of 96% v/v, producing hydrous ethanol (Karimi and Chisti 2017). To produce anhydrous ethanol (>99% v/v), used in gasoline blends, the alcohol undergoes dehydration via, e.g., pressure swing adsorption (PSA) (Karimi and Chisti 2017), hexane, or monoethylene glycol distillation. Both anhydrous and hydrous ethanol are stored in tanks and transported in tanker trucks to distributors and industries (Raízen 2022; NOVACANA 2024). Figure 12.1 presents the flowchart of the sugarcane ethanol production process.

The sugarcane industry generates solid, liquid, and gaseous waste, all of which have the potential for reuse. Among the solid residues, the most significant are straw and bagasse, used as bioenergy sources, and filter cake, applied as an organic fertilizer. Liquid waste includes vinasse, used as a liquid fertilizer, and molasses, employed in ethanol production. The primary gaseous waste is CO_2 released during fermentation, which is used in industries for producing carbonated beverages and refrigeration processes (Formann et al. 2020; Soares et al. 2022; OICS 2025).

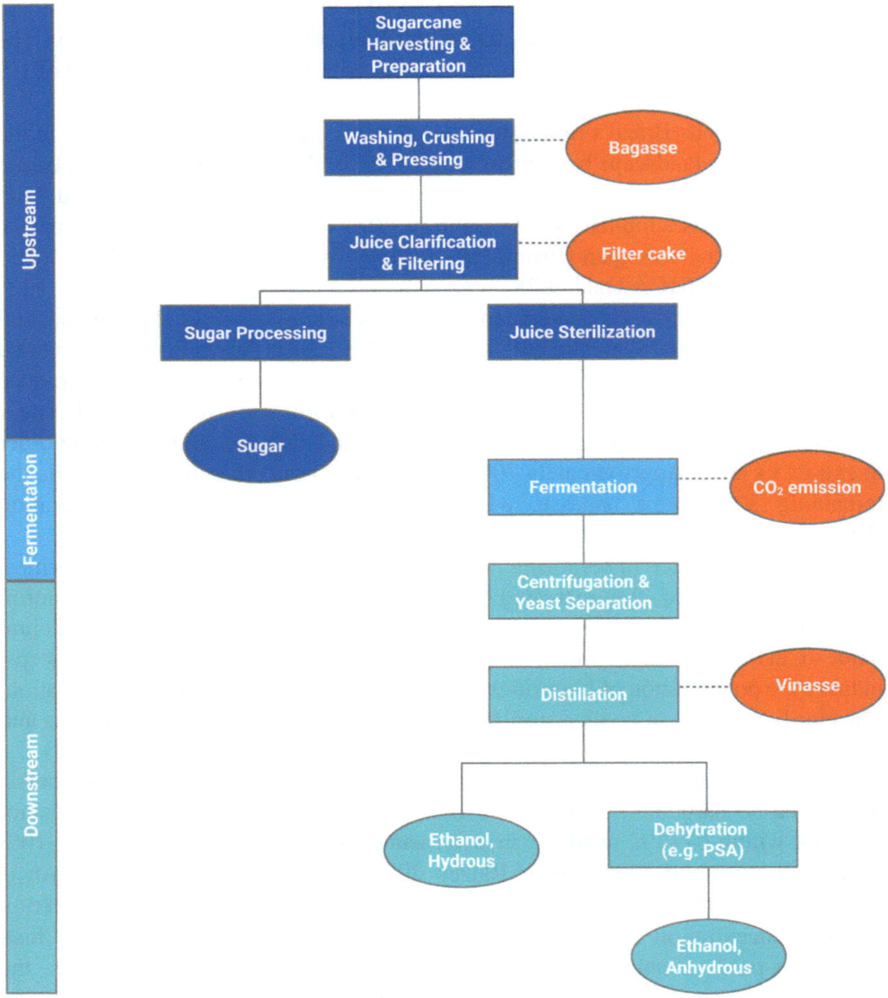

Fig. 12.1 Sugarcane ethanol production process

In addition to these residues, ethanol production generates important by-products, such as bioelectricity. Derived from the combustion of bagasse and straw, bioelectricity is currently the fourth largest source of electricity in Brazil, with a potential share of 15–55% of the energy matrix by 2030, driven by policies such as RenovaBio (UNICA data 2025). Furthermore, second-generation ethanol (2G) is produced from lignocellulosic waste, such as bagasse (Dias et al. 2011), enabling the biofuel production expansion without increasing cultivated areas.

By-products such as vinasse, filter cake, straw, and bagasse are increasingly used in biogas and biomethane production through anaerobic digestion. Biogas, consisting of approximately 60% methane and 40% CO_2, can generate electricity or be purified into biomethane (Soares et al. 2022; Sugarcane 2024).

In the cosmetics industry, an interesting example is the sustainable production of squalane from sugarcane biomass, replacing its traditional extraction from shark livers. Additionally, sugarcane-derived bioplastics are used in packaging as an alternative to fossil-based plastics, reinforcing sustainable practices in the sector (Sugarcane 2024).

Beyond producing sugar and ethanol, the sugarcane industry adopts a circular economy practice that enhances both profitability and sustainability. Sugarcane bagasse, the primary by-product, is utilized for steam and electricity cogeneration, fulfilling the energy demands of the processing plants while enabling surplus energy to be supplied to the national grid (ADECOAGRO 2023).

Waste such as boiler ash and filter cake are transformed into biofertilizers, applied to plantations to regenerate the soil and reduce the use of chemical inputs. Vinasse, the liquid by-product of ethanol distillation, besides being concentrated and utilized as a liquid organic fertilizer or for producing biogas and biomethane (Soares et al. 2022; Sugarcane 2024), could be used as a culture medium for microorganisms such as microalgae, yielding high-value-added products.

The implementation of these integrated sustainability strategies minimizes waste and greenhouse gas emissions while simultaneously promoting energy self-sufficiency, strengthening the economic and environmental resilience of the sugarcane ethanol production chain.

12.2.2 Corn Ethanol

Corn is the main raw material for ethanol production in the USA and the second most important in Brazil after sugarcane. Corn ethanol production can be carried out through two main routes: the dry process and the wet process. Each route has distinct characteristics in terms of stages, products generated, and operating costs. The dry process is the most widely used method, mainly due to its lower initial costs. Here, the whole corn grain is processed, and the separation of residual components occurs only in the final stages (Monlau et al. 2019; Assaf et al. 2024).

The wet process is characterized by a more complex approach, aiming at the separation of the different fractions of the corn grain in the initial stages, which favors obtaining a wider range of high-value products. Although the initial stages may be similar to the dry process, the wet process differs in its ability to extract multiple components from the corn before fermentation.

Figure 12.2 depicts both dry and wet milling processes. On the left, the dry route is illustrated, which involves steps such as cleaning, milling, liquefaction, saccharification, fermentation, distillation, and dehydration, resulting in ethanol as the main product and by-products such as corn oil and bran; on the right, the wet route is shown, which includes soaking the grain (steeping) and sequential separations of germ, fiber, and starch. This route generates several products, such as corn oil, corn

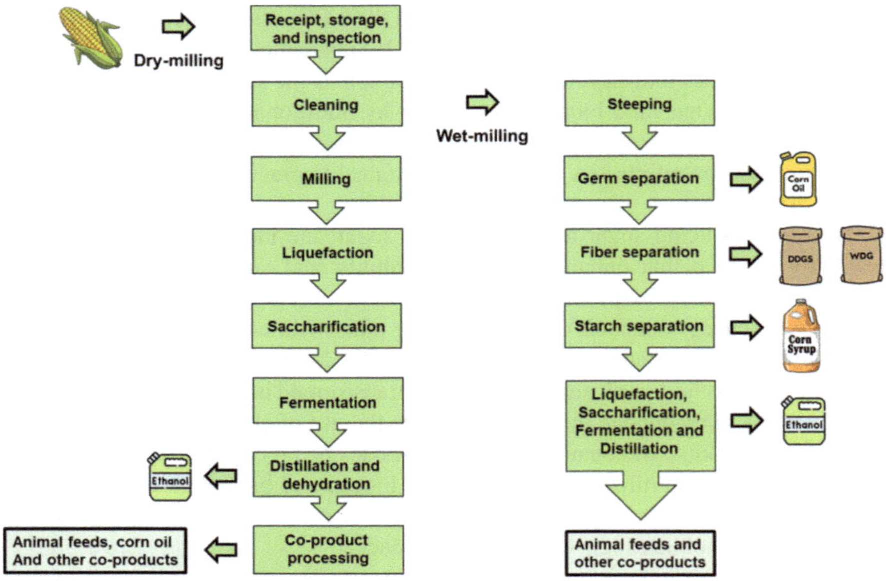

Fig. 12.2 Simplified flowchart of the corn ethanol production process

syrup, ethanol, and protein by-products (wet distillers grains (WDG) and dried distillers' grains with solubles (DDGS)), used as animal feed. Both routes demonstrate the full valorization of corn in the biofuels chain.

The dry process of corn ethanol production begins with pretreating and cleaning the corn to remove metal objects. The corn is then ground in hammer mills, resulting in corn flour. This flour is then subjected to liquefaction, where it is mixed with water and cooked to gelatinize the starch, and the enzyme α-amylase is added to break down the starch into dextrins and oligosaccharides. The resulting corn mash, now called "wort," undergoes saccharification, where the enzyme glucoamylase converts the dextrins into glucose, which makes the wort rich in fermentable sugars. The wort is then transferred to fermentation, where yeasts of the species *S. cerevisiae* convert the glucose into ethanol and carbon dioxide (CO_2) (Mohammadi Shad et al. 2021).

Fermentation times for corn ethanol are typically longer than for sugarcane, ranging from 38 to 45 h, and can be as long as 72 or even 96 h under high solids conditions (Assaf et al. 2024). A common process variation is the Simultaneous Saccharification and Fermentation (SSF), where enzymes and yeast are added together in a single step, which can reduce overall process time, reduce contamination risk, and save energy. The fermented must is then directed to distillation, where ethanol is separated and can be sold as hydrous or anhydrous. The distillation residue, the raw vinasse, is processed to separate the fine vinasse from the solids, which are later used to produce coproducts such as DDGS and WDG with solubles, as well as corn oil and CO_2 (Mohammadi Shad et al. 2021).

The wet process is characterized by the separation of the different fractions of the corn grain in the initial stages. The corn is kept in warm water for 24–48 h, and then wet milled, when the germ, fiber, starch, and gluten are separated. The high-value-added products and coproducts generated in the wet process include ethanol, starch, corn oil, corn steep liquor, corn gluten meal (rich in protein), corn gluten feed (low in protein), high-fructose corn syrup, carbon dioxide (CO_2), and others depending on the industrial facilities (Council USG 2023).

Diversification is a strategic and fundamental characteristic of the wet process. While dry milling is dominant due to lower upfront costs, wet milling offers a broader portfolio of high-value coproducts. This diversification provides greater market resilience because the plant is not solely dependent on ethanol prices. If ethanol prices fluctuate, the ability to generate significant revenue from products such as high-fructose corn syrup, starches, and various protein feeds can stabilize profitability. This implies a more complex but potentially more robust business model, especially for larger, integrated operations seeking to maximize the value of each component of the corn grain (Monlau et al. 2019).

The economic viability of corn ethanol production in Brazil has been attracting increasing interest, driven by the national energy scenario and the sector's growth prospects. This progress has been supported by business models that adapt to the reality and the profile of different investors. One example is the flex plant, which takes advantage of the existing structure of sugarcane mills during the off-season. With reduced initial investment and lower demand for labor, these plants have proven to be especially advantageous in regions with a large availability of corn, such as the state of Mato Grosso (Boschiero 2025). Integrated plants, which operate year-round with sugarcane and corn simultaneously, require greater investment but offer gains in productivity. Finally, full plants, which are entirely dedicated to corn, operate continuously and produce both ethanol and coproducts, adding more value to the chain, although requiring higher investments. This diversity of models has attracted different investors, from traditional mills to grain producers, and has been essential for the rapid expansion and consolidation of the industry, ensuring greater flexibility, competitiveness, and resilience over the years (Boschiero 2025).

From a circular economy and sustainability perspective, corn ethanol offers several opportunities. One of the main examples is the use of by-products such as DDGS, resulting from fermentation, in animal feed, especially in livestock and pig farming. This replaces inputs such as soybean meal, generating synergies between sectors and optimizing land use. Another important aspect is the recovery and use of corn oil, which can be applied in the food industry or for biodiesel production, adding value to the chain. In addition, the CO_2 generated during fermentation can be captured and used in industrial applications (Neves et al. 2021).

Choosing renewable biomass, such as eucalyptus chips or sugarcane bagasse, to generate energy in ethanol plants significantly reduces carbon emissions by replacing fossil fuels. The use of second-crop corn, grown in rotation with soybeans, also avoids agricultural expansion and improves the efficiency of the system (Neves et al. 2021). In summary, the bioeconomic performance of corn ethanol depends on maximizing resource utilization, diversifying revenue sources, and adopting low-carbon production practices.

12.3 First-Generation Biodiesel

Biodiesel is a liquid biofuel that can be produced through a chemical reaction called transesterification between vegetable oils or animal fats and an alcohol in the presence of a catalyst (U.S. Department of Energy 2025; Romano and Sorichetti 2010). Biodiesel is usually obtained from oily crops and is thus classified as a first-generation biofuel. Advantages of first-generation biodiesel are its low emission of greenhouse gases and its cost-efficient conversion technology. Drawbacks include an inadequate yield to meet demand, food shortages, and a high land footprint (Vignesh et al. 2021). Compared specifically with conventional diesel, the upsides are lower toxicity, faster degradation, lower emission of contaminants, lower health risk, no SO_2 emission, higher flash point, the possibility of blending with diesel, good lubricant properties, and no need for modifications in the diesel engine. Disadvantages include slightly higher fuel consumption and NOx emissions, a higher freezing point, less stability, potential degradation of plastic and rubber, and dissolving deposits from sediments and other contaminants. These limitations are reasonably overcome when mixing biodiesel and conventional diesel (Romano and Sorichetti 2010).

In general, the biodiesel production process can be divided into a few steps, as represented in Fig. 12.3. First, the feedstock is pretreated to ensure a better yield in the reaction. After that, the alcohol is mixed with the catalyst, and then the reaction occurs when the oil is added. Subsequently, glycerol, another product of the transesterification process, is separated from the biodiesel. Finally, the ester undergoes a purification process to ensure optimal performance as biodiesel (Romano and Sorichetti 2010).

Transesterification is the chemical reaction in which the conversion of one ester to another occurs by the alteration of alkoxy groups. For biodiesel, it usually involves the reaction of triglycerides with an alcohol (most often methanol or ethanol) in the presence of a catalyst. The result is methyl or ethyl esters (depending on which alcohol was used) and glycerol as a coproduct. Di- and monoglycerides, excess reactants, and free fatty acids are possible by-products of the reaction (Aktaş et al. 2020; Vignesh et al. 2021). For first-generation biodiesel production, rapeseed, soybean, palm, and sunflower oils are the most common feedstocks. Using soybeans results in other valuable by-products: soybean meal, pellets, and flour (Romano and Sorichetti 2010; Najafpour 2015).

Methanol is the preferred alcohol despite its high toxicity because it requires simpler technology, excess alcohol can be recovered at low cost, while higher reaction speeds are achieved. Methanol is used in excess to achieve a higher conversion efficiency. Excess alcohol as a reactant shifts the equilibrium to the right side of the equation, increasing the amount of products. Usually, a 100%

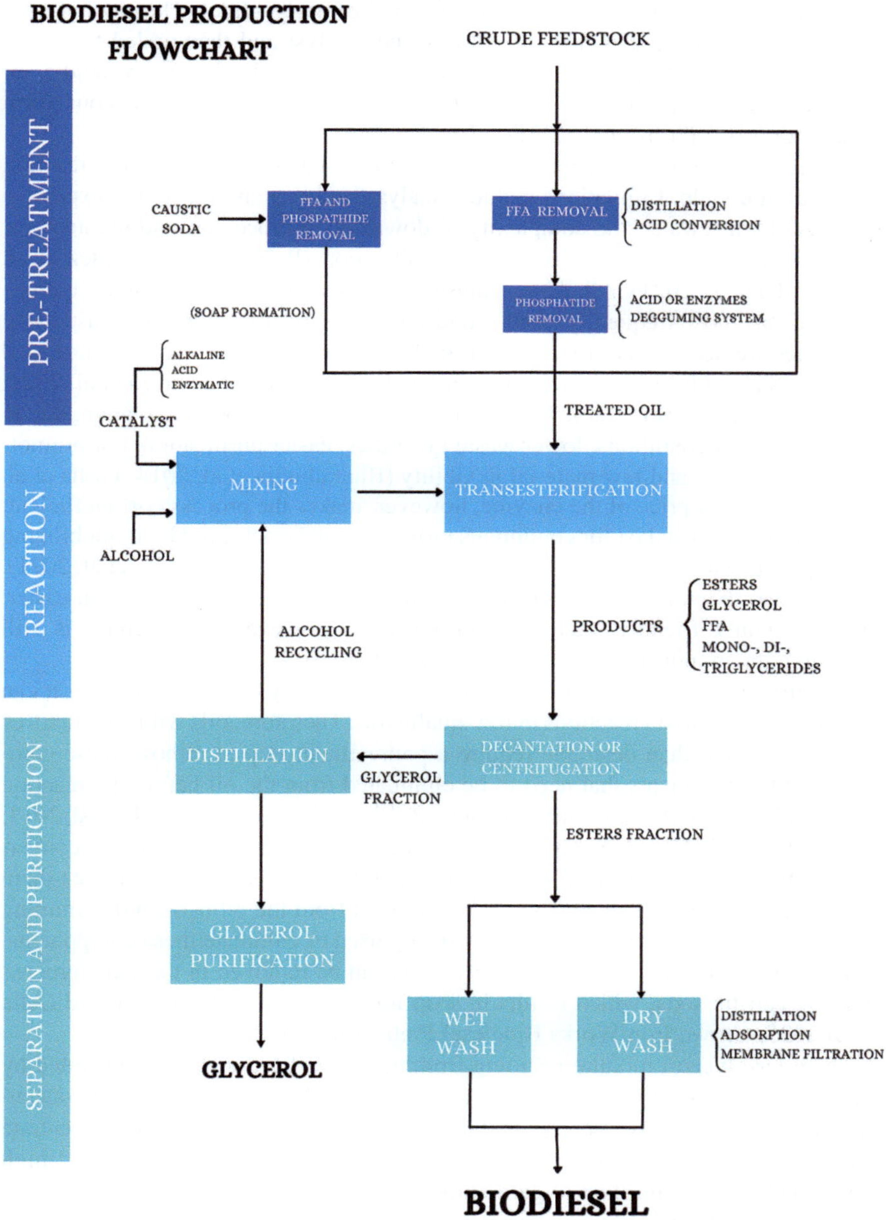

Fig. 12.3 Biodiesel production flowchart

alcohol excess is used (6 mol of alcohol per mol of triglyceride). The alcohol used for biodiesel production is first mixed with the catalyst and then added to the oil (Romano and Sorichetti 2010). Catalysts can be chemical (alkaline or acidic) or enzymatic. They can also be present in a homogeneous or heterogeneous form (Najafpour 2015; Pasha et al. 2021).

Chemical catalysts are the most commonly used route due to their shorter reaction time and high yield. Limitations include catalyst recovery and recycling, excessive amounts of wastewater, the complexity of downstream processes, and the need for high-quality raw materials to prevent saponification (Peñarrubia Fernandez et al. 2017; Pasha et al. 2021). Alkaline catalysts, such as sodium and potassium hydroxides, are the most frequently used. Acid catalysts are less commonly used and sometimes are used as the first stage for highly acidic raw materials, recommended for feedstocks with free fatty acid content levels above 1% (Sirajunnisa et al. 2021; Romano and Sorichetti 2010). The biocatalytic process is a good alternative, due to its mild reaction conditions, lower waste quantities, easier purification, not promoting side reactions, and raw material flexibility (Bharathiraja et al. 2014; Pasha et al. 2021). The higher price of the enzyme, however, makes the process cost inefficient. Reusable enzyme catalyst for continuous processes can be achieved by immobilizing lipases, which enables recycling, recovery, and lower costs (Sirajunnisa et al. 2021). There is also the non-catalytic supercritical methanol process, which has an easier, environmentally friendly, and rapid purification. However, this method is still unfeasible on an industrial scale (Bharathiraja et al. 2014).

Industrially, the alkaline catalyst is the most frequently used, even though it can promote saponification reactions in low-quality oil. Therefore, oils with a higher free fatty acid content than desirable require a pretreatment step to remove these components. Other elements that need to be eliminated from the oil before the reaction are phosphatides (Romano and Sorichetti 2010; Javidialesaadi and Raeissi 2013; Nordblad et al. 2016). One way to remove free fatty acids from the oil is to convert them to biodiesel using an acid catalyst, although it is also possible to separate them through distillation. Phosphatides can be removed from the oil using a degumming system in which acids or enzymes are typically used to eliminate these components. Furthermore, phosphatides and free fatty acids can be removed in the same process by using caustic soda, which results in soap formation that is then separated (Alfa Laval 2025; Crown Iron Works Biodiesel Pretreatment 2025).

The ester-glycerol mixture resulting from transesterification can be separated by decantation or centrifugation. Most of the catalyst and excess alcohol will concentrate in the lower phase (glycerol), and mono-, di-, and triglycerides will concentrate in the upper phase. The glycerin obtained in the chemical reaction is not of high quality and must be purified to gain commercial value. For large-scale production plants, besides high-quality glycerin, part of the methanol can be recovered for reutilization, thus lowering biodiesel production costs. Methanol can be recovered by the distillation of the glycerol portion (Romano and Sorichetti 2010).

After glycerol and ester separation, the fatty acid methyl ester (FAME) mixture contains impurities such as remnants of alcohol, catalyst, and mono-, di-, and triglycerides, which confer undesirable characteristics to FAME; therefore purification is necessary to reduce the concentration of contaminants to acceptable levels so that the biofuel complies with international standards. These components can be removed through wet washing, using multiple washing steps to remove methanol, the catalyst, and glycerin (these contaminants are water-soluble). Although this process works well, it is not cost-effective due to the high expenses associated with wastewater treatment and energy (Romano and Sorichetti 2010; Wan Osman et al. 2024).

Considering the limitations of the wet washing process, there are other approaches that do not use water, the so-called "dry washing" processes. One of them is distillation, which works by vaporizing the impurities. The most commonly used type of distillation in biodiesel purification is vacuum distillation, in which the pressure is lowered so that the components can vaporize at lower temperatures. This process can effectively remove the impurities, but its implementation cost is high, which can be a barrier for smaller facilities (Wan Osman et al. 2024).

Another dry washing approach is adsorption, in which a polar washing agent adsorbs the impurities. Although this process is easier to run than distillation, its implementation cost is still high, and the washing agent needs to be changed when it is saturated. Another disadvantage is that, depending on the legislation, this process may not achieve the quality requirements (Wan Osman et al. 2024).

The impurities can also be removed from biodiesel by membrane separation. This process is effective, and consumes less energy compared to distillation. In addition, it is versatile to be used in small to large facilities. One disadvantage is that the process cost can be high due to membrane fouling (Wan Osman et al. 2024).

12.4 First-Generation Bio-Jet Fuels

In 2016, the International Civil Aviation Organization (ICAO) established the Carbon Offsetting and Reduction Scheme for International Aviation (CORSIA), an initiative aimed at reducing carbon emissions in international aviation, which will become mandatory for airlines starting in 2027 (IATA 2016). This highlights the importance of seeking alternatives to traditional petroleum-based aviation fuels. Aviation biofuels are produced from renewable feedstocks, with the potential to reduce CO_2 emissions throughout their life cycle, making them a viable alternative for a more sustainable aviation industry (Staples et al. 2018). First-generation bio-jet fuels are derived from edible biomass, such as vegetable oils (e.g., soybean, palm, and canola oil), sugars (sugarcane and beet), and starch (corn and wheat) (Tiwari et al. 2023). There are four general categories of bio-jet fuel production pathways, differentiated by the feedstock used: alcohol-to-jet (ATJ), oil-to-jet (OTJ), sugar-to-jet (STJ), and gas-to-jet (GTJ) (Wang and Tao 2016). The latter does not use edible biomass in its production (Neuling and Kaltschmitt 2018; Wei et al. 2019).

There are two main conversion pathways for producing bio-jet fuels from oil. Hydroprocessed esters and fatty acids (HEFA) is a process certified by the ASTM International standards development organization that involves converting triglycerides, and saturated or unsaturated fatty acids from animal fats, vegetable oils, and waste cooking oil into fuel. The method is divided into the following steps: (1) catalytic hydrogenation using a catalyst (usually NiMo or Al_2O_3), which converts unsaturated fatty acids and triglycerides into saturated fatty acids; (2) depending on the process conditions, hydrodeoxygenation, decarboxylation, or decarbonylation reactions may occur, converting the saturated fatty acids into linear-chain alkanes (C8–C20), depending on the vegetable oil used; (3) isomerization, where linear-chain alkanes are cracked using a catalyst, breaking the molecules into short fragments while simultaneously saturating the open bonds with hydrogen and partially recombining them, thereby generating highly branched alkanes (Neuling and Kaltschmitt 2018; Wei et al. 2019). Another pathway for producing aviation biofuels from oil is catalytic hydrothermolysis (CH), which uses algae or oilseed plants as feedstocks. The process includes: (1) pretreatment of triglycerides, including conjugation, cyclization, and cross-linking to enhance molecular structures; (2) cracking and hydrolysis reactions; (3) catalytic decarboxylation and dehydration; and (4) post-refining hydrotreatment and fractionation to convert straight-chain, branched, and cyclo-olefins into alkanes (Neuling and Kaltschmitt 2018; Wei et al. 2019).

Aviation biofuels based on sugar can be produced via a biochemical pathway in a process called direct sugar to hydrocarbons (DSHC) or direct fermentation of sugar-to-jet (DFSTJ), or as named by ASTM, hydroprocessed fermented sugar synthetic isoparaffin (SIP). This technology involves the following steps: (1) fermentation and enzymatic hydrolysis to convert sugars into farnesene ($C_{15}H_{24}$); (2) farnesene recovery; (3) conversion of farnesene into farnesane ($C_{15}H_{32}$) via hydroprocessing (Ng et al. 2021). Studies on this type of process involve using genetically modified microorganisms to ferment sugars derived from feedstocks such as sugarcane, beet, and corn (Wei et al. 2019). In addition to these methods, there is also a thermochemical conversion pathway for sugar into fuel known as aqueous phase reforming (APR). Aqueous phase catalytic reforming is a technology for converting soluble plant sugars into chemical intermediates such as alcohols, ketones, aldehydes, acids, furans, and other oxygenated hydrocarbons, which can then be converted into hydrocarbons used as aviation fuel (Neuling and Kaltschmitt 2018; Wei et al. 2019).

The conversion process from alcohol to aviation fuels requires that feedstocks rich in sugars or starch (after enzymatic hydrolysis) first be converted into alcohol through fermentation. Then, the alcohol undergoes dehydration to generate olefins, which are then oligomerized in the presence of catalysts, and the product is separated by distillation. Finally, the middle distillates are hydrogenated to produce hydrocarbons within the jet fuel range (Neuling and Kaltschmitt 2018; Wei et al. 2019).

Lim et al. conducted a study on economic opportunities and challenges in bio-jet fuel production. The most frequently mentioned opportunities were the growing demand for aviation fuels, the lack of low-carbon alternatives, supportive government policies, the volatility of the fossil fuel market, and the importance of energy independence and diversification. The main challenges reported for the use of bio-jet fuels were the high total production cost including the high feedstock cost, the large capital investment, insufficient production scale, the availability of sustainable feedstocks, the lack of consistent government incentives, and financial uncertainties (Lim et al. 2023).

12.5 Conclusion and Future Perspectives

Although the global energy matrix is undergoing a significant transition, characterized by the increasing importance of electrification, liquid biofuels continue to play a substantial role, particularly within transportation systems. First-generation biofuels, which exhibit the highest degree of technological maturity, offer considerable potential for the optimization of their bioeconomic parameters, thereby contributing to the advancement of a more sustainable and low-carbon economy.

The bioethanol industry can benefit from the implementation of circular economy strategies aimed at improving both techno-economic and environmental performance. One promising approach involves the valorization of sugarcane vinasse and solid residues through their conversion into biogas, providing a surplus energy carrier within the production chain. Also, by-products of corn ethanol production offer significant nutritional value and can serve as a substitute for soybean in animal feed. Additionally, the recovery of carbon dioxide (CO_2) as a coproduct presents valuable opportunities from both economic and environmental perspectives.

In the biodiesel industry, advances in process performance and product quality are still a focus of interest, while enhancing the sustainability of the production chain is another goal. One concept strongly connected to the circular economy is the regenerative design: all materials or wastes are reintroduced into the process or transformed into new products (Geisendorf and Pietrulla 2018). This concept can be applied to biodiesel production in several ways, for example, through the recovery and subsequent reuse of alcohol and catalysts (Romano and Sorichetti 2010; Mandari and Devarai 2021). Glycerol, a by-product of biodiesel production, can be purified to achieve commercial value or alternatively be utilized as a source of carbon and hydrogen in both biochemical and thermochemical conversion processes. Furthermore, the free fatty acids separated from the oil during pretreatment can also be recovered and marketed as valuable coproducts (Contreras-Andrade et al. 2015).

In the case of bio-jet fuels, alternative feedstocks and conversion processes can be investigated to reduce the total production costs. Advancements in catalytic upgrading and hydroprocessing technologies may increase conversion efficiencies and product yields. Integrating bio-jet fuel production into existing biorefineries or colocating with other industrial operations can facilitate shared infrastructure and reduce capital expenditure. Additionally, optimizing supply chain logistics and implementing carbon capture and utilization strategies can further enhance the environmental and economic performance of bio-jet fuels. Continued research into microbial and enzymatic conversion pathways also holds promise for the development of more efficient and scalable production systems.

References

Abbasi M, Pishvaee MS, Mohseni S (2021) Third-generation biofuel supply chain: a comprehensive review and future research directions. J Clean Prod 323:129100. https://doi.org/10.1016/j.jclepro.2021.129100

Abdul Kareem Joyia M, Ahmad M, Chen YF et al (2024) Trends and advances in sustainable bioethanol production technologies from first to fourth generation: a critical review. Energy Convers Manag 321:119037. https://doi.org/10.1016/J.ENCONMAN.2024.119037

ADECOAGRO (2023) Circular economy. https://sustainability.adecoagro.com/en/certifications/environment/circular-economy/. Accessed 24 Feb 2025

Aktaş ES, Demir Ö, Uçar D (2020) A review of the biodiesel sources and production methods. Int J Energy Smart Grid 5:1. https://dergipark.org.tr/tr/download/article-file/941575

Alfa Laval (2025) FAME biodiesel pre-treatment systems | Alfa Laval. https://www.alfalaval.com/products/process-solutions/edible-oil-solutions/pretreatment-systems-for-biofuels/fame-biodiesel-pretreatment-systems/. Accessed 21 Feb 2025

Alternative Fuels Data Center: Ethanol Fuel Basics. https://afdc.energy.gov/fuels/ethanol-fuel-basics. Accessed 24 Mar 2025

Amorim HV, Lopes ML, de Castro Oliveira JV et al (2011) Scientific challenges of bioethanol production in Brazil. Appl Microbiol Biotechnol 91:1267–1275. https://doi.org/10.1007/s00253-011-3437-6

Ashokkumar V, Chandramughi VP, Kumar G et al (2024) Advancements in lignocellulosic biomass: a critical appraisal of fourth-generation biofuels and value-added bioproduct. Fuel 365:130751. https://doi.org/10.1016/j.fuel.2023.130751

Assaf JC, Mortada Z, Rezzoug SA et al (2024) Comparative review on the production and purification of bioethanol from biomass: a focus on corn. Processes 12. https://doi.org/10.3390/pr12051001

Audilakshmi S, Swarnalatha M (2019) Sorghum for starch and grain ethanol. Breed Sorghum Diverse End Uses:239–254. https://doi.org/10.1016/B978-0-08-101879-8.00015-2

Azizi K, Keshavarz Moraveji M, Abedini Najafabadi H (2018) A review on bio-fuel production from microalgal biomass by using pyrolysis method. Renew Sustain Energy Rev 82:3046–3059. https://doi.org/10.1016/j.rser.2017.10.033

Basso LC, Basso TO, Rocha SN (2011) Ethanol production in Brazil: the industrial process and its impact on yeast fermentation. In: Biofuel production-recent developments and prospects. InTech. https://www.intechopen.com/chapters/20058

Batten R, Galant O, Karanjikar M, Spatari S (2023) Meeting sustainable aviation fuel policy targets through first generation corn biorefineries. Fuel 33:126294. https://doi.org/10.1016/j.fuel.2022.1262943

EPE (2023) Balanço Energético Nacional 2023. https://www.epe.gov.br/pt/publicacoes-dados-abertos/publicacoes/balanco-energetico-nacional-2023. Accessed 20 Feb 2025

Bharathiraja B, Chakravarthy M, Kumar RR et al (2014) Biodiesel production using chemical and biological methods – a review of process, catalyst, acyl acceptor, source and process variables. Renew Sustain Energy Rev 38:368–382. https://doi.org/10.1016/j.rser.2014.05.084

Bhuiya MMK, Rasul MG, Khan MMK et al (2016) Prospects of 2nd generation biodiesel as a sustainable fuel – Part 1: Selection of feedstocks, oil extraction techniques and conversion technologies. Renew Sustain Energy Rev 55:1109–1128. https://doi.org/10.1016/j.rser.2015.04.163

Bioethanol – European Biomass Industry Association. https://www.eubia.org/cms/wiki-biomass/biofuels/bioethanol/. Accessed 24 Mar 2025

Boschiero BN (2025) Etanol de milho no Brasil: 3 razões que impulsionam o crescimento. Agroadvance. https://agroadvance.com.br/blog-etanol-de-milho-no-brasil-crescimento/

Carpio LGT, de Souza FS (2017) Optimal allocation of sugarcane bagasse for producing bioelectricity and second generation ethanol in Brazil: scenarios of cost reductions. Renew Energy 111:771–780. https://doi.org/10.1016/j.renene.2017.05.015

Chaos-Hernández D, Reynel-Ávila HE, Bonilla-Petriciolet A, Villalobos-Delgado FJ (2023) Extraction methods of algae oils for the production of third generation biofuels – a review. Chemosphere 341:139856. https://doi.org/10.1016/j.chemosphere.2023.139856

Contreras-Andrade I, Avella-Moreno E, Sierra-Cantor JF et al (2015) Purification of glycerol from biodiesel production by sequential extraction monitored by 1H NMR. Fuel Process Technol 132:99–104. https://doi.org/10.1016/J.FUPROC.2014.12.016

Correa DF, Beyer HL, Possingham HP et al (2017) Biodiversity impacts of bioenergy production: microalgae vs. first generation biofuels. Renew Sustain Energy Rev 74:1131–1146. https://doi.org/10.1016/j.rser.2017.02.068

Council USG (2023) Dry-grind production of ethanol, distillers corn oil, and corn co-products. Precision DDGS Nutrition, p 234

Crown Iron Works Biodiesel Pretreatment – Crown Iron Works. https://www.crowniron.com/ethanol/biodiesel-pretreatment/. Accessed 21 Feb 2025

de Almeida MA, Colombo R (2021) Production chain of first-generation sugarcane bioethanol: characterization and value-added application of wastes. BioEnergy Res 16:924–939. https://doi.org/10.1007/S12155-021-10301-4

Dias MOS, Modesto M, Ensinas AV et al (2011) Improving bioethanol production from sugarcane: evaluation of distillation, thermal integration and cogeneration systems. Energy 36:3691–3703. https://doi.org/10.1016/j.energy.2010.09.024

Sirajunnisa AR, Duraiarasan S, Baskar T et al (2021) Current and Future Perspectives on Lipid-Based Biofuels. In: Prospects of Renewable Bioprocessing in Future Energy Systems. Springer, pp 387–429

EIA (2024) Biofuels explained. Biodiesel, renewable diesel, and other biofuels. https://www.eia.gov/energyexplained/biofuels/biodiesel-rd-other-basics.php. Accessed 23 Feb 2025

EPE (2023) Análise de Conjuntura dos Biocombustíveis 2023. https://www.epe.gov.br/pt/publicacoes-dados-abertos/publicacoes/analise-de-conjuntura-dos-biocombustiveis-2023. Accessed 24 Feb 2025

Formann S, Hahn A, Janke L et al (2020) Beyond sugar and ethanol production: value generation opportunities through sugarcane residues. Front Energy Res:8. https://doi.org/10.3389/fenrg.2020.579577

Geisendorf S, Pietrulla F (2018) The circular economy and circular economic concepts – a literature analysis and redefinition. Thunderbird Int Bus Rev 60:771–782. https://doi.org/10.1002/TIE.21924

Grandis A, da Fortirer SJ, Pagliuso D, Buckeridge MS (2024) Scientific research on bioethanol in Brazil: history and prospects for sustainable biofuel. Sustainability (Switzerland) 16:4167. https://doi.org/10.3390/SU16104167/S1

Halimatussadiah A, Nainggolan D, Yui S et al (2021) Progressive biodiesel policy in Indonesia: does the Government's economic proposition hold? Renew Sustain Energy Rev:150. https://doi.org/10.1016/j.rser.2021.111431

IATA (2016) Carbon Offsetting and Reduction Scheme for International Aviation (CORSIA). https://www.icao.int/CORSIA. Accessed 20 Feb 2025

IEA Bioenergy (2022) WS28: Technology advances in liquid biofuels and renewable gas. https://www.ieabioenergy.com/blog/ieaevent/ws28-technology-advances-in-liquid-biofuels-and-renewable-gas/. Accessed 20 Feb 2025

Javidialesaadi A, Raeissi S (2013) Biodiesel production from high free fatty acid-content oils: experimental investigation of the pretreatment step. APCBEE Proc 5:474–478. https://doi.org/10.1016/J.APCBEE.2013.05.080

Jones F (2024) Brazil prepares to produce sustainable aviation fuel. Revista Pesquisa FAPESP. https://revistapesquisa.fapesp.br/en/brazil-prepares-to-produce-sustainable-aviation-fuel/#:~:text=More%20than%20600%20million%20liters,and%20Mexico%20currently%20manufacture%20SA. Accessed 24 Feb 2025

Karimi K, Chisti Y (2017) Bioethanol production and technologies. Encyclopedia of Sustainable Technologies, pp 273–284. https://doi.org/10.1016/B978-0-12-409548-9.10111-3

Kasegn MM, Simachew A, Redda YT et al (2024) Evaluation of sugar content and bioethanol production of Ethiopian local varieties "Nech Tinkish" and "Hawaye" sweet sorghum (*Sorghum bicolor* (L.)). Biotechnology for the Environment 1:1–13. https://doi.org/10.1186/S44314-024-00001-6

Li J, Zhao R, Xu Y et al (2022) Fuel ethanol production from starchy grain and other crops: an overview on feedstocks, affecting factors, and technical advances. Renew Energy 188:223–239. https://doi.org/10.1016/J.RENENE.2022.02.038

Lim M, Luckert MK (Marty), Qiu F (2023) Economic opportunities and challenges in biojet production: A literature review and analysis. Biomass and Bioenergy 170:106727. https://doi.org/10.1016/j.biombioe.2023.106727

Mall AK, Misra V, Santeshwari et al (2021) sugar beet cultivation in India: prospects for bio-ethanol production and value-added co-products. Sugar Tech 23:1218–1234. https://doi.org/10.1007/S12355-021-01007-0/TABLES/12

Mandari V, Devarai SK (2021) Biodiesel production using homogeneous, heterogeneous, and enzyme catalysts via transesterification and esterification reactions: a critical review. BioEnergy Res 15:935–961. https://doi.org/10.1007/S12155-021-10333-W

Marzo C, Díaz AB, Caro I, Blandino A (2019) Status and perspectives in bioethanol production from sugar beet. In: Bioethanol production from food crops: sustainable sources, interventions, and challenges, pp 61–79. https://doi.org/10.1016/B978-0-12-813766-6.00004-7

Mohammadi Shad Z, Venkitasamy C, Wen Z (2021) Corn distillers dried grains with solubles: production, properties, and potential uses. Cereal Chem 98:999–1019. https://doi.org/10.1002/CCHE.10445

Mohanty SK, Swain MR (2019) Bioethanol production from corn and wheat: food, fuel, and future. Bioethanol Prod Food Crops:45–59. https://doi.org/10.1016/B978-0-12-813766-6.00003-5

Monlau F, Sambusiti C, Barakat A (2019) Comparison of dry versus wet milling to improve bioethanol or methane recovery from solid anaerobic digestate. Bioengineering:6. https://doi.org/10.3390/bioengineering6030080

Müller C, Scapini T, Rempel A et al (2023) Challenges and opportunities for third-generation ethanol production: a critical review. Eng Microbiol 3:100056. https://doi.org/10.1016/j.engmic.2022.100056

Tabatabaei M, Najafpour G, Mollahosseini A (2015) Biofuel Production. In: Biochemical Engineering and Biotechnology. Elsevier, pp 597–630

Neuling U, Kaltschmitt M (2018) Conversion routes from biomass to biokerosene. In: Biokerosene: status and prospects, pp 435–473. https://doi.org/10.1007/978-3-662-53065-8_18

Neves MF, Milan P, Valério FR et al (2021) Etanol de Milho: Cenário Atual e Perspectivas para a Cadeia no Brasil. Ribeirão Preto

Ng KS, Farooq D, Yang A (2021) Global biorenewable development strategies for sustainable aviation fuel production. Renew Sustain Energy Rev 150:111502. https://doi.org/10.1016/J.RSER.2021.111502

Nordblad M, Pedersen AK, Rancke-Madsen A, Woodley JM (2016) Enzymatic pretreatment of low-grade oils for biodiesel production. Biotechnol Bioeng 113:754–760. https://doi.org/10.1002/BIT.25846

NOVACANA (2024) Processos de fabricação do etanol. https://www.novacana.com/noticias/fabricacao. Accessed 24 Feb 2025

OICS (2025) Captura e utilização de carbono da produção de Etanol na indústria de alimentos. https://oics.cgee.org.br/solucoes/captura-e-utilizacao-de-carbono-da-producao-de-etanol-na-industria-de-alimentos_63441808437eb1283e0a49f9. Accessed 24 Feb 2025

Padder SA, Khan R, Rather RA (2024) Biofuel generations: new insights into challenges and opportunities in their microbe-derived industrial production. Biomass Bioenergy 185:107220. https://doi.org/10.1016/j.biombioe.2024.107220

Pasha MK, Dai L, Liu D et al (2021) An overview to process design, simulation and sustainability evaluation of biodiesel production. Biotechnol Biofuels 14:129. https://doi.org/10.1186/s13068-021-01977-z

Patel A, Shah AR (2021) Integrated lignocellulosic biorefinery: gateway for production of second generation ethanol and value added products. J Bioresour Bioprod 6:108–128. https://doi.org/10.1016/j.jobab.2021.02.001

Pavlečić M, Novak M, Trontel A et al (2021) Mathematical modelling of bioethanol production from raw sugar beet Cossettes in a horizontal rotating tubular bioreactor. Fermentation 8:13. https://doi.org/10.3390/FERMENTATION8010013

Pelkmans L, Cowie A, Berndes G (2024) The role of bioenergy in the energy transition, and implications on the global use of biomass. https://www.ieabioenergy.com/wp-content/uploads/2025/01/Commentary_RoleBioenergy_Dec2024l.pdf. Accessed 20 Feb 2025

Peñarrubia Fernandez IA, Liu DH, Zhao J (2017) LCA studies comparing alkaline and immobilized enzyme catalyst processes for biodiesel production under Brazilian conditions. Resour Conserv Recycl 119:117–127. https://doi.org/10.1016/j.resconrec.2016.05.009

Propeq (2025) Biocombustíveis: Etanol de primeira geração. https://propeq.com/etanol-biocombustiveis/. Accessed 24 Feb 2025

Raízen (2022) Etanol: o que é, como é produzido e quais seus benefícios? https://www.raizen.com.br/blog/etanol. Accessed 24 Feb 2025

Ramos NP, Mendes NC, Da MI et al (2024) Final work report: Regionalization of "typical" sugarcane production profiles for use in Renovabio. https://www.gov.br/anp/pt-br/assuntos/renovabio/arq/arquivos-estudos-relatorio-e-seminarios/sugarcane-agricultural-profiles-renovabio-2024.pdf. Accessed 24 Feb 2025

Ray RC, Uppuluri KB, Trilokesh C, Lareo C (2019) Sweet sorghum for bioethanol production: scope, technology, and economics. In: Bioethanol production from food crops: sustainable sources, interventions, and challenges, pp 81–100. https://doi.org/10.1016/B978-0-12-813766-6.00005-9

Renewable Fuels Association (2025) 2025 Pocket Guide to Ethanol. https://ethanolrfa.org/file/2913/2025%20Pocket%20Guide.pdf. Accessed 24 Feb 2025

Romano SD, Sorichetti PA (2010) Introduction to biodiesel production. Green Energy Technol 29: 7–27. https://doi.org/10.1007/978-1-84996-519-4_2

Soares IP, Schultz EL, Gambetta R, Gonçalves SB (2022) Biogás e suas contribuições para os Objetivos de Desenvolvimento Sustentável. Embrapa Agroenergia

Stamenković OS, Siliveru K, Veljković VB et al (2020) Production of biofuels from sorghum. Renew Sustain Energy Rev 124:109769. https://doi.org/10.1016/J.RSER.2020.109769

Staples MD, Malina R, Suresh P et al (2018) Aviation CO_2 emissions reductions from the use of alternative jet fuels. Energy Policy 114:342–354. https://doi.org/10.1016/J.ENPOL.2017.12.007

Sugarcane (2024) Ethanol. https://www.sugarcane.org/sugarcane-products/ethanol/. Accessed 24 Feb 2025

Tanwar R, Panghal A, Chaudhary G et al (2023) Nutritional, phytochemical and functional potential of sorghum: a review. Food Chem Adv 3:100501. https://doi.org/10.1016/J.FOCHA.2023.100501

Tiwari R, Mishra R, Choubey A et al (2023) Environmental and economic issues for renewable production of bio-jet fuel: a global prospective. Fuel 332:125978. https://doi.org/10.1016/J.FUEL.2022.125978

U.S. Department of Energy (2025) Alternative Fuels Data Center: Biodiesel Production and Distribution. https://afdc.energy.gov/fuels/biodiesel-production. Accessed 21 Feb 2025

UNICA (2025) Ethanol. https://unicadata.com.br/listagem.php?idMn=4. Accessed 24 Feb 2025

UNICA Data (2025) Painel Histórico da Área de Cultivo, Moagem e Produção de Açúcar e Etanol Brasil. https://unicadata.com.br/listagem.php?idMn=4. Accessed 24 Feb 2025

Varghese G, Saeed K, Lu X, J Rutt K (2025) Measurements of the tailpipe emissions characteristics of algae biodiesel compared with the 1st and 2nd generation biodiesel fuels. Fuel 388. https://doi.org/10.1016/j.fuel.2025.134315

Vignesh P, Pradeep Kumar AR, Ganesh NS et al (2021) Biodiesel and green diesel generation: An overview. Oil Gas Sci Technol:76. https://doi.org/10.2516/ogst/2020088

Wan Osman WNA, Rosli MH, Mazli WNA, Samsuri S (2024) Comparative review of biodiesel production and purification. Carbon Capture Sci Technol 13:100264. https://doi.org/10.1016/J.CCST.2024.100264

Wang WC, Tao L (2016) Bio-jet fuel conversion technologies. Renew Sustain Energy Rev 53:801–822. https://doi.org/10.1016/J.RSER.2015.09.016

Wei H, Liu W, Chen X et al (2019) Renewable bio-jet fuel production for aviation: a review. Fuel 254:115599. https://doi.org/10.1016/J.FUEL.2019.06.007

Wirawan SS, Solikhah MD, Setiapraja H, Sugiyono A (2024) Biodiesel implementation in Indonesia: experiences and future perspectives. Renew Sustain Energy Rev:189. https://doi.org/10.1016/j.rser.2023.113911

Wu B, Wang YW, Dai YH et al (2021) Current status and future prospective of bio-ethanol industry in China. Renew Sustain Energy Rev 145:111079. https://doi.org/10.1016/J.RSER.2021.111079

Chapter 13
Second-Generation Biofuels

Natália Paes de Oliveira, Walter José Martinez-Burgos, Maísa Emanuelle Colodel, Laura Sebben Galarce, Israel Carvalho de Assis, and Adriane Bianchi Pedroni Medeiros

Abstract This chapter delves into the essential aspects of second-generation (2G) biofuels, highlighting the innovative conversion technologies and primary feedstocks that position them as sustainable alternatives to fossil fuels and first-generation biofuels. Unlike first-generation biofuels, which are derived from food crops such as corn and sugarcane, 2G biofuels utilize nonfood biomass. This includes a variety of sources, such as corn stover, sugarcane bagasse and vinasse, forest residues, vegetable oils, and animal fats that are not suitable for food consumption. This significant distinction not only mitigates the competition with food production but also addresses concerns related to food security and land use. Furthermore, the chapter outlines the key types of 2G biofuels, each with unique production processes and applications. These include cellulosic bioethanol, which is produced through the fermentation of lignocellulosic materials; biodiesel derived from nonedible oils, such as waste cooking oil; biomethane, which is generated through anaerobic digestion of organic matter; and biohydrogen, created from various biomass sources through thermochemical or biological processes. Each of these biofuels plays a crucial role in advancing sustainable energy solutions, reducing greenhouse gas emissions, and supporting the transition to a more circular economy.

13.1 Introduction

Globally, there is a growing demand for sustainable and renewable energy sources due to concerns about climate change, energy security, and the depletion of fossil fuel reserves (Antizar-Ladislao and Turrion-Gomez 2008; Nanda et al. 2018). In this context, second-generation (2G) biofuels have emerged as a promising alternative to

N. P. de Oliveira · W. J. Martinez-Burgos · M. E. Colodel · L. S. Galarce · I. C. de Assis
A. B. P. Medeiros (✉)
Federal University of Paraná, Department of Bioprocess Engineering and Biotechnology, Centro Politécnico, Curitiba, Paraná, Brazil
e-mail: adrianebpm@ufpr.br

A. L. Woiciechowski et al. (eds.), *Biomass for Bioenergy and Bioproducts*, Biofuel and Biorefinery Technologies 17, https://doi.org/10.1007/978-3-032-19898-3_13

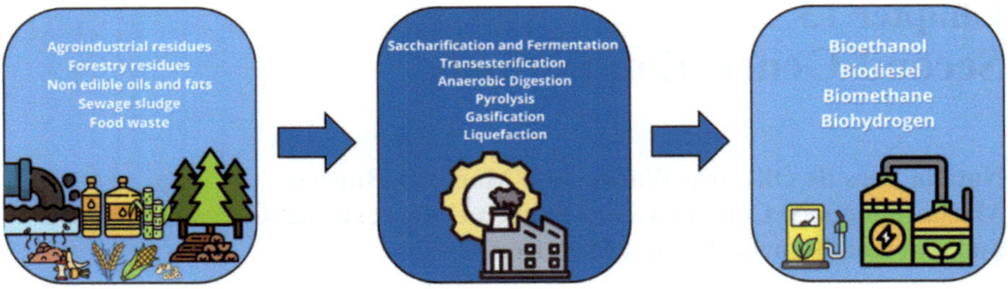

Fig. 13.1 Overview of second-generation (2G) biofuels production

transitioning to sustainable energy (Demirbas 2009). The global market for 2G biofuels was valued at USD 8.2 billion in 2022 and was projected to grow significantly, reaching USD 87.5 billion by 2032, with a compound annual growth rate (CAGR) of 26.8% (Prasad 2023).

Unlike first-generation biofuels, which are derived from food crops such as sugarcane and corn, second-generation biofuels are produced from nonfood biomass, such as lignocellulosic feedstocks, forest and agricultural residues, and municipal waste, making them more sustainable and less competitive with food supplies (Sims et al. 2010; Bhatia et al. 2017; Hirani et al. 2018).

Second-generation biofuels primarily include cellulosic bioethanol, biodiesel from nonedible oils, biomethane, and biohydrogen (Fig. 13.1) (Nanda et al. 2018). Lignocellulosic raw materials can be converted into biofuels through either the thermochemical or biochemical routes (Ganguly et al. 2020). In the thermochemical process, biomass conversion into biofuels involves applying heat, pressure, and catalysts during pyrolysis, gasification, and combustion (Zhang and Zhang 2019; Jha et al. 2022). The biochemical route involves enzymatic saccharification and fermentation, utilizing microorganisms and enzymes to break down biomass into biofuels (Foust et al. 2009; Nanda et al. 2023).

2G biofuels offer several advantages, making them a sustainable alternative to 1G biofuels and fossil fuels (Naik et al. 2010). Since these biofuels can be produced from a diverse range of lignocellulosic feedstocks and residual biomass, they minimize competition with food crops and the food industry, thereby increasing food security (Nanda et al. 2018; Hassan et al. 2019; Chowdhury 2021). Additionally, they have a more significant potential to reduce greenhouse gas emissions (GHG) and are considered more sustainable due to the use of waste materials as raw materials, integrating the concept of a circular economy (Foust et al. 2009; Hirani et al. 2018; Srivastava et al. 2020).

However, 2G biofuels have some disadvantages and face technological challenges because of the complexity of lignocellulosic biomass conversion, which increases production costs and causes logistical difficulties (Carriquiry et al. 2011; Bhatia et al. 2017). Even though research and development have advanced over the years, there are still some impediments to the development of cost-efficient, flexible, and scalable pretreatment and conversion technologies (Sims et al. 2010; Bhatia et al. 2017). Additionally, the logistics system providing a competitive biomass feedstock supply requires improvements to meet a commercial scale (Antizar-Ladislao and Turrion-Gomez 2008).

The use of 2G biofuels contributes to energy diversification and the decarbonization of the transport sector, improving energy security since 2G bioethanol and biodiesel can be integrated into conventional fuels and used in conventional engines and in aviation or maritime transport, and the biomethane can be applied as a substitute for natural gas in vehicles (Demirbas 2009; O. Kukoyi et al. 2015; Hirani et al. 2018; Srivastava et al. 2020; Mignogna et al. 2024). However, to fully realize the potential of 2G biofuels and make their production processes more feasible, continued research and government support are not only beneficial but also necessary. The urgency and importance of these actions cannot be overstated.

13.2 2G Biofuels Types and Conversion Processes

Among the main 2G biofuels are biomethane (bio-CH_4), 2G bioethanol, 2G biodiesel, and biohydrogen (bio-H_2). The most-produced 2G energy vectors in the world are bio-CH_4 with an annual production of 578 million metric tons (Statista 2024a), and 2G bioethanol with a production of approximately 885.6 million gallons, representing 3.0% of the total bioethanol production in 2023 (Sharma et al. 2020; Statista 2024b). In the case of 2G biodiesel, its production is still insignificant, mainly due to production costs, and in the case of bio-H_2, several challenges related to the sterilization of the fermentation medium still need to be overcome to avoid the production of other molecules such as CH_4 and H_2S (Martinez-Burgos et al. 2019, 2021a).

The technologies used to produce 2G biofuels depend on the type of fuel. Thus, in biogas production, although several types of bioreactors are known for anaerobic digestion, the most widely used on an industrial scale are: the continuous stirred tank reactor (CSTR). Due to its high mixing efficiency, which ensures uniform contact between microorganisms and the substrate, in addition to being easily scalable and having good microbial retention, this bioreactor is mainly suitable for effluents with high solids content (Van Lier et al. 2010).

Another bioreactor is the upflow anaerobic sludge blanket (UASB), which is mainly used for low-viscosity effluents such as domestic sewage and some industrial effluents (Brito et al. 2020). For biohydrogen, the same bioreactors can be used; however, either pure strains or hydrogen-producing microbial consortia must be used, such as some species of the genera *Clostridium*, *Enterobacter*, *Acetobacter*, *Rhodobacter*, *Clespiella*, *Escherichia*, *Pyrococcus*, and *Thermococcus* (Martinez-Burgos et al. 2020; Rosa et al. 2020; do Nascimento Junior et al. 2021).

Biogas undergoes a purification process that utilizes physical absorption systems, such as pressure water washing, to remove CO_2, H_2S, and traces of other volatile compounds, resulting in the production of bio-CH4. Some methods use amine solutions, like methyldiethanolamine and monoethanolamine, to enhance CO_2

capture efficiency (Law et al. 2018; Aghel et al. 2022). Another commonly used technique is pressure swing adsorption (PSA), which employs absorbent materials, primarily zeolites and activated carbon. In this process, biogas is compressed to high pressures between 4 and 10 bar and passed through a column containing selective materials that retain traces of CO_2, H_2S, and water vapor. After this phase, desorption is performed at low pressure. When the absorbent materials become saturated, the system pressure is reduced to release the retained contaminants, allowing the adsorbent to be regenerated for reuse in subsequent cycles.

The yields of biomethane production (measured as L CH_4/g chemical oxygen demand (COD)) vary based on the organic matter content of the effluent. For instance, average yields for different types of effluents are as follows: cassava processing wastewater yields approximately 0.35 L CH_4/g COD, vinasse wastewater yields around 0.23 L CH_4/g COD, and slaughterhouse effluents yield about 0.14 L CH_4/g COD (Martinez-Burgos et al. 2021b). The cost of producing biomethane ranges from €30 to €80/MWh (which is roughly equivalent to 100 m^3 of methane). This cost depends on several factors, including the type of substrate and the logistics involved in transportation, as well as the purification technology employed. One of the expensive but efficient purification methods is pressure swing adsorption (PSA). Additionally, the scale of production plays a significant role in influencing costs; larger plants typically have lower production expenses (European-Biogas-Association 2023).

In the case of 2G bioethanol, the lignocellulosic biomass is initially pretreated to remove the recalcitrant fraction of the biomass (lignin), either by steam explosion or chemical treatments (acid or base) at high pressures and temperatures (Kumar et al. 2020). Then, the biomass (cellulose + hemicellulose) is treated with cellulose-lytic enzymes to transform cellulose and hemicellulose into fermentable sugars, such as glucose and xylose. The sugars are fermented to ethanol using genetically modified yeasts of *Saccharomyces cerevisiae* or some bacteria such as *Zymomonas mobilis* capable of fermenting the aforementioned monomers. Finally, purification occurs by fractional distillation at a temperature of 78.5 °C. However, to obtain anhydrous ethanol, additional steps are required, such as dehydration with molecular sieves (the primary method used on an industrial scale) or azeotropic distillation using cyclohexane (Wang and Cheng 2024). The estimated production cost for 2G ethanol by 2025 is approximately €0.25 ($1.5BR) per liter of ethanol (Agencia-FAPESP 2017). The yield is significantly lower when compared to first-generation bioethanol, reaching conversions between 30% and 50% (Kang et al. 2014).

Finally, for second-generation biodiesel production, the process and technologies depend on the type of material used. When reused oils are used, they must initially be decanted and filtered to remove traces of impurities. In traditional biodiesel production processes, the transesterification reaction is carried out in stirred tank reactors. A technology that has been gaining momentum is the use of cavitation reactors due to their efficiency in mixing and intensifying chemical reactions. These reactors use hydrodynamic, acoustic, or controlled cavitation to improve the interaction between oil and methanol, reducing reaction time and increasing yield (Panda et al. 2020; Bizualem and Nurie 2024). Subsequently, the biodiesel is separated from glycerin through decantation and centrifugation, and washing and drying can even be used to remove traces of glycerin. According to (van Eijck et al. 2014), the production costs of 2G biodiesel can vary between $ USD 0.6–1.0/L.

13.2.1 Biofuels from the Biochemical Route

2G energy vectors are produced via microorganisms, except for biodiesel, which is obtained through a transesterification reaction after the extraction of microbial oil.

13.2.1.1 2G Bioethanol

For the production of 2G bioethanol, the lignocellulosic biomass is initially subjected to physical–chemical hydrolysis processes and then to enzymatic hydrolysis. The first hydrolysis is to separate lignin, which is a recalcitrant part of the biomass, from cellulose and hemicellulose. In enzymatic hydrolysis, cellulose and hemicellulose are hydrolyzed to their monomeric glucose and fructose, which are fermentable sugars (Martinez-Burgos et al. 2023; Savian et al. 2024).

Alcoholic fermentation is a process that occurs under anaerobic conditions and is mainly developed by yeasts and some bacteria (Maicas 2020). Although several species are known, the most widely used in the industry is *Saccharomyces cerevisiae*. Due to its high efficiency, high yields are obtained; this type of yeast also tolerates high concentrations of alcohol without undergoing premature inhibition processes (Parapouli et al. 2020). It is noteworthy that this yeast is highly resistant to environmental changes. In other words, it can grow in different conditions of temperature, osmotic pressure, and pH. *S. cerevisiae* has the ability to ferment different types of sugars, mainly monosaccharides such as glucose and fructose, and some disaccharides such as maltose. Another advantage is that *S. cerevisiae* grows quickly and competes effectively with other microorganisms (Parapouli et al. 2020).

Initially, sugar enters the cell through active or passive transport, depending on the type of sugar and its concentration in the growth medium. In the case of monomeric sugars in high concentrations, such as glucose and fructose, the cell uses proteins such as permeases that help sugar cross the membrane without expending energy. When sugar is in low concentrations, the cell uses proton pumps (H+) to transport sugar to the interior of the cell with energy expenditure (Bisson et al. 1993).

Ethanol synthesis occurs in three macro-stages: (i) Glycolysis, which occurs in the cell cytoplasm, where the glucose molecule is split into two molecules of pyruvic acid ($C_3H_4O_3$), in addition to generating energy in the form of adenosine triphosphate (ATP) and nicotinamide adenine dinucleotide (NADH). (ii) Decarboxylation of pyruvic acid, in the absence of oxygen, PA undergoes a decarboxylation process in which it loses a CO_2 molecule and is transformed into acetaldehyde (C_2H_4O).

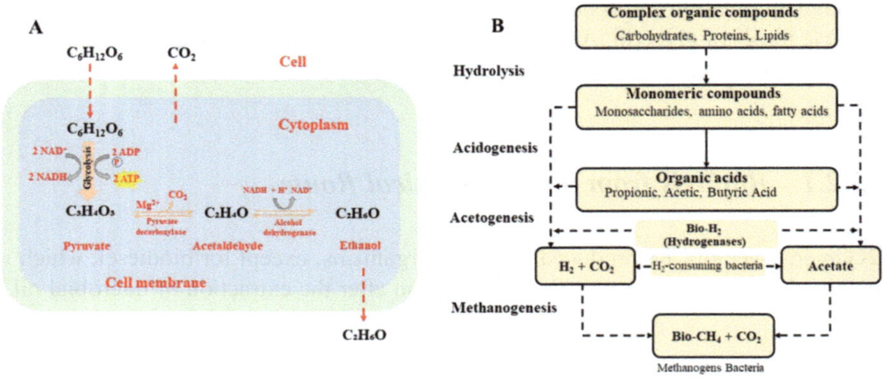

Fig. 13.2 (**a**) Metabolic pathway for ethanol synthesis. (**b**) Metabolic pathway for the production of bio-H_2 and bio-CH_4

(iii) Reduction of acetaldehyde to ethanol, in this last stage, acetaldehyde gains electrons from NADH, reducing it to ethanol (Fig. 13.2a). In addition, NAD^+ is regenerated, which allows the glycolysis process to continue (Ma and Liu 2010; Maicas 2020).

Finally, ethanol is released into the environment. According to Guijarro and Lagunas (1984), the release of ethanol from the yeast cytoplasm occurs through simple diffusion processes through the plasma membrane, since ethanol molecule, in addition to being small, is relatively polar and can therefore easily cross the lipid layers of the membrane without the need for transporters.

13.2.1.2 Biohydrogen and Biomethane

Biohydrogen (bio-H_2) and biomethane (bio-CH_4) are produced through anaerobic processes primarily facilitated by bacteria. Biohydrogen is generated as an intermediate metabolite during biodegradation, a process referred to as dark fermentation (Martinez-Burgos et al. 2020; do Nascimento Junior et al. 2021). In contrast, bio-CH_4 is the final product resulting from the anaerobic biodegradation of organic matter (Karp et al. 2022; de Carvalho et al. 2023). The biodegradation process occurs in four stages: hydrolysis, acidogenesis, acetogenesis, and methanogenesis (see Fig. 13.2).

In the first step, polysaccharides are hydrolyzed to their monomeric units: polysaccharides into monosaccharides, proteins to amino acids, and lipids to fatty acids. Bio-H_2 is produced in acetogenesis, along with other metabolites, for example, some short-chain fatty acids, such as acetic acid and butyric acid. Stoichiometrically, when the process follows the acetate pathway, 4 moles of bio-H_2 are produced for each mole of glucose, and in the butyrate pathway, only 2 moles of bio-H_2 are produced for each mole of glucose (Sydney et al. 2020).

In addition, some solvents, such as acetone and bioethanol, are also generated in small proportions (Martinez-Burgos et al. 2020). When methanogenic bacteria are not present in the system, the process ends only with the formation of bio-H_2 and CO_2 (dark fermentation). On the other hand, when methanogenic bacteria are present in the process, they develop methanogenesis, that is, they take the final products of acetogenesis, such as H_2, CO_2, acetate, etc., and are transformed into bio-CH_4, a process known as biodigestion of organic matter.

13.2.2 *Biofuels from the Thermochemical Route*

The increasing demand to decarbonize energy systems, reduce dependency on fossil fuels, and transition to circular economies has led to a growing interest in biomass thermochemical conversion technologies as paths to obtain renewable fuels, chemicals, and materials. In parallel to the biochemical methods, these processes align with global priorities such as climate change mitigation, energy and food security, and valorization of agricultural byproducts (Ibarra-Gonzalez and Rong 2019).

Thermochemical routes for biomass processing involve the application of heat and/or chemical reactions to convert biomass, typically lignocellulosic or lipid-based materials derived from agricultural and urban waste or algae, into biofuels or other valuable materials (Ong et al. 2019; Zhang and Zhang 2019). This allows for the production of lower-greenhouse-gas-emissions biofuels from nonfood resources, alleviating concerns about food security while also utilizing waste materials, reducing the carbon footprint of the whole chain significantly.

These methods include combustion, pyrolysis, gasification, and liquefaction, all conducted in relatively high temperatures (200–1200 °C) when compared to biochemical routes. The key points that set apart these thermochemical processes and define the characteristics of the final products are temperature, time of processing, and the presence or absence of oxygen (Zhang et al. 2011; Damartzis and Zabaniotou 2011). Biomass composition also significantly affects the resulting products (Jha et al. 2022).

13.2.2.1 2G Biofuels from Pyrolysis

Pyrolysis is the thermal decomposition of biomass in the absence of or with minimal quantities of oxygen, preventing combustion during the process and producing bio-oil (liquid fraction), biochar (solid fraction), and fuel gases (Demirbaş and Arin 2002; Rahimi et al. 2022; Escalante et al. 2022). Biochar is an energy-dense, highly stable, and resistant to biological decomposition material, used mainly for energy obtention, soil improvement, pollution remediation, and carbon sequestration (Fahmy et al. 2020). Bio-oil is a complex mixture of water, inorganic, and organic compounds, primarily containing oxygenated hydrocarbons, such as aldehydes, ketones, phenols, and acids (Xu et al. 2011). It serves as a versatile intermediate for producing fuels, such as gasoline, diesel, and jet fuel, as well as plastics and chemicals. Advances in refining technologies are still needed to maximize the bio-oil potential as a sustainable alternative to fossil oil (Jacobson et al. 2013; Hu and Gholizadeh 2020).

The gases produced during pyrolysis, primarily composed of methane (CH_4), carbon monoxide (CO), carbon dioxide (CO_2), and hydrogen (H_2), can be utilized for energy production or further chemical processing (Demirbaş and Arin 2002; Rahimi et al. 2022). They can be used to generate heat or power for the pyrolysis process itself, making it a more energy-efficient method (Ibarra-Gonzalez and Rong 2019; Rahimi et al. 2022).

This thermochemical process is operated in an inert atmosphere, usually nitrogen or argon is utilized, and can be operated under a wide range of conditions, with temperatures ranging from 300 to 1000 °C and reaction times varying from a few seconds to up to 12 h. These variations in operational conditions lead to different yields and compositions of the final products (Strugnell and Patrick 1995; Bridgwater and Peacocke 2000; Nanda et al. 2014; Tripathi et al. 2016). Based on these differences, pyrolysis can be categorized into three main types: slow pyrolysis, fast pyrolysis, and flash pyrolysis, each with specific characteristics: heating rate, temperature, residence time, and particle size (Brown 2015; Escalante et al. 2022).

Slow pyrolysis normally employs lower temperatures (300–500 °C), slower heating rates (0.1–1 °C/s), vapor residence time in the reactor ranging from 5 to 30 min, which can reach hours or days, and larger biomass particle sizes (5–50 mm) (Patra et al. 2021; Okolie et al. 2022; Igliński et al. 2023; Güleç et al. 2024). In general, the main goal of slow pyrolysis is to maximize biochar production with low liquid yields (Grande et al. 2021; Patra et al. 2021). In opposition, the fast pyrolysis is usually used to maximize bio-oil yields (60–70% of dry feed) with a fast heating rate (10–200 °C/s), moderate to high temperatures (400–800 °C), a shorter vapor residence time of 0.5–5 s, and thinner biomass particles (1–3 mm) (Bhaskar et al. 2011; Okolie et al. 2021; Makepa et al. 2023; Igliński et al. 2023; Güleç et al. 2024). However, to avoid secondary reactions, fast pyrolysis requires rapid cooling of the pyrolysis vapors to obtain the bio-oil product (Ibarra-Gonzalez and Rong 2019; Makepa et al. 2023).

Flash pyrolysis has extremely high heating rates (>1000 °C/s), the highest temperatures among the three types (650–1000 °C), the shortest vapor residence time (<1 s) and usually requires tiny biomass particle sizes (<0.2 mm) (Okolie et al. 2021, 2022; Igliński et al. 2023). These process conditions minimize heat and mass transfer limitations, maximizing bio-oil yields up to 75% of dry feed (Rahimi et al. 2022; Makepa et al. 2023). Slow pyrolysis is mainly performed in fixed bed, drum, rotatory kilns, and screw/auger reactors, whereas fast and flash pyrolysis are carried out in fluidized bed, rotating cones, vacuum, and ablative reactors, or transported beds (Brown 2015; Rahimi et al. 2022).

The overall efficiency and sustainability of pyrolysis are closely linked to optimizing the conversion of biomass into valuable products, with ongoing research focused on improving product yields, refining bio-oil quality, and enhancing gas utilization (Ibarra-Gonzalez and Rong 2019; Igliński et al. 2023).

13.2.2.2 2G Bio-Oil from Liquefaction

The liquefaction of biomass thermochemically converts the solid biomass into a liquid product (bio-oil) in the presence of a solvent, usually water or alcohols (Bhaskar et al. 2011; Okolie et al. 2022). It occurs under anaerobic conditions and high temperatures (250–500 °C) and diverges from pyrolysis primarily in its goal of maximizing bio-oil yield, with less focus on solid or gaseous byproducts (Gollakota et al. 2018). This process is conducted under higher pressure conditions, typically ranging from 5 to 25 MPa. Liquefaction often involves the use of solvents and catalysts to improve the breakdown of biomass (Wang et al. 2013; Biswas et al. 2021). These additives aid in the conversion of complex organic materials, particularly wet feedstocks such as algae, sewage sludge, and high-moisture agricultural residues, which are challenging to process using pyrolysis (Gollakota et al. 2018; Okolie et al. 2022).

The liquefaction process can be classified into direct liquefaction and indirect liquefaction. The direct liquefaction can be divided based on solvent choice into hydrothermal liquefaction (HTL) and solvolysis (Bhaskar et al. 2011; Baredar et al. 2020). Direct liquefaction converts biomass directly into liquid products employing catalysts and solvents (e.g., water, ethanol, acetone) at high temperatures (350–500 ° C) and moderate pressure (5–25 MPa), being suitable for dry biomass (Ibarra-Gonzalez and Rong 2019; Okolie et al. 2022). The indirect liquefaction occurs in two steps: First, the biomass is converted into a gaseous intermediate product, mainly synthesis gas (syngas), at high temperatures (<700 °C), without combustion and with a controlled amount of a gasifying agent. In the second step, this syngas is processed through catalytic synthesis processes, such as Fischer-Tropsch (F-T) synthesis or methanol synthesis, to produce long-chain liquid fuels, for example, synthetic diesel, aviation fuel, or synthetic oil (Sims et al. 2010; Bhaskar et al. 2011; Baredar et al. 2020).

Hydrothermal liquefaction (HTL) utilizes water as a solvent under high pressure and temperature (10–25 MPa, 300–400 °C), making it ideal for wet feedstocks like algae and sewage sludge and producing bio-oil with higher water content, alongside gases and an aqueous phase for further processing (Gollakota et al. 2018). Solvolysis uses organic solvents, commonly of low molecular weight, such as alcohols (methanol and ethanol), and ketones, and the operational conditions tend to be lighter than those using water as the solvent; normally temperature and pressure range from 250 to 350 °C and 1–25 MPa, respectively (Ibarra-Gonzalez and Rong 2019; Okolie et al. 2022; Syed et al. 2024; Meena et al. 2024). This method yields bio-oil with lower water content and higher energy density, ideal for fuel refining (Shui et al. 2010). There is also supercritical liquefaction that leverages supercritical fluids like water or CO_2, offering efficient breakdown of biomass into cleaner, more homogeneous bio-oil, being particularly effective for the use of plastics as a substrate (Mazaheri et al. 2010).

The solvent choice is crucial, as it affects the solubility of biomass, especially lignin, and influences the reaction mechanisms. Mixtures of solvents (cosolvents), such as water with methanol or ethanol, are being employed to combine the advantages of different polarities and properties (Okolie et al. 2022; Syed et al. 2024). The catalysts employed in biomass liquefaction may vary as well, depending on the process and feedstock. Common options include homogeneous catalysts such as alkali hydroxides (e.g., NaOH, KOH) or acids (e.g., H_2SO_4) (Biswas et al. 2021), heterogeneous catalysts such as metal oxides (e.g., Al_2O_3, TiO_2) or supported metals (e.g., Ni, Ru) (Wang et al. 2013), and bifunctional catalysts combining acid/base and metal sites (e.g., Pt/Al_2O_3) (Sekhon et al. 2021). In supercritical liquefaction, catalysts like Ru/C in supercritical water are often used (Miao et al. 2017).

13.2.2.3 2G Syngas from Gasification

Gasification focuses on producing a gaseous output known as syngas while minimizing solid or liquid byproducts (Ibarra-Gonzalez and Rong 2019). Syngas, also known as synthesis gas, is mainly a mixture of hydrogen (H_2) and carbon monoxide (CO) but it can also contain other gases in smaller quantities, such as carbon dioxide (CO_2), methane (CH_4), and other low molecular weight hydrocarbons (ethane, ethylene, propane), as well as traces of ammonia (NH_3), nitrogen oxides (NOx), sulphur oxides (SOx), and low molecular weight alcohols (Okolie et al. 2021; Rahimi et al. 2022; Enebe et al. 2023; Güleç et al. 2024). Unlike liquefaction, gasification operates at higher temperatures (700–1000 °C) and low to moderate pressures (1–5 MPa). The production of syngas occurs in a controlled environment in the presence of an oxidant agent (air, oxygen, or steam) and in a multi-step process (Sikarwar et al. 2017; Ibarra-Gonzalez and Rong 2019; Enebe et al. 2023).

Initially, biomass is dried at temperatures between 100 and 200 °C to remove moisture, which enhances energy efficiency in the subsequent stages. Biomass with 5–30% humidity is adequate for the gasification process (Güleç et al. 2024). The biomass then undergoes pyrolysis at temperatures of 200–500 °C, where it thermally decomposes in the absence of oxygen, resulting in the release of volatile gases (CO, CO_2, CH_4, H_2), tars, and solid char (Damartzis and Zabaniotou 2011). Next is the oxidation stage (700–1000 °C), where a limited amount of oxygen is introduced. This partially combusts the volatiles and char, generating heat that sustains the process. In the final reduction stage (also at 700–1000 °C), the remaining char reacts with air, O_2, CO_2, and H_2O in endothermic reactions such as the Boudouard reaction and the water-gas reaction to produce syngas (Damartzis and Zabaniotou 2011; Güleç et al. 2024; Meena et al. 2024).

The resulting syngas, composed of CO, H_2, CO_2, and CH_4, is then purified for various applications, including energy generation, fuel synthesis, or chemical production (e.g., methanol). The composition and yield of syngas are influenced by factors such as feedstock, temperature, and the type of oxidant used (Göransson et al. 2011; Sikarwar et al. 2017; Gao et al. 2023). Syngas can be utilized for electricity generation, combined heat and power (CHP) systems, and fuel production (e.g., hydrogen, synthetic natural gas, and liquid fuels via Fischer-Tropsch synthesis). Additionally, it serves as a feedstock for chemical synthesis, including methanol, ammonia, and oxo-chemicals (Sikarwar et al. 2017; Ibarra-Gonzalez and Rong 2019; Rahimi et al. 2022).

The gasification process can be classified according to the necessity of an external heat source to support the reactions into autothermal or direct gasification, when the gasifying agent already includes O_2 that promotes the oxidation of biomass inside the gasifier and generates heat, and allothermal or indirect gasification, when an external source needs to provide heat to the system (Pio and Tarelho 2021). Although conventional gasification is the main route, there are variations such as supercritical water gasification (SCWG), especially suitable for wet biomass, which operates under supercritical water conditions ($T > 374$ °C, $P > 218$ atm) to produce syngas rich in H_2 and/or CH_4 (Güleç et al. 2024).

Three main types of reactor designs are used for gasification: fixed bed, fluidized bed and entrained flow reactors. The fixed bed reactors are usually used for small-scale production and produce low-heating-valued syngas and can be divided into updraft and downdraft configurations (Bhaskar et al. 2011; Kaur et al. 2019; Pio and Tarelho 2021). Fluidized bed gasifiers are widely used and offer uniform temperature distribution, better gas-solid contact, and high heat transfer rates. They can have installed power over 100 MWth and are divided into bubbling fluidized bed (BFB) and circulating fluidized bed (CFB) (Damartzis and Zabaniotou 2011; Pio and Tarelho 2021; Meena et al. 2024). Entrained flow reactors operate at higher pressures (2–8 MPa) and temperatures (>1000 °C) and are more often used in large-scale production (Damartzis and Zabaniotou 2011; Meena et al. 2024).

13.2.2.4 2G Biodiesel

To produce 2G biodiesel, nonedible oils and fats, residual oils and fats, and waste products from households, restaurants, and slaughterhouses are used as raw materials, including cottonseed oil, animal tallow, fish oil, used cooking oil, jatropha oil, chicken fat, and other waste oils or oils derived from nonfood crops (Gashaw et al. 2015; Verma et al. 2016; Spanou et al. 2024; Praveena et al. 2024). However, when using this material, it is necessary to pretreat the oils and fats to remove high levels of free fatty acids or water content before the transesterification process converts the oil into biodiesel (Verma et al. 2016; Malode et al. 2021).

Biodiesel production primarily relies on transesterification, where triglycerides, usually from animal fat or vegetable oil source, react with alcohol (e.g., methanol) in the presence of a catalyst (e.g., NaOH) to produce fatty acid alkyl esters (biodiesel) and glycerol, in a process that breaks ester bonds in triglycerides, replacing glycerol with smaller alcohol molecules (Pinto et al. 2005). The catalyst can be a strong alkali such as NaOH or an acid, but recently it has also been used as an enzymatic catalyst with lipases because enzymes can increase efficiency and selectivity, resulting in a better-quality biodiesel (Gashaw et al. 2015; Spanou et al. 2024). Key factors affecting the transesterification products include the alcohol-to-oil ratio, catalyst type, temperature (50–65 °C), and reaction time (1–2 h) (Hamze et al. 2015; Ong et al. 2021).

Another process that can be used to produce 2G biodiesel is the hydrogenation process, which consists in the conversion of triacylglycerols into biodiesel in the presence of hydrogen gas and a metal-based catalyst at high pressure and medium temperature. (Praveena et al. 2024; Zhang et al. 2024). However, it faces some limitations in terms of the quality of the raw material, the catalysts, and the production processes. A new process that combines the suspended-bed hydrogenation and fixed bed hydrogenation arises as an alternative to overcome some of these limitations, allowing the use of a wider range of residual raw materials and aiming to efficiently decompose and remove impurities such as metals, sulphur and phosphorus. This combined process demonstrates a higher yield of 2G biodiesel (80–83%) than conventional hydrogenation in a fixed bed (usually 70–80%) (Zhang et al. 2024).

Biodiesel is carbon-neutral, significantly reducing greenhouse gas emissions and particulate matter compared to conventional diesel. However, its production and commercialization face challenges such as high feedstock costs, competition with the food supply, cold flow limitations, and oxidative instability (Saleh et al. 2023). Furthermore, the efficient management of byproducts such as glycerol, along with the energy-intensive nature of the production process, presents additional hurdles.

13.3 Feedstocks for 2G Biofuel Production

Sustainable economic and industrial growth requires safe, renewable energy resources that do not compromise food security, enable cost-effective processing and energy output, and align with the principles of a circular economy. These resources should also contribute to a carbon-neutral energy matrix, helping to address the global climate crisis.

Using nonfood biomass like agricultural residues, forestry byproducts, and waste oils as raw materials for energy production offers a promising solution to fulfill these requirements. Currently, the most relevant feedstocks globally are corn stover, sugarcane bagasse, and vinasse, forest residues, vegetable oils, and animal fats, as they are low-value, high-volume byproducts of agribusiness, being widely and heterogeneously distributed across different geographical regions (Naik et al. 2010). This includes countries with extensive agricultural and forestry industries, such as the United States, Brazil, and various European nations, as well as regions with large-scale livestock production like Argentina and Australia (FAO 2017).

13.3.1 Sugarcane Residues

Sugarcane (*Saccharum officinarum*) is one of the most widely produced agricultural commodities in the world, being extensively used for sugar and ethanol production. Brazil leads the production of sugarcane, generating 759 million tons in 2024, followed by India and China (FAO 2024). However, its production chain generates various byproducts that can be utilized for the production of second-generation (2G) biofuels.

The main byproducts used for 2G biofuel production are bagasse, straw, and vinasse (Gholami et al. 2021). Sugarcane bagasse (SCB) is a residue from juice extraction, comprising about 1/3 of the harvested plant and composed of 32–45% cellulose, 20–32% hemicellulose, 17–32% lignin, and 1–9% ash, making it a promising feedstock for the obtention of bio-oil, biogas, and biochar through thermochemical processes and 2G bioethanol through biochemical routes (Alokika et al. 2021; Shabbirahmed et al. 2022). Additionally, SCB is used for direct heat and energy production through combustion (Iwuozor et al. 2022). Despite its seasonal availability, SCB's potential is vast (Iwuozor et al. 2022). Its versatility supports sustainable energy, bioeconomy growth, and reduced environmental impact, highlighting its importance in advancing renewable energy and industrial applications. Straw, which consists of the dry leaves and tops of the sugarcane plant, has a similar composition, with approximately 33–40% cellulose, 26–30% hemicellulose, 19–31% lignin, and 2–8% ash (Szczerbowski et al. 2014).

The production of 2G bioethanol from bagasse and straw consists of four main stages (Fig. 13.3): (i) physical, chemical, or biological pretreatment for lignin removal; (ii) enzymatic hydrolysis, where polysaccharides are broken down into fermentable sugars; (iii) fermentation, where sugars are converted into ethanol by microorganisms such as *Saccharomyces cerevisiae*; (iv) distillation for product purification. There are different technological approaches for the hydrolysis and fermentation stages, the main ones being: separate hydrolysis and fermentation (SHF), simultaneous saccharification and fermentation (SSF), co-fermentation (CF) of pentoses and hexoses, and simultaneous saccharification and co-fermentation (SSCF). Recently, consolidated bioprocessing (CBP) has emerged as a promising alternative, as this process integrates cellulase production, biomass hydrolysis, and ethanol fermentation into a single step, reducing the need for external enzymatic inputs and improving substrate utilization, which can potentially increase process efficiency and lower costs (Beltrán-Ramírez et al. 2019; Pan et al. 2022).

Another significant byproduct of sugarcane processing is sugarcane vinasse. It is generated from the distillation of first-generation (1G) ethanol and contains organic acids (such as lactate and acetate), alcohols (mainly glycerol and ethanol), and carbohydrates (Moraes et al. 2015). Approximately 10–15 L of vinasse are produced per liter of ethanol (Carrilho et al. 2016), resulting in vast quantities globally. Traditionally, vinasse has been used as a fertilizer in sugarcane fields due to its high nutrient content, including potassium, nitrogen, and organic matter (Reis and Hu 2017). However, improper disposal or over-application can lead to environmental issues such as soil salinization and groundwater contamination. To address these challenges, vinasse is being increasingly explored for resource recovery, particularly for biogas production via anaerobic digestion, aiding renewable energy generation. It can also be processed to yield valuable compounds like organic acids, biofertilizers, and bio-based chemicals, promoting a circular bioeconomy (Naspolini et al. 2017). Advanced treatments, including thermochemical conversion and microbial fermentation, are being developed to enhance vinasse's value while reducing environmental risks.

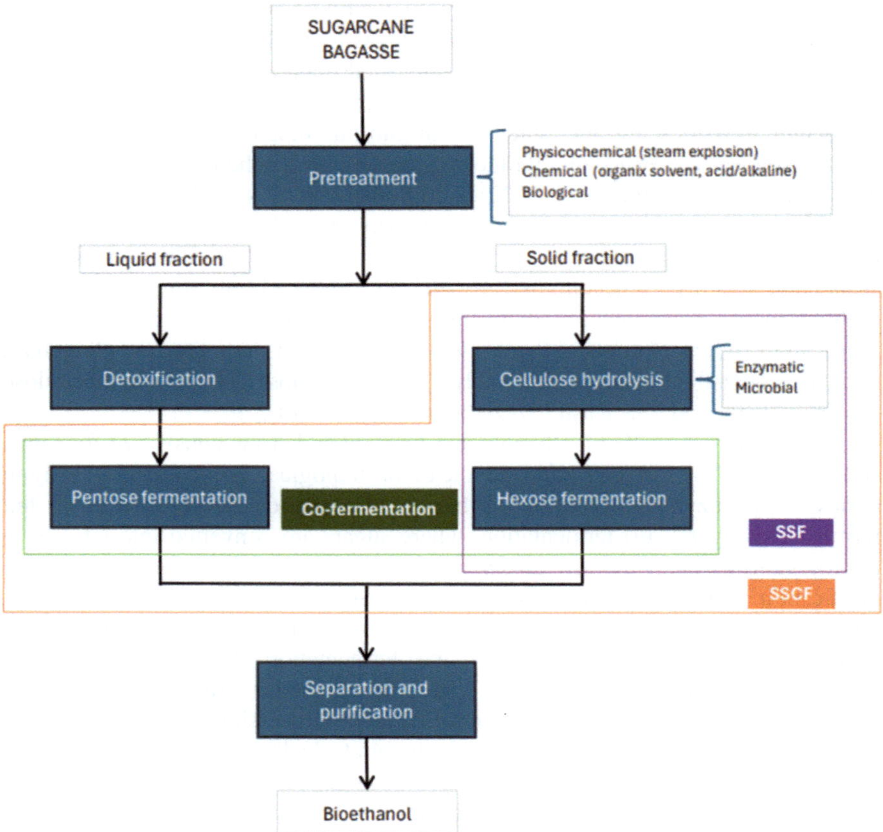

Fig. 13.3 Process schemes of bioethanol production from sugarcane bagasse

The production of biogas and biomethane through anaerobic digestion is a potential route for valorizing sugarcane residues. This process occurs in four stages: enzymatic hydrolysis, acidogenesis, acetogenesis, and methanogenesis, resulting in a biogas primarily composed of methane (CH_4) and carbon dioxide (CO_2) (Ojo-kupoluyi et al. 2024). Before these stages, straw and bagasse need to undergo pretreatment for lignin removal. Additionally, in some refineries, vinasse undergoes neutralization treatments with sulfate compounds, which can lead to the formation of hydrogen sulfide (H_2S) during anaerobic digestion, potentially causing microbial inhibition (Formann et al. 2020).

In addition to bioethanol and biogas, sugarcane residues can be used for biohydrogen production. Bagasse and vinasse can undergo dark fermentation or photofermentation, with dark fermentation requiring lower energy input. Hydrogen can also be produced through thermochemical processes such as pyrolysis, gasification, and hydrothermal liquefaction, which convert biomass into synthesis gas (H_2 and CO). However, these thermochemical approaches pose environmental challenges and have high energy costs (Pan et al. 2022).

Bagasse can also be used for biometanol production through thermochemical and biochemical pathways. The thermochemical route is based on the gasification or pyrolysis of bagasse, generating syngas, which is converted into methanol using catalysts such as Cu/ZnO or Ni/Al_2O_3; however, this process requires high energy input and advanced infrastructure for purification. The biochemical route utilizes methanotrophic bacteria to convert methane from biogas into methanol through the action of the enzyme methane monooxygenase (MMO), representing a promising field that is still in the experimental stage and that therefore requires a deeper understanding of the metabolic mechanisms involved (Pan et al. 2022).

Finally, biobutanol emerges as a viable alternative for utilizing straw and bagasse, being a biofuel with a higher calorific value, lower volatility, and lower corrosivity compared to ethanol (Pratto et al. 2020). Biobutanol production occurs through the acetone-butanol-ethanol (ABE) fermentation process by Clostridium bacteria, with a product ratio of 3:6:1. The process involves an acidogenic phase, where organic acids and gases are produced, followed by a solventogenic phase, which converts the acids into solvents. However, fermentation can be inhibited by compounds derived from biomass pretreatment, and the downstream process is more complex and expensive compared to ethanol (Pratto et al. 2020; Pan et al. 2022).

13.3.2 Corn Residues

The production and processing of corn grain, either for the food or the fuel industry, generate a great number of residues composed of lignocellulosic material (cellulose, hemicellulose, and lignin), such as corn stover and corn cob. Per kilogram of dry corn grain, one kilogram of corn stover and 0.25 kg of corn cob are produced, this material, when compared to corn stover, has a higher percentage of cellulose and a lower percentage of lignin (Ruan et al. 2019) Therefore, by estimation, in 2024–2025, when 1.21 billion metric tons of corn were produced worldwide, the production of corn stover and corn cob would be around 1.21 and 0.30 billion metric tons, respectively (Statista 2025). However, not all this stover can be used as a bioenergy feedstock, as some is left in the fields to protect and enhance soil quality, while another portion is reserved for animal feed or bedding. Its use needs to be carefully weighed against these competing purposes. When in depth about the conversion of those materials into biofuel, some references may merge the corn stover and corn cob as only corn stover, given that the process of conversion is pretty similar (Liu et al. 2022; Celignis Analytical 2024; Porninta et al. 2024).

Corn stover primarily consists of cellulose, hemicellulose, and lignin, with approximate weight percentages of 35%, 20%, and 12%, respectively (Kumar and Singh 2019). Regarding corn stover conversion, this material can be turned into ethanol by biological or thermochemical conversion, which will not be described here (Khan et al. 2021). The biological conversion encompasses the process of first submitting the stover to a pretreatment that may be of one or more of three natures to surpass the recalcitrance of the material, the complex of cellulose and hemicellulose with lignin, being those physical, such as milling and steam explosion; biological, using microorganisms or enzymes; or chemical, which uses mainly acid or alkali agents (Molaverdi et al. 2021; Yang et al. 2022; Sun et al. 2022; Zhang et al. 2023). After that, it passes through a saccharification (hydrolysis) process of a chemical or enzymatic nature, which converts cellulose and hemicellulose into suitable six or five-carbon sugars for fermentation, a process that is followed by the recovery of the biofuel, commonly by distillation (Aghaei et al. 2022).

In the process of fermentation, due to its nature and the previous process, a solid residue, mainly composed of lignin, is produced. This solid residue can be used to produce heat by burning the material or be transformed into coproducts that can be used in the chemical or pharmaceutical industry, such as phenolic and alcoholic compounds, carboxylic acids, benzoquinones, syngas, and bio-oil (Yu et al. 2021; Zhu et al. 2022; Fang et al. 2023; Brown 2023; Tong et al. 2024; Kim et al. 2025). Other possible coproducts include D-xylitol, which may be refined out of a five-carbon sugar by a biological process, instead of being converted into ethanol (Baptista et al. 2018).

Given the possible processes, it is clear that there are many alternatives for the production of biofuel with corn residues. However, those can only be feasible through a deep techno-economic analysis. It is common, in the studies of the implantation of biorefineries for this production, that the higher costs are related to the pretreatment stage and to the inputs on the process, such as enzyme cocktails and the main material (Kazi et al. 2010; Patel and Shah 2021). Furthermore, when using corn cobs, there is evidence in economic analysis that the production of D-xylitol is more economical than converting the five-carbon compound into ethanol (Gandam et al. 2022). Therefore, a deep study is needed to assess the possible problems and bottlenecks of the designed project.

13.3.3 Forest Feedstocks

Forest residues are also an important source of biomass for 2G biofuel production, with availability varying by region due to differences in forest cover, logging practices, and climate. In North America and Europe, forest residues are abundant, especially in the United States, Russia, Sweden, and France, with estimates suggesting hundreds of millions of tons annually, while tropical regions like South America and Southeast Asia also contribute substantially (FAO 2021).

Forest resources used in the production of lignocellulosic biofuels can be classified into three main categories: harvest residues (branches, tree tops, small or low-quality trunks), sawmill residues (sawdust, wood chips, bark, and hog fuel), and fast-growing woods, such as Salix L. sp. (willow), Populus L. sp. (poplar), Eucalyptus L'Her. sp. (eucalyptus), and Pinus L. sp. (pine) (Papini and Simeone 2010; Cambero et al. 2015). The conversion of forest biomass into biofuels can follow different technological routes, depending on factors such as biomass composition, economic feasibility, and process efficiency.

Gasification, for example, transforms biomass into synthesis gas (syngas), a mixture rich in hydrogen (H_2), carbon monoxide (CO), carbon dioxide (CO_2), and methane (CH_4). This gas can be used for electricity generation or converted into liquid fuels, such as methanol and synthetic diesel, through catalytic processes. For optimal performance, the feedstock used in gasification should have low ash and impurity content (Cambero et al. 2015). Another widely used technology is pyrolysis, a thermochemical process in which organic molecules are thermally degraded in the absence of oxygen. Fast pyrolysis enables the production of bio-oil, a biofuel of commercial interest. Its efficiency is higher when the biomass has low moisture content, making wood-based feedstocks suitable due to its low ash and moisture content (Cambero et al. 2015; Osman et al. 2023). Campbell et al. (2018) investigated the feasibility of a pyrolysis system based on forest feedstock for the coproduction of biochar and biofuel, as shown in Fig. 13.4. The study concluded that the main impact factor is the volatility of market prices for the products, posing a challenge to financial planning and the economic viability of production facilities (Campbell et al. 2018).

An alternative approach to converting forest residues into advanced biofuels is the integrated hydropyrolysis and hydroconversion (IH^2) process. This process involves three main stages: (i) biomass preparation (drying, grinding, and resizing), (ii) hydropyrolysis, where the biomass is subjected to a pressurized fluidized bed reactor (20–35 bar) with hydrogen input, resulting in the production of light hydrocarbons, C1–C3 gases, H_2O, COx, and biochar, and (iii) catalytic hydroconversion, in which vapors pass through catalysts for oxygen removal, promoting hydrodeoxygenation and hydrogenation reactions. The final product is then separated by distillation to obtain renewable gasoline and diesel. Additionally, C1–C3 gases can be reused for electricity generation or hydrogen gas production, reducing the need for external inputs. In the case of woody biomass, ammonia/ammonium sulfate can also be recovered as a byproduct, which can be used as a fertilizer or an industrial chemical feedstock, enhancing the valorization of process byproducts. Waste heat can also be recovered for biomass drying, increasing energy efficiency and system circularity (Zupko 2019). The process flowchart is presented in Fig. 13.5.

Beyond these routes, forest biomass can be converted into bioethanol through a process that includes pretreatment for lignin removal, enzymatic hydrolysis to convert cellulose and hemicellulose into monosaccharides, fermentation to transform sugars into ethanol, and distillation to purify the final product (Raihan 2023). Another alternative is anaerobic digestion, which enables the conversion of forest biomass into biogas, which is composed mainly of methane and CO_2 (Matsakas et al. 2016). However, the high lignin concentration in forest biomass presents a challenge in these processes, requiring more intensive pretreatments compared to lignocellulosic residues with lower lignin content. For this reason, research is being conducted to optimize yields and improve the efficient utilization of this biomass (Matsakas et al. 2016; Raihan 2023).

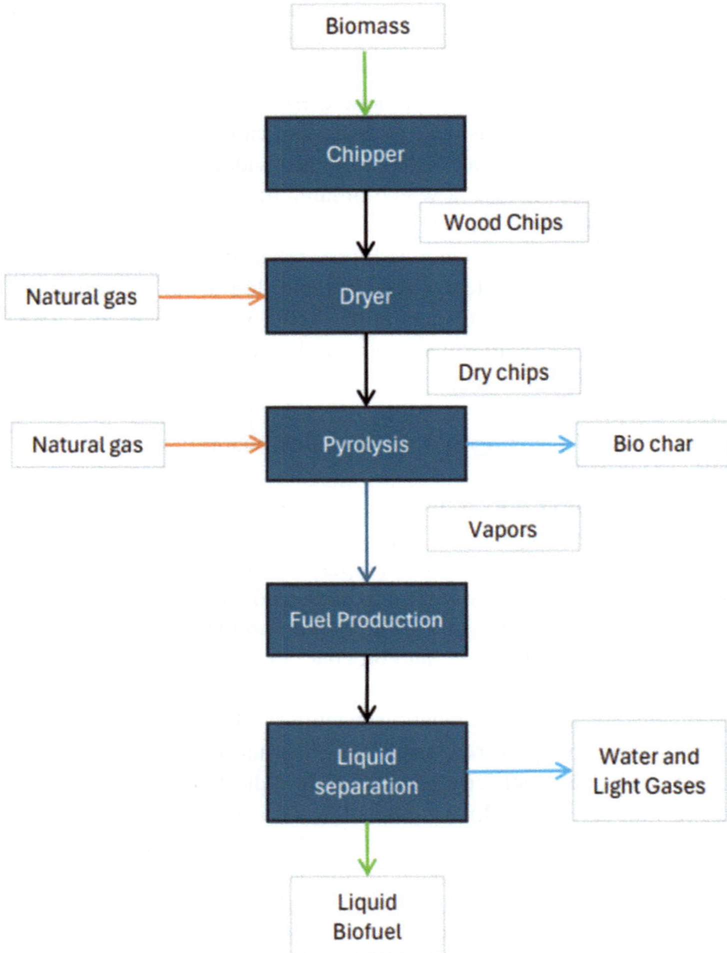

Fig. 13.4 Pyrolysis using forest raw material for the coproduction of biochar and biofuel

13.3.4 Residual Vegetable Oils and Animal Fats

Vegetable oils and animal fats hold significant potential as feedstocks for the production of renewable fuels. Regions with high oilseed agricultural output, such as Southeast Asia (palm oil), South America (soybean oil), and Europe (rapeseed oil), are particularly promising for large-scale production. In 2022, global production of palm oil reached approximately 77 million metric tons (FAO 2021), primarily sourced from Indonesia and Malaysia. Soybean oil production exceeded 60 million metric tons, mainly driven by Brazil and Argentina (Sobczak-Malitka and Sobczak 2023). Additionally, Europe produced around 27 million metric tons of rapeseed oil

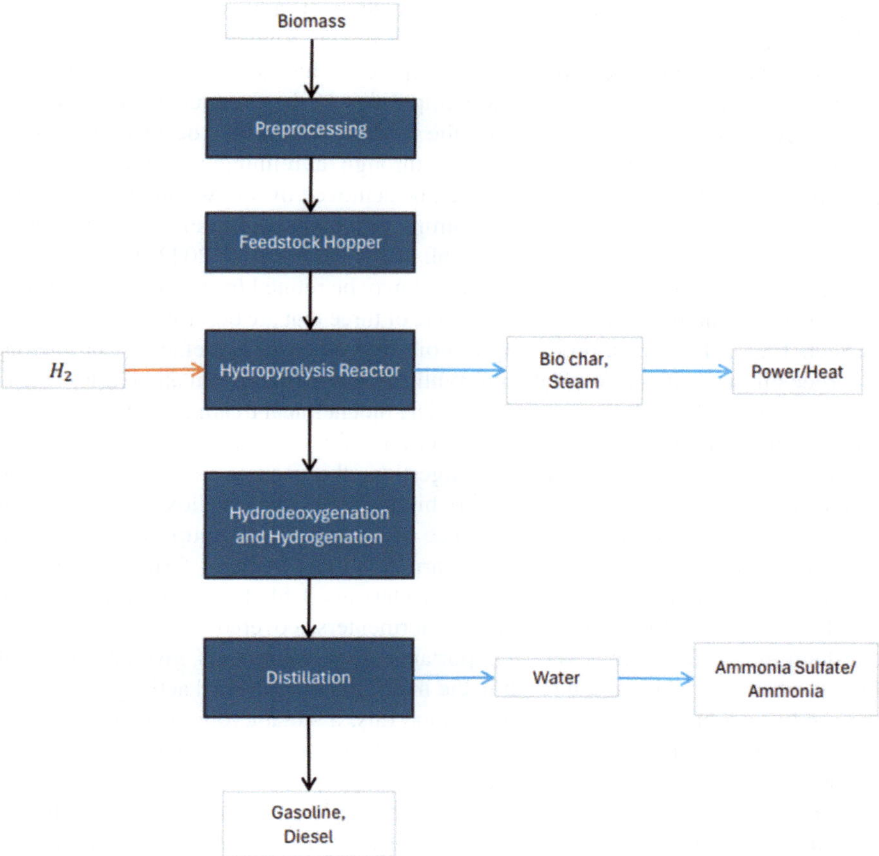

Fig. 13.5 Simplified flowchart of the IH^2 process

annually, contributing to a diverse pool of feedstocks. These vegetable oils can be converted into renewable diesel and sustainable aviation fuel (SAF) through processes such as hydroprocessing or transesterification. When targeting biofuel production to replace petroleum-derived fuels, particularly diesel, these renewable fuels can achieve up to a 90% reduction in life cycle greenhouse gas emissions compared to conventional fossil fuels.

Despite the challenge of designing conversion processes for more than 20 years (Vilas Bôas and Mendes 2022), the use of residual vegetable oils and animal fats as feedstocks remains a feasible alternative. There are many various pathways for converting these oils into biofuels, which include chemical, thermochemical, and biochemical treatments (Binhweel et al. 2023).

Chemical treatment involves the transesterification of the material and biodiesel production, which is by far the most common alternative. This process occurs with an alcohol reactant and a strong alkaline or acidic catalyst at controlled temperatures, so that the triglycerides from the oil are converted into biodiesel, which is composed of fatty acid monoalkyl esters (FAMEs), and glycerol, a coproduct of the process (Vignesh et al. 2021; Paladino and Neviani 2021). Another alternative for this process is an enzymatic catalyst, such as lipases, which requires less energy and results in higher conversion and fewer impurities in the product (Cerón Ferrusca et al. 2023; Mustafa et al. 2023). After the process, which may occur in a batch or continuous reactor, the biodiesel passes through a refining process to remove glycerol and other components, which can be achieved by wet washing the systems with water, distillation, or adsorption columns, or membrane filtration (Budžaki et al. 2020; Polishchuk et al. 2020; Arenas et al. 2021; Bashir et al. 2022; Reginio et al. 2022; Wan Osman et al. 2024). The glycerol may be refined for further applications.

The thermochemical treatment can be one of three that are described and analyzed in Sect. 13.2.2: Pyrolysis, producing bio-oil; hydrothermal liquefaction, producing biocrude oil; or gasification, producing synthesis gas (Shumovskii and Gorlov 2021; Kumar et al. 2021; Binhweel et al. 2023). The biochemical treatment involves either fermentation, producing ethanol, or anaerobic digestion, producing biomethane (Binhweel et al. 2023). In anaerobic digestion, the process revolves around the conversion of organic compounds into biomethane, carbon dioxide, water, and biomass, in a system without the presence of oxygen, by a group of well-adapted microorganisms. This process can be carried out in a single fermenter or in a combination of serial fermenters. The fermenters available for this process include: CSTR, plug, fixed film, garage-type dry fermenters, covered lagoon, and UASB (Rashama et al. 2019). However, an important part of this process, given the usage of lipidic substrates, is that it can result in the inhibition of microbial activity, hindering the production of biomethane. To overcome this, a suitable option is to add combined additives to the fermentation, such as calcium-modified or potassium carbonate biochar (Kobayashi and Kuramochi 2022; Thakur et al. 2024).

Those processes show that it is possible to reutilize residual vegetable oils and animal fats as a sources for biofuel production. However, when designing a biorefinery, a few topics need to be discussed. The first is the size, volume, and material of the necessary reactors, especially in the transesterification process. Once that, depending on the catalyst utilized, such as sulfuric acid, the use of stainless steel is mandatory (Feng et al. 2022). Other topics include the quantity of alcohol and catalyst needed, which may involve greater costs for the implantation and operation of the biorefinery. In the case of anaerobic digestion, one of the major points that needs to be taken into account is the yield of the process, which influences the type of reactor, the ratio of carbon/nitrogen, and the need to add other compounds to enhance the production, which may lead to a significant difference in costs, depending on the budget available (Rashama et al. 2019).

13.4 Conclusion and Perspectives

Second-generation biofuels are not just a local solution, but a global recognition of the need for a more sustainable energy future. The conversion of nonfood biomass, such as agricultural residues, forestry byproducts, and waste oils, into second-generation (2G) biofuels is a promising solution contributing to carbon neutrality and promoting a circular economy. The projected market growth is a reflection of this global recognition. Low-value byproducts like corn stover, sugarcane bagasse, and vinasse are widely available in agricultural regions, including the United States, Brazil, and parts of Europe and Australia. The most-produced 2G biofuel is ligno-cellulosic ethanol. Other technologies for converting lignocellulosic biomass are still developing and are not cost-competitive. However, challenges such as optimizing production processes, reducing costs, and developing efficient supply chains still need to be overcome for large-scale adoption. Future research should focus on more efficient conversion technologies, the exploration new biomass sources, and the integration of 2G biofuels with existing energy infrastructure to maximize their environmental and economic benefits.

References

Agencia-FAPESP (2017) Second-generation ethanol can be economically viable by 2025, pp 1–23. Retrieved from: https://agencia.fapesp.br/second-generation-ethanol-can-be-economically-viable-by-2025/26422. Accessed 4 Feb 2025. https://doi.org/10.1007/978-3-031-01241-9

Aghaei S, Karimi Alavijeh M, Shafiei M, Karimi K (2022) A comprehensive review on bioethanol production from corn stover: worldwide potential, environmental importance, and perspectives. Biomass Bioenergy 161:106447. https://doi.org/10.1016/j.biombioe.2022.106447

Aghel B, Janati S, Wongwises S, Shadloo MS (2022) Review on CO2 capture by blended amine solutions. Int J Greenh Gas Con 119:103715. https://doi.org/10.1016/j.ijggc.2022.103715

Alokika, Anu, Kumar A et al (2021) Cellulosic and hemicellulosic fractions of sugarcane bagasse: potential, challenges and future perspective. Int J Biol Macromol 169:564–582. https://doi.org/10.1016/j.ijbiomac.2020.12.175

Antizar-Ladislao B, Turrion-Gomez JL (2008) Second-generation biofuels and local bioenergy systems. Biofuels Bioprod Biorefin 2:455–469

Arenas E, Villafán-Cáceres SM, Rodríguez-Mejía Y et al (2021) Biodiesel dry purification using unconventional bioadsorbents. Processes 9:194. https://doi.org/10.3390/pr9020194

Baptista SL, Cunha JT, Romaní A, Domingues L (2018) Xylitol production from lignocellulosic whole slurry corn cob by engineered industrial Saccharomyces cerevisiae PE-2. Bioresour Technol 267:481–491. https://doi.org/10.1016/j.biortech.2018.07.068

Baredar P, Khare V, Nema S (2020) Biomass liquefaction. In: Design and optimization of biogas energy systems. Elsevier, pp 231–266

Bashir MA, Wu S, Zhu J et al (2022) Recent development of advanced processing technologies for biodiesel production: a critical review. Fuel Process Technol 227:107120. https://doi.org/10.1016/j.fuproc.2021.107120

Beltrán-Ramírez F, Orona-Tamayo D, Cornejo-Corona I et al (2019) Agro-industrial waste revalorization: the growing biorefinery. In: Biomass for bioenergy – recent trends and future challenges. IntechOpen

Bhaskar T, Bhavya B, Singh R et al (2011) Thermochemical conversion of biomass to biofuels. In: Biofuels: alternative feedstocks and conversion processes. Elsevier, pp 51–77

Bhatia SK, Kim SH, Yoon JJ, Yang YH (2017) Current status and strategies for second generation biofuel production using microbial systems. Energy Convers Manag 148:1142–1156

Binhweel F, Hossain MS, Ahmad MI (2023) Recent trends, potentials, and challenges of biodiesel production from discarded animal fats: a comprehensive review. Bioenergy Res 16:778–800. https://doi.org/10.1007/s12155-022-10527-w

Bisson LF, Coons DM, Kruckeberg AL, Lewis DA (1993) Yeast sugar transporters. Crit Rev Biochem Mol Biol 28:259–308. https://doi.org/10.3109/10409239309078437

Biswas B, Kumar A, Kaur R et al (2021) Catalytic hydrothermal liquefaction of alkali lignin over activated bio-char supported bimetallic catalyst. Bioresour Technol 337:125439. https://doi.org/10.1016/j.biortech.2021.125439

Bizualem YD, Nurie AG (2024) A review on recent biodiesel intensification process through cavitation and microwave reactors: yield, energy, and economic analysis. Heliyon 10:e24643. https://doi.org/10.1016/j.heliyon.2024.e24643

Bridgwater AV, Peacocke GVC (2000) Fast pyrolysis processes for biomass. Renew Sust Energ Rev 4:1–73. https://doi.org/10.1016/S1364-0321(99)00007-6

Brito MGSL, Nunes FCB, Magalhães HLF et al (2020) Hydrodynamics of UASB reactor treating domestic. Water (Basel) 1:1–20

Brown TR (2015) A techno-economic review of thermochemical cellulosic biofuel pathways. Bioresour Technol 178:166–176

Brown EE (2023) Minireview: recent efforts toward upgrading lignin-derived phenols in continuous flow. J Flow Chem 13:91–102. https://doi.org/10.1007/s41981-022-00248-8

Budžaki S, Zelić B, Tišma M et al (2020) Membranska filtracija kao ekološki prihvatljiva metoda pročišćavanja sirovog biodizela. Kem Ind 69:175–181. https://doi.org/10.15255/KUI.2019.049

Cambero C, Sowlati T, Marinescu M, Röser D (2015) Strategic optimization of forest residues to bioenergy and biofuel supply chain. Int J Energy Res 39:439–452. https://doi.org/10.1002/er.3233

Campbell RM, Anderson NM, Daugaard DE, Naughton HT (2018) Financial viability of biofuel and biochar production from forest biomass in the face of market price volatility and uncertainty. Appl Energy 230:330–343. https://doi.org/10.1016/j.apenergy.2018.08.085

Carrilho ENVM, Labuto G, Kamogawa MY (2016) Destination of vinasse, a residue from alcohol industry: resource recovery and prevention of pollution. In: Environmental materials and waste: resource recovery and pollution prevention. Elsevier, Amsterdam

Carriquiry MA, Du X, Timilsina GR (2011) Second generation biofuels: economics and policies. Energy Policy 39:4222–4234. https://doi.org/10.1016/j.enpol.2011.04.036

Celignis Analytical (2024) Feedstocks analysed at celignis – corn stover. https://www.celignis.com/feedstock.php?value=27. Accessed 12 Feb 2025

Cerón Ferrusca M, Romero R, Martínez SL et al (2023) Biodiesel production from waste cooking oil: a perspective on catalytic processes. Processes 11:1952. https://doi.org/10.3390/pr11071952

Chowdhury R (2021) From the Guest Editor's desk: biofuels from biomass and wastes. Indian Chem Eng 63:117–118

Damartzis T, Zabaniotou A (2011) Thermochemical conversion of biomass to second generation biofuels through integrated process design—a review. Renew Sust Energ Rev 15:366–378

de Carvalho JC, de Souza Vandenberghe LP, Sydney EB et al (2023) Biomethane production from sugarcane vinasse in a circular economy: developments and innovations. Fermentation 9:349. https://doi.org/10.3390/fermentation9040349

Demirbas A (2009) Biofuels securing the planet's future energy needs. Energy Convers Manag 50:2239–2249. https://doi.org/10.1016/j.enconman.2009.05.010

Demirbaş A, Arin G (2002) An overview of biomass pyrolysis. Energy Sources 24:471–482. https://doi.org/10.1080/00908310252889979

do Nascimento Junior JR, Zevallos Torres LA, Medeiros ABP et al (2021) Enhancement of biohydrogen production in industrial wastewaters with vinasse pond consortium using lignin-mediated iron nanoparticles. Int J Hydrog Energy 46:27431–27443. https://doi.org/10.1016/j.ijhydene.2021.06.009

Enebe NL, Chigor CB, Obileke K et al (2023) Biogas and syngas production from sewage sludge: a sustainable source of energy generation. Methane 2:192–217. https://doi.org/10.3390/methane2020014

Escalante J, Chen WH, Tabatabaei M et al (2022) Pyrolysis of lignocellulosic, algal, plastic, and other biomass wastes for biofuel production and circular bioeconomy: a review of thermogravimetric analysis (TGA) approach. Renew Sust Energ Rev 169:112914. https://doi.org/10.1016/j.rser.2022.112914

Eurpean-Biogas-Association (2023) New study shows that monetary value of whole-system benefits of biomethane far outweighs current production costs. Retrieved from: https://www.europeanbiogas.eu/new-study-shows-that-monetary-value-of-whole-system-benefits-of-biomethane-far-outweighs-c. https://doi.org/10.1007/978-3-031-01241-9

Fahmy TYA, Fahmy Y, Mobarak F et al (2020) Biomass pyrolysis: past, present, and future. Environ Dev Sustain 22:17–32

Fang Z, Li F, Wang M et al (2023) Selective electrocatalytic upgrading of lignin to aryl aldehydes and carboxylic acids over dodecyl sulfate-intercalated CoS nanocones. Appl Catal B 323: 122149. https://doi.org/10.1016/j.apcatb.2022.122149

FAO (2017) FAOSTAT database. FAO. Fisheries and Agriculture Organisation United Nations, Rome

FAO (2021) FAOSTAT crops and livestock production. FAOSTAT Analytical Brief 60 Agricultural production statistics 2000–2021. FAO

FAO (2024) FAOSTAT Data. Crop and livestock produts. http://www.fao.org/faostat/en/#data/QCL

Feng L, Liu J, Lu H et al (2022) Techno-economic and profitability analysis of plant for producing biodiesel from fresh vegetable oil and waste frying oil on large-scale. Fuel 323:124304. https://doi.org/10.1016/j.fuel.2022.124304

Formann S, Hahn A, Janke L et al (2020) Beyond sugar and ethanol production: value generation opportunities through sugarcane residues. Front Energy Res 8:579577. https://doi.org/10.3389/fenrg.2020.579577

Foust TD, Aden A, Dutta A, Phillips S (2009) An economic and environmental comparison of a biochemical and a thermochemical lignocellulosic ethanol conversion processes. Cellulose 16: 547–565. https://doi.org/10.1007/s10570-009-9317-x

Gandam PK, Chinta ML, Pabbathi NPP et al (2022) Second-generation bioethanol production from corncob—a comprehensive review on pretreatment and bioconversion strategies, including techno-economic and lifecycle perspective. Ind Crop Prod 186:115245. https://doi.org/10.1016/j.indcrop.2022.115245

Ganguly P, Sarkhel R, Das P (2020) The second- and third-generation biofuel technologies: comparative perspectives. In: Sustainable fuel technologies handbook. Elsevier, pp 29–50

Gao Y, Wang M, Raheem A et al (2023) Syngas production from biomass gasification: influences of feedstock properties, reactor type, and reaction parameters. ACS Omega 8:31620

Gashaw A, Getachew T, Teshita A (2015) A review on biodiesel production as alternative fuel. J For Prod Ind 4:80–85

Gholami M, Torreggiani D, Tassinari P, Barbaresi A (2021) Narrowing uncertainties in forecasting urban building energy demand through an optimal archetyping method. Renew Sust Energ Rev 148:111312. https://doi.org/10.1016/j.rser.2021.111312

Gollakota ARK, Kishore N, Gu S (2018) A review on hydrothermal liquefaction of biomass. Renew Sust Energ Rev 81:1378

Göransson K, Söderlind U, He J, Zhang W (2011) Review of syngas production via biomass DFBGs. Renew Sust Energ Rev 15:482

Grande L, Pedroarena I, Korili SA, Gil A (2021) Hydrothermal liquefaction of biomass as one of the most promising alternatives for the synthesis of advanced liquid biofuels: a review. Materials 14:5286

Guijarro JM, Lagunas R (1984) Saccharomyces cerevisiae does not accumulate ethanol against a concentration gradient. J Bacteriol 160:874–878. https://doi.org/10.1128/jb.160.3.874-878.1984

Güleç F, Parthiban A, Umenweke GC et al (2024) Progress in lignocellulosic biomass valorization for biofuels and value-added chemical production in the EU: a focus on thermochemical conversion processes. Biofuels Bioprod Biorefin 18:755–781

Hamze H, Akia M, Yazdani F (2015) Optimization of biodiesel production from the waste cooking oil using response surface methodology. Process Saf Environ Prot 94:1–10. https://doi.org/10.1016/j.psep.2014.12.005

Hassan SS, Williams GA, Jaiswal AK (2019) Moving towards the second generation of lignocellulosic biorefineries in the EU: drivers, challenges, and opportunities. Renew Sust Energ Rev 101:590–599

Hirani AH, Javed N, Asif M et al (2018) A review on first- and second- generation biofuel productions. In: Biofuels: greenhouse gas mitigation and global warming: next generation biofuels and role of biotechnology. Springer, pp 141–154

Hu X, Gholizadeh M (2020) Progress of the applications of bio-oil. Renew Sust Energ Rev 134: 110124

Ibarra-Gonzalez P, Rong BG (2019) A review of the current state of biofuels production from lignocellulosic biomass using thermochemical conversion routes. Chin J Chem Eng 27:1523–1535

Igliński B, Kujawski W, Kiełkowska U (2023) Pyrolysis of waste biomass: technical and process achievements, and future development—a review. Energies (Basel) 16:1829

Iwuozor KO, Emenike EC, Ighalo JO et al (2022) Valorization of sugar industry's by-products: a perspective. Sugar Tech 24:1052–1078. https://doi.org/10.1007/s12355-022-01143-1

Jacobson K, Maheria KC, Kumar Dalai A (2013) Bio-oil valorization: a review. Renew Sustain Energy Rev 23:91

Jha S, Nanda S, Acharya B, Dalai AK (2022) A review of thermochemical conversion of waste biomass to biofuels. Energies (Basel) 15:6352

Kang Q, Appels L, Tan T, Dewil R (2014) Bioethanol from lignocellulosic biomass: current findings determine research priorities. Sci World J 2014:1–13. https://doi.org/10.1155/2014/298153

Karp SG, Burgos WJM, Vandenberghe LPS et al (2022) Sugarcane: a promising source of green carbon in the circular bioeconomy. Sugar Tech 24:1230–1245. https://doi.org/10.1007/s12355-022-01161-z

Kaur R, Gera P, Jha MK, Bhaskar T (2019) Thermochemical route for biohydrogen production. In: Biomass, biofuels, biochemicals: biohydrogen, 2nd edn. Elsevier, pp 187–218

Kazi FK, Fortman JA, Anex RP et al (2010) Techno-economic comparison of process technologies for biochemical ethanol production from corn stover. Fuel 89:S20–S28. https://doi.org/10.1016/j.fuel.2010.01.001

Khan MFS, Akbar M, Xu Z, Wang H (2021) A review on the role of pretreatment technologies in the hydrolysis of lignocellulosic biomass of corn stover. Biomass Bioenergy 155:106276. https://doi.org/10.1016/j.biombioe.2021.106276

Kim M, Choi D, Kim JY et al (2025) Carbon dioxide-mediated catalytic pyrolysis of lignin in syngas production. Int J Biol Macromol 296:139786. https://doi.org/10.1016/j.ijbiomac.2025.139786

Kobayashi T, Kuramochi H (2022) Optimized production conditions and activation of biochar for effective promotion of long-chain fatty acid degradation in anaerobic digestion. Bioresour Technol 358:127393. https://doi.org/10.1016/j.biortech.2022.127393

Kumar D, Singh V (2019) Chapter 22: Bioethanol production from corn. In: Corn, 3rd edn. Elsevier

Kumar B, Bhardwaj N, Agrawal K et al (2020) Current perspective on pretreatment technologies using lignocellulosic biomass: an emerging biorefinery concept. Fuel Process Technol 199: 106244. https://doi.org/10.1016/j.fuproc.2019.106244

Kumar A, Pali HS, Kumar M (2021) Application of pyrolysis and hydrothermal liquefaction process to convert bio-fuel from biomass. J Phys Conf Ser 2062:012029. https://doi.org/10.1088/1742-6596/2062/1/012029

Law LC, Yusoff Azudin N, Syamsul SR (2018) Optimization and economic analysis of amine-based acid gas capture unit using monoethanolamine/methyl diethanolamine. Clean Techn Environ Policy 20:451–461. https://doi.org/10.1007/s10098-017-1430-1

Liu S, Yu Y, Xu Z et al (2022) Efficient corncob biorefinery for ethanol initiated by a novel pretreatment of densifying lignocellulosic biomass with sulfuric acid. Fermentation 8:661. https://doi.org/10.3390/fermentation8110661

Ma M, Liu ZL (2010) Mechanisms of ethanol tolerance in saccharomyces cerevisiae. Appl Microbiol Biotechnol 87:829–845. https://doi.org/10.1007/s00253-010-2594-3

Maicas S (2020) The role of yeasts in fermentation processes. Microorganisms 8:1–8. https://doi.org/10.3390/microorganisms8081142

Makepa DC, Chihobo CH, Musademba D (2023) Advances in sustainable biofuel production from fast pyrolysis of lignocellulosic biomass. Biofuels 14:529–550

Malode SJ, Prabhu KK, Mascarenhas RJ, Shetti NP, Aminabhavi TM (2021) Recent advances and viability in biofuel production. Energy Convers Manage X(10):100070. https://doi.org/10.1016/j.ecmx.2020.100070

Martinez-Burgos WJ, Sydney EB, Brar SK et al (2019) The effect of hydrolysis and sterilization in biohydrogen production from cassava processing wastewater medium using anaerobic bacterial consortia. Int J Hydrog Energy 44:25551–25564. https://doi.org/10.1016/j.ijhydene.2019.08.085

Martinez-Burgos WJ, Sydney EB, de Paula DR et al (2020) Biohydrogen production in cassava processing wastewater using microbial consortia: process optimization and kinetic analysis of the microbial community. Bioresour Technol 309:123331. https://doi.org/10.1016/j.biortech.2020.123331

Martinez-Burgos WJ, de Souza Candeo E, Pedroni Medeiros AB et al (2021a) Hydrogen: current advances and patented technologies of its renewable production. J Clean Prod 286:124970. https://doi.org/10.1016/j.jclepro.2020.124970

Martinez-Burgos WJ, Sydney Bittencourt E, Bianchi Pedroni Medeiros A et al (2021b) Agro-industrial wastewater in a circular economy: characteristics, impacts and applications for bioenergy and biochemicals. Bioresour Technol 341:125795. https://doi.org/10.1016/j.biortech.2021.125795

Martinez-Burgos WJ, Porto de Souza Vandenberghe L, Karp SG et al (2023) Microbial lipid production from soybean hulls using Lipomyces starkeyi LPB53 in a circular economy. Bioresour Technol 372:128650. https://doi.org/10.1016/j.biortech.2023.128650

Matsakas L, Rova U, Christakopoulos P (2016) Strategies for enhanced biogas generation through anaerobic digestion of forest material—an overview. Bioresources 11:5482–5499. https://doi.org/10.15376/biores.11.2.Matsakas

Mazaheri H, Lee KT, Bhatia S, Mohamed AR (2010) Sub/supercritical liquefaction of oil palm fruit press fiber for the production of bio-oil: effect of solvents. Bioresour Technol 101:7641–7647. https://doi.org/10.1016/j.biortech.2010.04.072

Meena PK, Awale SD, Kumar S et al (2024) Thermochemical conversion of agricultural residue for the production of hydrogen, methane, and biofuels: a comprehensive overview. Ind Crop Prod 221:119340

Miao R, Zhang Q, Shi Y et al (2017) Liquefaction of lignite with a Ru/C catalyst in supercritical ethanol. RSC Adv 7:5402. https://doi.org/10.1039/c6ra21818k

Mignogna D, Szabó M, Ceci P, Avino P (2024) Biomass energy and biofuels: perspective, potentials, and challenges in the energy transition. Sustainability 16:7036

Molaverdi M, Karimi K, Mirmohamadsadeghi S, Galbe M (2021) High efficient ethanol production from corn stover by modified mild alkaline pretreatment. Renew Energy 170:714–723. https://doi.org/10.1016/j.renene.2021.02.002

Moraes BS, Zaiat M, Bonomi A (2015) Anaerobic digestion of vinasse from sugarcane ethanol production in Brazil: challenges and perspectives. Renew Sust Energ Rev 44:888–903. https://doi.org/10.1016/j.rser.2015.01.023

Mustafa A, Faisal S, Ahmed IA et al (2023) Has the time finally come for green oleochemicals and biodiesel production using large-scale enzyme technologies? Current status and new developments. Biotechnol Adv 69:108275. https://doi.org/10.1016/j.biotechadv.2023.108275
Naik SN, Goud VV, Rout PK, Dalai AK (2010) Production of first and second generation biofuels: a comprehensive review. Renew Sust Energ Rev 14:578–597
Nanda S, Mohammad J, Reddy SN et al (2014) Pathways of lignocellulosic biomass conversion to renewable fuels. Biomass Convers Biorefinery 4:157–191
Nanda S, Rana R, Sarangi PK et al (2018) A broad introduction to first-, second-, and third-generation biofuels. In: Recent advancements in biofuels and bioenergy utilization. Springer, Singapore, pp 1–25
Nanda S, Pattnaik F, Patra BR et al (2023) A review of liquid and gaseous biofuels from advanced microbial fermentation processes. Fermentation 9:813
Naspolini BF, De Oliveira Machado AC, Cravo WB et al (2017) Bioconversion of sugarcane vinasse into high-added value products and energy. Biomed Res Int 2017:8986165. https://doi.org/10.1155/2017/8986165
O. Kukoyi T, Muzenda E, Mashamba A (2015) Biomethane and bioethanol as alternative transport fuels. In: Seventh international conference on latest trends in engineering & technology. International Institute of Engineers. https://doi.org/10.15242/IIE.E1115054
Ojo-kupoluyi OJ, Dele-Afolabi TT, Tahir SM (2024) Harnessing sugarcane bagasse for bioenergy production: current status, optimization, and future directions. Sustain Energy Technol Assess 72:104074. https://doi.org/10.1016/j.seta.2024.104074
Okolie JA, Mukherjee A, Nanda S et al (2021) Next-generation biofuels and platform biochemicals from lignocellulosic biomass. Int J Energy Res 45:14145–14169
Okolie JA, Epelle EI, Tabat ME et al (2022) Waste biomass valorization for the production of biofuels and value-added products: a comprehensive review of thermochemical, biological and integrated processes. Process Saf Environ Prot 159:323–344
Ong HC, Chen WH, Farooq A et al (2019) Catalytic thermochemical conversion of biomass for biofuel production: a comprehensive review. Renew Sustain Energy Rev 113:109266
Ong HC, Tiong YW, Goh BHH et al (2021) Recent advances in biodiesel production from agricultural products and microalgae using ionic liquids: opportunities and challenges. Energy Convers Manag 228:113647
Osman AI, Farghali M, Ihara I et al (2023) Materials, fuels, upgrading, economy, and life cycle assessment of the pyrolysis of algal and lignocellulosic biomass: a review. Environ Chem Lett 21:1419–1476. https://doi.org/10.1007/s10311-023-01573-7
Paladino O, Neviani M (2021) Sustainable biodiesel production by transesterification of waste cooking oil and recycling of wastewater rich in glycerol as a feed to microalgae. Sustainability 14:273. https://doi.org/10.3390/su14010273
Pan S, Zabed HM, Wei Y, Qi X (2022) Technoeconomic and environmental perspectives of biofuel production from sugarcane bagasse: current status, challenges and future outlook. Ind Crop Prod 188:115684. https://doi.org/10.1016/j.indcrop.2022.115684
Panda D, Saharan VK, Manickam S (2020) Controlled hydrodynamic cavitation: a review of recent advances and perspectives for greener processing. Processes 8:220. https://doi.org/10.3390/pr8020220
Papini A, Simeone MC (2010) Forest resources for second generation biofuel production. Scand J For Res 25:126–133. https://doi.org/10.1080/02827581.2010.485827
Parapouli M, Vasileiadis A, Afendra AS, Hatziloukas E (2020) Saccharomyces cerevisiae and its industrial applications. AIMS Microbiol 6:1–31. https://doi.org/10.3934/microbiol.2020001
Patel A, Shah AR (2021) Integrated lignocellulosic biorefinery: gateway for production of second generation ethanol and value added products. J Bioresour Bioprod 6:108–128. https://doi.org/10.1016/j.jobab.2021.02.001
Patra BR, Nanda S, Dalai AK, Meda V (2021) Slow pyrolysis of agro-food wastes and physicochemical characterization of biofuel products. Chemosphere 285:131431. https://doi.org/10.1016/j.chemosphere.2021.131431

Pinto AC, Guarieiro LLN, Rezende MJC et al (2005) Biodiesel: an overview. J Braz Chem Soc 16: 1313–1330
Pio DT, Tarelho LAC (2021) Industrial gasification systems (>3 MWth) for bioenergy in Europe: current status and future perspectives. Renew Sust Energ Rev 145:111108
Polishchuk V, Tarasenko S, Antypov I et al (2020) Investigation of the efficiency of wet biodiesel purification. E3S Web Conf 154:02006. https://doi.org/10.1051/e3sconf/202015402006
Porninta K, Mahakuntha C, Khemacheewakul J et al (2024) Cell recycling application in single-stage and sequential-stage co-production of xylitol and ethanol using corn cob hydrolysates. Agriculture 14:1062. https://doi.org/10.3390/agriculture14071062
Prasad E (2023) Second generation biofuels market. Global opportunity analysis and industry forecast, 2021-2030
Pratto B, Chandgude V, de Sousa R et al (2020) Biobutanol production from sugarcane straw: defining optimal biomass loading for improved ABE fermentation. Ind Crop Prod 148:112265. https://doi.org/10.1016/j.indcrop.2020.112265
Praveena V, Martin LJ, Matijošius J et al (2024) A systematic review on biofuel production and utilization from algae and waste feedstocks – a circular economy approach. Renew Sust Energ Rev 192:114178
Rahimi Z, Anand A, Gautam S (2022) An overview on thermochemical conversion and potential evaluation of biofuels derived from agricultural wastes. Energy Nexus 7:100125
Raihan A (2023) A concise review of technologies for converting forest biomass to bioenergy. J Technol Innov Energy 2:10–36. https://doi.org/10.56556/jtie.v2i3.592
Rashama C, Ijoma G, Matambo T (2019) Biogas generation from by-products of edible oil processing: a review of opportunities, challenges and strategies. Biomass Convers Biorefinery 9:803–826. https://doi.org/10.1007/s13399-019-00385-6
Reginio RR, Tongco AJ, Aguilar JCS (2022) Purification of raw biodiesel using modified fractional distillation. Key Eng Mater 937:201–210. https://doi.org/10.4028/p-mn7659
Reis CER, Hu B (2017) Vinasse from sugarcane ethanol production: better treatment or better utilization? Front Energy Res 5:7
Rosa D, Medeiros ABP, Martinez-Burgos WJ et al (2020) Biological hydrogen production from palm oil mill effluent (POME) by anaerobic consortia and Clostridium beijerinckii. J Biotechnol 323:17–23. https://doi.org/10.1016/j.jbiotec.2020.06.015
Ruan Z, Wang X, Liu Y, Liao W (2019) Corn. In: Integrated processing technologies for food and agricultural by-products. Elsevier, pp 59–72
Saleh AM, Alias AB, Saleh NM et al (2023) Production of first and second-generation biodiesel for diesel engine operation: a review. NTU J Renew Energy 5:8–23. https://doi.org/10.56286/ntujre.v5i1
Savian CM, Scapini T, Martínez-Burgos WJ et al (2024) Imidazole as a new strategy in the lignocellulosic pretreatment: an application in corn sub-products with a focus on bioenergy production. Biomass Bioenergy 183:107075. https://doi.org/10.1016/j.biombioe.2024.107075
Sekhon SS, Lee J, Park JS (2021) Biomass-derived bifunctional electrocatalysts for oxygen reduction and evolution reaction: a review. J Energy Chem 65:149–172. https://doi.org/10.1016/j.jechem.2021.05.052
Shabbirahmed AM, Haldar D, Dey P et al (2022) Sugarcane bagasse into value-added products: a review. Environ Sci Pollut Res 29:62785
Sharma B, Larroche C, Dussap CG (2020) Comprehensive assessment of 2G bioethanol production. Bioresour Technol 313:123630. https://doi.org/10.1016/j.biortech.2020.123630
Shui H, Cai Z, Xu CC (2010) Recent advances in direct coal liquefaction. Energies (Basel) 3:155–170
Shumovskii AV, Gorlov EG (2021) Preparation of biomass mixtures with residual oil for gasification to produce synthesis gas. Solid Fuel Chem 55:260–265. https://doi.org/10.3103/S0361521921040078
Sikarwar VS, Zhao M, Fennell PS et al (2017) Progress in biofuel production from gasification. Prog Energy Combust Sci 61:189

Sims REH, Mabee W, Saddler JN, Taylor M (2010) An overview of second generation biofuel technologies. Bioresour Technol 101:1570–1580

Sobczak-Malitka W, Sobczak E (2023) Changes in the vegetable oil market, with particular emphasis on market instability in relation to the war in Ukraine. Zeszyty Naukowe SGGW w Warszawie – Problemy Rolnictwa Światowego 23:46–57. https://doi.org/10.22630/prs.2023.23.1.4

Spanou A, Moschona A, Theodosiou E et al (2024) Novel concepts for the biocatalytic synthesis of second-generation biodiesel. Front Catal 4:1360702. https://doi.org/10.3389/fctls.2024.1360702

Srivastava RK, Shetti NP, Reddy KR, Aminabhavi TM (2020) Biofuels, biodiesel and biohydrogen production using bioprocesses. A review. Environ Chem Lett 18:1049–1072

Statista (2024a) Actual and potential biomethane production worldwide in 2022, pp 1–23. Retrieved from: https://www.statista.com/statistics/1296541/global-biomethane-production-and-potential-production/. Accessed 4 Feb 2025. https://doi.org/10.1007/978-3-031-01241-9

Statista (2024b) Fuel ethanol production worldwide in 2023, by leading country. Retrieved from: https://www.statista.com/statistics/281606/ethanol-production-in-selected-countries/. Accessed 4 Feb 2025. Bacterial cellulose: production, scale-up and applications, pp 7–21. https://doi.org/10.1201/9781003355434-2

Statista (2025) Global corn production from 2014/2015 to 2024/2025 (in million metric tons). https://www.statista.com/statistics/1156213/global-corn-production/#:~:text=In%20the%202023/24%20trade,over%201.23%20billion%20metric%20tons. Accessed 12 Feb 2025

Strugnell B, Patrick JW (1995) Hydropyrolysis yields in relation to coal properties. Fuel 74:481–486. https://doi.org/10.1016/0016-2361(95)98349-J

Sun W, Li X, Zhao J, Qin Y (2022) Pretreatment strategies to enhance enzymatic hydrolysis and cellulosic ethanol production for biorefinery of corn Stover. Int J Mol Sci 23:13163. https://doi.org/10.3390/ijms232113163

Sydney EB, Duarte ER, Martinez Burgos WJ et al (2020) Development of short chain fatty acid-based artificial neuron network tools applied to biohydrogen production. Int J Hydrog Energy 45:5175–5181. https://doi.org/10.1016/j.ijhydene.2019.10.128

Syed MW, Kazmi WW, Hussain A et al (2024) Lignin liquefaction: unraveling the effect of process conditions and sustainable pathways for biofuel production—a comprehensive review. Energy Convers Manag 313:118615

Szczerbowski D, Pitarelo AP, Zandoná Filho A, Ramos LP (2014) Sugarcane biomass for biorefineries: comparative composition of carbohydrate and non-carbohydrate components of bagasse and straw. Carbohydr Polym 114:95–101. https://doi.org/10.1016/j.carbpol.2014.07.052

Thakur N, Jalalah M, Alsareii SA et al (2024) Anaerobic digestion of fat, oil, and grease (FOG) under combined additives: enhanced digestibility, biogas production, and microbiome. Renew Sust Energ Rev 191:114155. https://doi.org/10.1016/j.rser.2023.114155

Tong Y, Yang T, Wang J et al (2024) A review on the overall process of lignin to phenolic compounds for chemicals and fuels: from separation and extraction of lignin to transformation. J Anal Appl Pyrolysis 181:106663. https://doi.org/10.1016/j.jaap.2024.106663

Tripathi M, Sahu JN, Ganesan P (2016) Effect of process parameters on production of biochar from biomass waste through pyrolysis: a review. Renew Sustain Energy Rev 55:467–481

van Eijck J, Batidzirai B, Faaij A (2014) Current and future economic performance of first and second generation biofuels in developing countries. Appl Energy 135:115–141. https://doi.org/10.1016/j.apenergy.2014.08.015

Van Lier JB, Vashi A, Van Der Lubbe J, Heffernan B (2010) Anaerobic sewage treatment using UASB reactors: engineering and operational aspects. In: Environmental anaerobic technology: applications and new developments. World Scientific Publishing, pp 1–30. https://doi.org/10.1142/9781848165434_0004

Verma D, Raj J, Pal A, Jain M (2016) A critical review on production of biodiesel from various feedstocks. J Sci Innov Res 5: 51–58. https://doi.org/10.31254/jsir.2016.5205

Vignesh P, Pradeep Kumar AR, Shankar Ganesh N et al (2021) A review of conventional and renewable biodiesel production. Chin J Chem Eng 40:1–17. https://doi.org/10.1016/j.cjche.2020.10.025

Vilas Bôas RN, Mendes MF (2022) A review of biodiesel production from non-edible raw materials using the transesterification process with a focus on influence of feedstock composition and free fatty acids. J Chil Chem Soc 67:5433–5444. https://doi.org/10.4067/S0717-97072022000105433

Wan Osman WNA, Rosli MH, Mazli WNA, Samsuri S (2024) Comparative review of biodiesel production and purification. Carbon Capture Sci Technol 13:100264. https://doi.org/10.1016/j.ccst.2024.100264

Wang X, Cheng B (2024) Integrating molecular dynamics simulations and experimental data for azeotrope predictions in binary mixtures. J Chem Phys 161:1–11. https://doi.org/10.1063/5.0217232

Wang Y, Wang H, Lin H et al (2013) Effects of solvents and catalysts in liquefaction of pinewood sawdust for the production of bio-oils. Biomass Bioenergy 59:158–167. https://doi.org/10.1016/j.biombioe.2013.10.022

Xu Y, Hu X, Li W, Shi Y (2011) Preparation and characterization of bio-oil from biomass. In: Progress in biomass and bioenergy production. IntechOpen

Yang J, Gao C, Yang X et al (2022) Effect of combined wet alkaline mechanical pretreatment on enzymatic hydrolysis of corn stover and its mechanism. Biotechnol Biofuels Bioprod 15:31. https://doi.org/10.1186/s13068-022-02130-0

Yu J, Wang D, Sun L (2021) The pyrolysis of lignin: pathway and interaction studies. Fuel 290: 120078. https://doi.org/10.1016/j.fuel.2020.120078

Zhang J, Zhang X (2019) The thermochemical conversion of biomass into biofuels. In: Biomass, biopolymer-based materials, and bioenergy: construction, biomedical, and other industrial applications. Elsevier, pp 327–368

Zhang X, Tu M, Paice MG (2011) Routes to potential bioproducts from lignocellulosic biomass lignin and hemicelluloses. Bioenergy Res 4:246–257

Zhang H, Zhang R, Song Y et al (2023) Enhanced enzymatic saccharification and ethanol production of corn stover via pretreatment with urea and steam explosion. Bioresour Technol 376:128856. https://doi.org/10.1016/j.biortech.2023.128856

Zhang F, Yi J, Pan W, Mei G (2024) A new process for the production of second-generation biodiesel from waste oils and fats. Biomass Convers Biorefinery 14:9013–9022

Zhu Y, Song W, Yao R et al (2022) In-situ catalytic hydropyrolysis of lignin for the production of aromatic rich bio-oil. J Energy Inst 101:187–193. https://doi.org/10.1016/j.joei.2022.01.011

Zupko R (2019) Life cycle assessment of the production of gasoline and diesel from forest residues using integrated hydropyrolysis and hydroconversion. Int J Life Cycle Assess 24:1793–1804. https://doi.org/10.1007/s11367-019-01616-8

Chapter 14
Third Generation Biofuels

Hissashi Iwamoto, Juliana Cardoso, Renato Canci Vieira, Denisse Tatiana Molina Aulestia, Carlos Ricardo Soccol, and Júlio Cesar de Carvalho

Abstract The third generation of biofuels uses algae as a raw material. Algae are a preferred raw material compared to agricultural source (i.e., corn). However, the algae industry struggles to meet the biofuel demand, and these biofuels are still mostly a promise. However, several factors are contributing to the growing study of production, including a low demand for algae cultivation areas; the possibility of cultivation avoiding agricultural areas already considered scarce; rapid growth of algae when compared to many other higher plants; the better use of biomass in the conversion of biofuels; and, probably the most important, possibility of various coproducts derived from a single biomass. For biofuel production, biomass can be subjected to biochemical, thermochemical, chemical, or direct conversion. Those techniques involve photobiological H_2 production, fermentation, anaerobic digestion, gasification, pyrolysis, liquefaction, transesterification, and direct power generation. Microalgae biomass can yield various important and valuable bioproducts, and the biomass can still be explored for biofuel production. This chapter discusses how new technologies are going to lead to the third-generation of biofuels, as well as their future market perspectives on a global scale.

14.1 Introduction

Biofuels are essential products for enabling and sustaining a variety of technologies fundamental to contemporary society due to their high-energy properties upon combustion and the variety of biological raw materials used during industrial production. These biofuels, including bioethanol, butanol, biogas, biohydrogen, and bio-oils, hold significant market value (Alam et al. 2015). Given the wide diversity of raw materials and products classified as biofuels, these products have

H. Iwamoto (✉) · J. Cardoso · R. C. Vieira · D. T. M. Aulestia · C. R. Soccol ·
J. C. de Carvalho (✉)
Bioprocess Engineering and Biotechnology Department, Federal University of Paraná–Polytechnic Center, Curitiba, Brazil
e-mail: jccarvalho@ufpr.br

A. L. Woiciechowski et al. (eds.), *Biomass for Bioenergy and Bioproducts*, Biofuel and Biorefinery Technologies 17, https://doi.org/10.1007/978-3-032-19898-3_14

often been categorized into four generations. This classification is based on the type of raw material used, the technology needed for production, and the environmental impact associated with each process (Patil et al. 2008).

First-generation biofuels are produced through fermentation mediated by bacteria and yeasts, using sugar extracts and vegetable oils obtained from the processing of food-based bioproducts rich in mono- and polysaccharides or lipids, such as soybeans, corn, and sugarcane. Second-generation biofuels, in contrast, are obtained from nonedible raw materials, including agricultural, forestry, or urban residues (Kiran et al. 2014). Their production is similar to that of first-generation biofuels, often using comparable bacterial and yeast strains in mainstream production. Third-generation biofuels, in turn, are classified as biofuels obtained through processes conducted by microalgae or macroalgae (Stephens et al. 2010). These biofuels can use agro-industrial or municipal wastes and carbon dioxide as raw materials, making algae a rich source of proteins, carbohydrates, and lipids for highly efficient biofuel production. However, despite their high productivity, algae present processing challenges as well. Lastly, fourth-generation biofuels rely on the genetic modification of microorganisms (such as algae, bacteria, and fungi) to increase their biofuel synthesis capabilities, allowing for the use of a broader range of raw materials, including carbon dioxide, as a potential substrate for biofuel production. A summarized description of the four different generations of biofuels is presented in Fig. 14.1.

Four key criteria can be used to evaluate third-generation biofuels: the final product quality, the potential for obtaining valuable coproducts, the biodiversity impact, and the favorable contribution to the circular economy. It is reported that third-generation feedstocks and algae biomass are economically more sustainable and have less environmental impact than other feedstocks used in biofuel production (Hebbale and Ramachandra 2022). However, when analyzing isolated processes,

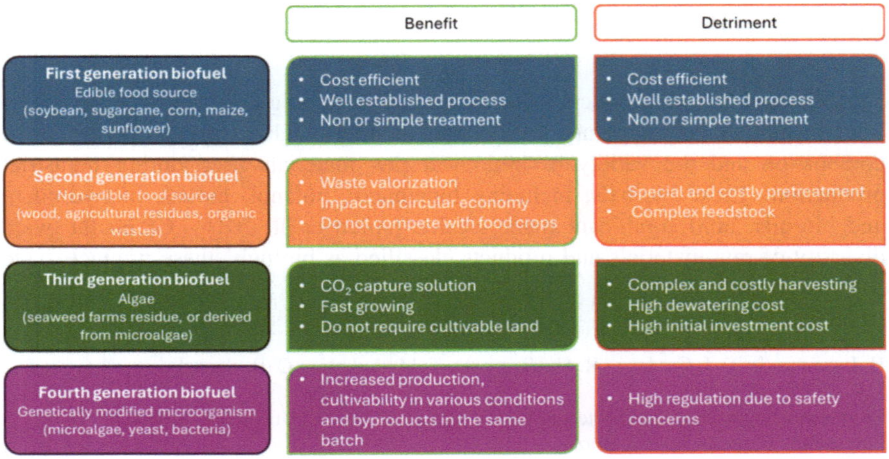

Fig. 14.1 Four generations of biofuels: definitions, and their strengths and weaknesses

these advantages may appear contradictory. For example, algae biomass pretreatment for bioethanol production may require chemical products that make the lignocellulosic bioethanol production pretreatment slightly more environmentally friendly.

Algae are photosynthetic and aquatic organisms that absorb carbon dioxide from the atmosphere and minerals from the environment, generating oxygen and a rich biomass. In comparison to higher plants (the basis for first-generation feedstock), algae are more efficient in the photosynthesis process and grow faster, resulting in a better capacity for fixing CO_2 (Hebbale and Ramachandra 2022).

Fossil energy is recognized to be limited and environmentally harmful, making the search for sustainable sources important. The truth is that the widespread consumption of fossil fuels is considered the main cause of anthropogenic greenhouse gas emissions, which is the primary cause of the contemporary climate crisis. Environmental programs like the European Green Deal and the Paris Agreement aim that around 80–90% of greenhouse gas emissions may be reduced by 2050 through the widespread consumption of biofuels. Fortunately, various countries around the world have initiated sustainable projects in biofuel production to avoid fossil source dependency, which currently accounts for around 92% of the demand for fuels today. It is important to notice that in the present period, electric technology (like electric vehicles) is a growing market and is contributing significantly to greenhouse gas emissions reduction. However, the emissions related to electricity production and battery production have often been overlooked (Meşe Erdoğan et al. 2024).

14.2 Biomass and Biofuel

14.2.1 What Are Algae and How Are They Produced?

Algae are a diverse group of unicellular and multicellular organisms capable of photosynthesis. They are relatively simple compared to plants, lacking sap-carrying vessels or a dense lignocellulosic body, but they are excellent at capturing CO_2 from the atmosphere and utilizing sunlight energy for growth and producing intracellular storage compounds. Some algae show greater photosynthetic efficiency than other plants, and some species even outperform the world's fastest-growing plants. In fact, photosynthetic efficiencies for algae have been reported to range from 3% to 8%, compared with 0.5% for many terrestrial crops. Algae can be classified into two categories: seaweed and microalgae (Alaswad et al. 2015).

Macroalgae, or "seaweed," consist of multicellular algae. They are generally similar to plants, showing differentiated cellular structures and functions. However, they do not have true roots, stems, leaves, or vascular tissue and have simple reproductive structures. They can grow in both salt and fresh water. As mentioned above, they grow rapidly and can reach sizes up to 60 meters in length. Most macroalgae are divided into three large groups, which are easily identified based on their pigmentation: brown algae (*Phaeophyceae*), red algae (*Rhodophyceae*), and green algae (*Chlorophyceae*). The main products currently obtained from the cultivation of a few marine algae are mostly the production of special foods and the extraction of hydrocolloids. In their natural environment, macroalgae grow attached to firm substrates, such as rocks and bottom sands, forming a stable multilayered vegetation to capture as much light as possible (Alaswad et al. 2015).

Microalgae, on the other hand, are unicellular organisms. Some microalgae may form colonies (for example, *Volvox*, *Scenedesmus*, or the cyanobacteria *Spirulina* sp. or *Nostoc* sp.). The term "microalgae" refers to a broad classification of microscopic organisms capable of performing photosynthesis, including both eukaryotes and prokaryotes from various classes, such as *Euglenoidea, Chlorophyceae*, and even *Cyanophyceae*. Microalgae are further categorized into three relevant groups: diatoms (Bacillariophyceae), green algae (*Chlorophyceae*), and golden algae (*Chrysophyceae*). Due to their presence in diverse groups of organisms, microalgae can grow under various conditions and synthesize a wide variety of biological and natural products, ranging from toxins to pigments, antioxidants, proteins, sugars, and different types of lipids. To be cultivated, they require water and need to be submerged. They do not have growth limitations when compared to plants, supporting a wide range of salinity and nutritional needs. Unlike plants and algae that are attached to a substrate, microalgae need to be cultivated in closed containers to control contamination and cultivation conditions. These containers, when used on a large scale, can be ponds, swimming pools, or closed tanks. Microalgae can be cultivated on any land type, whether arable or nonarable, or even in aquatic environments, as technologies exist for plastic-bag containers designed for this purpose. The biomass obtained from microalgae offers several advantages over marine plants and algae, including low fiber or lignin content, high protein richness, the potential to accumulate significant amounts of sugars and lipids, and the ability to produce a wide variety of value-added bioproducts for industries such as nutraceuticals, cosmetics, pharmaceuticals, and biorefineries (Alaswad et al. 2015).

The author Upadhyay et al. (2024) conducted an analysis based on a 2019 report published by Statista, which examined global consumption trends of various energy resources from 2000 to 2019, with projections extending to 2050. During the editing of this chapter, we saw that Statista updated its projections (Madhumitha 2025), presenting a profile of lower fuel consumption (Exajoules (EJ)) overall for 2050. Our most recent analysis, based on Statista's 2022 report, shows a 16.4% decrease (from 760 EJ to 635 EJ) in the total energy consumption expected for 2050, which can be explained by the growth of electric technologies (e.g., electric cars) in recent years. Between 2015 and 2019, approximately 85% of fuel sources were petroleum-based (Upadhyay et al. 2024). In 2022, Statista reported the following energy distribution: 32% oil, 24% natural gas, 28% coal, 2% nuclear energy, 3% hydroelectric power, and 12% renewable energy. Consumption of renewable energy alone grew from 56 EJ in 2015 to 70 EJ in 2022, representing a 25% increase (Madhumitha 2025). In Fig. 14.2, it is clear that during this same period, other energy sources saw an increase in consumption ranging from 3% to a maximum of 14.3%, indicating that renewable energy is contributing more to global energy demand, replacing conventional sources with significant economic and environmental impacts globally.

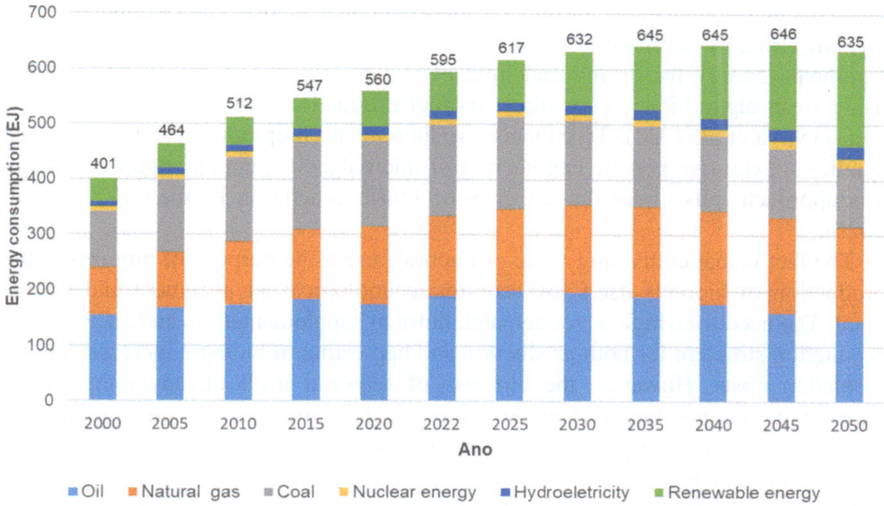

Fig. 14.2 Consumption rate of energy resources from 2000 to 2050 (prediction) (Madhumitha 2025)

Projections for the coming years suggest a decrease in the consumption of nonrenewable fuels. By 2050, renewable fuels are expected to increase, accounting for approximately 27.4% of the total energy (174 EJ out of 635 EJ). Compared to Statista's 2019 results, where this share was 33%, the increase in electric technologies and their impact on reduced fuel consumption suggests that renewable energies will make up around one-third of global energy consumption, as interpreted by Upadhyay et al.'s (2024) to achieve energy security in the coming decades, it is critical to shift the focus from fossil fuel dependence to renewable sources. Third-generation fuels are an alternative to oil and are being explored as practical options.

14.2.2 High Production Cost Is Still a Problem

Algae production still requires an economic solution for large-scale implementation. In the case of first-generation and second-generation biofuels, concerns arise over the use of arable land, fertilizers, and water, as these biofuels are derived from food crops. In contrast, third-generation biofuels do not require arable land but demand significant amounts of water for cultivation, which increase total production costs. To mitigate this issue, microalgae can be cultivated using nutrient-rich wastewater (Kant et al. 2025). One of the post-cultivation challenges is biomass harvesting and pretreatment. The colloidal characteristics, negative surface charge, and low sedimentation velocity of microalgae make harvesting difficult. It is estimated that microalgae harvesting accounts for approximately 20–30% of the total production costs. To reduce the cost of traditional harvesting methods (e.g., centrifugation, filtration, and flocculation), bio-flocculation may offer a cost-effective and environmentally friendly alternative (Tripathi et al. 2024). Biomass treatment also requires significant energy for drying and lipid extraction. Several studies have explored using fresh algae biomass for direct biofuel production (e.g., hydrothermal treatment) (Silva et al. 2024). To enhance production and bioproduct yields for biofuel synthesis, extensive genetic engineering of microalgae is gaining attention, though this approach falls under the category of fourth-generation biofuels (Kant et al. 2025).

The fact is that cultivating algae or microalgae for the purpose of biomass in the production of biofuels itself still requires technological advancement and investment. The need for a high water demand, nutrient supplementation, and an advanced production structure for both production and harvesting of biomass increases implementation costs. However, the biomass of seaweed itself already comes from existing production farms. The use of residual biomass for biofuel production, after the extraction of hydrocolloids, becomes relevant for macroalgae farms as an addition to their revenue. Several studies and adaptations have been studied and implemented in existing energy production systems or fuel plants, since algae can use various industrial wastes, although costly, require little space, and can therefore be coupled in the vicinity of these industries. One example is the modeling of hybrid energy generation systems with wind, solar, and biomass technologies. This combination has shown a considerably high cost benefit. Furthermore, the integration of algae production systems into industries perfectly meets the concerns in terms of the circular bioeconomy that aims to recycle, reuse, remanufacture, and move away from the traditional linear economy (Kant et al. 2025).

14.2.3 How Algae Biomass Could Be Competitive

The composition of biomass for third-generation biofuels is an important subject, involving carbohydrates, lipids, and proteins. One key difference in algae is their high protein content compared to plants. Some studies propose obtaining biofuels exclusively from proteins or amino acids as an innovative alternative. While algae's high protein content was not a focus in first-generation and second-generation fuels, it holds added value and importance as a nutraceutical, although it can also serve as a biofuel source. Both macroalgae and microalgae have considerable amounts of proteins, which could support the production of differentiated fuels. For example, the optimal condition for biomethane production involves macroalgal biomass with low lipid content and high carbohydrate and protein content (Ramaraj et al. 2022).

Lipids can accumulate at various sites in algal cells, including the endoplasmic reticulum, chloroplast, and cytosol. Microalgal lipids serve functions such as providing structural support and energy storage. Based on polarity, microalgal lipids are classified into (a) polar lipids, such as glycolipids and phospholipids, and (b) neutral lipids, such as triacylglycerols (TAGs). Neutral lipids, present as TAGs, form the bulk of algal oil, comprising 20–50% of the dry weight of the cell. The synthesis of

Table 14.1 Algae lipid and carbohydrate content in dried biomass

	Lipid content (% of dry wt.)	Starch	Reference
Seaweeds			
Ulva	2.1	42.0	Alaswad et al. (2015)
Enteromorpha	0.3	64.9	
Monostroma	1.2	63.9	
Laminaria	2.4	39.3	
Alaria	3.6	39.8	
Sargassum	2.9	33.0	
Padina	1.7	31.6	
Porphyra	4.5	45.1	
Rhodymenia	1.7	44.6	
Gracilaria	1.8	61.75	
Microalgae			
Acutodesmus dimorphus	33.40	53.30	Zhang (2023)
Botryococcus braunii	30.2	20	
Chlorella sp.	21.8	47.4	
C. vulgaris	9.81	16.8	
Dunaliella salina	6	32	
Scenedesmus obliquus	30.8	35	
Tetraselmis maculata	3	15	

TAGs within microalgae occurs in two stages: first, in the chloroplast, and then in the endoplasmic reticulum (Pandey et al. 2024).

Algae can be a potential source of various biofuels, including biogas from the anaerobic degradation of biomass and biodiesel from lipids accumulated in algal cells. Biogas and biodiesel can replace natural gas and diesel oil, respectively, and are the most common biofuels produced from algae. Microalgae typically produce substantial amounts of lipids, while seaweed produces more sugar and carbohydrates than lipids. The oil content of macroalgae is minimal, less than 5% of the dry weight, making macroalgae generally uneconomical as a source of biodiesel (Alaswad et al. 2015).

In summary, microalgae produce substantial amounts of lipids, while seaweed produces more sugars and carbohydrates than lipids. The oil content of macroalgae is very low, less than 5% of the dry weight, making macroalgae uneconomical as a source of biodiesel. Table 14.1 presents the lipid and starch content in several types of algae.

Microalgae have gained attention for their potential in biorefineries due to their ability to use nutrients from wastewater. Minimally processed algae biomass can generate large quantities of raw fractions, such as lipids and biodiesel. To improve efficiency, it is essential to focus on producing small quantities of high-value products, like vitamins, proteins, DHA, and carbohydrates (Kant et al. 2025). Both seaweed and microalgae biomass contain various compounds that have potential applications in the feed, food, nutraceutical, and pharmaceutical industries. Algae, as natural phototrophic organisms, can thrive in various conditions and have a range of cellular protective properties due to bioactive substances such as mycosporine-like amino acids, carotenoids, sulfated polysaccharides, sterols, polyphenols, and vitamins. These include photoprotective chemicals such as UV absorbers, antioxidants, matrix-metalloproteinase inhibitors, and compounds with anti-aging and immunomodulatory properties. As a result, algae can be used in cosmeceutical products to prevent various skin diseases (Nilam et al. 2023).

Seaweed is primarily cultivated as a raw material for the food industry. Their biomass is used as fertilizer and for the extraction of hydrocolloids. Hydrocolloids are long-chain polymers that form viscous solutions or gels when dispersed in water, playing a key role as ingredients for thickening, gelling, or stabilizing food and other materials. For example, ulvan, a sulfated polysaccharide hydrocolloid, can be extracted from the green macroalga *Ulva rigida*. Similarly, other types of sulfated polysaccharides, such as carrageenan from red algae and fucoidans from brown algae, can also be obtained (Llano et al. 2023).

Microalgae, including both cyanobacteria and eukaryotic microalgae, contain natural pigments with significant potential for biotechnological applications, such as feed, food colorants, cosmetics, and pharmaceuticals. These pigments include phycobiliproteins, primary carotenoids, and fucoxanthin. The primary source of phycocyanin is the cyanobacterium *Arthrospira platensis* (*Spirulina platensis*), though recently, the red algae *Galdieria sulphuraria* has also been identified as a source. Other microalgae, such as *Chlorella*, *Dunaliella*, *Nannochloropsis*, *Scenedesmus*, and *Haematococcus*, are promising for producing several light-harvesting and protective carotenoids. These microalgae can manage oxidative stress due to their free radical scavenging properties. The carotenoids lutein, zeaxanthin, beta-carotene, astaxanthin, and others are suitable sources of natural antioxidants (Chini Zittelli et al. 2023).

14.2.4 Technologies for Biofuel Conversion from Algae Biomass

Producing third-generation biofuels requires integrating knowledge from fields like biology, physiochemistry, and quality control. As such, applying various techniques from these areas is crucial for efficient synthesis (Singh 2024).

To ensure an effective process, the algal inoculum used for conversion must be carefully handled and properly stored through subculturing and cryopreservation in liquid nitrogen. For example, the widely used alga *Chlorella* should be cultivated in a specific medium (BG11) to enhance cellular concentration in solution, initiating the scale-up process. Afterward, the inoculum must adapt to a medium made from human-derived residues and undergo conditions like those at an industrial scale (Medipally et al. 2015).

The inoculum should be cultivated in a pilot-scale bioreactor under conditions similar to those of industrial environments. Microalgae can be grown in several types of bioreactors, such as ponds, raceway ponds, bubble columns, tubular, airlift, and plastic-bag photobioreactors for secondary energy production. Microalgae thrive under different conditions: in the presence of light, assimilating CO_2 (photoautotrophic); in the presence of organic carbon (mixotrophic); or even without light, relying solely on organic carbon (heterotrophic). This versatility gives microalgae an advantage in environmental remediation strategies. To ensure a mainstream process, cultivation can be analyzed using common methods, such as standard growth curves, optical density measurements, and dry weight determination. Finally, third-generation biofuels can also be produced using other cell types in cocultivation systems, particularly for carbohydrate fermentation or photobiological hydrogen (H_2) production (Alam et al. 2015).

Algae obtain the necessary nutrients directly from the water. To improve nutrient absorption efficiency, upstream processing techniques are often employed. At the laboratory scale, chemical treatments such as medium clarification, pH correction, and nutrient supplementation are common. However, for larger-scale applications, cost reduction and environmental impact minimization take precedence, leading to the use of physical management techniques such as temperature elevation, pressurization, and centrifugation—processes that, while effective, require sophisticated, high-cost equipment (Alam et al. 2015).

After the upstream and mainstream processes, various separation techniques are applied to isolate biofuel molecules from the reaction medium, increasing their purity and enhancing overall yields. These downstream techniques also allow for the reutilization of byproducts, supporting the development of a circular biorefinery system. Common conversion pathways in the literature include biochemical conversion, thermochemical conversion, esterification reactions, and direct combustion (Brigagão et al. 2019). In Fig. 14.3, a summary of these processes is presented.

Biochemical conversion involves anaerobic digestion or the direct production of bio-H_2. Anaerobic digestion degrades microalgal biomass generated during the mainstream phase, mediated by microorganisms in the absence of oxygen, producing biogas, primarily methane (CH_4) and carbon dioxide (CO_2) (Singh et al. 2011). This process can produce H_2; though cyanobacteria, green algae, and seaweed also have mechanisms for producing hydrogen by capturing light energy to break water into molecular oxygen, protons, and electrons (Zieliński et al. 2025).

Thermochemical conversion aims to increase reaction temperatures through various mechanisms, depending on the target biofuel molecule. Examples from the literature include gasification, pyrolysis, biomass liquefaction, transesterification reactions, and direct combustion of the feedstock. Thermogenic pyrolysis works at high temperatures and in the absence of oxygen, generating bio-oil, biochar, and gaseous byproducts from biomass carbonization. Thermogenic gasification involves heating algal biomass with a limited oxygen supply at high, controlled temperatures, avoiding carbonization and producing syngas (a mixture of H_2, CO_2, and CH_4) and liquid biofuels. Lastly, thermogenic liquefaction, conducted at lower temperatures and high pressures, is recommended for producing liquid biofuels like biocrude,

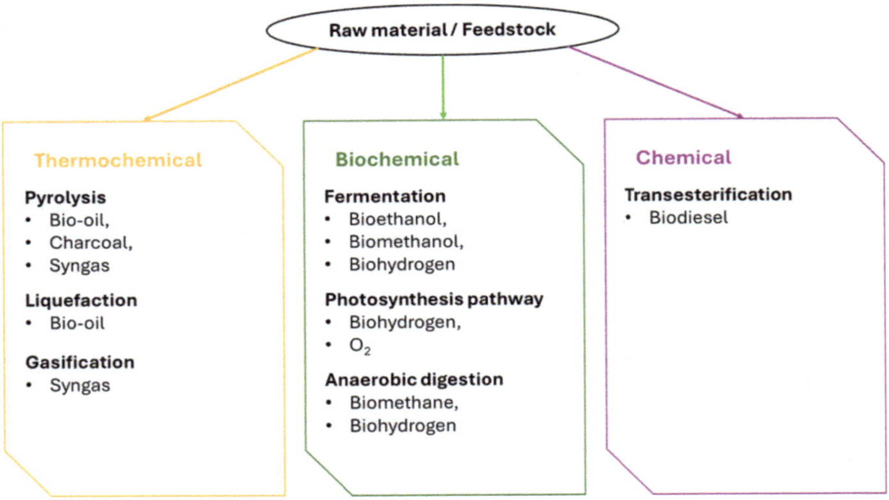

Fig. 14.3 Conventional process for biofuel production

which shares characteristics with petroleum (Patil et al. 2008; Stephens et al. 2010). Both the processes of liquefaction and gasification, also known as hydrothermal liquefaction and gasification, may be conducted in the presence of water. This condition takes advantage of using algal biomass without drying treatment, and presents high yield on producing syngas, biohydrogen, and biocrude (Vonk et al. 2025).

The transesterification reaction involves the direct chemical reaction of lipids extracted from algae with an alcohol in the presence of a catalyst (acid, base, enzyme, or biocatalyst), leading to biodiesel synthesis. The transesterification is crucial to the quality and quantity of the final biodiesel produced. Even more, the lipid extraction must be considered, as it may affect the maximum yield of lipids obtainable from the biomass. It is common for extraction and transesterification to be conducted in a single-step process, saving conversion costs. Catalysis is another important part of the biodiesel synthesis through this process, as it influences the rate of lipid reaction and the conversion rate into the final biodiesel. Enzymes and biocatalysts are gaining great attention, as they are cost-effective and promote sustainable use of waste (Neto et al. 2019; Pandey et al. 2024).

Lastly, an alternative approach for using residual biomass in downstream processing is the direct combustion of biomass. This method allows for generating electrical and thermal energy, which can be reintegrated into the process to improve energy efficiency (Neto et al. 2019).

Each of these processes is associated with the extraction of specific biofuels and high-value-added molecules. Given the variability of microalgal strains, process adaptations are always necessary, considering the microorganism utilized, the type of biochemical conversion necessary, the target biofuel that needs to be produced, potential contaminants that can be purified, and the corresponding analytical methods required for quantification and quality control (Alam et al. 2015). These methods will change depending on the type of biofuel being produced. The third-generation biofuels addressed in this capture were biodiesel, bioethanol, biobutanol, biosynthetic natural gas (SNG), biomethane, biohythane, hydrogen, biokerosene, and incomplete biomass combustion products such as syngas and biocrude.

14.3 Types of Third-Generation Biofuel

Biofuels should not present significant differences in their composition or quality, especially when it comes to ethanol or biogas, for example. Specifically, biodiesel composition may have the presence of metals and metalloids, making analysis important, as during combustion these fuels can adversely affect the performance of engines and shorten the lifetime of equipment. Also, some elements act as catalyst poisons, affecting the amount of unwanted gases and particulate matter emitted from engines. According to the American Society for Testing and Materials (ASTM), certain compounds such as calcium (Ca), magnesium (Mg), phosphorus (P), potassium (K), and sodium (Na) should be analyzed due to their higher concentrations in biodiesel compared to other biofuels such as ethanol or methane (ASTM D8351-22).

It is expected that produced fuels, both fossil fuels and biofuels, do not change in terms of energy conversion and quality. However, some physical and chemical properties are listed in Table 14.2.

Next, we describe various biofuels derived from third-generation feedstocks. These include biodiesel, bioethanol, biobutanol, biosynthetic natural gas (SNG), biomethane, biohythane (or hydrogen), biokerosene, and incomplete combustion (pyrolysis) biofuels (e.g., syngas and biocrude).

14.3.1 Biodiesel

Microalgae may be adequate for biodiesel production compared to seaweed due to lipid accumulation. Even more, their oil production could easily surpass plant-based production. Based on Rajpoot et al. (2025), it is expected that microalgae could yield

Table 14.2 Some differences between fossil fuels and biofuels properties (Rajpoot et al. 2025)

Particulates	Fossil fuels	Biofuels
Energy per unit mass	Relatively high	Relatively low
Viscosity	Relatively low	Relatively high
Freezing point	Relatively low	Relatively high
Potential products	Coal, petroleum, NG	Bioethanol, biodiesel, SNG
Pollution	Relatively high	Relatively low
Impact on health	Toxic	Low toxicity ingredients and bioproducts

more third-generation biodiesel than the average production from second generation, based on corn oil, coconut, palm oil, and others. Microalgae could yield 136,900 L/ha of biodiesel per land, which is 24 times more than palm oil (5950 L/ha), 215 times more than soybean oil (636 L/ha), and 790 times more than corn oil (172 L/ha).

As a coproduct, lipid or FAMEs accumulation may occur after inducing stress in microalgae cultivation. A *Chlorella* sp. strain has been found to have potential for lipid production. In the experiment, this microalgae under optimized conditions accumulated 21.21% of FAME content from its dry biomass. Cultivation mode was based on two steps: biomass production followed by nitrogen deprivation. Identified FAME on *Chlorella* sp. was palmitoleic acid (C16:1), heptadecanoic acid (C17:0), stearic acid (C18:0), oleic acid (C18:1), linolenic acid (C18:2), arachidonic acid (C20:4), and heneicosanoic acid (C21:0) (Bejarano Meza et al. 2025).

The use of waste materials is important for environmental remediation, which collaborates with the goal of using biofuels for pollution reduction. In Brazil, nutshell has been studied for *Tolypothrix* sp. CACIAM 22 cultivation. These cyanobacteria reach a significant amount of 12 g/L (10 days) of biomass when cultivated in media supplemented with hydrolysate of Brazil nut shells (namely "*Castanha do Pará*" in Brazil). This experiment also evaluated glycerol as a carbon source for *Tlypothrix* sp. and obtained 9.3 g/L of dry biomass after 10 days of cultivation (Terra de Oliveira et al. 2024). As told previously, microalgae, although being a photosynthetic organism, can also assimilate organic carbon sources; research shows biomass production increases on media supplemented with organic carbon. For instance, *Chlorella* sp. Y8–1 (Lin and Wu 2015) and *Spirulina* sp. LEB18 (de Morais et al. 2019) yielded 45 g/L and 3 g/L, respectively, when cultivated at mixotrophic conditions, supplemented with glycerol. This result was threefold and fivefold better than those obtained from phototrophic conditions, respectively, for *Chlorella* sp. Y8-1 and *Spirulina* sp. LEB18.

A cultivated microalgae consortium in wastewater showed a good yield of bio-oil production. The cultivation involved resulted in a biomass with 30% of *Chlorella vulgaris*, 69% of *Tetradesmus obliquus*, and 1% of cyanobacteria *Limnothrix planctonica*, treated with hydrothermal liquefaction, which converted the biomass into a bio-oil important for biofuel (Silva et al. 2024). Microalgae-based biodiesel is important not only as a renewable energy source but also because it generates up to 80% less toxic or polluted gases or particulate matter (Ofori-Boateng 2022).

14.3.2 Bioethanol

For bioethanol production, a better yield is expected using macroalgal biomass. Compared to microalgae, seaweed has higher starch and lignin content, making it a better fit to existing processes of first- and second-generation ethanol production. To start ethanol fermentation, the first step is to depolymerize the cell wall polysaccharides of the algae. The method most widely used for this purpose is methanolysis.

However, there are studies that recommend acid hydrolysis to enhance effectiveness (Llano et al. 2020). The hydrolyzed sugars may be used for fermentation. Traditionally, ethanol is obtained from alcoholic fermentation that uses yeast to convert sugars into ethanol and carbon dioxide. Other fermentation strategies are commonly employed for more efficient conversion and the production of other coproducts. For example, acetone-butanol-ethanol fermentation (ABE) (Llano et al. 2023).

Microalgae offer immense potential for large-scale, cost-effective recombinant protein production. Biofuels are primarily derived from carbohydrates and lipids present in these organisms. In contrast, proteins have not been used for fuel production due to their high value for nutraceuticals and animal feed, coupled with challenges in deaminating protein hydrolysates. Recombinant strategies, however, have proven practical for large-scale enzyme production for biofuel generation. Metabolic engineering in microorganisms can deaminate protein hydrolysates, enabling the conversion of proteins into C4 and C5 alcohols, with a theoretical yield of 56%. *S. cerevisiae*, *E. coli*, *Bacillus subtilis*, and microalgae can serve as protein sources, yielding up to 4 g/L of alcohols from biomass containing around 22 g/L of amino acids. These findings show the feasibility of using proteins in biorefineries, with high-protein microalgae as potential feedstocks to maximize algal growth and CO_2 fixation (Neto et al. 2019).

14.3.3 Biobutanol

Fermentation of first-generation biomass for biobutanol production is limited by high costs, making the process unfeasible. The second-generation process similarly incurs high costs for biomass pretreatment. However, third-generation biobutanol can be obtained through acetone-butanol-ethanol (ABE) fermentation, making macroalgae a practical alternative for clean, renewable gas production. In research, macroalgae biomass was analyzed for biobutanol production using *Clostridium beijerinckii* ATCC 10132. Seaweed biomass has the advantage of containing little to no lignin, which eliminates the need for extreme pretreatments such as lignocellulose pretreatment. Three types of biomass—*Gracilaria tenuistipitata*, *Ulva intestinalis,* and *Rhizoclonium* sp.—were used for biobutanol production. These biomasses underwent low-temperature hydrothermal pretreatment (HTP) to obtain reducing sugar. *Clostridium beijerinckii* produced 14 g/L of butanol from 60 g/L of glucose. The pretreated *Ulva intestinalis* and *Rhizoclonium* sp. biomasses did not show significant improvements in ABE fermentation. However, *Gracilaria tenuistipitata* exhibited a rapid increase in butanol concentration after 16 hours of fermentation. *G. tenuistipitata* resulted in 1,38 g/L of butanol, from optimized low-HTP that yielded a 57.6% of reductive sugar and a yield of 19.87% of ABE fermentation (Khaonuan et al. 2023).

It is expected that macroalgae contain 55–65% sugars, mostly complex carbohydrates, in their dry biomass. The pretreatment for the biomass is mild when compared to first-generation biogas production. For *Sargassum wightii* macroalgae biomass, Dubey et al. (2015) reported the conditions for the optimized pretreatment of the biomass as a 2 N H_2SO_4 acid treatment at 120 °C for 30 min. The hydrolyzed algae were used in ABE fermentation with *Clostridium saccharoperbutylacetonicum* (ATCC 27022), which resulted in 15 g/L of butanol from 70 g/L of hydrolyzed algae material, with a theoretical yield of 48% (Dubey et al. 2015).

For butanol production, it is important to note that this biofuel is toxic to both native and non-native butanol-producing organisms. High-yield butanol fermentation is limited by this factor. To avoid toxic issues, continuous butanol recovery during fermentation may be helpful. Selecting or manipulating genetic *Clostridium* strains is also considered an alternative. Lastly, fermentation process enhancement by agitation or aeration may affect cell metabolism. Xia et al. (2015) appointed the periodic-peristole agitation that enhanced *Clostridium acetobutylicum* ATCC 824 butanol fermentation. The mechanism enhances butanol production by 2.4-fold in comparison to the traditional agitator. In addition, the technology presented less shear driven (74%) compared to the traditional Rushton turbine (Xia et al. 2015).

14.3.4 Biosynthetic Natural Gas (SNG)

Biomass use has gained interest due to growing concerns about greenhouse gas emissions and the increasing use of renewables in the energy mix. Wood biomass is an important renewable energy source expected to play a significant role in reducing reliance on fossil-based resources for households, industry, and transportation. However, wood production and consumption are related to various problems of uneven distribution of the resource, insufficient availability of the resource, and difficulties of storing it (Mesfun et al. 2016). In 2022, global natural gas usage stood at 3.94 trillion m^3, which is unsustainable and contributes to greenhouse gas emissions. Synthetic natural gas (SNG) is gaining attention as an alternative source of renewable energy, and algae-derived SNG can significantly contribute to this scenario (Sun et al. 2024).

SNG can be produced from carbon-containing feedstocks using thermochemical or biological treatments, or from CO_2 through direct hydrogenation (power-to-SNG). Power-to-SNG offers a promising solution for storing excess renewable energy, especially from solar and wind power plants, to alleviate the disparity between supply and demand (Sun et al. 2024). It is expected that advanced biomass-based motors will become increasingly important to replace fossil energy sources within the coming decades. It is likely that the new biorefineries will evolve from existing forest and petroleum industry sites, as they already have the required biomass handling infrastructure in place (Mesfun et al. 2016).

The *Chlorella vulgaris* biomass and municipal solid waste (MSW) were used as alternative materials for SNG production. Each material was preprocessed and converted into pellets by the refuse-derived fuel technique. Initially, the pellets derived from a single biomass of microalgae resulted in a low yield of SNG production, as during the gasification phase, the oxygen carrier system was inhibited. The maximum SNG and hydrogen production was achieved when the pellets were composed of 40%:60% of *C. vulgaris* and MSW, respectively. In this condition, the oxygen carrier splitting was eliminated, as the thermal distribution was more balanced across the reactors. The study boosted the fuel-to-hydrogen efficiency to 77.8% and SNG thermal and exergy to 61.3% and 57.9%, respectively (Sun et al. 2024).

14.3.5 Biomethane

In general, methane is the primary component of natural gas, among others, such as ethane, propane, butane, and pentane. If not consumed as fuel, this gas itself is a greenhouse gas, having a relatively short lifespan in the atmosphere (about a 10-year half-life), but it is much more potent than CO_2 in trapping heat in the atmosphere. For methane production, the most recommended biomass is that with an elevated carbon and nitrogen ratio (C:N) in the composition. It is essential in anaerobic digestion, for cell growth and metabolism of various groups of microbial populations, and the toxicity related to ammonia is also drastically reduced by selecting algae biomass that is carbohydrate- and protein-rich (Ramaraj et al. 2022).

The use of *Spirogyra varians* biomass for biomethane production is proposed, as this filamentous macroalgae biomass is high in carbohydrates and protein but low in lipids, showing biochemicals to generate biomethane. The fresh and pretreated biomass of *S. varians* was fermented anaerobically. The pretreated biomass was subjected to a 1% NaOH alkaline hydrolysis. The result was 580 mg/L of biogas after 70 days of fermentation, in which 69.4% was composed of methane (Ramaraj et al. 2022). In another work, *Ulva* sp., *Codium* sp., and *Undaria pinnatifida* biomass were used to obtain biomethane. The anaerobic fermentation was optimized by an artificial neural network. After the fermentation, the biochemical methane potential (CH_4 mL/g of volatile solids) of the three biomasses was reported as 288 mL/g VS, 315 mL/g VS, and 442 mL/g VS for *Ulva* sp., *Codium* sp., and *Undaria pinnatifida*, respectively (Córdoba et al. 2024).

14.3.6 BioHythane or Hydrogen

Hydrogen has become a symbol of the transformation in the global energy matrix, often hailed as one of the most promising environmentally friendly fuels. The fact that its reaction consumes oxygen and generates water is undoubtedly surprising. However, as innovative technologies emerge, they also bring new challenges. While hydrogen production and consumption are gaining traction, numerous unresolved issues remain regarding safety-risk prevention and control (Yao et al. 2025). A critical challenge for hydrogen fuel is the need for high pressure in the storage of both gaseous and liquid hydrogen. Fortunately, solid-state hydrogen storage has advanced significantly in recent years and is now considered the safest mode of hydrogen storage (Yue et al. 2021).

In 2022, China alone consumed 35.2 million tons of hydrogen, accounting for 29% of global hydrogen consumption. Even today, most of the hydrogen consumed is obtained from electrochemical processes through electrolysis (Yue et al. 2021). On the other hand, third-generation biohydrogen can be directly obtained from algae through photosynthesis. Cyanobacteria, green algae, and seaweed have developed mechanisms for producing hydrogen while capturing light energy to break water into molecular oxygen, protons, and electrons. Biohydrogen (naturally occurring hydrogen) is a resource that requires fewer inputs and is clean, economical, and widely accepted (Kajla et al. 2025). The microalga *Tetraselmis subcordiformis* SAG 161-1a, cultivated under optimal conditions, has been recognized as an economically valuable biomass. This alga was used to capture CO_2-rich gas from microbial fuel cells. Maximum growth efficiency occurred when the CO_2 concentration was kept at 220 mg/L. The biomass proliferation rate reached 0.52 g VS/L/day, with a final biomass concentration of 2.68 g VS/L by the end of the cultivation cycle. The sedimented biomass was used for biohydrogen production in a fresh culture medium designed to stimulate bio-H_2 production. The highest bio-H_2 production reached 312 mL, corresponding to a production rate of 62.4 mL/day (Zieliński et al. 2025).

Hydrogen can still be obtained through anaerobic fermentation, producing biohythane, a mixture of hydrogen and methane gas. As previously mentioned, anaerobic digestion is used for biomass treatment, where complex macromolecules are broken down into simpler ones. Biohythane production occurs in two stages: The first involves hydrolysis and acidogenesis, which release volatile fatty acids and hydrogen, while the second stage consists of acetogenesis and methanogenesis. Methanogenic bacteria easily consume the hydrogen produced during the acidic stage, so it is recommended to physically separate the two stages. Several types of biomass, such as sludge, algae, and waste, can be processed through anaerobic fermentation for biohythane production. However, hydrogen yield from sludge is much lower (10–90 mL/g VS) compared to algae (29.5–158 mL/g VS), food waste (100–250 mL/g VS), or crude glycerol (29.2–219.1 mL/g VS) (Bolzonella et al. 2019). The hydrolyzed *Chlorella* sp. biomass was used for hydrogen and methane production via fermentation. The maximum hydrogen yield was achieved after dark fermentation followed by photo-fermentation, resulting in 125 mL/g VS. Dark fermentation followed by anaerobic digestion produced hydrogen and methane yields of 47 mL/g VS and 153 mL/g VS, respectively (Phanduang et al. 2019).

14.3.7 Biokerosene

Among all the third-generation biofuels, kerosene may be the most sought after. Kerosene is consumed in various areas besides as engine fuel, such as a solvent, a cleaning agent, for heating, cooking, or even in lamps. But its highest value lies in its use as aviation fuel, allowing the high-speed operation of jet engines. Biokerosene can be produced through biocrude purification. The thermochemical processing uses the entire algal biomass, including carbohydrates and proteins, to generate algal oils. However, this process is effective when consuming dry feedstocks; the thermochemical processes conducted in the presence of water, such as hydrothermal carbonization, hydrothermal gasification, or hydrothermal liquefaction (HTL), are more suitable for wet feedstocks. Moreover, from an economic standpoint, those techniques are more appealing due to the high quantity of water in algal biomass. Processing the whole algal biomass can offer significant energetic cost savings. To know, the HTL can be applied to use biomass at concentrations of 10–25% wet weight in water, reducing by almost 88% the energy consumption for biomass preparation (Raikova et al. 2016).

Bio-oil and kerosene were obtained through cleaner production by blending *Eucheuma cottonii* and low-rank coal, using microwave-assisted pyrolysis integrated with a distillation column. In a concentration of 1:1 macroalgae-coal ratio, the optimized process yielded 61.7% of bio-oil, of which 50.56% was kerosene. So, the expected kerosene yield from the feedstock is calculated to be 31.19%. From the same process, biodiesel, bio gasoline, and syngas could be obtained, with yields of 14.48% and 10.50% from the feedstock, respectively, to biodiesel and gasoline, and 23% of syngas from the total gas formed (Mahfud et al. 2025).

14.3.8 Incomplete Biomass Combustion Products (Syngas, Biocrude)

The truth is, there are many strategies for obtaining various types of fuels from different feedstock states, including solid-to-solid, solid-to-liquid, solid-to-gas, liquid-to-liquid, liquid-to-gas, and even reversed processes (e.g., gas-to-liquid—syngas to gasoline and biodiesel) (Huber et al. 2006). Fuel production generally results in the generation of residual gas or oil, which are considered incomplete combustion products, commonly known as biocrude and syngas. Specific biocrude and syngas production requires intermediate-temperature thermochemical processes. Hydrothermal liquefaction (HTL) at temperatures ranging from 250 °C to 375 °C can lead to biocrude production, and some yield of syngas can be obtained. This process can generate biocrude that can be upgraded to the entire distillate range of petroleum-derived products. If the process is conducted at a temperature above 375 °C, gasification reactions dominate, and the process is known as hydrothermal gasification (HTG), resulting in the production of syngas (Vonk et al. 2025).

In HTL, the algal biomass composition markedly affects the product. Lipids will be entirely fractionated into the biocrude as fatty acids and alkanes. Carbohydrates, on the other hand, tend to form char if the water is not kept alkaline to increase biocrude yields. Proteins are converted readily to biocrude, increasing pH because of ammonium formation. In contrast to anaerobic digestion for biomethane obtention

(Ramaraj et al. 2022), the HTL process is more suitable for protein-rich biomass, putting microalgae as a recommended feedstock for third generation biocrude production (Biller and Ross 2016).

The HTG for syngas production occurs in the high-temperature region above the supercritical point, where water is in the supercritical state. Under this condition, the primary product of algal biomass becomes syngas, high in combustible gases such as H_2, CH_4, CO, and light hydrocarbons (C_2–C_3). HTG has many promising attributes for the formation of syngas rich in H_2 and synthetic natural gas (CH_4), as it can use a variety of feedstocks of different carbohydrate-to-protein ratios. The fraction of H_2 in the syngas increases from 400 °C to 800 °C while the CH_4 drops in the HTG process (Biller and Ross 2016).

14.4 Algae Biorefineries in the Economy and Environmental Considerations

14.4.1 Sustainability

Algae cultures have been used to produce many valuable products, such as vitamins, feed, nutraceuticals, and others, on an industrial scale. In industrial microalgae farms, usually the production of high-yield biomass is attached to another industrial waste treatment plant. Since emission control and wastewater management are costly and technically demanding, the use of waste as a source of nutrients for algae production adds environmental and economic benefits (Patil et al. 2008).

The Energy Return on Investments (EROI) reported for third-generation biofuels is quite low compared to other generation types. For instance, first-generation biofuels reported EROI is about 8, second-generation biofuels production reported EROI can have around 4, while third-generation biofuels production can have EROI around 1. The EROI for the third generation biofuels excluded scenarios that were the result of the coproduction of electricity and heat or biogas/methane/hydrogen. It must be noted that, compared to lignocellulosic biomass, the theoretical advantages are slow advancement of technologies, and the challenges for third-generation biofuels to become competitive and available on a large scale are significant. Most of the research for biofuels production considered in this EROIs results were from modeling and bench scale experiments, with a small fraction, near 5% of those articles presented results from pilot scale or demonstration scale plants. The fact is, it raises the question of potential discrepancies between theoretical and actual values of biofuels derived from alternative and renewable feedstocks for high-yield, quality, and efficient biofuels production (Papagianni et al. 2024).

14.4.2 Life Cycle Assessment

In the production of butanol from cyanobacteria, the environmental impact and cumulative energy demand (CED) of consequential life cycle assessment (LCA) showed a significant reduction in greenhouse gas emissions. A hypothetical production plant in northern Sweden was considered, with an area of one hectare and an output of 5–85 m^3 of n-butanol per year. Various cultivation methods and productivity scenarios were evaluated, showing greenhouse gas emissions ranging from 16.9 to 58.6 $gCO_2eq/MJBuOH$ and CED between 3.8 and 13 MJ/MJBuOH. Only when a nearby paper factory provided waste heat and CO_2 was the sustainability requirement of at least 60% greenhouse gas emission savings compared to fossil fuels met. Although the location reduced cooling energy needs, a high CED in all scenarios indicated the necessity of significant metabolic engineering, such as a >90% carbon partition to butanol, as well as improved light utilization, to replace fossil fuels or first- and second-generation bioethanol effectively (Nilsson et al. 2020).

Life cycle assessment models also highlight the environmental benefits of biohythane production. It allows the assessment of impacts across a product's life cycle, showing key areas for process improvement. LCA studies emphasize that anaerobic digestion (AD) in single- or two-stage processes is an effective way to manage organic waste, significantly reducing methane and CO_2 emissions compared to landfill disposal. Biogas from AD is used for electricity generation, converting methane into neutral CO_2, while landfill disposal releases methane directly into the atmosphere. Two-stage AD enhances sustainability by simultaneously producing hydrogen, a carbon-free fuel, and reducing overall CO_2 emissions by an additional 10%. Furthermore, the digestate from two-stage AD is more stable, with lower acid, nitrogen, and carbon content, improving environmental performance by reducing acidification and eutrophication effects (Bolzonella et al. 2019).

14.5 Innovative Trends in Third-Generation Biofuels

Knowledge and innovation are strictly related. On this topic, the average publication of patents and articles in the last decade (2015 to 2025) was analyzed to determine the status of innovations on the theme of biofuels production from algae. The search was performed for patents using the Derwent Innovations database, while the articles were searched using the Scopus database. The patents and articles were organized by year, country, and related area of application. Scopus results were limited to review and experimental articles, so publications such as chapters, book sections, or reports were not selected for this analysis. The number of patents found was 374, and the number of articles was 5521. Figure 14.4a illustrates the evolution of published documents and articles. The results show that interest in third-generation biofuels has risen in the last 15 years, since 2010. Until then, research in this theme has grown in only 3 years and has remained in constant study and development since 2013

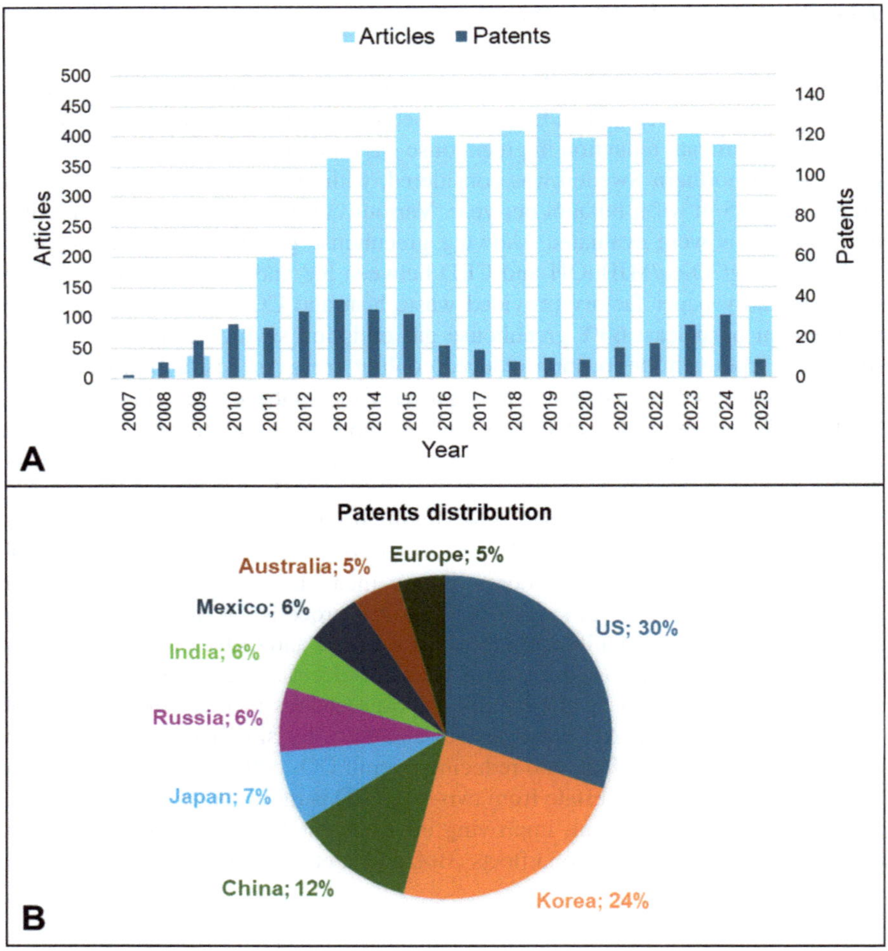

Fig. 14.4 Study growth around the theme of third generation biofuel: articles and patents

(with an average of 400 papers per year). This consistency shows that interest in the topic has not diminished. And although some institutions may be losing interest or ceasing to conduct research on this topic, others are taking their place by bringing new developments or advancing research, preventing the publication statistics from decreasing. Also, since 2008, there has been a surge in technological growth with some natural fluctuations. It has been almost two decades since third-gen biofuels emerged in technology, jointly with companies, research centers, and universities. According to World Intellectual Property Organization (WIPO), the United States of America has the most patents applied, followed by the Republic of Korea and China. Also, Japan and India remain among the top ten country applicants of patents, which indicates that the primary market is focused on Asia and North America (specifically the USA).

14.6 Global Economy in Biofuels Production

The demand for fuel is increasing, and biofuels may contribute to the current market. It is expected that, in various sectors requiring fuel, the use of renewable and environmentally safe alternatives may replace fossil-based fuels, not only from an environmental perspective but also because fossil sources are limited. The renewable energy market size (in the segments of solar, hydro, wind, and bioenergy) is estimated at 4.53 TW in 2025, and is expected to reach 6.38 TW by 2030, at a compound annual growth rate (CAGR) of 7.09% during the forecast period (2025–2030). The renewable energy industry is experiencing an unprecedented transformation driven by technological advancements and manufacturing scale-up. Based on analysis, a significant indicator of this transformation was found in the aggressive expansion of polysilicon production capacity in China, which is projected to increase from 530 gigatons by 2021 to 4000 gigatons by 2024 (Mordor Intelligence 2019).

The global next-generation biofuels market size, based on Allied Market Research (Mohite et al. 2021), was valued at $6.0 billion in 2020, and is projected to reach $59.4 billion by 2030, growing at a CAGR of 26.4% from 2021 to 2030. Biofuels are produced through contemporary biological processes, such as anaerobic digestion, rather than through fuels produced by geological/fossil processes. In the near future, it is expected that second-generation biofuels will be produced from nonfood crops such as organic waste, wood, biomass crops, and food crop waste, whereas third-generation biofuels are based on improvements in the production of biomass and employ engineered energy crops, such as algae, as their feedstock. Algae-based biofuels are expected to be manufactured into a wide range of fuels, such as petrol, diesel, and jet fuel (Mohite et al. 2021).

14.7 Conclusion

Third-generation biofuels, those derived from algae, are important because they offer a promising path to renewable energy production without competing with food crops and can be grown in nonarable land. Algae can also be grown in wastewater, saline, or brackish water, further reducing their environmental impact. Their rapid growth rate and high oil content make them a potentially efficient and sustainable source of biofuels. However, both seaweed and microalgae are producers of various valuable natural resources, and many studies have proposed alternative methods for extracting important bioproducts following the production of biofuels, making use of the residual biomass or oils. The fact is that third-generation biofuels are still in need of advancements and are an emergent sector. The many studies conducted so far have proven that many types of fuels may be produced by algae sources, and the microalgae production system can be adapted to already existing fuel plants. Whether these productions are more efficient or not, the results of publications and patent applications show that it may be a matter of time until large-scale algae farms emerge to produce renewable fuels, impacting the current concern about the environment and the state of the greenhouse effect.

References

Alam F, Mobin S, Chowdhury H (2015) Third generation biofuel from Algae. Procedia Eng 105: 763–768. https://doi.org/10.1016/j.proeng.2015.05.068

Alaswad A, Dassisti M, Prescott T, Olabi AG (2015) Technologies and developments of third generation biofuel production. Renew Sust Energ Rev 51:1446–1460. https://doi.org/10.1016/j.rser.2015.07.058

Bejarano Meza ME, Escalante Palacios AY, Jiménez Pacheco HG (2025) Analysis of biodiesel production (FAME) from microalgae Chorella SP. in a photobioreactor under optimum laboratory conditions. Case Stud Chem Environ Eng 11:101087. https://doi.org/10.1016/j.cscee.2024.101087

Biller P, Ross AB (2016) Production of biofuels via hydrothermal conversion. In: Handbook of biofuels production. Elsevier, pp 509–547. https://doi.org/10.1016/B978-0-08-100455-5.00017-5

Bolzonella D, Battista F, Cavinato C, Gottardo M, Micolucci F, Pavan P (2019) Biohythane production from food wastes. In: Biohydrogen. Elsevier, pp 347–368. https://doi.org/10.1016/B978-0-444-64203-5.00013-7

Brigagão GV, Wiesberg IL, Pinto JL, Araújo O d QF, de Medeiros JL (2019) Upstream and downstream processing of microalgal biogas: emissions, energy and economic performances under carbon taxation. Renew Sust Energ Rev 112:508–520. https://doi.org/10.1016/j.rser.2019.06.009

Chini Zittelli G, Lauceri R, Faraloni C, Silva Benavides AM, Torzillo G (2023) Valuable pigments from microalgae: phycobiliproteins, primary carotenoids, and fucoxanthin. Photochem Photobiol Sci 22(8):1733–1789. https://doi.org/10.1007/s43630-023-00407-3

Córdoba V, Bavio M, Acosta G (2024) Biomethane production modelling from third-generation biomass. Renew Energy 234:121211. https://doi.org/10.1016/j.renene.2024.121211

de Morais EG, Druzian JI, Nunes IL, de Morais MG, Costa JAV (2019) Glycerol increases growth, protein production and alters the fatty acids profile of spirulina (Arthrospira) sp LEB 18. Process Biochem 76:40–45. https://doi.org/10.1016/j.procbio.2018.09.024

Dubey KK, Dhingra AK, Rana S (2015) Optimisation of process parameters for enhanced biobutanol production from Sargassum wightii hydrolysate. Int J Energy Technol Policy 11(3):303. https://doi.org/10.1504/IJETP.2015.073806

Hebbale D, Ramachandra TV (2022) Third-generation bioethanol: status, scope, and challenges. In: Handbook of biofuels, pp 295–312. https://doi.org/10.1016/B978-0-12-822810-4.00015-4

Huber GW, Iborra S, Corma A (2006) Synthesis of transportation fuels from biomass: chemistry, catalysts, and engineering. Chem Rev 106(9):4044–4098. https://doi.org/10.1021/cr068360d

Kajla S, Kumari R, Kumari P, Patel VK (2025) Emerging trends and strategies involved in microbial biohydrogen production. Int J Hydrog Energy 103:797–819. https://doi.org/10.1016/j.ijhydene.2025.01.077

Kant, G., Hasan, A., Yadav, P., Pandey, A., & Srivastava, S. (2025). The generational shift in biofuels: a path toward sustainable energy solutions. In Biomass Bioenerg (Vol. 196). Elsevier Ltd. https://doi.org/10.1016/j.biombioe.2025.107757

Khaonuan S, Jariyaboon R, Usmanbaha N, Cheirsilp B, Birkeland N, Kongjan P (2023) Potential of butanol production from Thailand marine macroalgae using *Clostridium beijerinckii* ATCC 10132-based ABE fermentation. Biotechnol J 18(10). https://doi.org/10.1002/biot.202300026

Kiran B, Kumar R, Deshmukh D (2014) Perspectives of microalgal biofuels as a renewable source of energy. Energy Convers Manag 88:1228–1244. https://doi.org/10.1016/j.enconman.2014.06.022

Lin TS, Wu JY (2015) Effect of carbon sources on growth and lipid accumulation of newly isolated microalgae cultured under mixotrophic condition. Bioresour Technol 184:100–107. https://doi.org/10.1016/j.biortech.2014.11.005

Llano TA, Sánchez CM, Trancho Bedoya A, Robles Chomón J, Arroyo Hailuoto NL, Coz Fernández A (2020) Acid depolymerization of cell wall polysaccharides from Ulvan-rich extracts of green seaweeds. In: II international congress on biorefineries and renewable energies supported in ICTs, pp 51–54. http://hdl.handle.net/10902/23803 ISSN: 2539-1925

Llano T, Arce C, Gallart LE, Perales A, Coz A (2023) Techno-economic analysis of macroalgae biorefineries: a comparison between ethanol and butanol facilities. Fermentation 9(4). https://doi.org/10.3390/fermentation9040340

Madhumitha J (2025) Energy consumption worldwide from 2000 to 2022, with a forecast until 2050, by energy source. https://www.statista.com/statistics/222066/projected-global-energy-consumption-by-source/

Mahfud M, Ramadhana AH, Novita NA, Kurniawansyah F, Sardi B, Mirzan M, Mahmudin L, Ali AA, Indrawan N (2025) Cleaner production of bio-oil from macroalgae and low-rank coal mixture by pyrolysis in a microwave reactor integrated with a distillation column. Energy 314: 134309. https://doi.org/10.1016/j.energy.2024.134309

Medipally SR, Yusoff FM, Banerjee S, Shariff M (2015) Microalgae as sustainable renewable energy feedstock for biofuel production. BioMed Res Int 2015). Hindawi Publishing Corporation. https://doi.org/10.1155/2015/519513

Meşe Erdoğan E, Yılmaz H, Topuz E, Özkan M (2024) Decision making with analytical hierarchy process approach to select feedstock type for bioethanol production: is 3G more sustainable than others? Renew Energy 236:121464. https://doi.org/10.1016/j.renene.2024.121464

Mesfun S, Anderson J-O, Umeki K, Toffolo A (2016) Integrated SNG production in a typical Nordic sawmill. Energies 9(5):333. https://doi.org/10.3390/en9050333

Mohite S, Danekar R, Prasad E (2021) Next generation biofuels market size, share, competitive landscape and trend analysis report, by biofuel type, process, raw material and application: global opportunity analysis and industry forecast, 2021–2030. Allied Market Research. https://www.alliedmarketresearch.com/second-third-generation-biofuels-market

Mordor Intelligence (2019) Renewable Energy Market – Growth Trends & Forecasts Analysis (2025–2030). https://www.mordorintelligence.com/industry-reports/global-renewable-energy-market-industry?network=g&source_campaign=&utm_source=google&utm_medium=cpc&matchtype=p&device=c&gad_source=1&gclid=Cj0KCQjwkZm_BhDrARIsAAEbX1F1T9MnkSP8HSDm9KpBBvlol01SsY2OYtCUuI7VB6RvpQ5PGOfLGK0aAvOKEALw_wcB

Neto JM, Komesu A, da Silva Martins LH, Gonçalves VOO, de Oliveira JAR, Rai M (2019) Third generation biofuels: an overview. In: Sustainable bioenergy. Elsevier, pp 283–298. https://doi.org/10.1016/B978-0-12-817654-2.00010-1

Nilam N, Pal A, Panda P, Birje MS (2023) Algae: prospective and futuristic component for cosmeceuticals. Int J Algae 26(1):29–52. https://doi.org/10.1615/InterJAlgae.v26.i1.30

Nilsson A, Shabestary K, Brandão M, Hudson EP (2020) Environmental impacts and limitations of third-generation biobutanol: life cycle assessment of *n* -butanol produced by genetically engineered cyanobacteria. J Ind Ecol 24(1):205–216. https://doi.org/10.1111/jiec.12843

Ofori-Boateng C (2022) Global profile and market potentials of the third-generation biofuels. In: 3rd generation biofuels. Elsevier, pp 745–756. https://doi.org/10.1016/B978-0-323-90971-6.00018-8

Pandey S, Narayanan I, Selvaraj R, Varadavenkatesan T, Vinayagam R (2024) Biodiesel production from microalgae: a comprehensive review on influential factors, transesterification processes, and challenges. Fuel 367:131547. https://doi.org/10.1016/j.fuel.2024.131547

Papagianni S, Capellán-Pérez I, Adam A, Pastor A (2024) Review and meta-analysis of energy return on investment and environmental indicators of biofuels. Renew Sust Energ Rev 203: 114737. https://doi.org/10.1016/j.rser.2024.114737

Patil V, Tran KQ, Giselrød HR (2008) Towards sustainable production of biofuels from microalgae. Int J Mol Sci 9(7):1188–1195. https://doi.org/10.3390/ijms9071188

Phanduang O, Lunprom S, Salakkam A, Liao Q, Reungsang A (2019) Improvement in energy recovery from Chlorella sp. biomass by integrated dark-photo biohydrogen production and dark fermentation-anaerobic digestion processes. Int J Hydrog Energy 44(43):23899–23911. https://doi.org/10.1016/j.ijhydene.2019.07.103

Raikova S, Le CD, Wagner JL, Ting VP, Chuck CJ (2016) Towards an aviation fuel through the hydrothermal liquefaction of algae. In: Biofuels for aviation. Elsevier, pp 217–239. https://doi.org/10.1016/B978-0-12-804568-8.00009-3

Rajpoot AS, Shende V, Chelladurai HM, Dwivedi G, Verma TN, Choudhary T (2025) Exploring the potential and progress of microalgae biomass production, harvesting, and pretreatment for sustainable biofuel production: a comprehensive review. Environ Dev Sustain. https://doi.org/10.1007/s10668-025-05984-0

Ramaraj R, Junluthin P, Dussadee N, Unpaprom Y (2022) Potential evaluation of biogas production through the exploitation of naturally growing freshwater macroalgae Spirogyra varians. Environ Dev Sustain. https://doi.org/10.1007/s10668-021-02051-2

Silva TA, do Couto E d A, Assemany PP, Costa PAC, Marques PASS, Paradela F, dos Reis AJD, Calijuri ML (2024) Biofuel from wastewater-grown microalgae: a biorefinery approach using hydrothermal liquefaction and catalyst upgrading. J Environ Manage 368:122091. https://doi.org/10.1016/j.jenvman.2024.122091

Singh R (2024) Domestically produced bioethanol contributing to India's pursuit of energy self-reliance and de-carbonization of the road transport sector. https://doi.org/10.4271/2024-28-0126

Singh A, Olsen SI, Nigam PS (2011) A viable technology to generate third-generation biofuel. J Chem Technol Biotechnol 86(11):1349–1353. https://doi.org/10.1002/jctb.2666

Stephens E, Ross IL, Mussgnug JH, Wagner LD, Borowitzka MA, Posten C, Kruse O, Hankamer B (2010) Future prospects of microalgal biofuel production systems. Trends Plant Sci 15(10): 554–564. https://doi.org/10.1016/j.tplants.2010.06.003

Sun Z, Chen X, Kuo P-C, Ding L, Aziz M (2024) Co-feeding biomass and municipal solid waste for enhanced hydrogen and synthetic natural gas yields employing chemical looping process. Chem Eng J 493:152487. https://doi.org/10.1016/j.cej.2024.152487

Terra de Oliveira D, de Jesus Paiva R, Albuquerque de Mescouto V, Ferreira da Silva SR, Farias Da Costa AA, Santos AV, Gonçalves EC, Narciso da Rocha Filho G, Rodrigues Noronha RC, Santos do Nascimento LA (2024) The potential of third-generation biodiesel from Tolypothrix sp. CACIAM22 as a feedstock. Heliyon 10(16):e36343. https://doi.org/10.1016/j.heliyon.2024.e36343

Tripathi G, Dubey P, Shamim A, Farooqui A, Mishra V (2024) Bio-flocculation: a cost effective and energy efficient harvesting technique for algal biofuel production and wastewater treatment. Bioresour Technol Rep 28:101969. https://doi.org/10.1016/j.biteb.2024.101969

Upadhyay N, Kumar K, Kumar Das R, Kumar Ghosh S (2024) A thermodynamic approach to energy, exergy, exergoeconomic, enviroeconomic, and sustainability assessments involving an VCR diesel engine employing third-generation biodiesel with TiO2 NPs and n-heptane. Energy Convers Manage 321. https://doi.org/10.1016/j.enconman.2024.119064

Vonk G, Boy V, Lanoisellé J-L, Lendormi T (2025) The gasification of marine and coastal resources for syngas production: a review. Energies 18(3):616. https://doi.org/10.3390/en18030616

Xia M, Wang L, Yang Z, Chen H (2015) Periodic-peristole agitation for process enhancement of butanol fermentation. Biotechnol Biofuels 8(1):225. https://doi.org/10.1186/s13068-015-0409-6

Yao Y, Tan L, Chen F, Pan A, Ma Q, Zhao J, Kang R (2025) Hydrogen energy industry in China: the current status, safety problems, and pathways for future safe and healthy development. Saf Sci 186:106808. https://doi.org/10.1016/j.ssci.2025.106808

Yue M, Lambert H, Pahon E, Roche R, Jemei S, Hissel D (2021) Hydrogen energy systems: a critical review of technologies, applications, trends and challenges. Renew Sustain Energy Rev 146). Elsevier Ltd. https://doi.org/10.1016/j.rser.2021.111180

Zhang C (2023) Dynamics of energy and biofuel markets in the context of rising oil prices. Agribusiness. https://doi.org/10.1002/agr.21849

Zieliński M, Barczak Ł, Rusanowska P, Nowicka A, Dębowski M (2025) Microbial fuel cells as CO2 source in the autotrophic cultivation of the green microalgae Tetraselmis subcordiformis: impact on biomass growth, nutrient removal, and hydrogen production. Energies 18(4):874. https://doi.org/10.3390/en18040874

Chapter 15
Biochar: Production Processes, Applications, Importance and Advances

Matheus Cavali, Aline Perin Dresch, Thamarys Scapini, Igor Marcon Belli, Walter José Martinez-Burgos, Nelson Libardi Junior, Adenise Lorenci Woiciechowski, and Armando Borges de Castilhos Junior

Abstract In the face of mounting concerns regarding climate change, the circular economy has emerged as a strategic model for reducing the extraction of natural resources and minimizing environmental impacts. One of the pillars of this concept is waste valorization, which converts discarded by-products into materials with high added value. In this context, biochar production stands out as an innovative and sustainable solution, converting wastes through either pyrolysis, hydrothermal carbonization, gasification, or torrefaction into a stable carbonaceous material. The potential applications of biochar are diverse, spanning environmental, agricultural, and energy sectors. Biochar production, therefore, offers a multifaceted approach to addressing the challenges of climate change and promoting the development of a circular economy. In addition, biochar production can be integrated into the context of biorefineries. Therefore, this chapter, through an introductory approach, addresses biochar production processes, possible biochar applications, the potential of biochar production in a biorefinery context, and the advances in biochar research.

15.1 Introduction

Climate change represents a pressing environmental challenge, primarily driven by the escalating emissions of greenhouse gases (GHG) resulting from anthropogenic activities. The consequences of global warming have been significant, given the extreme weather events, rising sea levels, and the loss of biodiversity faced

M. Cavali (✉) · I. M. Belli · N. Libardi Junior · A. B. de Castilhos Junior
Department of Sanitary and Environmental Engineering, Federal University of Santa Catarina, Florianópolis, Santa Catarina, Brazil

A. P. Dresch
Department of Environmental Engineering and Technology, Federal University of Paraná, Palotina, Paraná, Brazil

T. Scapini · W. J. Martinez-Burgos · A. L. Woiciechowski
Department of Bioprocess Engineering and Biotechnology, Federal University of Paraná, Curitiba, Paraná, Brazil

A. L. Woiciechowski et al. (eds.), *Biomass for Bioenergy and Bioproducts*, Biofuel and Biorefinery Technologies 17, https://doi.org/10.1007/978-3-032-19898-3_15

nowadays (United Nations 2025). Therefore, it is imperative to implement strategies to reduce pressure on ecosystems and foster a transition to a more sustainable economic model. In this regard, the circular economy has emerged as a novel and strategic approach to replace the conventional linear model of production, which is based on the extraction, utilization, and disposal of materials (Ellen MacArthur Foundation 2017). A transition to a circular economy has the potential to markedly curtail waste generation and the demand for new raw materials. This paradigm shift would contribute to climate change mitigation by reducing emissions associated with the extraction and processing of natural resources. Furthermore, it would generate new economic and social opportunities by promoting more resilient and sustainable production chains (International Energy Agency 2022).

From that perspective, waste valorization is a crucial component for establishing and consolidating a circular economy, as it involves the transformation of discarded materials into products with added value (Bamigbade et al. 2025; Letti et al. 2021). In this approach, wastes, once regarded as an environmental liability, can be considered a resource for novel production processes, thereby reducing pollution, preserving natural resources, and generating income. Innovative technologies, such as the thermochemical conversion of organic waste, are promising routes for the utilization of these materials (Joshi et al. 2024).

An interesting application of this concept is the production of biochar from wastes (Zhang et al. 2019a). Biochar is produced through the thermal decomposition of organic matter under conditions of limited oxygen (Cavali et al. 2022; Cha et al. 2016). This material has the potential for a variety of applications, including its use as a soil conditioner, adsorbent, carbon sequestration, and even remediation of contaminated areas (Biswal et al. 2022; Dissanayake et al. 2020; Kazemi Shariat Panahi et al. 2020; Sajjadi et al. 2019). Notably, biochar production can contribute to the mitigation of GHG emissions by capturing and sequestering carbon that would otherwise be released into the atmosphere through the natural decomposition of biomass (Sri Shalini et al. 2020). Consequently, this technology emerges as a promising solution for waste treatment and valorization. By integrating concepts such as the circular economy and waste valorization, biochar production has the potential to transform environmental challenges into opportunities for sustainable development.

15.2 Biochar

Biochar has a long history as a soil conditioner, with origins dating back millennia. Its utilization in agricultural practices can be traced back more than 2500 years. The use of biochar can be traced back to the indigenous communities of the Amazon Basin. The soils of the Amazon Basin, known as Terra Preta, meaning “black soil,” were developed by ancient indigenous cultures, suggesting evidence of the extensive use of biochar. These soils are distinguished by their elevated concentrations of essential nutrients, including potassium, phosphorus, and nitrogen, underscoring the importance of biochar in enhancing soil fertility (Kamarudin et al. 2022).

More recently, certain organizations have been engaged in the promotion of biochar production and market development. The most prominent examples include the International Biochar Initiative (IBI) and the United States Biochar Initiative (USBI). The mission of these organizations is to promote the utilization of biochar as a means of waste valorization to mitigate climate change. For IBI, biochar is a "*solid material obtained from thermochemical conversion of biomass in an oxygen-limited environment,*" while USBI has a more zero-waste-oriented definition, defining biochar as a "*fine-grained charcoal made by pyrolysis, the process of heating biomass with limited to no oxygen in a specially designed furnace capturing all emissions, gases, and oils for reuse as energy*" (International Biochar Initiative 2015; United States Biochar Initiative 2025). Nonetheless, despite the organization, it is understood that biochar is the carbon-rich (low O/C and H/C atomic ratios) solid product resulting from the oxygen-limited thermal decomposition of organic matter, preferably residues.

15.3 Biochar Production Processes

Biochar is obtained through the thermochemical conversion of biomass waste, using either pyrolysis (fast and slow), hydrothermal carbonization (HTC), gasification, or torrefaction (Prochnow et al. 2024). Table 15.1 reports some examples of biochar production from these processes, presenting the biochar yield and characteristics. The physicochemical characteristics of biochar depend on both the production method and conditions (e.g., temperature, residence time, and heating rate), as well as the type of biomass used (Rajput et al. 2024). However, biochar characteristics, such as carbon concentration, surface area, and higher heating value, can be improved by physical, chemical, or biological modification techniques to enhance potential biochar applications (Antor et al. 2023; Rajapaksha et al. 2016).

15.3.1 Pyrolysis

Pyrolysis is the thermal decomposition of organic materials under an oxygen-free environment, occurring generally at temperatures between 250 and 900 °C (Liu et al. 2014). This process converts waste biomass into valuable products, such as biochar, bio-oil, and synthesis gas (syngas). Bio-oil, the liquid product of pyrolysis, can be stored and processed for energy generation, while the volatile fraction contains non-condensable gases, such as CO, CO_2, H_2, CH_4, and heavier hydrocarbons, known as pyrolysis gas or syngas (Kung and Mu 2019). Biochar, the solid product, is rich in carbon and can be used as an energy source or for soil amelioration (Foong et al. 2020). Factors such as the type of biomass used and the temperature of the

Table 15.1 Examples of biochar production processes using different biomass residues and the respective biochar yields and characteristics

Method of production	Waste	Temperature (°C)	Biochar yield (wt. %)	Ultimate analysis (wt. %)			Proximate analysis (wt. %)			S_{BET} ($m^2\ g^{-1}$)	Higher Heating Value (HHV) - HHV ($MJ\ kg^{-1}$)	pH	Ref.
				C	O	H	Ash	Moisture	VM/FC				
Slow pyrolysis	Cocopeat	500	62.9	84.4	11.7	2.9	15.9	2.5	0.21	13.7	–	10.3	Lee et al. (2013)
	Rice husks	300	38	46.1	23.3	3.8	22.7	3.6	1.4	–	17.6	4.1	Vieira et al. (2020)
	Coffee husks	350	40	69.9	22.5	3.6	9.8	4.6	0.4	–	26.7	–	Setter et al. (2020)
Fast pyrolysis	Rice husks	550	39	44.7	7.7	1.8	45	–	0.3	117	17.7	–	Zhang and Xiong (2016)
	Used tires	500	40	85.1	–	0.53	11.7	1.7	0.2	–	29.7	–	Raclavská et al. (2015)
Hydrothermal carbonization (HTC)	Pine saw-dust and sewage sludge	250	64	55.8	16.2	4.3	18.4	1.9	2.6	22.7	20.3	5.8	Cavali et al. (2023a)
	Municipal solid waste (MSW)	220	54.8	66.6	16.8	7.0	5.8	–	2.8	–	29.9	–	Merzari et al. (2018)
	Wood meal	245	55	69.9	22.7	5.4	1.4	–	1.1	–	28.4	–	Kang et al. (2012)

Gasification	Grape pomace	1200	15	53	41	3.9	42.3	5.43	0.3	76.1	18.05	11.8	Hernández et al. (2020)
	Miscanthus plant	800	22.6	92.5	5.2	1.6	12.3	12.3	0.1	629.3	–	–	Tian et al. (2021)
	Bamboo guadua	850	13	60.8	6.6	0.8	31.5	–	–	–	–	10.9	Romero Millán et al. (2021)
Torrefaction	Microalgae residue	250	71.6	36.5	49.5	6.1	23.8	1.8	4.9	–	12.6	–	Zhang et al. (2018)
	Rice husk	250	52	57.2	22.9	4.5	14.3	–	0.9	–	22.6	–	Nakason et al. (2019)
	Camellia shell	240	70	55.6	31.9	4.8	6.2	7.5	1.5	1.03	20.8	–	Tu et al. (2022)

process significantly affect the yield and quality of the biochar, with higher temperatures tending to increase the production of synthesis gas and reduce the yield of biochar (Yaashikaa et al. 2020). The process can be conducted in different types of reactors, such as bubbling fluidized beds, circulating beds, and rotating cone reactors, as well as agitated drums, rotary sand kilns, and blade pyrolysis kilns (Garcia-Nunez et al. 2017). In addition, pyrolysis can be classified as fast or slow, based on the temperature, heating rate, and residence time.

Generally, slow pyrolysis occurs at temperatures lower than 500 °C (Desjardins et al. 2024; Prochnow et al. 2024) and is ideally suited to obtaining biochar with a high fixed carbon content (Tan et al. 2021). This process involves long residence times (>1 h) and a moderate heating rate (5–7 °C/min). Fast pyrolysis, however, due to the high pyrolysis temperature and heating rate, produces a lower biochar yield and releases volatile gaseous matter. This process is indicated, therefore, when the objective is to obtain bio-oil (Tan et al. 2021). Its conditions include fast heating rates (>100 °C/min), low residence time (0.5–2 s), and moderate operating temperatures (400–600 °C). Over the past decade, there has been a significant increase of more than 100% in the number of publications related to biochar produced by pyrolysis, demonstrating the growing interest of the scientific community in this method of waste valorization (Prochnow et al. 2024).

Pyrolysis is a simpler and more economical conversion process when compared to other thermochemical methods (Kavan Kumar and Panwar 2024). For example, the construction of a pyrolysis reactor can be realized relatively quickly (Hoang et al. 2021). However, pyrolysis requires relatively dry feedstocks, usually with a moisture content of less than 30% (Amalina et al. 2022), and its disadvantages include the difficulty of controlling the properties of the biochar, especially during fast pyrolysis, and the high production of tar (Afshar and Mofatteh 2024). Nevertheless, slow pyrolysis at low temperatures is an efficient approach to generating bioenergy, and the use of biochar in soil reduces greenhouse gas emissions more significantly (Gaunt and Lehmann 2008). Overall, pyrolysis has been suggested as the most efficient, sustainable, and preferable technique for biochar production (Amalina et al. 2022).

15.3.2 Hydrothermal Carbonization (HTC)

The HTC process is considered an attractive approach to producing biochar by performing at relatively low temperatures (180–250 °C) and under autogenous pressure in a medium containing water (Cavali et al. 2023a). The material from HTC is called hydrochar, a type of biochar whose properties differ from those obtained by pyrolysis and gasification (Prochnow et al. 2024). Commonly, HTC produces hydrochar by using temperatures lower than 250 °C. At higher temperatures (250–400 °C), bio-oil is obtained, and the process comes to be named hydrothermal liquefaction. Under even higher temperatures (>400 °C), gases (CO, CO_2, H_2, and CH_4) are the main products, and the process is called hydrothermal gasification (Yaashikaa et al. 2020). In HTC, the process conditions and biomass composition influence hydrochar yield and characteristics (Cavali et al. 2025b; Oliveira et al. 2013).

Biomass degradation is controlled by temperature, as well as pressure, reaction time, and solid–to–liquid ratio, which have a direct influence on the yield and properties of the hydrochar (Rajput et al. 2024). Under high-temperature conditions, the aromaticity of the hydrochar is increased, which improves both the thermal stability and the energy content of the material (Zhang et al. 2019a). However, synthesis at high temperatures reduces the hydrochar yield (Ferrentino et al. 2020). By varying the temperature and pressure, hydrochars generated at temperatures and pressures below 150 °C and 24 MPa present minimal structural changes compared to those produced under stricter conditions (Güleç et al. 2021). The reaction time in the HTC process generally ranges from 1 to 16 h (Zhang et al. 2019b), and for the solid–to-liquid ratio, wide ratios can be applied between 1/40 and 1/1 (Ahmed Khan et al. 2019). When high solid–liquid ratios are used, the impact of the reaction time tends to be irrelevant, since the amount of water may not be enough to hydrolyze the feedstock completely. However, with lower solid-to-liquid ratios, the time of the HTC process can have a greater influence on determining the final distribution of the products (Cavali et al. 2023a).

HTC is a process capable of transforming, without needing prior drying, waste biomass into hydrochar (Cavali et al. 2023b) with a higher energy density than the original material (Afshar and Mofatteh 2024). The application possibilities of HTC range from the production of biofuels to the development of doped porous materials for adsorption, energy storage, and catalysis (Román et al. 2018). In addition, HTC can work as a Co-HTC process when using two types of waste. By applying a moisture-rich and a moisture-poor waste, Co-HTC can avoid the addition of water, thereby making the process more sustainable (Cavali et al. 2023a; Wang et al. 2022). However, HTC still faces the challenges of high investment costs due to the need for specialized equipment (Afshar and Mofatteh 2024).

15.3.3 Gasification

Gasification is a thermochemical process that converts carbon-rich raw materials, whether organic or fossil, into syngas at high temperatures (>700 °C) and in the presence of oxidizing agents (oxygen, air, water vapor, or combinations thereof) (Qian et al. 2015). In this process, the reaction temperature is the most significant factor in determining synthesis gas production; as the temperature rises, CO and H_2 production increase, while the content of other gases, such as CH_4, CO_2, and hydrocarbons, decreases (Yaashikaa et al. 2020). The main products of gasification is syngas, and biochar is considered a lower-yield by-product (Gale et al. 2021; Sharma et al. 2023). Therefore, the interest in gasification to produce biochar is still less studied compared to pyrolysis or HTC, but it has increased over the last decade (Prochnow et al. 2024).

The gasification mechanism can be subdivided into four stages: (i) drying (100–150 °C), a stage in which the moisture content of the biomass is completely evaporated without energy recovery, being a separate process when the biomass has a high moisture content; (ii) pyrolysis (200–500 °C); (iii) oxidation/combustion (800–1200 °C), a phase during which the oxidation and combustion reactions of the gasification agents are the main sources of energy for the process, reacting with the combustible species present in the gasifier to produce CO_2, CO, and water; and (iv) gasification (>700 °C) (Rajput et al. 2024; Yaashikaa et al. 2020).

At high temperatures and with a controlled amount of oxygen, the biochar yield in gasification is generally between 15% and 30% of the initial biomass (Barbhuiya et al. 2024). This process necessitates a specific reactor with complex technology, requiring specific capital, operation/maintenance, and availability (Afshar and Mofatteh 2024; Barbhuiya et al. 2024). The biggest disadvantage associated with this technique is the smaller amount of biochar obtained in comparison with pyrolysis, as well as the generation of GHG (You et al. 2018). One of the problems with steam gasification is the formation of tar (Cay et al. 2019). However, it is highly efficient at producing syngas and has the capacity to extract the energy contained in lignin, the hardest structural part of biomass (Afshar and Mofatteh 2024).

15.3.4 Torrefaction

Torrefaction is a process that employs a low heating rate; therefore, it can be referred to as soft pyrolysis. The process is characterized by operating conditions that include a temperature range of 200–300 °C, a residence time of less than 30 minutes, a heating rate of less than 50 °C/min, and oxygen deprivation (Yaashikaa et al. 2020). Torrefaction modifies biomass properties, including but not limited to particle size, moisture content, surface area, and energy density (Zakaria et al. 2023). This process is organized into distinct stages: preheating (the temperature is raised to 100 °C until all the moisture is removed); heating (the biomass is heated until it reaches and stabilizes the ideal temperature, ensuring total evaporation of the moisture); post-drying (the temperature is raised to 200 °C); torrefaction (conducted at 200 °C, maintaining a constant temperature); and cooling (Yaashikaa et al. 2020).

In addition, the torrefaction process can be realized in three ways: (i) steam torrefaction, where the biomass is treated with steam at a temperature of no more than 260 °C and a residence time of approximately 10 minutes; (ii) wet torrefaction, also known as HTC, which occurs by contacting the biomass with water at a temperature of between 180 °C and 260 °C and a residence time of between 5 minutes and 240 minutes; and (ii) oxidative torrefaction, which involves treating the biomass with oxidizing agents, such as gases, used in the combustion process to generate thermal energy. These approaches aim to convert the raw material efficiently, using variations in temperature and residence time to optimize the process (Rajput et al. 2024; Yaashikaa et al. 2020).

15.4 Biochar Applications

The stable, carbon-rich matrix and beneficial chemical and physical properties make biochar a valuable and sustainable material for applications in agriculture, environmental management, energy production, and industry (Fig. 15.1).

15.4.1 Environmental Applications

Due to increasing industrialization and urbanization, soil and water pollution have become major environmental concerns in recent years (Hama Aziz and Kareem 2023). Pollutants in water and soil pose significant risks to human health because of their high toxicity, persistence, non-biodegradability, and carcinogenicity (Hama Aziz and Kareem 2023; Qiu et al. 2022). Thus, biochar has emerged as a sustainable

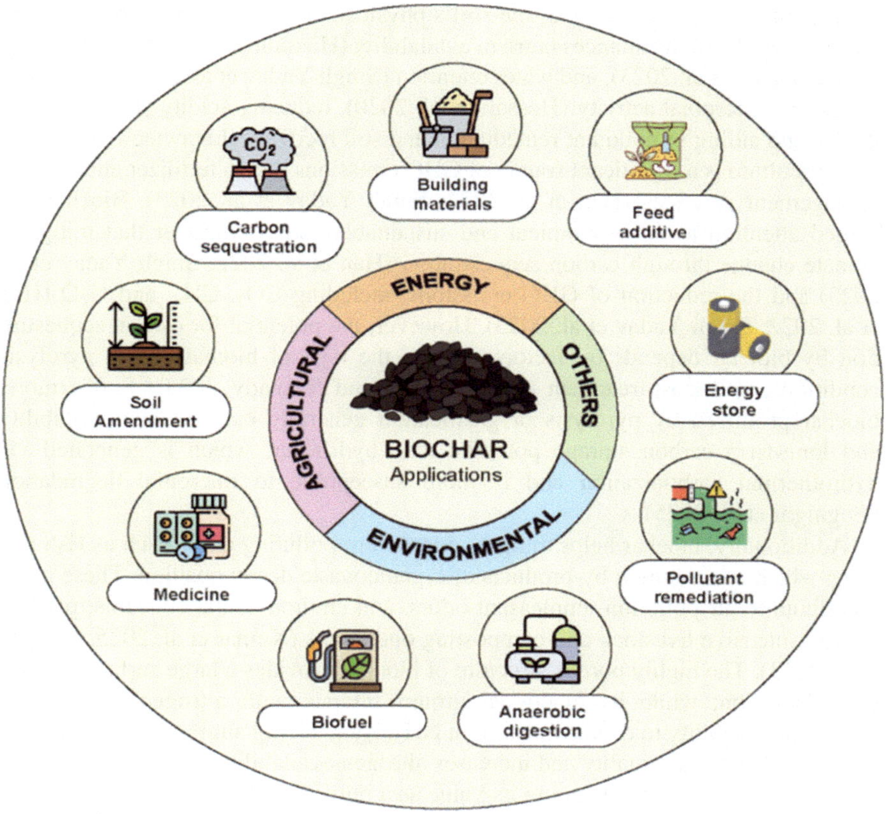

Fig. 15.1 Overview of potential applications of biochar

remediation material due to its highly porous structure, large surface area, ion exchange capacity, and abundance of functional groups. It can effectively remove pollutants such as dyes (Senapati et al. 2025), pharmaceuticals (Du et al. 2023; Liu et al. 2025), phenols (Guo et al. 2024), pesticides (Holanda et al. 2023; Rambabu et al. 2021), micro- and nanoplastics (Kumar et al. 2023; Zhu et al. 2022), and oils (Sun et al. 2022; Urgel et al. 2024) from soil and water. Additionally, it shows strong adsorption potential for heavy metals, including nickel (Ni) (Ibrahim et al. 2022), copper (Cu) (Ibrahim et al. 2022; Wang et al. 2020), cadmium (Cd) (Bakr et al. 2025; Ibrahim et al. 2022), and lead (Pb) (Ibrahim et al. 2022). Ongoing research into biochar's mechanisms of action and its production from various raw materials may further establish it as a low-cost, sustainable solution for soil and water decontamination, offering an eco-friendly alternative to synthetic adsorbents (Qiu et al. 2022).

15.4.2 Agricultural Applications

In agriculture, biochar improves the soil's physical, chemical, and biological properties. Its application enhances nutrient availability (Hossain et al. 2020), soil fertility (Singh Yadav et al. 2023), and water retention (Singh Yadav et al. 2023), while also promoting microbial activity (Hossain et al. 2020), reducing acidity (Hossain et al. 2020), and aiding in pollutant remediation and soil recovery (Latawiec et al. 2019).

Agriculture is a significant source of GHG emissions due to fertilizer and manure management practices (Han et al. 2022; Singh Yadav et al. 2023). Biochar has gained attention as an economical and sustainable soil conditioner that mitigates climate change through carbon sequestration (Han et al. 2022; Singh Yadav et al. 2023) and the reduction of GHG emissions, including CO_2, CH_4, and N_2O (Han et al. 2022; Singh Yadav et al. 2023). However, the potential for carbon sequestration by biochar depends on factors such as the type of biomass used, pyrolysis conditions, and post-treatment processes (Li and Tasnady 2023). Furthermore, biochar produced by pyrolysis or gasification generally exhibits higher stability and long-term carbon storage potential than hydrochar, which is generated via hydrothermal carbonization and is more susceptible to microbial degradation (Cognigni et al. 2025).

Additionally, biochar helps mitigate odors from polluting gases such as H_2S and NH_3, which are common by-products of organic waste decomposition. These gases contribute to air pollution, unpleasant odors, and environmental concerns, particularly in intensive livestock and composting operations (Kumar et al. 2025; Nguyen et al. 2023). The highly porous structure of biochar provides a large surface area for gas adsorption, while its functional groups interact with nitrogen compounds, enhancing its ability to capture and retain NH_3. By reducing ammonia volatilization, biochar improves air quality and increases nitrogen availability in the soil, minimizing nutrient losses and enhancing its value as a nutrient-rich fertilizer (Kumar et al. 2025; Nguyen et al. 2023).

15.4.3 Energy Production Applications

Biochar has emerged as a versatile and sustainable material in the transition to renewable energy sources. Its high chemical stability, porous structure, and fixed carbon content enable various energy applications, including its use as an adsorbent, catalyst, and fuel (Cardarelli and Barbanera 2024; Kant Bhatia et al. 2021).

As a solid fuel, biochar's high carbon content and relatively high calorific value make it a potential substitute for fossil coal. Its lower moisture content and higher energy density improve combustion efficiency while reducing pollutant emissions (Qian et al. 2020). However, its effectiveness as a fuel largely depends on its higher heating value (HHV), which is influenced by the type of feedstock and production conditions (Table 15.1) (Awad et al. 2024; Qian et al. 2020). Typically, biochars with higher fixed carbon and lower ash content exhibit superior HHVs (Awad et al. 2024; Qian et al. 2020). To enhance biochar's viability as an alternative energy source, optimizing production parameters is crucial to maximizing its energy potential and sustainability (Awad et al. 2024; Qian et al. 2020).

In biodiesel production, biochar has been explored as a heterogeneous catalyst in transesterification and esterification reactions (Kant Bhatia et al. 2021). Its advantages include reusability, thermal stability, and reduced waste generation. Additionally, chemical modifications can enhance its catalytic activity, making biodiesel conversion more efficient and cost-effective (Foroutan et al. 2021; Kant Bhatia et al. 2021).

Biochar is also suggested for microbial fuel cells (MFCs), a clean technology that converts organic matter into electricity using microorganisms (Kant Bhatia et al. 2021). Its high electrical conductivity and large surface area make it an ideal electrode material, promoting microbial biofilm growth and enhancing electricity generation efficiency (Kant Bhatia et al. 2021; Zha et al. 2020).

For biogas production, biochar serves as an additive in anaerobic digestion, improving process stability, increasing CH_4 yields, and mitigating inhibitory compounds, such as NH_3 and volatile fatty acids (Cavali et al. 2022; Lim et al. 2022). Its highly porous structure provides a favorable surface for microbial immobilization, promoting syntrophic interactions and facilitating direct interspecies electron transfer (DIET), which accelerates methanogenesis and improves overall digestion efficiency (Lim et al. 2022; Wang et al. 2021a).

In hydrogen production, biochar serves multiple roles: (i) as a catalyst in biohydrogen production, where its porous structure provides a suitable environment for microbial immobilization and electron transfer, improving process efficiency (Kant Bhatia et al. 2021); (ii) as a catalyst in methane-to-hydrogen conversion, where metal-modified biochar (e.g., with nickel, iron, or cobalt) enhances catalytic activity, facilitating CH_4 breakdown and hydrogen formation while reducing CH_4 emissions (Mishra et al. 2024; Zhao et al. 2023); and (iii) as an additive in anaerobic digestion, where it stabilizes the process, supports microbial attachment, and boosts hydrogen yields (Sugiarto et al. 2021; Zhang et al. 2021). However, despite challenges related to scalability and initial costs, biochar's application in energy production presents a promising, sustainable alternative for a cleaner energy transition.

15.4.4 Other Applications

In addition to its applications in environmental remediation, agriculture, and energy production, biochar has been explored in advanced materials and industrial processes. For example, it has been proposed as an electrode material for rechargeable lithium-ion batteries (Kane et al. 2022; Seroka et al. 2024), sodium-ion batteries (Bartoli et al. 2024), and zinc-air batteries (Zhang et al. 2023). However, its performance and storage capacity depend on key factors such as porosity, electrical conductivity, graphitic structure, and chemical composition (Kane et al. 2022). In supercapacitors, biochar-based materials offer high capacitance, sustainability, cost-effectiveness, and good electrochemical performance. However, optimizing pore structure and thermal stability is crucial for their effective use (Rashidi et al. 2022).

In pharmaceutical formulations, biochar's adsorbent and biocompatible properties make it valuable for drug delivery systems, particularly in controlled release and toxin absorption applications (Wang et al. 2021b). In construction materials, biochar can be incorporated into concrete (Mobili et al. 2023), bricks, and insulating materials (Osman et al. 2023), increasing their mechanical strength and thermal properties while also contributing to CO_2 emission reduction (Mobili et al. 2023). Another application that is still little-explored is its use as a feed additive. Biochar has recently been applied to ruminant feed at concentrations between 0.5% and 3%, showing potential benefits for animal health (Rajpoot et al. 2024).

Biochar is also utilized in the production of reinforced plastics, enhancing mechanical strength, thermal stability, and biodegradability in packaging materials (Vidakis et al. 2024). This contributes to reducing dependence on petroleum-based plastics. Additionally, biochar plays a crucial role in food preservation, helping to extend storage life (Feng et al. 2024). For example, a study developed composite films incorporating silver nanoparticles and biochar, prolonging the shelf life of blueberries while exhibiting antimicrobial and antioxidant properties, thereby contributing to the reduction of food waste (Feng et al. 2024).

In filter systems for drinking water treatment, biochar has demonstrated superior performance compared to conventional filter materials such as gravel, sand, and anthracite, making it a sustainable and promising strategy for contaminant removal (García-Ávila et al. 2023). In the metallurgical industry, biochar has been explored as a sustainable alternative to coke and coal in metal smelting processes, helping to reduce carbon emissions (Yang et al. 2025; Ye et al. 2019). However, economic barriers still limit its large-scale application in the steel industry, requiring further research and technological advancements.

Despite some challenges, the versatility of biochar across various sectors underscores its potential as a valuable raw material for driving sustainable innovations and supporting the circular economy.

15.5 Biochar in a Biorefinery Context

The need for an economically viable, environmentally friendly, and highly efficient approach to managing large volumes of solid waste drives the development of biochar-based strategies, offering multiple benefits across applications ranging from agriculture to environmental remediation. Biochar produced from waste contributes to the circular economy by transforming residual biomass into a stable resource that can be applied in various fields, adding economic value, reducing environmental impact, and supporting the achievement of the Sustainable Development Goals (Cavali et al. 2025a; Neogi et al. 2022). Within this concept, waste is converted into high-value products, promoting an efficient resource utilization cycle and reinforcing the circular economy.

The use of abundant and low-cost residual biomass for biochar conversion offers an effective solution for managing agro-industrial waste, with potential integration into biorefinery processes. From an environmental sustainability perspective, biochar production, storage, and application play a crucial role in the global carbon cycle, with estimated carbon sequestration potential ranging from 0.3 to 2 Gt of CO_2 per year by 2050. This represents a viable strategy for large-scale waste management and bioproduct generation, with carbon removal costs ranging from 52 to 131 GBP/ton CO_2 (Fawzy et al. 2021).

Recent studies have demonstrated that integrating biochar into advanced biorefineries has enabled biofuels to achieve negative emissions of up to −20.2 gCO_2e/MJ for Fischer-Tropsch (FT) fuels and −4.2 gCO_2e/MJ for hydroprocessed pyrolysis oils (HPO). In a 100 MW plant, this approach resulted in an additional removal of 13.3 kton CO_{2eq}/year for HPO and 6.8 kton CO_{2eq}/year for FT, demonstrating that biochar can be an efficient mechanism for carbon sequestration and decarbonization in energy biorefineries (Buffi et al. 2024).

Integrated into the management of agro-industrial waste, sludge, and other by-products, biochar conversion for multiple applications can be a versatile solution for stabilizing organic materials. Biochar produced from lignocellulosic and agricultural residues has demonstrated a high adsorption capacity for odorous compounds in wastewater treatment plants, efficiently removing NH_3, SO_2, and H_2S, acting as a catalyst in the oxidation of sulfur compounds, becoming an alternative for atmospheric emission control in sanitation facilities, reducing odors, and promoting the circular economy (Lee et al. 2023). Another example is biochar derived from sewage sludge, which accelerated aerobic granulation in a sequential batch reactor, enhancing biomass retention and aggregation. This process resulted in the removal of 80% of COD and 70% of total phosphorus, in addition to contributing to integrated waste management through the conversion of treated sludge by hydrothermal liquefaction (Paul et al. 2025).

In the concept of an integrated biorefinery, where different products are obtained from the same biomass, Rahul and coworkers conducted studies with *Madhuca indica* for conversion into biodiesel, bio-oil, and biochar, promoting a sustainable model. The biochar, obtained through the pyrolysis of de-oiled cake, was applied as a support for lipase immobilization, achieving an enzyme retention rate of 57%. Applied in enzymatic catalysis during transesterification, it resulted in a maximum conversion of 93.4% of fatty acid methyl esters (FAME) (Rahul et al. 2024).

Recently, a study evaluated the incorporation of biochar into batteries. Biochar derived from the hydrolysis of cashew nutshells exhibited high electrical conductivity and was combined with MnO_2 and PEDOT to form an electrochemically active composite, demonstrating lower electrical resistance and greater structural stability, allowing the battery to reach 101.2 mW/m^2 and maintain a stable discharge for 9.6 hours, with 16-hour cycles without significant variation in potential. The scalability of production was assessed, with the potential to manufacture 9540 cells from 1 ton of waste (Gupta et al. 2025).

The economic viability of using biochar derived from agricultural residues was evaluated for environmental applications in dye removal. The results were promising, with peanut shell biochar exhibiting an adsorption capacity of 46.3 mg/g and the lowest cost per unit of Basic Red 09 dye removal proving to be more economically viable than commercial activated carbon (Praveen et al. 2021). The economic analysis of a *Madhuca indica* biorefinery for conversion into biodiesel, bio-oil, and biochar indicated a payback period of 3.15 years, thereby suggesting the economic feasibility of this approach. (Rahul et al. 2024).

It is worth noting that the efficiency of biorefineries depends on optimizing biomass energy use and controlling the conversion of thermochemical and biochemical intermediates. Reducing biomass moisture and energy consumption in biochar processing can result in savings of up to 80 MW, directly impacting emission balances and system competitiveness. The conversion of biochar and non-condensable gases into advanced fuels (ethanol and biohydrogen) reinforces the viability of these units as integrated platforms for energy transition, expanding the application spectrum of biomass in sustainable value chains. This model allows biofuels derived from biochar to reach a reduced emission intensity of 13.54 gCO_{2eq}/MJ, while the replacement of fossil fuels in plant heat generation further reduces emissions by up to 53% (Giwa et al. 2022).

Thus, biochar stands out as a valuable bioproduct within biorefinery processes, adding value to the system and expanding potential applications as a catalyst, adsorbent, and environmental improvement component. Soil applications, for example, reduce the need for chemical fertilizers and increase crop yields. In this same scenario, biochar retains carbon in the soil for decades, as the conversion processes make the carbon more stable and recalcitrant in the soil than other organic fertilizers, such as animal waste, thereby minimizing GHG emissions into the atmosphere. For example, it was reported that the mean residence time of the labile fraction of biochar was 108 days, while the stable fraction remained for 556 days, contributing to long-term carbon sequestration in the soil (Wang et al. 2016). Moreover, recent studies show that biochar modification can optimize its efficiency in mitigating CO_2 and N_2O emissions, as observed in magnesium-modified biochar, which raised soil pH above 8.5 and reduced CO_2 emissions by 58.8% (15,652 μg/kg vs. 38,001 μg/kg in untreated soil) and N_2O emissions by 52.9% (288.3 μg/kg vs. 613.25 μg/kg in untreated soil) (Dong et al. 2024).

Furthermore, converting biorefinery by-products into biochar has proven effective in removing contaminants (e.g., heavy metals). Specifically, biochar derived from acidified lignin, a by-product of the cellulose and hemicellulose extraction process, has demonstrated selective Pb^{2+} adsorption capabilities of up to 126.6 mg/g, exhibiting notable chemical stability (Lee et al. 2024). Beyond its application in the remediation of wastewater (Trivedi et al. 2025; Xiang et al. 2020), biochar production can be an economically feasible process, facilitating the implementation of large-scale treatment units and promoting positive impacts on waste management in biorefineries.

15.6 Advances in the Biochar Field

In order to assess scientific progress and innovation in the biochar domain, both for production and application, it is possible to search for patents, as these documents may represent technological developments and innovations. In this topic, a patent search was carried out in the Derwent Innovation database using the algorithm "*(((((TS=(biochar)) OR TS=(biochar derivatives)) OR TS=(Biochar-based materials)) OR TS=(Biochar composites)) OR TS=(Biochar products)) OR TS=(Activated biochar),*" and the analysis period was between 2020 and 2025. During the evaluation period, 139 patents were found. In Fig. 15.2, it is worth noting that in

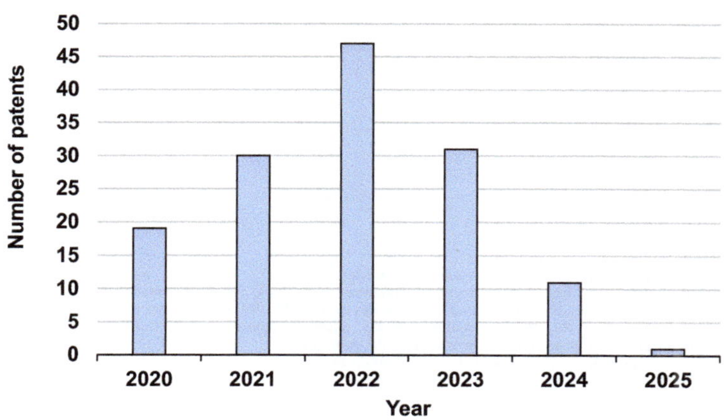

Fig. 15.2 Number of patents filed between 2020 and 2025

2022, 47 documents were deposited, and for the years 2023 and 2024, a significant decrease was observed, i.e., 31 and 11 patents, respectively. This can be explained by the fact that the publication of a patent takes around 18 months after the deposit of the document (Vandenberghe et al. 2021). Furthermore, it was observed that around 95% of the patents were oriented toward processes or equipment to produce biochar and that the main holders of this technology were China and South Korea, with 65% and 17% of the patents, respectively.

The main problems of biochar production devices are generally the large amounts of energy required in the process, mainly due to the size of the devices and the types of conformation traditionally used. For instance, patent CN112430468A presents a device or chamber for biochar production, focusing on enhancing the system's energy efficiency. This is achieved by optimizing the energy expenditure of the carbonization chamber, the combustion chamber, and the conduction layer. The energy expenditure problem is solved by optimally arranging the carbonization chamber, the combustion chamber, and the heat-conducting layer in a circular distribution, as well as detachably arranging the main components of the device (Fig. 15.3a) (Huimin et al. 2020). According to the inventors, the system has high efficiency in the production of coal, high energy conservation due to the insulation layer, and the production process is simplified, taking place in a short period. Likewise, the structure is simple, easy to maintain, and safe. A limitation of this system is that it only works in batch mode.

Patent CN115287084A reports a solution to one of the problems of the previous utility model: its operation in discontinuous or batch form. The utility model CN115287084A presents a continuous production system of environmentally

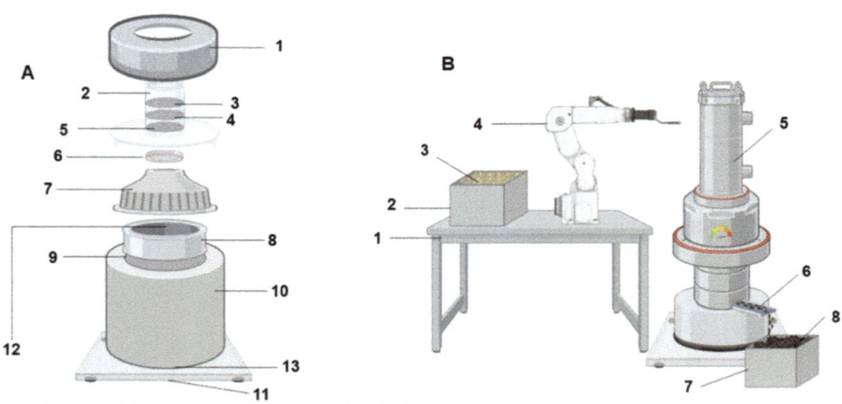

Outer barrel cover (1), Tail gas treatment system (2), Third layer of physical dust removal device (3), Second layer of chemical dust removal devices (4), First layer of physical dust removal device (5), Regionalized isolation plate (6), Built-in reaction fixing device (7),Carbon chamber cover (8), Heat conduction interlayer (9), Aerogel thermal insulation layer (10), Universal wheel (11), Carbonization chamber (12), Combustion chamber (13), Outer barrel body (14).

Table (1), Raw material box (2), Containment box (3), Mechanical arm (4), Carbonization furnace (5), Conveyor belt (6), Containment box (7), Biochar (8)

Fig. 15.3 Chamber for biochar production (**a**) continuous process for biochar production (**b**). (Source: Chunyan et al. (2022); Huimin et al. (2020))

responsible biochar. The system is composed of several accessories, as shown in Fig. 15.3b: table (1), raw material box (2), containment box (3), mechanical arm (4), carbonization furnace (5), conveyor belt (6), containment box (7), and biochar (8) (Chunyan et al. 2022). The system operates automatically. Initially, the raw material box is filled. The arm places the raw material on top of the carbonization furnace. The furnace is activated, and the raw materials are carbonized. The biochar generated is then transferred to the receiving box via the conveyor belt at the discharge port at the bottom of the furnace to complete the production process. According to the inventors, the system not only speeds up biochar production but also saves labor and optimizes energy resources. Table 15.2 presents other patents related to biochar production processes.

In the domain of biochar modification, patent CN118652705A, for example, presented a biochar produced at high pressures in the presence of alkaline activators, zinc oxide, ferric oxide, and carbon nanotubes. According to the inventors, the alkaline activators act to increase the pores and specific surface area of the biochar; zinc oxide acts as a structure-directing agent to help form ordered pores and increase the specific surface area of the biochar. Also, zinc oxide can improve the distribution and size of the pores when interacting with carbon materials, thus affecting the adsorption properties of the biochar. Ferric oxide has high surface energy and a special electronic structure, which can help increase the adsorption capacity of the biochar for certain pollutants. According to the inventors, the prepared biochars present higher performance in terms of specific surface area, pore volume, and pore size. They can remove 90% and 94% of cadmium and methylene blue, respectively, from a solution with a concentration of 800 ppm (Yongqiang and Yongming 2024).

The main advances in biochar regarding its application are mainly focused on its use as an absorbent or an agent to remove pollutants. For example, patent CN114225902A showed the use of biochar modified with iron oxide for the remediation of soils contaminated by heavy metals, such as copper, cadmium, and lead. According to the inventors, for efficient removal of pollutants, between 0.5% and 1.5% of biochar per mass of contaminated soil should be used, and the water content should preferably be adjusted between 60% and 70%. The modified biochar, in addition to being efficient in removing heavy metals, also significantly improves soil pH. The remediation time should vary between 7 and 110 days (Hua et al. 2021).

In patent CN119175075A, it was pointed out that the biochar obtained by acid pretreatment and carbonization has high performance in the removal of different organic pollutants. The equilibrium adsorption amounts of methylene blue and Congo red reached 251.03 and 226.44 mg/g, respectively, and the equilibrium adsorption amount of tetracycline, which is a typical antibiotic, reached 154.61 mg/g. These equilibrium values are higher compared with other types of absorbents (Jun et al. 2024). In another example, patent CN115999505, poplar biochar was suggested for efficiently removing xenobiotic compounds, such as antibiotics. It was determined that the removal capacity of the antibiotics tetracycline, doxycycline, and oxytetracycline of biochar ranges between 70.28 and 89.58 mg/g. In the case of fluoroquinolone antibiotics, enrofloxacin, and enoxacin, the absorption capacity of biochar was significantly lower (35.54–60.31 mg/g) (Qingrong et al. 2022).

Table 15.2 Some examples of patented biochar production processes

Patent code	Patent title	Novelty	Inventors
CN118652705A	A method and application of preparing biochar by pyrolysis of agricultural, forestry and livestock and poultry waste	They presented a method or process for preparing biochar from agricultural, forestry, livestock, and poultry waste. The process begins with the separation and grinding of the waste using a ball mill at a speed of 300 rpm. The ratio of agricultural and forestry waste is 5: 2, and livestock and poultry waste is 4:0.08. The conditions of the hydrothermal carbonization treatment should be between 200 and 250 °C and the pressure between 10 and 15 MPa for 3–4 hours. Some additives should also be added, such as NaOH or KOH (30 to 40 parts), zinc oxide (10 to 25 parts), ferric oxide (5 to 8 parts), carbon nanotubes with a diameter of 50–100 nm (1 to 3 parts), and a dispersing agent (0.1 to 0.3 parts). The method enables the preparation of biochar, having effectively solved problems of an underdeveloped pore structure and a limited specific surface area in the biochar preparation process.	Yongqiang and Yongming (2024)
CN119175075A	A process for preparing biochar with high adsorption performance by acid pretreatment and coordinated carbonization of coking sludge	The novelty establishes a biochar preparation technique with high absorption performance by acid pretreatment and synergistic carbonization of coke sludge. The coke sludge is pretreated by acid immersion using hydrochloric acid solution (2.3–2.7 M). The mass ratio of coke sludge to hydrochloric acid should be in the range from 1:20 to 25:3. The treatment time varies between 7 hours and 9 hours. Then the sludge is separated from the liquid by centrifugation, and carbonization is carried out by pyrolysis under anaerobic conditions, at temperatures between 700 and 900 °C in a period between 2 and 3 hours.	Jun et al. (2024)

CN117866648A	Method for preparing biochar by collaborative pyrolysis of oily sludge and agricultural and forestry waste	The patent presents a biochar production process using oily sludge (oil content can vary between 5% and 30%) and agricultural and forestry waste. The production process begins with drying the waste until it reaches a maximum moisture content of 10%. The waste is then ground and sieved (40–100 mesh). The powder is soaked in an acid solution and placed in a chamber with an inert gas (N_2 gas). Then, it is treated with ultrasound (800–5000 W) in anaerobic conditions and at a controlled pressure between 7 and 8 MPa for 15 minutes. The acidic biochar was heated until the pH is between 6.5 and 7, and finally the product is dried at temperatures between 100 and 105 °C.	Lili et al. (2023)
CN115999505A	Preparation method of poplar biochar for adsorbing antibiotics	The patent presents a process for producing active biochar from poplar. The process begins with the pulverization and sieving of the poplar biomass, followed by activation in an alkaline solution, the mass fraction of which is between 20% and 35% of the mass of the solution, for a period of 30–90 minutes. Then, it is dried and carbonized under vacuum. Next, the biochar is ground and sieved with a 200-mesh sieve and treated with an HCl solution (1.5 M). Finally, the biochar is mixed with a solution of divalent iron $(NH_4)_2Fe(SO_4)_2$ in the proportion (1.1–1.3).	Qingrong et al. (2022)

(continued)

Table 15.2 (continued)

Patent code	Patent title	Novelty	Inventors
CN114225902A	Alpha-iron oxide modified biochar and preparation method and application thereof	The novelty of the patent is the production of a biochar modified with iron oxide. The raw materials used are cattle and poultry manure, which are placed in controlled atmospheres where it is calcined at temperatures between 500–700 °C for a period between 1 and 3 hours. the biochar is modified with a potassium ferrocyanide solution, in a mass ratio of biochar: Ferrous ion between 4:1 and 10:1. The time to soak the biochar with the solution should be between 20 and 40 minutes, followed by a second calcination at 250–550 °C.	Hua et al. (2021)
CN114029035A	Preparation method of modified sludge biochar, obtained biochar and application	The patent presents a process for producing a modified sludge biochar. Initially, the sludge is dried and sieved to obtain a powder with a homogeneous particle diameter. The sludge is then mixed with calcium chloride and magnesium chloride (ratio 1:3) and an alkaline solution (pH: 10–11), at 15–25 mL solution/g powder. The sludge is then washed and dried, to obtain a modified sludge powder. The modified powder is placed in a tubular furnace for pyrolysis at a temperature between 400 and 800 °C for 1.5–2.5 hours.According to the inventors, this type of modified biochar has a high capacity to absorb phosphorus, which is why it can be widely used in effluent treatment.	Fang et al. (2021)

Accordingly, there have been technological developments in both the production and application of biochar. A predominant focus was observed on the optimization of production processes, with emphasis on energy efficiency and automation. Furthermore, advancements in biochar modification have rendered it a promising alternative for environmental remediation and pollutant removal. Notwithstanding these advancements, challenges persist, including energy consumption and the necessity for optimization. These challenges point to potential areas for future research and technological development in this field.

15.7 Conclusion

Biochar has emerged as a novel and sustainable solution for waste management and valorization. Its production by either pyrolysis, HTC, gasification, or torrefaction can have an essential role in the development and consolidation of a circular economy, thereby reducing environmental impacts and promoting the efficient use of residues. However, each technique has specific advantages, and the choice of the most suitable process is influenced by the characteristics of the waste used. For instance, when the waste material exhibits a high moisture content, HTC can be a particularly efficient option by avoiding the necessity for pre-drying.

Biochar production aligns with the growing global demand for more efficient and sustainable processes, making this material an attractive alternative given its possible applications. For example, biochar can remove pollutants from soil and water, improve soil properties, and produce energy. Therefore, in addition to contributing to waste management, biochar production yields a material that can play an important role in addressing some environmental issues, such as soil and water contamination and dependence on fossil fuels. Moreover, the benefits can even increase when considering the integration of biochar production into the biorefinery concept, reinforcing the significance of this material in sustainability. Once implemented, this approach would agree with circular economy principles.

In light of these considerations, biochar emerges as a promising alternative for waste management and mitigating environmental impacts. The versatility of this material in different applications positions biochar as a pivotal element in the transition to a more sustainable and efficient future. Despite substantial progress achieved in the domains of biochar production, activation, and application, continued research and technological development are imperative to further boost its adoption and maximize its benefits.

References

Afshar M, Mofatteh S (2024) Biochar for a sustainable future: environmentally friendly production and diverse applications. Results Eng 23:102433. https://doi.org/10.1016/j.rineng.2024.102433

Ahmed Khan T, Kim H, Gupta A, Jamari SS, Jose R (2019) Synthesis and characterization of carbon microspheres from rubber wood by hydrothermal carbonization. J Chem Technol Biotechnol 94:1374–1383. https://doi.org/10.1002/jctb.5867

Amalina F, Razak ASA, Krishnan S, Sulaiman H, Zularisam AW, Nasrullah M (2022) Biochar production techniques utilizing biomass waste-derived materials and environmental applications—a review. J Hazard Mater Adv 7:100134. https://doi.org/10.1016/j.hazadv.2022.100134

Antor NH, Mia S, Hasan MM, Lipi NJ, Jindo K, Sanchez-Monedero MA, Rashid MH (2023) Chemically and biologically activated biochars slow down urea hydrolysis and improve nitrogen use efficiency. Pedosphere 33:659–669. https://doi.org/10.1016/j.pedsph.2022.06.050

Awad MI, Makkawi Y, Hassan NM (2024) Yield and energy modeling for biochar and bio-oil using pyrolysis temperature and biomass constituents. ACS Omega 9:18654–18667. https://doi.org/10.1021/ACSOMEGA.4C01646/ASSET/IMAGES/LARGE/AO4C01646_0011.JPEG

Bakr MMA, Wang Y, Hao P, Dawoud MMA, Peng L, Wang Y (2025) Biochar from Al (OH)3-activated sugarcane bagasse for efficient removal of cadmium: optimization using the response surface method (RSM). Biomass Convers Biorefinery 2025:1–18. https://doi.org/10.1007/S13399-025-06594-6

Bamigbade GB, Oyelami OI, Babalola OO, Adewolu A, Omemu AM, Ogunsanya TF, Sanusi JOF, Daniel OM (2025) An updated comprehensive review on waste valorization: Informetric analysis, current insights and future perspectives on cereal waste and byproduct utilization for sustainable industrial applications. Bioresour Technol 418:131868. https://doi.org/10.1016/J.BIORTECH.2024.131868

Barbhuiya S, Bhusan Das B, Kanavaris F (2024) Biochar-concrete: a comprehensive review of properties, production and sustainability. Case Stud Constr Mater 20:e02859. https://doi.org/10.1016/j.cscm.2024.e02859

Bartoli M, Piovano A, Elia GA, Meligrana G, Pedraza R, Pianta N, Tealdi C, Pagot G, Negro E, Triolo C, Gomez LV, Comisso N, Tagliaferro A, Santangelo S, Quartarone E, Di Noto V, Mustarelli P, Ruffo R, Gerbaldi C (2024) Pristine and engineered biochar as Na-ion batteries anode material: a comprehensive overview. Renew Sust Energ Rev 194:114304. https://doi.org/10.1016/J.RSER.2024.114304

Biswal BK, Vijayaraghavan K, Tsen-Tieng DL, Balasubramanian R (2022) Biochar-based bioretention systems for removal of chemical and microbial pollutants from stormwater: a critical review. J Hazard Mater. https://doi.org/10.1016/j.jhazmat.2021.126886

Buffi M, Hurtig O, Prussi M, Scarlat N, Chiaramonti D (2024) Energy and GHG emissions assessment for biochar-enhanced advanced biofuels value chains. Energy Convers Manag 309:118450. https://doi.org/10.1016/j.enconman.2024.118450

Cardarelli A, Barbanera M (2024) Hydrochar from winemaking industry wastes as solid biofuel: a thermal and kinetic analysis of pyrolysis and combustion. Fuel 372:132256. https://doi.org/10.1016/J.FUEL.2024.132256

Cavali M, Junior NL, Mohedano RDA, Filho PB, Helena R, Borges A, Junior DC (2022) Biochar and hydrochar in the context of anaerobic digestion for a circular approach: an overview. Sci Total Environ 822:153614. https://doi.org/10.1016/j.scitotenv.2022.153614

Cavali M, Benbelkacem H, Kim B, Bayard R, Libardi Junior N, Gonzaga Domingos D, Woiciechowski AL, de Castilhos Junior AB (2023a) Co-hydrothermal carbonization of pine residual sawdust and non-dewatered sewage sludge—effect of reaction conditions on hydrochar characteristics. J Environ Manage 340:117994. https://doi.org/10.1016/j.jenvman.2023.117994

Cavali M, Libardi Junior N, de Sena JD, Woiciechowski AL, Soccol CR, Belli Filho P, Bayard R, Benbelkacem H, de Castilhos Junior AB (2023b) A review on hydrothermal carbonization of potential biomass wastes, characterization and environmental applications of hydrochar, and biorefinery perspectives of the process. Sci Total Environ. https://doi.org/10.1016/j.scitotenv.2022.159627

Cavali M, Dresch AP, Belli IM, Libardi Junior N, Woiciechowski AL, Soares SR, de Castilhos Junior AB (2025a) Hydrothermal carbonization—a critical overview of its environmental and economic sustainability. J Clean Prod 491:144838. https://doi.org/10.1016/J.JCLEPRO.2025.144838

Cavali M, Hennig TB, Libardi Junior N, Kim B, Garnier V, Benbelkacem H, Bayard R, Woiciechowski AL, Matias WG, de Castilhos Junior AB (2025b) Co-hydrothermal carbonization of sawdust and sewage sludge: assessing the potential of the Hydrochar as an adsorbent and the Ecotoxicity of the process water. Appl Sci (Switz) 15. https://doi.org/10.3390/app15031052

Cay H, Duman G, Yanik J (2019) Two-step gasification of biochar for hydrogen-Rich gas production: effect of the biochar type and catalyst. Energy Fuel 33:7398–7405. https://doi.org/10.1021/acs.energyfuels.9b01354

Cha JS, Park SH, Jung SC, Ryu C, Jeon JK, Shin MC, Park YK (2016) Production and utilization of biochar: a review. J Ind Eng Chem 40:1–15. https://doi.org/10.1016/j.jiec.2016.06.002

Chunyan Z, Fengjun Z, Xu L, Zeyuan H, Yinhu Q, Feiyue L, Yue X, Yunlei B (2022) A new type of biochar continuous environmental protection production equipment. CN Patent 115287084A

Cognigni P, Leonelli C, Berrettoni M (2025) A bibliographic study of biochar and hydrochar: differences and similarities. J Anal Appl Pyrolysis 187:106985. https://doi.org/10.1016/J.JAAP.2025.106985

Desjardins SM, Ter-Mikaelian MT, Chen J (2024) Cradle-to-gate life cycle analysis of slow pyrolysis biochar from forest harvest residues in Ontario, Canada. Biochar 6:58. https://doi.org/10.1007/s42773-024-00352-z

Dissanayake PD, You S, Igalavithana AD, Xia Y, Bhatnagar A, Gupta S, Kua HW, Kim S, Kwon JH, Tsang DCW, Ok YS (2020) Biochar-based adsorbents for carbon dioxide capture: a critical review. Renew Sust Energ Rev 119:1–14. https://doi.org/10.1016/j.rser.2019.109582

Dong Y, Zhang X, Wang X, Xie C, Liu J, Cheng Y, Yue Y, You X, Li Y (2024) Modified biochar affects CO2 and N2O emissions from coastal saline soil by altering soil pH and elemental stoichiometry. Sci Total Environ 954:176283. https://doi.org/10.1016/j.scitotenv.2024.176283

Du L, Ahmad S, Liu L, Wang L, Tang J (2023) A review of antibiotics and antibiotic resistance genes (ARGs) adsorption by biochar and modified biochar in water. Sci Total Environ 858:159815. https://doi.org/10.1016/J.SCITOTENV.2022.159815

Ellen MacArthur Foundation (2017) The circular economy in detail [WWW document]. The Circular Economy in Detail. https://www.ellenmacarthurfoundation.org/explore/the-circular-economy-in-detail. Accessed 2.7.21

Fang F, Jialing Z, Lu T, Jingyang L, Qian F, Jiashun C (2021) Preparation method of modified sludge biochar, obtained biochar and application. CN Patent 114029035A 15:37–48

Fawzy S, Osman AI, Yang H, Doran J, Rooney DW (2021) Industrial biochar systems for atmospheric carbon removal: a review. Environ Chem Lett 19:3023–3055. https://doi.org/10.1007/s10311-021-01210-1

Feng Q, Fan B, He YC, Ma C (2024) Antibacterial, antioxidant and fruit packaging ability of biochar-based silver nanoparticles-polyvinyl alcohol-chitosan composite film. Int J Biol Macromol 256:128297. https://doi.org/10.1016/J.IJBIOMAC.2023.128297

Ferrentino R, Ceccato R, Marchetti V, Andreottola G, Fiori L (2020) Sewage sludge Hydrochar: an option for removal of methylene blue from wastewater. Appl Sci 10:3445. https://doi.org/10.3390/app10103445

Foong SY, Liew RK, Yang Y, Cheng YW, Yek PNY, Wan Mahari WA, Lee XY, Han CS, Vo D-VN, Van Le Q, Aghbashlo M, Tabatabaei M, Sonne C, Peng W, Lam SS (2020) Valorization of biomass waste to engineered activated biochar by microwave pyrolysis: progress, challenges, and future directions. Chem Eng J 389:124401. https://doi.org/10.1016/j.cej.2020.124401

Foroutan R, Mohammadi R, Razeghi J, Ramavandi B (2021) Biodiesel production from edible oils using algal biochar/CaO/K2CO3 as a heterogeneous and recyclable catalyst. Renew Energy 168:1207–1216. https://doi.org/10.1016/J.RENENE.2020.12.094

Gale M, Nguyen T, Moreno M, Gilliard-AbdulAziz KL (2021) Physiochemical properties of biochar and activated carbon from biomass residue: influence of process conditions to adsorbent properties. ACS Omega 6:10224–10233. https://doi.org/10.1021/acsomega.1c00530

García-Ávila F, Galarza-Guamán A, Barros-Bermeo M, Alfaro-Paredes EA, Avilés-Añazco A, Iglesias-Abad S (2023) Integration of high-rate filtration using waste-derived biochar as a potential sustainable technology for drinking water supply. Biochar 5:1–23. https://doi.org/10.1007/S42773-023-00256-4/TABLES/18

Garcia-Nunez JA, Pelaez-Samaniego MR, Garcia-Perez ME, Fonts I, Abrego J, Westerhof RJM, Garcia-Perez M (2017) Historical developments of pyrolysis reactors: a review. Energy Fuel 31: 5751–5775. https://doi.org/10.1021/acs.energyfuels.7b00641

Gaunt JL, Lehmann J (2008) Energy balance and emissions associated with biochar sequestration and pyrolysis bioenergy production. Environ Sci Technol 42:4152–4158. https://doi.org/10.1021/es071361i

Giwa T, Akbari M, Kumar A (2022) Assessment of the greenhouse gas emission footprint of a biorefinery over its life cycle. Energy Convers Manage 271:116327. https://doi.org/10.1016/j.enconman.2022.116327

Güleç F, Riesco LMG, Williams O, Kostas ET, Samson A, Lester E (2021) Hydrothermal conversion of different lignocellulosic biomass feedstocks—effect of the process conditions on hydrochar structures. Fuel 302:121166. https://doi.org/10.1016/j.fuel.2021.121166

Guo J, Xiao H, Zhang JB, Dai C, Li T, Gao MT, Hu J, Li J (2024) Characterization of highly stable biochar and its application for removal of phenol. Biomass Convers Biorefin 14:13311–13321. https://doi.org/10.1007/S13399-022-03375-3/FIGURES/7

Gupta M, Dorn E, Nguyen L, McFarlan A (2025) Techno-economic evaluation of pulp and paper mill derived biochar, liquid and gaseous biofuel precursors: a British Columbia case study. Biomass Bioenergy 195:107677. https://doi.org/10.1016/j.biombioe.2025.107677

Hama Aziz KH, Kareem R (2023) Recent advances in water remediation from toxic heavy metals using biochar as a green and efficient adsorbent: a review. Case Stud Chem Environ Eng 8: 100495. https://doi.org/10.1016/J.CSCEE.2023.100495

Han L, Chen L, Li D, Ji Y, Feng Y, Feng Y, Yang Z (2022) Influence of polyethylene terephthalate microplastic and biochar co-existence on paddy soil bacterial community structure and greenhouse gas emission. Environ Pollut 292:118386. https://doi.org/10.1016/J.ENVPOL.2021.118386

Hernández JJ, Saffe A, Collado R, Monedero E (2020) Recirculation of char from biomass gasification: effects on gasifier performance and end-char properties. Renew Energy 147:806–813. https://doi.org/10.1016/j.renene.2019.09.063

Hoang AT, Ong HC, Fattah IMR, Chong CT, Cheng CK, Sakthivel R, Ok YS (2021) Progress on the lignocellulosic biomass pyrolysis for biofuel production toward environmental sustainability. Fuel Process Technol 223:106997. https://doi.org/10.1016/j.fuproc.2021.106997

Holanda MAS, Coelho Menezes JM, Coutinho HDM, Teixeira RNP (2023) Effectiveness of biochar as an adsorbent for pesticides: systematic review and meta-analysis. J Environ Manag 345:118719. https://doi.org/10.1016/J.JENVMAN.2023.118719

Hossain MZ, Bahar MM, Sarkar B, Donne SW, Ok YS, Palansooriya KN, Kirkham MB, Chowdhury S, Bolan N (2020) Biochar and its importance on nutrient dynamics in soil and plant. Biochar 2:379–420. https://doi.org/10.1007/S42773-020-00065-Z

Hua Y, Yingchao L, Zhì D, Xin Q (2021) Alpha-iron oxide modified biochar and preparation method and application thereof. CN Patent CN114225902A 6

Huimin X, Biying L, Liuzhen R, Chunyu L, Ranqi F, Chengzhi L, Jianjun F, Hui-zhen M, Lei Z (2020) Biochar preparation device. CN112430468A. China: Guangzhou University of Chinese Medicine. https://patents.google.com/patent/CN112430468A

Ibrahim EA, El-Sherbini MAA, Selim EMM (2022) Effects of biochar on soil properties, heavy metal availability and uptake, and growth of summer squash grown in metal-contaminated soil. Sci Hortic 301:111097. https://doi.org/10.1016/J.SCIENTA.2022.111097

International Biochar Initiative (2015) Standardized product definition and product testing guidelines for biochar that is used in soil (aka IBI biochar standards)

International Energy Agency (2022) Task 42: biorefining in a circular economy [WWW document]. Activities: Tasks. https://www.ieabioenergy.com/blog/task/biorefining-sustainable-processing-of-biomass-into-a-spectrum-of-marketable-biobased-products-and-bioenergy/. Accessed 6.25.22

Joshi NC, Sinha S, Bhatnagar P, Nath Y, Negi B, Kumar V, Gururani P (2024) A concise review on waste biomass valorization through thermochemical conversion. Curr Res Microb Sci 6: 100237. https://doi.org/10.1016/J.CRMICR.2024.100237

Jun Z, Xiaofei Y, Jinglong L, Houhu Z, Susan, Yingying W (2024) A process for preparing biochar with high adsorption performance by acid pretreatment and coordinated carbonization. 15, 37–48

Kamarudin NS, Dahalan FA, Hasan M, An OS, Parmin NA, Ibrahim N, Hamdzah M, Zain NAM, Muda K, Wikurendra EA (2022) Biochar: a review of its history, characteristics, factors that influence its yield, methods of production, application in wastewater treatment and recent development. Biointerface Res Appl Chem. https://doi.org/10.33263/BRIAC126.79147926

Kane S, Storer A, Xu W, Ryan C, Stadie NP (2022) Biochar as a renewable substitute for carbon black in lithium-ion battery electrodes. ACS Sustain Chem Eng 10:12226–12233. https://doi.org/10.1021/ACSSUSCHEMENG.2C02974/SUPPL_FILE/SC2C02974_SI_001.PDF

Kang S, Li X, Fan J, Chang J (2012) Characterization of hydrochars produced by hydrothermal carbonization of lignin, cellulose, d-xylose, and wood meal. Ind Eng Chem Res:9023–9031. https://doi.org/10.1021/ie300565d

Kant Bhatia S, Palai AK, Kumar A, Kant Bhatia R, Kumar Patel A, Kumar Thakur V, Yang YH (2021) Trends in renewable energy production employing biomass-based biochar. Bioresour Technol 340:125644. https://doi.org/10.1016/J.BIORTECH.2021.125644

Kavan Kumar V, Panwar NL (2024) Pyrolysis technologies for biochar production in waste management: a review. Clean Energy 8:61–78. https://doi.org/10.1093/ce/zkae036

Kazemi Shariat Panahi H, Dehhaghi M, Ok YS, Nizami AS, Khoshnevisan B, Mussatto SI, Aghbashlo M, Tabatabaei M, Lam SS (2020) A comprehensive review of engineered biochar: production, characteristics, and environmental applications. J Clean Prod 270:122462. https://doi.org/10.1016/J.JCLEPRO.2020.122462

Kumar R, Verma A, Rakib MRJ, Gupta PK, Sharma P, Garg A, Girard P, Aminabhavi TM (2023) Adsorptive behavior of micro(nano)plastics through biochar: co-existence, consequences, and challenges in contaminated ecosystems. Sci Total Environ 856:159097. https://doi.org/10.1016/J.SCITOTENV.2022.159097

Kumar T, Ansari SA, Sawarkar R, Agashe A, Singh L, Nidheesh PV (2025) Bamboo biochar: a multifunctional material for environmental sustainability. Biomass Convers Biorefin 2025:1–25. https://doi.org/10.1007/S13399-025-06608-3

Kung C-C, Mu JE (2019) Prospect of China's renewable energy development from pyrolysis and biochar applications under climate change. Renew Sust Energ Rev 114:109343. https://doi.org/10.1016/j.rser.2019.109343

Latawiec AE, Strassburg BBN, Junqueira AB, Araujo E, Luiz LF, Pinto HAN, Castro A, Rangel M, Malaguti GA, Rodrigues AF, Barioni LG, Novotny EH, Cornelissen G, Mendes M, Batista N, Guerra JG, Zonta E, Jakovac C, Hale SE (2019) Biochar amendment improves degraded pasturelands in Brazil: environmental and cost-benefit analysis. Sci Rep 9:1–12. https://doi.org/10.1038/s41598-019-47647-x

Lee Y, Park J, Ryu C, Gang KS, Yang W, Park Y-K, Jung J, Hyun S (2013) Comparison of biochar properties from biomass residues produced by slow pyrolysis at 500°C. Bioresour Technol 148:196–201. https://doi.org/10.1016/j.biortech.2013.08.135

Lee J, Lee S, Lin K-YA, Jung S, Kwon EE (2023) Abatement of odor emissions from wastewater treatment plants using biochar. Environ Pollut 336:122426. https://doi.org/10.1016/j.envpol.2023.122426

Lee J-M, Kim H-B, Wang JJ, Zhou B, Seo D-C, Park J-H (2024) Conversion of acidified lignin containing sulfur discharged from a biorefinery process into neutralized biochar: characterization and metal adsorption. Sci Total Environ 954:176438. https://doi.org/10.1016/j.scitotenv.2024.176438

Letti LAJ, Woiciechowski AL, Medeiros ABP, de Carvalho JC, Vandenberghe LP d S, Karp SG, Torres LAZ, Guarnizo AFC, Colonia BSO, Siqueira JGW, Diestra KV, Cavali M, Burgos WJM, Soccol CR (2021) Valorization of solid and liquid wastes from palm oil industry. In: Bhaskar T, Pandey A, Varjani S, Rene ER (eds) waste biorefinery: value addition through resource utilization. Elsevier, pp 235–265. https://doi.org/10.1016/B978-0-12-821879-2.00009-0

Li S, Tasnady D (2023) Biochar for soil carbon sequestration: current knowledge, mechanisms, and future perspectives. C 9:67. https://doi.org/10.3390/C9030067

Lili W, Dan W, Lichen L, Houhu Z, Haifeng T, Yuanshun X, Hao J, Wen Y (2023) Method for preparing biochar by collaborative pyrolysis of oily sludge and agricultural and forestry waste. CN Patent 117866648A

Lim EY, Lee JTE, Zhang L, Tian H, Ong KC, Tio ZK, Zhang J, Tong YW (2022) Abrogating the inhibitory effects of volatile fatty acids and ammonia in overloaded food waste anaerobic digesters via the supplementation of nano-zero valent iron modified biochar. Sci Total Environ 817:152968. https://doi.org/10.1016/J.SCITOTENV.2022.152968

Liu X, Li Z, Zhang Y, Feng R, Mahmood IB (2014) Characterization of human manure-derived biochar and energy-balance analysis of slow pyrolysis process. Waste Manag 34:1619–1626. https://doi.org/10.1016/j.wasman.2014.05.027

Liu D, Cai Y, Yu X, Wang Q (2025) Removal of tetracycline with grape leaves–based biochar: adsorption properties and mechanism. Biomass Convers Biorefinery:1–14. https://doi.org/10.1007/S13399-025-06507-7/FIGURES/1

Merzari F, Lucian M, Volpe M, Andreottola G, Fiori L (2018) Hydrothermal carbonization of biomass: design of a Bench- Scale Reactor for evaluating the heat of reaction. Chem Eng Trans 65:43–48. https://doi.org/10.3303/CET1865008

Mishra R, Shu CM, Ong HC, Gollakota ARK, Kumar S (2024) Progress and development of biochar as a catalyst for hydrogen production. J Clean Prod 477:143853. https://doi.org/10.1016/J.JCLEPRO.2024.143853

Mobili A, Cosoli G, Giulietti N, Chiariotti P, Bellezze T, Pandarese G, Revel GM, Tittarelli F (2023) Biochar and recycled carbon fibres as additions for low-resistive cement-based composites exposed to accelerated degradation. Constr Build Mater 376:131051. https://doi.org/10.1016/J.CONBUILDMAT.2023.131051

Nakason K, Pathomrotsakun J, Kraithong W, Khemthong P, Panyapinyopol B (2019) Torrefaction of agricultural wastes: influence of lignocellulosic types and treatment temperature on fuel properties of biochar. Int Energy J 19:253–266

Neogi S, Sharma V, Khan N, Chaurasia D, Ahmad A, Chauhan S, Singh A, You S, Pandey A, Bhargava PC (2022) Sustainable biochar: a facile strategy for soil and environmental restoration, energy generation, mitigation of global climate change and circular bioeconomy. Chemosphere 293:133474. https://doi.org/10.1016/j.chemosphere.2021.133474

Nguyen MK, Lin C, Hoang HG, Bui XT, Ngo HH, Le VG, Tran HT (2023) Investigation of biochar amendments on odor reduction and their characteristics during food waste co-composting. Sci Total Environ 865:161128. https://doi.org/10.1016/J.SCITOTENV.2022.161128

Oliveira I, Blöhse D, Ramke H-G (2013) Hydrothermal carbonization of agricultural residues. Bioresour Technol 142:138–146. https://doi.org/10.1016/j.biortech.2013.04.125

Osman AI, Farghali M, Dong Y, Kong J, Yousry M, Rashwan AK, Chen Z, Al-Fatesh A, Rooney DW, Yap PS (2023) Reducing the carbon footprint of buildings using biochar-based bricks and insulating materials: a review. Environ Chem Lett 22:71–104. https://doi.org/10.1007/S10311-023-01662-7

Paul T, Yee J-J, Park SH (2025) Biochar-assisted aerobic granulation in enhanced biological treatment of landfill leachate with municipal wastewater for sustainable resource recovery: a biorefinery approach. J Environ Chem Eng 13:115330. https://doi.org/10.1016/j.jece.2025.115330

Praveen S, Gokulan R, Pushpa TB, Jegan J (2021) Techno-economic feasibility of biochar as biosorbent for basic dye sequestration. J Indian Chem Soc 98:100107. https://doi.org/10.1016/j.jics.2021.100107

Prochnow FD, Cavali M, Dresch AP, Belli IM, Libardi N, de Castilhos AB (2024) Biochar: from laboratory to industry scale—an overview of scientific and industrial advances, opportunities in the Brazilian context, and contributions to sustainable development. Processes 12:1006. https://doi.org/10.3390/pr12051006

Qian K, Kumar A, Zhang H, Bellmer D, Huhnke R (2015) Recent advances in utilization of biochar. Renew Sust Energ Rev 42:1055–1064. https://doi.org/10.1016/j.rser.2014.10.074

Qian C, Li Q, Zhang Z, Wang X, Hu J, Cao W (2020) Prediction of higher heating values of biochar from proximate and ultimate analysis. Fuel 265:116925. https://doi.org/10.1016/J.FUEL.2019.116925

Qingrong P, Xianzhao Z, Fengmao L, Rui C, Xuehui C (2022) Preparation method of poplar biochar for adsorbing antibiotics. CN 115999505A Patent 15, pp 37–48

Qiu B, Shao Q, Shi J, Yang C, Chu H (2022) Application of biochar for the adsorption of organic pollutants from wastewater: modification strategies, mechanisms and challenges. Sep Purif Technol 300:121925. https://doi.org/10.1016/J.SEPPUR.2022.121925

Raclavská H, Corsaro A, Juchelková D, Sassmanová V, Frantík J (2015) Effect of temperature on the enrichment and volatility of 18 elements during pyrolysis of biomass, coal, and tires. Fuel Process Technol 131:330–337. https://doi.org/10.1016/j.fuproc.2014.12.001

Rahul S, Dhanuprabha D, Prabakaran S, Arumugam A (2024) An integrated biorefinery of Madhuca indica for co-production of biodiesel, bio-oil, and biochar: towards a sustainable circular bioeconomy. Ind Crop Prod 221:119409. https://doi.org/10.1016/j.indcrop.2024.119409

Rajapaksha AU, Chen SS, Tsang DCW, Zhang M, Vithanage M, Mandal S, Gao B, Bolan NS, Ok YS (2016) Engineered/designer biochar for contaminant removal/immobilization from soil and water: potential and implication of biochar modification. Chemosphere 148:276–291. https://doi.org/10.1016/j.chemosphere.2016.01.043

Rajpoot SK, Sharma PR, Singh J, Kumar A, Vijayakumar S, Chaudhary R, Kumar D (2024) Biochar as a novel feed additive for ruminants. In: Feed additives and supplements for ruminants, pp 423–435. https://doi.org/10.1007/978-981-97-0794-2_18

Rajput V, Saini I, Parmar S, Pundir V, Kumar V, Kumar V, Naik B, Rustagi S (2024) Biochar production methods and their transformative potential for environmental remediation. Discover App Sci 6:408. https://doi.org/10.1007/s42452-024-06125-4

Rambabu K, AlYammahi J, Bharath G, Thanigaivelan A, Sivarajasekar N, Banat F (2021) Nano-activated carbon derived from date palm coir waste for efficient sequestration of noxious 2,4-dichlorophenoxyacetic acid herbicide. Chemosphere 282:131103. https://doi.org/10.1016/J.CHEMOSPHERE.2021.131103

Rashidi NA, Chai YH, Ismail IS, Othman MFH, Yusup S (2022) Biomass as activated carbon precursor and potential in supercapacitor applications. Biomass Convers Biorefin 1:1–15. https://doi.org/10.1007/S13399-022-02351-1/FIGURES/3

Román S, Libra J, Berge N, Sabio E, Ro K, Li L, Ledesma B, Álvarez A, Bae S (2018) Hydrothermal carbonization: modeling, final properties design and applications: a review. Energies (Basel) 11:216. https://doi.org/10.3390/en11010216

Romero Millán LM, Sierra Vargas FE, Nzihou A (2021) Characterization of steam gasification biochars from Lignocellulosic Agrowaste towards soil applications. Waste Biomass Valoriz 12: 4141–4155. https://doi.org/10.1007/s12649-020-01241-9

Sajjadi B, Chen WY, Egiebor NO (2019) A comprehensive review on physical activation of biochar for energy and environmental applications. Rev Chem Eng 35:735–776. https://doi.org/10.1515/revce-2017-0113

Senapati S, Giri J, Mallick L, Biswal P, Mohapatra S, Behera D, Rath P, Panda AK (2025) Ultra-fast adsorption of the industrial cationic dye pollutant using nitric acid-activated rice straw biochar: insights into adsorption mechanisms. Biomass Convers Biorefinery:1–19. https://doi.org/10.1007/S13399-025-06540-6/FIGURES/1

Seroka NS, Luo H, Khotseng L (2024) Biochar-derived anode materials for lithium-ion batteries: a review. Batteries (Basel) 10:144. https://doi.org/10.3390/BATTERIES10050144

Setter C, Silva FTM, Assis MR, Ataíde CH, Trugilho PF, Oliveira TJP (2020) Slow pyrolysis of coffee husk briquettes: characterization of the solid and liquid fractions. Fuel 261:116420. https://doi.org/10.1016/j.fuel.2019.116420

Sharma AK, Ghodke PK, Goyal N, Bobde P, Kwon EE, Lin K-YA, Chen W-H (2023) A critical review on biochar production from pine wastes, upgradation techniques, environmental sustainability, and challenges. Bioresour Technol 387:129632. https://doi.org/10.1016/j.biortech.2023.129632

Singh Yadav SP, Bhandari S, Bhatta D, Poudel A, Bhattarai S, Yadav P, Ghimire N, Paudel P, Paudel P, Shrestha J, Oli B (2023) Biochar application: a sustainable approach to improve soil health. J Agric Food Res 11:100498. https://doi.org/10.1016/J.JAFR.2023.100498

Sri Shalini S, Palanivelu K, Ramachandran A, Raghavan V (2020) Biochar from biomass waste as a renewable carbon material for climate change mitigation in reducing greenhouse gas emissions—a review. Biomass Convers Biorefin:280. https://doi.org/10.1007/s13399-020-00604-5

Sugiarto Y, Sunyoto NMS, Zhu M, Jones I, Zhang D (2021) Effect of biochar in enhancing hydrogen production by mesophilic anaerobic digestion of food wastes: the role of minerals. Int J Hydrog Energy 46:3695–3703. https://doi.org/10.1016/J.IJHYDENE.2020.10.256

Sun X, Fu H, Bao M, Liu W, Luo C, Li Y, Li Y, Lu J (2022) Development of a new hydrophobic magnetic biochar for removing oil spills on the water surface. Biochar 4:1–17. https://doi.org/10.1007/S42773-022-00184-9/FIGURES/10

Tan H, Lee CT, Ong PY, Wong KY, Bong CPC, Li C, Gao Y (2021) A review on the comparison between slow pyrolysis and fast pyrolysis on the quality of Lignocellulosic and lignin-based biochar. IOP Conf Ser Mater Sci Eng 1051, 012075. https://doi.org/10.1088/1757-899X/1051/1/012075

Tian H, Hu Q, Wang J, Chen D, Yang Y, Bridgwater AV (2021) Kinetic study on the CO2 gasification of biochar derived from Miscanthus at different processing conditions. Energy 217: 119341. https://doi.org/10.1016/j.energy.2020.119341

Trivedi Y, Sharma M, Mishra RK, Sharma A, Joshi J, Gupta AB, Achintya B, Shah K, Vuppaladadiyamd AK (2025) Biochar potential for pollutant removal during wastewater treatment: a comprehensive review of separation mechanisms, technological integration, and process analysis. Desalination 600:118509. https://doi.org/10.1016/J.DESAL.2024.118509

Tu R, Sun Y, Wu Y, Fan X, cheng S, Jiang E, Xu X (2022) The fuel properties and adsorption capacities of torrefied camellia shell obtained via different steam-torrefaction reactors. Energy 238. https://doi.org/10.1016/j.energy.2021.121969

United Nations (2025) What is climate change? [WWW document]. https://www.un.org/en/climatechange/what-is-climate-change. Accessed 3.4.25

United States Biochar Initiative (2025) Biochar Introduction | US Biochar Initiative [WWW Document]. https://biochar-us.org/biochar-introduction. Accessed 3.4.25

Urgel JJDT, Briones JMA, Diaz EB, Dimaculangan KMN, Rangel KL, Lopez ECR (2024) Removal of diesel oil from water using biochar derived from waste banana peels as adsorbent. Carbon Res 3:1–17. https://doi.org/10.1007/S44246-024-00100-9/FIGURES/9

Vandenberghe LPS, Pandey A, Carvalho JC, Letti LAJ, Woiciechowski AL, Karp SG, Thomaz-Soccol V, Martínez-Burgos WJ, Penha RO, Herrmann LW, Rodrigues AO, Soccol CR (2021) Solid-state fermentation technology and innovation for the production of agricultural and animal feed bioproducts. Syst Microbiol Biomanuf 1:142–165. https://doi.org/10.1007/s43393-020-00015-7

Vidakis N, Petousis M, Kalderis D, Michailidis N, Maravelakis E, Saltas V, Bolanakis N, Papadakis V, Spiridaki M, Argyros A (2024) Reinforced HDPE with optimized biochar content for material extrusion additive manufacturing: morphological, rheological, electrical, and thermomechanical insights. Biochar 6:1–21. https://doi.org/10.1007/S42773-024-00314-5/FIGURES/5

Vieira FR, Romero Luna CM, Arce GLAF, Ávila I (2020) Optimization of slow pyrolysis process parameters using a fixed bed reactor for biochar yield from rice husk. Biomass Bioenergy 132: 105412. https://doi.org/10.1016/j.biombioe.2019.105412

Wang J, Xiong Z, Kuzyakov Y (2016) Biochar stability in soil: meta-analysis of decomposition and priming effects. GCB Bioenergy 8:512–523. https://doi.org/10.1111/gcbb.12266

Wang RZ, Huang DL, Liu YG, Zhang C, Lai C, Wang X, Zeng GM, Zhang Q, Gong XM, Xu P (2020) Synergistic removal of copper and tetracycline from aqueous solution by steam-activated bamboo-derived biochar. J Hazard Mater 384:121470. https://doi.org/10.1016/J.JHAZMAT.2019.121470

Wang G, Li Q, Yuwen C, Gong K, Sheng L, Li Y, Xing Y, Chen R (2021a) Biochar triggers methanogenesis recovery of a severely acidified anaerobic digestion system via hydrogen-based syntrophic pathway inhibition. Int J Hydrog Energy 46:9666–9677. https://doi.org/10.1016/J.IJHYDENE.2020.03.115

Wang F, Song Y, Huang J, Wu B, Wang Y, Pang Y, Zhang W, Zhu Z, Ma F, Wang X, Zhang X, Wang F, Song Y, Wang Y, Pang Y, Ma F, Zhang X, Huang J, Wu B, Zhang W, Zhu Z, Wang X (2021b) Lollipop-inspired multilayered drug delivery hydrogel for dual effective, long-term, and NIR-defined glaucoma treatment. Macromol Biosci 21:2100202. https://doi.org/10.1002/MABI.202100202

Wang Q, Wu S, Cui D, Zhou H, Wu D, Pan S, Xu F, Wang Z (2022) Co-hydrothermal carbonization of organic solid wastes to hydrochar as potential fuel: a review. Sci Total Environ. https://doi.org/10.1016/j.scitotenv.2022.158034

Xiang W, Zhang X, Chen J, Zou W, He F, Hu X, Tsang DCW, Ok YS, Gao B (2020) Biochar technology in wastewater treatment: a critical review. Chemosphere 252:126539. https://doi.org/10.1016/J.CHEMOSPHERE.2020.126539

Yaashikaa PR, Kumar PS, Varjani S, Saravanan A (2020) A critical review on the biochar production techniques, characterization, stability and applications for circular bioeconomy. Biotechnol Rep 28:e00570. https://doi.org/10.1016/j.btre.2020.e00570

Yang Y, Xu R, Zhang J, Wang W, Wang R, Shi J, Dong Y, Huang Y, Ye Y (2025) Effect of biochar addition on the strength properties of metallurgical dust briquettes. J Sustain Metall:1–13. https://doi.org/10.1007/S40831-024-01003-0/FIGURES/8

Ye L, Peng Z, Wang L, Anzulevich A, Bychkov I, Kalganov D, Tang H, Rao M, Li G, Jiang T (2019) Use of biochar for sustainable ferrous metallurgy. JOM 71:3931–3940. https://doi.org/10.1007/S11837-019-03766-4/FIGURES/7

Yongqiang F, Yongming H (2024) A method and application of preparing biochar by pyrolysis of agricultural, forestry and livestock and poultry waste CN 118652705A patent Zhong 15, 37–48

You S, Ok YS, Tsang DCW, Kwon EE, Wang C-H (2018) Towards practical application of gasification: a critical review from syngas and biochar perspectives. Crit Rev Environ Sci Technol 48:1165–1213. https://doi.org/10.1080/10643389.2018.1518860

Zakaria MR, Ahmad Farid MA, Andou Y, Ramli I, Hassan MA (2023) Production of biochar and activated carbon from oil palm biomass: current status, prospects, and challenges. Ind Crop Prod 199:116767. https://doi.org/10.1016/j.indcrop.2023.116767

Zha Z, Zhang Z, Xiang P, Zhu H, Zhou B, Sun Z, Zhou S (2020) One-step preparation of eggplant-derived hierarchical porous graphitic biochar as efficient oxygen reduction catalyst in microbial fuel cells. RSC Adv 11:1077–1085. https://doi.org/10.1039/D0RA09976G

Zhang S, Xiong Y (2016) Washing pretreatment with light bio-oil and its effect on pyrolysis products of bio-oil and biochar. RSC Adv 6:5270–5277. https://doi.org/10.1039/C5RA22350D

Zhang C, Ho S-H, Chen W-H, Xie Y, Liu Z, Chang J-S (2018) Torrefaction performance and energy usage of biomass wastes and their correlations with torrefaction severity index. Appl Energy 220:598–604. https://doi.org/10.1016/j.apenergy.2018.03.129

Zhang S, Zhu X, Zhou S, Shang H, Luo J, Tsang DCW (2019a) Hydrothermal carbonization for Hydrochar production and its application. In: Biochar from biomass and waste. Elsevier, pp 275–294. https://doi.org/10.1016/B978-0-12-811729-3.00015-7

Zhang Z, Zhu Z, Shen B, Liu L (2019b) Insights into biochar and hydrochar production and applications: a review. Energy 171:581–598. https://doi.org/10.1016/J.ENERGY.2019.01.035

Zhang J, Yang M, Zhao W, Zhang J, Zang L (2021) Biohydrogen production amended with nitrogen-doped biochar. Energy Fuels 35:1476–1487. https://doi.org/10.1021/ACS.ENERGYFUELS.0C03405/ASSET/IMAGES/LARGE/EF0C03405_0009.JPEG

Zhang P, Sun K, Liu Y, Zhou B, Li S, Zhou J, Wang A, Xie L, Li B, Jiang J (2023) Improving bifunctional catalytic activity of biochar via in-situ growth of nickel-iron hydroxide as cathodic catalyst for zinc-air batteries. Biochar 5:1–12. https://doi.org/10.1007/S42773-023-00259-1/FIGURES/6

Zhao S, Zhang L, Hu H, Jin L (2023) Catalytic decomposition of methane over enteromorpha prolifera-based hierarchical porous biochar. Int J Hydrog Energy 48:3328–3339. https://doi.org/10.1016/J.IJHYDENE.2022.10.215

Zhu N, Yan Q, He Y, Wang X, Wei Z, Liang D, Yue H, Yun Y, Li G, Sang N (2022) Insights into the removal of polystyrene nanoplastics using the contaminated corncob-derived mesoporous biochar from mining area. J Hazard Mater 433:128756. https://doi.org/10.1016/J.JHAZMAT.2022.128756

Chapter 16
Bio-oil: Production Processes, Applications, Importance, and Advances

Matheus Cavali, Aline Perin Dresch, Thamarys Scapini, Luiza Loss, Walter José Martinez-Burgos, Nelson Libardi Junior, Adenise Lorenci Woiciechowski, and Armando Borges de Castilhos Junior

Abstract It is well established that the transition from the current linear economy model to a circular one requires a novel approach for waste management. In other words, residues must be seen as feedstock for other processes instead of valueless materials. Within the domain of biofuels, for example, waste conversion into energetic molecules (e.g., ethanol, biodiesel, and biogas) is a suitable alternative to fossil fuels, and it has been developed successfully. Similarly, other value-added molecules can also be obtained from residues and avoid producing such molecules from nonrenewable feedstocks. Accordingly, if an industrial facility produces energetic and other marketable molecules from waste materials, i.e., a spectrum of bio-based products, it can be called a "biorefinery." Considering this context, bio-oil production emerges as a promising technology, capable of providing biofuels and high-value-added molecules while contributing to waste management. In that regard, this chapter, through an introductory approach, addresses bio-oil production processes, possible bio-oil applications, the potential of this waste valorization process in a biorefinery context, and the advances in bio-oil research.

16.1 Introduction

Climate change represents a significant contemporary challenge, predominantly attributed to greenhouse gas emissions resulting from the combustion of fossil fuels and industrial activities. Rising global temperatures have provoked extreme

M. Cavali (✉) · L. Loss · N. Libardi Junior · A. B. de Castilhos Junior
Department of Sanitary and Environmental Engineering, Federal University of Santa Catarina, Florianópolis, Santa Catarina, Brazil

A. P. Dresch
Department of Environmental Engineering and Technology, Federal University of Paraná, Paraná, Brazil

T. Scapini · W. J. Martinez-Burgos · A. L. Woiciechowski
Department of Bioprocess Engineering and Biotechnology, Federal University of Paraná, Paraná, Brazil

A. L. Woiciechowski et al. (eds.), *Biomass for Bioenergy and Bioproducts*, Biofuel and Biorefinery Technologies 17, https://doi.org/10.1007/978-3-032-19898-3_16

weather events, endangering ecosystems, economies, and food security (Intergovernmental Panel on Climate Change 2022). In light of these circumstances, transitioning to a more sustainable production model is paramount to mitigate environmental impacts and reduce reliance on nonrenewable resources. In this context, the circular economy has emerged as a promising approach to rethinking the way resources are extracted, used, and disposed of. This strategy is based on the minimization of environmental impacts while concomitantly generating new economic opportunities by transforming waste into valuable resources (Ellen MacArthur Foundation 2017).

A cornerstone of the circular economy is waste valorization, which aims to convert waste into value-added products. For instance, thermochemical (e.g., pyrolysis and gasification) and biological (e.g., fermentation) processes have been extensively studied for converting organic waste into bioenergy. Consequently, waste valorization plays a pivotal role in the diversification of the global energy matrix (Durak 2023; Joshi et al. 2024; Okolie et al. 2022).

A promising example of the application of these concepts is the development of biorefineries—industrial facilities designed to convert biomass into a wide range of products (Cherubini and Strømman 2011; Scapini et al. 2023). Among the products obtained in biorefineries, biofuels stand out as viable alternatives to replace fossil fuels (Cherwoo et al. 2023). In addition to ethanol, biodiesel, and biogas, which have already achieved notable applications in the transportation sector and energy generation, another noteworthy product is bio-oil. This product is notable for its high energy content, which can be harnessed through refinement to produce sustainable liquid fuels or utilized as a raw material in the chemical industry. Bio-oil utilization represents a strategic approach to reducing reliance on fossil fuels, concurrently valorizing discarded waste (Hu and Gholizadeh 2020; Machado et al. 2022).

Accordingly, the integration of the circular economy, waste recovery, and development of biorefineries emerges as a compelling solution to address the challenges posed by climate change. In this regard, the production of bio-oil from waste can be an alternative to transitioning from a linear economic model to one that is based on renewable resources. Therefore, a discussion of bio-oil production processes and their applications, as well as the significance of this waste-recovery route for sustainable development, is paramount.

16.2 Bio-oil

Since it is imperative to reduce dependence on oil and make efficient utilization of biomass residues, bio-oil has emerged as a promising alternative in the transition to a more sustainable energy matrix. As conversion and refining technologies have advanced, bio-oil has garnered significant attention as a renewable alternative to fossil fuels. Its applications include energy generation, biofuel production, and the production of chemical compounds of industrial interest (Hu and Gholizadeh 2020; Machado et al. 2022).

The composition of bio-oil depends on the feedstock utilized, feedstock pretreatment, production process, operating conditions, and upgrading techniques. Resulting from the thermochemical degradation of industrial, animal, forestry, and agricultural waste, bio-oil is a dark-colored, highly complex organic liquid containing various oxygenated organic compounds (Ilyin and Makarova 2022). Its composition includes organic acids, alcohols, ketones, sugars, anhydro sugars, phenols, esters, aldehydes, pyrolytic lignin, and furans (Das 2022; Drugkar et al. 2022; Ilyin and Makarova 2022; Machado et al. 2022), along with a significant fraction of water, which can account for 20–30% of its total volume (Hu and Gholizadeh 2020). The high oxygen content of bio-oil can result in corrosiveness, non-volatility, thermal instability, and immiscibility with fossil fuels, in addition to a tendency to polymerize when exposed to air (Kumar et al. 2025). Therefore, to improve bio-oil characteristics, upgrading processes are employed to make it more stable, less corrosive, and with a higher calorific value, increasing its compatibility with engines and fossil fuels (Jacobson et al. 2013; Kumar et al. 2025). For example, some methods have been investigated for upgrading bio-oil, such as emulsification, hydrodeoxygenation, catalytic cracking, esterification, solvent addition, hot vapor filtration, and supercritical fluid (Singh et al. 2023). Thus, the bio-oil production chain may require an upgrading step in addition to the thermochemical process used to obtain raw bio-oil.

16.3 Bio-oil Production Processes

Bio-oil, a renewable liquid fuel, can be produced through various thermochemical processes, with the main techniques being fast pyrolysis, hydrothermal liquefaction, and solvent liquefaction. These processes convert solid biomass into bio-oil but differ significantly in their operating conditions and the characteristics of the final product (Chang 2018). Each method has distinct operational parameters, advantages, and challenges, which influence the yield, quality, and potential applications of bio-oil. The key factors affecting the production process include temperature, reaction time, particle size, and biomass composition. A detailed comparison of these processes, including feedstock types, bio-oil yields, main characteristics, and applications, is presented in Table 16.1.

16.3.1 Pyrolysis

Pyrolysis is one of the most widely used methods for bio-oil production, being a thermochemical conversion process of biomass. This process occurs in an oxygen-free environment under controlled temperature, pressure, and residence time conditions, leading to the breakdown of complex hydrocarbon molecules in biomass into simpler gaseous, liquid, and solid fractions (Basu 2013).

Table 16.1 Examples of bio-oil production processes

Process	Feedstock	Bio-oil yield (%)	Main characteristics	Proposed application	By-products	References
Fast pyrolysis	*Sorghum bicolor*	24–25	High carbon content (46.4 wt%) Low water content (9.9 wt%) Presence of acids, phenols, ketones, and furans Heating value of ~24 MJ/kg	Fuel for energy generation Precursor for chemical platform production	Biochar Gases (CO, CO_2, H_2, CH_4, and light hydrocarbons)	Ajikashile et al. (2023)
	Pennisetum glaucum	24.4	High carbon content (45.5 wt%) Water content of 14 wt% Presence of acids, phenols, ketones, and furans Heating value of ~24 MJ/kg	Fuel for energy generation Precursor for chemical platform production	Biochar Gases (CO, CO_2, H_2, CH_4, and light hydrocarbons)	Ajikashile et al. (2023)
	Pine cone, tea bush, walnut shell	69.8	Heating value of 24.66 MJ/kg Low ash content (1.02 wt%) Presence of oxygenates (ketones, furans, phenols)	Liquid fuel for diesel engines Raw material for chemical products	Biochar Gases	Kar et al. (2023)
	Soybean hulls	38–45	Presence of oxygenates (ketones, furans, phenols) High acid content Low heating value	Liquid fuel after additional treatment to reduce oxygenates and acids	Biochar	Toro-Trochez et al. (2022)
	Microalgae (e.g., *Nannochloropsis Chlorella*, *Spirulina*)	17–72	Presence of hydrocarbons, aromatic compounds, phenols, ethers, and nitrogenous compounds Variable heating value (20–35 MJ/kg)	Liquid fuel after additional treatment to reduce oxygenates and nitrogenous compounds	Biochar Gases (CO, CO_2, CH_4, H_2)	Brindhadevi et al. (2021)

Hydrothermal liquefaction	Orange peels	28.4	Heating value of 32 MJ/kg Presence of hydrocarbons (alkanes, alkenes, aromatics), esters, ketones, and alcohols	Marine fuel or furnace oil Substitute/blend for diesel and biodiesel after additional treatment	Biochar Gases (mainly CO_2 and CO)	Divyabharathi and Subramanian (2021)
	Microalgae (e.g., *Nannochloropsis*, *Chlorella*)	20–66	Variable heating value (27–36 MJ/kg) Presence of hydrocarbon content, phenols, amides, esters, and nitrogenous compounds	Marine fuel or furnace oil Substitute/blend for diesel and biodiesel after additional treatment	Gases (CO_2, CO, CH_4, H_2)	Brindhadevi et al. (2021)
	Cassava pulp	23.6	High carbon (30.38%) and hydrogen (5.68%) content Low sulfur content (0.042%) H/C ratio comparable to biodiesel and petroleum	Renewable liquid fuel Potential substitute for diesel or biodiesel	Solid residues Wastewater	Nonchana and Pianthong (2020)
Solvent liquefaction	Lignocellulose (Glucose, Xylan, Lignin)	63.6	High level of ester content Low hydrocarbon content Presence of phenolic compounds from lignin	Production of ester- and phenol-derived chemicals	Solid residues Wastewater	Du et al. (2023)
	Oak wood	51.8	High phenolic compound content Variable heating values (22.1–35 MJ/kg)	Production of phenol-formaldehyde resins and adhesives	Solid residues Wastewater	Öcal and Yüksel (2023)
	Microalgae	51	Oxygenated compounds (~50%) and nitrogenous compounds (~18%) Low nitrogen content (1.2–1.5 wt%) Heating value of 32–33 MJ/kg	Liquid fuel with potential for refining into advanced biofuels	Solid residue	González-Gálvez et al. (2020)

Depending on the variation of parameters, pyrolysis can be categorized into slow, fast, and flash pyrolysis. However, fast pyrolysis is the predominant method for bio-oil production (Dincer and Ishaq 2022). Fast pyrolysis of biomass occurs at temperatures around 500 °C, with a residence time ranging from 0.5 to 5 s, typically within 1 s. It produces vapors that, after condensation, maximize bio-oil yield. The rapid heating, followed by immediate cooling, decomposes the biopolymeric fractions (hemicellulose, cellulose, and lignin) and condenses the volatile intermediates into bio-oil. Fast pyrolysis yields approximately 60% bio-oil, making it an efficient conversion method (Imtiaz Anando et al. 2023). Although wood is the most studied raw material, a wide variety of agricultural, forestry, and even industrial residues, such as straw, nut shells, sugarcane bagasse, and sewage sludge, have been successfully tested as fast pyrolysis feedstocks (Gautam and Chaurasia 2020).

Fast pyrolysis offers advantages, such as rapid biomass conversion into liquid products, achieving higher yields than slow pyrolysis. However, some challenges arise because bio-oil is not miscible with petroleum oils, contains water in the range of 20%–30%, has a lower heating value, is acidic, and has high viscosity. Thus, bio-oil requires upgrading processes to improve its stability (Santos et al. 2024).

Various types of reactors are employed in fast pyrolysis, each with its particularities. The fluidized bed reactor is one of the most common, using fine biomass particles to ensure rapid heat transfer and high bio-oil yields (Ly et al. 2015). The circulating fluidized bed reactor offers gains in energy efficiency and has the advantage of keeping the catalyst continuously activated (Mendes et al. 2023). Finally, the auger reactor is compact and does not require a transport gas, making it ideal for small scales. In this system, biomass is transported through a screw, ensuring precise control of temperature and residence time (Brassard et al. 2017).

Regarding catalytic pyrolysis, it comprises the utilization of catalysts in the process to improve bio-oil quality, reducing the formation of undesirable compounds and increasing product stability. Fast catalytic pyrolysis utilizes catalysts that promote cracking, dehydration, and deoxygenation reactions, resulting in bio-oil with lower oxygen content and higher heating value. This method is particularly promising for producing high-quality biofuels but still faces challenges related to catalyst cost and durability (Rahman et al. 2018).

Industrially, fast pyrolysis of woody biomass is rapidly advancing toward commercial scale, with operational plants in Europe and North America. The use of fast pyrolysis bio-oil is already a viable alternative to replace fossil fuels in heating applications, particularly in boilers, engines, and turbines (Oasmaa et al. 2015). Technologies for producing bio-oil from pyrolysis are at technology readiness levels (TRL) of 3–7, demonstrating that there are already functioning prototypes (Vincenzo et al. 2022). Examples of pre-commercial applications include the Empyro and Ensyn plants, which operate based on this technology.

16.3.2 Hydrothermal Liquefaction

Hydrothermal liquefaction (HTL) is a thermochemical conversion process that transforms wet biomass into liquid fuel, gas, and char. In this process, biomass is subjected to a hot, pressurized water environment, with controlled temperature and pressure ranging from moderate to high levels for a predetermined period. This leads to the breakdown of complex polymers into bio-oil. The temperature typically ranges from 280 to 370 °C, and the pressure varies between 10 and 25 MPa, with a reaction time of 5–120 min. Since HTL occurs in a liquid medium, one of its key advantages is that the biomass does not require drying, making it particularly suitable for wet feedstocks (Mahadevan et al. 2025). A significant advantage of HTL is its ability to utilize wet biomass, and various feedstocks have been extensively studied. This process can effectively convert wet biomass, such as algae, sewage sludge, wood, food waste, and other organic wastes, into bio-oil (Cronin et al. 2022).

HTL converts biomass into bio-oil through a series of chemical reactions. Initially, partial hydrolysis and depolymerization occur, breaking down long carbon chains into smaller monomers. Subsequently, dehydration and decarboxylation reactions play a crucial role in bio-oil production by removing oxygen in the form of water and carbon dioxide (Gong et al. 2024). The resulting bio-oil exhibits a relatively low oxygen content (Mahadevan et al. 2025) and a high heating value, typically ranging from 28 to 38 MJ/kg (Nagappan et al. 2021). The yield of products from hydrothermal technologies depends on several factors, including feedstock type, catalyst, temperature, and pressure conditions. According to Akhtar and coworkers, bio-oil yield can vary from 23.3% to 68.4% (Akhtar and Amin 2011).

Currently, HTL is advancing from laboratory scale (TRL 4-5) to pilot-industrial scale (TRL 5-6), with some projects nearing demonstration. However, challenges remain, including reactor optimization and high capital costs (Vincenzo et al. 2022). Additionally, there are technological gaps that need to be addressed for full commercialization. Examples of practical applications include technologies developed by Licella, Steeper Energy, and Genifuel, which have explored HTL at commercial scales.

16.3.3 Solvent Liquefaction

Solvent liquefaction employs organic solvents or solvent mixtures to achieve results similar to other bio-oil production methods. This process utilizes high temperatures, elevated pressures, and the presence of a solvent to break down the polymeric structure of biomass, converting it into liquid components. Solvent liquefaction is particularly advantageous for treating wet biomass, where water removal would be economically unfeasible, and it serves as a promising alternative to conventional methods, with the added benefit of reducing oxygenated compounds in the bio-oil (González-Gálvez et al. 2020; Öcal and Yüksel 2023).

Catalysts, such as alkalis or specific additives, can be incorporated into the process to improve depolymerization efficiency or direct the formation of desired products (Du et al. 2023). For instance, González-Gálvez and coworkers demonstrated that the conversion of wild microalgae from wastewater treatment into bio-oil using tetralin as a hydrogen-donor solvent achieved yields of up to 58%. Tetralin stabilized free radicals and prevented lignin repolymerization (González-Gálvez et al. 2020). In another study, two hydrocarbon liquids were blended and employed as solvents (CAS 64742-94-5 and CAS 64741-59-9) for the continuous liquefaction of pine waste, yielding 51.2% bio-oil with low oxygen content (23.2%) and high compatibility with petroleum streams (Haverly et al. 2018). Liu and coworkers investigated solvothermal liquefaction of kitchen waste using bimetallic Fe-Ni/Al_2O_3 catalysts and recycled ethanol, achieving a high bio-oil yield of 64.9% with enhanced hydrocarbon (13.52%) and ester (69.86%) content (Liu et al. 2025). However, despite these promising results, key challenges persist, including the optimization of continuous processes (Haverly et al. 2018), the reduction of operational costs (Liu et al. 2025), and the improvement of solvent recovery efficiency (Mullen et al. 2022).

16.4 Bio-oil Applications

Considering its characteristics, bio-oil has broad application potential, as depicted in Fig. 16.1. However, its exact composition depends on the type of raw material and the conditions of the thermochemical process used, which must be optimized to enhance its use as a biofuel or as a precursor in the production of chemicals and advanced materials.

One of the most studied applications of bio-oil is its use as fuel. Bio-oil has traditionally been used as a fuel mainly in static artifacts, such as boilers, furnaces, engines, and electric generators (Ilyin and Makarova 2022). Nonetheless, some of its properties—acidity, density, viscosity, pH, molecular weight, ash content, calorific value, and the presence of oxygenated organic compounds derived from biomass breakdown—pose challenges to its direct use in fuel production (Drugkar et al. 2022; Ilyin and Makarova 2022). To convert bio-oil into fuel, the complete removal of oxygenated compounds is necessary. Technologies such as hydrotreating, deoxygenation, catalytic cracking, catalytic hydrogenation, steam reforming, esterification, and emulsification can be employed to enable this conversion (Drugkar et al. 2022; Hu and Gholizadeh 2020). For instance, a significant improvement was reported in pyrolysis bio-oil characteristics using the thermo-catalytic reforming method (Khan et al. 2022). However, to make bio-oil a viable liquid fuel, further research into production technologies is essential to enhance its cost-competitiveness with conventional fuels, ultimately generating a new sustainable and profitable energy source (Hu and Gholizadeh 2020; Kumar et al. 2023).

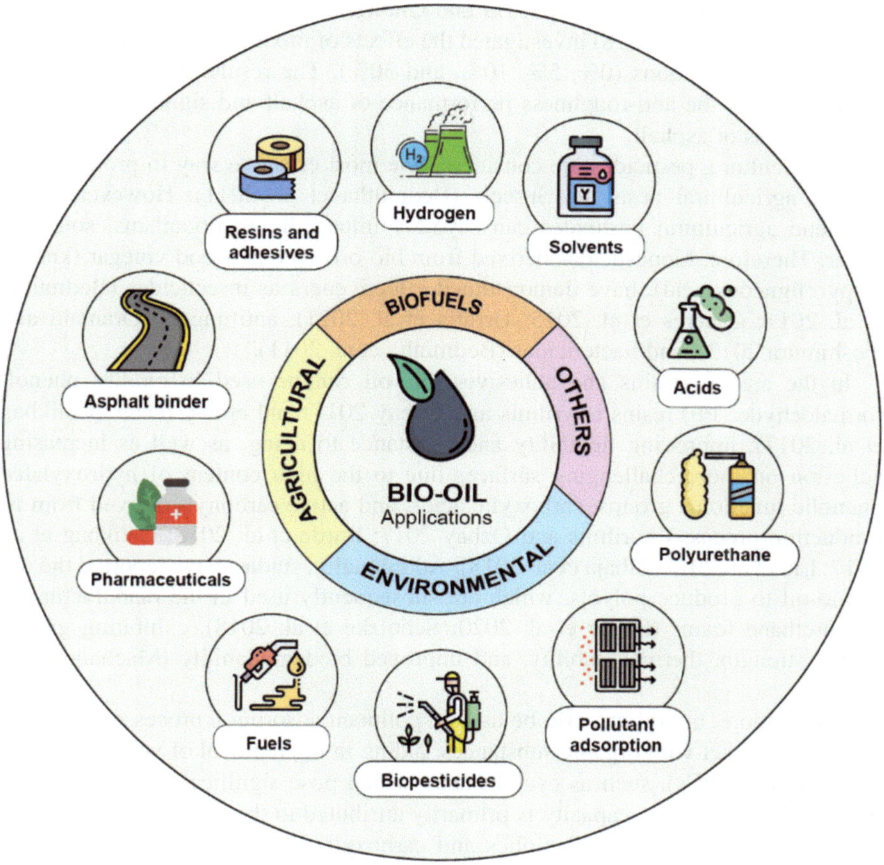

Fig. 16.1 Overview of potential applications of bio-oil

In contrast, bio-oil can also be used as a feedstock for hydrogen production through processes such as steam gasification of biomass and steam reforming of pyrolysis bio-oil (Kumar et al. 2017; Li et al. 2012; Sarkar and Kumar 2010). Despite its potential, challenges remain, including the high chemical complexity of bio-oil, coke formation, and the need for deactivation-resistant catalysts, which must be addressed to make this technology economically viable (Hu and Gholizadeh 2020). Given the ongoing technological advancements, the conversion of bio-oil into hydrogen could play a crucial role in transitioning to a more sustainable energy matrix.

In addition to its application in the fuel and energy sector, bio-oil can also serve as a renewable alternative to petroleum-based asphalt binders (Bhavinlal and Venudharan 2024; Flávia Justino Uchoa et al. 2021; Wang et al. 2021). Compared with conventional binders, bio-oil-based binders have a higher oxygen content due to the presence of phenolic compounds and pyrolytic lignin, which result from lignin degradation (Athparia et al. 2023; Hu and Gholizadeh 2020). A study by Gao and coworkers (Gao et al. 2018) investigated the effects of mixing bio-oil with asphalt at different concentrations (0%, 5%, 10%, and 30%). The results demonstrated that bio-oil reduces the anti-roughness performance of asphalt and significantly affects the mass loss of asphalt.

In agriculture, pesticides are considered the most effective way to protect crops against agricultural pests and insects (Bedmutha et al. 2011). However, using synthetic agricultural pesticides can severely impact living organisms, soil, and water. Therefore, biopesticides derived from bio-oil, such as wood vinegar (known as pyroligneous acid), have demonstrated effectiveness as insecticides (Bedmutha et al. 2011; Cáceres et al. 2015; Urrutia et al. 2021), antifungals (Oramahi and Yoshimura 2013), and bactericides (Bedmutha et al. 2011).

In the area of resins and adhesives, bio-oil can be used to modify phenol-formaldehyde (PF) resins (Ayrilmis and Özbay 2017) and epoxy resins (Celikbag et al. 2017), improving flexibility and resistance to aging, as well as increasing adhesion on more challenging surfaces due to the high content of hydroxylated phenolic functional groups, carboxylic acids, and active carbonyls derived from its production process (Ayrilmis and Özbay 2017; Barde et al. 2018; Celikbag et al. 2017; Liu et al. 2017; Sibaja et al. 2018). Additionally, studies have reported the use of bio-oil to produce polyols, which are subsequently used in the manufacture of polyurethane foams (Saffar et al. 2020; Schulzke et al. 2018), exhibiting greater tensile strength, thermal stability, and improved biodegradability (Machado et al. 2022).

Furthermore, bio-oil can also be used in pollutant adsorption processes due to its ability to interact with organic substances, aiding in the removal of volatile organic compounds (VOCs), such as cyclohexane, which pose significant risks to human health. This adsorption capacity is primarily attributed to the complex composition of bio-oil, which includes phenolics and carboxylic acids—such as pentadecylic acid, myristic acid, lauric acid, oleic acid, margaric acid, and tricosanoic acid—that enhance its affinity for various pollutants. Additionally, due to its physicochemical properties, bio-oil can play a crucial role in the development of sustainable strategies for pollutant mitigation, reducing toxic effects on human health, and contributing to environmental recovery (Azizi et al. 2023).

Although the applications of bio-oil in the pharmaceutical area are still little explored, the polar fraction generated in the extraction of pyrolytic lignin has great potential for application as an antimicrobial (Adfa et al. 2020; Dias et al. 2023; Firouzbehi et al. 2021) and antioxidant (Dias et al. 2023; Valois et al. 2024) agent due to the presence of phenolic compounds, carboxylic acids, and anhydrous sugars (Dias et al. 2023; Zadeh et al. 2020). With further studies to explore the mechanisms of action behind its antioxidant and antimicrobial effects, bio-oil may present a sustainable and economical alternative to synthetic antioxidants.

Regarding acids and solvents, bio-oil has the potential to generate a variety of valuable chemicals, such as acetic acid, phenols, styrene, ethanol, and xylose (Athparia et al. 2023; Drugkar et al. 2022; Machado et al. 2022; Park et al. 2016; Zadeh et al. 2020), which serve as important monomers for the synthesis of various plastics. This contributes to reducing environmental issues caused by nondegradable plastics while enabling the production of sustainable chemicals and materials. The resulting environmental, economic, and social benefits support a more ecological and circular approach to material production.

Overall, bio-oil has been demonstrated to possess versatile and promising properties that render it suitable for a range of applications. Its complex chemical composition and capacity for modification give bio-oil significant potential for application across various industrial sectors. However, challenges remain regarding its stability and compositional structure, requiring the development of new technologies to enhance its viability. Addressing these challenges is essential to ensuring the economic competitiveness and large-scale applicability of bio-oil, thereby positioning it as a strategic resource for the energy transition and environmental sustainability.

16.5 Bio-oil in a Biorefinery Context

Processes to convert waste biomass into bio-oil can serve as a strategy to develop cascading systems within a circular economy framework. Residues can supply feedstocks for promising sustainable alternatives in biorefineries, such as fuels and industrial chemicals (Selvarajoo et al. 2022). From the perspective of integrating waste management strategies with decarbonization efforts, this approach becomes even more significant, as bio-oil can be considered a renewable and low-carbon fuel (Njoku et al. 2024).

The carbon footprint of bio-oil production demonstrates the potential of this process. Bio-oil production processes using reed biomass showed a negative carbon footprint of up to -1301.06 kg CO_{2eq}. The combination of torrefaction and pyrolysis reduced emissions by 176.13 kg CO_{2eq} compared with simple pyrolysis (-1124.93 kg CO_{2eq}) and improved the energy efficiency of the process, making bio-oil a viable solution for reducing energy consumption and emissions in biorefineries (Zhan et al. 2025).

This reinforces the potential of bio-oil as a renewable and low-carbon resource, aligned with global decarbonization goals, and it can be incorporated into cascade processes within a sustainable and economical waste management system to reduce landfill disposal (Biney et al. 2025). Considering global scenarios that emphasize the relevance of searching and integrating technologies for waste utilization, ambitious goals aim to recycle or reuse approximately 65% of urban waste by 2035 (Royuela et al. 2024). Recently, polystyrene waste has been explored for conversion into bio-oil, synthesis gas, and biochar through catalytic co-pyrolysis (ZSM-5) with sewage sludge, showing improvements in the quality of bio-oil due to the addition of the catalyst (Xu et al. 2024). In this context, a model was proposed for electricity generation (20,863 kWh/d) from 10 tons/day of waste (Xu et al. 2024). Another study demonstrated the recovery of bio-oil from cardboard, with a yield of approximately 47%, in addition to biochar production (~10%) (Sotoudehnia et al. 2020).

From that perspective, there are also opportunities for exploration of other biomass wastes as bio-oil feedstocks, such as cassava peel (Lopes et al. 2024), sugarcane bagasse (Kopperi and Venkata Mohan 2023), sewage sludge (O'Boyle et al. 2024; Xu et al. 2024), wheat straw (O'Boyle et al. 2024), and fruit peels (Selvarajoo et al. 2022). For instance, in a cascade biorefinery concept, after exploring sugarcane or ethanol, the residue from sugarcane bagasse was processed via catalytic hydrothermal liquefaction to obtain bio-oil (~28% of the fraction with 11.6% hydrocarbons), biochar, and bio-crude, with the latter being converted into biohydrogen via dark fermentation (Kopperi and Venkata Mohan 2023). In another scenario, cassava peels were subjected to pyrolysis with nickel nanoparticles as catalysts, resulting in the recovery of biochar and bio-oil. The bio-oil was then evaluated in comparison to conventional diesel, showing favorable results in terms of viscosity, density, pH, flash point, and heating value (Egbosiuba 2022).

Exploring other wastes and different applications can also be a strategy for a sustainable route that integrates waste management with the production of high-value products, enhancing the economic viability of development (Royuela et al. 2024; Xu et al. 2024). However, for industrial implementation, the economic analysis of converting waste into bio-oil and other bioproducts is crucial. The economic feasibility of the process depends on a careful evaluation of the market prices of bioproducts, as well as on the reduction of operating costs through the valorization of all the fractions generated. Thus, considering the integration of environmental sustainability and economic viability, pyrolysis was proposed as a promising technology within the circular economy, with long-term benefits for waste reduction and energy generation (Biney et al. 2025). For instance, the economic analysis of pyrolysis and co-pyrolysis of dehydrated sewage sludge demonstrated the relevance of each stage of the process to ensure scalability. The choice of co-substrate and the quality of products such as biochar and bio-oil directly impacted profitability. For example, in the co-pyrolysis of sludge with sawdust, the annual biochar revenue ranged between 0.5 and 1.3 million Canadian dollars (CAD), while bio-oil generated between 0.3 and 0.4 million CAD. In contrast, co-pyrolysis with wheat straw showed a lower contribution of biochar and bio-oil to the total revenue, making the process economically unfeasible (O'Boyle et al. 2024). Such a variation highlights the importance of adjusting operational conditions to maximize the production of high-value bioproducts and ensure the financial sustainability of the project. These details are critical for optimizing large-scale processes, requiring careful analysis to maximize yields and ensure economic feasibility over time.

Within the optimization issue, the use of catalysts in bio-oil production to improve conversion yield and bio-oil characteristics has been a research topic of growing interest. For instance, the development of catalyzed pyrolysis plants can yield good returns if the quality of products, such as bio-oil, is optimized to meet market demands. Because of the fragmentation of cellulose, hemicellulose, and lignin during thermochemical conversion, the composition of bio-oil directly depends on the pyrolysis process conditions and the feedstock composition. Temperatures above 560 °C, coupled with the use of catalysts, tend to produce a higher phenolic content in bio-oil, which is desirable for chemical applications. For pharmaceutical applications, biomass with higher carbon content and volatile matter and lower ash content may be preferable, generating higher levels of ketones and aldehydes (Chen et al. 2025). Thus, to produce hydrocarbon-rich bio-oil, it is essential to develop multifunctional catalysts.

Another topic to be considered is the bio-oil upgrade. While processes such as catalytic cracking, hydrodeoxygenation, and supercritical fluids show potential for upgrading bio-oil, there are technological challenges, such as high cost, catalyst deactivation, low durability, and reactor clogging, which remain a research area likely to increase interest and consequently improve economic viability with technological maturity (Kumar et al. 2025). The use of bio-oil as aviation fuel, for example, has been explored, highlighting the importance of searching for upgrading unit operations to improve bio-oil quality (Lahijani et al. 2022).

Accordingly, the shift from a linear to a circular economy inevitably involves technological development and maturation. The transition from a waste management paradigm focused on disposal to one that is centered on sustainability, intending to maximize utilization, is a strategy for environmental impact reduction that is being promoted with increasing frequency in both academic and corporate settings. In this scenario, the potential for bio-oil production is evident, especially because it allows the use of a range of feedstocks and, within the same unit operation flow, the production of other high-value bioproducts. This process is therefore a strong candidate for implementation within the biorefinery concept.

16.6 Advances in the Bio-oil Field

As discussed in previous sections, bio-oil is one of the main products of biomass pyrolysis or hydrothermal liquefaction, and its quantitative and qualitative properties depend on several factors (Srifa et al. 2019). The main advances in bio-oil production are focused on processes, pretreatments, and some utility models used to obtain bio-oil. To assess scientific advances and innovation in the bio-oil field, a patent search was carried out in the Derwent database for the period 2020–2025 using the algorithm "*((TS=(bio-oil)) and =(pyrolysis)) or (liquefaction)*". A total of 85 documents were found. The main countries holding the technology are China, India, Korea, and the United States, with 55%, 13%, 9.4%, and 6% of deposits, respectively. However, around 90% of the patents were filed by Universities and Research Institutes, indicating that this is a developing technology. In Fig. 16.2, it is possible to see that the largest number of patents was filed between 2022 and 2023, with 21 patents in each year. In 2024 and 2025, 19 and 2 patents have already been filed, respectively. This can be explained by the fact that the time for patent publication after submission is approximately 18 months (Vandenberghe et al. 2021).

Considering pretreatment processes, patent CN114806616A is an example that proposes a biomass pretreatment method to improve the quality of bio-oil, called "low co-solvent assisted freeze–thaw." The pretreatment begins by freezing the solid–liquid mixture (biomass:water) in a 1:15 ratio in a liquid nitrogen environment

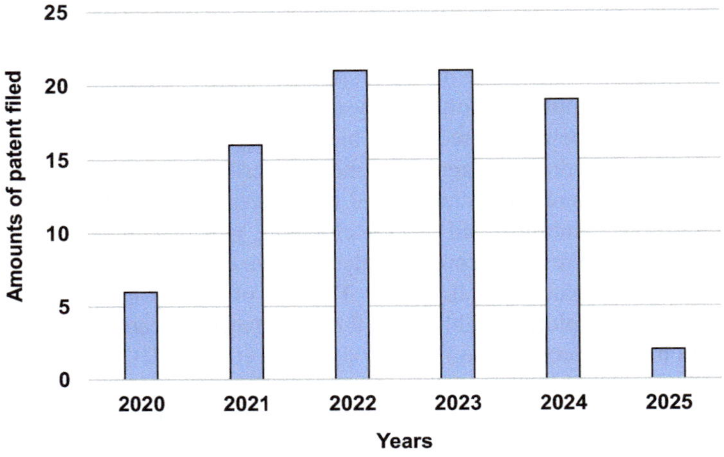

Fig. 16.2 Number of patents filed between 2020 and 2025

at −196 °C for 1–4 h, after which the mixture is thawed, and the process is carried out 4 times. The instantaneous freezing and thawing process makes the biomass extremely brittle, facilitating its grinding or pulverization. Next, pretreatment with the eutectic mixture (imidazole and choline chloride, 1:1-4) is carried out at a temperature between 105 and 125 °C. Imidazole helps in the separation of lignin since it interacts with lignin through hydrogen bonds and interactions with aromatic rings, which weakens the lignin bond network (Savian et al. 2025). These interactions increase the solubility of lignin, facilitating its extraction from biomass. When combined, these compounds can synergistically improve the solubilization of lignocellulosic biomass, reducing the amount of lignin and making cellulose and hemicellulose more accessible for bioconversion. Finally, the solids are separated, washed with a 50% ethanol solution, and dried to be used in pyrolysis. According to the inventors, the pretreatment significantly increases the levoglucan content in the bio-oil. Levoglucan contributes to greater efficiency in biomass conversion. In addition, levoglucan can be transformed into value-added compounds (Xiwei et al. 2022).

Patent CN117568060A also presents an electrochemical pretreatment method of dissolved salt to improve bio-oil quality. The pretreatment starts grinding with the raw materials (approximately 7 h), and then the biomass is mixed with a solution of molten salts (lithium nitrate, sodium nitrate, and potassium nitrate). The electrochemical process is carried out with a graphite cathode, a platinum anode, a voltage of 5–15 V, at a temperature of 160 ± 5 °C, and for a period of 1–5 h. The ratio of biomass to molten salt solvent system is 1.5:20. After the reaction is complete, the solid and liquid are separated under the conditions (1000 °C for 30 min), and the solid is washed with alcohol solution and subsequently dried at temperatures between 80 and 105 °C for 24–48 h. According to the inventors, the electrochemical molten salt pretreatment method promotes pyrolysis and saccharification of biomass feedstock and improves the levorotatory glucan content in bio-oil. This increases chemical stability, reducing the degradation of bio-oil over time. It also decreases acidity and the presence of undesirable compounds, making bio-oil more suitable for energy use. Since glucans are rich in carbon and hydrogen, they can increase the calorific value of bio-oil. Moreover, fast pyrolysis can thermally break down biomass instantly and convert it into high-value-added chemicals (Xiwei et al. 2023).

Regarding bio-oil production, patent CN118931575A presented a process for producing bio-oil by microwave pyrolysis from residual fat. Initially, fat, microwave absorbent (silicon carbide), copper salt, and the porous carbon material are mixed and subjected to a hydrothermal reaction at a temperature between 150 and 180 °C for 2–5 h. After completing the reaction, the mixture is filtered to separate the solid. This is calcined at a temperature between 500 and 700 °C for 2–5 h under a modified atmosphere to obtain a porous copper/biomass carbon composite material. The pyrolysis is carried out in a microwave with a power of 1000–2000 W and a temperature of 300–500 °C for 10–30 min. According to the inventors, microwave-assisted pyrolysis is one of the most attractive thermochemical processes developed in recent years. These processes significantly reduce pollutant emissions and simultaneously provide uniform and rapid heating. The preparation of bio-oil by microwave pyrolysis of waste oils is rarely reported, and it is necessary to build an efficient process (Junjie et al. 2024). Other patents are summarized in Table 16.2, reporting processes for obtaining bio-oil using different types of biomass and conditions.

Concerning companies that own bio-oil production technology, one can cite: Abundia Biomass-To-Liquids Limited, Green Carbon Shanghai Technology, Steeper, Yingna Green Carbon Shanghai Technology, Valtmet, and Fortum. It is worth noting that Valmet built the first commercial-scale bio-oil plant in Joensuu, Finland, for the company Fortum in 2015. The plant has an annual production capacity of 50,000 tons of bio-oil. The plant employs integrated pyrolysis technology and fluidized bed boilers. The plant mainly uses lignocellulosic biomass as raw material, although used oils can also be used (Valmet-Forward 2015). Another industry worth mentioning is the Bioenergy Association, which produces bio-oil using Hydrofaction technology, which was developed by the company Steeper Energy (Bioenergy-Association 2021). Another industry that uses Hydrofaction® technology is the multinational Statkraft, with a presence in more than 20 countries. The technology was implemented in 2022 at Statkraft Tofte in Norway, and the plant can produce around 4000 L/day of bio-oil (Statkraft 2015).

Accordingly, advances in bio-oil production continue to evolve, driven by innovations in processes, pretreatments, and new technologies. The patents registered between 2020 and 2025 demonstrate the global interest in developing this technology, with countries such as China, India, Korea, and the United States standing out. In addition, companies and research institutes continue to invest in more efficient and sustainable methods to improve the quality and viability of bio-oil as an alternative energy source.

Table 16.2 Some examples of patented bio-oil production processes

Patent code	Patent title	Novelty	References
CN112852457A	Method for rapidly evaluating the quality of lignin pyrolysis bio-oil based on free radical detection and application	Method for rapid quality assessment of lignin pyrolysis bio-oil based on free radical detection. The solid pyrolysis product is collected, washed with organic solvent (methanol, ethanol, propanol, isopropanol, acetone, ether, ethyl acetate, n-hexane) to remove the attached bio-oil, and dried to obtain lignin carbon. The lignin carbon is ground into powder with a particle size of less than 0.5 mm. The lignin carbon powder after grinding is placed in a quartz tube, and electron paramagnetic resonance spectroscopy is used to detect the spin content of free radicals. The method allows noncontact detection and reliable, economical, and rapid quality assessment of lignin pyrolysis bio-oil without an extraction and separation process by a simple process	Rui et al. (2020)
BR102021005872-A2	Production processes of bio-coal, bio-oil, and briquettes from residual biomass of baru and Jatobá fruits, products obtained, and uses	The patent presents a process to produce bio-oil from residual biomass of baru (*Dipteryx alata* Vog) and Jatobá (*Hymenaea stigonocarpa* Mart). The process begins with drying the biomass until it reaches 10–12% moisture content. Pyrolysis is then carried out at a temperature of between 500 and 550 °C for 8 h. The products obtained have applications in different niches of the energy market. According to the inventors, the process provides products with significant energy value, superior to woody and lignocellulosic biomass, and with energy equivalent to the energy value of some fossil fuels	Gouvêa et al. (2021)

MY193194A	A continuous process of decomposing biomass to produce bio-oil by an improved fluidized bed pyrolysis system	The patent presents a continuous process of biomass decomposition for bio-oil production by an improved fluidized bed pyrolysis system. This consists of a feeding system comprising a speed-adjustable hopper operated by a motor. The fluidized bed pyrolyzer was equipped with four electric jacket heaters with PID (proportional–integral–derivative) controllers, a coal separation system with two cyclones to collect fine coal particles, a fractional condensation system comprising three stages of fractional condensation of a series of six fractional condensers, and a filter system with plate filters. The decomposition system improves the fluidized bed pyrolysis system	Yoshimitsu et al. (2022)
IN202241050026-A	Producing bio-oil from seed pods	The patent presented the process for producing bio-oil from Cassia fistula seed pods. The process begins with drying biomass, followed by treatment of the biomass in a phosphoric acid solution at 60 °C and pyrolysis of the biomass at 450 °C with a nitrogen flow of 3 L/min for 30 min. According to the inventors, the method ensures the effective removal of metallic species in the C. fistula seed pods, alters the degradation pathway to increase the yield of pyrolysis oil, and improves the quality of the bio-oil	Muniyappan et al. (2024)
CN114613612-A	Carbon material prepared from bio-oil obtained by catalytic pyrolysis of biomass and a method thereof	The method uses catalytic pyrolysis of biomass to obtain bio-oil and prepare carbon material. First, seaweed char is modified with chemical activators and activated at a high temperature. This modified char catalyzes the fast pyrolysis of biomass, generating bio-oil, which undergoes dry distillation to separate light and heavy components. The heavy components are then carbonized under a nitrogen-enriched atmosphere to form doped carbon materials. The process involves temperatures between 450 and 900 °C and activators such as NaOH and KOH, resulting in carbon materials with specific properties	Yijie et al. (2022)

16.7 Conclusion

As a current and urgent environmental issue, proper waste management requires the implementation of efficient waste valorization processes. In this context, bio-oil production from organic residues can be a promising route. By using thermochemical processes, such as pyrolysis and liquefaction, those worthless materials, which would otherwise be an environmental burden if mismanaged, can serve as a feedstock to produce bio-oil, which can be converted into molecules of industrial interest (e.g., biofuels). Bio-oil production therefore has two main benefits: waste management and avoiding the use of nonrenewable resources (e.g., fossil fuels).

The production of bio-oil from residues is also in line with the concept of the circular economy, where residues are not a problem but a resource. In addition, bio-oil production has the potential to function as a biorefinery. Such an approach would extend the process from a biofuel-only technology to one capable of producing a variety of bio-based products. However, while the technological advances in bio-oil production have been significant, there are still challenges to overcome to consolidate the technology. The efficiency of bio-oil production can vary depending on the composition and characteristics of the waste used, the thermochemical process used and its operating conditions, and the product of interest, which may require an upgrading step to improve the bio-oil characteristics. Therefore, it is essential to continuously search for strategies that make bio-oil production sustainable and economically feasible.

References

Adfa M, Romayasa A, Kusnanda AJ, Avidlyandi A, Yudha S. S, Banon C, Gustian I (2020) Chemical components, antitermite and antifungal activities of Cinnamomum parthenoxylon wood vinegar. J Korean Wood Sci Technol 48:107–116. https://doi.org/10.5658/WOOD.2020.48.1.107

Ajikashile JO, Alhnidi MJ, Parku GK, Funke A, Kruse A (2023) A study on the fast pyrolysis of millet and sorghum straws sourced from arid and semi-arid regions of Nigeria in a twin-screw mixing reactor. Mater Sci Energy Technol 6:388–398. https://doi.org/10.1016/j.mset.2023.03.007

Akhtar J, Amin NAS (2011) A review on process conditions for optimum bio-oil yield in hydrothermal liquefaction of biomass. Renew Sust Energ Rev 15:1615–1624. https://doi.org/10.1016/j.rser.2010.11.054

Athparia M, Bora N, Deka A, Sohtun P, Padhi P, Bhuyan N, Bordoloi NJ, Gogoi L, Kataki R (2023) Non-fuel applications of bio-oil for sustainability in management of bioresources. Environ Sci Pollut Res 2023(1):1–25. https://doi.org/10.1007/S11356-023-31449-W

Ayrilmis N, Özbay G (2017) Technological properties of plywood bonded with phenol-formaldehyde resol resin synthesized with bio-oil. Cerne 23:493–500. https://doi.org/10.1590/01047760201723042351

Azizi M, Abdulrahman YJ, Abdessamad NEH, Azzaz AA, Naguib DM (2023) Valorization and characterization of bio-oil from Salvadora persica seed for air pollutant adsorption. Environ Sci Pollut Res 30:53397–53410. https://doi.org/10.1007/S11356-023-25566-9/FIGURES/9

Barde M, Adhikari S, Via BK, Auad ML (2018) Synthesis and characterization of epoxy resins from fast pyrolysis bio-oil. Green Mater 6:76–84. https://doi.org/10.1680/JGRMA.17.00038/ASSET/IMAGES/SMALL/JGRMA6-0076-F6.GIF

Basu P (2013) Chapter 5—Pyrolysis. In: Basu P (ed) Biomass gasification, pyrolysis and torrefaction, 2nd edn. Academic, Boston, pp 147–176. https://doi.org/10.1016/B978-0-12-396488-5.00005-8

Bedmutha R, Booker CJ, Ferrante L, Briens C, Berruti F, Yeung KKC, Scott I, Conn K (2011) Insecticidal and bactericidal characteristics of the bio-oil from the fast pyrolysis of coffee grounds. J Anal Appl Pyrolysis 90:224–231. https://doi.org/10.1016/J.JAAP.2010.12.011

Bhavinlal K, Venudharan V (2024) Exploring production and performance of popular bio-oil modified asphalts: a state-of-the-art research review. Innov Infrastruct Solut 9:1–23. https://doi.org/10.1007/S41062-024-01716-X/TABLES/6

Biney MA, Gusiatin MZ, Trakal L, Moško J, Pohořelý M (2025) Utilization of co-substrates in municipal sewage sludge co-pyrolysis: yields and characterization of biochar, bio-oil, and syngas, with economic feasibility analysis. J Anal Appl Pyrolysis:107091. https://doi.org/10.1016/j.jaap.2025.107091

Bioenergy-Association (2021) Bronze member in categories Technology developer. Retrieved from: <https://www.bioenergy.org.nz/member/steeper-energy> (March 4, 2025). 1296541

Brassard P, Godbout S, Raghavan V (2017) Pyrolysis in auger reactors for biochar and bio-oil production: a review. Biosyst Eng 161:80–92. https://doi.org/10.1016/j.biosystemseng.2017.06.020

Brindhadevi K, Anto S, Rene ER, Sekar M, Mathimani T, Thuy Lan Chi N, Pugazhendhi A (2021) Effect of reaction temperature on the conversion of algal biomass to bio-oil and biochar through pyrolysis and hydrothermal liquefaction. Fuel 285. https://doi.org/10.1016/j.fuel.2020.119106

Cáceres LA, McGarvey BD, Briens C, Berruti F, Yeung KKC, Scott IM (2015) Insecticidal properties of pyrolysis bio-oil from greenhouse tomato residue biomass. J Anal Appl Pyrolysis 112:333–340. https://doi.org/10.1016/J.JAAP.2015.01.003

Celikbag Y, Meadows S, Barde M, Adhikari S, Buschle-Diller G, Auad ML, Via BK (2017) Synthesis and characterization of bio-oil-based self-curing epoxy resin. Ind Eng Chem Res 56:9389–9400. https://doi.org/10.1021/ACS.IECR.7B02123/SUPPL_FILE/IE7B02123_SI_001.PDF

Chang SH (2018) Bio-oil derived from palm empty fruit bunches: fast pyrolysis, liquefaction and future prospects. Biomass Bioenergy 119:263–276. https://doi.org/10.1016/j.biombioe.2018.09.033

Chen X, Shafizadeh A, Shahbeik H, Nadian MH, Golvirdizadeh M, Peng W, Lam SS, Tabatabaei M, Aghbashlo M (2025) Enhanced bio-oil production from biomass catalytic pyrolysis using machine learning. Renew Sust Energ Rev 209:115099. https://doi.org/10.1016/j.rser.2024.115099

Cherubini F, Strømman AH (2011) Principles of biorefining. Biofuels:3–24. https://doi.org/10.1016/B978-0-12-385099-7.00001-2

Cherwoo L, Gupta I, Flora G, Verma R, Kapil M, Arya SK, Ravindran B, Khoo KS, Bhatia SK, Chang SW, Ngamcharussrivichai C, Ashokkumar V (2023) Biofuels an alternative to traditional fossil fuels: a comprehensive review. Sustain Energy Technol Assess 60:103503. https://doi.org/10.1016/j.seta.2023.103503

Cronin D, Schmidt AJ, Billing J, Hart TR, Fox SP, Fonoll X, Norton J, Thorson MR (2022) Comparative study on the continuous flow hydrothermal liquefaction of various wet-waste feedstock types. ACS Sustain Chem Eng 10:1256–1266. https://doi.org/10.1021/acssuschemeng.1c07214

Das P (2022) Pyrolytic bio-oil—production and applications. In: Advanced biofuel technologies: present status, challenges and future prospects, pp 243–304. https://doi.org/10.1016/B978-0-323-88427-3.00016-7

Dias IA, Horta RP, Matos M, Helm CV, Magalhães WLE, de Lima EA, da Silva BJG, de Muniz GIB, de Cademartori PHG (2023) Exploring the antioxidant and antimicrobial properties of the water-soluble fraction derived from pyrolytic lignin separation in fast-pyrolysis bio-oil. Biomass Convers Biorefinery 14:24333–24344. https://doi.org/10.1007/S13399-023-04561-7/TABLES/2

Dincer I, Ishaq H (2022) Chapter 8—Biomass energy-based hydrogen production. In: Dincer I, Ishaq H (eds) Renewable hydrogen production. Elsevier, pp 249–287. https://doi.org/10.1016/B978-0-323-85176-3.00002-0

Divyabharathi R, Subramanian P (2021) Biocrude production from orange (Citrus reticulata) peel by hydrothermal liquefaction and process optimization. Biomas Convers Biorefin 12:183–194. https://doi.org/10.1007/s13399-021-01383-3

Drugkar K, Rathod W, Sharma T, Sharma A, Joshi J, Pareek VK, Ledwani L, Diwekar U (2022) Advanced separation strategies for up-gradation of bio-oil into value-added chemicals: a comprehensive review. Sep Purif Technol 283:120149. https://doi.org/10.1016/J.SEPPUR.2021.120149

Du C, Yang T, Li B, Cao H, Liu Z, Huang S (2023) Effect of alkali and alkaline earth metals on the liquefaction of lignocellulosic model compounds to prepare bio-oil in ethanol solvent. Energy 278. https://doi.org/10.1016/j.energy.2023.127614

Durak H (2023) Comprehensive assessment of thermochemical processes for sustainable waste management and resource recovery. Processes 11:2092. https://doi.org/10.3390/pr11072092

Egbosiuba TC (2022) Biochar and bio-oil fuel properties from nickel nanoparticles assisted pyrolysis of cassava peel. Heliyon 8:e10114. https://doi.org/10.1016/j.heliyon.2022.e10114

Ellen MacArthur Foundation (2017) The circular economy in detail [WWW Document]. The Circular Economy In Detail. https://www.ellenmacarthurfoundation.org/explore/the-circular-economy-in-detail. Accessed 6 Feb 21

Firouzbehi F, Efhamisisi D, Hamzeh Y, Tarmian A, Oladi R (2021) Pyrolysis acid as sustainable wood preservative against rot fungi. Biofuels Bioprod Biorefin 15:74–84. https://doi.org/10.1002/BBB.2149

Flávia Justino Uchoa A, da Silva Rocha W, Peter Macedo Feitosa J, Lopes Nogueira R, Hellen Almeida de Brito D, Barbosa Soares J, de Aguiar Soares S (2021) Bio-based palm oil as an additive for asphalt binder: chemical characterization and rheological properties. Constr Build Mater 285:122883. https://doi.org/10.1016/J.CONBUILDMAT.2021.122883

Gao J, Wang H, You Z, Hasan MRM, Lei Y, Irfan M (2018) Rheological behavior and sensitivity of wood-derived bio-oil modified asphalt binders. Appl Sci 8(6):919. https://doi.org/10.3390/APP8060919

Gautam N, Chaurasia A (2020) Study on kinetics and bio-oil production from rice husk, rice straw, bamboo, sugarcane bagasse and neem bark in a fixed-bed pyrolysis process. Energy 190. https://doi.org/10.1016/j.energy.2019.116434

Gong M, Jiang W, Wang S, Liu P, Xu F, Wang W, Fan Y (2024) Bio-oil production from hydrothermal liquefaction of algal biomass: effects of feedstock properties and reaction parameters. J Environ Chem Eng 12(5):114010. https://doi.org/10.1016/j.jece.2024.114010

González-Gálvez OD, Nava Bravo I, Cuevas-García R, Velásquez-Orta SB, Harvey AP, Caero LC, Ledesma MTO (2020) Bio-oil production by catalytic solvent liquefaction from a wild microalgae consortium. Biomass Convers. Biorefin 11:2627–2639. https://doi.org/10.1007/s13399-020-00716-y/Published

Gouvêa ADFG, Costa EB, Carvalho AMML, Carneiro ADCO (2021) Production processes of biocoal, bio-oil and briquettes from residual biomass of baru and jatobá fruits, products obtained and uses. BR Patent 102021005872A2:2024–2025

Haverly MR, Schulz TC, Whitmer LE, Friend AJ, Funkhouser JM, Smith RG, Young MK, Brown RC (2018) Continuous solvent liquefaction of biomass in a hydrocarbon solvent. Fuel 211:291–300. https://doi.org/10.1016/j.fuel.2017.09.072

Hu X, Gholizadeh M (2020) Progress of the applications of bio-oil. Renew Sust Energ Rev 134:110124. https://doi.org/10.1016/j.rser.2020.110124

Ilyin SO, Makarova VV (2022) Bio-oil: production, modification, and application. Chem Technol Fuels Oils 58:29–44. https://doi.org/10.1007/S10553-022-01348-W

Imtiaz Anando A, Ehsan MM, Karim MR, Bhuiyan AA, Ahiduzzaman M, Karim A (2023) Thermochemical pretreatments to improve the fuel properties of rice husk: a review. Renew Energy 215:118917. https://doi.org/10.1016/j.renene.2023.118917

Intergovernmental Panel on Climate Change (2022) IPCC — Intergovernmental Panel on Climate Change [WWW Document]. URL https://www.ipcc.ch/. Accessed 24 July 22

Jacobson K, Maheria KC, Kumar Dalai A (2013) Bio-oil valorization: a review. Renew Sust Energ Rev 23:91–106. https://doi.org/10.1016/j.rser.2013.02.036

Joshi NC, Sinha S, Bhatnagar P, Nath Y, Negi B, Kumar V, Gururani P (2024) A concise review on waste biomass valorization through thermochemical conversion. Curr Res Microb Sci 6: 100237. https://doi.org/10.1016/J.CRMICR.2024.100237

Junjie X, Cong W, Zesang L, Hongbin L (2024) Method for preparing bio-oil by microwave pyrolysis of waste oil. CN Patent 118931575A:2024–2025

Kar T, Kaygusuz Ö, Güney MŞ, Cuce E, Keleş S, Shaik S, Owolabi AB, Nsafon BEK, Ogunsua JM, Huh JS (2023) Fast pyrolysis of tea bush, walnut Shell, and pine cone mixture: effect of pyrolysis parameters on pyrolysis crop yields. Sustainability (Switzerland) 15. https://doi.org/10.3390/su151813718

Khan MS, Min Y, Broumand M, Yun S, Hong Z, Thomson MJ (2022) Advancing the application of pyrolysis liquid (bio-oil) by the improvement of its fuel properties by Thermo-catalytic reforming. Energy Fuels 36:4381–4395. https://doi.org/10.1021/ACS.ENERGYFUELS.1C04129/SUPPL_FILE/EF1C04129_SI_001.PDF

Kopperi H, Venkata Mohan S (2023) Catalytic hydrothermal deoxygenation of sugarcane bagasse for energy dense bio-oil and aqueous fraction acidogenesis for biohydrogen production. Bioresour Technol 379:128954. https://doi.org/10.1016/j.biortech.2023.128954

Kumar A, Chakraborty JP, Singh R (2017) Bio-oil: the future of hydrogen generation. Biofuels 8: 663–674. https://doi.org/10.1080/17597269.2016.1141276

Kumar R, Strezov V, Weldekidan H, He J, Singh S, Kan T, Dastjerdi B (2023) Lignocellulose biomass pyrolysis for bio-oil production: biomass pre-treatment methods for production of drop-in fuels. In: Biowaste and biomass in biofuel applications, pp 123–196. https://doi.org/10.1201/9781003265597-8/LIGNOCELLULOSE-BIOMASS-PYROLYSIS-BIO-OIL-PRODUCTION-KUMAR-STREZOV-WELDEKIDAN-SINGH-KAN-DASTJERDI

Kumar V, Sharma N, Abdelaal AS, Chakraborty P, Thomas J, Duhan L, Pasrija R, Dogra S, Jayaraj I (2025) Bio-oil production and catalytic upgrade to value added product: a review on recent technologies. J Energy Inst 118:101880. https://doi.org/10.1016/j.joei.2024.101880

Lahijani P, Mohammadi M, Mohamed AR, Ismail F, Lee KT, Amini G (2022) Upgrading biomass-derived pyrolysis bio-oil to bio-jet fuel through catalytic cracking and hydrodeoxygenation: a review of recent progress. Energy Convers Manag 268:115956. https://doi.org/10.1016/j.enconman.2022.115956

Li Z, Hu X, Zhang L, Lu G (2012) Renewable hydrogen production by a mild-temperature steam reforming of the model compound acetic acid derived from bio-oil. J Mol Catal A Chem 355: 123–133. https://doi.org/10.1016/J.MOLCATA.2011.12.006

Liu Y, Via BK, Pan Y, Cheng Q, Guo H, Auad ML, Taylor S (2017) Preparation and characterization of epoxy resin cross-linked with high wood pyrolysis bio-oil substitution by acetone pretreatment. Polymers 9:106. https://doi.org/10.3390/POLYM9030106

Liu Y, Luo C, Hantoko D, Sun Y, Ye J, Cao H, Yan M (2025) Catalytic solvothermal liquefaction of kitchen waste for the production of high-quality bio-oil: effects of reusing ethanol phase as the reaction medium. J Environ Manag 376:124549. https://doi.org/10.1016/j.jenvman.2025.124549

Lopes T d S, Alves JLF, Delmiro TM, Calixto GQ, de Oliveira KFS, Barbosa A d S, Voigt EL, Melo DM d A, Braga RM (2024) From cassava peel (Manihot esculenta) to hydrocarbon-rich bio-oil: catalytic flash pyrolysis as a new valorization route. Biomass Bioenergy 190:107432. https://doi.org/10.1016/j.biombioe.2024.107432

Ly HV, Kim SS, Woo HC, Choi JH, Suh DJ, Kim J (2015) Fast pyrolysis of macroalga Saccharina japonica in a bubbling fluidized-bed reactor for bio-oil production. Energy 93:1436–1446. https://doi.org/10.1016/j.energy.2015.10.011

Machado H, Cristino AF, Orišková S, Galhano dos Santos R (2022) Bio-oil: the next-generation source of chemicals. Reactions 3:118–137. https://doi.org/10.3390/reactions3010009

Mahadevan V, Subbaiyan N, Kannappan Panchamoorthy G, Jayaseelan A, Palaniappan SK, Siengchin S (2025) Hydrothermal liquefaction of confused waste to bio-oil: a study on elemental and energy recovery. Energy Convers Manag 325. https://doi.org/10.1016/j.enconman.2024.119353

Mendes FL, de Pinho AR, de Almeida MBB, Caramão EB (2023) Biomass pyrolysis in a circulating fluidized bed reactor: evaluating two strategies for the reduction of oxygen content in bio-oil. J Anal Appl Pyrolysis 175. https://doi.org/10.1016/j.jaap.2023.106150

Mullen CA, Strahan GD, Elkasabi Y (2022) A comparison of the solvent liquefaction of lignin in ethanol and 1,4-butanediol. J Anal Appl Pyrolysis 164. https://doi.org/10.1016/j.jaap.2022.105522

Muniyappan D, Gaikwad BH, Ramanathan A (2024) Producing bio-oil from seed pods, by extracting seed pods from Cassia fistula fruit, drying and communicating of seed pods, soaking seed pod particles in aqueous phase bio-oil by-product, filtering, washing, drying, pyrolyzing dried biomass, performing ch 2024–2025

Nagappan S, Bhosale RR, Nguyen DD, Chi NTL, Ponnusamy VK, Woong CS, Kumar G (2021) Catalytic hydrothermal liquefaction of biomass into bio-oils and other value-added products—a review. Fuel 285. https://doi.org/10.1016/j.fuel.2020.119053

Njoku CN, Emori W, Ekerenam OO, Anorondu CC, Innocent CS, Etim I-IN, Ozioko FC, Obike AI, Orga AC, Nkuzinna OC (2024) Blending bio-oil with fossil fuels: opportunities for low-carbon fuel production. Energy Sources Part A Recover Util Environ Eff 46:45–64. https://doi.org/10.1080/15567036.2024.2428393

Nonchana T, Pianthong K (2020) Bio-oil synthesis from cassava pulp via hydrothermal liquefaction: effects of catalysts and operating conditions. Int J Renew Energy Dev 9:329–337. https://doi.org/10.14710/ijred.9.3.329-337

O'Boyle M, Mohamed BA, Li LY (2024) Co-pyrolysis of sewage sludge and biomass waste into biofuels and biochar: a comprehensive feasibility study using a circular economy approach. Chemosphere 350:141074. https://doi.org/10.1016/j.chemosphere.2023.141074

Oasmaa A, Van De Beld B, Saari P, Elliott DC, Solantausta Y (2015) Norms, standards, and legislation for fast pyrolysis bio-oils from lignocellulosic biomass. Energ Fuels 29:2471–2484. https://doi.org/10.1021/acs.energyfuels.5b00026

Öcal B, Yüksel A (2023) Liquefaction of oak wood using various solvents for bio-oil production. ACS Omega 8:40944–40959. https://doi.org/10.1021/acsomega.3c06419

Okolie JA, Epelle EI, Tabat ME, Orivri U, Amenaghawon AN, Okoye PU, Gunes B (2022) Waste biomass valorization for the production of biofuels and value-added products: a comprehensive review of thermochemical, biological and integrated processes. Process Saf Environ Prot 159: 323–344. https://doi.org/10.1016/j.psep.2021.12.049

Oramahi HA, Yoshimura T (2013) Antifungal and antitermitic activities of wood vinegar from Vitex pubescens Vahl. J Wood Sci 59:344–350. https://doi.org/10.1007/S10086-013-1340-8/TABLES/6

Park LKE, Ren S, Yiacoumi S, Ye XP, Borole AP, Tsouris C (2016) Separation of Switchgrass bio-oil by water/organic solvent addition and pH adjustment. Energy Fuels 30:2164–2173. https://doi.org/10.1021/ACS.ENERGYFUELS.5B02537/SUPPL_FILE/EF5B02537_SI_001.PDF

Rahman MM, Liu R, Cai J (2018) Catalytic fast pyrolysis of biomass over zeolites for high quality bio-oil—a review. Fuel Process Technol. https://doi.org/10.1016/j.fuproc.2018.08.002

Royuela D, Martínez JD, Callén MS, López JM, García T, Murillo R, Veses A (2024) Pyrolysis of polystyrene using low-cost natural catalysts: production and characterisation of styrene-rich pyro-oils. J Anal Appl Pyrolysis 182:106690. https://doi.org/10.1016/j.jaap.2024.106690

Rui X, Yuyang F, Chao L, Xiangchen K, Weicong X (2020) Method for rapidly evaluating quality of lignin pyrolysis bio-oil based on free radical detection and application. CN patent 112852457A. Wang 1296541

Saffar T, Bouafif H, Braghiroli FL, Magdouli S, Langlois A, Koubaa A (2020) Production of bio-based polyol from oxypropylated pyrolytic lignin for rigid polyurethane foam application. Waste Biomass Valorization 11:6411–6427. https://doi.org/10.1007/S12649-019-00876-7/FIGURES/8

Santos BSDF, Palacios-Bereche MC, Gallego AG, Nebra SA, Palacios-Bereche R (2024) Energy assessment and heat integration of biofuel production from bio-oil produced through fast pyrolysis of sugarcane straw, and its upgrading via hydrotreatment. Renew Energy 232: 121021. https://doi.org/10.1016/j.renene.2024.121021

Sarkar S, Kumar A (2010) Large-scale biohydrogen production from bio-oil. Bioresour Technol 101:7350–7361. https://doi.org/10.1016/J.BIORTECH.2010.04.038

Savian CM, Woiciechowski AL, Scapini T, Martinez-Burgos WJ, Zevallos Torres LA, Filho AZ, Ramos LP, Soccol CR (2025) Imidazole as a new strategy in the lignocellulosic pretreatment: an application in corn sub-products with a focus on bioenergy production. Biomass Bioenergy 183: 107075

Scapini T, Klanovicz N, Camargo AF, Bazoti SF, Cavali M, Bonatto C, Paliga L, Warken A, Alves SL Jr, Woiciechowski AL, Treichel H (2023) Future of biorefineries from a circular economy perspective. In: Nyström W (ed) The future of biorefineries. Nova Science Publishers

Schulzke T, Iakovleva A, Cao Q, Conrad S, Zabelkin S, Grachev A (2018) Polyurethane foams produced from pyrolysis oil—production and possible application. Biomass Bioenergy 115: 195–202. https://doi.org/10.1016/J.BIOMBIOE.2018.04.006

Selvarajoo A, Wong YL, Khoo KS, Chen W-H, Show PL (2022) Biochar production via pyrolysis of citrus peel fruit waste as a potential usage as solid biofuel. Chemosphere 294:133671. https://doi.org/10.1016/j.chemosphere.2022.133671

Sibaja B, Adhikari S, Celikbag Y, Via B, Auad ML (2018) Fast pyrolysis bio-oil as precursor of thermosetting epoxy resins. Polym Eng Sci 58:1296–1307. https://doi.org/10.1002/PEN.24694

Singh S, Pant KK, Krishania M (2023) Current perspective for bio-oil production from agricultural residues in commercialization aspect: a review. J Anal Appl Pyrolysis 175:106160. https://doi.org/10.1016/j.jaap.2023.106160

Sotoudehnia F, Baba Rabiu A, Alayat A, McDonald AG (2020) Characterization of bio-oil and biochar from pyrolysis of waste corrugated cardboard. J Anal Appl Pyrolysis 145:104722. https://doi.org/10.1016/j.jaap.2019.104722

Srifa A, Chaiwat W, Pitakjakpipop P, Anutrasakda W, Faungnawakij K (2019) Advances in bio-oil production and upgrading technologies. In: Sustainable bioenergy: advances and impacts, pp 167–198. https://doi.org/10.1016/B978-0-12-817654-2.00006-X

Statkraft (2015) Silva Green Fuel. Retrieved from: < https://www.statkraft.com/about-statkraft/where-we-operate/norway/silva-green-fuel/> (March 4, 2025) 1296541

Toro-Trochez JL, Haro Del Río DA, De Sandoval-Rangel L, Bustos-Martínez D, García-Mateos FJ, Ruiz-Rosas R, Rodríguez-Mirasol J, Cordero T, Carrilo-Pedraza ES (2022) Catalytic fast pyrolysis of soybean hulls: focus on the products. J Anal Appl Pyrolysis 163. https://doi.org/10.1016/j.jaap.2022.105492

Urrutia RI, Yeguerman C, Jesser E, Gutierrez VS, Volpe MA, Werdin González JO (2021) Sunflower seed hulls waste as a novel source of insecticidal product: pyrolysis bio-oil bioactivity on insect pests of stored grains and products. J Clean Prod 287:125000. https://doi.org/10.1016/J.JCLEPRO.2020.125000

Valmet-Forward (2015) Smart integration of bio-oil production. Retrieved from: <https://www.valmet.com/more-industries/bio/bio-oil/> (March 4, 2025). 1296541

Valois FP, Bezerra KCA, da Costa Assunção FP, Bernar LP, da Paz SPA, Santos MC, Feio WP, Silva RMP, Mendonça NM, de Castro DAR, Duvoisin S Jr, Gomes ARQ, Sousa VRC, Monteiro MC, Machado NT (2024) Improving the antioxidant activity, yield, and hydrocarbon content of bio-oil from the pyrolysis of Açaí seeds by chemical activation: effect of temperature and molarity. Catalysts 14:44. https://doi.org/10.3390/CATAL14010044/S1

Vandenberghe LPS, Pandey A, Carvalho JC, Letti LAJ, Woiciechowski AL, Karp SG, Thomaz-Soccol V, Martínez-Burgos WJ, Penha RO, Herrmann LW, Rodrigues AO, Soccol CR (2021) Solid-state fermentation technology and innovation for the production of agricultural and animal feed bioproducts. Syst Microbiol Biom 1:142–165. https://doi.org/10.1007/s43393-020-00015-7

Vincenzo M, Nicolae S, Oliver H, Marco B, Aliki G, Simon L, Aikaterini M, Geraldine JO (2022) Clean energy technology observatory: bioenergy in the European Union—2022 status report on technology development, trends, value chains and markets. https://doi.org/10.2760/577104. (online)

Wang H, Jing Y, Zhang J, Cao Y, Lyu L (2021) Preparation and performance evaluation of swine manure bio-oil modified rubber asphalt binder. Constr Build Mater 294:123584. https://doi.org/10.1016/J.CONBUILDMAT.2021.123584

Xiwei X, Fan Z, Yan S, Hong W, Haipeng Y, Linghao L, Zhiwen J, Yujian W, Xudong F (2022) Method for improving pyrolysis oil quality by pretreating biomass raw material with freeze-thaw auxiliary low cosolvent and application. CN patent 114806616A 1296541

Xiwei X, Fan Z, Péng H, Linghao L, Yan S, (2023) Method for ball milling assisted molten salt electrochemical pretreatment of biomass to improve pyrolysis oil quality and application thereof. CN patent 117568060A 1296541

Xu F, Xia X, Luo J, Luo D, Xu J (2024) Maximized energy recovery by catalytic co-pyrolysis of dewatered sewage sludge and polystyrene to contribute in bio-circular economy: synergistic compositional analysis of bio-oil and syngas through artificial neural networking. Process Saf Environ Prot 186:1253–1262. https://doi.org/10.1016/j.psep.2024.04.072

Yijie Z, Shuang W, Chuan Y, Bin C, Hao C, Yanyun C (2022) Carbon material prepared from bio-oil obtained by catalytic pyrolysis of biomass and method thereof. CN patent 114613612A Muniyappan 2024–2025

Yoshimitsu U, Norizan BA, Wissam NO, Noridah BO, Mohammad AFBM, (2022) Firdaus Bin Mazlan Mohammad. 2022. A continuous process of decomposing biomass for the production of bio-oil by an improved fluidized bed pyrolysis system. MY patent 193194A 2022–2023

Zadeh ZE, Abdulkhani A, Saha B (2020) Characterization of fast pyrolysis bio-oil from hardwood and softwood lignin. Energies 13:887. https://doi.org/10.3390/EN13040887

Zhan L, Zhou S, Yan Y, Jiang D, Li R, Wu Y (2025) Multistage thermal-chemical transformation of reed to produce phenol bio-oil and biochar: process exploration and life cycle assessment. Ind Crop Prod 226:120609. https://doi.org/10.1016/j.indcrop.2025.120609

Chapter 17
Biogas from Industrial Effluents: Production and Applications

Nadia Maria do Valle Ramos, Victor Brisk, Giuliana Varela Garcia Lesak, and Marcos Lucio Corazza

Abstract This chapter explores the potential of biogas production from industrial effluents as a sustainable and versatile energy source. It provides a comprehensive overview of the global biogas landscape, emphasizing regional developments, policies, and technological advancements. The chapter details the anaerobic digestion (AD) process and characterizes various industrial feedstocks—such as livestock manure, sewage sludge, food and crop waste, and residues from the food, beverage, pulp, and paper industries—highlighting their biogas yields, challenges, and opportunities. Emphasis is placed on co-digestion strategies, biodigester technologies, and biogas upgrading techniques that enhance methane content and usability. Furthermore, the chapter addresses the valorization of byproducts, particularly digestate and captured CO_2, within the framework of circular economy principles. By integrating case studies, technical data, and economic considerations, the chapter illustrates how biogas production from industrial waste streams can support climate change mitigation, promote energy security, and foster sustainable industrial development.

17.1 Introduction

Biogas represents a sustainable energy alternative, particularly in areas where access to national grids is limited or where there is a high demand for heat that cannot be met by renewable electricity (IEA 2020). It is composed primarily of methane, CO_2, and small amounts of other compounds, with its composition varying according to the type of feedstock used and the production pathway (IEA 2020). This renewable fuel can be produced from any biodegradable feedstock rich in organic carbon, and anaerobic digestion (AD) is typically the technology employed for this purpose, as it is a cost-effective and sustainable approach (Bhatt and Tao 2020). Biogas has a wide range of applications (Fig. 17.1), including heat and electricity generation, acting as a substitute for natural gas and heating oil, and being upgraded to biomethane or used as a transport fuel (Bhatt and Tao 2020).

N. M. do Valle Ramos · V. Brisk · G. V. G. Lesak · M. L. Corazza (✉)
Department of Chemical Engineering, Federal University of Paraná, Curitiba, PR, Brazil

A. L. Woiciechowski et al. (eds.), *Biomass for Bioenergy and Bioproducts*, Biofuel and Biorefinery Technologies 17, https://doi.org/10.1007/978-3-032-19898-3_17

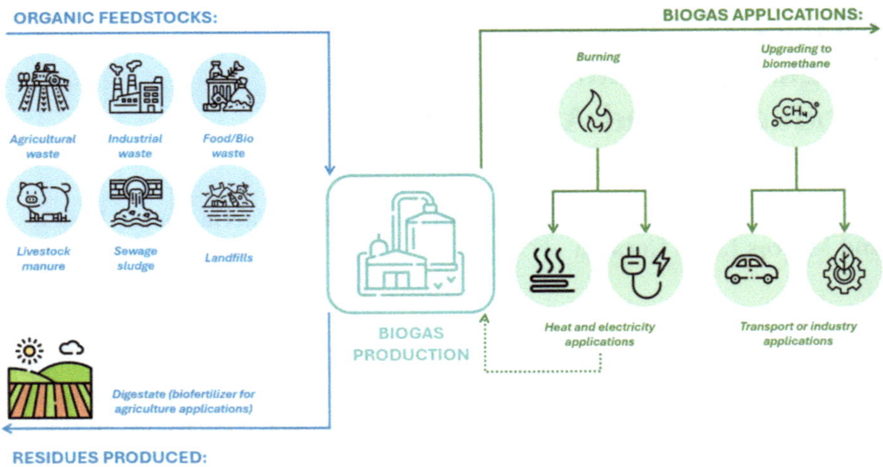

Fig. 17.1 Schematic diagram of biogas production and its applications

In response to the global imperative to diminish dependency on fossil fuels, the production of biogas has experienced substantial growth, with regions such as Europe, the Americas, and Asia playing pivotal roles in this expansion. Nonetheless, the development of the biogas sector varies significantly across these key players due to differences in policy frameworks, technological capabilities, and resource availability.

A comprehensive report by Gustafsson et al. (2024) details updates on the biogas sector across nine European countries: Denmark, Finland, France, Germany, Ireland, Norway, Sweden, Switzerland, and the United Kingdom. A total of 13,856 biogas plants operate across these nations. However, not all countries report actual biogas production figures. In several cases, production estimates have been derived from available data on electricity and heat production using an assumed efficiency of 38%. As a result, the estimated annual energy generated from biogas is approximately 96,693.8 GWh. This significant production capacity is closely linked to the European Union's robust policy frameworks and infrastructural investments. The Renewable Energy Directive (RED II), for example, sets ambitious targets for renewable energy consumption, thereby encouraging member states to adopt supportive measures for biogas production (European Commission 2025a). Additionally, the expansion of infrastructure, including facilities that upgrade biogas to biomethane suitable for injection into natural gas grids, has further broadened biogas applications, supporting the European Union (EU)'s objectives for energy diversification and security (López et al. 2024). The REPowerEU plan demonstrates policymakers' proactive efforts to bolster internal energy security in the aftermath of the 2022 crisis and to advance climate objectives. A key strategy involves promoting the sustainable production of biomethane, striving to achieve 35 billion cubic meters of biomethane by 2030. This initiative serves as a cost-effective approach to reducing natural gas imports from Russia (European Commission 2025b).

Over the past 20 years, the biogas sector in China has advanced significantly. The government has invested over €6.2 billion to advance rural biogas initiatives, positioning the country as one of the global leaders in the field. By 2017, China was generating an estimated 12.37 Gm^3 of biogas annually, and 2021 data revealed that more than 108,555 active biogas plants were generating 13.48 Gm^3/year of biogas, with nearly 94% located in agricultural areas (Gustafsson et al. 2024; Lu and Gao 2021). Another major player in Asia is India. Recently, the Ministry of New and Renewable Energy (MNRE) announced a new biogas production initiative under its National Bioenergy Programme with an initial phase budget of approximately $114.4 million USD. This initiative is designed to tap into India's enormous biogas potential, largely driven by its vast livestock population of about 535.78 million animals, including roughly 302 million bovines (Ministry of New and Renewable Energy 2022). Despite this impressive potential, current utilization remains far from adequate. Presently, India produces only about 2.07 billion m^3 of biogas annually (Mittal et al. 2019). A study by Singh and Kalamdhad (2021) estimates that less than 5% of the country's total biogas potential is being harnessed. This significant gap highlights the urgent need for increased investments in technology, infrastructure, and supportive policies to boost biogas production and fully capitalize on this renewable energy resource.

Across North America, biogas is increasingly recognized as a vital element of the renewable energy landscape, significantly contributing to environmental sustainability and energy security. In the United States, around 2300 facilities are currently harnessing biogas. These include approximately 530 anaerobic digesters on farms, over 1100 wastewater treatment plants, about 110 facilities dedicated to processing biowaste, and nearly 580 projects that capture landfill gas. Collectively, these operations have attracted roughly $37.5 billion in capital investment (American Biogas Council 2025). Estimates show that these facilities could collectively produce nearly 6.21 Gm^3/year of biogas (Thunder Said Energy 2023). Canada is also making notable progress in this sector. The country currently operates 282 biogas plants, with nearly 39% of these deriving energy from wastewater treatment processes and 35% capturing gas from landfills. These facilities collectively produce close to 22,000 TJ/year, underscoring the growing importance of biogas in Canada's renewable energy portfolio (Gustafsson et al. 2024).

In Brazil, the biogas sector is gradually expanding despite the absence of a dedicated federal policy regarding subsidies. Currently, there are 1365 biogas plants nationwide, with most of these operating on agricultural sites. These facilities collectively produce approximately 4.15 Gm^3 of biogas per year, highlighting the country's significant potential in renewable energy generation. Moreover, 50 biogas upgrading plants are in operation, producing approximately 580 million m^3 of high-quality biomethane annually (CIBiogas 2024). Although Brazil lacks a comprehensive federal biogas framework, a significant development occurred in 2015. The National Oil, Natural Gas and Biofuels Agency (ANP) introduced regulations that govern the use of biomethane derived from agriculture, the food industry, urban solid waste, and sanitation as fuel (Kanda et al. 2022). This regulatory move marks an important step toward bolstering the institutional framework for biogas in the country, paving the way for further growth in this renewable energy sector.

Overall, biogas production trends over the past decade highlight significant advancements in some regions while underscoring the need for further development in others. The continued expansion of biogas as a renewable energy source will likely depend on policy support, investment in infrastructure, and technological innovation. This chapter explores the potential of industrial effluents as feedstocks for biogas production, highlighting their role in mitigating the climate crisis. It provides an in-depth analysis of biogas generation from various industrial waste streams, covering their chemical composition and production processes. Additionally, it discusses biogas characterization, biodigester technologies, and purification techniques. The chapter also examines biogas applications and the management of process residues within the framework of the circular economy, concluding with key insights into the sustainability of biogas production.

17.2 In-Depth Description of the Effluent from Industrial Sectors

This chapter presents an overview of the main effluents with potential for biogas production, focusing on their origin, biogas generation potential, and the associated challenges and opportunities. For this discussion, the term industrial is used in a broader sense, encompassing the entire industrial chain and major effluent-generating sectors. Therefore, residues such as crop waste, livestock manure, and food waste are considered within the scope of the food processing industry, while wastewater sludge is included due to its significant availability and relevance as a biogas feedstock. Landfill gas is also briefly addressed, as it represents a considerable source of methane emissions and forms part of the life cycle of industrialized products. Table 17.1 summarizes the biogas yields of these different effluents, providing a comparative view of their energy generation potential.

17.2.1 Livestock Manure

Livestock manure is an abundant byproduct of animal husbandry that, if not managed properly, can lead to significant greenhouse gas emissions and water pollution. Traditionally used as a fertilizer, its natural decomposition releases a potent greenhouse gas known as methane. Alternatively, anaerobic digestion (AD) offers a sustainable solution by converting manure into biogas and producing a stabilized digestate that can enhance soil fertility when applied properly (Jain et al. 2019).

Table 17.1 Comparison of biogas yield and electricity produced from different potential feedstocks

Feedstock	Biogas yield (m^3/ton of fresh biomass)	Electricity[a] (kWh/ton of fresh biomass)
Cattle dung	55–68	122.5
Chicken dung	126	257.3
Pig slurry	11–25	23.5
Horse manure	56	114.3
Sewage sludge	47	96.0
Municipal solid waste	101.5	207.2
Food waste (disinfected)	110	224.6
Fruit wastes	74	151.6

Adapted from Achinas et al. (2017)
[a]Considering 35% of electrical efficiency combined heat power, biogas heating value 21 MJ/m^3 with 55% methane content, 3.6 MJ/kWh

Different manure types exhibit different biogas potentials. For example, as per Table 17.1, cattle dung typically yields between 55 and 68 m^3 of biogas per ton of fresh matter, while chicken dung can produce up to 126 m^3. Pig slurry, on the other hand, generates approximately 11–25 m^3 and horse manure about 56 m^3 per ton. These figures underscore the importance of manure composition in determining energy recovery potential in AD systems.

AD systems rely on a consistent supply of feedstock, and livestock manures from cattle, pigs, and poultry are particularly suitable due to their availability and ease of collection. However, the digestion of poultry manure often poses greater challenges compared to cattle or pig manure, sometimes requiring additional pretreatment or co-digestion strategies (Liebetrau et al. 2021). Moreover, manure is a promising substrate for AD because it naturally contains many of the microbial groups essential for the process, particularly in cattle manure. This reduces the need for extensive pretreatment and contributes to a stable digestion process with good buffering capacity (Luque et al. 2016).

Farm size, husbandry methods, and manure management practices, which include bedding, storage, and transportation, are critical factors in the design and profitability of AD facilities. Smaller farms may struggle to achieve economic viability without governmental incentives, whereas larger operations can optimize manure handling to maximize biogas production and minimize methane leakage (Jain et al. 2019; Brahmi et al. 2024). Effective manure management through anaerobic digestion not only mitigates environmental impacts but also promotes a circular bioeconomy by recycling nutrients, reducing greenhouse gas emissions, and generating renewable energy and biofertilizer.

17.2.2 Sewage Sludge

The growing world population, combined with industrialization and excessive consumerism, has led to a significant increase in wastewater production, especially in urban areas. A byproduct of the wastewater treatment process, sewage sludge is generated in large quantities worldwide—approximately ten million tons annually in Europe, eight million tons in the United States, and four million tons in China (Yesil and Tugtas 2019). In 2017, global sewage sludge production reached approximately 45 million dry tons per year (Di Giacomo and Romano 2022). This increasing volume of sewage sludge, which contains heavy metal pollutants, a high proportion of organic matter, and toxins, not only poses environmental challenges but also offers opportunities for sustainable use (Enebe et al. 2023).

Anaerobic digestion (AD) is one of the most effective methods for processing sewage sludge, as it stabilizes the residue, reduces pathogens and odors, and generates biogas for energy production. On average, each ton of sewage sludge produces approximately 47 m^3 of biogas, which can generate around 96 kWh of energy (Achinas et al. 2017). Additionally, AD produces digestate, which, when properly treated, can be used as a soil amendment, contributing to nutrient recycling. In areas lacking sewage treatment infrastructure, AD can be applied in decentralized systems, such as community or farm-based digesters linked to toilets, preventing pollution and enabling on-site biogas utilization (Jain et al. 2019).

Although the energy potential of sewage sludge is lower than that of food waste or crop residues, its continuous availability makes it a reliable feedstock. The biogas generated can partially or fully supply the high energy demands of wastewater treatment plants. Moreover, the high nutrient content of sewage sludge, particularly phosphorus—a critical element for plant growth—enhances the value of digestate as a fertilizer. However, appropriate regulations and standards are necessary to ensure its safe application and prevent contamination (Jain et al. 2019).

17.2.3 Food Waste

In 2022, approximately 1.05 billion tons of food waste, including both edible and inedible components, removed from the human food supply chain, were generated worldwide across household, retail, and food service sectors. Although reducing this waste is crucial, as highlighted by UN Sustainable Development Goal 12, a portion of it remains unavoidable, particularly in urban areas where food waste management can be challenging and exacerbates landfill pressures (UNEP 2024). To mitigate these issues, various strategies are employed, including reintroducing food waste into food or feed processing chains, extracting valuable components, composting, and biodigesting.

Food waste, especially that originating from households and catering establishments, is notably heterogeneous yet generally easy to digest. According to

Table 17.1, there is a high methane potential when using this substrate, with biogas yields reaching about 110 m^3 per ton of fresh food waste. In contrast, fruit waste, for example, produces a slightly lower biogas yield of approximately 74 m^3 per ton of fresh matter.

However, the nitrogen-rich protein fraction present in food waste can lead to ammonia accumulation during the anaerobic digestion (AD) process, potentially inhibiting performance when concentrations become too high (Kunz et al. 2019). Addressing these challenges, one promising approach for food waste digestion involves modifying the dominant archaea and metabolic pathways by altering the digestion media composition through trace element supplementation. This method, however, demands careful analysis due to the complex interactions within the AD process, variations in food waste composition, and the risk of elemental dispersion into the environment (Banks et al. 2018).

Other challenges associated with the AD of food waste include acidification, ammonia nitrogen ($NH4^+$-N) inhibition, foaming, and the presence of emerging contaminants. To overcome these issues, several strategies are employed, such as pretreatment, co-digestion, and the addition of specific additives. Pretreatment techniques, for instance, can be particularly beneficial for food waste that is difficult to degrade, such as waste high in lipids. Nevertheless, the selection of pretreatment and other enhancement techniques must be approached with caution since they may introduce new challenges that can overshadow their benefits (He et al. 2024).

Emerging contaminants, including antibiotics (along with antibiotic-resistant genes), plasticizers, and microplastics, represent an additional concern. These substances can adversely affect both the AD process and the broader environment, posing risks to human health. Although there is some evidence that AD can reduce emerging contaminants such as antibiotics and antibiotics resistance genes (ARGs), further research is needed to fully understand their impact on the AD process and the efficacy of contaminant removal, thereby preventing their release into the environment (He et al. 2024).

17.2.4 Crop Residues

Crop residues—including stalks, husks, leaves, and bagasse—offer substantial potential for energy generation due to the global diversity of crops. However, only 40–70% of these materials can be diverted to anaerobic digestion (AD) because maintaining soil quality requires reintegrating 30–60% of the residues to preserve carbon content, structure, and water retention (Jain et al. 2019; Prasad et al. 2020; Scarlat et al. 2010).

When processed via AD, crop residues yield biogas and a nutrient-rich digestate that can be returned to the fields, promoting a circular nutrient cycle. In addition to energy recovery, sustainably managed crop residues serve other valuable purposes—ranging from animal feed and bedding to heating—while avoiding detrimental practices like open-field burning, which degrades soil quality, causes nutrient loss, and contributes to air pollution (Jain et al. 2019; Prasad et al. 2020).

A recent study illustrates the variability in biogas production among different feedstocks. For example, pineapple waste produced the highest specific biogas production (SBP) at 0.817 m^3/kg volatile solids (VS) added, followed by orange peels at 0.771 m^3/kg VS. In contrast, lignocellulosic residues, such as coffee husks, yielded only 0.366 m^3/kg VS, underscoring the challenges associated with their biological conversion and the need for pretreatment (Suhartini et al. 2021).

The lignocellulosic nature of many crop residues means they have a complex structure that requires pretreatment (whether physical, chemical, or biological) to enhance biodegradability. Additionally, their high carbon-to-nitrogen ratio can disrupt AD performance unless balanced through co-digestion or nutrient supplementation (Sukhesh and Rao 2018). Seasonal availability also complicates matters, necessitating effective collection, processing, and storage strategies. Often, residues from multiple farms must be combined, and operational techniques like liquid recycling or solid-state digestion may be required to overcome issues such as biomass stratification and pipe clogging.

17.2.5 Industrial Residues

Organic residues deriving from manufacturing processes are largely varied depending on the nature of the raw materials and process technologies. The main industries pointed out in the literature with potential for biogas generation are those related to food, beverage, and animal residue processing (e.g., slaughterhouses) (IEE-Project and FABbiogas [s.d.]). Other industries, such as pulp and paper and sugar processing for fuel, also represent a great potential for biogas generation due to the high loading of organic residues. The potential of other industrial residues for biogas generation is also explored in the literature since, in an industrial context that usually presents relatively high heat and power demands, retrofitting energy produced by residues in the process may present economic advantages. It should be noted that residue treatment and disposal are costly, and converting residues into useful products can help in this sense. The need to mitigate the climate crisis also puts pressure on the utilization of fossil fuels in an industrial context, and biogas can be an appealing alternative for converting the already existing natural gas utilization structure and equipment.

Regarding the implementation of AD systems in industrial processes, case-by-case detailed analysis and tailored systems should be considered according to the specific process and feedstock under consideration, integrating the AD plant as part of the process and extracting as much value in the form of residue treatment and energy recovery. Potential industries for AD implementation are those with continuous energy demand (power and/or heat) and constant generation of considerable amounts of waste streams with high organic load and microbial degradability potential, with characteristics that comply with local regulations (IEE-Project and FABbiogas [s.d.]).

In the food industry, dairy presents a pivotal role in residue generation, which mainly consists of cheese whey and wastewater, requiring treatment before disposal. A part of these residues can be further processed into food products and animal feed to reuse their fat, sugar, and protein content, and the process water can be reutilized. However, the remaining residue presents a great potential for environmental pollution, such as water eutrophication, due to its protein, sugar, salt, and phosphor contents, and reduction of its organic load is needed before release into the environment. AD of whey and wastewater presents a great potential for energy production in dairies, with large-scale plants already implemented. Another potential generator of residues suitable for AD related to the food industries is slaughterhouses. However, abattoir residues (comprising wastewater and animal byproducts such as fats, rumen content, blood, among others) are subject to strict control and regulation due to the possibility of spreading diseases. Thus, they are categorized according to risk, and not all categories can be digested. In this sense, thermal pretreatment may also be required by some regulations prior to AD of some animal byproducts. The high nitrogen concentration in slaughterhouse residues (such as blood) makes them potential sources of environmental contamination and can also be detrimental to AD processes, causing inhibition, for which trace element addition can be helpful (Bochmann et al. 2020).

Breweries also produce multiple organic residues, such as wastewater, malt dust, hot and cold break, among others, and the most notable for AD are brewers' spent grains (BSG). Despite having other uses such as animal feed, its quick degradation rate and the high energy demand of brewing processes raise attention to energy recovery through AD. Co-digestion or trace elements supplementation may be needed, and the implementation of AD must be accompanied by an energy efficiency analysis and energy consumption reduction measures, allowing for full exploitation of the energy recovery potential (Bochmann et al. 2020).

Another industry of interest for AD that can be related to food, beverage, and fuel production is the sugar industry, which is mostly based on sugarcane and sugar beet. Among the organic residues from sugarcane are bagasse, filter cake, and vinasse. Bagasse is the lignocellulosic residue obtained after juice extraction, and it is utilized for heat and electricity production by combustion and to produce second-generation biofuels such as ethanol. Bagasse can also be utilized for biogas production but requires pretreatment to increase digestibility or long retention times. Molasses containing relatively high amounts of sugar are also obtained as a residue in sugar production. Molasses can be utilized for animal feeding, in the chemical, food, and pharmaceutical industries, and its fermentation leads to the production of alcohol. This fermentation, in turn, generates vinasse, another residue that has applications such as agricultural use. Both vinasse and molasses can be utilized in AD processes. Filter cake is another residue that can serve as animal feed and is utilized in AD. The residues of sugarcane processing have remarkably versatile applications, and sugar production is usually associated with alcohol production. However, since great amounts of sugar and alcohol are produced and utilized worldwide, with proper management of sugarcane processing residues, AD can be a potential opportunity. Sugar beet processing industries produce sugar beet pulp and molasses as residues that can also be utilized for biogas production (Bochmann et al. 2020).

Paper mill residues are relatively frequently investigated in the literature. These residues present challenging treatment properties due to the large amounts of fibers, toxic compounds, and organic load, presenting a recalcitrant nature. Furthermore, paper mill effluents present hard-to-remove chemical oxygen demand (COD) in traditional activated sludge treatment, and the final effluent usually contains high concentrations of high molecular weight organic compounds. Thus, utilization of anaerobic digestion can be an interesting option that combines the capability of dealing with high organic load effluents and recovering energy from waste (Zhang 2018). In this sense, the study of Zhang (2018) investigated, on a pilot scale, different effluents from real industries with high organic loadings that present challenges in traditional treatment processes and are strictly regulated. Some anaerobic–aerobic biological treatment systems were evaluated. Different types of waste streams from real paper-making processes (such as bleaching and pulping effluents) were tested on a pilot scale, along with small amounts of wasted sugar water from food processing, coupled with an aerobic treatment. A good treatability (COD removal and substrate utilization) was obtained for paper mill effluents, showing the potential of AD for effluent treatment and energy recovery in the pulp and paper industry.

In the same study (Zhang 2018), the AD of brown grease in a high-rate anaerobic digestion system was also investigated. This is a residue containing trapped grease from restaurants, food industries, and sewage facilities, being costly to dispose of and presents a very high organic load. Due to its high lipid content, the AD of this residue (alone or in high amounts) is prone to long-chain fatty acid inhibition, fat clogging, and scum and foam formation, being commonly studied in co-digestion with other residues. Biogas with high methane yield and content was produced from diluted brown grease, showing itself as a potentially profitable alternative for biogas production. Corn stillage from real ethanol industries, which presents a pollution potential and demands high energy consumption for treatment, impacting the process economy, was also studied. The results also showed that AD is an interesting approach to treat corn thin stillage, which can be integrated into dry mill ethanol plants' process, saving money and energy, corroborating the idea of utilizing residues from the sugar/ethanol industries in AD (Zhang 2018).

As previously stated, the literature regarding AD of industrial wastes frequently mentions as potential residues those related to food, beverage, and animal residue processing. Some examples are those previously detailed from dairy, slaughterhouses, sugar, and breweries. Other potential waste streams are those of meat and fish industries, fruits, vegetables, coffee, tea, olives, wine, and pet food processing, to mention some of the examples (IEE-Project and FABbiogas [s.d.]). Thus, the residues will have different compositions and biogas/digestate generation potentials, requiring the previously mentioned tailored solutions. It should be noted, however, that the residues discussed in this chapter are fairly different in nature, and those other industrial residues will likely be similar to some of the discussed ones in terms of water content, lignocellulosic and organic composition and degradability, potentially requiring similar AD approaches such as pretreatment to increase degradability, co-digestion and supplementation to enhance nutritional balance, pretreatment to reduce pathogens, among others, with proper adaptations to different extents. Industrial residues can also have high concentrations of toxic compounds that can cause inhibition and pollution by digestate utilization, and in those cases, specific operations need to be considered. Ultimately, this chapter summarizes some of the most relevant residues for AD generated on an industrial scale, providing insights into the main topics and building a conceptual basis for other industrial residues of a similar nature.

17.2.6 Landfill Gas

Landfilling is a widely utilized solid waste disposal method in which layers of compacted solid residue are interspersed with soil, while implementing pollution control measures to minimize environmental and health risks. These measures help reduce waste volume, control disease vectors, mitigate odors, and prevent water contamination.

Once deposited in landfills, waste residues undergo four different decomposition stages, the first being aerobic and the others with anaerobic conditions. Because waste is continuously added, multiple decomposition stages occur simultaneously, leading to the gradual production of landfill gas (LFG). Vertical and horizontal piping is installed within the landfill to collect the generated gases. The collected gas has approximately 50% CO_2 and 50% methane, along with traces of other organic components and impurities (EPA 2024).

Without proper management, landfills can become a potential source of methane emissions, contributing to climate change. However, the collection and treatment of LFG through impurity removal and advanced upgrading techniques can enhance its energy content, allowing it to be refined into high-quality biogas. This process not only prevents methane emissions but also enables the use of landfill gas as a renewable energy source (EPA 2024).

17.2.7 Co-digestion

The digestion of a single substrate often faces long-term challenges associated with its composition, such as ammonia or acid accumulation, low organic content, and nutrient imbalances. Co-digestion, which combines different feedstocks to be anaerobically digested, helps overcome these issues by improving nutritional balance, buffering capacity, and microbial synergy, reducing inhibition and enhancing process stability (Luque et al. 2016).

Due to these advantages, co-digestion can improve methane yield, increase biogas production rate, stabilize the system, and allow for higher organic loadings, among other benefits. In fact, biogas production can increase by 25–400% compared to mono-digestion of the same substrates (Hagos et al. 2017). Commonly co-digested feedstocks include sewage sludge and the organic fraction of municipal solid waste. However, the proper substrate ratio is essential, considering the carbon-to-nitrogen (C/N) ratio, nutrient composition, pH, and the presence of inhibitory compounds. Properly formulated mixtures with balanced composition and biodegradation potential can optimize biogas yields and produce high-quality digestate (Luque et al. 2016).

17.3 Biogas Characterization and Purification Techniques

17.3.1 Biogas Characterization

Biogas is a gaseous mixture primarily consisting of methane and carbon dioxide. However, its exact composition varies depending on the type of feedstock used in the anaerobic digestion process and the operational conditions under which it occurs. Factors such as temperature, retention time, and the biochemical characteristics of the feedstock influence the proportion of methane, CO_2, and other trace gases. Table 17.2 provides a summary of the typical biogas composition derived from various feedstocks.

Anaerobic digestion is the most common process for biogas production. It is a complex, oxygen-free biological system in which specialized microorganisms transform organic matter into biogas. This conversion unfolds through four interconnected phases (hydrolysis, acidogenesis, acetogenesis, and methanogenesis), each crucial for breaking down intricate organic polymers into

Table 17.2 Biogas composition from different feedstocks

Component	Sewage sludge (%)	Food waste/ biowaste (%)	Agricultural residues (%)	Livestock manure (%)[a]	Landfills (%)	Industrial residues (%)
CH_4	55–75	55–70	50–80	60–75	50–80	50–70
CO_2	25–45	35–45	30–50	30–35	20–50	30–50
H_2S	0–1	0–1	0.7	0.1–0.6	0.1	0.8
H_2	0–1	0–1	0–2	–	0–5	0–2
N_2	0–5	–	0–1	–	0–3	0–1
O_2	0–2	–	0–1	–	0–1	0–1
NH_3	–	–	Traces	0–0.1	Traces	Traces
Siloxanes	–	–	Traces	–	Traces	Traces
Reference	Demirbas et al. (2016)	Silva et al. (2022)	Chen et al. (2015)	Sihlangu et al. (2024)	Chen et al. (2015)	Chen et al. (2015)

[a]Livestock manure considered: cattle manure, chicken manure, and pig manure

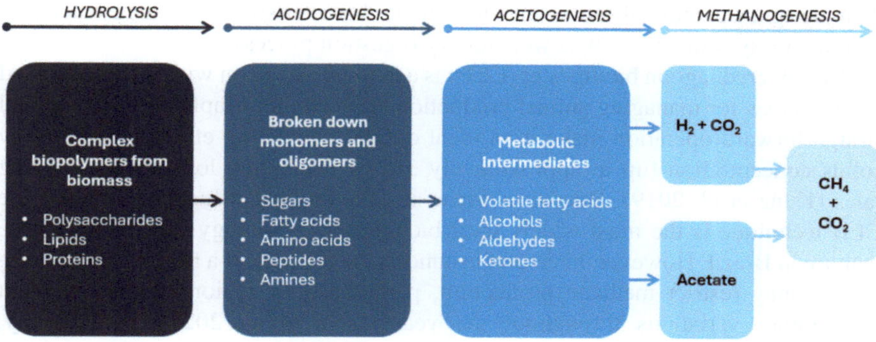

Fig. 17.2 Schematic anaerobic digestion pathway

simpler, energy-rich compounds (Fig. 17.2). Not only does this process offer a renewable source of energy, but it also contributes to waste reduction and produces a nutrient-dense digestate that can serve as a sustainable biofertilizer, thereby closing the loop in waste management and environmental conservation (Kunz et al. 2019).

During the initial hydrolysis phase, complex organic polymers—such as proteins, lipids, and polysaccharides—are enzymatically cleaved into simpler, water-soluble monomeric units through the action of extracellular enzymes produced by hydrolytic bacteria. The efficiency of hydrolysis is critical, especially when the organic matter is structurally complex, as it can limit the overall digestion rate. The monomers released during hydrolysis are then fermented by a diverse group of acidogenic bacteria. During this stage, the compounds are converted into short-chain organic acids (such as acetic, butyric, and propionic acids) along with the production of other byproducts like hydrogen and CO_2. This transformation lays the groundwork for subsequent reactions (Kunz et al. 2019).

Next, acetogenic bacteria act on organic acids, predominantly converting them into acetic acid (acetate), while releasing additional hydrogen and CO_2. This phase is essential because the formed acetate serves as a substrate for methane production and relies on a close syntrophic relationship between microorganisms to maintain favorable hydrogen partial pressures. In the final phase, methanogenic archaea utilize the acetate, hydrogen, and CO_2 to synthesize methane and CO_2 through exothermic reactions. Depending on the metabolic pathway, these archaea may either convert acetate directly (acetolactic methanogenesis) or use hydrogen and CO_2 (hydrogenotrophic methanogenesis). The methane produced is the main component of biogas, which can be harnessed for electricity, thermal energy, or further refined into biomethane for vehicular use (Kunz et al. 2019).

17.3.2 Biodigesters

Anaerobic digestion is typically carried out in biodigesters, which are specifically designed systems that facilitate the microbial breakdown of organic matter in the absence of oxygen. These biodigesters can be classified into different types based on their design, operational mode, and feedstock characteristics, including batch and continuous systems, as well as wet and dry digestion processes.

The covered lagoon biodigester (CLB) is a low-tech solution widely implemented in rural areas for managing animal production effluents. Its simple construction and straightforward operation make it an ideal option for treating effluents with a low solids concentration (up to approximately 3% by mass) and low organic loading rates (Kunz et al. 2019). Due to its lower installation and management costs, the CLB technique is the most widely used biodigester technology for treating swine manure in Brazil. However, in certain instances, the absence of a heating and stirring system may restrict methane production, particularly in regions with significant temperature variations throughout the year (Silva et al. 2024). Additionally, depending on the substrate and several other parameters, biogas productivity can reach up to 0.15 m^3 per cubic meter of reactor volume per day, providing a cost-effective and sustainable method for waste management (Kunz et al. 2019).

Among the various anaerobic wastewater treatment technologies, upflow anaerobic sludge blanket (UASB) reactors have achieved considerable success and are widely applied to treat effluents from industries such as sugar, pulp and paper, dairy, chemical, potato starch, soft drinks, fish processing, noodle processing, yeast production, slaughterhouses, and coffee processing (Daud et al. 2018). The UASB biodigester is particularly valued for its high biomass retention and ability to operate at low hydraulic retention times (from 4 to 72 h, typically). Its stability under fluctuating wastewater characteristics and capacity to handle high organic loading rates make it well-suited for treating solubilized organic matter. While UASB systems are primarily designed for wastewater with low total solids content (less than 2%), pretreatment is often necessary when processing effluents from animal production (Kunz et al. 2019).

The continuous stirred tank reactor (CSTR) is the most widely used biodigester configuration in biogas plants, particularly for co-digesting diverse substrates and processing effluents with high solid concentrations, typically ranging from 5% to 10% by mass (Shah et al. 2024). Its design ensures constant agitation, promoting homogenization, efficient heat transfer, and uniform suspension of solids. Additionally, CSTRs require a minimum hydraulic retention time (HRT) of 15–40 days, allowing sufficient contact time for microbial activity and maximizing digestion efficiency (Kunz et al. 2019). A pilot-scale CSTR with a 500 L capacity, operating with a feedstock of 80% food waste and 20% garden waste, achieved a methane yield of approximately 0.47 L g VS^{-1} (Perin et al. 2020).

The solid-state (or dry digestion) biodigester requires a smaller reactor volume for a specific content of solids and efficiently processes substrates with high solid concentrations (20–40%) and larger particle sizes without extensive dilution. Typically operated in batch mode, this system produces biogas with a high methane concentration (around 80%). However, digestion time ranges from 2 to 4 weeks, depending on the substrate, and biogas productivity is 15% to 40% lower than in wet digestion. Additionally, batch operation often demands more frequent manual handling and monitoring, increasing operational complexity and labor requirements (Kunz et al. 2019).

Special Content: How to Size a General Biodigester?

The design and sizing of biodigesters are essential for optimizing anaerobic digestion performance, ensuring efficient organic matter conversion and biogas production. The following exercise provides a general approach to sizing a biodigester, considering key parameters such as feedstock availability, organic loading rate, and retention time. This simplified methodology aims to give an overall understanding of the fundamental principles involved in biodigester design.

Consider a scenario in which an unstirred biodigester is to be installed in a small coffee shop serving 200 customers daily and employing 25 staff members, located in a tropical climate. The production of wet food waste is estimated at 0.2 kg per person per day, with a total solids (TS) content of 20%, of which 80% consists of volatile solids (VS). The biogas generated is intended for use in a kitchen stove for heat production.

Estimate the required biodigester volume and the total biogas production based on the following parameters:

- The mass ratio of water to food waste is 2:1.
- The hydraulic retention time is 30 days.
- The organic loading rate is defined as: Flowrate * concentration/reactor volume.
- The biogas yield from food waste is 0.84 m^3/kg of volatile solids.

Feedstock

First, the total daily food waste generation is determined as follows:

$$225 \text{ people} * 0.2 \frac{\text{kg food waste}}{\text{person} * \text{day}} = 45 \frac{\text{kg food waste}}{\text{day}}$$

Given a mass ratio of water to waste of 2:1, the total feedstock input consists of:

$$45 \text{ kg food waste} + 90 \text{ kg of water} = 135 \text{ kg (or liter) of feedstock per day}$$

Retention Time and Reactor Volume

Considering the retention time of 30 days:

$$135 \frac{\text{kg feedstock}}{\text{day}} * 30 \text{ days} = 4050 \text{ kg (or liters)} = 4.05 \text{ m}^3$$

Thus, the estimated biodigester volume required for feedstock retention is 4.05 m^3.

Feedstock Quality

The total solid (TS) content in the biomass is calculated as:

$$45 \frac{\text{kg food waste}}{\text{day}} * 20\% = 9 \text{ kg dry matter}$$

The volatile solids (VS) fraction is then determined:

$$9 \text{ kg dry matter} * 80\% = 7.2 \text{ kg VS per day}$$

To express this value per cubic meter of feedstock:

$$\frac{7.2 \text{ kg VS}}{135 \text{ L}} * 1000 = 53.3 \text{ kg VS/m}^3$$

Thus, the volatile solids concentration is 53.3 kg.

Organic Loading Rate
Using the defined equation for organic loading rate:

$$\frac{\frac{0.135 \text{ m}^3}{\text{day}} * 53.3 \text{ kg} \frac{\text{VS}}{\text{m}^3}}{4.05 \text{ m}^3} = 1.78 \text{ kg} \frac{\text{VS}}{\text{m}^3}$$

The calculated organic loading rate is 1.78 kg VS/m^3.

Biogas Production Estimation
Given that the biogas yield from food waste is 0.84 m^3/kg VS, the expected biogas production is:

$$1.78 \text{ kg} \frac{\text{VS}}{\text{m}^3} * 0.84 \frac{\text{m}^3}{\text{kg VS}} * 4.05 \text{ m}^3 = 6.05 \text{ m}^3/\text{day}$$

Total Biodigester Volume Consideration
Assuming that the organic slurry occupies approximately 75% of the reactor volume, the total biodigester volume, including space for gas storage, is calculated as:

$$\frac{6.05\text{m}^3}{0.75} = 8.06 \text{ m}^3$$

Thus, the final required biodigester volume is 8.06 m^3.

17.3.3 Purification Techniques

The composition of biogas is influenced by the specific substrate used in the anaerobic digestion process and the operational conditions under which it takes place (Ramos et al. 2024). Certain compounds, such as NH_3 and H_2S, can have harmful effects on biogas applications. These impurities are toxic and highly corrosive, causing equipment damage through the emission of SO_2 during combustion (Angelidaki et al. 2018). Additionally, gases like CO_2 and N_2 reduce the energy

Table 17.3 Comparison of different upgrading technologies

	Membrane separation	PSA	Chemical absorption	Water scrubbing	Organic scrubbing	Cryogenic separation
Consumption for raw biogas (kWh/Nm^3)	0.18–0.20	0.23–0.30	0.05–0.15	0.25–0.30	0.20–0.30	0.76
Heat consumption (kWh/Nm^3)	None	None	0.50–0.75	None	< 0.20	–
Process cost	High	Medium	High	Medium	Medium	High
CH_4 losses	< 0.6%	< 4%	< 2%	3–5%	< 2%	< 1%
CH_4 final content	> 96%	95–99%	99%	97%	> 97%	> 97%
Pre-purification	Suggested	Yes	Yes	Suggested	Suggested	Yes
H_2S removal	Possible	Possible	Yes	Yes	Possible	Yes
Biogas cost (€/Nm^3)	0.12	0.40	0.17	0.13	–	0.44

Adapted from Chen et al. (2015), Angelidaki et al. (2018), Ahmed et al. (2021), Awe et al. (2017), Baker (2012), Bauer et al. (2013), Zhao et al. (2010)

potential of biogas, hinder its conversion into other chemical compounds, such as syngas, and lower its efficiency in generating heat and electricity (Ramos et al. 2024). Therefore, to enhance biogas quality and optimize its use, these impurities must be removed through upgrading techniques.

In recent years, several biogas upgrading technologies have been developed, utilizing diverse processes and operating conditions. Their primary objective is to remove impurities and, consequently, increase the methane content of the biogas. The unit operations include membrane systems (both high and low pressure), cryogenic separation, pressure swing adsorption (PSA) or temperature swing adsorption (TSA), and absorption techniques (either physical or chemical) (Gustafsson et al. 2024). These methods vary in terms of heat consumption and demand, costs, methane (CH_4) loss and recovery, H_2S removal efficiency, operational conditions, environmental impact, and other critical parameters, as shown in Table 17.3.

17.3.3.1 Membrane Separation

Membrane separation is an alternative process to conventional absorption or adsorption methods for biogas purification. This process relies on the selective permeability of the membrane concerning the components of the gas mixture. Depending on the primary objective, the process can be carried out using either the dry method (gas–gas separation, operating at high pressures) or the wet method (gas–liquid separation, operating at low pressures) (Angelidaki et al. 2018).

Polymeric membranes are the most used in biogas upgrading processes, since the diffusion coefficient and CO_2 solubility are higher compared to CH_4. As a result, CO_2 permeates toward the lower-pressure side of the membrane—potentially carrying approximately 10–15% of CH_4—while methane remains on the higher-pressure side (Angelidaki et al. 2018; Awe et al. 2017). The typical membrane polymeric material able to separate CO_2/CH_4 are cellulose acetate and polyimide (Baker 2012).

Both glassy and rubbery polymeric membranes can perform CO_2 and H_2S removal from biogas. However, glassy membranes are more effective in permeating CO_2 by using molecular size as the selectivity criterion. In contrast, rubbery membranes rely on sorption-based selectivity and are more effective in permeating H_2S due to its larger molecular size and higher condensability (greater critical temperature) compared to CO_2 (Chen et al. 2015; Baker 2012).

In practical applications, it is recommended to remove H_2S from biogas beforehand to prevent system corrosion. Additionally, single-stage membrane systems exhibit lower CH_4 recovery rates compared to two- or three-stage systems, which can achieve 98% CH_4 purity. Another significant disadvantage of membrane separation is its high installation cost and the fragility of the system, with a typical lifespan ranging between 5 and 10 years (Angelidaki et al. 2018).

17.3.3.2 Pressure Swing Adsorption

Adsorption processes can be categorized as pressure swing adsorption (PSA), vacuum swing adsorption (VSA), temperature swing adsorption (TSA), and electrical swing (ESA) (Chen et al. 2015). Pressure swing adsorption (PSA) is the most widely used adsorption technology for biogas purification in industrial applications, separating biogas components based on their molecular properties and affinity for the adsorbent material (Angelidaki et al. 2018). Adsorbents such as zeolite, silica gel, activated carbon, activated alumina, and polymeric materials provide highly selective surfaces effectively capturing CO_2 over CH_4 (Ahmed et al. 2021). The process relies on the ability of pressurized gases to adhere to solid surfaces, enabling selective separation. At high pressure, CO_2 and H_2S are preferentially adsorbed, increasing the methane concentration. When the pressure is reduced, these gases are released, allowing the recovery of purified methane (Chen et al. 2015).

In the PSA process, compressed biogas (4–10 bar) is introduced into an adsorption column where the adsorbent selectively captures undesirable impurities (CO_2, N_2, O_2, H_2O, and H_2S), allowing methane to pass through. Once the adsorbent becomes saturated, the gas flow is directed to a subsequent column while the saturated column undergoes regeneration. During regeneration, the pressure is reduced, releasing the trapped gases. Since the desorbed gas contains residual methane, it is typically recycled back to the PSA inlet to minimize losses (Angelidaki et al. 2018; Awe et al. 2017).

Since H_2S adsorption is typically irreversible, it must be removed before biogas enters the PSA system to prevent adsorbent degradation. This technology offers several advantages, such as a compact equipment design and the ability to upgrade raw biogas to a methane concentration of 96–98%. However, it also presents challenges, including high investment and operational costs, the need for process control, and methane losses of approximately 4% in the off-gas stream (Chen et al. 2015; Zhao et al. 2010).

17.3.3.3 Chemical Absorption

Chemical absorption is a widely used method for upgrading biogas by CO_2 and H_2S through chemical reactions. This process typically employs aqueous solutions of alkanolamines, such as diethanolamine (DEA), monoethanolamine (MEA), and methyl-diethanolamine (MDEA). Additionally, other alkaline solutions like sodium, potassium, and calcium hydroxides can also be used to facilitate the chemical reaction with CO_2 (Chen et al. 2015; Angelidaki et al. 2018).

The process involves an absorption column in which the amine solution captures CO_2 and H_2S from biogas, producing a purified gas. The amine solution, now loaded with acid gases, is sent to a regeneration unit consisting of a stripper and a reboiler, where heat is applied to release the absorbed gases and regenerate the amine solution for reuse. The separated gases in the overhead stream are rich in CO_2 and H_2S (Awe et al. 2017).

One of the major advantages of chemical absorption is its high efficiency, with methane concentrations exceeding 99% in the purified gas. This method also allows for complete H_2S removal, operates effectively at low pressures, and has higher reaction rates compared to physical absorption methods like water scrubbing. Furthermore, methane losses are often below 2% in efficiency processes due to the high selectivity of the chemical reactions (Ahmed et al. 2021). However, the process has notable drawbacks, including high capital costs, the need for thermal energy during amine regeneration, and operational challenges such as corrosion, amine degradation due to oxygen and other impurities, salt precipitation, and foaming (Chen et al. 2015). Moreover, significant chemical inputs are required; for instance, capturing one ton of CO_2 theoretically requires 1.39 tons of MEA or 0.9 tons of sodium hydroxide (Yoo et al. 2013). Despite these challenges, chemical absorption remains a prevalent choice for biogas upgrading due to its superior performance and reliability in achieving high methane purity.

17.3.3.4 Water Scrubbing

Water scrubbing is a physical absorption process that separates and purifies methane from biogas by exploiting the different solubility levels of gas components in water. This method is particularly effective for industrial-scale biogas upgrading and is most suitable for higher flow rates, typically between 500 and 2000 Nm^3/h (Ahmed et al. 2021). The main advantage of water scrubbing is its ability to simultaneously remove both CO_2 and H_2S due to their significantly higher solubility in water compared to methane. At 20 °C, the solubility of CO_2 in water is 26 times greater than CH_4 (901 mL/L), while H_2S is even more soluble (2830 mL/L) (Angelidaki et al. 2018; Ahmed et al. 2021).

The process begins by pressurizing the biogas and introducing it at the bottom of the absorption column, while water flows counter-currently from the top. As the biogas ascends, CO_2 and H_2S dissolve into the water, and the purified biomethane exits from the top of the scrubber. The water, now enriched with the impurities removed, is sent to a flush column where the pressure is reduced to recover dissolved CH_4 (Bauer et al. 2013).

Two primary water scrubbing methods are used: “single pass scrubbing” and “regenerative absorption.” Single pass scrubbing is applied when water from external sources, such as sewage treatment plants, is available. In regenerative absorption, water is regenerated through desorption by reducing pressure to atmospheric levels, often using air stripping to remove absorbed gases. This regeneration step is essential due to the high-water consumption; for example, upgrading 1000 Nm^3/h of raw biogas typically requires 180–200 m^3/h of water (Angelidaki et al. 2018).

Although the CO_2 released during water regeneration is usually not captured, its purity can reach 80–90% through air stripping (Ahmed et al. 2021). The biomethane produced by water scrubbing is saturated with water, necessitating a drying step to achieve final methane purity levels of up to 99% (Angelidaki et al. 2018). Methane loss during the process generally ranges from 3% to 5%, although some equipment suppliers claim losses can be controlled to below 2% (Ahmed et al. 2021). Despite its higher investment and operating costs, water scrubbing remains an environmentally friendly technology that effectively upgrades biogas without the use of hazardous chemicals. It achieves approximately 97% methane recovery while simultaneously removing both CO_2 and H_2S (Chen et al. 2015; Bauer et al. 2013).

17.3.3.5 Organic Scrubbing

Organic solvent scrubbing technology operates on the same principles as water scrubbing, but it uses organic solvents—such as dimethyl ethers of polyethylene glycol (e.g., Selexol® or Genosorb®)—instead of water. These solvents absorb CO_2 and H_2S more efficiently than water, requiring less volume. Their low vapor pressure reduces losses, minimizing consumption and allowing for a more compact, high-performance system (Ahmed et al. 2021).

The process involves pressurizing biogas and introducing it into an absorption column, where CO_2 and H_2S dissolve in the organic solvent. Like water scrubbing, the methane is released from the top of the scrubber, while the solvent enriched with acid gases is sent to a regeneration unit. The regeneration process typically involves pressure reduction, vacuum stripping, or boiling to release the absorbed gases and recover the solvent for reuse (Angelidaki et al. 2018). However, the high solubility of CO_2 in organic solvents makes complete regeneration challenging. This problem is amplified when H_2S concentrations are high, demanding higher temperatures for effective separation during solvent regeneration. To mitigate this energy demand, it is advisable to pre-remove H_2S using an activated carbon filter if its concentration is significant (Ahmed et al. 2021).

Organic solvent scrubbing achieves high methane recovery (>97%) and removes multiple impurities with lower solvent consumption. However, it requires high investment, complex management, and significant energy for solvent regeneration. Water dilution can reduce efficiency, and its adoption remains limited (6% of the market) due to high costs and energy demands, especially for biogas with high H_2S levels (Chen et al. 2015; Ahmed et al. 2021; Zhao et al. 2017).

17.3.3.6 Cryogenic Separation

Cryogenic separation purifies biogas by gradually reducing its temperature, enabling the liquefaction and separation of methane from CO_2 and other components. The process involves drying and compressing the raw biogas to approximately 80 bar, followed by a controlled, stepwise temperature reduction to −170 °C (Chen et al. 2015). During this cooling process, impurities such as water, hydrogen sulfide (H_2S), siloxanes, and halogens are removed. Subsequently, CO_2, the second most abundant component of biogas, is separated, allowing the recovery of high-purity biomethane (>97%) (Angelidaki et al. 2018).

Although cryogenic separation facilitates biomethane liquefaction and is an environmentally friendly process that does not require chemicals, it remains energy-intensive and is not yet widely adopted. This technology represents only 0.4% of the biogas upgrading market, with limited commercial-scale facilities (Ahmed et al. 2021). To prevent CO_2 from solidifying below −78 °C, which can lead to pipeline blockages and temperature fluctuations, maintaining high pressure throughout the process is essential, thereby increasing energy costs. Additionally, combining cryogenic separation with other upgrading technologies to pre-remove water and H_2S can help reduce operational challenges. The system also relies on turbines and distillation columns, further contributing to its high capital and operational costs (Ahmed et al. 2021).

17.4 Biogas Applications

Biogas is very versatile, and it can be utilized either directly or after being upgraded to biomethane. Direct biogas utilization involves its application as a fuel, for example, that may occur after a cleaning step to remove major impurities. When upgraded to biomethane, its energy content is increased, and nobler applications may be aimed.

17.4.1 Direct Biogas Application

Biogas can serve as a fuel after the removal of some impurities such as H_2S that cause corrosion in system components and water vapor to prevent condensation. It can be applied to generate electricity in power plants, in boilers to produce heat and steam, or in combined cycle power plants to produce both heat and power, increasing the efficiency of energy recovery. The scales of utilization differ. Biogas can be utilized on site after its generation, to produce heat and electricity to meet the demands of farms or industries, for example (Alengebawy et al. 2024).

However, there are some cases in which the amount of residue generated is not enough for feasible/profitable biogas generation or utilization. In those cases, cooperatives can be set up to collect the residues of multiple generators and to digest it into centralized equipment. Alternatively, if biogas generation is practical but the utilization is not required or even feasible, centralized schemes for biogas utilization or upgrading can be established, by constructing pipelines between nearby biogas generators. Centralized biogas utilization can provide district heating, for example (Alengebawy et al. 2024). The economic feasibility of those schemes needs to be thoroughly investigated, and a case-by-case life cycle assessment is needed to ensure environmental benefits. If economically and environmentally attractive, this type of approach can greatly contribute to waste-to-energy conversion, the circular economy, and local development.

Biogas can also be used in smaller-scale applications such as household heating and as a fuel in stoves in rural areas, for example. However, due to its lower energy content when compared to natural gas, biogas cannot be directly used as fuel in vehicles utilizing compressed natural gas (CNG) (Mertins and Wawer 2022). For these purposes, it needs to be upgraded to biomethane, as will be discussed in the following.

17.4.2 Upgrading to Biomethane

Although biogas can be directly used in some applications, high CO_2 concentration lowers its energy content and limits its applicability. For some purposes, such as the use in CNG-driven vehicles, biogas needs to be upgraded to biomethane, removing CO_2 and allowing for a CH_4 concentration of $>95\%$ (v/v) (Luque et al. 2016). Biomethane is analogous to natural gas and can be utilized in virtually all natural gas applications. Thus, it can take advantage of the already installed infrastructure for natural gas transportation, storage, and utilization. Furthermore, it can also be utilized in the applications cited for biogas, such as in power and heat generation, and despite being more expensive than biogas due to upgrading costs, it is also more versatile (Mertins and Wawer 2022).

By adding biomethane to the natural gas grid, this renewable raw material can be processed in different types of industries. One important application is the generation of syngas, a precursor of high-added-value products such as hydrogen, fuels, and chemicals. The generation of syngas from biomethane or biogas can be performed by reforming reactions, which are catalytic reactions, in which methane and other reactants are converted to a mixture of carbon monoxide and hydrogen under high temperatures. Multiple reforming reactions can be performed/favored, depending on the reactants and reaction conditions, and either biomethane or biogas (considering the utilization of CO_2) can be utilized. It should be mentioned, however, that not all types of methane reforming are utilized in commercial operations, and some of them are prone to coke deposition and catalyst deactivation and oxidation, which pose challenges to the operation (da Silva Pinto et al. 2022; Zhao et al. 2020).

The obtained syngas is a high-added-value product, and multiple applications can be aimed at. It can be purified to green hydrogen, for example, and this gas can be used in the multiple industrial applications that already exist for it and as a clean fuel that only emits water as combustion residue. Syngas can also be utilized as a feedstock to produce valuable products such as synthetic hydrocarbons, methanol, and dimethyl ether, among others, and can be applied in fuel cells (Zhao et al. 2020).

17.5 Residues Produced

Biogas production and processing can generate byproducts that can be repurposed for various applications, enhancing the sustainability of the process. Anaerobic digestion for biogas production typically produces a residual solid byproduct known as digestate. Due to its high organic content and nutrient-rich composition, digestate can be used as an organic fertilizer to improve soil quality and support agricultural activities. This sustainable reuse not only reduces waste but also promotes a circular economy within the biogas production cycle.

Additionally, the biogas upgrading process generates a gaseous byproduct primarily composed of CO_2 after biomethane is separated. This CO_2 stream can be captured and utilized in several industrial applications, including methanol synthesis, carbon capture and storage (CCS), and the food and beverage industry for carbonation. Effective management and utilization of these byproducts improve both the environmental and economic performance of biogas production.

17.5.1 Digestate and Agriculture

Digestate, the residual biomass produced through anaerobic digestion in biogas production, is emerging as a promising organic fertilizer and soil amendment in sustainable agriculture. In this oxygen-free process, bacteria decompose organic matter, yielding a nutrient-rich residue that not only supplies essential elements, such as nitrogen, phosphorus, and potassium in plant-accessible forms but also fosters a thriving soil microbiota (Du et al. 2018). This functionality allows digestate to act as an organic fertilizer, enhancing nutrient uptake, promoting plant growth, increasing organic matter content, enhancing soil structure, providing concentrated nutrients, and improving moisture retention. The performance and specific role of digestate depend on factors such as the composition of the feedstock and the treatment processes applied during and after anaerobic digestion. By contributing to a circular nutrient cycle and reducing reliance on conventional synthetic fertilizers, digestate supports soil preservation and regeneration, advancing sustainable land management practices (Sharma et al. 2023).

The application of digestate has shown positive effects on different crops. In wheat cultivation, 3 years of digestate application under various treatment regimens led to an increase in grain yield compared to the control. Additionally, compared to the NPK treatment, the grain exhibited higher protein, wet gluten, and phenol content (Różyło et al. 2017). In tomato cultivation, the use of digestate and concentrated digestate increased soil NPK levels and electrical conductivity (EC), while slightly reducing soil pH from 5.4 to 5.2 (Yu et al. 2010). For switchgrass, 3 years of digestate application at rates of 30 and 60 m^3/ha improved soil sorption properties, reduced acidity, and increased zinc, potassium, and organic matter content, leading to higher switchgrass yields (Głowacka et al. 2020). The presence of heavy metals is a significant concern regarding the safety of biofertilizers. Commonly found in food waste, particularly municipal solid waste (MSW), these metals include cadmium (Cd), chromium (Cr), copper (Cu), lead (Pb), mercury (Hg), nickel (Ni), and zinc (Zn) (Abdullah and Greetham 2016). Applying biofertilizers with elevated metal concentrations can result in the contamination of agricultural soil, with metals potentially accumulating in plant roots, persisting in the soil, and leaching into groundwater. So, to prevent this, the quality of digestate must be assessed through rigorous testing to ensure it meets safety standards and minimizes environmental risks (Du et al. 2018).

The quality of digestate and its agronomic potential depend on several interrelated factors. Firstly, the composition and variability of residues—such as animal waste, carcasses, agro-industrial byproducts, and plant biomass—directly influence nutrient availability. Secondly, the type of biodigester and the specific biodigestion technology employed determine the overall efficiency and characteristics of the process. Furthermore, proper segregation and minimization of nutrient losses in

Table 17.4 Characterization of digestate from biodigesters treating liquid swine manure

Farm type	Animals housed	TS (g/L)	N (mg/L)	$N\text{-}NH_4$ (mg/L)	P_2O_5 (mg/L)	K_2O (mg/L)
Piglet production unit (UPL)	400 sows	14.8	2008	1527	850	576
Growing and finishing unit (UCT)	750 pigs	4.2	987	954	31	919
Farrow-to-finish unit (CC)	150 sows	1.7	125	94	29	447
Nursery unit (UC)	1500 piglets	19.4	2376	1843	352	1438

Adapted from Kunz et al. (2019)

storage systems for both substrates and digestate are critical. Additionally, the effectiveness of substrate pretreatment systems (e.g., phase separation before biodigestion) and subsequent digestate treatment plays a vital role. Finally, the dilution of substrates and digestate with water further impacts the final quality. Table 17.4 displays different digestate characteristics, influenced by all these variables (Kunz et al. 2019).

While recycling digestate in agricultural systems can help offset some operational costs associated with treatment facilities, the design and implementation of such projects must also address several technical and economic challenges. These include ensuring a reliable supply of nutrients via digestate, matching nutrient availability with the agricultural demand in target areas, and managing the logistics of fertilizer distribution (Du et al. 2018). Despite these challenges, the growing interest in sustainable agricultural input has driven demand for fertilizers, including digestate, positioning them as valuable alternatives.

The market for biofertilizers, including digestate, is shaped by multiple factors that influence its economic value. These include the consistency of nutrient composition, overall product quality, the geographical location of production, seasonal demand variations, and public perception of bio-based fertilizers. A compelling economic advantage of biofertilizers lies in their potential to reduce agricultural input costs. Research indicates that substituting chemical fertilizers with biofertilizers can yield cost savings ranging from £84 to £118 per hectare, making them an attractive option for farmers seeking sustainable practices (Barzee et al. 2019).

In 2019, the global biofertilizer market was valued at approximately USD 1.0 billion (Ma et al. 2017). Projections suggest robust growth, with the market expected to reach USD 1.55 billion by 2025 and further expand to USD 2.83 billion by 2030 (Grand View Research 2025). This upward trajectory reflects increasing global demand for sustainable agricultural inputs and the rising adoption of organic farming practices.

Within this global context, South America stands out as a region with significant growth potential. The digestate market in South America is anticipated to achieve the second-highest compound annual growth rate (CAGR) during the next few years (Grand View Research 2025). This growth is driven by economic development in countries such as Argentina and Brazil, where market expansion is expected to outpace the regional average. The availability of arable land and ongoing farmland expansion, particularly in Argentina, Brazil, and Chile, further bolsters this trend. Additionally, crop production in South America has surged, with Brazil notably increasing its cultivated areas, thereby heightening the demand for fertilizers (Grand View Research 2025).

In Europe, comprehensive data on digestate production remains elusive, yet estimates provide valuable insights into its scale. It is estimated that the European Union (EU28) generates around 180 million metric tons of digestate annually. Of this total, 68% originates from agricultural feedstock, highlighting the agricultural sector's substantial contribution to digestate output (Catenacci et al. 2022). Additional projections indicate that digestate production is linked to the energy infrastructure of anaerobic digestion facilities. Specifically, it is estimated that approximately 20 cubic meters of digestate are produced per year for each kilowatt of electric capacity installed in downstream combined heat and power (CHP) units (Catenacci et al. 2022).

China has emerged as a major producer of digestate, with annual output surpassing 200 million tons. This significant volume is a direct result of the rapid expansion of anaerobic digestion projects across the country, reflecting China's strategic focus on renewable energy and sustainable waste management. The proliferation of these projects has led to a dramatic increase in the production of both solid digestate and liquid digestate, presenting opportunities for agricultural applications as well as challenges related to effective utilization and management (Fu et al. 2025).

This nutrient-rich byproduct of anaerobic digestion can offer promising opportunities for enhancing agricultural sustainability. Its application as an organic fertilizer can improve soil health, increase crop yields, and contribute to a circular nutrient economy. However, careful management is essential to mitigate potential environmental risks, such as nutrient leaching and soil salinity issues. Ongoing research and technological advancements are crucial to optimizing digestate utilization, ensuring that its benefits are maximized while minimizing any adverse effects. By addressing these challenges, digestate can play a vital role in promoting sustainable agricultural practices and supporting global food security.

17.5.2 CO_2

The removal of CO_2 from biogas is a well-established field, with several advanced technologies employed to achieve the desired biomethane quality. These methods, discussed in previous sections, utilize the distinct physical and chemical properties of CO_2 to effectively separate it from methane. The selection of a particular technique is often influenced by factors such as energy efficiency, cost-effectiveness, and scalability, since each method involves trade-offs related to energy consumption, capital investment, and operational complexity, with the appropriate choice depending on local conditions, system scale, and regulatory considerations. This upgrading process not only purifies the gas for use in various applications but also presents opportunities for utilizing the separated CO_2, such as in the production of methanol or through carbon capture and utilization technologies, enhancing sustainability.

As of 2021, the global production capacity of methanol (CH_3OH) was approximately 164 million tons per year, with a projected annual growth rate of around 10% over the next decade. Traditionally, methanol synthesis has been achieved using syngas derived from fossil sources, typically through steam reforming of natural gas or coal gasification. However, the utilization of captured CO_2 for methanol production is gaining significant attention worldwide. In this context, the CO_2 captured during the biogas upgrading process presents a promising opportunity for methanol production (Campos et al. 2022).

There are two prominent methodologies for this transformation: CO_2 hydrogenation and direct CO_2 electrolysis to methanol. The hydrogenation process involves the catalytic reaction of CO_2 with H_2 to synthesize methanol (CH_3OH). This process typically requires elevated temperatures (200–300 °C) and pressures (3–10 MPa), along with effective catalysts to overcome the high energy barrier associated with CO_2 reduction. Recent advancements have focused on developing catalysts that enhance activity and selectivity, aiming to make the process more efficient and economically viable (Cui et al. 2024). Direct CO_2 electrolysis refers to the electrochemical reduction of CO_2 to methanol, offering a potentially more sustainable route by integrating renewable energy sources. This method can be categorized into three primary approaches (Li et al. 2022): electrolysis of water to produce hydrogen, followed by CO_2 hydrogenation; direct electrochemical reduction of CO_2 to methanol; and co-electrolysis of CO_2 and H_2O to produce syngas, followed by methanol synthesis.

Using CO_2 to produce methanol involves advanced catalyst development and process optimization, offering a sustainable utilization pathway. Both methods face economic and policy challenges, with ongoing research aiming to enhance efficiency and increase viability. As the global demand for cleaner and more sustainable chemical production increases, further advancements in these technologies are critical to overcoming these challenges and enabling large-scale adoption.

17.6 Final Remarks

Biogas production offers a sustainable alternative to fossil fuels by converting a wide range of organic feedstocks into renewable energy and nutrient-rich digestate. Anaerobic digestion processes, when optimized through appropriate biodigester design and operating conditions, enable the efficient conversion of livestock manure, sewage sludge, food waste, crop residues, and industrial residues into biogas. Advanced purification techniques further upgrade biogas to biomethane, enhancing its applicability in various energy and industrial sectors. Overall, the sector's growth is supported by robust policy frameworks, technological innovation, and integrated waste management practices, contributing to energy security and environmental sustainability.

17.7 Compliance with Ethical Standards

The authors declare that this chapter complies with the ethical standards of research and publication. No studies involving human participants or animals were conducted by the authors for the preparation of this manuscript. Ethical approval and informed consent were not required. The authors declare that there are no conflicts of interest related to the content of this chapter, and that all sources of funding, if any, have been appropriately acknowledged.

References

Abdullah JJ, Greetham D (2016) Optimizing cellulase production from municipal solid waste (MSW) using solid state fermentation (SSF). J Fundam Renew Energy Appl 6(3):206. https://doi.org/10.4172/2090-4541.1000206

Achinas S, Achinas V, Euverink GJW (2017) A technological overview of biogas production from biowaste. Engineering 3(3):299–307. https://doi.org/10.1016/J.ENG.2017.03.002

Ahmed SF, Mofijur M, Tarannum K et al (2021) Biogas upgrading, economy and utilization: a review. Environ Chem Lett 19(6):4137–4164. https://doi.org/10.1007/s10311-021-01292-x

Alengebawy A, Ran Y, Osman AI et al (2024) Anaerobic digestion of agricultural waste for biogas production and sustainable bioenergy recovery: a review. Environ Chem Lett 22:2641–2668. https://doi.org/10.1007/s10311-024-01789-1

American Biogas Council (2025) Biogas market snapshot. In: American Biogas Council. Available at https://americanbiogascouncil.org/biogas-market-snapshot/. Accessed 03 Apr 2025

Angelidaki I, Treu L, Tsapekos P et al (2018) Biogas upgrading and utilization: current status and perspectives. Biotechnol Adv 36(2):452–466. https://doi.org/10.1016/j.biotechadv.2018.01.011

Awe OW, Zhao Y, Nzihou A et al (2017) A review of biogas utilisation, purification and upgrading technologies. Waste Biomass Valoriz 8(2):267–283. https://doi.org/10.1007/s12649-016-9826-4

Baker RW (2012) Membrane technology and applications, 3rd edn. Wiley, Newark. https://doi.org/10.1002/9781118359686

Banks C, Heaven S, Zhang Y, Baier U (2018) Food waste digestion: anaerobic digestion of food waste for a circular economy. IEA Bioenergy Task 37(2018):12

Barzee TJ, Edalati A, El-Mashad H et al (2019) Digestate biofertilizers support similar or higher tomato yields and quality than mineral fertilizer in a subsurface drip fertigation system. Front Sustain Food Syst 3:58. https://doi.org/10.3389/fsufs.2019.00058

Bauer F, Persson T, Hulteberg C, Tamm D (2013) Biogas upgrading—technology overview, comparison and perspectives for the future. Biofuels Bioprod Biorefin 7(5):499–511. https://doi.org/10.1002/bbb.1423

Bhatt AH, Tao L (2020) Economic perspectives of biogas production via anaerobic digestion. Bioengineering 7(3):74. https://doi.org/10.3390/bioengineering7030074

Bochmann G, Pesta G, Rachbauer L et al (2020) Anaerobic digestion of pretreated industrial residues and their energetic process integration. Front Bioeng Biotechnol 8:487. https://doi.org/10.3389/fbioe.2020.00487

Brahmi M, Bruno B, Dhayal KS et al (2024) From manure to megawatts: navigating the sustainable innovation solution through biogas production from livestock waste for harnessing green energy for green economy. Heliyon 10(14):e34504. https://doi.org/10.1016/j.heliyon.2024.e34504

Campos BLO, John K, Beeskow P et al (2022) A detailed process and techno-economic analysis of methanol synthesis from H2 and CO2 with intermediate condensation steps. PRO 10:1535. https://doi.org/10.3390/pr10081535

Catenacci A, Boniardi G, Mainardis M et al (2022) Processes, applications and legislative framework for carbonized anaerobic digestate: opportunities and bottlenecks. A critical review. Energy Convers Manag 263:115691. https://doi.org/10.1016/j.enconman.2022.115691

Chen XY, Vinh-Thang H, Ramirez AA et al (2015) Membrane gas separation technologies for biogas upgrading. RSC Adv 5(31):24399–24448. https://doi.org/10.1039/C5RA00666J

CIBiogas (2024) Biogas Outlook 2023. Available at https://abiogas.org.br/wp-content/uploads/protectedfiles/cms_files_54738_1716811508PANORAMA_DO_BIOGS_2023.pdf. Accessed 13 Mar 2025

Cui G, Lou Y, Zhou M et al (2024) Review of mechanism investigations and catalyst developments for CO_2 hydrogenation to alcohols. Catalysts 14(4):232. https://doi.org/10.3390/catal14040232

da Silva Pinto RL, Vieira AC, Scarpetta A et al (2022) An overview on the production of synthetic fuels from biogas. Bioresour Technol Rep 18:101104. https://doi.org/10.1016/j.biteb.2022.101104

Daud MK, Rizvi H, Akram MF et al (2018) Review of upflow anaerobic sludge blanket reactor technology: effect of different parameters and developments for domestic wastewater treatment. J Chem 2018:1596319. https://doi.org/10.1155/2018/1596319

Demirbas A, Taylan O, Kaya D (2016) Biogas production from municipal sewage sludge. Energy Sources Part A 38(20):3027–3033. https://doi.org/10.1080/15567036.2015.1124944

Di Giacomo G, Romano P (2022) Evolution and prospects in managing sewage sludge resulting from municipal wastewater purification. Energies 15(15):5633. https://doi.org/10.3390/en15155633

Du C, Abdullah JJ, Greetham D et al (2018) Valorization of food waste into biofertiliser and its field application. J Clean Prod 187:273–284. https://doi.org/10.1016/j.jclepro.2018.03.211

Enebe NL, Chigor CB, Obileke K et al (2023) Biogas and syngas production from sewage sludge: a sustainable source of energy generation. Methane 2(2):192–217. https://doi.org/10.3390/methane2020014

EPA (2024) Basic information about landfill gas. In: Landfill Methane Outreach Program (LMOP). Available at https://www.epa.gov/lmop/basic-information-about-landfill-gas. Accessed 03 Apr 2025

European Commission (2025a) Renewable energy—recast to 2030 (RED II). In: The Joint Research Centre: EU Science Hub. Available at https://joint-research-centre.ec.europa.eu/welcome-jec-website/reference-regulatory-framework/renewable-energy-recast-2030-red-ii_en. Accessed 13 Mar 2025

European Commission (2025b) REPowerEU: affordable, secure and sustainable energy for Europe. In: Directorate-General for Communication. Available at https://commission.europa.eu/strategy-and-policy/priorities-2019-2024/european-green-deal/repowereu-affordable-secure-and-sustainable-energy-europe_en. Accessed 13 Mar 2025.

Fu Z, Zhao J, Zhong Z et al (2025) Novel insights into liquid digestate-derived hydrochar enhances volatile fatty acids production from anaerobic co-digestion of sludge and food waste. Chem Eng J 507:160635. https://doi.org/10.1016/j.cej.2025.160635

Głowacka A, Szostak B, Klebaniuk R (2020) Effect of biogas digestate and mineral fertilization on the soil properties and yield and nutritional value of switchgrass forage. Agronomy 10(4):490. https://doi.org/10.3390/agronomy10040490

Grand View Research (2025) Biofertilizers market size & trends. In: Biofertilizers market size, share & trends analysis report by product (nitrogen fixation, phosphate solubilizing), by application (seed treatment, oil treatment), by crop type, by region, and segment forecasts, 2025–2030. Available at https://www.grandviewresearch.Com/Industry-Analysis/Biofertilizers-Industry. Accessed 21 Mar 2025

Gustafsson M, Meneghetti R, Souza Marques F et al (2024) IEA bioenergy: a perspective on the state of the biogas industry from selected member countries. IEA Bioenergy Task 37(2024):2. https://task37.ieabioenergy.com/country-reports.html

Hagos K, Zong J, Li D et al (2017) Anaerobic co-digestion process for biogas production: progress, challenges and perspectives. Renew Sustain Energy Rev 76:1485–1496. https://doi.org/10.1016/j.rser.2016.11.184

He K, Liu Y, Tian L et al (2024) Review in anaerobic digestion of food waste. Heliyon 10(7): e28200. https://doi.org/10.1016/j.heliyon.2024.e28200

IEA (2020) OECD Publishing. https://doi.org/10.1787/040c8cd2-en

IEE-Project, & FABbiogas ([s.d].) Biogas production and biogas potentials from residues of the European food and beverage industry. European Biogas Association. Available at https://www.europeanbiogas.eu/biogas-production-and-biogas-potentials-from-residues-of-the-european-food-and-beverage-industry/. Accessed 01 Apr 2025.

Jain S, Newman D, Nzihou A et al. (2019) Global potential of biogas. Available at https://www.worldbiogasassociation.org/wp-content/uploads/2019/09/WBA-globalreport-56ppa4_digital-Sept-2019.pdf. Accessed 24 Mar 2025.

Kanda W, Zanatta H, Magnusson T et al (2022) Policy coherence in a fragmented context: the case of biogas systems in Brazil. Energy Res Soc Sci 87:102454. https://doi.org/10.1016/j.erss.2021.102454

Kunz A, Steinmetz RLR, Amaral AC (2019) Fundamentals of anaerobic digestion, biogas purification, use and treatment of digestate, vol 1, 1st edn. Embrapa

Li P, Gong S, Li C, Liu Z (2022) Analysis of routes for electrochemical conversion of CO_2 to methanol. Clean Energy 6(1). https://doi.org/10.1093/ce/zkac007

Liebetrau J, Wellisch M, Lyng KA et al. (2021) Potential and utilization of manure to generate biogas in seven countries. IEA Bioenergy Task 37. Available at https://www.ieabioenergy.com/wp-content/uploads/2021/07/Potential-utilization_WEB_END_NEW.pdf. Accessed 24 Mar 2025.

López AF, Lago Rodríguez T, Faraji Abdolmaleki S et al (2024) From biogas to biomethane: an in-depth review of upgrading technologies that enhance sustainability and reduce greenhouse gas emissions. Appl Sci 14(6):2342. https://doi.org/10.3390/app14062342

Lu J, Gao X (2021) Biogas: potential, challenges, and perspectives in a changing China. Biomass Bioenergy 150:106127. https://doi.org/10.1016/j.biombioe.2021.106127

Luque R, Lin CSK, Wilson K, Clark J (2016) Handbook of biofuels production: processes and technologies, 2nd edn. Elsevier. https://doi.org/10.1016/C2014-0-04240-2

Ma Y, Yin Y, Liu Y (2017) New insights into co-digestion of activated sludge and food waste: biogas versus biofertilizer. Bioresour Technol 241:318–326. https://doi.org/10.1016/j.biortech.2017.05.154

Mertins A, Wawer T (2022) How to use biogas? A systematic review of biogas utilization pathways and business models. Bioresour Bioprocess 9:59. https://doi.org/10.1186/s40643-022-00545-z

Ministry of New and Renewable Energy, Government of India (2022) Biogas programme. Government of India, Ministry of New and Renewable Energy. Available at https://biogas.mnre.gov.in/about-the-programmes. Accessed 13 Mar 2025

Mittal S, Ahlgren EO, Shukla PR (2019) Future biogas resource potential in India: a bottom-up analysis. Renew Energy 141:379–389. https://doi.org/10.1016/j.renene.2019.03.133

Perin JKH, Biesdorf PLB, Torrecilhas AR et al (2020) Optimization of methane production parameters during anaerobic co-digestion of food waste and garden waste. J Clean Prod 272:123130. https://doi.org/10.1016/j.jclepro.2020.123130

Prasad S, Singh A, Korres NE et al (2020) Sustainable utilization of crop residues for energy generation: a life cycle assessment (LCA) perspective. Bioresour Technol 303:122964. https://doi.org/10.1016/J.BIORTECH.2020.122964

Ramos NMV, Del-Mazo-Alvarado O, Bonilla-Petriciolet A et al (2024) Multi-objective optimization of syngas production for Fischer-Tropsch synthesis based on biogas catalytic reforming and upgrading. Chem Eng Process Process Intensif 199:109758. https://doi.org/10.1016/j.cep.2024.109758

Różyło K, Świeca M, Gawlik-Dziki U et al (2017) Phytochemical properties and heavy metal accumulation in wheat grain after three years' fertilization with biogas digestate and mineral waste. Agric Food Sci 26(3):148–159. https://doi.org/10.23986/afsci.63156

Scarlat N, Martinov M, Dallemand JF (2010) Assessment of the availability of agricultural crop residues in the European Union: potential and limitations for bioenergy use. Waste Manag 30(10):1889–1897. https://doi.org/10.1016/J.WASMAN.2010.04.016

Shah SV, Yadav LB, Tiwari AK, Chen W-H (2024) Sustainable biogas production via anaerobic digestion with focus on CSTR technology: a review. J Taiwan Inst Chem Eng 162:105575. https://doi.org/10.1016/j.jtice.2024.105575

Sharma P, Bano A, Verma K et al (2023) Food waste digestate as biofertilizer and their direct applications in agriculture. Bioresour Technol Rep 23:101515. https://doi.org/10.1016/j.biteb.2023.101515

Sihlangu E, Luseba D, Regnier T et al (2024) Investigating methane, carbon dioxide, ammonia and hydrogen sulphide content in agricultural waste during biogas production. Sustainability 16(12):5145. https://doi.org/10.3390/su16125145

Silva CEV, Marder M, Hickman EV et al (2022) Qualitative aspects of biogas production in a small-scale biodigester by anaerobic digestion of food waste from university environment. Fronteiras 11(1):31–38. https://doi.org/10.21664/2238-8869.2022v11i1.p31-38

Silva JFF, Tápparo DC, Steinmetz RLR et al (2024) Anaerobic co-digestion of swine manure and enzymatically hydrolyzed carcasses at two different biodigesters. Biomass Conv Biorefin. https://doi.org/10.1007/s13399-024-06327-1

Singh P, Kalamdhad AS (2021) A comprehensive assessment of state-wise biogas potential and its utilization in India. Biomass Convers Biorefinery 13(14):12557–12579. https://doi.org/10.1007/s13399-021-02001-y

Suhartini S, Nurika I, Paul R et al (2021) Estimation of biogas production and the emission savings from anaerobic digestion of fruit-based agro-industrial waste and agricultural crops residues. Bioenergy Res 14:844–859. https://doi.org/10.1007/s12155-020-10209-5

Sukhesh MJ, Rao PV (2018) Anaerobic digestion of crop residues: technological developments and environmental impact in the Indian context. Biocatal Agric Biotechnol 16:513–528. https://doi.org/10.1016/J.BCAB.2018.08.007

Thunder Said Energy (2023) Global biogas production by country. In: Thunder Said Energy. Available at https://thundersaidenergy.com/downloads/global-biogas-production-by-country/. Accessed 13 Mar 2025.

UNEP (2024) Food waste index report 2024: think eat save: tracking progress to halve global food waste. United Nations Environment Programme. Available at https://www.unep.org/resources/publication/food-waste-index-report-2024. Accessed 25 Mar 2025

Yesil H, Tugtas AE (2019) Removal of heavy metals from leaching effluents of sewage sludge via supported liquid membranes. Sci Total Environ 693:133608. https://doi.org/10.1016/j.scitotenv.2019.133608

Yoo M, Han SJ, Wee JH (2013) Carbon dioxide capture capacity of sodium hydroxide aqueous solution. J Environ Manag 114:512–519. https://doi.org/10.1016/j.jenvman.2012.10.061

Yu FB, Luo XP, Song CF, Shan SD (2010) Concentrated biogas slurry enhanced soil fertility and tomato quality. Acta Agric Scand B 60(3):262–268. https://doi.org/10.1080/09064710902893385

Zhang P (2018) Biogas recovery from anaerobic digestion of selected industrial wastes. In: Advances in biofuels and bioenergy. IntechOpen. https://doi.org/10.5772/intechopen.72292

Zhao Q, Leonhardt E, MacConnell C et al. (2010) Purification technologies for biogas generated by anaerobic digestion. In: Climate friendly farming: improving the carbon footprint of agriculture in the Pacific northwest. CSANR research report 2010-00:24. Available at https://www.build-a-biogas-plant.com/PDF/BiogasPurificationTech2010.PDF. Accessed 19 Mar 2025.

Zhao X, Zhou H, Sikarwar VS et al (2017) Biomass-based chemical looping technologies: the good, the bad and the future. Energy Environ Sci 10(9):1885–1910. https://doi.org/10.1039/c6ee03718f

Zhao X, Joseph B, Kuhn J, Ozcan S (2020) Biogas reforming to syngas: a review. iScience 23(5): 101082. https://doi.org/10.1016/j.isci

Chapter 18
Integrating Biorefinery in the Pulp Industry: Advanced Technologies and Circular Economy

Graciela Ines Bolzon de Muniz, Rilton Alves de Freitas, Mayara Elita Braz Carneiro, Andrey Pereira Acosta, and Pedro Henrique Gonzalez de Cademartori

Abstract The incorporation of the biorefinery concept into the pulp and paper industry represents a strategic evolution toward a circular bioeconomy, proposing the full valorization of lignocellulosic biomass. This chapter presents a critical analysis of conventional pulp production processes (e.g., mechanical, sulphite, and kraft methods), highlighting their limitations in terms of selectivity, component recovery, and environmental sustainability. In contrast, the application of emerging pretreatment technologies, such as autohydrolysis, vapor explosion, ionic liquids, and organosolv processes, is explored, capable of promoting the selective fractionation of biomass and expanding the range of products obtained. In this context, emphasis is placed on the production of high-added-value biocomposites, including nanocellulose (crystalline and microfibrillated), second-generation bioethanol, regenerated fibers (such as lyocell), carbon fibers from lignin, vanillin, lignin

G. I. B. de Muniz (✉) · M. E. B. Carneiro · P. H. G. de Cademartori
Postgraduate Program in Forestry Engineering (PPGEF), Federal University of Paraná, Curitiba, Brazil

Postgraduate Program in Materials Science and Engineering (PIPE), Federal University of Paraná, Curitiba, Brazil

R. A. de Freitas
Postgraduate Program in Materials Science and Engineering (PIPE), Federal University of Paraná, Curitiba, Brazil

BioPol, Department of Pharmacy, Health Sciences Sector, Federal University of Paraná, Curitiba, Brazil

A. P. Acosta
Postgraduate Program in Forestry Engineering (PPGEF), Federal University of Paraná, Curitiba, Brazil

Postgraduate Program in Materials Science and Engineering (PIPE), Federal University of Paraná, Curitiba, Brazil

Human Resources Training Programme of the National Agency for Petroleum, Natural Gas and Biofuels (ANP), Federal University of Paraná, Curitiba, Brazil

A. L. Woiciechowski et al. (eds.), *Biomass for Bioenergy and Bioproducts*, Biofuel and Biorefinery Technologies 17, https://doi.org/10.1007/978-3-032-19898-3_18

adhesives, and functional materials with conductive properties. Case studies are presented from leading companies in the sector (e.g., Suzano, Borregaard, Stora Enso, and Lenzing) that have demonstrated the technical and economic viability of biorefinery integration, with a positive impact on portfolio diversification, reducing carbon emissions, and valorizing historically underused by-products. Finally, it can be concluded that the pulp industry's transition to a multifunctional biorefinery model is a viable way of promoting greater efficiency in the use of biomass, reducing environmental impacts, and making the sector a key player in consolidating the global bioeconomy.

Graphical Abstract

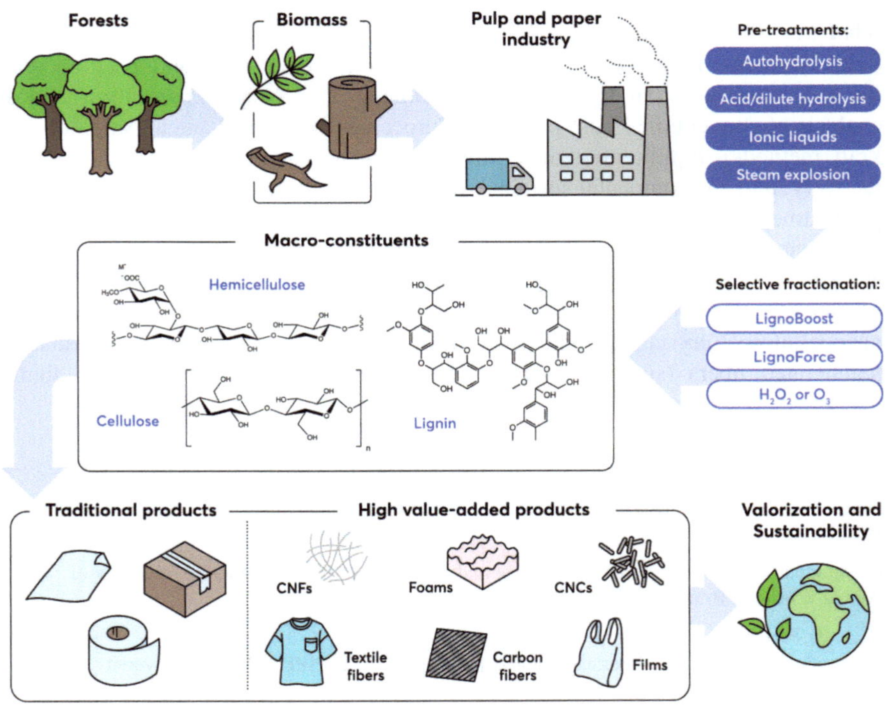

18.1 Introduction

The integration of biorefineries into the pulp and paper industry (P&PI) represents a transformative approach for valorizing lignocellulosic biomass, aligning industrial processes with the principles of the circular economy. Over the last century, the pulp and paper industry has undergone significant evolution, from the initial mechanical processes and acid sulfite pulping methods to the widespread adoption of the alkaline kraft process. Each advance has been driven by the need to improve pulp quality, increase fiber yield, and reduce operating costs. However, these traditional processes (such as mechanical pulping, the acid sulphite process, and, above all, the kraft process) were designed primarily to maximize pulp recovery, often neglecting the full recovery of hemicellulose and lignin fractions. This linear and resource-intensive model, although successful for large-scale production, ended up revealing critical limitations in terms of sustainability (such as high generation of liquid and solid effluents, intensive use of chemicals, high atmospheric emissions, low efficiency in biomass utilization, and dependence on fossil fuels). These limitations laid the foundations for the emergence of integrated biorefineries, which seek to optimize the use of all biomass components.

Lignocellulosic biomass, composed of cellulose, hemicellulose, and lignin, offers vast potential for producing high-value bioproducts such as biofuels, biochemicals, and advanced materials. Traditional pulp production, dominated by the kraft and sulfite processes, generates significant waste streams like black liquor and sludge, which are now being reimagined as feedstocks for biorefineries. Innovative technologies, including LignoBoost/LignoForce for lignin extraction (yields: 80–90%, purity >95%), autohydrolysis, and ionic liquid pretreatments, enable efficient fractionation and conversion of biomass components (Tomani 2010; Silveira et al. 2015; Casanova et al. 2023). These advancements have brought renewed attention to the limitations of conventional pulping methods and highlight the need for systemic change within the industry.

While traditional pulp and paper production processes, such as the kraft and sulfite methods, have long dominated the industry due to their efficiency in cellulose extraction, they are inherently linear, resource-intensive, and environmentally taxing. These methods primarily focus on pulp production, often discarding or combusting valuable components like lignin and hemicellulose, and generating high volumes of wastewater, sludge, and air pollutants. In contrast, integrated biorefineries adopt a circular, multi-product model that maximizes the use of all biomass fractions. By transforming side streams—like black liquor and hemicellulosic hydrolysates—into marketable biochemicals, fuels, and advanced materials, biorefineries significantly reduce environmental burdens while creating diversified revenue streams. This shift enables enhanced resource efficiency, greater energy self-sufficiency through internal residue valorization, and alignment with global sustainability goals (Kamali and Khodaparast 2015). Consequently, integrated biorefineries not only reconfigure the operational and environmental paradigm of the pulp and paper industry but also position it as a critical player in the emerging bioeconomy. Among these technologies, pretreatment stands out as a crucial step in unlocking the full potential of lignocellulosic feedstocks, directly influencing the quality and yield of downstream products.

Efficient pretreatment is a critical step in the valorization of lignocellulosic biomass for producing cellulose and high-value bioproducts. Methods such as autohydrolysis, acid or dilute acid hydrolysis, steam explosion, and ionic liquid pretreatment selectively break down biomass components, enhancing cellulose accessibility and downstream conversion efficiency (Galbe and Wallberg 2019). Among these, autohydrolysis stands out for its low environmental impact and cost-effectiveness, while acid hydrolysis, although efficient, generates inhibitory by-products and requires effluent treatment. Steam explosion offers high sugar yields but demands significant energy input, and ionic liquids demonstrate excellent selectivity and recyclability, though still limited by high costs (Martin-sampedro et al. 2014). Each method presents trade-offs between efficiency, cost, and sustainability, and their integration or hybridization is often pursued to optimize industrial-scale biorefinery performance. Following pretreatment, purification of cellulose is essential to remove residual lignin and hemicelluloses, improving its crystallinity and reactivity. This purified cellulose can be valorized into nanocellulose, bioethanol, regenerated fibers, or biodegradable films. Lignin, often separated during pretreatment, is increasingly used in advanced applications such as carbon fibers, adhesives, aromatic chemicals, and even graphene-like conductive materials (De Aguiar et al. 2020). Together, these valorized streams contribute to a biorefinery model that transforms waste into high-value products, supporting a circular and sustainable bioeconomy.

Companies such as Suzano (Brazil) and Borregaard (Norway) exemplify this shift by transforming lignin into valuable products like vanillin, carbon fibers, and adhesives. Additionally, nanocellulose applications are expanding into various industries, including packaging, textiles, and electronics. These companies are not alone; Stora Enso (Sweden/Finland), Lenzing (Austria), and CelluComp (United Kingdom) are also pioneering innovations in microfibrillated cellulose (MFC), cellulose nanocrystals (CNC), and sustainable textile fibers. For example, Suzano has made significant investments in MFC for reinforcing specialty papers and lightweight automotive composites, while Borregaard leads in biodegradable packaging and high-strength nanocellulose-based materials. Stora Enso's Cellulose Foam, Kruger's CNC applications in paints and composites, and Lenzing's lyocell fiber Tencel™ reflect the sector's growing commitment to replacing fossil-based inputs. This paradigm shift not only reduces reliance on fossil fuels but also minimizes waste, cuts CO_2 emissions, and enhances economic viability. Based on this, in this study, the key technologies, successful case studies, and the global impact of biorefinery integration in the pulp and paper industry were examined.

18.2 Lignocellulosic Biomass: Composition and Opportunities

18.2.1 Lignocellulosic Biomass Structure

The composition of lignocellulosic biomass is complex and varies depending on the source. However, generally, the main constituents of any lignocellulosic material include cellulose, hemicelluloses, lignin, organic extractives, and inorganic minerals

(Chen et al. 2016). Understanding the intricate structure and properties of each component is crucial to optimizing biomass conversion processes and maximizing the utilization of this valuable resource (Díez et al. 2020). The most abundant biomacromolecule on Earth, cellulose, is a linear "polymer" $(C_6H_{10}O_5)n$ formed by the linkage of monomeric β-D-glucose and dimeric cellobiose via 1,4-glycosidic bonds (Wang et al. 2022). This unique arrangement results in long, crystalline chains that provide structural support and rigidity to plant cell walls (Zhu et al. 2023). Although cellulose is generally crystalline, a combination of amorphous and crystalline regions can be observed in plants or biomass. The degree of polymerization depends on the type of lignocellulosic biomass. The degree of polymerization, which refers to the number of glucose units in a cellulose chain, can vary significantly depending on the source and processing conditions. Intra-chain and inter-chain hydrogen bonds within cellulose chains contribute to its high tensile strength and resistance to degradation. The ordered crystalline structure of cellulose presents a challenge for enzymatic breakdown, requiring pretreatment methods to disrupt the structure and enhance accessibility (Pereira et al. 2021).

The second main component of biomass, hemicelluloses, is a combination of several monomers, consisting of five different C_5 (xylose and arabinose) and C_6 (galactose, glucose, and mannose) sugars. Unlike cellulose, hemicellulose has a branched and amorphous structure in nature, making it more susceptible to hydrolysis. The composition and structure of hemicellulose vary significantly depending on the plant species and tissue type, with xylan being the dominant component in hardwoods. One of the main functions of hemicelluloses is to create bonds between cellulose and lignin. Cellulose and hemicellulose, combined with the third main component of biomass, lignin, make up more than 90% of lignocellulosic biomass (Segers et al. 2024).

Lignin is a three-dimensional aromatic heteropolymer composed of three phenylpropanoid monomers (coniferyl, sinapyl, and p-coumaryl alcohols). Within the "polymer" structure of lignin, the three monolignols are designated as guaiacyl, syringyl, and p-hydroxyphenyl subunits, respectively. Similar to hemicelluloses, the exact composition of lignin also varies with the type of lignocellulosic biomass. It acts as a cementing agent, offering structural support, impermeability, and resistance to microbial attack. Unlike cellulose and hemicellulose, lignin is an aromatic polymer and does not contain any sugar monomers. Its complex and irregular structure contributes to its high resistance to degradation, posing a significant challenge for biomass conversion processes (Segers et al. 2024). The chemical structures of all three constituents are shown in Fig. 18.1.

18.2.2 Planted Forests as Raw Material

Planted forests are gaining prominence as sustainable sources of feedstock for biorefineries, driven by the increasing demand for renewable resources and the need to mitigate climate change (Bušić et al. 2018). The use of these forests offers

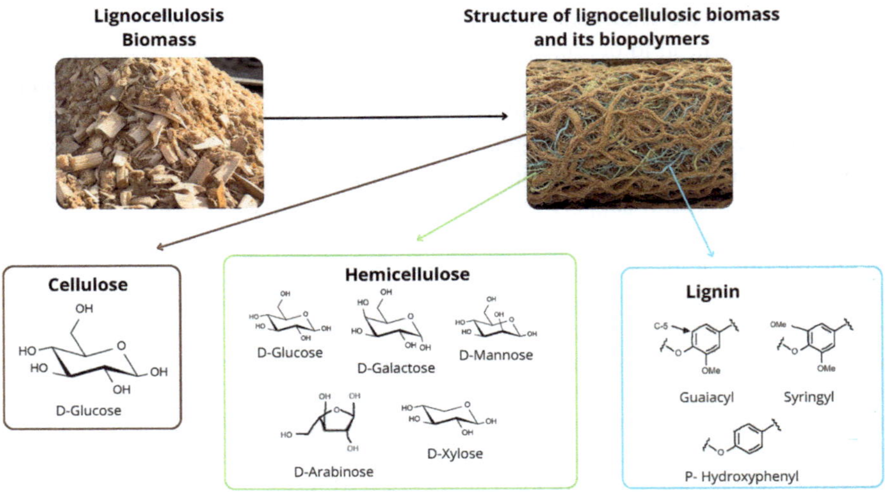

Fig. 18.1 Constituents of lignocellulosic biomass and their chemical structure

several advantages, including the abundance and renewability of forest resources, the potential for carbon sequestration, and reduced dependence on fossil fuels (Kalntremtziou et al. 2023). Fast-growing species such as eucalyptus and pine are particularly attractive due to their rapid biomass accumulation and adaptability to various environmental conditions (Vasudev et al. 2021). Furthermore, planted forests can be strategically managed to optimize biomass production, increase carbon storage (Brown et al. 2024), and provide better utilization of agricultural and forestry residues (Li et al. 2025). Globally, planted forests cover about 131 million ha, which represents 3% of the global forest (FAO 2022). The distribution of these planted forests, however, is not uniform. Brazil and China stand out as two of the most significant countries in terms of planted forest area and production (McEwan et al. 2020). Brazil has vast expanses of land suitable for silvicultural plantations, making it a prime location for the expansion of this sector. Eucalyptus plantations have witnessed rapid growth in Brazil, creating the largest cultivated area of this species worldwide.

18.2.3 Waste from the Pulp and Paper Industry

The pulp and paper industry, while essential for producing various paper products, generates substantial waste streams that pose environmental challenges (Saranya et al. 2020). These waste streams, however, also represent a significant opportunity for biorefineries, which can transform these underutilized resources into valuable products (Jin et al. 2018). The integration of biorefineries with the pulp and paper industry presents a compelling model for the circular economy, where waste is minimized, and resources are used efficiently (Branco et al. 2019). The waste generated from the pulp and paper industry includes materials like black liquor, sludge, and unused wood chips, which can be repurposed as feedstocks for biorefineries (Adu et al. 2018).

Black liquor, a by-product of the kraft pulping process, is a complex mixture of lignin, hemicelluloses, and inorganic chemicals (Chen and Wan 2017). Sludge, another significant waste stream, consists of organic matter, mainly cellulose fibers, and inorganic compounds such as kaolin, talc, and calcium carbonate (da Rosa et al. 2021). Unused wood chips, generated during wood processing, also constitute a substantial waste stream that requires proper management. However, instead of being discarded as waste, these materials can be viewed as valuable resources for biorefinery applications, offering opportunities for sustainable production of biofuels, biochemicals, and biomaterials (Azhar Zakir et al. 2015). The integration of biorefineries into existing pulp and paper mills can lead to diversification of products, increased profitability, and environmental benefits by valorizing these waste streams (Branco et al. 2019).

The selection and integration of biorefinery technologies should be based on waste characterization, with various technologies such as fermentation, anaerobic digestion, pyrolysis, incineration, and gasification (Nizami et al. 2017). By implementing effective waste management practices, the pulp and paper industry can minimize its environmental footprint and contribute to a more sustainable future. Integrating biorefineries with existing pulp and paper industrial plants can lead to product diversification and increased profitability, along with environmental benefits (Branco et al. 2019).

18.3 Traditional Paper Production Processes Versus Integrated Biorefineries

Pulp and paper production is fundamental to our society, but its process presents several environmental and economic drawbacks, which have led to a shift toward integrated biorefineries, which promise more sustainable and efficient manufacturing processes (Kamali and Khodaparast 2015). Conventional production methods, such as the kraft and sulfite processes, although effective in releasing cellulose fibers from wood, generate significant amounts of pollutants, including toxins characterized by high biological oxygen demand, chemical oxygen demand, and suspended solids (Saranya et al. 2020). These pollutants can contaminate water bodies, disrupt aquatic ecosystems, and pose risks to human health, creating a pressing need for cleaner and more sustainable alternatives (Jaiswal et al. 2021).

The kraft process, also known as the sulfate process, is the most widely used pulping method due to its ability to produce strong and durable paper from a wide range of wood species (Tolvaj et al. 2012). This process involves cooking wood chips in a chemical solution, known as white liquor, containing sodium hydroxide and sodium sulfide, which dissolves the lignin (Malik et al. 2020). The resulting pulp is then washed, screened, and bleached to achieve the desired brightness and purity for papermaking (Wang et al. 2012). Despite its widespread use, the kraft process has significant environmental limitations. Reduced total sulfur emissions, sulfur dioxide emissions, and hazardous air pollutants are some of the environmental effects of the process (Das and Jain 2001). In addition, the kraft process generates a dark-colored wastewater known as black liquor, which contains dissolved lignin, carbohydrates, and other organic compounds. Recovery and treatment of black liquor is energy-intensive and can generate additional pollutants if not managed properly (Vives et al. 2022).

The sulfite process, an alternative to the kraft process, utilizes sulfurous acid and its salts to dissolve lignin from wood chips. The specific chemicals used in the sulfite process, such as calcium, magnesium, sodium, or ammonium bisulfite, determine the characteristics of the resulting pulp and the process's environmental impact. Sulfite pulping generally operates under acidic conditions, which can lead to the degradation of cellulose fibers and the formation of unwanted by-products (Nada et al. 1998). While the sulfite process can produce pulps with high brightness and good printability, it is less versatile than the kraft process in terms of the wood species it can process. Sulfite pulping also faces challenges related to the recovery and reuse of chemicals, as well as the treatment of wastewater containing dissolved lignin and other organic compounds (Vásquez-Garay et al. 2021).

Both the kraft and sulfite processes share common limitations, including high water consumption, energy intensity, and the generation of solid waste, such as sludge from wastewater treatment. The use of harsh chemicals and the formation of toxic by-products also raise environmental concerns about the sustainability of these processes. These processes consume high levels of energy for cooking, bleaching, and drying, contributing to greenhouse gas emissions and dependence on energy sources such as fossil fuels. Disposal of sludge and other solid waste poses additional challenges and costs for these processes (Liu et al. 2020). Table 18.1 summarizes the yields of each traditional and modern process, as well as the pulp yields, advantages, and disadvantages.

Compared to traditional pulp production methods, integrated biorefineries offer a more sustainable approach to paper production, maximizing resource utilization and minimizing waste generation (Mongkhonsiri et al. 2021). By integrating different conversion technologies and valorizing waste streams, biorefineries can minimize waste generation, reduce reliance on fossil fuels, and increase the overall efficiency and profitability of the pulp and paper industry (Branco et al. 2019). The biorefinery concept, inspired by petroleum refineries, aims to convert biomass into a spectrum of valuable products, including biofuels, biochemicals, and biomaterials. This approach addresses the growing demand for sustainable alternatives to fossil resources and offers a pathway towards a circular bioeconomy (Bušić et al. 2018). Compared to traditional pulp production methods, integrated biorefineries offer a more sustainable approach to paper production, maximizing resource utilization and minimizing waste generation (Mongkhonsiri et al. 2021). By integrating different

Table 18.1 Advantages, disadvantages, and yields of each industrial process used for biomass pulping

Category	Process	Pulp yield (%)	Advantages	Disadvantages	References
Traditional	Mechanical pulping	90–98	High yield; relatively low cost	Low pulp strength; high energy consumption	Bajpai (2018)
Traditional	Acid sulfite process	45–55	Better bleachability than kraft	Requires effluent treatment; corrosive chemicals	Sixta (2008)
Traditional	Classic kraft process	45–55	High pulp strength; efficient chemical recovery	Low yield; dark pulp without bleaching	Cheremisinoff (2010)
Modern	Modified kraft (MCC/EMCC)	50–60	Higher selectivity and yield with fewer rejects	Requires advanced process control; higher operational cost	Antes and Joutsimo (2015)
Modern	Integrated autohydrolysis	60–70	Eco-friendly pretreatment; good hemicellulose recovery	Requires effluent conditioning; energy-intensive	Rodriguez et al. (2019)
Modern	Ionic liquid pulping	45–55	High selectivity; low residual lignin	High cost of ionic liquids; recovery still challenging	Song et al. (2016)
Modern	Organosolv process	45–55	Solvent recovery is feasible; lower environmental impact	High cost; industrial scale-up still in development	Nitsos et al. (2018)

conversion technologies and valorizing waste streams, biorefineries can minimize waste generation, reduce dependence on fossil fuels, and increase the overall efficiency and profitability of the pulp and paper industry (Branco et al. 2019).

Cost-effective biorefinery synthesis and design strategies and effective evaluation tools play a key role in the development of a practical and sustainable integrated biorefinery network for the pulp and paper industry. The selection of biorefinery technology for implementation depends on several factors such as processing conditions, local market supply chains, raw material product price, price of utility and fuels, and possible mass and energy integration. A systematic hierarchical methodology should be suitable for addressing all these factors in a step-wise manner (Babi et al. 2015). Cascade biorefinery represents an innovative strategy to maximize the utilization of woody biomass, moving beyond traditional single-product approaches to a more holistic and sustainable model (Jarre et al. 2020).

This approach seeks to extract high-value products by following a sequential process where residues from one stage become the raw materials for the next (Cozier 2014). This integrated approach promotes resource efficiency, reduces environmental impact, and improves the economic prospects of lignocellulosic biorefineries through the simultaneous generation of bio-based materials, chemicals, and energy, thereby decreasing dependence on fossil fuels (Mongkhonsiri et al. 2021). Implementing a cascade biorefinery requires meticulous design and optimization to ensure compatibility between different processing stages and maximize the yield and quality of the desired products. Furthermore, the cascade biorefinery concept aligns with the principles of the circular economy, where waste streams are seen as valuable resources that can be reintegrated into the production cycle (Branco et al. 2019). The gradual conversion of a kraft mill into a comprehensive biorefinery allows companies to gradually master novel technologies, evaluate various options, and formulate an appropriate business plan, offering a strategic path to adapt to evolving market demands and adopt sustainable practices (Moshkelani et al. 2013).

The conversion of biomass into bioenergy and bioproducts can be achieved through several pathways (Borole 2015). Biochemical conversions are particularly relevant for the production of fuels and products, involving several key steps that transform the complex structure of biomass into simpler and more usable compounds (Benavides et al. 2024). Pretreatment is essential to break down the complex structure of lignocellulosic biomass, increasing the accessibility of cellulose and hemicellulose to subsequent enzymatic or chemical processes (Yin et al. 2024). The effectiveness of pretreatment greatly impacts the overall efficiency of the biorefinery process, affecting the yield and quality of downstream products (Galbe and Wallberg 2019). Enzymatic hydrolysis uses cellulase and hemicellulase enzymes to hydrolyze polysaccharides into fermentable sugars, offering a specific and efficient route for sugar production. Fermentation then converts these sugars into biofuels, such as ethanol and butanol, using microorganisms such as yeast or bacteria. Biogas production stands out as an established method for renewable energy generation, simultaneously treating waste and recycling organic matter into fertilizer (Adnan et al. 2019). Anaerobic digestion, gasification, direct combustion, ethanol production, and biodiesel production are technologies to convert waste biomass into a renewable energy source, as well as protect human health and the environment (Schneemann et al. 2009). Ultimately, the integration of these biochemical conversion processes into a cascade biorefinery framework offers a promising pathway for the sustainable production of fuels and chemicals from woody biomass, aligning with the principles of the circular economy and resource efficiency.

Organosolv pretreatment represents an attractive strategy within the broader context of lignocellulosic biorefineries, offering a pathway to fractionate biomass. Unlike conventional pretreatment methods that may rely on chemicals or energy-intensive processes, organosolv pretreatment employs organic solvents such as ethanol, methanol, acetone, or glycerol, often in combination with water and catalysts, to selectively solubilize lignin (Silveira et al. 2015). Integrating organosolv pretreatment into a biorefinery setup offers several advantages over conventional methods (Brosse 2025). The effectiveness of organosolv pretreatment depends on several factors, including the choice of solvent, reaction temperature, residence time, and the presence of catalysts, all of which can be tailored to optimize fractionation

(Bušić et al. 2018). Although organosolv pretreatment holds considerable promise, it is crucial to address potential drawbacks, such as the high costs associated with solvent recovery and the risk of fire explosions if adequate safety measures are not implemented (Guo et al. 2019). The integration of organosolv pretreatment into a biorefinery setup offers several advantages over conventional methods (Brosse 2025). High cellulose digestibility can be achieved by effectively removing lignin (Chaturvedi and Verma 2013). Furthermore, the recovery of a relatively pure lignin stream opens up opportunities for its conversion into value-added products such as phenolic resins, adhesives, or carbon fibers, thus increasing the overall economic viability of the biorefinery (Silveira et al. 2015). The selection of an appropriate organic solvent is key to the success of organosolv pretreatment, including its ability to solubilize lignin, its boiling point, its cost, and its environmental impact. Optimization of the pretreatment process is necessary to maximize sugar yields, minimize solvent loss, and produce a high-quality lignin stream. Process simulations and techno-economic analyses are essential to assess the feasibility and economic competitiveness of organosolv pretreatment in an industrial-scale biorefinery (Zhang et al. 2016). Research has shown that organosolv pretreatment can efficiently fractionate biomass, leading to higher yields of enzymatic hydrolysis and bioethanol production (Brosse 2025).

Biological pretreatment methods, which use fungi, offer an environmentally friendly alternative, operating at mild temperatures and avoiding the creation of inhibitory compounds but are currently at the laboratory scale (Tsegaye et al. 2019). Green liquor pretreatment stands out for generating fewer toxic by-products, such as furfural and acetic acid, that can hinder fermentation and damage equipment; in addition, its chemicals can be fully recycled, improving its eco-economic profile (Alam et al. 2020). Advanced pretreatment methodologies, including those employing ionic liquids, deep eutectic solvents, and biological approaches, present viable solutions to overcome biomass recalcitrance and improve downstream processing. However, the need to maintain high temperature in sizeable reactors can require high capital and operating costs (Ahmed et al. 2019). Nonthermal plasma emerges as an environmentally conscious method, arousing interest for both the treatment and pretreatment of lignocellulosic materials (Pereira et al. 2021).

A biorefinery based on advanced pretreatment technologies requires careful consideration of several factors to ensure economic viability and environmental sustainability. These factors include feedstock selection, pretreatment optimization, enzyme selection, fermentation strategies, and product recovery methods. The ultimate goal of advanced pretreatment technologies is to increase the overall efficiency and sustainability of lignocellulosic biorefineries, paving the way for the production of biofuels and other bio-based products from renewable resources (Bhatia et al. 2020). Furthermore, pretreatment processing conditions must be tailored to the specific chemical and structural composition of the various lignocellulosic biomass sources (Mosier et al. 2005).

18.4 Integrated Biorefinery Paradigm

In contrast to traditional pulp and paper processes, the integrated biorefinery paradigm is driven by the efficient conversion of lignocellulosic biomass into different value-added products. Inspired by the logic of petroleum refineries, biorefineries aim to fractionate and upgrade all major components of biomass—cellulose, hemicellulose, and lignin—in terms of maximizing the efficiency and diversity of products. Thus, this systemic approach allows new economic opportunities through the coproduction of biofuels, biochemicals, bioplastics, and advanced materials, especially based on the circular economy concepts.

However, the paradigm involves the development and implementation of advanced fractionation technologies able to separate chemicals from lignocellulosic biomass under optimal and scalable conditions, preserving their chemical functionalities. Scientific reports have been published demonstrating the technical possibilities to advance on the production of valuable materials by the application of technologies like steam explosion, enzymatic hydrolysis, pyrolysis, and organosolv pulping.

The integrated biorefinery model also emphasizes process synergies and energy integration, where heat and power demands are met through internal recycling of residues, and where chemical strategies are closed to minimize waste. Specifically, by transforming pulp and paper mills into multi-bioproducts spots, this approach facilitates economic viability and diversification of the markets. Consequently, it may reduce the dependency on volatile raw material prices. Also, it improves the positive environmental aspects related to biorefinery, such as low greenhouse gas emissions and low water usage.

Therefore, the transition of conventional pulp and paper industries to integrated biorefineries comprehends significant changes in technological orientation, followed by economic and sustainability directions. The implementation of an integrated biorefinery can collaborate to avoid the traditional linear models in conventional pulp and paper mills with a single focus on products like kraft lignin, hemicelluloses, and cellulose pulp.

Table 18.2 provides a comparative overview of these two paradigms across several critical dimensions, from raw material utilization and product diversification to environmental performance and integration potential.

The contrast between the conventional pulp and paper industry and the integrated biorefinery model observed in Table 18.1 shows clear differences that can drive the future of the sector. In summary, the coproduction of pulp, energy, and different biochemicals and bioproducts from these raw materials widely available in pulp and paper mills makes it a key factor to really implement the concepts of biorefinery and bioeconomy in the sector, considering the technical and economic aspects.

Table 18.2 Traditional processes versus integrated biorefinery in the pulp and paper industries

Criteria	Conventional pulping process (Kraft and/or Sulfite)	Integrated biorefinery
Main goal	Pulp and paper production	Full valorization of lignocellulosic biomass
Typical application of lignin	Thermal energy production	Extraction and conversion in high-added-value products
Typical destination of hemicellulose	Partial recovery or discard	Conversion into fermentable sugars or chemical derivatives
Diversification of the products	Low	High—bioenergy, biochemicals, and bioplastics, for instance
Economic model	Linear (extraction, production, by-products)	Circular (recovery, reuse, valorization)
Sustainability	Limited due to the inefficient use of biomass and the degree of emissions	Elevated due to efficiency, reduction of residues, and low environmental impact
Integration with well-known processes	High, but with a focus on cellulose pulp production	Potential application with synergy and retrofit of units in operation
Current limitations	Well-established but not very flexible technologies	High initial investment with technical and logistical challenges

18.5 Key Technologies for Biomass Processing

18.5.1 Pretreatment

The production of cellulose and bioproducts from lignocellulosic biomass critically depends on efficient pretreatment methods, such as autohydrolysis, acid/dilute hydrolysis, ionic liquids, and steam explosion, each with different mechanisms for deconstructing the biomass structure. Autohydrolysis, for example, uses water under subcritical conditions and is economical and has a low environmental impact, while acid hydrolysis uses dilute acids for high efficiency, but at moderate cost and with a significant environmental impact. The use of ionic liquids, on the other hand, is a relatively new subject in the area of pulp production and stands out for its selectivity and sustainability, but with high costs, while steam explosion combines steam and rapid depressurization, offering efficiency and moderate costs, but with considerable energy consumption. This subtopic deals with the main pretreatment methods used for the production of cellulose and bioproducts derived from biomass, with a special focus on their efficiency, costs, and environmental impacts, showing the most recent studies on the progress of these methods in the industry. The main stages of each pretreatment are shown in Fig. 18.2.

Autohydrolysis is a thermochemical process that uses water under subcritical conditions (150–230 °C) to promote the partial hydrolysis of hemicellulose and the selective solubilization of lignocellulosic biomass components. This method does

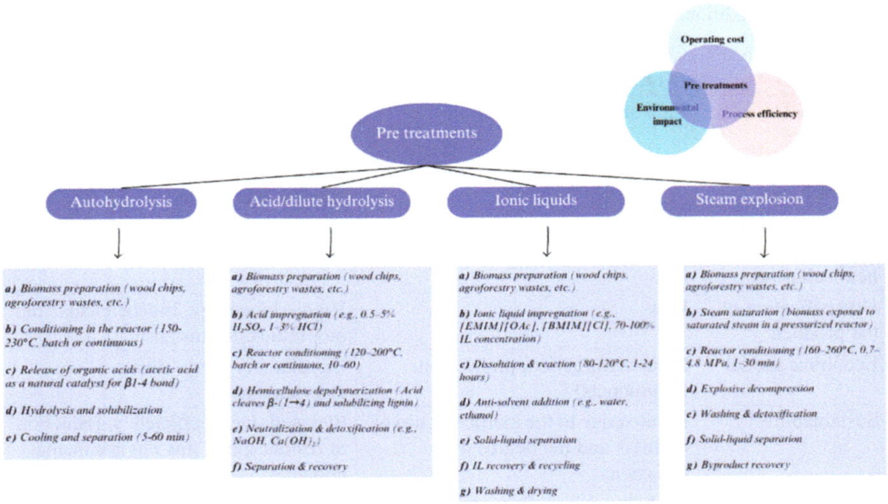

Fig. 18.2 Main stages of each pretreatment

not require external catalysts, as the necessary hydrogen ions (H^+) are naturally released from the biomass at high temperatures. Autohydrolysis preserves the cellulose, increasing its accessibility for enzymatic hydrolysis, and generates by-products such as xylooligosaccharides, which have commercial value. It is considered to have a low environmental impact and moderate costs, making it a sustainable option for pretreating biomass. Some studies mention autohydrolysis as the most promising approach to cellulose extraction, producing 63.50% of a solid fraction with around 70% cellulose (Casanova et al. 2023).

Autohydrolysis has emerged as one of the most promising and studied methods for producing high-purity cellulose, consolidating itself as a green alternative to conventional techniques. Several recent studies have explored not only its intrinsic efficiency but also its ability to be integrated with other technologies, expanding its potential in biorefineries. A paradigmatic example is the work of (Casanova et al. 2023), which compared ecological and traditional routes for extracting cellulose from sugarcane bagasse. While traditional methods (acid/alkali hydrolysis) showed lower yields (below ~50%) and high waste generation, autohydrolysis stood out with a 63.50% yield in cellulose, a 22.90% increase in purity, and an 11.20% reduction in hemicellulose, using only water and heat. This operational simplicity eliminated chemical and effluent neutralization costs, reducing the environmental impact by up to 40% compared to conventional techniques. In addition, (Gan et al. 2017) explored autohydrolysis to improve cellulose solubility and produce regenerated materials. Treating biomass with heated water (130–230 °C) under high pressure allowed selective solubilization of hemicelluloses and enrichment of the solid phase in cellulose (18–25% increase in purity). The technique also increased the viscosity of the cellulose by 30%, crucial for film and fiber applications, although temperatures above 200 °C reduced its molar mass by 15%, an effect that can be modulated to adjust rheological properties. In contrast to enzymatic or organosolv methods, which require additional costs for enzymes or solvent recovery, autohydrolysis proved to be economically superior, with operating costs 25% lower. The versatility of the method was further demonstrated by Joseph et al. (2024), who applied autohydrolysis to *Elaeis guineensis* (oil palm) fibers, achieving a 35% reduction in kappa number (an indicator of residual lignin) and a 28% increase in cellulose crystallinity. These results reinforce its applicability to non-timber biomass, expanding opportunities for tropical regions. Despite these advances, studies such as the one by Casanova et al. point out gaps, such as the lack of research into direct industrial applications of autohydrolyzed cellulose. For example, the integration of residual lignin into biomacromolecules or the conversion of xylooligosaccharides into prebiotics still lack pilot-scale prototypes. However, the synergy between autohydrolysis and complementary methods, such as alkaline pulping or enzymatic bleaching, has been explored to optimize processes, since autohydrolysis used alone as a pretreatment, despite its advantages, is already well studied in the literature.

Acid/dilute hydrolysis is a classic biomass pretreatment method that uses dilute inorganic acids, such as sulfuric acid (H_2SO_4) or hydrochloric acid (HCl) (0.5–5%), under moderate thermochemical conditions (100–200 °C), to promote the hydrolysis of hemicellulose and the partial dissolution of lignin. Unlike autohydrolysis, which uses water as the only reagent, this process depends on the addition of external acid catalysts, which accelerate the breaking of glycosidic bonds and esters, releasing sugars such as xylose and glucose. Although effective in depolymerizing hemicellulose (reaching conversions of over 80%), acid hydrolysis tends to partially degrade cellulose, reducing its yield (~45–55%) and generating inhibitory by-products such as furfural and 5-hydroxymethylfurfural (HMF), which compromise later stages such as fermentation. It also requires effluent neutralization and acid recovery steps, increasing operating costs and environmental impacts.

Throughout history, the first studies using dilute inorganic acids date back to the mid-1930s. Scholler–Tornesch, developed in Germany, was one of the first processes to use dilute sulfuric acid (0.4–0.8%) to hydrolyze wood and obtain fermentable sugars, although it was an industrial approach, not an academic one. Academically speaking, one of the first studies to use the hydrolysis method using dilute inorganic acids was carried out by Saeman (1945) (Saeman 1945). The authors investigated the kinetics of wood saccharification, specifically analyzing the hydrolysis of cellulose and the subsequent decomposition of sugars under treatment with dilute acid (H_2SO_4 in dilute concentrations [0.4–1.6%]) at high temperatures (160–220 °C). The results showed that increasing the acid concentration by 100% increased the hydrolysis rate by 153%, demonstrating an exponential relationship between catalyst concentration and reaction efficiency. At the same time, a 10 °C increase in temperature resulted in a ~186% increase in hydrolysis speed, reinforcing the critical role of thermal conditions in the process. The authors also explored the theory of consecutive reactions, concluding that the efficiency of converting cellulose into reducing sugars depends on the ratio between the rates of formation and degradation of the sugars. It was observed that increases in acid concentration or temperature accelerate hydrolysis more than glucose destruction, thus optimizing the overall yield. Although the experiments were carried out on a reduced scale, the principles identified were validated for application in larger industrial operations, such as single or multi-stage hydrolysis, indicating the practical relevance of the findings for lignocellulosic biomass saccharification processes. This pioneering work not only established essential kinetic bases but also guided the development of subsequent technologies in the area of biorefineries.

Currently, there have been a number of advances in the use of different dilute acids. Chimentão et al. (2014), for example, investigated the hydrolysis of cellulose pre-treated with dilute acids (sulfuric, H_2SO_4, and oxalic, $C_2H_2O_4$) under mild hydrothermal conditions, in order to compare the efficiency of glucose production and the impact on the purity of the residual cellulose. Pretreatment consisted of impregnating the cellulose with concentrations of 3% and 6% by mass of each acid. Hydrolysis was carried out using two methods, conventional oven heating (at 120 °C and 140 °C) and microwave irradiation, the latter being more efficient in speeding up the reaction. The results showed that sulfuric acid promoted greater release of total organic carbon (TOC $\approx$ 900 mg/L after 6 h), indicative of more intense biomass degradation. However, its action was less selective, generating only 30% glucose and by-products such as HMF. In contrast, oxalic acid, although with a lower TOC, showed 85% selectivity for glucose, making it the most efficient catalyst. This difference is linked to the mechanisms of action: H_2SO_4 partially disorganized the crystalline regions of cellulose, facilitating overall hydrolysis, but compromising the purity of the residual material. On the other hand, $C_2H_2O_4$ preferentially attacked the amorphous zones, preserving the crystalline structure and increasing the quality of the remaining cellulose. Li et al. (2017) proposed an innovative method for hydrolyzing cellulose into glucose using sulfonated carbon catalysts prepared via plasma treatment in dilute sulfuric acid (H_2SO_4 0.1 M, 0.5 M, and 1 M). The sulfonated carbon acted as a solid acid, whose sulfonic groups ($-SO_3H$) confer acidic properties essential for the reaction. The process combined mechanical pretreatment of the cellulose (ball milling for 48 h) with catalytic hydrolysis in an autoclave (150 ° C, 24 h), using the acid catalyst. The concentration of acid diluted in the sulfonation directly influenced efficiency: cellulose conversion increased from 10% (0.1 M H_2SO_4) to 40% (1 M H_2SO_4) in 30 min, with selectivity for glucose >80% in the case of the catalyst sulfonated in 1 M H_2SO_4. The sulfonic acid in the carbon remained stable, allowing 98% efficiency after 3 cycles, which reinforces its viability as a sustainable alternative to conventional liquid acids. This method stands out for integrating a dilute acid into the catalyst preparation, avoiding the direct use of concentrated acids in the reactor, reducing corrosion and environmental impacts, while maintaining high efficiency in converting biomass into sugars. It should be noted that most of the processes cited for dilute acid hydrolysis are currently focused on combining processes, mainly to increase yields.

The steam explosion method, first developed by William H. Mason in 1926, revolutionized biomass pretreatment by combining saturated steam at high pressure (0.7–4.8 MPa) and temperature (160–260 °C) with an abrupt depressurization, fragmenting the lignocellulosic structure by means of autocatalytic acid hydrolysis (via the release of acetic acid from the biomass itself) and mechanical forces (Sarker et al. 2021). This thermomechanical approach eliminates the need for aggressive chemical reagents and is widely adopted for its efficiency in increasing the accessibility of cellulose for enzymatic hydrolysis. Its advantages include the ability to solubilize up to 80% of hemicellulose and reduce the crystallinity of cellulose, increasing glucose yields to 85–90% in biomass such as sugarcane bagasse, as well as moderate operating costs due to the simplicity of the process and the absence of chemicals (Auxenfans et al. 2017).

Environmentally, the steam explosion method is considered sustainable, as it generates less toxic effluents and allows by-products such as xylooligosaccharides to be recovered in biorefineries. However, the method has significant disadvantages, such as the need for temperatures above 200 °C, which can degrade cellulose (reduction of up to 30% in molar mass) and generate inhibitory compounds (furfural and HMF), which affect later stages such as fermentation. In addition, energy consumption is high (20% higher than methods such as autohydrolysis), and lignin removal is limited (40–60%), requiring complementary steps such as alkaline bleaching for applications requiring high purity.

Serrano-Martínez et al. (2024), for example, propose the use of the steam explosion technique for the sustainable extraction of cellulose and lignin from rice straw waste, optimizing operating conditions and reducing environmental impacts. The method consists of a high-temperature (pressurized) steam cooking stage followed by abrupt decompression, which fragments the biomass structure, increasing its surface area and porosity, facilitating the release of cellulose and lignin. The ideal condition identified was an explosion time of 33 min, dispensing with the use of supercritical carbon dioxide ($ScCO_2$), which simplifies the process and eliminates costs associated with handling CO_2. The results showed 70% efficiency in the extraction of cellulose, with a yield of 20% from rice straw, indicating technical feasibility for the use of agricultural waste. The analysis also revealed that extending the explosion time beyond 33 min did not significantly increase cellulose concentration or yield, suggesting that the process can be optimized for maximum efficiency without wasting energy. Among the economic advantages is the reduction in operating costs, since limiting the oxidation time to 3 h did not compromise the yield, making the method faster and cheaper than conventional techniques. Environmentally, the absence of $ScCO_2$ minimizes the carbon footprint, while the recovery of rice straw contributes to the sustainable management of agricultural waste. The technique also demonstrated high purity of the extracted products, essential for applications in high-value-added biomaterials.

Another, but more recent, method is the use of ionic liquids as a pretreatment. The use of ionic liquids (ILs) as a method for pretreating lignocellulosic biomass has been consolidated as an innovative approach since the 2000s, with pilot studies such as that by (Swatloski et al. 2002), which for the first time demonstrated the ability of solvents such as 1-Ethyl-3-Methylimidazolium Acetate ([EMIM][OAc]) to dissolve cellulose at moderate temperatures (100 °C). These low-melting-point organic salts (<100 °C) act as biomass deconstruction agents, breaking hydrogen bonds and hydrophobic interactions between cellulose, hemicellulose, and lignin. The process involves immersing the ground biomass in LI (e.g., 1-butyl-3-methylimidazolium chloride ([BMIM][Cl]), 1-allyl-3-methylimidazolium chloride ([AMIM][Cl])) under stirring and heating (80–120 °C) for 1–24 h, followed by the addition of an antisolvent (water or ethanol) to precipitate the purified cellulose (>85%), while lignin and hemicellulose remain dissolved for later recovery.

The efficiency of the method using ILs has been extensively demonstrated in studies such as that by (Hou et al. 2017) which addressed the use of ionic liquids, such as 1-ethyl-3-methylimidazolium acetate ([Emim]Ac) and 1-butyl-3-methylimidazolium acetate ([Bmim]Ac), in the pretreatment of lignocellulosic biomass, highlighting their superiority over conventional methods. For example, by combining [Emim]Ac with ammonia (NH_3) and oxygen (O_2), the authors achieved a reduction of lignin to less than ~10% in Miscanthus biomass, surpassing treatments in an air or nitrogen (N_2) atmosphere. Another advance was the efficient dissolution of cellulose at room temperature (25 °C) using a mixture of [Bmim]Ac and dimethylsulfoxide (DMSO), a method that maintained the structural integrity of the cellulose and facilitated its application in regenerated materials. The results showed that cellulose with only 6.2% residual lignin was obtained from spruce wood (*Picea abies*), as well as significant energy efficiency, since processes under mild conditions reduced thermal consumption compared to traditional methods. Economically, despite the high cost of ionic liquids (US$ 50–100/kg), their recyclability (over 95% after multiple cycles) compensates for initial investments, while environmentally, their low toxicity and corrosiveness minimize hazardous waste. However, challenges such as the energy required to recover solvents and the cost of synthesis still limit industrial scalability.

Another interesting study was proposed by Usmani et al. (2020), which looked at the use of ionic liquids as a method of pretreating lignocellulosic biomass, with the aim of optimizing bioconversion processes for the production of biofuels and biochemicals. The study highlighted the ability of these compounds to deconstruct the lignocellulosic matrix, increasing the accessibility of cellulose for subsequent stages, such as enzymatic hydrolysis. Specific ionic liquids were evaluated, such as 1-butyl-3-methylimidazolium chloride ([Bmim]Cl), 1-allyl-3-methylimidazolium acetate ([Amim]OAc), and 1-butyl-3-methylimidazolium hydrogen sulfate, with a focus on their efficiency in dissolving lignin and increasing the availability of cellulose. The research showed that these ILs act through the interaction between their cations and anions with the structural components of the biomass, breaking hydrogen bonds and hydrophobic interactions. In the case of *Eucalyptus* sp., for example, the combination of an ionic liquid with sodium hydroxide (NaOH) resulted in a glucose production rate of 90.53%, showing synergy between chemical methods and ionic solvents. In addition, it was observed that the dissolved lignin maintained a consistent yield even after multiple cycles of IL recycling, indicating process stability and reuse viability. The authors pointed out that economically, the recyclability of ionic liquids reduces operating costs associated with solvent replacement and waste disposal, making them attractive alternatives for biorefineries. Environmentally, they stand out as less toxic and nonvolatile substitutes compared to traditional organic solvents, aligning themselves with the principles of green chemistry by minimizing the generation of hazardous by-products.

18.5.2 Selective Fractionation

This subtopic will detail the steps following biomass pretreatment, which are essential for isolating and purifying the cellulose and lignin-rich fractions. After the initial deconstruction of the lignocellulosic matrix (via methods such as autohydrolysis or steam explosion, mentioned above), processes such as delignification (e.g., LignoBoost or LignoForce techniques) and bleaching (with oxidizing agents such as H_2O_2 or O_3) are used to remove lignin and hemicellulose residues, respectively.

Hemicellulose can be extracted using methods such as enzymatic hydrolysis or hydrothermal processes, each with different mechanisms and advantages. In enzymatic hydrolysis, specific enzymes (e.g., xylanases and mannanases) depolymerize hemicellulose into oligomers or monomers (xylose, mannose, and arabinose), preserving its structure and functionality. This method is highly selective and operates under mild conditions (neutral pH, 40–60 °C), avoiding thermal degradation, but requires high enzyme costs and long times. Hydrothermal processes, such as autohydrolysis and steam explosion, use water at high temperatures (160–220 °C) and pressure to hydrolyze hemicellulose via the autocatalytic action of organic acids (e.g., acetic acid) released from the biomass itself. These techniques are faster and more scalable but can generate unwanted by-products (furfural, HMF) under severe conditions.

Both methods represent a critical intermediate step in biomass processing, preceding advanced delignification processes such as LignoBoost or LignoForce, which focus on selective lignin removal. The prior extraction of hemicellulose facilitates access to lignin and cellulose, optimizing the efficiency of subsequent stages. For example, in integrated biorefineries, after hemicellulose removal by autohydrolysis, the residual biomass (rich in cellulose and lignin) is subjected to methods such as LignoBoost to isolate high-purity lignin, followed by cellulose bleaching. The choice between enzymatic and hydrothermal hydrolysis depends on the balance between efficiency, cost, and sustainability, the former being preferred for high-value-added bioproducts (e.g., prebiotics) and the latter for industrial-scale applications such as the production of biofuels or lignocellulosic materials.

Sun et al. (2022) investigated the alkaline extraction of hemicellulose from *Dactylis glomerata* to optimize enzymatic hydrolysis (HE) of cellulose for the production of bioethanol. The method included sample preparation (grinding and removal of waxes/chlorophyll), delignification with sodium chlorite ($NaClO_2$) in an acidic medium, and alkaline treatment with sodium hydroxide (NaOH) or potassium hydroxide (KOH) to extract hemicelluloses. The results showed a significant increase in cellulose content (58.4–1.9%) and enzymatic hydrolysis efficiency (90.3–95.1% vs. 74.3% in the untreated sample), proving that alkaline extraction breaks ester bonds and reduces structural interferences. However, the use of alkaline reagents (NaOH/KOH) entails high operating costs and environmental risks associated with effluent disposal. In conclusion, the method proved to be effective in optimizing the conversion of biomass into bioethanol but requires adjustments to balance efficiency, costs, and sustainability.

The extraction of purified lignin from black liquor (waste from kraft pulp production) is made possible by technologies such as LignoBoost and LignoForce, which have revolutionized the valorization of this biomacromolecule. Both processes involve acidifying the lye to precipitate lignin, followed by stages of filtration, washing (to remove hemicelluloses and salts), and drying, resulting in lignin with a purity of over 95% and a low sulfur content. LignoBoost (developed in Sweden and marketed by Valmet) uses CO_2 to adjust pH, while LignoForce (Canadian, from FPInnovations) uses sulfuric acid, optimizing energy efficiency. These methods not only reduce the load on recovery boilers (increasing the plants' production capacity) but also turn a by-product into raw materials for bioplastics, chemical additives, and carbon fibers. Companies such as Stora Enso and Domtar have already adopted these technologies, integrating them into biorefineries to promote a circular economy and reduce CO_2 emissions. Isolated lignin thus becomes a pillar for replacing oil derivatives in sectors such as construction, green chemistry, and advanced materials.

18.5.3 Conversion and Purification

Cellulose, the most abundant biomacromolecule in nature, stands out as a renewable and sustainable resource with the potential to be transformed into high-value-added applications. Its conversion into products such as nanocellulose (nanometer-scale fibers) or high-purity fibers requires processes that balance efficiency, control of impurities, and preservation of structural properties. The production of nanocellulose, for example, involves deconstructing the lignocellulosic matrix using methods such as acid hydrolysis, enzymatic treatment, or intensive mechanization (e.g., microfluidization), with the aim of isolating crystals or fibrils with diameters of less than 100 nm (Phanthong et al. 2018). High-purity fibers, on the other hand, require rigorous purification steps—such as bleaching with oxidizing agents (e.g., hydrogen peroxide) or selective extraction of lignin and hemicelluloses—to eliminate contaminants that compromise performance in critical applications.

Purification is a central challenge, since organic residues, heavy metals, or by-products of chemical processes can reduce the thermal and mechanical stability of the final materials. Recent advances include green techniques, such as the use of deep eutectic solvents (DES) (Chen et al. 2019) or enzymatic processes (Pala et al. 2007), which minimizes environmental impacts. In addition, the integration of pretreatments (e.g., steam explosion) optimizes accessibility of cellulose, facilitating later stages, as previously mentioned. After the purification and conversion processes of cellulose into nanocellulose, the applications are vast. Nanocellulose, with its high mechanical strength (modulus of elasticity ~150 GPa) and high surface area, is used in advanced composites, biodegradable packaging, and even controlled drug release systems (Patel et al. 2019).

High-purity cellulose fibers are essential components in applications that demand physical and mechanical performance and reliability. In the textile sector, these fibers are used in the production of technical fabrics, such as lyocell and modal (Lyocell 2019), which combine softness, durability, and absorption capacity, as well as being sustainable alternatives to synthetic fibers derived from petroleum. In high-efficiency filters (High-Efficiency Particulate Air (HEPA) and Ultra-Low Penetration Air (ULPA)) (95–99.99%) (Wang et al. 2005), purified cellulose forms porous membranes capable of retaining microscopic particles, viruses, and even nanopollutants, making it essential in air and water purification systems and industrial processes. In flexible electronics, ultra-thin, conductive fibers (treated with metallic coatings or conductive polymers) are integrated into wearable devices, biodegradable sensors and foldable screens, where cellulose's flexibility and lightness are unmatched.

The synergy between sustainable methods and purification technologies is at the heart of this revolution. Processes such as enzymatic hydrolysis, hydrogen peroxide bleaching, and the use of DES reduce energy consumption and eliminate toxic waste while preserving fiber integrity. Advanced techniques, such as membrane fractionation and selective precipitation, guarantee the removal of residual lignin, hemicelluloses, and metals, raising purity to levels of over 99%. This combination not only meets stringent standards (e.g., Forest Stewardship Council [FSC] and ISO 14001 certifications) but also enables the circular economy: discarded fibers can be recycled or biodegraded without environmental risks, closing the material's life cycle.

Hemicellulose, a heterogeneous biomacromolecule present in plant cell walls, is emerging as a versatile raw material for the synthesis of high-value products such as xylitol, levulinic acid, and biomacromolecules (PHA), through sustainable biochemical routes. Unlike cellulose, its branched, pentose-rich structure (e.g., xylose) allows conversion via microbial fermentation or catalytic processes but requires efficient pretreatment and purification steps. Acid or enzymatic hydrolysis of hemicellulose releases fermentable sugars such as xylose (Yuan et al. 2021), which are transformed by yeasts (e.g., *Candida tropicalis*) into xylitol—a low-glycemic-index sweetener with applications in food and pharmaceuticals. For levulinic acid, a precursor to biofuels and plastics, hemicellulose can be deconstructed via acid hydrolysis (H_2SO_4, HCl), followed by catalytic dehydration, while the synthesis of PHA (polyhydroxyalkanoates)—biodegradable biomacromolecules—involves bacteria (e.g., *Cupriavidus necator*) that metabolize sugars into intracellular polymers (Jakob et al. 2024). Purification generates by-products, such as furfurals or organic acids, that inhibit fermentation, requiring techniques such as adsorption on activated carbon, ultrafiltration, or liquid–liquid extraction to isolate the target products.

Challenges remain, such as low conversion efficiency and the complexity of the lignocellulosic matrix. Advances in integrated biorefineries and metabolic engineering (e.g., inhibitor-tolerant microorganisms) seek to optimize yields (Velvizhi et al.

2022). Xylitol, for example, is reaching global markets in the food industry, while PHA is gaining ground as an alternative to conventional plastics. By combining low-impact processes (e.g., eutectic solvents and reusable catalysts) with effective fermentation routes, hemicellulose is consolidating itself as a pillar of the circular bioeconomy, transforming agro-industrial waste into strategic inputs.

Lignin, a complex polymer abundant in lignocellulosic waste, is a strategic resource for the production of bio-oils, vanillin, and composite materials, in line with the transition to a sustainable economy. Through fast pyrolysis (450–600 °C in the absence of oxygen), lignin is converted into bio-oil, a liquid rich in phenolic compounds and guaiacols, used as a biofuel precursor or chemical input (Baloch et al. 2018). Purification of this bio-oil, which is essential to remove carboxylic acids and impurities, involves techniques such as fractional distillation or hydrotreating, ensuring thermal stability and commercial viability. In catalytic oxidation, lignin reacts with metal catalysts (e.g., CuO) in an alkaline medium, generating vanillin—a highly valuable compound in food and pharmaceuticals. Purification of vanillin requires steps such as solvent extraction (e.g., ethyl acetate) or chromatography to isolate it from unwanted by-products.

In the context of composite materials, lignin is incorporated into polymers such as polylactic acid (PLA), polycaprolactone (PCL), polyethylene, polypropylene, or rubbers to improve mechanical properties, UV resistance, and biodegradability (Kun and Pukánszky 2017). Pretreatments such as sulphonation or acetylation increase their compatibility with the polymer matrix, while purification (e.g., removal of residual sugars via acid washing) prevents embrittlement of the final material. Despite the potential, challenges remain: the structural heterogeneity of lignin and the presence of impurities (metals, by-products) require advanced fractionation methods (ultrafiltration, selective precipitation) and selective catalysis.

18.6 Valorization of Biomass Fractions

18.6.1 Cellulose

Cellulose, traditionally associated with paper production, is proving to be a versatile and revolutionary biomacromolecule in applications that go beyond its conventional use. In the textile industry, fibers such as viscose (derived from regenerated cellulose) offer a sustainable alternative to polyester and are widely used in biodegradable clothing and fabrics. In packaging, cellulose is gaining prominence as a raw material for biodegradable films and containers, such as those based on cellulose acetate or nanocellulose, which replace fossil-based plastics in applications ranging from food packaging to pharmaceutical capsules. In addition, nanocellulose-reinforced films are emerging as high-performance materials, combining transparency, mechanical strength, and gas barrier, ideal for smart packaging, flexible sensors, and even medical dressings. These innovations not only extend the reach of cellulose but also align with the global demand for renewable materials, reducing carbon footprint and driving the transition to a circular economy.

In a recent study, Zhu et al. (2024), for example, synthesized transparent cellulose films to replace plastics via water-induced controllable deswelling, the authors pointed out that the films manufactured in this study stand out as a sustainable alternative to conventional plastics due to complete biodegradation in 14 days in the soil (vs. plastics that persist for decades). They have superior mechanical strength (30 MPa wet, 65.7% retention) and efficient oxygen barrier (2.31 cm^3 $\mu m/m^2$/day/ kPa), outperforming petroleum-based plastics in food preservation. High water vapor transmission (678.4 g/m^2/day) helps reduce spoilage, and ultra-fast production (less than 1 min) makes industrial scale feasible.

Agumba et al. (2024) manufactured advanced hybrid cellulose films that are hydro-stable, recyclable, and degradable as renewable alternatives to synthetic plastics, the authors mentioning that the hybrid cellulose films showed superior properties compared to conventional plastic films. Mechanically, the cellulose nanofibers (CNF)/esterified-polyvinyl alcohol-citric acid-lignin (E-PCL) film had a tensile strength of 375.1 MPa and an elastic modulus of 11.16 GPa, values much higher than those of ordinary plastics. Water absorption was reduced to less than 57% in 24 h, whereas common plastics can be highly permeable. Thermally, the hybrid films showed degradation only between 238.8 and 270.1 °C, ensuring greater stability than many synthetic plastics. In addition, they exhibited excellent UV-blocking capacity, transmitting less than 15% of radiation at 600 nm, whereas conventional films have low protection. In addition to these studies, there are a number of innovations in the use of cellulose to manufacture films, such as Xia et al. (2022) who produced transparent cellulose-based bio-hybrid films with improved anti-ultraviolet, antioxidant, and antibacterial performance and Chen et al. (2021) who produced high-strength, self-cleaning, and biodegradable packaging based on cellulose.

Cellulose is often converted into nanocellulose (crystalline or fibrillated), which, in its nature, is still considered cellulose and can be used for various purposes, such as polymer reinforcement. Some studies have highlighted the impact of nanocellulose on improving the properties of different polymeric matrices. Bras et al. (2010) showed that adding 10% CNF to natural rubber increased tensile strength by up to 10×, but made the matrix more fragile, indicating significant mechanical improvements. Zhang et al. (2015) successfully developed a nonwoven CNF/polyethylene terephthalate (PET) composite separator using a papermaking process. The material showed excellent wettability, high mechanical and thermal resistance, as well as superior electrochemical performance. The porous structure was obtained by coating the PET with CNF, resulting in a porosity of 70%, significantly higher than the 40% of polypropylene (PP) separators. The composite also showed superior hydrophilicity, being completely wetted by the electrolyte in a few seconds, while the PP separator was not completely wetted after a minute.

18.6.2 Hemicellulose

Hemicellulose, a key component of lignocellulosic biomass, is proving to be a strategic resource for innovative applications beyond its structural function in plants. In the production of second-generation biofuels (2G ethanol), sugars such as xylose and arabinose, released via enzymatic or acid hydrolysis, are fermented into ethanol, offering a renewable alternative to gasoline and reducing emissions by up to 85% compared to fossil fuels (Milessi et al. 2020). In the area of human and animal nutrition, oligosaccharides derived from hemicellulose, such as xylooligosaccharides (XOS), act as prebiotics, stimulating the growth of beneficial bacteria in the intestine and being incorporated into food supplements and functional foods (Rathamat et al. 2021). In green chemistry, hemicellulose serves as a raw material for the synthesis of biomacromolecules, such as furfural and HMF, used in the production of biodegradable plastics, epoxy resins, and even pharmaceuticals, replacing petrochemical derivatives (Luo et al. 2019). These applications not only diversify the use of biomass but also leverage the circular economy, transforming agricultural and forestry waste into high-value-added products, while reducing dependence on nonrenewable resources and promoting low-carbon industrial systems.

18.6.3 Lignin

Lignin, a biomacromolecule abundant in plant biomass, stands out as a versatile raw material for applications that transcend its traditional role in the cellulose industry. Currently, there are numerous studies using lignin as a concrete additive to improve durability and reduce cement consumption, while it acts as a natural dispersant in paints and agrochemicals, replacing synthetic surfactants (Silva et al. 2023). Transformed into activated carbon, it becomes effective in purifying water and air, and, in the form of bio-based resins, it replaces petroleum-derived phenolic compounds in adhesives and coatings, reducing emissions and toxicity. These applications reinforce its potential in the transition to sustainable industrial processes, aligning productivity and the circular economy.

In a recent study, (Barbosa et al. 2024) investigated the acid-catalyzed liquefaction of kraft lignin to produce polyols and subsequently bio-based polyurethane (PU) foams. Kraft lignin, a by-product of the pulp industry, was subjected to a liquefaction process using sulfuric acid (H_2SO_4) as a catalyst in different concentrations (0%, 3%, and 6% by weight). The aim was to assess how the addition of the catalyst affects the properties of the polyols and the resulting PU foams. The polyurethane foams in the study, derived from kraft lignin catalyzed with H_2SO_4, have a higher density (~1.26 g.cm^{-3}) compared to conventional foams (usually 0.05–0.2 g.cm^{-3}), reflecting a more compact cell structure. Their thermal stability (degradation >300 °C) is similar to commercial foams with flame retardant additives, while the stiffness (E′ +35%) surpassed many bio-based foams in the literature.

There are also several studies using lignin as a natural plasticizer for cement, hydrophobic agent, etc., using alkaline sulphonated lignin as a dispersant in cement (Ouyang et al. 2009). Takahashi et al. (2014) evaluated the performance of softwood soda-anthraquinone lignin as a water-reducing chemical additive in cement (Takahashi et al. 2014). Huang et al. (2018) explored the possibility of preparing lignosulfonates from lignin by sulfomethylation for application as a water reducer for cement (Huang et al. 2018). Lou et al. (2013) evaluated the possibility of preparing lignin-based superplasticizer by graft sulfonation and investigating the dispersive performance and mechanism in cement (Lou et al. 2013).

Lignin, due to its aromatic structure with a high carbon content of 60%, has been considered a qualified source for the production of carbon fibers, although it is not the only natural polymer that can be converted into CF. Carbonization yields of around 40–50% compared to carbonization yields of 10–30% for cellulose, a potential competitor, make lignin the most attractive option for economic processing. In 2002, Kadla et al. (2002) proposed the innovative manufacture of carbon fibers based on kraft lignin. The authors added poly(ethylene oxide) (PEO) to kraft lignin to facilitate the spinning of fibers in concentrations above 5%. The carbon fibers obtained had a total yield of 45%, with a tensile strength of 400–550 MPa and a modulus of 30–60 GPa. A few years later, several studies have focused on the production of lignin-based carbon fibers, such as Bengtsson et al. (2022), who studied the stabilization and continuous carbonization of a lignin-cellulose precursor for carbon fibers; Qu et al. (2022) who studied the manufacture of lignin-based carbon fibers as a renewable and low-cost substitute for fiber-shaped pseudo-capacitor electrodes; or Jiang et al. (2018) who suggested the application of lignin-based carbon fibers for the production of Ablative material, among many other applications.

18.7 Success Cases

The global bioeconomy is advancing with companies transforming biomass into high-value products, going beyond lignin to include derivatives such as cellulose microfibrils (MFC), nanocellulose crystals (CNC), and innovative textile fibers. Companies like Borregaard, Suzano, Stora Enso, and Lenzing are leading this revolution, showing that cellulose and its derivatives are pillars for a sustainable future. Borregaard (Norway), in addition to its expertise in lignin (bio-based vanillin and concrete additives), is exploring nanocellulose for applications in biodegradable packaging and high-strength composites. Suzano (Brazil), meanwhile, in addition to extracting kraft lignin for fertilizers and biomacromolecules, invests in MFC to reinforce special papers and lightweight materials for the automotive industry. In the MFC field, Stora Enso (Sweden/Finland) stands out with its Cellulose Foam product, a microfibril-based material used in sustainable packaging to replace plastic. CelluComp (United Kingdom) is innovating with Curran®, an MFC extracted from root vegetable waste (such as beet), applied in composites for construction and anticorrosion coatings. For CNC, Kruger (Canada), through the CelluForce joint venture, produces cellulose nanocrystals on an industrial scale, used in automotive paints (to increase durability) and aeronautical composites. American Process Inc. (USA) developed BioPlus™ CNC, used in biodegradable films for packaging and polymer reinforcement in medical prostheses. In the textile fibers sector, Lenzing AG (Austria) revolutionized the market with Tencel™, a lyocell fiber produced from wood cellulose, combining softness, biodegradability, and water efficiency. Spinnova (Finland) created a 100% cellulose textile fiber without the use of toxic chemicals, adopted by brands such as Marimekko and Adidas for sustainable collections. Renewcell (Sweden) innovated with Circulose®, a fiber made from recycled textiles (such as used jeans), closing the circular fashion loop in partnership with H&M and Levi's. These companies illustrate how the synergy between innovation and sustainability drives diversified markets. While lignin paves the way for green chemicals, MFC and CNC revolutionize structural materials, and cellulose-derived textile fibers reduce fashion's environmental footprint.

18.8 The Global Importance of the Paper Sector

18.8.1 Macroeconomic Data

The pulp and paper sector is one of the most dynamic pillars of the global economy, with world production reaching ~500 million tons by 2023 (Charles et al. 2024). China is the main player in this sector, with production of ~85 million tons per year, an increase of ~1.3% over the previous year (China Briefing 2023). The USA is one of the main players in this area, with production of ~65 million tons in 2023 (Charles et al. 2024). Brazil is the fourth largest global producer of pulp, generating 2.6 million direct and indirect jobs, achieving gross revenues of R$ 260 billion and breaking production records by reaching 25 million tons of pulp and 11 million tons of paper, with a growth of 10.2% in pulp production and 3.5% in paper production with a direct impact on the country's GDP ~6.3% of Brazil's GDP is directly or indirectly related to the paper production chain (IBÁ – Indústria Brasileira de Árvores 2019). The transition to the bioeconomy, with investments in nanocellulose and lignin to replace plastics, reinforces the sector's strategic role in the circular economy, integrating industrial productivity, technological innovation, and environmental responsibility. The indirect chain ranges from research into forest biotechnology, such as the eucalyptus clones developed by Embrapa in Brazil (Embrapa 2025), to the production of biochemicals in Finland, consolidating the sector as a central axis for sustainable development and generation of value on a global scale.

18.8.2 Sustainability as a Driver

The global transition to a low-carbon economy and the growing demand for sustainable practices have placed the pulp and paper industry (P&PI) at the center of a significant transformation. In this context, forest certifications, such as FSC (Forest Stewardship Council) and PEFC (Programme for the Endorsement of Forest Certification), and the pressure for production chains aligned with ESG (Environmental, Social, Governance) criteria have become key elements in guaranteeing the environmental, social, and economic viability of the sector. At the same time, Industrial Conversion and Products (IC & P) is playing a strategic role in building the bioeconomy by replacing products derived from fossil sources with innovative bioproducts, reducing emissions, and promoting circular production models.

The FSC and PEFC certifications are international systems that ensure responsible forest management, guaranteeing that the extraction of wood and fibers does not degrade ecosystems, respects labor rights, and benefits local communities. By 2023, the FSC had certified 238 million hectares of forests, while PEFC reached 330 million hectares, covering 20% of forests managed for industrial purposes. In 2023, around 14 million hectares of forests were certified in Brazil under the FSC and PEFC certification schemes, including both planted and native forests (Florestas do Brasil 2023). A practical example of this impact in Brazil is Suzano, the largest global producer of eucalyptus pulp, which maintains 100% of its forests certified by the FSC (Suzano 2025), guaranteeing access to premium markets and avoiding non-tariff barriers. These certifications are essential to mitigate the risks of illegal deforestation and to meet the growing pressure from institutional investors such as BlackRock and Vanguard, who make financing conditional on adherence to ESG standards.

The rise of ESG criteria has redefined IC&P's priorities. Companies in the sector are required to reduce their CO_2 emissions, with the sector accounting for 1.2 billion tons of CO_2eq/year. In order to meet neutrality targets by 2050, the sector is investing in renewable energy sources, such as replacing fuel oil with black liquor. In addition, it is essential to guarantee social equity, with programs such as Stora Enso's, which trains thousands of rural communities in Finland in forest management. Transparency in governance is also a growing requirement, with regulations, such as the European Union's Zero Deforestation, penalizing companies that do not present full traceability of the production chain. Non-compliance with these requirements can result in heavy financial losses, as happened with Asia Pulp & Paper (APP), which lost contracts worth US$500 million in Europe after allegations of illegal deforestation in Sumatra.

In addition to its contribution to sustainability, IC&P has played an important role in the bioeconomy, replacing plastics, chemicals, and fossil fuels with bioproducts derived from cellulose, lignin, and hemicellulose. Examples include bioplastics, such as the one produced by Metsä Spring, which replaces 30% of PET in bottles, and biokerosene, with the collaboration between Raízen and Klabin to produce sustainable aviation fuel. Another important advance is the production of nanocellulose, which already has a projected market of US$ 1.5 billion by 2030, with applications in high-tech sectors such as automobiles and antimicrobial medical dressings. In addition, the circular economy is gaining momentum, with biorefineries such as Suzano's, which convert biomass waste into products such as prebiotics and petroleum terephthalate substitutes. IC&P also contributes to carbon capture. Forests planted for pulp in Brazil sequester around 150 million tons of CO_2 per year, which is equivalent to 10% of national emissions. However, the sector faces challenges, such as high certification costs, which can be an obstacle for small producers, as well as the limited scalability of bioproducts, with only 5% of global lignin being commercialized. Competition with virgin fibers also represents a difficulty, since the global paper recycling rate is 58%, but developing countries still rely on virgin fibers to a large extent.

The pulp and paper Industry is undergoing a process of reinvention, where it is no longer limited to paper production but is positioning itself as one of the protagonists of the bioeconomy. By integrating forest certifications, ESG practices, and innovation in bioproducts, the sector not only reduces dependence on fossil fuels and mitigates climate change but also creates resilient and sustainable value chains. The next decade will be decisive in determining whether IC&P will be remembered as an industry of the past or as a key player in the green revolution, where sustainability and profitability go hand in hand.

18.9 Conclusion

The transition from traditional pulp and paper production to integrated biorefineries represents a transformative shift in biomass utilization, driven by the need for sustainable and efficient resource management. Lignocellulosic biomass, composed of cellulose, hemicellulose, and lignin, offers immense potential for producing high-value products such as nanocellulose, biofuels, biochemicals, and advanced materials. Technologies like LignoBoost and LignoForce have revolutionized lignin extraction, achieving yields of 80–90% with high purity (>95%), while processes such as autohydrolysis, steam explosion, and ionic liquid pretreatment optimize cellulose and hemicellulose recovery. The success of companies like Borregaard (vanillin from lignin), Suzano (kraft lignin for composites), and Lenzing (Tencel™ fibers) underscores the economic and environmental viability of biorefinery models. These innovations align with global sustainability goals, reducing reliance on fossil fuels and minimizing waste through circular economy principles.

However, challenges remain, including the high costs of advanced pretreatment methods (e.g., ionic liquids), the need for scalable enzymatic processes, and the heterogeneity of biomass feedstocks. Future advancements in metabolic engineering, solvent recycling, and cascade biorefining will be critical to improving efficiency and reducing costs. The pulp and paper industry, contributing ~6.3% to Brazil's GDP and 1.2 billion tons of CO_2eq/year globally, must continue integrating ESG criteria and certifications (FSC/PEFC) to meet decarbonization targets. By leveraging lignocellulosic biomass as a renewable feedstock, the sector can drive the bioeconomy forward, transforming waste into wealth while addressing climate change and resource scarcity. The integration of biorefineries is not just an industrial evolution—it is a necessity for a sustainable future.

References

Adnan AI, Ong MY, Nomanbhay S et al (2019) Technologies for biogas upgrading to biomethane: a review. Bioengineering 6:1–23. https://doi.org/10.3390/bioengineering6040092

Adu C, Jolly M, Thakur VK (2018) Exploring new horizons for paper recycling: a review of biomaterials and biorefinery feedstocks derived from wastepaper. Curr Opin Green Sustain Chem 13:21–26. https://doi.org/10.1016/j.cogsc.2018.03.003

Agumba DO, Kumar B, Kim J (2024) Advanced hydrostable, recyclable and degradable cellulose hybrid films as renewable alternatives to synthetic plastics. Int J Biol Macromol 260:129370. https://doi.org/10.1016/j.ijbiomac.2024.129370

Ahmed B, Aboudi K, Tyagi VK et al (2019) Improvement of anaerobic digestion of lignocellulosic biomass by hydrothermal pretreatment. Appl Sci 9:1–17. https://doi.org/10.3390/app9183853

Alam A, Wang Y, Liu F et al (2020) Modeling of optimal green liquor pretreatment for enhanced biomass saccharification and delignification by distinct alteration of wall polymer features and biomass porosity in Miscanthus. Renew Energy 159:1128–1138. https://doi.org/10.1016/j.renene.2020.06.013

Antes R, Joutsimo OP (2015) Fiber surface and paper technical properties of Eucalyptus globulus and Eucalyptus nitens pulps after modified cooking and bleaching. BioResources 10:1599–1616. https://doi.org/10.15376/biores.10.1.1599-1616

Auxenfans T, Crônier D, Chabbert B, Paës G (2017) Understanding the structural and chemical changes of plant biomass following steam explosion pretreatment. Biotechnol Biofuels 10:1–16. https://doi.org/10.1186/s13068-017-0718-z

Azhar Zakir MJ, Ramalingam S, Balasubramanian P et al (2015) Innovative material from paper and pulp industry for leather processing. J Clean Prod 104:436–444. https://doi.org/10.1016/j.jclepro.2015.05.001

Babi DK, Holtbruegge J, Lutze P et al (2015) Sustainable process synthesis-intensification. Comput Chem Eng 81:218–244. https://doi.org/10.1016/j.compchemeng.2015.04.030

Bajpai P (2018) Biermann's handbook of pulp and paper: raw material and pulp making

Baloch HA, Nizamuddin S, Siddiqui MTH et al (2018) Recent advances in production and upgrading of bio-oil from biomass: a critical overview. J Environ Chem Eng 6:5101–5118. https://doi.org/10.1016/j.jece.2018.07.050

Barbosa KT, Fuentes Da Silva SH, Magalhães WLE et al (2024) Acid-catalyzed Kraft lignin liquefaction for producing polyols and polyurethane foams. J Wood Chem Technol 44:9–21. https://doi.org/10.1080/02773813.2024.2303036

Benavides PT, Balchandani S, Gracida-Alvarez UR (2024) Environmental analysis of biotechnologies for biofuels, bioplastics, and bioproducts: a greenhouse gas (GHG) emissions review. Biotechnol Environ 1:1–16. https://doi.org/10.1186/s44314-024-00010-5

Bengtsson A, Bengtsson J, Jedvert K et al (2022) Continuous stabilization and carbonization of a lignin−cellulose precursor to carbon fiber. ACS Omega 7:16793–16802. https://doi.org/10.1021/ACSOMEGA.2C01806

Bhatia SK, Jagtap SS, Bedekar AA et al (2020) Recent developments in pretreatment technologies on lignocellulosic biomass: effect of key parameters, technological improvements, and challenges. Bioresour Technol 300:122724. https://doi.org/10.1016/j.biortech.2019.122724

Borole AP (2015) Sustainable and efficient pathways for bioenergy recovery from low-value process streams via bioelectrochemical systems in biorefineries. Sustainability 7:11713–11726. https://doi.org/10.3390/su70911713

Branco RHR, Serafim LS, Xavier AMRB (2019) Second generation bioethanol production: on the use of pulp and paper industry wastes as feedstock. Fermentation 5:1–30. https://doi.org/10.3390/fermentation5010004

Bras J, Hassan ML, Bruzesse C et al (2010) Mechanical, barrier, and biodegradability properties of bagasse cellulose whiskers reinforced natural rubber nanocomposites. Ind Crops Prod 32:627–633. https://doi.org/10.1016/j.indcrop.2010.07.018

Brosse N (2025) Organosolv processes references. Biorefineries

Brown J, Lindstrom JK, Ghosh A et al (2024) Production of sugars from lignocellulosic biomass via biochemical and thermochemical routes. Front Energy Res 12:1–34. https://doi.org/10.3389/fenrg.2024.1347373

Bušić A, Mardetko N, Kundas S et al (2018) Bioethanol production from renewable raw materials and its separation and purification: a review. Food Technol Biotechnol 56:289–311. https://doi.org/10.17113/ftb.56.03.18.5546

Casanova F, Freixo R, Pereira CF et al (2023) Comparative study of green and traditional routes for cellulose extraction from a sugarcane by-product. Polymers (Basel) 15. https://doi.org/10.3390/polym15051251

Charles M, Narayan KB, Edmonds J, Yu S (2024) The role of the pulp and paper industry in achieving net zero U.S. CO2 emissions in 2050. Energy Clim Chang 5:100160. https://doi.org/10.1016/j.egycc.2024.100160

Chaturvedi V, Verma P (2013) An overview of key pretreatment processes employed for bioconversion of lignocellulosic biomass into biofuels and value added products. 3 Biotech 3:415–431. https://doi.org/10.1007/s13205-013-0167-8

Chen Z, Wan C (2017) Biological valorization strategies for converting lignin into fuels and chemicals. Renew Sustain Energy Rev 73:610–621. https://doi.org/10.1016/j.rser.2017.01.166

Chen X, Li H, Sun S et al (2016) Effect of hydrothermal pretreatment on the structural changes of alkaline ethanol lignin from wheat straw. Sci Rep 6:1–9. https://doi.org/10.1038/srep39354

Chen YL, Zhang X, You TT, Xu F (2019) Deep eutectic solvents (DESs) for cellulose dissolution: a mini-review. Cellulose 26:205–213. https://doi.org/10.1007/s10570-018-2130-7

Chen Q, Chang C, Zhang L (2021) Surface engineering of cellulose film with myristic acid for high strength, self-cleaning and biodegradable packaging materials. Carbohydr Polym 269:118315. https://doi.org/10.1016/j.carbpol.2021.118315

Cheremisinoff NP (2010) Handbook of pollution prevention and cleaner production

Chimentão RJ, Lorente E, Gispert-Guirado F et al (2014) Hydrolysis of dilute acid-pretreated cellulose under mild hydrothermal conditions. Carbohydr Polym 111:116–124. https://doi.org/10.1016/j.carbpol.2014.04.001

China Briefing (2023) China's pulp and paper industry – opportunities for foreign investors

Cozier M (2014) Business highlights: collaboration: bigger and beta. Biofuels Bioprod Biorefin 8:743. https://doi.org/10.1002/BBB

da Rosa TS, Trianoski R, Michaud F et al (2021) Sludge fiber waste and kraft lignin powder as fillers in polylactic acid biocomposites: physical, mechanical, and thermal properties. Polymers (Basel) 13:1–14. https://doi.org/10.3390/polym13050672

Das TK, Jain AK (2001) Pollution prevention advances in pulp and paper processing. Environ Prog 20:87–92. https://doi.org/10.1002/ep.670200211

De Aguiar J, Bondancia TJ, Claro PIC et al (2020) Enzymatic deconstruction of sugarcane bagasse and straw to obtain cellulose nanomaterials. ACS Sustain Chem Eng 8:2287–2299. https://doi.org/10.1021/acssuschemeng.9b06806

Díez D, Urueña A, Piñero R et al (2020) Determination of hemicellulose, cellulose, and lignin content in different types of biomasses by thermogravimetric analysis and pseudocomponent kinetic model (TGA-PKM Method). Processes 8. https://doi.org/10.3390/pr8091048

Embrapa (2025) Embrapa. https://www.embrapa.br

FAO (2022) Food and Agriculture Organization of the United Nations. Faostat: database – crop production. In: Datos sobre Aliment. y Agric. http://www.fao.org/faostat/en/#data/QC
Florestas do Brasil (2023) Certificação Florestal
Galbe M, Wallberg O (2019) Pretreatment for biorefineries: a review of common methods for efficient utilisation of lignocellulosic materials. Biotechnol Biofuels 12:1–26. https://doi.org/10.1186/s13068-019-1634-1
Gan S, Zakaria S, Chen RS et al (2017) Autohydrolysis processing as an alternative to enhance cellulose solubility and preparation of its regenerated bio-based materials. Mater Chem Phys 192:181–189. https://doi.org/10.1016/j.matchemphys.2017.01.012
Guo H, Zhao Y, Chen X et al (2019) Pretreatment of Miscanthus with biomass-degrading bacteria for increasing delignification and enzymatic hydrolysability. Microb Biotechnol 12:787–798. https://doi.org/10.1111/1751-7915.13430
Hou Q, Ju M, Li W et al (2017) Pretreatment of lignocellulosic biomass with ionic liquids and ionic liquid-based solvent systems. Molecules 22:1–24. https://doi.org/10.3390/molecules22030490
Huang C, Ma J, Zhang W et al (2018) Preparation of lignosulfonates from biorefinery lignins by sulfomethylation and their application as a water reducer for concrete. Polymers (Basel) 10. https://doi.org/10.3390/polym10080841
IBÁ – Indústria Brasileira de Árvores (2019) Relatório 2019 report 2019. Relatório:1–80
Jaiswal S, Kumar Gupta G, Panchal K et al (2021) Synthetic organic compounds from paper industry wastes: integrated biotechnological interventions. Front Bioeng Biotechnol 8:1–13. https://doi.org/10.3389/fbioe.2020.592939
Jakob A, Likozar B, Grilc M (2024) Model-assisted optimization of xylose, arabinose, glucose, mannose, galactose and real hemicellulose streams dehydration to (hydroxymethyl)furfural and levulinic acid. ChemSusChem:202400962. https://doi.org/10.1002/cssc.202400962
Jarre M, Petit-Boix A, Priefer C et al (2020) Transforming the bio-based sector towards a circular economy – what can we learn from wood cascading? For Policy Econ 110:101872. https://doi.org/10.1016/j.forpol.2019.01.017
Jiang X, Ouyang Q, Liu D et al (2018) Preparation of low-cost carbon fiber precursors from blends of wheat straw lignin and commercial textile-grade polyacrylonitrile (PAN). Holzforschung 72: 727–734. https://doi.org/10.1515/hf-2017-0191
Jin Q, Yang L, Poe N, Huang H (2018) Integrated processing of plant-derived waste to produce value-added products based on the biorefinery concept. Trends Food Sci Technol 74:119–131. https://doi.org/10.1016/j.tifs.2018.02.014
Joseph N, Mohamad Kassim MH, Ibrahim M et al (2024) The effects of autohydrolysis pretreatment on the properties of OPT pulps for the production of dissolving pulp. Pertanika J Sci Technol 32:27–39. https://doi.org/10.47836/pjst.32.S3.03
Kadla JF, Kubo S, Venditti RA et al (2002) Lignin-based carbon fibers for composite fiber applications. Carbon N Y 40:2913–2920. https://doi.org/10.1016/S0008-6223(02)00248-8
Kalntremtziou M, Papaioannou IA, Vangalis V et al (2023) Evaluation of the lignocellulose degradation potential of Mediterranean forests soil microbial communities through diversity and targeted functional metagenomics. Front Microbiol 14:1–16. https://doi.org/10.3389/fmicb.2023.1121993
Kamali M, Khodaparast Z (2015) Review on recent developments on pulp and paper mill wastewater treatment. Ecotoxicol Environ Saf 114:326–342. https://doi.org/10.1016/j.ecoenv.2014.05.005
Kun D, Pukánszky B (2017) Polymer/lignin blends: interactions, properties, applications. Eur Polym J 93:618–641. https://doi.org/10.1016/j.eurpolymj.2017.04.035
Li OL, Ikura R, Ishizaki T (2017) Hydrolysis of cellulose to glucose over carbon catalysts sulfonated via a plasma process in dilute acids. Green Chem 19:4774–4777. https://doi.org/10.1039/c7gc02143g
Li Y, Sun H, Mu T, Garcia-Vaquero M (2025) Sustainable closed-loop biorefinery of γ-valerolactone from lignocellulosic biomass: pretreatments of multiple biomass and synthesis

of γ-valerolactone from multiple biomass-derived feedstocks. Biomass Bioenergy 193:107594. https://doi.org/10.1016/j.biombioe.2025.107594
Liu Y, Man Y, Ren J (2020) Waste-to-wealth by sludge-to-energy: a comprehensive literature reviews. INC
Lou H, Lai H, Wang M et al (2013) Preparation of lignin-based superplasticizer by graft sulfonation and investigation of the dispersive performance and mechanism in a cementitious system. Ind Eng Chem Res 52:16101–16109. https://doi.org/10.1021/ie402169g
Luo Y, Li Z, Li X et al (2019) The production of furfural directly from hemicellulose in lignocellulosic biomass: a review. Catal Today 319:14–24. https://doi.org/10.1016/j.cattod.2018.06.042
Lyocell (2019) Discover the fabric of the future: Lyocell. https://lyocell.info
Malik S, Rana V, Joshi G et al (2020) Valorization of wheat straw for the paper industry: pre-extraction of reducing sugars and its effect on pulping and papermaking properties. ACS Omega 5:30704–30715. https://doi.org/10.1021/acsomega.0c04883
Martin-sampedro R, Eugenio ME, Moreno JA et al (2014) Bioresource Technology Integration of a kraft pulping mill into a forest biorefinery: pre-extraction of hemicellulose by steam explosion versus steam treatment. Bioresour Technol 153:236–244. https://doi.org/10.1016/j.biortech.2013.11.088
McEwan A, Marchi E, Spinelli R, Brink M (2020) Past, present and future of industrial plantation forestry and implication on future timber harvesting technology. J For Res 31:339–351. https://doi.org/10.1007/s11676-019-01019-3
Milessi TS, Perez CL, Zangirolami TC et al (2020) Repeated batches as a strategy for high 2G ethanol production from undetoxified hemicellulose hydrolysate using immobilized cells of recombinant Saccharomyces cerevisiae in a fixed-bed reactor. Biotechnol Biofuels 13:1–12. https://doi.org/10.1186/s13068-020-01722-y
Mongkhonsiri G, Anantpinijwatna A, Charoensuppanimit P et al (2021) Novel biorefinery-Integrated-Kraft-pulping network for sustainable development. Chem Eng Process Process Intensif 163:108373. https://doi.org/10.1016/j.cep.2021.108373
Moshkelani M, Marinova M, Perrier M, Paris J (2013) The forest biorefinery and its implementation in the pulp and paper industry: energy overview. Appl Therm Eng 50:1427–1436. https://doi.org/10.1016/j.applthermaleng.2011.12.038
Mosier N, Wyman C, Dale B et al (2005) Features of promising technologies for pretreatment of lignocellulosic biomass. Bioresour Technol 96:673–686. https://doi.org/10.1016/j.biortech.2004.06.025
Nada AAMA, El-Sakhawy M, Kamel SM (1998) Infra-red spectroscopic study of lignins. Polym Degrad Stab 60:247–251. https://doi.org/10.1016/s0141-3910(97)00072-4
Nitsos C, Rova U, Christakopoulos P (2018) Organosolv fractionation of softwood biomass for biofuel and biorefinery applications. Energies 11. https://doi.org/10.3390/en11010050
Nizami AS, Rehan M, Waqas M et al (2017) Waste biorefineries: enabling circular economies in developing countries. Bioresour Technol 241:1101–1117. https://doi.org/10.1016/j.biortech.2017.05.097
Ouyang X, Ke L, Qiu X et al (2009) Sulfonation of alkali lignin and its potential use in dispersant for cement. J Dispers Sci Technol 30:1–6. https://doi.org/10.1080/01932690802473560
Pala H, Mota M, Gama FM (2007) Enzymatic depolymerisation of cellulose. Carbohydr Polym 68:101–108. https://doi.org/10.1016/j.carbpol.2006.07.015
Patel DK, Dutta SD, Lim KT (2019) Nanocellulose-based polymer hybrids and their emerging applications in biomedical engineering and water purification. RSC Adv 9:19143–19162. https://doi.org/10.1039/c9ra03261d
Pereira GN, Cesca K, Vieira Cubas AL, de Oliveira D (2021) Use of non-thermal plasma in lignocellulosic materials: a smart alternative. Trends Food Sci Technol 109:365–373. https://doi.org/10.1016/j.tifs.2021.01.047
Phanthong P, Reubroycharoen P, Hao X et al (2018) Nanocellulose: extraction and application. Carbon Resour Convers 1:32–43. https://doi.org/10.1016/j.crcon.2018.05.004

Qu W, Hu P, Liu J et al (2022) Lignin-based carbon fiber: a renewable and low-cost substitute towards featured fiber-shaped pseudocapacitor electrodes. J Clean Prod 343:131030. https://doi.org/10.1016/j.jclepro.2022.131030

Rathamat Z, Choorit W, Chisti Y, Prasertsan P (2021) Two-step isolation of hemicellulose from oil palm empty fruit bunch fibers and its use in production of xylooligosaccharide prebiotic. Ind Crops Prod 160:113124. https://doi.org/10.1016/j.indcrop.2020.113124

Rodriguez JML, Colodette JL, García JC, López F (2019) Autohydrolysis, pulping, and bleaching of Eucalyptus urograndis in a biorefinery framework. BioResources 14:5467–5487. https://doi.org/10.15376/biores.14.3.5467-5487

Saeman JF (1945) Forest products Lab. 8 bept Agr. BUU 81446:1

Saranya, Kumar SL, Jawahar RR (2020) Prediction of paper mill wastewater treatment process parameters in sequencing batch reactor using fuzzy logic technique. Int J Recent Technol Eng 8: 1929–1937. https://doi.org/10.35940/ijrte.e6004.018520

Sarker TR, Pattnaik F, Nanda S et al (2021) Hydrothermal pretreatment technologies for lignocellulosic biomass: a review of steam explosion and subcritical water hydrolysis. Chemosphere 284:131372. https://doi.org/10.1016/j.chemosphere.2021.131372

Schneemann J, Miller S, Liu YS, Safferman S (2009) Waste biomass inventory to support renewable energy development. Am Soc Agric Biol Eng Annu Int Meet 2009, ASABE 2009 6:3592–3599. https://doi.org/10.13031/2013.27356

Segers B, Nimmegeers P, Spiller M et al (2024) Lignocellulosic biomass valorisation: a review of feedstocks, processes and potential value chains and their implications for the decision-making process. RSC Sustain:3730–3749. https://doi.org/10.1039/d4su00342j

Serrano-Martínez VM, Pérez-Aguilar H, Carbonell-Blasco MP et al (2024) Steam explosion-based method for the extraction of cellulose and lignin from rice straw waste. Appl Sci 14. https://doi.org/10.3390/app14052059

Silva DM, Innocentini MDM, Marin CP et al (2023) Valorization of lignin residue in mortars and concretes. Ceramica 69:364–370. https://doi.org/10.1590/0366-69132024703923503

Silveira MHL, Morais ARC, Da Costa Lopes AM et al (2015) Current pretreatment technologies for the development of cellulosic ethanol and biorefineries. ChemSusChem 8:3366–3390. https://doi.org/10.1002/cssc.201500282

Sixta H (2008) Handbook of pulp. Handb Pulp 1–2:1–1352. https://doi.org/10.1002/9783527619887

Song W, Deng Y, Zhu H (2016) Research on wheat straw pulping with ionic liquid 1-ethyl-3-methylimidazole bromide. Kem u Ind Chem Chem Eng 65:579–585. https://doi.org/10.15255/KUI.2016.032

Sun SF, Yang HY, Yang J, Shi ZJ (2022) The effect of alkaline extraction of hemicellulose on cocksfoot grass enzymatic hydrolysis recalcitrance. Ind Crops Prod 178:114654. https://doi.org/10.1016/j.indcrop.2022.114654

Suzano (2025) Suzano certificações. https://www.suzano.com.br/suzano/transparencia/certificacoes

Swatloski RP, Spear SK, Holbrey JD, Rogers RD (2002) Dissolution of cellose with ionic liquids. J Am Chem Soc 124:4974–4975. https://doi.org/10.1021/ja025790m

Takahashi S, Hattori M, Morimoto M et al (2014) Performance of softwood soda-anthraquinone lignin aswater-reducing chemical admixture in concrete. J Wood Chem Technol 34:31–38. https://doi.org/10.1080/02773813.2013.820322

Tolvaj L, Papp G, Varga D, Lang E (2012) Effect of steaming on the colour change of softwoods. BioResources 7:2799–2808. https://doi.org/10.15376/biores.7.3.2799-2808

Tomani P (2010) The lignoboost process. Cellul Chem Technol 44:53–58

Tsegaye B, Balomajumder C, Roy P (2019) Microbial delignification and hydrolysis of lignocellulosic biomass to enhance biofuel production: an overview and future prospect. Bull Natl Res Cent 43. https://doi.org/10.1186/s42269-019-0094-x

Usmani Z, Sharma M, Gupta P et al (2020) Ionic liquid based pretreatment of lignocellulosic biomass for enhanced bioconversion. Bioresour Technol 304:123003. https://doi.org/10.1016/j.biortech.2020.123003

Vásquez-Garay F, Carrillo-Varela I, Vidal C et al (2021) A review on the lignin biopolymer and its integration in the elaboration of sustainable materials. Sustain 13:1–15. https://doi.org/10.3390/su13052697

Vasudev L, Chavan RL, Devagiri G (2021) Carbon sequestration potential in total biomass of Melia dubia cav. under semi-arid region of Karnataka. Int J Chem Stud 9:898–902. https://doi.org/10.22271/chemi.2021.v9.i2m.11933

Velvizhi G, Balakumar K, Shetti NP et al (2022) Integrated biorefinery processes for conversion of lignocellulosic biomass to value added materials: paving a path towards circular economy. Bioresour Technol 343:126151. https://doi.org/10.1016/j.biortech.2021.126151

Vives MB, Thuvander J, Arkell A, Lipnizki F (2022) Low-molecular-weight lignin recovery with nanofiltration in the kraft pulping process. Membranes (Basel) 12. https://doi.org/10.3390/membranes12030310

Wang W, Winters P, Read M (2005) Statistically significant e iciency testing of HEPA filters. J IEST 4321

Wang X, Cui X, Zhang L (2012) Preparation and characterization of lignin-containing nanofibrillar cellulose. Procedia Environ Sci 16:125–130. https://doi.org/10.1016/j.proenv.2012.10.017

Wang D, Kao MR, Li J et al (2022) Novel two-step process in cellulose depolymerization: hematite-mediated photocatalysis by lytic polysaccharide monooxygenase and fenton reaction. J Agric Food Chem 70:9941–9947. https://doi.org/10.1021/acs.jafc.2c02445

Xia G, Ji X, Xu Z, Ji X (2022) Transparent cellulose-based bio-hybrid films with enhanced anti-ultraviolet, antioxidant and antibacterial performance. Carbohydr Polym 298:120118. https://doi.org/10.1016/j.carbpol.2022.120118

Yin T, Huhe T, Li X et al (2024) Research on life cycle assessment and performance comparison of bioethanol production from various biomass feedstocks. Sustainability 16. https://doi.org/10.3390/su16051788

Yuan Q, Liu S, Ma MG et al (2021) The kinetics studies on hydrolysis of hemicellulose. Front Chem 9:1–12. https://doi.org/10.3389/fchem.2021.781291

Zhang H, Wang X, Liang Y (2015) Preparation and characterization of a Lithium-ion battery separator from cellulose nanofibers. Heliyon 1:1–16. https://doi.org/10.1016/j.heliyon.2015.e00032

Zhang Z, Harrison MD, Rackemann DW et al (2016) Organosolv pretreatment of plant biomass for enhanced enzymatic saccharification. Green Chem 18:360–381. https://doi.org/10.1039/c5gc02034d

Zhu Y, Zhang J, Wang D et al (2023) Preparation of anti-bacterial cellulose nanofibrils (CNFs) from bamboo pulp in a reactable citric acid–choline chloride deep eutectic solvent. Polymers (Basel) 15. https://doi.org/10.3390/polym15010148

Zhu P, Vo A, Sun X et al (2024) Water-induced controllable deswelling strategy enabled rapid fabrication of transparent cellulose film for plastics replacement. Chem Eng J 492:152200. https://doi.org/10.1016/j.cej.2024.152200

Chapter 19
Silk Industry Waste Biomass: Repurposing for Bioenergy and Bioproducts

Reema Kumar, Seyedehsetayesh Kazemi, and Satinder Kaur Brar

Abstract The global silk industry, while centuries old, continues to produce significant amounts of waste at various stages, from sericulture to postproduction processes. These waste streams, rich in biomaterials, offer immense potential for sustainable valorization but remain underexploited. This chapter investigates the different types of silk biomass generated, exploring their chemical properties and suitability for converting into bioenergy, particularly biogas through anaerobic digestion. It also examines innovative pathways for transforming silk waste into valuable bioproducts such as bioplastics, biomedical materials, cosmetics, and animal feed, highlighting the technological and economic opportunities involved. Despite promising advancements, several barriers, including extraction challenges, technological limitations, and awareness gaps, hinder widespread implementation. By reviewing successful case studies and current challenges, the chapter underscores the vital role of silk biomass in advancing circular economy practices. Ultimately, it emphasizes the importance of future research, innovation, and policy support to unlock the full potential of silk waste for sustainable development.

19.1 Introduction

The silk industry is among the most ancient and esteemed sectors within the global textile market, with origins traceable to ancient China over 3000 years ago. Today, it remains a significant economic contributor, with the global silk market valued at approximately USD 20 billion in 2024 and projected to grow at a compound annual growth rate of around 6.62% over 2025–2033 (Silk Market Size, Share, Trends, Report, and Forecast 2033). Major producers include China, which accounts for

R. Kumar · S. K. Brar (✉)
Department of Civil Engineering, Lassonde School of Engineering, York University, Toronto, ON, Canada
e-mail: satinder.brar@lassonde.yorku.ca

S. Kazemi
Department of Biology, York University, Toronto, ON, Canada

A. L. Woiciechowski et al. (eds.), *Biomass for Bioenergy and Bioproducts*, Biofuel and Biorefinery Technologies 17, https://doi.org/10.1007/978-3-032-19898-3_19

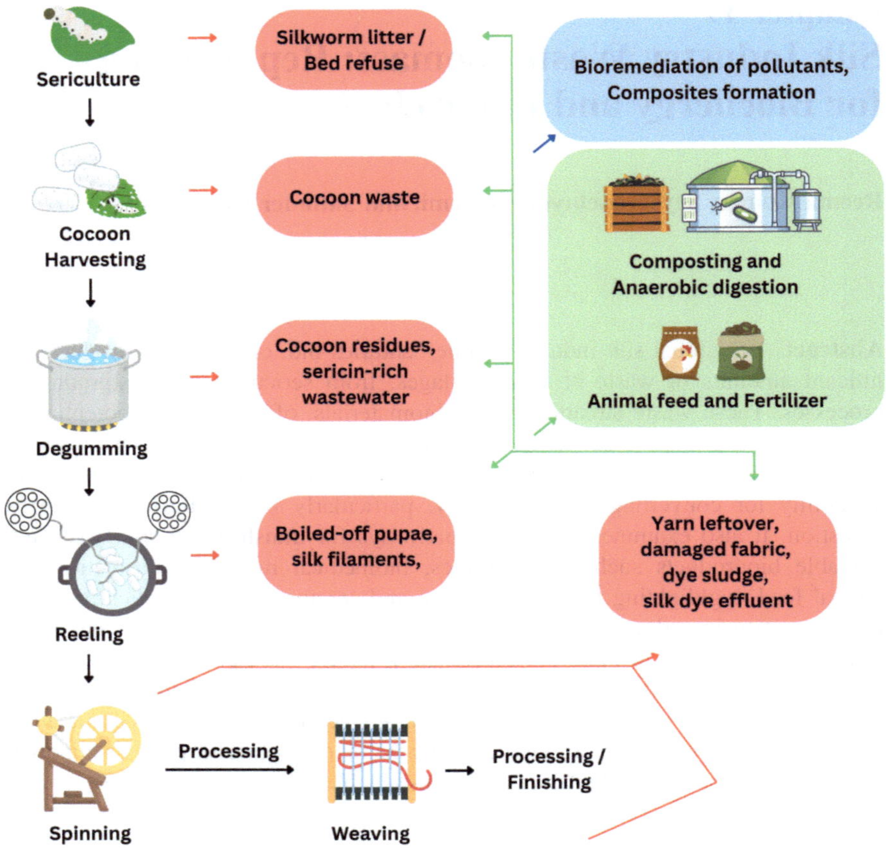

Fig. 19.1 Different stages of the silk industry, subsequent waste types, and their applications

roughly 70% of the world's silk output, followed by India, Uzbekistan, and Thailand (Statistics | International Sericultural Commission 2025). Silk's unique properties, such as its luster, strength, and smoothness, have sustained its demand across the luxury fashion, home décor, and technical sectors. Beyond its economic impact, silk holds considerable cultural, artistic, and symbolic importance, serving as a valuable resource for employment and technological development in numerous countries. Despite its longstanding prominence, the industry faces modern challenges related to environmental sustainability, resource consumption, and waste management, necessitating innovative approaches to ensure continued relevance and responsible production practices.

The silk industry encompasses a series of intricate processes (Fig. 19.1) that transform raw silkworm cocoons into high-quality silk textiles. The initial stage, sericulture, involves the domestication and cultivation of *Bombyx mori* silkworms, which are primarily reared on mulberry (*Morus alba*) leaves. This phase requires meticulous attention to environmental conditions, such as temperature, humidity, and hygiene, to optimize silkworm health and cocoon production. Once the silkworms reach the spinning stage, they produce a protective cocoon composed of raw silk fibers enveloped in sericin, a natural protein matrix. The subsequent step involves harvesting of cocoons, typically carried out when the silkworms have fully spun their cocoons, approximately 2–3 days before they emerge as moths to prevent damage and ensure fiber integrity. The cocoons are then subjected to a degumming process, which entails boiling in hot water or steaming to soften and dissolve sericin, thereby facilitating the unwinding of the silk filaments. This step is crucial for preserving the length and strength of the silk fibers and eliminating impurities. Following degumming, the silk threads are reeled, a process whereby multiple filaments are unwound from individual cocoons and combined to form a continuous filament yarn. This reeling process requires precision to maintain fiber uniformity and quality. The reeled silk is then spun into yarn using traditional hand-spinning or modern mechanized spinning methods, each influencing the final characteristics of the textile (Silk: Processing, Properties and Applications – K. Murugesh Babu – Google Books 2025).

The spun yarn undergoes further processing, including cleaning, stretching, and sometimes dyeing, to achieve the desired aesthetic and functional qualities. Dyeing can be performed at various stages, either on the raw yarn, the warp, or the finished fabric, using diverse techniques such as piece dyeing, yarn dyeing, or printing (Chakraborty and Chakraborty 2015). The weaving phase involves interlacing the silk yarns on different types of looms to produce fabrics with varied textures and designs, ranging from plain weaves to intricate brocades (Basu 2015). Post-weaving, the silk fabric undergoes finishing processes that may include bleaching to enhance brightness, mechanical or chemical smoothing, and decorative treatments. These finishing stages are essential for imparting the desired texture, appearance, and additional properties such as luster or water resistance (Basu 2015).

In the context of escalating environmental concerns, waste valorization has emerged as a critical strategy for promoting sustainability within the silk industry. The production process generates various waste streams, including sericin-rich wastewater, nonwoven cocoon residues, and leftover fibers, which, if improperly disposed of, pose environmental risks such as water pollution and increased landfill burden (Javali et al. 2015). Recent estimates suggest that approximately 11 million tonnes of silk waste are produced annually worldwide, emphasizing the need for effective waste management solutions (Hassan et al. 2025). Converting these wastes into value-added products, such as bio-based materials, organic fertilizers, or biofuels, significantly reduces the ecological footprint of silk manufacturing. Moreover, waste valorization aligns with circular economy principles, fostering resource efficiency and economic resilience by transforming industrial residues into profitable commodities. This approach not only mitigates environmental impacts but also opens pathways for innovation, sustainable development, and competitive advantage in the global marketplace. This chapter explores the valorization of silk industry waste and its bioproducts.

19.2 Waste Generation in the Silk Industry

The sericulture industry generates substantial amounts of waste throughout its production cycle, from silkworm rearing to silk reeling and post-fabrication processes. Effective waste management within this sector not only contributes to environmental sustainability but also opens new avenues for resource recovery and value addition. Sericulture waste primarily originates during the rearing of silkworms, where the most common byproducts include silkworm litter, dead or diseased larvae, uneaten mulberry leaves, and pupal remains. Silkworm litter, often called bed refuse, is a heterogeneous mix of excreta, leaf debris, and unconsumed feed that accumulates rapidly across rearing cycles (Tilahun et al. 2015). Rich in organic carbon and essential nutrients such as nitrogen and potassium, this waste can be repurposed into high-quality compost through aerobic or vermicomposting methods (Kalaiyarasan et al. 2015). Studies have shown that combining silkworm litter with cattle dung or microbial inoculants accelerates decomposition and enriches nutrient composition, making it an excellent organic fertilizer for agricultural use (Kumar et al. 2020). Dead larvae and diseased worms, which pose potential biohazards, are also biodegradable and protein-rich, thus serving as promising feedstock for anaerobic digestion (Viswanath and Nand 1994). When subjected to biogas production processes, particularly co-digestion with cow dung or food waste, these proteinaceous materials can yield high volumes of methane (Łochyńska and Frankowski 2018). Research from various Asian sericulture hubs, including China and India, demonstrates the feasibility of using enzymatic pretreatment or thermal conditioning to improve biogas yields from silkworm biomass (Ngabala and Emmanuel 2024; Kim et al. 2024). Furthermore, silk cocoon waste, such as pierced or flimsy cocoons, which are not viable for reeling, can be blended with other agro-residues for similar bioenergy applications.

The silk reeling process, which extracts raw silk threads from cocoons by boiling and unwinding, produces another layer of waste. A major byproduct is the boiled-off pupae, which are typically discarded in large quantities. However, these pupae are rich in proteins and fats, making them suitable for conversion into high-protein animal feed, organic fertilizer, or even for oil extraction (Hu et al. 2017). Pupal oil, comprising omega-3 and omega-6 fatty acids, has applications in cosmetic formulations and is being explored for biodiesel production (Halada Nandakrishnan et al. 2023). In addition to pupae, the reeling process also generates waste silk threads and degumming water effluents. Waste threads, usually composed of short filament lengths or broken fibers, are often converted into spun silk yarn or used in handloom sectors to produce low-cost fabrics. Meanwhile, the effluents from degumming and reeling processes, often containing soaps, proteins, and fine particulate matter, can burden local water bodies if left untreated (Wu et al. 2025). Advanced biological treatment methods, such as the use of microbial consortia or constructed wetlands, are gaining traction for managing these effluents in an eco-friendly manner (Shi et al. 2015).

Postproduction waste in the sericulture value chain arises during the finishing stages of silk processing, including dyeing, weaving, and tailoring. These wastes include yarn leftovers, defective or damaged fabrics, dye sludge, and cutting remnants. Among these, silk dye effluents pose one of the most critical challenges due to the presence of complex dye molecules, heavy metals, and high chemical oxygen demand (COD) (Mia et al. 2022). Conventional treatment plants often struggle with the complete removal of these pollutants. However, recent innovations in bio-remediation, particularly involving dye-degrading bacteria such as *Pseudomonas putida* and *Bacillus subtilis*, offer promising solutions (Rima et al. 2022). Additionally, the sludge generated from effluent treatment plants has found secondary use in the production of eco-bricks and cement composites, thereby supporting the construction sector while reducing waste disposal burdens (Limami et al. 2021). Fabric remnants and off-cuts from garment manufacturing are increasingly being upcycled into artisanal products such as patchwork quilts, fashion accessories, and home décor items. In some parts of India, women's cooperatives and self-help groups have built successful micro-enterprises around silk fabric upcycling, adding both economic and ecological value to postproduction waste (Kamble and Behera 2021).

19.3 Composition and Characteristics of Silk Industry Waste

Silk waste is a significant biomass material derived from sericulture, primarily consisting of fibroin and sericin, the two major protein components of silk. The utilization of silk waste for sustainable applications requires an understanding of its chemical composition and physicochemical properties.

19.3.1 Chemical Composition of Silk Waste

Silk waste is predominantly composed of fibroin (70–80%) and sericin (20–30%), with minor constituents including waxes, carbohydrates, and mineral salts (Yuan et al. 2024). Fibroin is an insoluble fibrous protein with a highly ordered structure, consisting mainly of glycine (Gly), alanine (Ala), and serine (Ser) residues. These amino acids contribute to the formation of β-sheet structures, providing silk with its remarkable tensile strength and biocompatibility (Vepari and Kaplan 2007). Sericin, the water-soluble protein surrounding fibroin fibers, contains a higher proportion of polar amino acids, such as serine, aspartic acid, and glutamic acid. This characteristic makes sericin highly hydrophilic, playing a crucial role in moisture retention and biomedical applications (Seo et al. 2023). The removal of sericin from silk fibers, commonly achieved through degumming processes, enhances the mechanical properties of silk fibroin and facilitates its use in textiles and biomaterials (Yuan et al. 2024).

19.3.2 Physicochemical Properties of Silk Waste

Silk waste exhibits unique physicochemical properties that make it suitable for various industrial applications. The molecular weight of silk fibroin varies between 200 and 350 kDa, and its crystallinity influences its degradation and mechanical behavior (Nguyen et al. 2019). Additionally, the presence of β-sheet structures in fibroin enhances its thermal stability, with degradation temperatures exceeding 250 °C (Madappura and Madduri 2023). Silk waste also contains small amounts of lipid and carbohydrate impurities, which affect its solubility and processing efficiency (Cao and Zhang 2016). The pH of silk fibroin solutions significantly affects their solubility and self-assembly, which is crucial for optimizing conditions in biomedical and textile applications. Research has shown that adjusting the pH can induce conformational changes in silk fibroin, leading to the formation of hydrogels with varying properties. For instance, at certain pH levels, silk fibroin transitions from a random coil to a β-sheet structure, promoting gelation through self-aggregation (Zheng and Zuo 2021). The chemical composition and characteristics of silk waste play a crucial role in its processing and application potential.

19.4 Repurposing Silk Biomass for Bioenergy

Repurposing silk biomass for bioenergy production presents a sustainable strategy to mitigate silk industry waste while contributing to renewable energy solutions. With increasing concerns over waste generation from textile industries and the need for circular economy models, silk waste, particularly degumming residues, fibroin-rich byproducts, and sericin, has emerged as a promising feedstock for bioenergy applications such as biogas production (Samanta et al. 2015). Biogas, primarily composed of methane (CH_4) and carbon dioxide (CO_2), is generated through anaerobic digestion of organic matter by a consortium of microorganisms. Silk biomass, rich in proteinaceous compounds like sericin and fibroin, offers a high biochemical oxygen demand (BOD) and COD, making it conducive for methane generation (Dey and Thomson 2023). A study by Narendra et al. reported the successful use of silkworm larval litter as a co-substrate with cashew nut fruit waste, yielding about 73% methane due to the nitrogen-rich profile of silk-derived compounds, which supports microbial growth during AD processes (Narendra et al. 2022). Silk biomass, including degumming liquor and residual cocoon matter, can serve as a valuable substrate in anaerobic digestion. The high protein and low lignin content of silk residues make them easier to biodegrade compared to lignocellulosic materials. Priyanga et al. explored the digestion of silk reeling and cocoon waste and found a significant increase in cumulative biogas yield when it was blended with rumen fluid. This co-digestion approach helped balance the carbon-to-nitrogen (C/N) ratio, optimizing microbial digestion pathways (Priyanga et al. 2019). Moreover, some silk waste streams can help maintain a favorable pH in digesters, counteracting the acidification often seen in single-substrate digestion (Azcona et al. 2023). One study assessed the biogas production potential from silkworm waste, finding that silkworm excreta produce 167.32 m^3/Mg TS of methane and 331.97 m^3/Mg TS of biogas, while silkworm breeding waste yields 256.59 m^3/Mg TS of methane and 489.24 m^3/Mg TS of biogas (Łochyńska and Frankowski 2018). It emphasized that breeding waste was characterized by a higher energy potential than silkworm excrement. Another study explored the anaerobic biodegradation of natural and synthetic fabrics, including silk, assessing the mesophilic biochemical methane potential at different total solid concentrations. The highest methane yields were attained for silk fabrics at the lowest TS concentrations, with conformational changes in fibroin (Azcona et al. 2023).

Despite promising research outcomes, large-scale application of silk biomass for biogas production remains limited by supply consistency, transportation logistics, and the need for integrated waste management systems. However, with increasing pressure on the textile sector to transition to sustainable models, silk-producing regions like India, China, and Thailand have strong potential to lead innovations in bioconversion technologies.

19.5 Bioproducts from Silk Biomass

The silk industry generates significant amounts of waste biomass, including sericin, fibroin scraps, and pupal residues. These byproducts, often discarded as waste, have gained increasing attention due to their potential applications in sustainable bioproducts. Repurposing silk biomass can contribute to the development of biodegradable materials, pharmaceutical and cosmetic formulations, and nutrient-rich animal feed, minimizing environmental impact while creating value-added products (Bari et al. 2020). As a renewable biomass resource, silk waste offers sustainable solutions in biotechnology, healthcare, and material science.

19.5.1 Biopolymers and Bioplastics

Silk-derived proteins, primarily fibroin and sericin, have remarkable structural and functional properties that make them suitable for biopolymer production. Silk fibroin, known for its mechanical strength and biocompatibility, has been investigated for applications in biodegradable materials. Research has demonstrated that fibroin can be blended with other biopolymers, such as chitosan and chitin, to enhance flexibility and water resistance, making it a promising alternative to petroleum-based plastics (Kim et al. 2025). Sericin, a hydrophilic silk protein often removed during silk processing, has been explored for its role in bioplastic formulations. Studies indicate that sericin can be crosslinked with natural polymers to improve the mechanical and barrier properties of biofilms, enhancing their biodegradability (Nam et al. 2021). These applications highlight the potential of silk industry waste in reducing plastic pollution while promoting sustainable material development.

19.5.2 Cosmetics and Pharmaceuticals

Silk-derived compounds have found valuable applications in the cosmetic and pharmaceutical industries due to their bioactive properties. Sericin, another silk-derived protein, is produced in the middle silk gland of the silkworm and is known for its hydrophilic nature and high serine content. Its molecular weight and biological activity can vary depending on the extraction method employed. For instance, Dragojlov et al. investigated the extraction and characterization of sericin, emphasizing its application in sustainable and biodegradable cosmetic formulations. The study found that sericin's molecular structure contributes to its effectiveness in skin health applications (Dragojlov et al. 2025). In pharmaceuticals, both fibroin and sericin have been explored for their potential as drug delivery carriers. Silk fibroin, derived from the cocoons of *Bombyx mori* silkworms, is renowned for its superior mechanical characteristics and compatibility with the human body. Its controllable degradation rate and ability to be processed into various formats, such as films, gels, particles, and scaffolds, make it a versatile biomaterial. Recent studies have highlighted sericin's potential in drug delivery, gene delivery, and tissue engineering applications (Wani et al. 2022). Research has explored its applications in various fields, including tissue engineering, due to its beneficial properties (Jo et al. 2020). Advancements in nanotechnology have facilitated the development of silk-based nanoparticles and hydrogels designed for targeted drug delivery. These systems aim to improve therapeutic outcomes by ensuring controlled and sustained release of drugs at specific sites, thereby enhancing efficacy and reducing side effects. The inherent properties of silk proteins, such as their biodegradability and minimal immunogenicity, further support their suitability in these advanced drug delivery applications.

19.5.3 Animal Feed

Silk industry byproducts, notably silkworm pupae and sericin, have garnered attention as potential high-protein animal feed sources. Silkworm pupae, a residual product from silk cocoon processing, are particularly rich in essential amino acids, lipids, and minerals, making them a promising alternative to traditional protein sources in livestock and aquaculture feeds. Research indicates that silkworm pupae meal (SWPM) is a protein-rich feed ingredient of animal origin with high nutritional value. On a dry matter basis, its crude protein content ranges from 50% to more than 80% (defatted meal). The lysine (6–7% of the protein) and methionine (2–3% of the protein) contents are particularly high. Studies have shown that silkworm pupae can partially replace conventional protein sources in animal diets without adverse effects on growth performance (Banday et al. 2018). Additionally, sericin, the glue protein that binds fibroin fibers together in silkworm cocoons, has gained attention for its wide range of properties and possible applications. Recent studies have explored its potential in various fields, including its use in animal feed (Aad et al. 2024). Incorporating these silk industry byproducts into animal feed not only offers a sustainable approach to waste management but also provides a cost-effective and nutritious alternative to traditional feed ingredients.

19.6 Challenges and Opportunities

The valorization of silk biomass presents both significant challenges and promising opportunities, particularly in the context of sustainable material innovation and waste management. One of the primary barriers lies in the complex processes required for the extraction and separation of useful components from silk waste (Hu et al. 2017). Silk biomass, especially from postindustrial or postconsumer waste, is often a heterogeneous mixture comprising fibroin, sericin, and various impurities. Efficiently isolating high-purity fibroin and sericin suitable for biomedical, cosmetic, or textile applications demands advanced chemical or enzymatic processes (Biswal et al. 2022). These methods, while effective, can be resource-intensive and environmentally taxing if not optimized, presenting a major technical hurdle for scalable valorization. Compounding this issue is the limited awareness and slow adoption of new technologies in many silk-producing regions. Despite the growing body of research highlighting the utility of silk-derived materials in high-value applications, ranging from drug delivery systems to biodegradable films, knowledge transfer to industry and local stakeholders remains insufficient (Bari et al. 2020). Many sericulture communities, particularly in developing countries, lack access to technical training and infrastructure that could enable them to process silk waste beyond conventional methods (Jaiswal et al. 2021). This technological gap restricts the integration of innovative, circular economy practices within traditional silk industries.

However, several successful case studies demonstrate the viability and socioeconomic potential of silk biomass valorization. For example, initiatives in India and China have explored the conversion of degummed silk waste into sericin-based hydrogels and fibroin sponges for biomedical use (Babu and Suamte 2024; Zhang et al. 2022; Liu et al. 2022). These projects not only reduced environmental burden but also created new revenue streams for local industries. Another noteworthy implementation involves the use of silk fibroin for developing sustainable packaging materials, which could replace petroleum-based plastics while adding value to what would otherwise be discarded as waste (Low et al. 2022). Such examples underscore the broader economic opportunity of embedding silk waste processing into bioeconomy frameworks. With proper policy support, industry-academic partnerships, and access to sustainable processing technologies, silk biomass could transition from an underutilized byproduct to a cornerstone of value-added manufacturing. Integrating these practices into rural economies could foster innovation, generate employment, and contribute to national goals of sustainable development and waste reduction.

19.7 Conclusion

Waste generated from sericulture, whether during rearing, reeling, or postproduction, encompasses a broad spectrum of biodegradable and industrial byproducts. With strategic intervention, including composting, bioenergy production, material reuse, and eco-friendly treatments, the sericulture sector holds vast potential to transition into a circular economy model. Such integrated waste management not only mitigates environmental impact but also enhances the overall value chain, turning liabilities into assets and promoting sustainability in silk production ecosystems.

References

Aad R, Dragojlov I, Vesentini S (2024) Sericin protein: structure, properties, and applications. J Funct Biomater 15(11):322. https://doi.org/10.3390/JFB15110322

Azcona J, Olguín C, Durán A, Fernández-Rodríguez J (2023) Approach to anaerobic bio-degradation of natural and synthetic fabrics: physico-chemical study of the alteration processes. J Environ Manag 342:118366. https://doi.org/10.1016/J.JENVMAN.2023.118366

Babu PJ, Suamte L (2024) Applications of silk-based biomaterials in biomedicine and biotechnology. Eng Regen 5(1):56–69. https://doi.org/10.1016/J.ENGREG.2023.11.002

Banday IM et al (2018) Division of Livestock production and management, faculty of veterinary sciences utilization of silkworm pupae meal as an alternative source of protein in the diet of livestock and poultry: a review IU Sheikh, MT Banday, IA Baba, S Adil. J Entomol Zool Stud 6(4):1010

Bari E, Perteghella S, Torre ML, Sorlini M, Menato S (2020) From textile to pharmaceutical and cosmetic industry: circular economy applied to silk manufacturing wastes. In: Silk-based drug delivery systems, pp 205–224. https://doi.org/10.1039/9781839162664-00205

Basu A (2015) Advances in the spinning, weaving, and knitting of silk. Adv Silk Sci Technol:41–53. https://doi.org/10.1016/B978-1-78242-311-9.00003-3

Biswal B et al (2022) Extraction of silk fibroin with several sericin removal processes and its importance in tissue engineering: a review. J Polym Environ 30(6):2222–2253. https://doi.org/10.1007/S10924-022-02381-W

Cao TT, Zhang YQ (2016) Processing and characterization of silk sericin from Bombyx mori and its application in biomaterials and biomedicines. *Mater Sci Eng C* 61:940–952. https://doi.org/10.1016/J.MSEC.2015.12.082

Chakraborty JN, Chakraborty JN (2015) Dyeing of silk. In: Fundamentals and practices in colouration of textiles. WPI Publishing, pp 284–292. https://doi.org/10.1201/B18243-25

Dey A, Thomson RC (2023) The biomethane generation potential of wastes and wastewaters from the sericulture, fisheries, and agro-industrial sectors in India. Energy Sustain Dev 75:40–59. https://doi.org/10.1016/J.ESD.2023.05.001

Dragojlov I, Aad R, Ami D, Mangiagalli M, Natalello A, Vesentini S (2025) Silk Sericin-based electrospun nanofibers forming films for cosmetic applications: preparation, characterization, and efficacy evaluation. Molecules 30, no. 3:715. https://doi.org/10.3390/MOLECULES30030715/S1

Halada Nandakrishnan M, Dhanavel N, Mohan CB, Radhakrishna PG, Maharaddi VH, Reddy N (2023) Spent silkworm pupae as a renewable and sustainable source for biodiesel. Biofuels Bioprod Biorefin 17(1):167–177. https://doi.org/10.1002/BBB.2436;CTYPE:STRING:JOURNAL

Hassan R, Acerbi F, Terzi S, Rosa P (2025) Transitioning the silk industry towards circularity: a thematic analysis of sustainable value chain practices. Sustain Prod Consump 56:94–111. https://doi.org/10.1016/J.SPC.2025.03.009

Hu B et al (2017) Microwave-assisted extraction of silkworm pupal oil and evaluation of its fatty acid composition, physicochemical properties and antioxidant activities. Food Chem 231:348–355. https://doi.org/10.1016/J.FOODCHEM.2017.03.152

Jaiswal KK, Banerjee I, Mayookha VP (2021) Recent trends in the development and diversification of sericulture natural products for innovative and sustainable applications. Bioresour Technol Rep 13:100614. https://doi.org/10.1016/J.BITEB.2020.100614

Javali UC, Padaki NV, Das B, Malali KB (2015) Developments in the use of silk by-products and silk waste. Adv in Silk Sci Technol:261–270. https://doi.org/10.1016/B978-1-78242-311-9.00013-6

Jo YY, Kweon H, Oh JH (2020) Sericin for tissue engineering. Appl Sci 10(23):8457. https://doi.org/10.3390/APP10238457

Kalaiyarasan V, Nandhini DU, Udhayakumar K (2015) Seriwaste vermicompost-A trend of new sustainable generation-A review. Agri Rev 36(2):159–163. https://doi.org/10.5958/0976-0741.2015.00019.7

Kamble Z, Behera BK (2021) Upcycling textile wastes: challenges and innovations. Text Prog 53(2):65–122. https://doi.org/10.1080/00405167.2021.1986965;WEBSITE:WEBSITE:TFOPB;REQUESTEDJOURNAL:JOURNAL:TTPR20;PAGEGROUP:STRING:PUBLICATION

Kim JY, Kwon D, Jung S, Tsang YF, Kwon EE (2024) Thermochemical conversion of silkworm by-product into syngas. Int J Biol Macromol 265:130956. https://doi.org/10.1016/J.IJBIOMAC.2024.130956

Kim Y et al (2025) Multifunctional fructose-crosslinked fibroin film with a developed β-sheet structure for advanced food packaging. Int J Biol Macromol 286:138370. https://doi.org/10.1016/J.IJBIOMAC.2024.138370

Kumar KA, Rajesh G, Rani DS (2020) Utilizing seri waste to produce valuable bio-gas and production of bio compost. Gorteria J 33:12

Limami H, Manssouri I, Cherkaoui K, Khaldoun A (2021) Recycled wastewater treatment plant sludge as a construction material additive to ecological lightweight earth bricks. Cleaner Eng Technol 2:100050. https://doi.org/10.1016/J.CLET.2021.100050

Liu J et al (2022) Silk sericin-based materials for biomedical applications. Biomaterials 287:121638. https://doi.org/10.1016/J.BIOMATERIALS.2022.121638

Łochyńska M, Frankowski J (2018) The biogas production potential from silkworm waste. Waste Manag 79:564–570. https://doi.org/10.1016/J.WASMAN.2018.08.019

Low JT, Yusoff NISM, Othman N, Wong TW, Wahit MU (2022) Silk fibroin-based films in food packaging applications: a review. Compr Rev Food Sci Food Saf 21(3):2253–2273. https://doi.org/10.1111/1541-4337.12939

Madappura AP, Madduri S (2023) A comprehensive review of silk-fibroin hydrogels for cell and drug delivery applications in tissue engineering and regenerative medicine. *Comput Struct Biotechnol J* 21:4868–4886. https://doi.org/10.1016/J.CSBJ.2023.10.012

Mia MS, Yao P, Zhu X, Zhang J, Xing T, Chen G (2022) Removal of heavy metals from aqueous solutions by modified waste silk. Text Res J 92(11–12):1952–1965. https://doi.org/10.1177/00405175221076026/ASSET/29820372-2849-4F2A-9DDA-4FA326CE2E5E/ASSETS/IMAGES/LARGE/10.1177_00405175221076026-FIG15.JPG

Nam J, Hyun Y, Oh S, Park J, Jin HJ, Kwak HW (2021) Effect of cross-linkable bacterial cellulose nanocrystals on the physicochemical properties of silk sericin films. Polym Test 97:107161. https://doi.org/10.1016/J.POLYMERTESTING.2021.107161

Narendra N, Krishnamurthy N, Sagar MB, Murali MS, Arunkumar T (2022) Characterisation and feasibility study of potential energy for biogas yield from co-digestion of silkworm larval litter and cashew nut fruit. Int J Ambient Energy 43(1):2759–2766. https://doi.org/10.1080/01430750.2020.1772874;SUBPAGE:STRING:FULL

Ngabala FJ, Emmanuel JK (2024) Potential substrates for biogas production through anaerobic digestion-an alternative energy source. Heliyon 10(23):e40632. https://doi.org/10.1016/J.HELIYON.2024.E40632/ASSET/49562339-79BC-4FF6-BAC4-2FB010A6D413/MAIN.ASSETS/GR1.JPG

Nguyen TP et al (2019) Silk fibroin-based biomaterials for biomedical applications: a review. Polymers 11(12):1933. https://doi.org/10.3390/POLYM11121933

Priyanga S, Ramesh PT, Maheswari M, Karthikeyan S (2019) Biogas production from sericulture wastewater as feed stocks in batch process using rumen fluid. Accessed 04 May 2025. [Online]. Available: www.tjprc.org

Rima SAJ et al (2022) Efficacy of Pseudomonas sp. and Bacillus sp. in textile dye degradation: a combined study on molecular identification, growth optimization, and comparative degradation. J Hazard Mater Lett 3:100068. https://doi.org/10.1016/J.HAZL.2022.100068

Samanta KK, Basak S, Chattopadhyay SK (2015) Recycled fibrous and nonfibrous biomass for value-added textile and nontextile applications. In: Muthu S (ed) Environmental footprints and eco-design of products and processes. Springer, Singapore, pp 167–212. https://doi.org/10.1007/978-981-287-643-0_8

Seo SJ, Das G, Shin HS, Patra JK (2023) Silk Sericin protein materials: characteristics and applications in food-sector industries. Int J Mol Sci 24(5):4951. https://doi.org/10.3390/IJMS24054951

Shi N et al (2015) A novel protection and utilization pattern for urban wetland—as an example of Hemu wetland in Hangzhou, China. J Water Resour Prot 7(4):300–311. https://doi.org/10.4236/JWARP.2015.74024

Silk Market Size, Share, Trends, Report & Forecast (2033) Accessed: May 04, 2025. [Online]. Available: https://www.imarcgroup.com/silk-market

Silk: Processing, Properties and Applications – K. Murugesh Babu – Google Books. Accessed 04 May 2025. [Online]. Available: https://books.google.ca/books?hl=en&lr=&id=oAx-DwAAQBAJ&oi=fnd&pg=PP1&dq=silk+industry+processes&ots=ns94Rrl2nq&sig=W734HstHT31VqgtYxHrP8w4Qe8M&redir_esc=y#v=onepage&q=silk%20industry%20processes&f=false

Statistics | International Sericultural Commission. Accessed 04 May 2025. [Online]. Available: https://inserco.org/en/statistics

Tilahun A, Shifa K, Ibrahim A, Terefe M (2015) Study on silkworm bed cleaning frequency during larval growth period. Ethiopian J Anim Prod 15(1):107–123. Accessed: May 04, 2025. [Online]. Available: https://www.ajol.info/index.php/ejap/article/view/255682

Vepari C, Kaplan DL (2007) Silk as a biomaterial. Prog Polym Sci 32(8–9):991–1007. https://doi.org/10.1016/J.PROGPOLYMSCI.2007.05.013

Viswanath P, Nand K (1994) Anaerobic digestion of silk industry wastes. Bioresour Technol 49(3):273–276. https://doi.org/10.1016/0960-8524(94)90052-3

Wani SUD et al (2022) Silk fibroin as an efficient biomaterial for drug delivery, gene therapy, and wound healing. Int J Mol Sci 23(22):14421. https://doi.org/10.3390/IJMS232214421

Wu Z, Xu W, Wu X, Ding X (2025) Life cycle environmental and economic assessment of raw silk production in China. Sustain Prod Consum 55:11–23. https://doi.org/10.1016/J.SPC.2025.02.002

Yuan Y et al (2024) Sericin coats of silk fibres, a degumming waste or future material? Mater Today Bio 29:101306. https://doi.org/10.1016/J.MTBIO.2024.101306

Zhang H, Xu D, Zhang Y, Li M, Chai R (2022) Silk fibroin hydrogels for biomedical applications. Smart Med 1(1):e20220011. https://doi.org/10.1002/SMMD.20220011

Zheng H, Zuo B (2021) Functional silk fibroin hydrogels: preparation, properties and applications. *J Mater Chem B* 9(5):1238–1258. https://doi.org/10.1039/D0TB02099K

Chapter 20
Utilization of Wood Sawdust and Tree Pruning into Bioproducts

Richa Parashar, Saurabh Jyoti Sarma, Deepa Khare, and Satinder Kaur Brar

Abstract Wood sawdust and tree pruning residues are underutilized bioresources that can be converted into valuable bioproducts, addressing raw material shortage and reducing municipal pruning costs. These residues are an abundant source of lignocellulosic biomass (cellulose, hemicellulose, lignin), making them an ideal source for a range of applications, including composting, fuel production, and animal bedding, offering a versatile and sustainable solution for various industries. Cellulose can be converted into biopolymers, natural fabrics, and biomedical devices, while hemicellulose can be converted into fermentable carbohydrates for biofuels and biochemicals. Lignin has potential applications in energy, chemicals, pharmaceuticals, and environmental pollution, and promotes sustainable development.

The increasing demand for sustainable solutions in waste management and renewable energy has highlighted the potential of lignocellulosic biomass as a valuable resource generated from wood sawdust and tree pruning residues, generally considered as waste, which offer an abundant and renewable feedstock to produce biopolymers, biofuels, nano compounds, aromatic compounds, and other high-value bioproducts. This book chapter explores the potential applications of these residues, their conversion into valuable bioproducts, and their significance in the circular economy. Wood sawdust and tree pruning, rich in lignocellulosic biomass, can be transformed into bioethanol, biodiesel, biogas, biochar, compost, biofilms, hydrogels, and other industrially and agriculturally relevant materials.

R. Parashar · S. J. Sarma (✉) · D. Khare
Department of Biotechnology, School of Engineering and Applied Sciences, Bennett University, Greater Noida, India
e-mail: saurabh.sarma@bennett.edu.in

S. K. Brar
Lassonde School of Engineering, York University, Toronto, ON, Canada

A. L. Woiciechowski et al. (eds.), *Biomass for Bioenergy and Bioproducts*, Biofuel and Biorefinery Technologies 17, https://doi.org/10.1007/978-3-032-19898-3_20

20.1 Introduction

During regular maintenance and trimming of trees that are planted along streets, sidewalks, and other public areas, tree pruning residues are produced, which consist of bark, leaves, and branches. Wood sawdust is a fine wood particle by-product generated during wood cutting and processing, while woodworking processes generate tiny wood shavings and dust-like particles (Guo et al. 2024). It is a waste product commonly produced by sawmills and woodworking industries, at carpentry and woodworking shops, by chainsaws, and power tools. These residues are underutilized bioresources that can help address biomass shortage and reduce municipal pruning costs. Wood sawdust and tree pruning waste are substantial sources of biomass residues that can be converted into valuable bioproducts, but in many regions, sawdust and tree pruning are often disposed of in large heaps and burned, contributing to environmental pollution. These cost-effective raw materials are rich in cellulose, hemicellulose, and lignin, making them ideal for various industrial applications. The chemical composition of wood sawdust contains 40%–50% cellulose (Muley et al. 2016), 25%–35% hemicellulose (Ebringerová et al. 2005), and 20%–30% lignin (Li et al. 2020). Tree pruning consists of bark (45% lignin), leaves (28% lignin), and branches (14% lignin) (Vangeel et al. 2021).

Wood sawdust and tree pruning materials are valuable resources with various applications, including fuel production, composting, absorbent material, bioplastics, and hydrogels. The key components of these materials, cellulose, hemicellulose, and lignin, can be converted into various products. Cellulose can be transformed into natural biopolymers, natural fabrics, food additives, and biomedical devices. Hemicellulose can be converted into fermentable sugars for biofuels, biochemicals, and animal feeds. Lignin has potential applications in energy, chemicals, pharmaceuticals, cosmetics, and environmental sustainability, offering a wide range of possibilities for these underutilized resources (Sun et al. 2021).

The implementation of biomass residues, such as wood sawdust and tree pruning waste, offers numerous benefits, contributing to a more sustainable future. Converting waste into valuable products mitigates environmental pollution, minimizes landfill waste, and reduces greenhouse gas emissions. This approach supports sustainable development, efficient waste management, and circular economy principles. Leveraging biomass residues ensures sustainable utilization of renewable energy sources. Ultimately, this practice fosters sustainable industries, innovations, economic growth, and environmental stewardship, while reducing pollution and supporting sustainable development.

Lignocellulosic biomass is vital in mitigating the use of fossil-based resources and promoting a sustainable industrial framework (Garg et al. 2023). Reusing and repurposing biomass residues reduces waste, carbon footprints, and promotes ecological balance. Valorizing these materials fosters innovations in renewable materials and energy production, creating economic opportunities in the bioeconomy sector and contributing to a more sustainable future (Guo et al. 2024).

20.2 Wood Sawdust Components

20.2.1 Cellulose

Cellulose is one of the most abundant polymers and makes the primary structural component of cell walls of plants, which contributes up to 40% of plant biomass (Fig. 20.1). Cellulose has its widespread use in textile industries to produce rayon, lyocell, and other cellulose-based fibers, in paper production as a primary component of paper, provides strength and durability, in the food industry it is used as food additive which provides texture and structure to various products, and in pharmaceuticals as it is used as an excipient in some pharmaceutical products (Heinze 2016).

The structure of cellulose is a complex carbohydrate that is composed of long chains of glucose molecules linked together by beta-1,4-glycosidic bonds. This creates a long chain-like structure that provides strength, rigidity, and resilience to plant cell walls.

Cellulose plays a pivotal role in various industries like textiles, paper production, and other bio-based products. Extracting cellulose from lignocellulosic biomass, such as wood sawdust and tree pruning, provides a sustainable approach to meet industrial demands while reducing environmental waste. This essay explores the processes involved in cellulose production, technological advancements, derived bioproducts, potential applications, and the significance of sustainable biomass utilization (Marchessault and Sundararajan 1983).

The rising interest in renewable and sustainable materials has highlighted the potential of cellulose as a versatile biopolymer. Derived from plant cell walls,

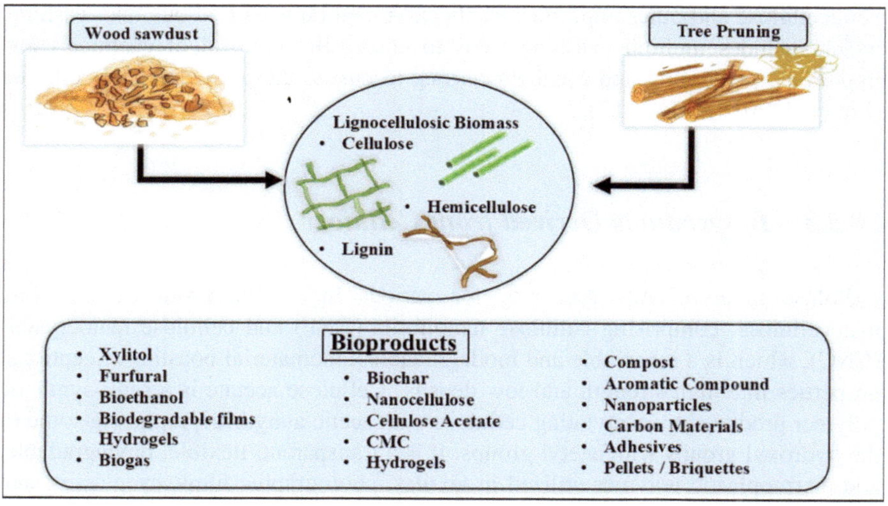

Fig. 20.1 Bioproducts derived from wood sawdust and tree pruning

cellulose is a primary component of lignocellulosic biomass, including wood sawdust and tree pruning.

Wood processing industries generate large amounts of sawdust as a by-product, while tree pruning from urban landscapes is often discarded or burned, contributing to environmental pollution. These lignocellulosic residues are valuable feedstocks for cellulose production, reducing waste and supporting the circular economy.

This essay delves into the processing technologies for cellulose extraction, its derived bioproducts, applications in various industries, and the environmental and economic benefits of utilizing wood sawdust and tree pruning (Béguin and Aubert 1994).

20.2.2 Processing Technologies for Cellulose Production

The process of transforming cellulose from wood sawdust and tree pruning involves a multistep process. (1) The first step, pretreatment, is a crucial step in breaking down the complex structure of lignocellulosic biomass, making cellulose more accessible for subsequent processing. This can be achieved through various methods, including physical techniques such as milling or grinding, chemical methods using acids and alkalis, and biological methods. Following pretreatment, the biomass undergoes delignification to remove lignin (Postek et al. 2013). (2) Delignification removes lignin, a phenolic polymer that encases cellulose fibers, like Alkaline Delignification (Uses sodium hydroxide (NaOH) to solubilize lignin), Organosolv Process (Utilizes organic solvents such as ethanol or acetone), Kraft Process (A widely used industrial method involving sodium sulphide and sodium hydroxide) (Kumar and Murthy 2011). (3) Purification, to remove residual hemicellulose and other impurities purification method is carried out, and washing, which includes multiple washing steps to ensure the removal of chemicals and impurities. (4) Drying and Final Processing to ensure the purification of cellulose (Liu et al. 2019) (Fig. 20.2).

20.2.3 Bioproducts Derived from Cellulose

Cellulose is a versatile precursor for various high-value products, including nanocellulose, comprising cellulose nanofibrils (CNF) and cellulose nanocrystals (CNC), which is a renewable and biodegradable nanomaterial boasting exceptional properties like high strength and low density. Cellulose acetate is a semi-synthetic polymer produced by acetylating cellulose with acetic anhydride, replacing some of the hydroxyl groups with acetyl groups. It is a transparent, flexible, biodegradable, and thermoplastic polymer utilized in textiles, photographic films, eyeglasses, and cigarette filters. Carboxymethyl cellulose (CMC) is a cellulose derivative with a carboxymethyl group attached to its cellulose backbone, and it is a water-soluble

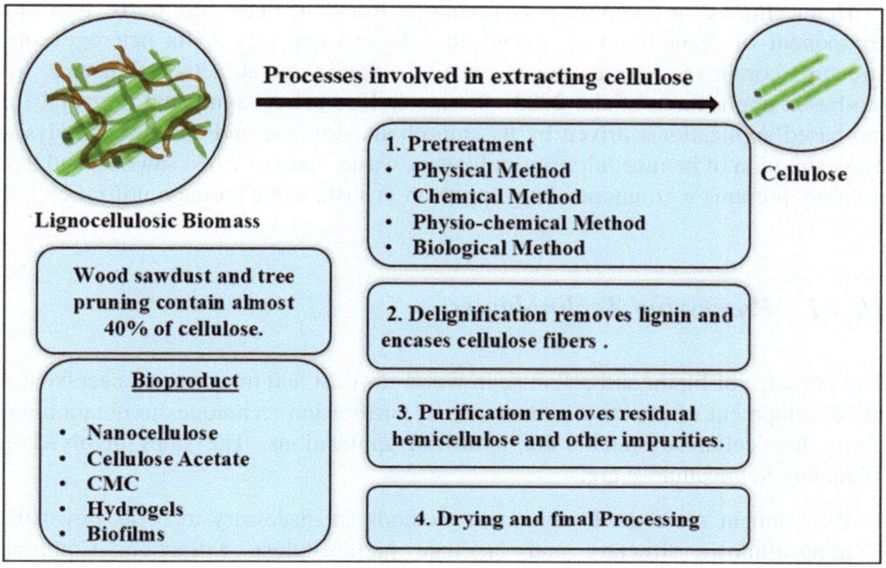

Fig. 20.2 Overview of the process involved in extracting cellulose from lignocellulosic biomass

cellulose polymer with applications in the food industry as a thickener and stabilizer, pharmaceuticals as excipients, and cosmetics (Bindu Sharmila et al. 2024). Hydrogels, three-dimensional networks of hydrophilic polymers, have the ability to absorb and retain large amounts of water, making them suitable for various applications, including contact lenses, wound dressings, and drug delivery systems, particularly when derived from cellulose (Sidek et al. 2022).

The production of cellulose from wood sawdust and tree pruning represents a sustainable approach to biomass utilization. The adoption of cutting-edge processing technologies enables industries to convert waste materials into high-value bioproducts, fostering environmental sustainability and economic prosperity (Wang et al. 2023).

20.3 Hemicellulose

The utilization of renewable biomass resources such as wood sawdust and tree pruning has gained significance in the production of hemicellulose, a valuable polysaccharide with diverse industrial applications. Hemicellulose, primarily composed of pentoses and hexoses, plays a crucial role in bio-based material production, food additives, pharmaceuticals, and biofuels. By optimizing processing methods and improving scalability, hemicellulose extraction from wood waste can significantly contribute to sustainable industrial development (Rao et al. 2023).

Hemicellulose, a complex polysaccharide found in plant cell walls, is a key component of plant biomass, alongside cellulose and lignin. Its heterogeneous structure comprises various sugar units, making it an attractive feedstock for bio-based applications (Saha 2003). Hemicellulose offers significant potential for bio-based applications, driven by its amorphous structure and ease of hydrolysis. The extraction of hemicellulose from forestry waste, such as wood sawdust and tree pruning, presents a groundbreaking approach to sustainable biomass utilization.

20.3.1 Processing Technologies

The presence of lignin and cellulose in wood sawdust and tree pruning necessitates the development of effective extraction and purification techniques to obtain high-purity hemicellulose suitable for industrial applications. The steps involved in extracting hemicellulose are:

1. Pretreatment methods: Pretreatment methods are necessary to break down the lignocellulosic structure and facilitate hemicellulose extraction. Common pretreatment methods include (a) Physical pretreatment, which includes grinding and milling of biomass. (b) Chemical pretreatment, which includes acid and alkaline treatments to dissolve hemicellulose while minimizing cellulose degradation. (c) Biological pretreatment in which fungal or bacterial degradation is used to selectively remove lignin and enhance hemicellulose recovery (Carvalheiro et al. 2008).
2. Hemicellulose extraction techniques: These include (a) alkaline extraction in which sodium hydroxide or potassium hydroxide solutions are commonly used to solubilize hemicellulose while preserving its polymer integrity. (b) Autohydrolysis, in which steam explosion or hot water treatment is used, which releases hemicellulose without the need for harsh chemicals. (c) Enzymatic hydrolysis comprises lignocellulolytic enzymes that selectively hydrolyze hemicellulose into fermentable sugars. (d) Organosolv treatment in which organic solvents are used to dissolve hemicellulose, facilitating its separation from cellulose and lignin (Preece 1944).
3. Purification: After the extraction of hemicellulose, it undergoes purification through ultrafiltration and precipitation (Liao et al. 2024) (Fig. 20.2).

20.3.2 Bioproducts

Hemicellulose serves as a precursor for various bio-based products; these include xylitol, which is a polyol derived from xylose, a pentose sugar found in hemicellulose. Xylitol is used as a low-calorie sweetener in food products, such as sugar-free gum, candy, and baked goods, produced through a multistep process including hydrolysis, purification, and hydrogenation. It has applications in the food industry as a low-calorie sweetener, in pharmaceuticals as an excipient, and in mouthwash and toothpaste (Vallejos and Area 2017). Furfural is an important chemical intermediate derived from hemicellulose. It is used to produce resins, solvents, and other chemicals. Furfural is produced through hydrolysis and dehydration steps. Furfural has diverse industrial applications, including the production of phenol-formaldehyde resins and solvents like furfuryl alcohol, and serves as a crucial intermediate in the manufacture of certain pharmaceuticals (Machado et al. 2016). Bioethanol is a second-generation biofuel produced through the fermentation of hemicellulose-derived sugars. It is produced mainly through hydrolysis and fermentation. Bioethanol has its major applications as a transportation fuel, either as neat fuel or blended with gasoline, and as a solvent and intermediate in various industrial applications (Weusthuis et al. 2011). Hemicellulose-based biodegradable films offer an eco-friendly alternative to synthetic plastics, mainly produced through hydrolysis, fermentation, and extrusion. It is used in agriculture as mulch films and greenhouse films, and used in packaging (Mathura et al. 2024). Hydrogels can also be derived from hemicellulose through a series of processes, including hydrolysis, polymerization, and cross-linking. These have extensive applications in both biomedical fields (Wang et al. 2020) (Fig. 20.3).

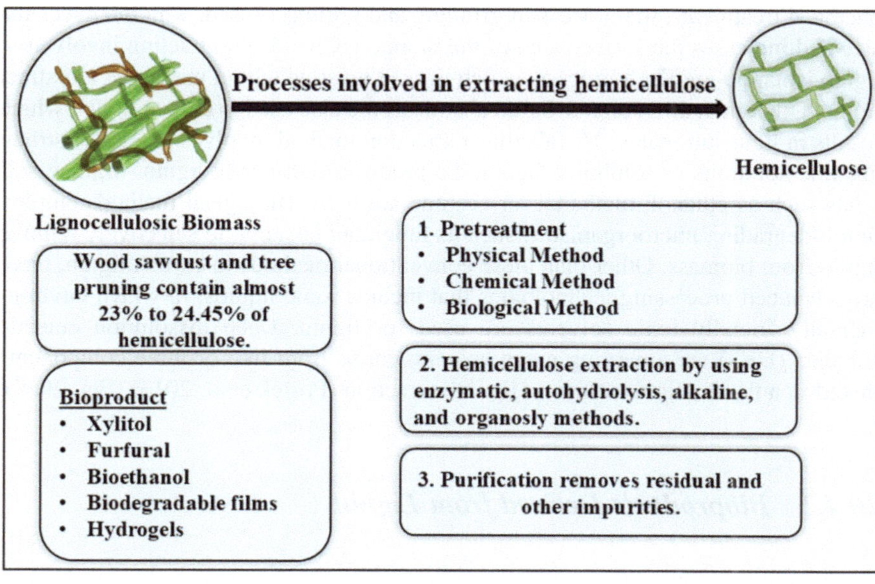

Fig. 20.3 Overview of the process involved in extracting hemicellulose from lignocellulosic biomass

20.4 Lignin

Lignin is a complex, aromatic organic polymer that provides structural support and rigidity to plant cell walls, constituting 15%–30% of lignocellulosic biomass (Boerjan et al. 2003). Unlike cellulose and hemicellulose, which are primarily carbohydrate-based, lignin has a complex phenolic structure. It plays a crucial role in resisting microbial degradation but is often discarded as waste during the production of cellulose-based material. Wood sawdust and tree pruning, generated as by-products in urban landscaping and mainly in cities during the winter, and the management of public gardens, are valuable sources of lignin.

The extraction of lignin from these renewable resources can lead to the development of high-value bioproducts, contributing to a more sustainable bio-based economy.

20.4.1 Processing Technologies for Cellulose Production

The processing technologies involved in breaking down the complex plant structure to separate lignin from lignocellulosic biomass. Methods employed for this purpose are: Pretreatment to make the lignocellulosic biomass more accessible for chemical treatments, the process of grinding and milling is used, which reduces the size and increases the surface area of the biomass. Chemical extraction involved in lignin isolation are Kraft process, which is used most widely in industries to extract lignin. Other than this, sulphur-based chemicals are also used to extract lignin, which results in lignosulfonates. The alkaline extraction method involves the use of strong alkaline solutions to solubilize lignin. To produce high-purity lignin, organic solvents such as ethanol, methanol, or acetone are used. Biological methods employ lignin-degrading microorganisms, such as fungi and bacteria, to selectively remove lignin from biomass. Other than these conventional methods to extract lignin, there are advanced processing technologies that include ionic liquids, in which environmentally friendly ionic solvents are used for lignin. Deep dissolution eutectic solvents (DES) are emerging green solvents made from two or three components mixed in a fixed ratio for efficient lignin extraction (Postek et al. 2013) (Fig. 20.4.).

20.4.2 Bioproducts Derived from Lignin

Contrary to its traditional classification as a waste product, lignin can be transformed into numerous bioproducts of substantial industrial worth. For example, lignin-based adhesives are used in the production of wood composites and construction materials. Compared to traditional petroleum-based adhesives, these adhesives offer various advantages, including lower volatile organic compound (VOC) emissions, enhanced

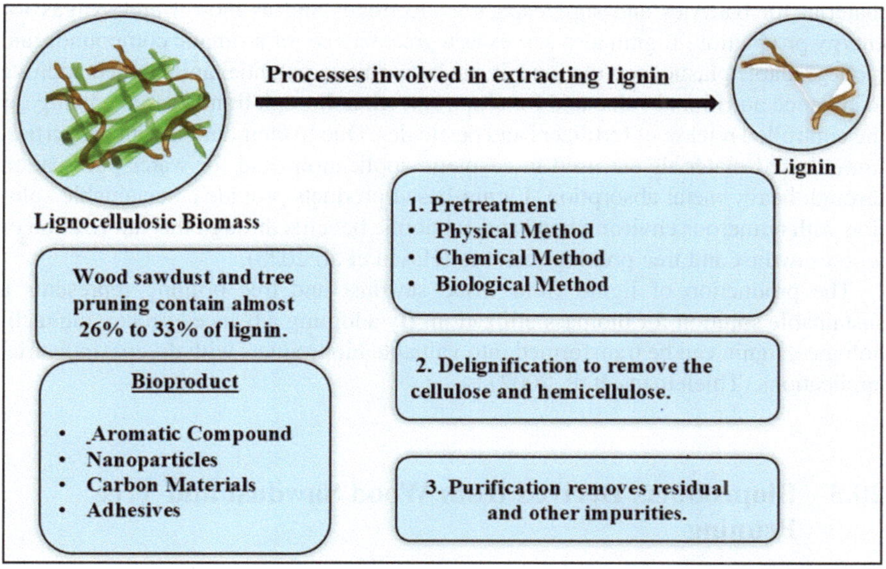

Fig. 20.4. Overview of the process involved in extracting lignin from lignocellulosic biomass

bonding strength, improved thermal stability, and renewable and sustainable sources (Zhou et al. 2022). Through the process of pyrolysis or gasification, lignin-derived carbon materials are produced, and these materials have a high surface area, which offers energy storage. Lignin has high absorption capacity, improved electrochemical performance, and potential applications in air purification and water treatment. Lignin can be used as a replacement for petroleum-based phenol in resin production, which offers improved thermal stability and enhanced bonding strength. Lignin nanoparticles are formulated through mechanical or chemical breakdown of lignin. Lignin has a high surface area, making it ideal for applications in drug delivery and functional coating; these nanoparticles have improved solubility and bioavailability. Lignin can also be converted into aromatic compounds such as vanillin and phenols. These compounds have numerous uses in food and flavoring, pharmaceuticals, cosmetics, plastics and polymers, agrochemicals, and pesticides (Li and Wilkins 2021).

20.4.3 Applications of Lignin

Lignin has a wide range of applications across multiple industries, including construction and building materials, where lignin-based adhesives and binders are used for plywood and particleboard manufacturing, as well as concrete additives to improve workability and strength. In energy and biofuels, lignin is used as a carbon material for batteries and supercapacitor electrodes and as bio-oil and syngas for energy production. Lignin also serves as a great source for aromatic compounds and biodegradable plastic formulations. It can be used as a potential animal feed additive to enhance nutritional value and for crop protection through lignin-based coating for the controlled release of fertilizers and pesticides. Due to their antioxidant properties, lignin-based materials are used in cosmetic applications and for water purification through heavy metal absorption. Lignin-based products provide a sustainable solution with numerous environmental and economic benefits through the valorization of wood sawdust and tree pruning residues (Hasan et al. 2023).

The production of lignin from wood sawdust and tree pruning represents a sustainable solution for biomass utilization. By adopting advanced processing technologies, lignin can be transformed into valuable bioproducts with diverse industrial applications (Thielemans et al. 2002).

20.5 Bioproducts Derived from Wood Sawdust and Tree Pruning

20.5.1 Bioethanol

The increasing global energy demands and the environmental consequences of fossil fuels have intensified the quest for alternative, sustainable sources. Bioethanol, an eco-friendly alternative to gasoline, is primarily produced from carbohydrate-rich biomass, particularly lignocellulosic materials, like wood sawdust and tree pruning, and offers a promising solution for bioethanol production. These abundant, renewable residues can be converted into bioethanol through advanced biotechnological processes, contributing to energy security and environmental sustainability (Wetzel et al. 2006).

The lethal environmental effects of fossil fuels have ignited the need for sustainable and alternative energy sources. Lignocellulosic biomass, specifically wood sawdust and pruning of trees, holds the potential for a sustainable alternative to produce bioethanol. The renewable biomass sources can easily be converted into bioethanol with the assistance of modern biotechnological processes, hence improving energy security and promoting environmental sustainability. Bioethanol is a renewable and eco-friendly substitute for gasoline, primarily derived from biomass with high carbohydrate content. With dwindling fossil fuel resources and growing concern about climate change, bioethanol has emerged as a promising, eco-friendly alternative to decrease the dependence on nonrenewable energy sources (Wetzel et al. 2006).

Wood sawdust and tree pruning generated as waste from wood processing and urban land management are of large potential as feedstocks for bioethanol due to material availability and high carbohydrate content. Use of these wastes not only saves the environment from pollution but also promotes the generation of a circular bioeconomy towards a greener future (Turgut and Algin 2007).

20.5.1.1 Processing Technologies for Bioethanol Production

The conversion of wood sawdust and tree pruning into bioethanol involves multiple stages: pretreatment, hydrolysis, fermentation, and product recovery. Pretreatment is a critical step that breaks down lignocellulose, making cellulose and hemicellulose accessible for enzymatic hydrolysis, and can be achieved through physical, chemical, physicochemical, or biological methods. Hydrolysis then breaks down complex carbohydrates into simple fermentable sugar through enzymatic or acid hydrolysis, with enzymatic hydrolysis being preferred due to its high specificity and lower by-product formation. Following fermentation, where simple sugars are converted into ethanol by microorganisms, product recovery and purification involve distillation and dehydration to achieve fuel-grade ethanol purity (Susmozas et al. 2020).

20.5.1.2 Applications of Bioethanol

Bioethanol, a renewable energy source, offers various benefits and applications. As a transportation fuel, it is blended with gasoline to produce ethanol–gasoline mixtures, reducing greenhouse gas emissions and decreasing reliance on fossil fuels. Industrially, bioethanol serves as a solvent and feedstock for chemical production, pharmaceuticals, and renewable energy generation. Its production from lignocellulosic biomass contributes to waste management, pollution reduction, and carbon sequestration. The environmental benefits of bioethanol production include reduced carbon dioxide emissions, mitigation of air pollution, and contributions to waste management and resource conservation. Economically, it creates new value chains and job opportunities, and diversifies income sources for rural communities.

However, bioethanol production faces technological limitations, including high pretreatment and enzyme production costs, and inefficient fermentation of mixed sugars. Despite these challenges, bioethanol production from wood sawdust and tree pruning has the potential to address energy and environmental issues.

20.5.2 Biodiesel

Biodiesel production from renewable materials has picked up pace as an environmentally friendly alternative to fossil fuels. Lignocellulosic biomass, such as wood sawdust and tree pruning, is a promising avenue for biodiesel production. These widespread and untapped residues can be transformed into high-value biofuels through advanced chemical and biochemical transformations. Biodiesel is a renewable, biodegradable, and clean fuel compared to petroleum diesel. Conventionally made from vegetable oils and animal fat, new developments have investigated lignocellulosic biomass as a possible feedstock since it is inexpensive and readily available. Wood sawdust and tree pruning, usually considered waste products, are produced in large amounts from forestry activities, woodworking mills, and urban landscaping.

20.5.2.1 Processing Technologies for Biodiesel Production

The conversion of wood sawdust and tree pruning into biodiesel involves several steps, including pretreatment of biomass, which is essential for breaking down the lignocellulosic structure and enhancing the accessibility of lignocellulosic components for subsequent processing. Physical methods, such as grinding and milling, increase the surface area of biomass. Chemical methods involve acid or alkaline treatments that remove lignin and release fermentable sugars. Biological methods involve the use of lignin-degrading microorganisms. Efficient pretreatment ensures higher yields of fermentable intermediates for biodiesel production (Bashir et al. 2022). Extraction of lipids and intermediate compounds from wood sawdust and tree pruning contains minimal lipids. However, intermediate compounds, like sugars, can be converted into lipids by oleaginous microorganisms. Microbial lipid production via fungi and yeasts, such as *Cryptococcus curvatus* and *Yarrowia lipolytica*, can accumulate lipids when grown on lignocellulosic hydrolysates. Subsequently, through the fermentation process, the hydrolysates from pretreated biomass are fermented to produce microbial oils (Shahidi and Wanasundara 2002). Then, the process of transesterification converts lipids into biodiesel through a chemical reaction with alcohol in the presence of acidic, alkaline, or enzymatic catalysts. Optimal temperature and reaction time are important for achieving better yield. The resulting biodiesel consists of fatty acid methyl esters (FAMEs) that meet fuel product standards. The final step involves purification, in which biodiesel is separated and purified to remove impurities and by-products, such as glycerol, followed by washing to remove residual catalysts and soaps. Finally, the drying process is carried out to ensure the removal of water content (Otera 1993).

20.5.2.2 Applications of Biodiesel

Biodiesel is a sustainable, eco-friendly alternative fuel that is used as a direct substitute or blended with petroleum to reduce carbon emissions. Its benefits include improved engine lubrication, compatibility with existing diesel engines, and the advantages of being renewable, biodegradable, and nontoxic. In industries, biodiesel serves as a solvent, lubricant, and feedstock for chemical industries for the production of fatty acid derivatives. It is used in machinery to reduce wear and friction. In energy generation, it can be utilized in power plants and generators for electricity production. It is a renewable energy source with lower sulphur emissions compared to fossil fuels. The utilization of wood sawdust and tree pruning for biodiesel production contributes to environmental conservation (Rastegari et al. 2019). The production of biodiesel from wood sawdust and tree pruning presents a sustainable approach to addressing energy and environmental challenges. By transforming these lignocellulosic residues into valuable biofuels, industries can reduce waste, mitigate pollution, and contribute to a greener economy (Xing et al. 2021). The use of these lignocellulosic residues to produce biodiesel not only solves waste management problems but also helps in energy security and environmental sustainability (Ma and Hanna 1999).

20.5.2.3 Biogas

The growing demand for sustainable energy sources has driven research into renewable alternatives such as biogas. Wood sawdust and tree pruning, commonly considered waste materials, offer potential feedstocks for biogas production through anaerobic digestion. This essay explores the feasibility of utilizing these lignocellulosic biomass sources for biogas generation. The book chapter delves into various processing technologies, their efficiency, applications, and environmental benefits. With proper pretreatment, optimized digestion conditions, and technological advancements, wood-based biomass can significantly contribute to sustainable energy production (Scarlat et al. 2018).

Biogas production has emerged as a vital component of the global transition, offering a clean, sustainable, and environmentally friendly alternative energy source. It is primarily composed of methane (CH_4) and carbon dioxide (CO_2), produced via anaerobic digestion of organic matter by microorganisms. Traditional feedstocks include agricultural residues, animal manure, and food waste; however, woody biomass, such as sawdust and tree pruning, presents an underutilized alternative with immense potential.

Wood-based biomass is rich in lignocellulosic compounds, making its digestion challenging. Lignin, cellulose, and hemicellulose form a complex structure that hinders microbial access, necessitating effective pretreatment methods. Despite these challenges, advancements in enzymatic, thermal, and chemical treatments have made anaerobic digestion of such materials increasingly viable (Weiland 2010).

Processing wood sawdust and tree pruning into biogas involves the following techniques: pretreatment of wood sawdust and tree pruning. Since wood-based biomass is recalcitrant to microbial degradation, pretreatment is crucial for enhancing biogas yield. The major pretreatment methods include size reduction (grinding, milling) to increase surface area and improve microbial accessibility. Acid or alkali treatments break down lignin and hemicellulose, making cellulose more available. Steam explosion or hydrothermal treatment helps to disintegrate the lignocellulosic matrix, improving digestibility.

Following pretreatment, anaerobic digestion of sawdust and tree pruning occurs through four key stages. First, hydrolysis, in which complex organic polymers are depolymerized. Second, acidogenesis, in which monomers are transformed into volatile fatty acids (VFAs), hydrogen, and CO_2. Then, acetogenesis, in which VFAs are converted into acetic acid, CO_2, and hydrogen. Last, methanogenesis, in which methanogenic archaea convert acetic acid and hydrogen into methane and CO_2.

A balanced mix of sawdust, pruning, and other co-substrates (e.g., manure) can improve biogas yield. Co-digestion with other biomass sources is beneficial, as wood biomass has a high carbon-to-nitrogen (C/N) ratio, which can hinder microbial activity, and co-digestion with nitrogen-rich materials enhances microbial efficiency and methane yield. Studies indicate that optimal C/N ratios of 20–30:1 provide the best conditions for anaerobic digestion.

Applications of Biogas from Wood-Based Biomass
Biogas provides a sustainable alternative to fossil fuels, and its by-product, digestate, serves as an environmental issue by reducing waste accumulation, landfill emissions, and carbon footprints. It contributes to sustainable forestry management by utilizing tree waste effectively (Vignesh et al. 2024).

20.5.3 Compost

Composting is a centuries-old, sustainable practice that recycles organic materials into a rich soil amendment. This natural process improves soil structure, water retention, and fertility while reducing the need for chemical fertilizers. Wood sawdust and tree pruning, often discarded or incinerated, can be composted to harness their carbon and nutrients. By utilizing these lignocellulosic residues, composting manages organic waste, promotes circular economy principles, and contributes to environmental sustainability, offering a valuable solution for waste reduction and sustainable resource management.

The production of compost from wood sawdust and tree pruning involves several stages. First, pretreatment prepares the raw materials through size reduction, moisture adjustment, and balancing the carbon-to-nitrogen ratio. Next, various composting methods like windrow, aerated static pile, and in-vessel composting. The decomposition process then occurs, involving thermophilic, mesophilic, and curing phases. Monitoring and control of temperature, moisture, aeration, and pH levels are crucial to ensure high-quality compost. In the final stage, the compost is allowed to mature and undergoes screening and quality testing to meet agricultural standards (Clark 2019).

Compost has various applications across different industries. In agriculture and horticulture, compost enhances soil health and productivity by providing essential nutrients, improving soil structure, and controlling erosion. It is also used in landscaping and urban green spaces for mulching and as a soil amendment. Compost is used in industrial and commercial applications, such as in compost-based products like potting mixes and soil conditioners, and in erosion control products like erosion blankets (Shao et al. 2017).

20.5.4 Bioplastics

Bioplastics, derived from renewable biomass, are gaining attention as a sustainable alternative to petroleum-based plastics. Wood sawdust and tree pruning, rich in cellulose, hemicellulose, and lignin, offer a promising raw material for bioplastic synthesis.

The lignocellulosic nature of wood biomass makes its conversion into bioplastics challenging. However, advancements in extraction, chemical modification, and polymer blending techniques have significantly improved the feasibility of producing bioplastics from these materials. This essay examines the production process, different processing technologies, potential applications, and the environmental and economic implications of bioplastics derived from wood-based biomass (Arikan and Ozsoy 2015).

The conversion of wood sawdust and tree pruning into bioplastics involves a multistep process. First, pretreatment of the lignocellulosic biomass is carried out, achieved by mechanical methods. This is followed by the extraction of cellulose and hemicellulose through the breakdown of lignin by chemical or biological methods. Subsequently, the extraction of biopolymers is carried out using alkali and bleaching processes. The extracted biopolymers are then modified into bioplastics through thermoplastic processing.

Bioplastics from wood sawdust offer a sustainable solution to environmental challenges with applications in various industries. In packaging, biodegradable materials replace petroleum-based plastics, reducing waste. The automotive and construction sectors benefit from high-strength, eco-friendly materials. Medical and biodegradable products, such as surgical implants and wound dressings, can be made from cellulose-based bioplastics. Textiles, consumer goods, and household products also utilize bioplastics. Integrating bioplastic production with wood waste management can help create a greener, circular economy, reducing reliance on fossil fuels and plastic pollution (Arikan and Ozsoy 2015).

20.5.5 Biochar

It is a kind of charcoal derived from the thermal decomposition of organic matter in a low-oxygen environment. It offers benefits like improved water retention and enhances soil fertility and carbon content (Weber and Quicker 2018). Biochar produced from lignocellulosic biomass provides multiple applications in energy, agriculture, and provides environmental significance.

The production of biochar involves processes like pyrolysis, in which the thermal decomposition of biomass occurs in an oxygen-limited environment. It involves slow pyrolysis, which operates at lower temperatures ranging from 300 to 600 °C, and it yields high-quality biochar. Fast pyrolysis, which is conducted at high temperatures ranging from 400 to 800 °C, produces biochar along with bio-oil and syngas. Then, torrefaction is a mild form of pyrolysis that enhances biomass energy density at temperatures between 200 and 300 °C. Gasification converts biomass into syngas, leaving behind biochar as a by-product, operating at high temperatures of 700–1000 °C, and is typically used for energy production. Hydrothermal carbonization (HTC) processes biomass in water under high temperatures and pressure, producing hydrochar, a precursor to biochar, suitable for wet biomass and minimizing energy consumption. Additionally, carbonization with chemical activation enhances biochar's surface area and porosity using chemical activators like phosphoric acid or potassium hydroxide, making it suitable for advanced applications such as adsorption and catalysis (Higman 2008).

Biochar has diverse applications across various industries. In agriculture, it enhances soil fertility by increasing water retention and nutrient availability, reducing soil acidity, and promoting microbial activity. It also acts as a slow-release fertilizer when combined with compost. Additionally, biochar plays a crucial role in carbon sequestration and climate mitigation by storing carbon for centuries, reducing CO_2 emissions, and offsetting industrial carbon footprints. It is also used in waste management and pollution control, converting agricultural and forestry waste into valuable resources, and adsorbing heavy metals and contaminants from wastewater and polluted soils. Furthermore, biochar is utilized in energy production as a solid fuel or processed into activated carbon, with syngas produced during pyrolysis serving as a renewable energy source. In livestock and animal husbandry, biochar is used as a feed additive to improve digestion and nutrient absorption, reducing methane emissions from ruminants.

20.5.6 Briquettes

Briquettes offer an eco-friendly, sustainable, and efficient energy solution, and support waste management. They not only provide an eco-friendly substitute for conventional fuels like firewood and coal but also minimize the waste and pollution by repurposing wood sawdust. The potential sources to produce briquettes are wood wastes, including sawdust, tree pruning residues, wood chips, agricultural residues, industrial and municipal waste, and plant-based materials like coffee grounds, bamboo waste, and coconut husks (Dinesha et al. 2019).

The process of manufacturing briquettes involves the transformation of raw biomass into compact, efficient fuel sources by following several steps, including the collection of raw materials, which is followed by drying to reduce the moisture content below 12% using solar or mechanical dryers. If required, grinding and shredding of biomass into finer particles are then performed, and the briquettes are formed by compressing the material under high pressure using briquetting machines and binding it together with agents like starch and clay. This technique offers a game-changing solution for converting biomass waste into a valuable energy resource, promoting energy independence, sustainable development, and environmental stewardship (Richards 1990).

20.5.7 Biofilms

Biofilms, derived from cellulose and hemicellulose, offer a promising sustainable solution for various industries, which leverage the multidisciplinary and sustainable approach. The sources for cellulose and hemicellulose are mainly from agricultural waste, forestry residues like sawdust, wood chips, industrial by-products such as fruit peels, industry waste, and processing residues, and bacterial cellulose produced through fermentation by microorganisms (Sidek et al. 2022).

Production of biofilms from cellulose and hemicellulose involves a multistep process and modification techniques, including extraction and purification to extract the cellulose and hemicellulose by breaking the lignocellulosic biomass, using grinding, milling, and sieving. To remove lignin and other impurities, chemical treatment or the enzymatic method is used. To enhance the properties of biofilm, techniques such as surface functionalization, polymer blending, and nanocellulose incorporation are used. This provides enhanced barrier properties, improved mechanical properties, and increased structural integrity (Kumar and Murthy 2011).

Biofilms derived from cellulose and hemicellulose have comprehensive uses across multiple industries, offering sustainable and innovative solutions. Biofilms have diverse applications in biomedical fields, including tissue engineering scaffolds, and in promoting wound healing and tissue repair. Biofilms are used as antimicrobial coatings, as they reduce spoilage and contamination, and as biodegradable coatings. Biofilms are also used as water filtration membranes, mulching films, and conductive materials (Marchessault and Sundararajan 1983).

20.5.8 Xylitol

To provide a sustainable, environmentally friendly, and economically viable health-conscious alternative to conventional sugars, xylitol produced from wood sawdust and tree pruning has driven interest. Wood sawdust and tree pruning contain substantial amounts of hemicellulose that can be hydrolyzed to yield xylose—the precursor for xylitol production (Vallejos and Area 2017).

Production of xylitol from wood sawdust and tree pruning involves several steps, as mentioned below:

(1) Biomass pretreatment, in which chemical, mechanical, or steam methods are used to break down the lignocellulosic biomass. (2) Hydrolysis involves the breakdown of hemicellulose into its constituent sugars, primarily xylose. Acid Hydrolysis: Sulfuric acid is commonly used to hydrolyze hemicellulose. Enzymatic Hydrolysis: Xylanase enzymes selectively degrade hemicellulose into xylose with fewer by-products. (3) Xylose Purification is done from the hydrolysate is critical to achieving high xylitol yields. Filtration: Removal of insoluble residues. Ion Exchange Chromatography: Separation of sugars from impurities. Activated Carbon Treatment: Adsorption of colored and inhibitory compounds. (4) Production of xylitol: There are two primary methods for converting xylose into xylitol. Chemical Reduction: Catalytic hydrogenation of xylose using nickel or ruthenium catalysts under high pressure. Microbial fermentation: Use of yeast strains such as *Candida tropicalis* and *Pichia stipitis* for bioconversion of xylose to xylitol under controlled conditions. (5) The final step involves the separation and purification of xylitol from the reaction mixture. Crystallization: Xylitol is crystallized from the solution. Drying and Packaging: The purified xylitol is dried and packaged for commercial use.

20.5.8.1 Applications of Xylitol

Xylitol has diverse applications across multiple industries due to its unique properties. In the food and beverage industry, it is used to manufacture sugar-free gums, candies, and baked goods. It provides a low-calorie alternative to conventional sugars and maintains sweetness even at high temperatures. In oral healthcare, it is used in toothpaste and mouthwash to inhibit the growth of cavity-causing bacteria such as *Streptococcus mutans*, and in chewing gums, as it promotes remineralization of tooth enamel. In the pharmaceutical industry, it is used as a sweetener in syrups and lozenges to produce sugar-free medications. In nutritional supplements, it is incorporated into diabetic-friendly health products. In the cosmetics industry, it is used to produce moisturizers and skincare products, as xylitol exhibits antimicrobial and hydrating benefits. Other than this, it is used in industries for polymer production.

20.5.9 Hydrogels

Due to their exceptional water absorption properties, hydrogels have gained significant attention. To address the growing demand for biodegradable and sustainable materials, the utilization of natural waste resources, such as wood sawdust and tree pruning, is being addressed. As the conventional production of hydrogels involves the use of nonrenewable, nonbiodegradable petroleum-based polymers, it contributes to environmental pollution.

By harnessing the potential of wood sawdust and tree pruning, we can promote the development of sustainable, cost-effective, and biodegradable hydrogels. The production stages of hydrogels include biomass pretreatment, which breaks down the lignocellulosic structures to separate the polysaccharides. After that, the extraction of cellulose and hemicellulose is carried out for hydrogel formation. Modification of the extracted polysaccharides is done using techniques such as grafting polymerization, cross-linking, and freeze–thaw cycles.

Hydrogels derived from wood sawdust and tree pruning have a wide range of applications in agriculture, biomedical devices, including drug delivery systems, wound dressings, and tissue engineering; as well as oil spill cleaning, biodegradable packaging, smart fabrics for moisture regulation, and cosmetic products, such as moisturizing creams and face masks.

20.6 Conclusions

The utilization of wood sawdust and tree pruning residue represents a transformative approach to sustainable biomass management. These feedstocks are rich in cellulose, hemicellulose, and lignin, and provide a diverse range of industrial applications, from biofuels and biopolymers to composting and soil amendments. Advancements in biomass conversion technologies offer practical solutions for repurposing these residues into high-value products, instead of their underutilization and often improper disposal.

Beyond energy applications, the transformation of lignocellulosic biomass into bioproducts such as xylitol, hydrogels, and pharmaceuticals enhances industrial sustainability. The extraction of cellulose and hemicellulose from wood feedstock facilitates the development of bio-based materials, providing eco-friendly alternatives to synthetic polymers, and lignin valorization presents opportunities for biochemical and bio-composites, expanding the scope of sustainable manufacturing.

The adoption of biomass valorization strategies aligns with circular economy principles, reducing waste accumulation while fostering economic and environmental benefits. By promoting efficient waste management and resource recovery, industries can contribute to a low-carbon economy and mitigate the ecological impacts. The implementation of innovative biomass processing technologies, coupled with policy support and industry collaboration, is essential for scaling up sustainable bioeconomy solutions.

The conversion of wood sawdust and tree pruning residues into biofuels and bioproducts is a crucial step towards a greener future. As research and technology continue to evolve, biomass resources are poised to become a crucial component in the fight against climate change, driving sustainable industrial growth and preserving the environment for future generations. Investing in biomass utilization is a sole responsibility toward ecological stewardship and a sustainable global economy.

References

Arikan EB, Ozsoy HD (2015) A review: investigation of bioplastics. J Civ Eng Arch 9(1):188–192

Bashir MA, Wu S, Zhu J, Krosuri A, Khan MU, Aka RJN (2022) Recent development of advanced processing technologies for biodiesel production: a critical review. Fuel Process Technol 227: 107120

Béguin P, Aubert JP (1994) The biological degradation of cellulose. FEMS Microbiol Rev 13(1): 25–58

Bindu Sharmila TK, Julie Chandra CS, Sasi S, Arundhathi CK (2024) Modification of cellulose. In: Handbook of biomass. Singapore, Springer Nature Singapore, pp 535–571

Boerjan W, Ralph J, Baucher M (2003) Lignin biosynthesis. Annu Rev Plant Biol 54(1):519–546

Carvalheiro F, Duarte LC, Gírio F (2008) Hemicellulose biorefineries: a review on biomass pretreatments. J Sci Ind Res 67:849–864

Clark G (2019) Technologies for compost production from plant byproducts. In: Byproducts from agriculture and fisheries: adding value for food, feed, pharma, and fuels. Wiley, pp 545–562

Dinesha P, Kumar S, Rosen MA (2019) Biomass briquettes as an alternative fuel: a comprehensive review. Energ Technol 7(5):1801011

Ebringerová A, Hromádková Z, Heinze T (2005) Hemicellulose. In: Polysaccharides I: structure, characterization and use. Springer, Berlin, pp 1–67

Garg R, Sabouni R, Ahmadipour M (2023) From waste to fuel: challenging aspects in sustainable biodiesel production from lignocellulosic biomass feedstocks and role of metal organic framework as innovative heterogeneous catalysts. Ind Crop Prod 206:117554

Guo J, Zhang Y, Fang J, Ma Z, Li C, Yan M, Qiao N, Liu Y, Bian M (2024) Reduction and reuse of forestry and agricultural bio-waste through innovative green utilization approaches: a review. Forests 15(8):1372

Hasan MS, Sardar MR, Shafin AA, Rahman MS, Mahmud M, Hossen MM (2023) A brief review on applications of lignin. J Chem Rev 5(1):56–82

Heinze T (2016) Cellulose: structure and properties. In: Cellulose chemistry and properties: fibers, nanocelluloses and advanced materials. Springer, pp 1–52

Higman C (2008) Gasification. In: Combustion engineering issues for solid fuel systems. Academic Press, pp 423–468

Kumar D, Murthy GS (2011) Impact of pretreatment and downstream processing technologies on economics and energy in cellulosic ethanol production. Biotechnol Biofuels 4:1–19

Li M, Wilkins M (2021) Lignin bioconversion into valuable products: fractionation, depolymerization, aromatic compound conversion, and bioproduct formation. Syst Microbiol Biomanuf 1: 166–185

Li W, Jiang Y, Yin X (2020) Characterization of hydrolysis lignin bonding properties during the pelletization of eucalyptus sawdust. Waste Biomass Valorization 11:995–1003

Liao G, Sun E, Kana EG, Huang H, Sanusi IA, Qu P, Jin H, Liu J, Shuai L (2024) Renewable hemicellulose-based materials for value-added applications. Carbohydr Polym 341:122351

Liu CG, Xiao Y, Xia XX, Zhao XQ, Peng L, Srinophakun P, Bai FW (2019) Cellulosic ethanol production: progress, challenges and strategies for solutions. Biotechnol Adv 37(3):491–504

Ma F, Hanna MA (1999) Biodiesel production: a review. Bioresour Technol 70(1):1–15

Machado G, Leon S, Santos F, Lourega R, Dullius J, Mollmann ME, Eichler P (2016) Literature review on furfural production from lignocellulosic biomass. Nat Resour 7(3):115–129

Marchessault RH, Sundararajan PR (1983) Cellulose. In: The polysaccharides. Academic Press, pp 11–95

Mathura SR, Landázuri AC, Mathura F, Sosa AGA, Orejuela-Escobar LM (2024) Hemicelluloses from bioresidues and their applications in the food industry–towards an advanced bioeconomy and a sustainable global value chain of chemicals and materials. Sustain Food Technol 2(5): 1183–1205

Muley PD, Henkel C, Abdollahi KK, Marculescu C, Boldor D (2016) A critical comparison of pyrolysis of cellulose, lignin, and pine sawdust using an induction heating reactor. Energy Convers Manag 117:273–280

Otera J (1993) Transesterification. Chem Rev 93(4):1449–1470

Postek MT, Moon RJ, Rudie AW, Bilodeau MA (2013) Production and applications of cellulose. Tappi Press, Peachtree Corners

Preece IA (1944) Some hemicellulose extraction methods. Biochem J 38(5):402

Rao J, Lv Z, Chen G, Peng F (2023) Hemicellulose: structure, chemical modification, and application. Prog Polym Sci 140:101675

Rastegari H, Jazini H, Ghaziaskar HS, Yalpani M (2019) Applications of biodiesel by-products. In: Biodiesel: from production to combustion. Springer, pp 101–125

Richards SR (1990) Physical testing of fuel briquettes. Fuel Process Technol 25(2):89–100

Saha BC (2003) Hemicellulose bioconversion. J Ind Microbiol Biotechnol 30(5):279–291

Scarlat N, Dallemand JF, Fahl F (2018) Biogas: developments and perspectives in Europe. Renew Energy 129:457–472

Shahidi F, Wanasundara PKJPD (2002) Extraction and analysis of lipids. In: Food lipids. CRC Press, pp 152–187

Shao B, Liu Z, Zhong H, Zeng G, Liu G, Yu M, Liu Y, Yang X, Li Z, Fang Z, Zhang J (2017) Effects of rhamnolipids on microorganism characteristics and applications in composting: a review. Microbiol Res 200:33–44

Sidek IS, Draman SFS, Rozaimah S, Abdullah S, Anuar N (2022) Chemical composition and thermal analysis of cellulose extracted from senduduk (*Melastoma malabathricum* L.). J Sustain Sci Manag 17(4):141–150

Sun SF, Yang J, Wang DW, Yang HY, Sun SN, Shi ZJ (2021) Enzymatic response of ryegrass cellulose and hemicellulose valorization introduced by sequential alkaline extractions. Biotechnol Biofuels 14:1–12

Susmozas A, Martín-Sampedro R, Ibarra D, Eugenio ME, Iglesias R, Manzanares P, Moreno AD (2020) Process strategies for the transition of 1G to advanced bioethanol production. Processes 8(10):1310

Thielemans A, Can E, Morye SS, Wool RP (2002) Novel applications of lignin in composite materials. J Appl Polym Sci 83(2):323–331

Turgut P, Algin HM (2007) Limestone dust and wood sawdust as brick material. Build Environ 42(9):3399–3403

Vallejos ME, Area MC (2017) Xylitol as bioproduct from the agro and forest biorefinery. In: Food bioconversion. Academic Press, pp 411–432

Vangeel T, Neiva DM, Quilho T, Costa RA, Sousa V, Sels BF, Pereira H (2021) Tree bark characterization envisioning an integrated use in a biorefinery. Biomass Convers Biorefinery 13:1–15

Vignesh NS, Vimali E, Rajkeerthana S, Kaleeshwari R, Kiruthika M, Ashokkumar B, Varalakshmi P, Yesodharan V, Vigneswaran VS, Rajendran K (2024) Utilization of woody biomass for biogas production. In: Sustainable biorefining of woody biomass to biofuels and biochemicals. Woodhead Publishing, pp 103–123

Wang Y, Zhao M, Liu T (2020) Journal of bioresources and bioproducts. J Bioresour Bioprod 5: 286–293

Wang J, Zhao H, Larter SR, Kibria MG, Hu J (2023) One-pot sequential cascade reaction for selective gluconic acid production from cellulose photobiorefining. Chem Commun 59(23): 3451–3454

Weber K, Quicker P (2018) Properties of biochar. Fuel 217:240–261

Weiland P (2010) Biogas production: current state and perspectives. Appl Microbiol Biotechnol 85: 849–860

Wetzel S, Duchesne LC, Laporte MF (2006) Forests as a source of bioproducts. In: Bioproducts from Canada's forests: new partnerships in the bioeconomy. Springer, pp 1–33

Weusthuis RA, Aarts JM, Sanders JP (2011) From biofuel to bioproduct: is bioethanol a suitable fermentation feedstock for synthesis of bulk chemicals? Biofuels Bioprod Biorefin 5(5): 486–494

Xing Y, Zheng Z, Sun Y, Agha Alikhani M (2021) A review on machine learning application in biodiesel production studies. Int J Chem Eng 2021(1):2154258

Zhou N, Thilakarathna WW, He QS, Rupasinghe HV (2022) A review: depolymerization of lignin to generate high-value bio-products: opportunities, challenges, and prospects. Front Energy Res 9:758744

Chapter 21
A Strategic Overview of Conversion of Biomass into Bioproducts

Lachi Wankhede, Khyati Joshi, Gaurav Bhardwaj, and Satinder Kaur Brar

Abstract In recent years, the pressing global challenges of environmental sustainability, resource scarcity, and waste management have led to increased interest in converting waste biomass into high-value products aligned with the principles of a circular bioeconomy. Such an approach utilizes renewable resources, reduces dependence on fossil fuels, and mitigates greenhouse gas emissions, thereby offering viable pathways to address environmental concerns and focusing on economic development. This chapter explains the possibilities, challenges, and real-world considerations involved in transforming waste biomass into valuable bioproducts. This chapter first examines the abundance and diversity of waste biomass sources, highlighting their global distribution and potential for valorization. It then explores how chemistry, biology, engineering, and data analytics converge to develop both conventional and emerging conversion technologies. Further, the focus is on ensuring product quality standards and keeping pace with changing market needs, highlighting economic feasibility, scale-up hurdles, and environmental performance measures. Case studies illustrate real-world applications and insights, while discussions on policy frameworks and prospects offer strategic guidance for stakeholders. By providing an overview of these interconnected elements, the chapter seeks to empower researchers, industry professionals, and policymakers to make decisions that advance waste biomass conversion toward large-scale, sustainable implementation.

21.1 Introduction

The global paradigm shift toward sustainable development and environmental awareness has catalyzed significant interest in converting waste biomass into valuable bioproducts. This transformation represents not merely a technological

L. Wankhede · K. Joshi · G. Bhardwaj · S. K. Brar (✉)
Department of Civil Engineering, Lassonde School of Engineering, York University, North York, Toronto, ON, Canada
e-mail: satinder.brar@lassonde.yorku.ca

A. L. Woiciechowski et al. (eds.), *Biomass for Bioenergy and Bioproducts*, Biofuel and Biorefinery Technologies 17, https://doi.org/10.1007/978-3-032-19898-3_21

advancement but a fundamental reimagining of renewable resource utilization, positioning biomass conversion as a cornerstone of the circular bioeconomy. The urgency of transitioning from fossil-based resources to renewable alternatives has become increasingly apparent, with recent global environmental assessments emphasizing the critical need to reduce fossil fuel dependence (Pintér et al. 2022; Sertolli et al. 2022). According to the International Energy Agency, sustainable biomass could potentially contribute 100–300 EJ annually by 2050, representing a substantial portion of global energy and material needs (Popp et al. 2013). The complexity of biomass conversion is reflected in the diverse range of feedstock sources, from agricultural residues and forest waste to industrial by-products and municipal solid waste. Current statistics indicate that agricultural residues alone generate approximately 140 billion metric tons annually, while industrial processes contribute over 50 million tons of potential feedstock (Correa and Kruse 2018; Abbie 2024). Municipal solid waste, containing 40%–50% organic fraction, further enhances this resource pool (Turner et al. 2022). This abundance of waste biomass presents significant opportunities for valorization through various conversion pathways.

These statistics show not only the magnitude of the available resources but also the immense potential for their valorization. The complexity of biomass conversion is highlighted by the process of transforming waste into valuable bioproducts, which requires a coordinated integration of diverse scientific fields and technological methods. Traditional conversion methods have been complemented by emerging technologies, including microwave-assisted conversion, nanocatalysis, and plasma technology (Kataya et al. 2023; Lanjekar et al. 2024). This technological evolution has expanded the spectrum of products, ranging from biofuels and biochemicals to high-value materials and specialty products. However, the translation of these technologies from the laboratory scale to industrial implementation remains challenging, with only select strategies achieving commercial success (Makepa et al. 2023; Wankhede et al. 2024, 2025). This implementation challenge has prompted a deeper examination of the factors influencing success or failure in biomass conversion strategies. Technical challenges, such as process efficiency and product quality consistency, intertwine with economic considerations like capital investment requirements and market competitiveness. Environmental sustainability, which offers a primary driver for biomass conversion, must be balanced against economic viability and social acceptance. This complex web of factors necessitates a thorough understanding of the various conversion pathways and their implications (Yamakawa et al. 2018; Andiappan et al. 2021). Recent technological advancements, particularly in process engineering, catalysis, and biotechnology, have begun addressing these challenges. The incorporation of advanced digital tools, such as artificial intelligence and machine learning, has enhanced process optimization and control (Ambaye et al. 2021; Makepa et al. 2023; Kasoju et al. 2023). However, regional variations add complexity to implementation strategies. Developed nations typically focus on high-efficiency processes and premium products, while developing countries prioritize cost-effective solutions addressing local waste management challenges (Brodny and Tutak 2020; Szarka et al. 2021; Sertolli et al. 2022; Ranjbari et al. 2022).

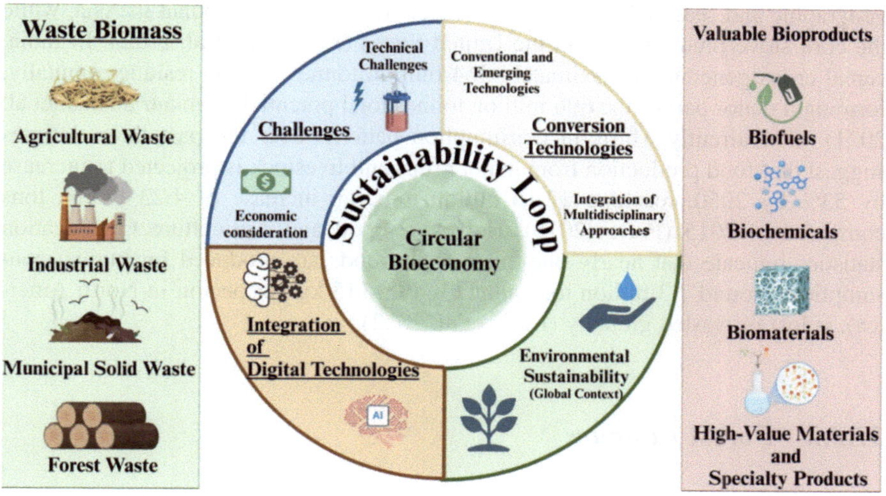

Fig. 21.1 Circular bioeconomy framework: waste biomass to valuable bioproducts

This economic landscape reflects these regional disparities, with market dynamics shaped by fluctuating oil prices, environmental regulations, and changing consumer preferences. The global bio-based products market is projected to reach USD 396.01 billion by 2027, growing at a CAGR (Compound Annual Growth Rate) of 25.40% (LLP 2024). Showing increasing market acceptance rate of sustainable products across industries. This chapter navigates through this complex landscape, analyzing various conversion strategies and their practical implications. By examining process efficiency, usability, sustainability, scalability, and cost-time effectiveness, it provides insights into factors determining the successful implementation of different conversion approaches as shown in Fig. 21.1.

21.2 Sources of Waste Biomass

The diversity and abundance of waste biomass sources present opportunities and challenges for converting them into valuable bioproducts. Understanding these sources, their characteristics, and their distribution patterns is crucial for developing effective conversion strategies and ensuring sustainable resource utilization.

21.2.1 Agricultural Residues

Agricultural residues represent one of the most abundant and widely distributed sources of waste biomass globally, including wheat and rice straw, corn stover, sugarcane bagasse, and various husks, with their availability varying according to geographic and seasonal factors. For instance, rice straw is abundant in Asia, while the corn stove predominates in the United States (Kaniapan et al. 2022). In India, cereal crops generate approximately 364 million tonnes of crop residues annually, forming a major part of the 696 million tonnes total potential (Venkatramanan et al. 2021). Concurrently, shifts in consumption patterns over the past three decades suggest that food production from major crops and livestock is projected to increase by 53% by 2050, reaching 12.16 billion tons, an increase of 4.23 billion tons compared to 2015 (Porter 2016). However, Food and Agriculture Organization statistics indicate that nearly one-third of the foods are produced for human consumption, around 1.3 billion tons annually (95–115 kg per person in North America), is lost or wasted globally (Karić et al. 2022).

21.2.2 Forest Residues

Forest residues constitute a significant biomass resource, primarily generated from logging, wood processing industries, and forest management activities. These materials consist of bark, branches, leaves, sawdust, and wood chips. The sustainable extraction of forest residues requires careful consideration of ecological impacts, including soil fertility, biodiversity, and carbon sequestration (Pergola et al. 2022). Recent studies have shown that proper forest residue management can enhance forest health while providing valuable biomass resources. For instance, in North America, sustainable forest residue collection could yield a considerable amount of biomass annually without compromising forest ecosystem services (Kukrety et al. 2015).

21.2.3 Industrial Waste Biomass

Industrial waste biomass emerges from various manufacturing and processing industries, including paper and pulp mills, food processing industries, textile industries, and agricultural processing units. This category of biomass is particularly attractive due to its concentrated availability and relatively consistent composition. For example, the paper industry alone generates over 100–500 million tons of waste annually worldwide, including black liquor, sludge, and fiber rejects (Haile et al. 2021). Similarly, food processing industries contribute significant quantities of organic waste, estimated at 1.4 billion tons annually (Secretariat 2000; Batool et al. 2024). These industrial residues often have well-defined compositions and are already collected at centralized locations, making them ideal candidates for bioconversion processes.

21.2.4 Municipal Solid Waste

Municipal solid waste (MSW) represents a growing source of biomass, particularly in urban areas. The organic fraction of MSW typically constitutes 40% of the total waste volume in developed countries and up to 90% in developing nations (Yadav and Samadder 2018). Global MSW generation is expected to reach 3.4 billion tons by 2050, presenting both a challenge and an opportunity for biomass conversion (Sharma et al. 2021). The heterogeneous nature of MSW poses unique challenges in collection, sorting, and processing. However, advanced sorting technologies and integrated waste management systems have improved the feasibility of utilizing MSW for bioproduct generation. Recent innovations in mechanical-biological treatment have enhanced the recovery of organic fractions from MSW, making it a more viable feedstock for bioconversion (Karthikeyan et al. 2024).

21.2.5 Aquatic Biomass

Aquatic biomass, including macro and microalgae, represents an emerging and promising biomass source. Marine and freshwater systems can produce significant quantities of biomass without competing for agricultural land or freshwater resources (Namasivayam et al. 2023). Microalgae, in particular, show remarkable potential with biomass yield higher than traditional energy crops (Bora et al. 2024). Recent technological advances in cultivation systems, including photobioreactors and open pond systems, have improved the feasibility of large-scale aquatic biomass production. Marine macroalgae (seaweed) cultivation has also gained attention, with global production valued at USD 5.9 billion and an expected compound annual growth rate of 9.1% until 2027, and showing potential for significant expansion (Khan et al. 2024).

21.3 Interdisciplinary Strategies in Biomass Conversion

The conversion of waste biomass into high-value products demands an integrated approach across multiple scientific disciplines, particularly when considering industrial-scale implementation challenges. While numerous laboratory-scale technologies show promise, their successful translation into commercial applications remains limited, highlighting the complexity of scale-up processes. The interdisciplinary integration of biology, chemistry, engineering, and information technology has become crucial in addressing key industrial challenges, including feedstock variability, process stability, and economic viability (Sulis et al. 2025). Despite their simplicity, the physical treatment method faces significant scaling challenges due to high energy requirements and limited effectiveness in lignin removal (Millati et al. 2020). Chemical pre-treatment approaches, though effective at the laboratory scale, encounter hurdles in industrial implementation due to equipment corrosion, waste handling, and environmental concerns (Galbe and Wallberg 2019). Biological conversion pathways, while offering environmental advantages, struggle with long processing times and sensitivity to feedstock variations in large-scale operations (Millati et al. 2020). Modern digital technologies and process integration strategies have emerged as critical tools in bridging the gap between laboratory success and industrial implementation, enabling real-time optimization and control of complex biorefinery operations (Mohammad et al. 2024; Makepa and Chihobo 2024). Recent advances in genetic engineering and computational biology have improved process robustness and efficiency, though many promising strains still face challenges in industrial environments (Abbate et al. 2023; Wankhede et al. 2024). The implementation of digital twins and advanced control systems, while showing potential in improving operational efficiency, requires significant investment and technical expertise, often lacking in traditional biomass processing industries (Sheik et al. 2024). Despite these challenges, the successful integration of multiple disciplines has led to some notable commercial achievements, particularly in lignocellulosic bioethanol production, demonstrating the potential for overcoming scale-up barriers through coordinated multidisciplinary approaches.

21.4 Conventional Conversion Technologies

Conventional conversion technologies represent established pathways for transforming waste biomass into valuable products, though their industrial implementation success varies significantly. Among these, thermochemical conversion methods have shown varying degrees of commercial viability, each presenting unique challenges in large-scale operations (Zhang and Zhang 2019). Pyrolysis technology, operating under oxygen-free conditions at 300–600 °C, offers flexibility in processing diverse waste streams but faces significant hurdles in product quality consistency and upgrading requirements, limiting its widespread industrial adoption (Saravanan et al. 2021). While gasification shows promise for producing versatile syngas products at 700–1200 °C, its industrial implementation remains constrained by substantial capital investments and complex gas purification systems, factors critical for commercial success (AlNouss et al. 2019). Traditional combustion, despite being the most widely implemented technology, continues to struggle with efficiency optimization and emission control, particularly when processing heterogeneous waste biomass feedstocks (Yang et al. 2021). Hydrothermal liquefaction, though capable of producing higher-quality bio-crude oil, faces considerable challenges in scaling up due to severe operating conditions and associated equipment costs, limiting its commercial applications (Zhang et al. 2023). While technologies offer more environmentally sustainable approaches in the biochemical conversion domain, they encounter different implementation barriers. Fermentation processes, successfully commercialized for bioethanol production, require substantial optimization for emerging high-value products to achieve economic feasibility (Broda et al. 2022). Similarly, despite its proven waste management benefits, anaerobic digestion faces significant challenges in maintaining process stability and product quality at industrial scales (Kataya et al. 2023). Enzymatic processes, though crucial for efficient biomass breakdown, continue to face economic barriers in large-scale applications, primarily due to high enzyme costs and efficiency limitations (Jain and Kumar 2024). The selection and implementation of these technologies at an industrial scale requires careful consideration of technical feasibility, economic viability, and environmental impact, factors that have historically limited the commercial success of many promising laboratory-scale developments.

21.5 Advanced Technologies and Product Diversification in Biomass Conversion

The landscape of waste biomass conversion has been revolutionized by emerging technologies that enhance process efficiency and expand the spectrum of valuable products. While conventional conversion methods have established a foundation for biomass processing, the increasing demand for process efficiency and product diversity has driven the development of advanced technologies. These innovative approaches address many limitations of traditional methods while opening new pathways for product development and market opportunities. For example, microwave-assisted conversion technologies have shown significant advantages in biomass processing, reducing reaction times by up to 70% while improving product selectivity through precise energy transfer mechanisms (Lozano Pérez et al. 2024; Ke et al. 2024). Ultrasound-assisted processes have shown remarkable success in biomass pre-treatment and extraction, with case studies reporting enhanced sugar yields compared to conventional methods (Sindhu et al. 2013; Nakashima et al. 2016; Perrone et al. 2016). Similarly, supercritical fluid technology, particularly using CO_2 and water, offers environmentally benign extraction and conversion pathways, achieving higher selectivity in bioactive compound isolation and biomass fractionation (Essien et al. 2020). On the other hand, plasma technology has emerged as a promising tool for surface modification and biomass activation, improving downstream conversion efficiency, though scaling challenges persist (Wright et al. 2020; Jain and Kumar 2024). Nanocatalysis has transformed reaction specificity and efficiency, with engineered nano catalysts showing up to 50% improvement in conversion rates, although cost and stability remain critical concerns (Pandit et al. 2023). These advanced technologies have enabled the production of a diverse range of high-value products. In the biofuels sector, improved processes have led to enhanced bioethanol yields exceeding 90% of the theoretical maximum, while biodiesel production has achieved better quality control through optimized catalytic systems (Jain and Kumar 2024). Other biochemicals, such as lactic acid and succinic acid, have seen increased production efficiency through hybrid technologies combining multiple conversion methods, with market demand growing at 19.6% annually (Kumar et al. 2024; Market Research Future 2026). These high-value products, such as specialty chemicals and functional materials, represent a growing market segment with premium pricing potential, though production scale optimization remains crucial for commercial viability (Ambaye et al. 2021). The synergistic integration of these advanced technologies has improved process efficiencies and expanded the range of economically viable products from waste biomass. However, challenges in scaling and cost-effectiveness continue to influence industrial implementation decisions.

21.6 Process Efficiency Analysis

The successful industrial implementation of waste biomass conversion technologies heavily depends on comprehensive process efficiency analysis, which encompasses multiple interconnected parameters crucial for commercial viability. Process efficiency evaluation has become increasingly critical as industries strive to optimize resource utilization while minimizing environmental impact and operational costs.

21.6.1 Energy Balance

Energy balance analysis reveals that most current biomass conversion processes require significant energy inputs, with typical energy efficiency ratios ranging from 45% to 75%, depending on the technology employed (Huang and Zhang 2011). Advanced heat integration systems and improved process designs have shown potential energy savings in industrial-scale operations, though implementation challenges persist due to high capital investment requirements. For example, waste heat boilers, featuring parallel water tubes aligned with the exhaust flow, capture medium- to high-temperature heat to produce steam for power generation or energy recovery (Jouhara et al. 2018). Therefore, optimizing energy inputs versus outputs remains crucial for economic viability, with successful industrial implementations and enhanced energy efficiency measures.

21.6.2 Mass Balance

Mass balance assessments indicate that material flows in conversion processes often face efficiency losses during various transformation stages, highlighting the need for improved process control and optimization strategies (Rissman et al. 2020). Recent industrial implementations have achieved yield improvements of up to 20% through better material flow management and advanced monitoring systems (Rissman et al.

2020). The analysis of conversion yields across different process stages has become essential for identifying bottlenecks and optimization opportunities in commercial operations.

21.6.3 Carbon Footprint

Carbon footprint analysis has become increasingly crucial. Recent studies have shown that optimized biomass conversion processes can achieve net carbon reductions of 60%–80% compared to conventional fossil-based processes, though these benefits vary significantly with feedstock type and processing methods (Huovila et al. 2022). Implementing carbon capture and storage technologies in industry has shown the potential for further reducing greenhouse gas emissions by 15%–25%, though economic feasibility remains a key consideration (Hanson et al. 2025).

21.6.4 Resource Utilization

Resource utilization efficiency remains a significant challenge, as current industrial processes do not fully approach their theoretical yield potential and concurrently generate various by-products that necessitate additional treatment or disposal. Advanced separation and purification technologies have enabled better resource recovery and utilization of input biomass through integrated biorefinery approaches. For instance, A recent syngas fermentation study showed that an integrated downstream process using vacuum distillation and advanced separation techniques can recover 99.5% of ethanol at 99.8 wt% purity from a 6 wt% feed stream. This approach reduced annual costs by 54.2%, energy consumption by 66.1%, and CO_2 emissions by 82.6% (Janković et al. 2023). These findings underscore the potential of integrated biorefinery approaches for enhanced resource recovery and biomass utilization. Minimizing waste streams and optimizing by-product utilization is, therefore, crucial for enhancing overall process economics.

21.6.5 Process Optimization

Process optimization efforts have focused on developing integrated approaches that combine multiple efficiency enhancement strategies, including advanced catalysis, improved reactor designs, and control systems, resulting in overall efficiency improvements at the pilot scale. The implementation of real-time monitoring systems and artificial intelligence-based control strategies has shown promise in addressing scale-up challenges, though economic feasibility often determines the extent of their industrial adoption. Recent process advances have reduced equipment footprints and enhanced operational flexibility, essential factors for commercial-scale implementation (Ramírez-Márquez et al. 2023).

21.7 Usability and Product Quality

The commercial success of waste biomass-derived products heavily depends on their ability to meet stringent quality standards and functional requirements while gaining market acceptance. This critical aspect of biomass conversion has become increasingly important as bioproducts compete with well-established conventional products in various market segments.

21.7.1 Product Standards and Specifications

The implementation of rigorous quality control measures in industrial-scale biomass conversion processes has become essential for ensuring consistent product quality. Some industries maintain strict product quality parameters through comprehensive monitoring and control systems, meeting or exceeding established industry standards. Quality benchmarks for bioproducts have evolved significantly, with international standards organizations establishing specific criteria for different product categories. For instance, the biodiesel production industry must now achieve high purity levels with strict limits on contaminants to comply with American Society for Testing and Material (ASTM) and European Standards (EN) standards, requiring quality management systems and advanced analytical capabilities (Wan Osman et al. 2024).

21.7.2 Market Acceptance and Demand

Consumer preferences for bio-based products have grown consistently, driven by increasing environmental awareness and sustainability concerns (Fletcher et al. 2024). However, market uptake varies considerably across different product categories. While bio-based packaging materials have achieved a significant market presence in some regions, specialized biochemicals face more challenges due to supply chains and technical specifications (Reichert et al. 2020). Price considerations remain a critical factor, with market success often depending on competitive pricing while maintaining product quality standards.

21.7.3 Functional Properties of Bioproducts

The performance characteristics of biomass-derived products have improved substantially through technological advancements in conversion processes. These improvements are particularly evident in bio-based polymers, where recent developments have successfully bridged the performance gap with petroleum-based counterparts while offering additional advantages in biodegradability and environmental performance (Atiwesh et al. 2021). Similarly, in the biofuel sector, extensive research and process optimization have resulted in products showing competitive energy densities and performance characteristics in standard engine tests (Ambaye et al. 2021). Success in achieving these enhanced functional properties relies on a comprehensive approach to product development. This approach carefully balances parameters such as stability, durability, and application-specific requirements while also ensuring cost-effective production processes.

21.8 Sustainability Assessment

The comprehensive assessment of sustainability in waste biomass conversion processes encompasses environmental, economic, and social dimensions, forming a critical framework for evaluating the long-term viability of industrial implementations.

21.8.1 Environmental Impacts

Life cycle assessment (LCA) studies of waste biomass conversion show variations in environmental impacts across different technologies and scales of operation. Recent analyses indicate that optimized conversion processes can reduce greenhouse gas emissions by 60%–80% compared to conventional fossil-based alternatives, though these benefits heavily depend on feedstock sourcing and processing efficiency (Jeswani et al. 2020). The ecological footprint of the industrial scale extends beyond direct emissions, encompassing water consumption, land use changes, and biodiversity impacts. Advanced manufacturing processing implementing closed-loop systems has shown potential reductions in water consumption and waste generation compared to traditional processing methods (Karkou et al. 2024).

21.8.2 Economic Viability

The economic sustainability of biomass conversion processes is a critical factor for successful industrial implementation. Techno-economic analyses (TEA) reveal that integrated biorefineries often require higher initial capital investments compared to conventional biorefineries, while achieving optimal process efficiency can significantly lower operational costs. Market competitiveness varies across product categories, with high-value biochemicals typically enjoying robust profit margins and bulk biofuels operating on thinner margins. For instance, TEA studies have reported competitive Minimum selling price for biofuels such as bioethanol (US$0.5–1.8/L) and biobutanol (US$0.5–2.2/kg), whereas biohydrogen remains comparatively expensive (US$9–33/kg); similarly, high-value chemicals like xylitol (US $1.5–3.1/kg), succinic acid (US$1.5–6.9/kg), 5-HMF (US$1–2/kg), and lactic acid (US$0.5–1.9/kg) are priced competitively (Patel et al. 2025). Moreover, the synergistic integration of multiple product streams and the valorization of by-products have emerged as crucial strategies for enhancing overall economic viability.

21.8.3 Social Implications

The social aspect of sustainability has become increasingly important in biomass conversion projects. Large-scale facilities create direct jobs based on their processing capacity and additional indirect jobs in feedstock collection and logistics. Energy is essential for social and economic growth (Demırbas 2017), and converting biomass to bioenergy adds value by using existing resources. Biofuel production also strengthens energy security and supports rural and regional development. For example, it is reported that over 2 million biofuel-related jobs were created worldwide in 2018 (Ambaye et al. 2021). Community support improves with early stakeholder involvement and clear communication. In contrast, local benefits such as higher tax revenues and the growth of supporting industries further highlight the socio-economic advantages of these projects.

21.8.4 Resource Conservation

Sustainable sourcing of biomass feedstock represents a fundamental aspect of resource conservation in conversion processes. Successful processes have implemented comprehensive feedstock management systems that ensure a reliable supply while maintaining ecological balance. The adoption of circular economy principles has led to the development of integrated biorefinery concepts, where waste streams from one process become feedstock for another (Leong et al. 2021). Advanced tracking systems and certification schemes have enhanced transparency in biomass sourcing, with some industries achieving full traceability of their feedstock supply chains. (Arias et al. 2023).

21.9 Scalability and Industrial Implementation

The transition from laboratory-scale developments to successful industrial implementation represents a critical challenge in waste biomass conversion technologies. This complex journey encompasses various interconnected factors that determine the commercial viability of these processes.

21.9.1 Scale-Up Challenges

The scale-up of biomass conversion processes presents numerous technical hurdles that often become apparent only during implementation. Common challenges include maintaining process stability with heterogeneous feedstocks, managing heat and mass transfer limitations in larger reactors, and ensuring consistent product quality at increased production volumes. For example, scaling up the production of single-cell oils for biodiesel using oleaginous yeasts remains particularly challenging despite significant technological advances. High capital costs are driven by relatively low lipid yields, with feedstock expenses accounting for 50%–80% of overall costs and downstream processing comprising about 13%. Although waste-based feedstocks like volatile fatty acids could reduce costs, their toxicity at high concentrations poses a barrier to industrial implementation (Robles-Iglesias et al. 2023). These insights show the importance of pilot-scale validation and careful process optimization to mitigate the risks of premature scaling.

21.9.2 Technical Barriers

Technical barriers in industrial implementation often emerge at various process stages, requiring innovative solutions. Critical bottlenecks include feedstock handling and preprocessing limitations, reactor design constraints, and separation efficiency challenges. Recent innovations addressing these challenges include advanced pre-treatment technologies, modular reactor designs, and improved separation systems (Okoro et al. 2022). The development of robust control systems and real-time monitoring capabilities has proven essential in overcoming operational challenges. (Sharma et al. 2023).

21.9.3 Economic Barriers

Economic feasibility remains a fundamental concern in scaling up biomass conversion processes. Investment challenges include high capital requirements, extended payback periods, and market uncertainties. Some cases have employed various strategies to enhance economic viability, including strategic partnerships, government incentives, and innovative financing models (Wan Osman et al. 2024).

21.9.4 Policy and Regulatory Frameworks

Government policies and regulatory support play vital roles in facilitating industrial implementation. Effective policy frameworks include renewable energy mandates, carbon pricing mechanisms, and investment incentives. Regulatory support through streamlined permitting processes and clear environmental guidelines has significantly impacted project success rates. The alignment of regional policies with national and international sustainability goals has created favorable conditions for industrial-scale implementations.

21.10 Cost-Effectiveness

The journey from concept to commercial reality in waste biomass conversion technologies begins with a critical examination of cost-effectiveness, where success depends on carefully balancing multiple economic factors. At the foundation lies the challenge of capital investment, where project developers must navigate varying requirements across different conversion technologies. While thermochemical processes often demand substantial initial investments in robust processing equipment, biochemical routes might require specialized fermentation and purification systems (Manikandan et al. 2023). These initial decisions influence the subsequent operational landscape, where daily running costs become the next crucial consideration. Different findings show that strategic approaches to managing operating costs, including innovative process integration and smart automation systems, can substantially improve overall economic viability. As these move into full operation, the focus shifts to monitoring and optimizing return on investment, with experience showing that carefully planned implementations can achieve financial sustainability through balanced product portfolios and efficient operations. The path to profitability, however, is intimately linked with time-to-market considerations, where shorter development cycles can provide competitive advantages but must be balanced against technical readiness and market dynamics. Throughout this journey, risk assessment and management play important roles, requiring project developers to anticipate and address potential challenges ranging from feedstock supply uncertainties to market fluctuations. This interconnected web of economic factors ultimately shapes the commercial viability of waste biomass conversion projects, where success depends not just on individual cost elements but on their harmonious integration within a broader strategic framework.

21.11 Recent Progress and Advances

Recent advancements in waste biomass conversion technologies have shown breakthroughs through various successful industrial implementations. These developments showcase the integration of innovative technologies with practical applications, which is particularly evident in several notable case studies from different global contexts.

For example, Novozyme's adoption of enzymatic transesterification for biodiesel production has transformed conventional methods, reducing greenhouse gas emissions by 40% and energy consumption by 30% compared to traditional techniques. Moreover, the ability to reuse enzymatic catalysts up to 50 times with minimal loss of activity marks a significant step forward in process sustainability (Novozymes 2025). Similarly, Algenol's commercial-scale microalgae-based biodiesel production highlights feedstock innovations, achieving yields of 10,000 gallons per hectare annually while capturing 2.5 tons of CO_2 per ton of biodiesel produced (Algenol 2025). Additionally, integrated biorefinery approaches have advanced processing techniques, as seen in Neste's large-scale operations, where advanced hydrotreatment processes enable a production capacity of 2.7 million tons of renewable diesel per year while cutting greenhouse gas emissions by up to 90% compared to fossil diesel (Neste 2025). Their capability to process diverse feedstocks, including vegetable oils, animal fats, and waste materials, further demonstrates significant improvements in processing flexibility (Wan Osman et al. 2024). Similarly, Eni's conversion of traditional oil refineries into biorefineries showcases innovative hydroprocessing technology, achieving an 80% reduction in greenhouse gas emissions while maintaining an annual production capacity of 360,000 tons of biofuel (Eni 2019). Moreover, modern biorefineries have embraced smart manufacturing and data-driven optimization. ADM's large-scale biodiesel production, for instance, successfully integrates continuous processing with advanced control systems to achieve a 20% reduction in energy usage and a 95% recovery of glycerol through sophisticated monitoring and control (Max Manturov 2025). These digital innovations extend to supply chain management and process optimization, enabling real-time adjustments and improved resource utilization across the production cycle.

21.12 Future Directions and Conclusion

The advancement of waste biomass conversion technologies requires addressing multiple interconnected challenges while using emerging opportunities across research, industry, and policy domains. Further research must focus on process optimization and scalability, with particular emphasis on catalyst efficiency improvements and heterogeneous feedstock processing capabilities. The development of novel catalytic systems for enhanced selectivity and reduced energy requirements represents a critical research direction, alongside innovations in process integration and control systems. Industrial implementations by companies like Novozymes and Neste have shown commercial viability through the alignment of technological innovation, market demand, and supportive policies, providing valuable insights for future developments. Further, the establishment of comprehensive policy frameworks, including standardized sustainability criteria, carbon pricing mechanisms, and streamlined regulatory pathways, also plays a crucial role in facilitating industrial scale. These frameworks must address both environmental performance metrics and economic viability considerations to enhance industry growth. Emerging market opportunities in high-value biochemicals and specialized materials present significant potential for bio-based alternatives, though market penetration requires strategic product positioning and showing performance advantages. The development of bio-based platform chemicals, advanced materials, and specialty products offers promising routes for market expansion and value creation. The research must prioritize developing robust technologies for feedstock handling variability and efficient separation processes, while industrial stakeholders should implement phased approaches and integrated biorefinery concepts. This includes advancing separation technologies, improving process monitoring systems, and developing flexible manufacturing platforms capable of handling diverse feedstock sources. Policymakers have a fundamental role in establishing consistent regulatory frameworks and supporting demonstration projects that bridge laboratory developments to commercial applications. These frameworks should include incentive structures for sustainable practices, standards for product quality, and mechanisms for technology validation and scale-up support. The integration of technological innovation, practical implementation strategies, clear policy frameworks, and market development initiatives establishes a clear pathway for waste biomass conversion technologies. Advanced process control systems, artificial intelligence applications, and digital twin technologies offer additional tools for optimizing operations and reducing implementation risks. This comprehensive approach, encompassing process efficiency improvements, cost optimization, and environmental performance, supported by-product quality, charts the course toward a sustainable bioeconomy based on waste biomass conversion. The future of this field lies in the synergistic development of advanced technologies, supportive policies, and market acceptance, driving the transition toward sustainable bio-based production systems.

Acknowledgements This research was funded by the Natural Sciences and Engineering Research Council of Canada (NSERC) CREATE Program, Training in Advanced Biotechnology for Environmental Sustainability (TABES) Grant no: 554777-2021, and the James and Joanne Love Chair in Environmental Engineering at York University for the support provided.

Declaration of Competing Interests The authors have no relevant financial or non-financial interests to disclose.

References

Abbate E, Andrion J, Apel A et al (2023) Optimizing the strain engineering process for industrial-scale production of bio-based molecules. J Ind Microbiol Biotechnol 50:kuad025. https://doi.org/10.1093/jimb/kuad025

Abbie (2024) From farm to fuel: the potential of agricultural waste for bioenergy|waste managed. Waste Manag Serv Recycl WasteManaged. https://www.wastemanaged.co.uk/our-news/environment/from-farm-to-fuel-the-potential-of-agricultural-waste-for-bioenergy/. Accessed 4 Feb 2025

Algenol (2025) Making algal biofuel production more efficient, less expensive. Energy.gov. https://www.energy.gov/eere/articles/making-algal-biofuel-production-more-efficient-less-expensive. Accessed 25 July 2025

AlNouss A, McKay G, Al-Ansari T (2019) Superstructure optimization for the production of fuels, fertilizers and power using biomass gasification. In: Computer aided chemical engineering. Elsevier, pp 301–306

Ambaye TG, Vaccari M, Bonilla-Petriciolet A et al (2021) Emerging technologies for biofuel production: a critical review on recent progress, challenges and perspectives. J Environ Manag 290:112627. https://doi.org/10.1016/j.jenvman.2021.112627

Andiappan V, How BS, Ngan SL (2021) A perspective on post-pandemic biomass supply chains: opportunities and challenges for the new norm. Process Integr Optim Sustain 5:1003–1010. https://doi.org/10.1007/s41660-021-00176-5

Arias A, Feijoo G, Moreira MT (2023) Biorefineries as a driver for sustainability: key aspects, actual development and future prospects. J Clean Prod 418:137925. https://doi.org/10.1016/j.jclepro.2023.137925

Atiwesh G, Mikhael A, Parrish CC et al (2021) Environmental impact of bioplastic use: a review. Heliyon 7:e07918. https://doi.org/10.1016/j.heliyon.2021.e07918

Batool F, Kurniawan TA, Mohyuddin A et al (2024) Environmental impacts of food waste management technologies: a critical review of life cycle assessment (LCA) studies. Trends Food Sci Technol 143:104287. https://doi.org/10.1016/j.tifs.2023.104287

Bora A, Thondi Rajan AS, Ponnuchamy K et al (2024) Microalgae to bioenergy production: recent advances, influencing parameters, utilization of wastewater—a critical review. Sci Total Environ 946:174230. https://doi.org/10.1016/j.scitotenv.2024.174230

Broda M, Yelle DJ, Serwańska K (2022) Bioethanol production from lignocellulosic biomass—challenges and solutions. Molecules 27:8717

Brodny J, Tutak M (2020) Analyzing similarities between the European Union countries in terms of the structure and volume of energy production from renewable energy sources. Energies 13:913. https://doi.org/10.3390/en13040913

Correa CR, Kruse A (2018) Biobased functional carbon materials: production, characterization, and applications—a review. Materials 11:1568. https://doi.org/10.3390/ma11091568

Demırbas A (2017) The social, economic, and environmental importance of biofuels in the future. Energy Sources Part B Econ Plan Policy 12:47–55. https://doi.org/10.1080/15567249.2014.966926

Eni (2019) Eni opens second biorefinery in Italy, including pilot waste-to-fuel plant. Biofuels International Magazine. https://biofuels-news.com/news/eni-opens-second-biorefinery-in-italy-including-pilot-waste-to-fuel-plant/. Accessed 25 July 2025

Essien SO, Young B, Baroutian S (2020) Recent advances in subcritical water and supercritical carbon dioxide extraction of bioactive compounds from plant materials. Trends Food Sci Technol 97:156–169. https://doi.org/10.1016/j.tifs.2020.01.014

Fletcher CA, Aureli S, Foschi E et al (2024) Implications of consumer orientation towards environmental sustainability on the uptake of bio-based and biodegradable plastics. Curr Res Environ Sustain 7:100246. https://doi.org/10.1016/j.crsust.2024.100246

Max Manturov (2025) Investment analysis of Archer-Daniels-Midland Co Investment analysis of Archer-Daniels-Midland Co. https://freedom24.com/ideas/details/13691. Accessed 25 July 2025

Galbe M, Wallberg O (2019) Pretreatment for biorefineries: a review of common methods for efficient utilisation of lignocellulosic materials. Biotechnol Biofuels 12:294. https://doi.org/10.1186/s13068-019-1634-1

Haile A, Gelebo GG, Tesfaye T et al (2021) Pulp and paper mill wastes: utilizations and prospects for high value-added biomaterials. Bioresour Bioprocess 8:35. https://doi.org/10.1186/s40643-021-00385-3

Hanson E, Nwakile C, Hammed VO (2025) Carbon capture, utilization, and storage (CCUS) technologies: evaluating the effectiveness of advanced CCUS solutions for reducing CO2 emissions. Results Surf Interfaces 18:100381. https://doi.org/10.1016/j.rsurfi.2024.100381

Huang W-D, Zhang Y-HP (2011) Energy efficiency analysis: biomass-to-wheel efficiency related with biofuels production, fuel distribution, and powertrain systems. PLoS One 6:e22113. https://doi.org/10.1371/journal.pone.0022113

Huovila A, Siikavirta H, Antuña Rozado C et al (2022) Carbon-neutral cities: critical review of theory and practice. J Clean Prod 341:130912. https://doi.org/10.1016/j.jclepro.2022.130912

Jain S, Kumar S (2024) Advances and challenges in pretreatment technologies for bioethanol production: a comprehensive review. Sustain Chem Clim Action 5:100053. https://doi.org/10.1016/j.scca.2024.100053

Janković T, Straathof AJJ, Kiss AA (2023) Advanced downstream processing of bioethanol from syngas fermentation. Sep Purif Technol 322:124320. https://doi.org/10.1016/j.seppur.2023.124320

Jeswani HK, Chilvers A, Azapagic A (2020) Environmental sustainability of biofuels: a review. Proc Math Phys Eng Sci 476:20200351. https://doi.org/10.1098/rspa.2020.0351

Jouhara H, Khordehgah N, Almahmoud S et al (2018) Waste heat recovery technologies and applications. Therm Sci Eng Prog 6:268–289. https://doi.org/10.1016/j.tsep.2018.04.017

Kaniapan S, Pasupuleti J, Patma Nesan K et al (2022) A review of the sustainable utilization of rice residues for bioenergy conversion using different valorization techniques, their challenges, and techno-economic assessment. Int J Environ Res Public Health 19. https://doi.org/10.3390/ijerph19063427

Karić N, Maia AS, Teodorović A et al (2022) Bio-waste valorisation: agricultural wastes as biosorbents for removal of (in)organic pollutants in wastewater treatment. Chem Eng J Adv 9:100239. https://doi.org/10.1016/j.ceja.2021.100239

Karkou E, Angelis-Dimakis A, Parlapiano M et al (2024) Process innovations and circular strategies for closing the water loop in a process industry. J Environ Manag 370:122748. https://doi.org/10.1016/j.jenvman.2024.122748

Karthikeyan PK, Bandulasena HCH, Radu T (2024) A comparative analysis of pre-treatment technologies for enhanced biogas production from anaerobic digestion of lignocellulosic waste. Ind Crop Prod 215:118591. https://doi.org/10.1016/j.indcrop.2024.118591

Kasoju N, Remya NS, Sasi R et al (2023) Digital health: trends, opportunities and challenges in medical devices, pharma and bio-technology. CSI Trans ICT 11:11–30. https://doi.org/10.1007/s40012-023-00380-3

Kataya G, Cornu D, Bechelany M et al (2023) Biomass waste conversion technologies and its application for sustainable environmental development—a review. Agronomy 13:2833. https://doi.org/10.3390/agronomy13112833

Ke L, Zhou N, Wu Q et al (2024) Microwave catalytic pyrolysis of biomass: a review focusing on absorbents and catalysts. Npj Mater Sustain 2:1–20. https://doi.org/10.1038/s44296-024-00027-7

Khan N, Sudhakar K, Mamat R (2024) Macroalgae farming for sustainable future: navigating opportunities and driving innovation. Heliyon 10:e28208. https://doi.org/10.1016/j.heliyon.2024.e28208

Kukrety S, Wilson DC, D'Amato AW, Becker DR (2015) Assessing sustainable forest biomass potential and bioenergy implications for the northern Lake States region, USA. Biomass Bioenergy 81:167–176. https://doi.org/10.1016/j.biombioe.2015.06.026

Kumar V, Kumar P, Maity SK et al (2024) Recent advances in bio-based production of top platform chemical, succinic acid: an alternative to conventional chemistry. Biotechnol Biofuels Bioprod 17:72. https://doi.org/10.1186/s13068-024-02508-2

Lanjekar PR, Panwar NL, Agrawal C et al (2024) A review on environmental assessment of thermochemical conversion technologies for energy generation from crop residue. Earth Energy Sci. https://doi.org/10.1016/j.ees.2024.07.001

Leong HY, Chang C-K, Khoo KS et al (2021) Waste biorefinery towards a sustainable circular bioeconomy: a solution to global issues. Biotechnol Biofuels 14:87. https://doi.org/10.1186/s13068-021-01939-5

LLP SI (2024) Global bio-based materials market size to worth USD 396.01 billion by 2033|CAGR of 25.40%. GlobeNewswire News Room. https://www.globenewswire.com/news-release/2024/07/23/2916929/0/en/Global-Bio-Based-Materials-Market-Size-To-Worth-USD-396-01-Billion-By-2033-CAGR-Of-25-40.html. Accessed 5 Feb 2025

Lozano Pérez AS, Lozada Castro JJ, Guerrero Fajardo CA (2024) Application of microwave energy to biomass: a comprehensive review of microwave-assisted technologies, optimization parameters, and the strengths and weaknesses. J Manuf Mater Process 8:121. https://doi.org/10.3390/jmmp8030121

Makepa DC, Chihobo CH (2024) Barriers to commercial deployment of biorefineries: a multifaceted review of obstacles across the innovation chain. Heliyon 10:e32649. https://doi.org/10.1016/j.heliyon.2024.e32649

Makepa DC, Chihobo CH, Ruziwa WR, Musademba D (2023) A systematic review of the techno-economic assessment and biomass supply chain uncertainties of biofuels production from fast pyrolysis of lignocellulosic biomass. Fuel Commun 14:100086. https://doi.org/10.1016/j.jfueco.2023.100086

Manikandan S, Vickram S, Sirohi R et al (2023) Critical review of biochemical pathways to transformation of waste and biomass into bioenergy. Bioresour Technol 372:128679. https://doi.org/10.1016/j.biortech.2023.128679

Market Research Future (2026) MRF bio-based succinic acid market size, share, growth|report, 2030. https://www.marketresearchfuture.com/reports/bio-based-succinic-acid-market-7779. Accessed 10 Feb 2025

Millati R, Wikandari R, Ariyanto T et al (2020) Pretreatment technologies for anaerobic digestion of lignocelluloses and toxic feedstocks. Bioresour Technol 304:122998. https://doi.org/10.1016/j.biortech.2020.122998

Mohammad REA, Abdullahi SS, Muhammed HA et al (2024) Recent technical and non-technical biorefinery development barriers and potential solutions for a sustainable environment: a mini review. Case Stud Chem Environ Eng 9:100586. https://doi.org/10.1016/j.cscee.2023.100586

Nakashima K, Ebi Y, Kubo M et al (2016) Pretreatment combining ultrasound and sodium percarbonate under mild conditions for efficient degradation of corn Stover. Ultrason Sonochem 29:455–460. https://doi.org/10.1016/j.ultsonch.2015.10.017

Namasivayam SKR, Prakash P, Babu V et al (2023) Aquatic biomass cellulose fabrication into cellulose nanocomposite and its application in water purification. J Clean Prod 396:136386. https://doi.org/10.1016/j.jclepro.2023.136386

Neste (2025) Neste Renewable diesel. Neste. https://www.neste.com/products-and-innovation/neste-my-renewable-diesel. Accessed 25 July 2025

Novozymes (2025) Novozymes enzymes-biodiesel. Accessed 12 Feb 2025—Google Search. https://www.novonesis.com/en/biosolutions/bioenergy/biodiesel. Accessed 25 July 2025

Okoro V, Azimov U, Munoz J (2022) Recent advances in production of bioenergy carrying molecules, microbial fuels, and fuel design—a review. Fuel 316:123330. https://doi.org/10.1016/j.fuel.2022.123330

Pandit C, Banerjee S, Pandit S et al (2023) Recent advances and challenges in the utilization of nanomaterials in transesterification for biodiesel production. Heliyon 9:e15475. https://doi.org/10.1016/j.heliyon.2023.e15475

Patel R, Rajaraman TS, Rana PH et al (2025) A review on techno-economic analysis of lignocellulosic biorefinery producing biofuels and high-value products. Results Chem 13:102052. https://doi.org/10.1016/j.rechem.2025.102052

Pergola MT, Saulino L, Castellaneta M et al (2022) Towards sustainable management of forest residues in the southern Apennine Mediterranean mountain forests: a scenario-based approach. Ann For Sci 79:14. https://doi.org/10.1186/s13595-022-01128-w

Perrone OM, Colombari FM, Rossi JS et al (2016) Ozonolysis combined with ultrasound as a pretreatment of sugarcane bagasse: effect on the enzymatic saccharification and the physical and chemical characteristics of the substrate. Bioresour Technol 218:69–76. https://doi.org/10.1016/j.biortech.2016.06.072

Pintér G, Zsiborács H, Baranyai NH (2022) Aspects of determining the energy storage system size linked to household-sized power plants in Hungary in accordance with the regulatory needs of the electric energy system. Sustainability 14. https://doi.org/10.3390/su14052622

Popp J, Harangi-Rákos M, Pető K, Nagy A (2013) Bioenergy: risks to food-, energy- and environmental security. Appl Stud Agribus Commer 7:121–130. https://doi.org/10.19041/APSTRACT/2013/4-5/17

Porter C (2016) World agricultural prospects the road to 2050. Supply Intell Ltd

Ramírez-Márquez C, Al-Thubaiti MM, Martín M et al (2023) Processes intensification for sustainability: prospects and opportunities. Ind Eng Chem Res 62:2428–2443. https://doi.org/10.1021/acs.iecr.2c04305

Ranjbari M, Shams Esfandabadi Z, Quatraro F et al (2022) Biomass and organic waste potentials towards implementing circular bioeconomy platforms: a systematic bibliometric analysis. Fuel 318:123585. https://doi.org/10.1016/j.fuel.2022.123585

Reichert CL, Bugnicourt E, Coltelli M-B et al (2020) Bio-based packaging: materials, modifications. Ind Appl Sustain Polym 12:1558. https://doi.org/10.3390/polym12071558

Rissman J, Bataille C, Masanet E et al (2020) Technologies and policies to decarbonize global industry: review and assessment of mitigation drivers through 2070. Appl Energy 266:114848. https://doi.org/10.1016/j.apenergy.2020.114848

Robles-Iglesias R, Naveira-Pazos C, Fernández-Blanco C et al (2023) Factors affecting the optimisation and scale-up of lipid accumulation in oleaginous yeasts for sustainable biofuels production. Renew Sust Energ Rev 171:113043. https://doi.org/10.1016/j.rser.2022.113043

Saravanan A, Kumar PS, Khoo KS et al (2021) Biohydrogen from organic wastes as a clean and environment-friendly energy source: production pathways, feedstock types, and future prospects. Bioresour Technol 342:126021

Secretariat O (2000) United Nations environment programme. Synth Rep Sci Environ Eff Technol Econ Assess Panels Montr Protoc

Sertolli A, Gabnai Z, Lengyel P, Bai A (2022) Biomass potential and utilization in worldwide research trends—a bibliometric analysis. Sustainability 14:5515. https://doi.org/10.3390/su14095515

Sharma HB, Vanapalli KR, Samal B et al (2021) Circular economy approach in solid waste management system to achieve UN-SDGs: solutions for post-COVID recovery. Sci Total Environ 800:149605. https://doi.org/10.1016/j.scitotenv.2021.149605

Sharma V, Tsai M-L, Chen C-W et al (2023) Advances in machine learning technology for sustainable biofuel production systems in lignocellulosic biorefineries. Sci Total Environ 886:163972. https://doi.org/10.1016/j.scitotenv.2023.163972

Sheik AG, Kumar A, Ansari FA et al (2024) Reinvigorating algal cultivation for biomass production with digital twin technology—a smart sustainable infrastructure. Algal Res 84:103779. https://doi.org/10.1016/j.algal.2024.103779

Sindhu R, Kuttiraja M, Elizabeth Preeti V et al (2013) A novel surfactant-assisted ultrasound pretreatment of sugarcane tops for improved enzymatic release of sugars. Bioresour Technol 135:67–72. https://doi.org/10.1016/j.biortech.2012.09.050

Sulis DB, Lavoine N, Sederoff H et al (2025) Advances in lignocellulosic feedstocks for bioenergy and bioproducts. Nat Commun 16:1244. https://doi.org/10.1038/s41467-025-56472-y

Szarka N, Haufe H, Lange N et al (2021) Biomass flow in bioeconomy: overview for Germany. Renew Sust Energ Rev 150:111449. https://doi.org/10.1016/j.rser.2021.111449

Turner T, Wheeler R, Oliver IW (2022) Evaluating land application of pulp and paper mill sludge: a review. J Environ Manag 317:115439. https://doi.org/10.1016/j.jenvman.2022.115439

Venkatramanan V, Shah S, Prasad S et al (2021) Assessment of bioenergy generation potential of agricultural crop residues in India. Circ Econ Sustain 1:1335–1348. https://doi.org/10.1007/s43615-021-00072-7

Wan Osman WNA, Rosli MH, Mazli WNA, Samsuri S (2024) Comparative review of biodiesel production and purification. Carbon Capture Sci Technol 13:100264. https://doi.org/10.1016/j.ccst.2024.100264

Wankhede L, Bharadwaj G, Saini R et al (2024) Recent advances in genetic and metabolic engineering for drop-in biofuel production. In: Brar SK, Osorio Gonzalez CS, Soccol CR, Saini R (eds) The microbiology of the drop-in biofuel production. Springer Nature Switzerland, Cham, pp 345–367

Wankhede L, Bhardwaj G, Pereira GVDM et al (2025) Raw material selection for sustainable fermentation-derived alternative protein production: a review. Syst Microbiol Biomanuf 5:1–14. https://doi.org/10.1007/s43393-024-00294-4

Wright A, Rollinson A, Yadav D et al (2020) Plasma-assisted pre-treatment of lignocellulosic biomass for anaerobic digestion. Food Bioprod Process 124:287–295. https://doi.org/10.1016/j.fbp.2020.09.005

Yadav P, Samadder SR (2018) Environmental impact assessment of municipal solid waste management options using life cycle assessment: a case study. Environ Sci Pollut Res 25:838–854. https://doi.org/10.1007/s11356-017-0439-7

Yamakawa CK, Qin F, Mussatto SI (2018) Advances and opportunities in biomass conversion technologies and biorefineries for the development of a bio-based economy. Biomass Bioenergy 119:54–60. https://doi.org/10.1016/j.biombioe.2018.09.007

Yang W, Pudasainee D, Gupta R et al (2021) An overview of inorganic particulate matter emission from coal/biomass/MSW combustion: sampling and measurement, formation, distribution, inorganic composition and influencing factors. Fuel Process Technol 213:106657

Zhang J, Zhang X (2019) The thermochemical conversion of biomass into biofuels. In: Biomass, biopolymer-based materials, and bioenergy. Elsevier, pp 327–368

Zhang W, Chen Q, Chen J et al (2023) Machine learning for hydrothermal treatment of biomass: a review. Bioresour Technol 370:128547

MIX
Papier aus verantwortungsvollen Quellen
Paper from responsible sources
FSC® C105338

If you have any concerns about our products,
you can contact us on
ProductSafety@springernature.com

In case Publisher is established outside the EU,
the EU authorized representative is:
Springer Nature Customer Service Center GmbH
Europaplatz 3, 69115 Heidelberg, Germany

Printed by Libri Plureos GmbH
in Hamburg, Germany